Environmental Biology

Environmental Biology offers a fresh, problem-solving treatment of the topic for students requiring a biology background before further study in environmental science, sustainable development or environmental engineering.

The text begins with an environmental theme that carries throughout, using three major case studies with a regional focus. Key foundational knowledge in biology is introduced and developed as the text progresses, with students encouraged to integrate their accumulating learning to reach solutions. A comprehensive coverage of scientific method, including field experimentation and field techniques, is an important part of the approach. While emphasising the environmental theme, the book introduces all facets of the discipline of biology, including cell biology, evolution, ecology, conservation and restoration.

With over 500 line drawings, diagrams and photos throughout, including full-colour sections, each chapter includes:

- chapter summaries
- comprehension questions
- activities that reinforce learning and encourage scientific analysis
- topics for debate with other students
- lists of further reading.

An online Instructors' Resource offers multiple-choice questions, 'Test your knowledge' solutions, video footage, a full repository of text-based and supplementary photos, and a vast list of relevant journal articles.

Mike Calver is Associate Professor in the School of Biological Sciences and Biotechnology at Murdoch University.

Alan Lymbery is Associate Professor of Parasitology in the School of Veterinary and Biomedical Sciences at Murdoch University.

Jen McComb is Emeritus Professor in the School of Biological Sciences and Biotechnology at Murdoch University.

Mike Bamford is Consulting Ecologist at Bamford Consulting.

Environmental Biology

Edited by
Mike Calver
Alan Lymbery
Jen McComb
Mike Bamford

Illustrated by Belinda Cale and Mike Bamford

CAMBRIDGE
UNIVERSITY PRESS

477 Williamstown Road, Port Melbourne, VIC 3207, Australia

Published in the United States of America by Cambridge University Press, New York

Cambridge University Press is part of the University of Cambridge.

It furthers the University's mission by disseminating knowledge in the pursuit of education, learning and research at the highest international levels of excellence.

www.cambridge.edu.au
Information on this title: www.cambridge.org/9780521679824

First published 2009
Reprinted 2012, 2014

Designed by Pier Vido
Typeset by Aptara Corp

A catalogue record for this publication is available from the British Library

A Cataloguing-in-Publication entry is available from the catalogue of the National Library of Australia at www.nla.gov.au

978-0-521-67982-4 paperback

Contents

Contributors

Mike Bamford is Consulting Ecologist at Bamford Consulting

Mike Calver is Associate Professor in the School of Biological Sciences and Biotechnology at Murdoch University

Alan Lymbery is Associate Professor of Parasitology in the School of Veterinary and Biomedical Sciences at Murdoch University

Jen McComb is Emeritus Professor in the School of Biological Sciences and Biotechnology at Murdoch University

David Ayre is Professor in the School of Biological Sciences at the University of Wollongong.

Barbara Bowen is Lecturer in the School of Biological Sciences and Biotechnology at Murdoch University.

Stuart Bradley is Professor and Dean of the Faculty of Sustainability, Environmental and Life Sciences at Murdoch University.

Treena Burgess is Postdoctoral Fellow in the School of Biological Sciences and Biotechnology at Murdoch University.

Jane Chambers is Senior Lecturer in the School of Environmental Science at Murdoch University.

Mathew Crowther is Lecturer in the School of Biological Sciences at the University of Sydney.

Jenny Davis is Professor in the School of Biological Sciences at Monash University

Bernie Dell is Professor in the School of Biological Sciences and Biotechnology at Murdoch University.

Chris Dickman is Professor in the School of Biological Sciences at the University of Sydney.

Mark Garkaklis is Adjunct Lecturer in the School of Biological Sciences and Biotechnology at Murdoch University.

Howard Gill is Senior Lecturer in the School of Biological Sciences and Biotechnology at Murdoch University.

Richard Hobbs is Professor in the School of Plant Biology at the University of Western Australia

Stephen Hopper is Director, Royal Botanic Gardens Kew, United Kingdom

Pierre Horwitz is Associate Professor in the School of Natural Sciences at Edith Cowan University

Carolyn Jones is Senior Lecturer in the School of Biological Sciences and Biotechnology at Murdoch University.

Annette Koenders is Senior Lecturer in the School of Natural Sciences at Edith Cowan University.

Philip Ladd is Senior Lecturer in the School of Environmental Science at Murdoch University.

Dan Lunney is Principal Research Scientist in the Department of Environment and Climate Change New South Wales

Arthur McComb is Emeritus Professor in the School of Environmental Science at Murdoch University.

Graham O'Hara is Associate Professor in the School of Biological Sciences and Biotechnology at Murdoch University.

Eric Paling is Associate Professor in the School of Environmental Science at Murdoch University.

Harry Recher is Emeritus Professor in the School of Natural Sciences at Edith Cowan University.

Luke Twomey is Principal Scientist of the Swan River Trust, Western Australia

Mike van Keulen is Senior Lecturer in the School of Biological Sciences and Biotechnology at Murdoch University.

Grant Wardell-Johnson is Associate Professor in the School of Environmental Biology at Curtin University of Technology.

Preface

There are many excellent introductory biology textbooks available, so why write another? The answer lies partly in the rapid expansion of modern biology and partly in the needs and aspirations of modern students.

The second half of the 20th century and the early 21st century have seen such major developments as the unravelling of the structure of DNA, the complete cataloguing of the genome of humans and other species, and the first successful cloning. These developments are reflected in university biology curricula, which offer new units and courses in subjects such as molecular genetics and biotechnology and a much greater prominence for molecular biology in introductory textbooks. Simultaneously, other biologists have noted with concern the impacts of climate change, increasing human populations and changing technologies on natural environments and other species. They note that the rate of extinction in species at present is well above the background extinction rate shown in the fossil record, suggesting that the world is in a period of human-caused mass extinction that is reducing our biological heritage. These realisations are reflected in the curricula too, with new courses and units in conservation biology and restoration biology, as well as chapters on conservation in introductory textbooks.

Students majoring in biology at university need a thorough grounding in all these new areas as well as the more traditional aspects of the discipline. They are well served by existing textbooks, but many non-majors lack the space in their crowded timetables to cover all the topics in such detail. Instead, they need to emphasise the biology of direct relevance to their major field of study. Unfortunately, for many students it may not be clear how basic biology is relevant to their varying majors. This has long been recognised in biomedical education, where biology textbooks for physicians, dentists and other health professionals take an explicit human emphasis in their examples. Similarly, we believe that there is a need for a text with an environmental emphasis for those students needing a semester of biology as background for their specialist studies in environmental science, conservation biology, sustainable development, environmental engineering and related fields.

Environmental Biology is our attempt to meet that need. It begins with an account of the human species and our impact on the environment, before developing the biological knowledge and skills necessary to solve environmental problems through a consideration of scientific method (including the major unifying theories of evolution and the cell), biodiversity and the interactions of organisms with each other and with the physical environment. The final chapters integrate this background material in the applied disciplines of conservation biology and environmental restoration. The specialist chapter authors are all experienced researchers and accomplished teachers, and they illustrate their points with theoretical and practical environmental examples. We hope that this approach will enable students with interests in environmental science or sustainable development to see immediately the relevance of biology to their major discipline and integrate biological knowledge and skills into solving pressing environmental problems.

Acknowledgements

A project such as this is possible only because of the generosity, assistance and hard work of many people. We are grateful to the chapter authors who have endured our questioning and nagging and accepted with good grace the editorial adjustments necessary to ensure a uniform style across all chapters. Many also kindly provided excellent photographs to illustrate their chapters.

Individual copyright holders are acknowledged in the text for permission to include figures, tables and quotes, but we would particularly like to thank Rodney Armistead, Richard Calver, Jane Chambers, Jenny Davis, Bernie Dell, Bill Dunstan, Hugh Finn, Ray Froend, Alex George, Richard Hobbs, John Huisman, N. Insalud, Manfred Jusaitis, Philip Ladd, Dan Lunney, Jenny Lawrence, David Macey, Neville Marchant, John Martin, Brett Mawbey, Ron Mawbey, Martina Müller, Jim Negus, Eric Paling, H. Patterson, John Plaza, Michael Shane, Laurie Twigg, Grant Wardell-Johnston, Maria Waters and Robert Whyte for generously providing photographs from their own collections. Jiri and Marie Lochman quickly and efficiently met our urgent requests for photographs to cover gaps in our requirements. Belinda Cale drew most of the original illustrations and adaptations, impressing everyone with her knack for turning rough conceptions into polished images.

We wore out many people at Cambridge University Press with the demands and mistakes that only novice editors can make. Thuong Du, Zoe Hamilton, Jill Henry, Karen Hildebrandt, Jodie Howell, Debbie Lee and Joy Window all gave valued advice and encouragement. Finally, we thank our families, friends and colleagues for their boundless patience during the book's long gestation.

Theme 1

What is environmental biology?

About 65 million years ago, life on Earth was stressed by global climate change and rising sea levels. At this time, two large asteroids struck the planet simultaneously in North and South America. The impacts threw a huge dust cloud into the atmosphere and blocked the sun for at least several months. They also coincided with the extinction of the large dinosaurs and profound changes in other life.

Biologist Peter Ward has called attention to similarities between that time and our own. Climate change and rising sea levels are stressing the world's biota again and, to use Ward's analogy, another 'asteroid' struck in Africa about 100 000 years ago – the evolution of the human species. Ward argues that the rise of humans has meant the diminishment of the great diversity of life on Earth, in just the same way as the impacts of the earlier asteroids altered the direction of evolution.

In this first theme of *Environmental Biology* we examine why an understanding of our species is essential to understanding and conserving the diversity of life.

1

Environmental biology and our time

Mike Calver, Alan Lymbery and Jen McComb

Setting the scene

About 40 years ago, the first comprehensive biology textbook written specifically for Australian students opened with the photograph in Figure 1.1, showing a flock of sheep in a paddock.

Figure 1.1 An Australian sheep paddock. (Source: Morgan, D. (supervising editor) (1973). *Biological science: the web of life*. Australian Academy of Science, Canberra)

The scene was typical of many agricultural areas in Australia then, and remains so today. The authors commented on the questions a biologist might ask when viewing the picture: why do the sheep prefer to stand in the shade? Why are there no sheep under the far tree? Why are there no young trees in the paddocks? Today, those questions seem less relevant. A contemporary biologist might ask: what was the landscape like before the establishment of European agriculture? What plants and animals have been lost from this area? Are there signs of land degradation and, if so, how could they be reversed? Is the agricultural production sustainable? If not, what are the implications for local human communities? These new questions reveal a growing concern about the impacts of expanding human populations and the application of new technologies on the natural environment.

Chapter aims

This chapter describes how the success of the world's dominant animal species, humans, has severely altered biodiversity and natural ecosystems. Three in-depth examples of environmental problems are introduced, together with an explanation of the knowledge and skills biologists need to reverse or mitigate such problems.

Humans and environmental problems

The environment can be regarded as a series of linked ecosystems, each of which includes all the organisms of a given area and the physical surroundings with which they interact. Collectively, the ecosystems of the Earth form the biosphere. They provide an enormous range of 'goods and services' for human populations, including food production, cycling of nutrients important for agricultural productivity, water purification, oxygen production, preservation of topsoil, and a wide range of renewable materials such as drugs, fibres and timber. Humans are now placing great demands on these services as a result of our population growth and increasing resource consumption.

The human population density a million years ago has been estimated at about $0.004/km^2$. By 10 000 years ago it had increased to $0.04/km^2$, by 2000 years ago it had reached $1.0/km^2$ and by the early 19th century it was $6.2/km^2$, rising further to $46.0/km^2$ by the turn of the 21st century. Current projections suggest a global population of about 11 billion, or over $90.0/km^2$, by 2050. Over the same period, technology has advanced from simple tools of stone, bone and wood powered by human muscle, to modern sophisticated metalworking, electronics and machines powered by fossil fuels. Thus the capacity for rapid modification of the environment has grown along with the population. For example, at present humans divert about 50% of accessible fresh water on the Earth for our own use. Every extinction of an organism on the Earth in the last 200 years has probably involved human activity. About 7% of the productive capacity of the Earth's terrestrial ecosystems is lost annually as a result of human-caused habitat degradation. We could mention many other examples.

Why should we care? Much concern about the environment stems from two utilitarian reasons: the usefulness of many products that the environment provides and the essential need to maintain the ecosystem services upon which human life depends. Aside from these, people are concerned about the aesthetic value of natural environments and the organisms within them, and argue that their preservation enriches human life. The yearning for wildlife and wild places is now reflected in a burgeoning environmental tourism industry, which seeks to provide people with such experiences. A further issue is one of ethics. If any one human generation destroys a non-renewable environmental asset, then that asset is unavailable to future generations and their quality of life might be reduced as a result. Each generation has a custodial responsibility for the quality of the environment.

Overall, the human species is the critical factor in understanding environmental biology. It is both the problem, because of its impacts, and the solution, because the impacts are preventable. Therefore this chapter begins with a discussion of the human species, exploring the reasons for its success and for its effects on the Earth's ecosystems. Then three case studies of environmental problems are introduced:

1. the conflict between timber production and conservation of Leadbeater's possum (*Gymnobelideus leadbeateri*) in the mountain highlands of Victoria, Australia
2. concern as to whether damaging outbreaks of crown-of-thorns starfish (*Acanthaster planci*) on tropical reefs have a human cause, and
3. conservation of a rare plant species, the Corrigin grevillea (*Grevillea scapigera*), following extensive land clearing for agriculture in the wheat belt of Western Australia.

Each case study highlights the background knowledge necessary to solve environmental problems, and thus the relevance of the background studies presented in this book. In later chapters we will revisit these problems to demonstrate progress in developing solutions.

The human animal

Evolutionary origins of modern humans

Many of the steps in the evolution of humans are open to different interpretations because the incomplete nature of the fossil record makes it difficult to build a conclusive case. However, it appears that the human lineage separated from that of the great apes about 5–7 million years ago. The stimulus may have been climate change turning the tropical forests of southern and eastern Africa into open woodland with large gaps in the canopy and, still later, into wooded savanna. Walking upright on the hind limbs – bipedalism – aided the mobility of ancestral humans in the changed environment, increased their range of vision and also freed their forelimbs for manipulating objects. This in turn may have encouraged brain development and opened new opportunities for finding and using environmental resources by means of tools, which are external objects that extend the body's functions to achieve an immediate goal. Steady increases in brain size and, with them, increasing skill in tool making and environmental modification have characterised human evolution ever since.

Overall, the major evolutionary trends in the development of humans involved:

1. bipedal posture (walking on two legs)
2. a considerable increase in brain size, both in absolute volume and also in proportion to overall body size (a modern human has a brain volume of about 1450 mL, compared with 400 mL in a chimpanzee)
3. a shortening of the jaws and flattening of the face (perhaps the increasing use of tools to prepare food and for defence reduced the need for large jaws and teeth)
4. reduced sexual dimorphism, with males being only slightly larger than females and
5. a long period of infant dependence on the parents, which in turn may have promoted both long-term pair-bonding and increased opportunities for teaching infants.

Modern human characteristics

The substantial development of the brain in modern humans was accompanied by significant increases in using and making tools. Tool use is not a uniquely human trait; for example, Egyptian vultures (*Neophron percnopterus*) may break ostrich eggs by striking them with stones, while New Caledonian crows (*Corvus moneduloides*) strip the spiny edges of plant leaves to make hooks to lever insects out of their hiding places under bark. However, humans are unusual in the range of tools they use and in their ingenuity in fashioning or modifying tools for special tasks. These tools, especially those powered by fossil-fuel energy sources, give humans a great capacity to modify their environment.

Modern humans also display an outstanding facility with language, both spoken and written, and this is an important basis for learning to be passed on between generations. Language aids communication about the physical world and also expresses abstract ideas such as ethics, justice or evil. Throughout their lifetimes, but especially in their protracted period of dependence on their parents, humans learn both by observing others and also by sharing their experiences via language. As the playwright George Bernard Shaw observed, if you and I each have an apple and we swap apples, we each still have only one apple. However, if we each have an idea and we swap ideas, then each of us will have two ideas. The accumulation of this shared experience and its transmission from generation to generation is what makes up culture. It is a flexible, rapidly changing means of adjusting human responses to environmental change that considerably outpaces biological evolution. Language, both written and spoken, is a powerful medium for its transmission. Box 1.1 gives common characteristics of all human cultures.

Science and technology are important parts of human culture. Science involves solving questions raised by observations of nature, using reasoned evaluation of evidence and careful testing of ideas. Good science has a high success in predicting the outcomes of

Box 1.1 Common features of all human cultures

A detailed anthropological study published in 1945 listed 67 characteristics common to all known human cultures at that time. Culture was taken to mean the social behaviours and institutions of the societies. The list included:

> Age-grading, athletic sports, bodily adornment, calendar, cleanliness training, community organization, cooking, cooperative labour, cosmology, courtship, dancing, decorative art, divination, division of labour, dream interpretation, education, eschatology (myths about the beginning and end of time), ethics, ethno-botany, etiquette, faith healing, family feasting, fire-making, folklore, food taboos, funeral hospitality, housing, hygiene, incest taboos, inheritance rules, joking, kin groups, kinship nomenclature, language, law, luck superstitions, personal names, population policy, postnatal care, pregnancy usages, property rites, propitiation of supernatural beings, puberty customs, religious ritual, residence rules, sexual restrictions, soul concepts, status differentiation, surgery, tool-making, trade, visiting, weather control and weaving.[1]

More recently, other anthropologists suggested adding 'marriage customs' (not implying monogamy) to the list. Thus despite the unique features of individual cultures, there is still much in common between them.

[1] From Wilson, E.O. (1998). *Consilience: the unity of knowledge.* Vintage Books, New York and Little, Brown Book Group, London.

new situations, because of its sound understanding of underlying processes. Technology is the practical application of science in daily life, in which the understandings gained through science are used to solve problems or design new tools.

Human cultural development

Despite the diversity of human cultures, three broad stages of cultural development can be recognised: pre-agricultural, agricultural and urban. The transition from each caused a surge in human population numbers associated with improved food production (Table 1.1), but created a greater capacity to damage the environment.

In pre-agricultural cultures humans survived by gathering plants and hunting or scavenging animals for food. The domestication of plants for agriculture profoundly altered human lifestyles and their interactions with their environment by enabling protracted settlement in one place. In particular, urban living with very high population

Table 1.1 Human population growth and cultural development over the last 1 million years.

Years before present	Cultural development	Approximate density per square kilometre	Approximate population (millions)
1000000	Pre-agricultural	0.00425	0.6
30000–40000	Pre-agricultural	0.02	2.5
10000	Pre-agricultural, first storage of wild grain	0.04	5
2000–9000	Village farming, early urban development	1.0	133
160	Intensive agriculture and industrialisation	6.2	906
Present	Modern technological	46	6300

densities became possible. Its potential was fully realised during the Industrial Revolution, when the replacement of cottage industries with large-scale, machine-based production concentrated workers in large urban centres. The new production techniques and living conditions created enormous demands for energy. This was supplied initially by timber and coal, and more recently by oil, natural gas and nuclear energy. The environmental interactions characterising these three stages of cultural development are shown in Figure 1.2.

Pre-agricultural ecosystems

Pre-agricultural ecosystems were characterised by solar energy input and by the large-scale recycling of matter. Humans lived entirely within such ecosystems for at least 100000 years, subsisting by collecting plant foods, scavenging remains of animals killed by other predators and limited hunting. About 50000 years ago humans developed more sophisticated weapons, enabling active hunting of large game. A detailed example of such a lifestyle is given in Box 1.2.

The environmental impacts of pre-agricultural lifestyles are controversial. Some anthropologists believe that aggressive scavenging by armed people could have driven large carnivores from their kills and depleted their food supply. Hunting methods such as driving herds of animals over cliffs wastefully killed large numbers, and the use of fire to flush out game altered plant ecosystems. It is argued that the development of more sophisticated weapons led to the hunting and ultimate extermination of a range of large animals including woolly rhinoceros, mammoths and giant deer in Europe, a range of large mammals in North America and large marsupials in Australia. Proponents of this view argue that human migrants carried the new hunting skills into previously unoccupied environments where the animals, unadapted to human predation, succumbed quickly. This is called the blitzkrieg hypothesis, after Nazi Germany's aggressive use of

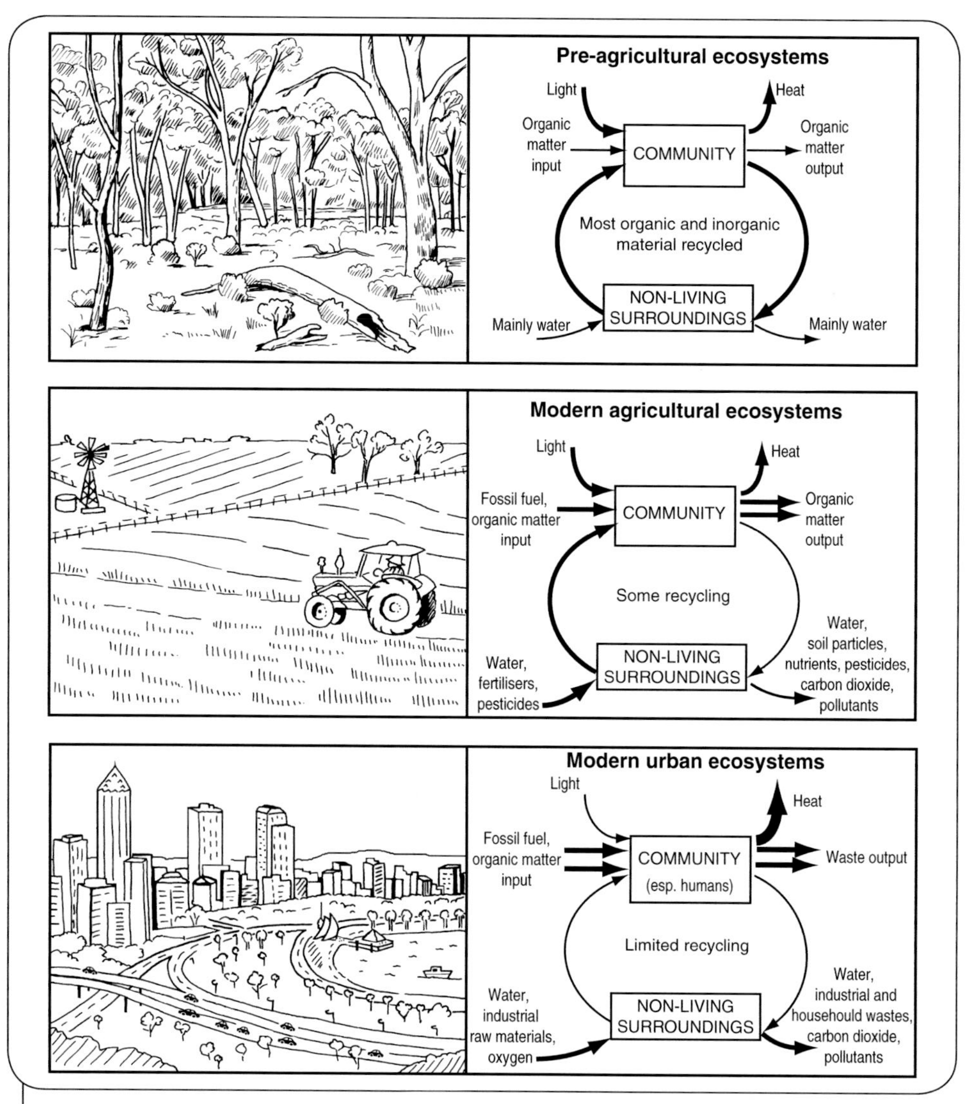

Figure 1.2 Characteristics of pre-agricultural, agricultural and urban ecosystems. (Source: Adapted from Morgan, D. (supervising editor) (1973). *Biological science: the web of life*. Australian Academy of Science, Canberra)

a novel combination of tanks and infantry to overrun opponents in early World War II. Some of these conclusions are debated fiercely, with alternative explanations such as climate change and disease proposed as sole or contributory causes of the extinctions. However, one case is reasonably free of controversy. The 11 species of New Zealand moas, flightless birds some of which stood up to 2 metres high at the shoulder, were hunted to extinction within 100 years of Polynesian settlement of New Zealand. Furthermore, the Polynesians introduced rodents that may have contributed to the extinctions by preying on moa eggs.

Box 1.2 The Martu of Western Australia – an example of a pre-agricultural culture

The traditional lands of the Martu people lie in the Pilbara, Western Australia. The average annual rainfall is 306 mm, with localised heavy falls. Many water sources are ephemeral, so the location and permanency of water is the major constraining resource.

Foods are widely dispersed and separately of poor quality. People choosing where to search for foods probably used a hierarchy of scales. In order from the smallest to the largest, these are:

1. selection within micropatches
2. targeting of some plant communities
3. use of particular landscapes
4. habitation of general regions.

Plant food comprised 70–80% of the diet. The people ate at least 106 of the 330 plants from the region, with 20–40 being staple foods. Their choice depended on:

1. abundance, predictability and access
2. the energy costs of processing
3. the energy and nutrient returns of species, and
4. cultural preferences.

Animals were eaten too – at least 12 species of mammals. Reptiles were a reliable food source.

The people used biological indicators such as cricket calls to predict resource availability. They recognised specific vegetation communities to gather plant resources, while hunters ranged more widely. Group size and pattern of movement over the countryside fluctuated with resource availability, and there was some evidence for species management (e.g. sowing seeds on burned ground).

These practices were specific to the region and we would expect variations elsewhere. Such ecosystems have solar energy input, heat output, and a little input and output of matter, with a high degree of recycling.

Agricultural ecosystems

The domestication of plants about 10 000 to 15 000 years ago improved the reliability of the human food supply and enabled people to settle in a single place for long periods. The sun remained the energy source, but inputs of matter such as manure or other fertilisers were needed to optimise crop production. The food and fibre produced were not necessarily consumed on site, some being traded with other communities. Thus, although extensive recycling took place, it was not as great as that occurring in pre-agricultural ecosystems. Early agriculture could have severe localised effects when vegetation was cleared to grow crops. Furthermore, the associated increases in population and the activities of domestic animals placed great demands on food production, leading

Box 1.3 Slash-and-burn farming – an example of agricultural ecosystems

(Contributed by Philip Ladd)

Slash-and-burn agriculture is an evocative term for an agricultural practice, perhaps less emotively termed 'shifting agriculture', first recorded about 8000 years ago in China. Today it is associated with clearing patches in rainforest in countries such as Papua New Guinea, Indonesia and Africa. However, it can also be practised in savanna and grasslands. Natural vegetation is felled, dried and burned to provide ash as a fertiliser. Crops are planted in the ash bed for several years afterwards.

Productivity is initially high because of the nutrients provided in the ash bed. However, many agricultural crops (especially the annuals, such as maize, taro and wheat) have high nutrient requirements and harvesting removes a large proportion of the nutrients accumulated in the plants, so they are lost to the system. Nutrients are also lost in erosion when heavy rains fall on cleared plots. Ultimately, depletion of soil nutrients, especially phosphorus, and increases in the acidity of the soil render it increasingly unproductive for crops, and the cleared areas are abandoned. The local native plants are adapted to the lower nutrient availability, reinvade the cleared areas and increase the biomass production when people move on. In some ways it is similar to a crop–fallow cycle in a more conventional farming system where crop paddocks are rested for 1 to 3 years between crops. If only small plots are disturbed in large tracts of forest, the environment copes with the disturbance, but when the proportion of disturbed ground to forest increases problems arise.

Agricultural systems such as this have inputs of solar energy, water and natural fertilisers. Organic matter and mineral nutrients are removed by both crop plants and erosion. Modern intensive agriculture greatly increases inputs of chemical fertilisers, fossil fuels for energy, water from irrigation and pesticides to give greater crop yields.

in some cases to land degradation and loss of productive capacity. An example of such an agricultural culture is given in Box 1.3.

Despite these problems, the food stores generated by agriculture allowed greater division of labour in human societies. Fewer people were involved directly in food production, freeing others for different activities. Cultures were enriched and the pace of innovation increased.

Urban ecosystems

The concentration of people in cities characteristic of the modern world was stimulated by the Industrial Revolution that began in the late 18th century. The development of machines ended the localised cottage industries of earlier times and shifted production of goods into the mechanised factories of the cities. Machines also transformed agriculture and, because of the lower demand for agricultural labour, people migrated to cities in search of work. Thus, before describing the features of urban ecosystems it is necessary to understand the implications of mechanisation for agriculture.

Mechanisation increased agricultural productivity, but required considerably greater energy inputs to drive the machines and to manufacture the synthetic fertilisers needed to produce ever-larger volumes of food. At first, energy came from burning wood, but this was replaced quickly by coal and, in more recent times, by oil fuels burned in internal combustion engines. Outputs of heat, carbon dioxide and a range of pollutants grew. Increasing proportions of the organic matter produced were transported well away from the agricultural areas, consuming more fuel in transportation and reducing possibilities for local recycling.

In cities, inputs of energy (now mainly in the form of fossil fuels) and inputs of raw materials and oxygen are huge. For example, the per capita energy consumption of a modern Australian is between 650 and 1000 MJ/day, compared with about 15 MJ/day for a human living a pre-agricultural lifestyle. Much of this energy is consumed in transport and industrial production, because food production occurs largely outside the urban area, and a very high proportion of the population is free to pursue other activities. Volumes of waste outputs such as carbon monoxide, carbon dioxide, water vapour, sulfur dioxide, hydrocarbons, oxides of nitrogen, solid waste and manufacturing waste are also extremely high and their safe disposal is a major problem for urban planning. Problems are not necessarily confined to the urban area itself nor even to particular countries, because wind and tides may disperse pollutants well beyond national borders. Box 1.4 describes urban ecology in large modern cities.

Box 1.4 Problems in urban ecosystems – air pollution in Indian cities

(Contributed by Frank Murray)

In this first decade of the 21st century more people live in cities than in rural areas for the first time in human history. The globalisation of production has led to explosive economic and population growth in the large cities of developing countries. The cities attract poor rural people fleeing the hardships of poverty-stricken villages for the potential economic opportunities of the large cities. The accompanying acceleration in industrial, transport and energy production emissions is exposing growing numbers of people to increasing levels of air pollution and damaging their health in many cities.

As an example, India's urban population was 285 million in 2001 and is projected to be 473 million in 2021 and 820 million in 2051. Concentrations of suspended particulate matter such as dust and smoke in the air in all Indian cities greatly exceed the Indian National Ambient Air Quality Standard for residential areas, with Delhi having the highest concentrations during the period 1997–2005. In some cities the annual average is twice the air quality standard. Actions to reduce smoke emissions from vehicles and industry caused decreases in concentrations in some cities, but not others.

The potential environmental consequences of increasing numbers of humans living such lifestyles places many organisms at risk through the habitat destruction associated with food production, diversion of water to human needs and the sheer space taken by growing human populations and their infrastructure. Environmental degradation may ultimately compromise the capacity of the environment to provide the ecological services that still underpin human life, regardless of our scientific and technological advances.

Studying human impacts

Humans have always affected the natural world of which we are part. It is the scale of such impacts that has increased so dramatically through pre-agricultural to urban stages of development and that now threatens ecosystems throughout the world.

How, though, can one identify an environmental impact before it becomes irreversible? What remedial actions can be taken when a problem is identified? What precautionary measures can be taken to anticipate impacts before they arise? Clearly, important background knowledge and specialist skills are needed to answer these questions. One way to identify them is to consider specific examples of environmental impacts and observe how biologists approached each problem.

Three such examples are given below. One deals with successfully combining timber production with nature conservation, the second concerns the sudden increase of a destructive organism in a marine environment and the third explores strategies for the conservation of a rare plant. In each case, the basic problem is described and the critical background knowledge and skills involved are summarised. These examples demonstrate that the fundamental knowledge and skills that a good environmental biologist needs to study putative human impacts can be organised into a few major categories or themes. These themes become the content of the rest of this book.

Wildlife or wood? Conservation of Leadbeater's possum

Leadbeater's possum is a small (120 g) marsupial restricted to parts of the mountain ash eucalypt forests of the Victorian Central Highlands, Australia (Figure 1.3).

It was believed extinct in the early 20th century, but was rediscovered in 1961. It lives on insects and plant exudates, mainly from wattle trees, and shelters in the hollows of large, old trees. Its optimum habitat is a forest with unevenly aged trees, with both wattles for food and old trees with suitable nest hollows. Such habitat may be created by regrowth following wildfire, or by logging operations in which some old hollow trees are retained. However, old trees may collapse and wattles diminish in number as the forest matures, so areas of optimum habitat continually change in response to changes in forest structure. It takes nearly 200 years for tree hollows suitable for Leadbeater's possum to develop, so foresight and planning are needed to ensure an ongoing supply of suitable habitat. Studies of the possum and its habitat were necessary before reserves that would effectively conserve the species could be established. It was clearly not possible to reserve

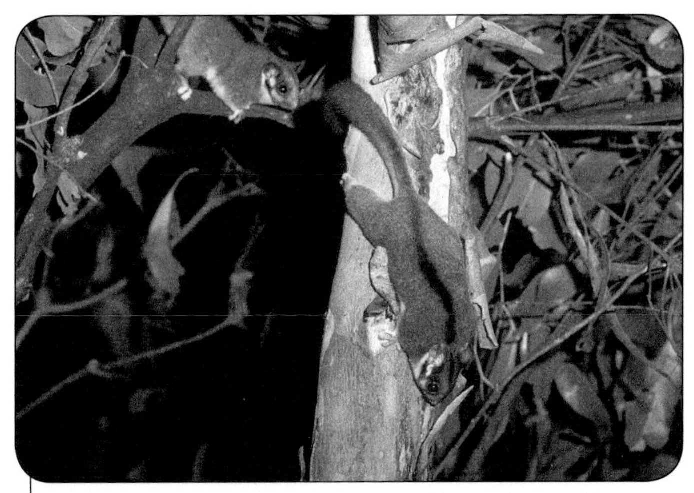

Figure 1.3 Leadbeater's possum, an endangered marsupial from the mountain ash forests of Victoria. (Source: Lochman Transparencies)

all the territory in which the possum was found, because the area is also valuable for timber. The types of questions that biologists needed to answer included:

1. What is the possum's basic biology, including its social structure, ideal population density, preferred habitats and social behaviour?
2. How does the possum interact with other animal and plant species in its environment?
3. What determines the cycle of tree growth and senescence?
4. Should salvage logging (the removal of fire-damaged trees after a wildfire) be continued?
5. How much logging is compatible with possum conservation and what should be the time interval between logging rotations?
6. How many habitat trees with nesting hollows should be left after logging?
7. Are there adequate reserves protecting the possum from disturbance?
8. Are there enough animals left for the population to survive and expand?

Solving the problem also required the skills of:

1. designing and implementing field survey techniques for both the vegetation and the animal populations, and
2. mathematical and statistical techniques for applying population viability analysis (a computer-based modelling method of predicting the outcome of various management techniques).

Clearly, these should be part of the environmental biologist's toolkit.

The starfish, the reef and the scientists

The crown-of-thorns starfish is a specialist coral predator (Figure 1.4).

Corals are colonies of tiny polyps, joined to each other by a thin membrane and supported and protected by a skeleton of calcium carbonate. Each soft polyp has its own

Figure 1.4 Crown-of-thorns starfish, a possible threat to the Great Barrier Reef. (Source: Lochman Transparencies)

protective hole in the communal skeleton. Because the ratio of soft tissue to skeleton is low, and the polyps are hard to get at, most predators do not find coral an energetically attractive food source. Crown-of-thorns starfish have a simple solution. Instead of bringing the polyps to their stomach, as do most predators, the starfish bring their stomachs to the polyps. The membranous stomach of a crown-of-thorns starfish is everted through the mouth over the coral. Digestive juices pour into the coral skeleton and liquefy the polyps. The starfish then sucks in the polyp 'soup', leaving only the limestone skeleton. The stomach is retracted and the starfish moves on to a new patch of coral. Crown-of-thorns starfish usually feed twice a day for several hours, and each individual can eat from 2 to 6 m^2 of coral a year.

Few biologists had ever seen a crown-of-thorns starfish before 1960. Then, a large specimen was found at Green Island, a popular tourist spot in the Great Barrier Reef, and became an object of scientific curiosity. Not only are the starfish large (adults grow to about 60 cm in diameter), but the sharp spines that cover their bodies contain a poison that can cause intense pain, nausea and vomiting in people unlucky enough to be pierced by them. Special demonstration visits were made to Green Island so that the starfish could be observed. More starfish were soon discovered. By 1962, large numbers of crown-of-thorns were reported at Green Island, moving across the reef in a front, leaving dead, white coral in their wake. By the end of the decade, about 80% of the coral on Green Island reef had been destroyed. The crown-of-thorns starfish plague continued on the Great Barrier Reef throughout the 1970s and, after a brief pause, during the 1980s. The Great Barrier Reef stretches along the coast of Queensland for 2000 km and is composed of about 2000 individual reefs. Since 1986, when regular surveys of starfish numbers began, the percentage of reefs with crown-of-thorns starfish outbreaks (defined as densities at which the starfish are consuming coral tissue faster than the coral can regrow) has fluctuated between 1 and 15%.

The crown-of-thorns starfish plagues of the 1970s and 1980s intrigued both scientists and the public. They were often portrayed in the media as heralds of impending environmental disaster – an unnatural consequence of our exploitation of the planet, which would destroy coral reefs throughout the Indo-Pacific. Many marine scientists also viewed the starfish population explosions as new occurrences, resulting from some environmental disturbance that needed to be controlled so that coral reef communities could recover. Others, however, believed that the outbreaks were natural, cyclical occurrences, which would have no long-term deleterious effects and may even be beneficial to coral reefs by feeding only on fast-growing coral species and promoting a greater diversity of life forms. These issues were debated fiercely in scientific journals and at conferences around the world.

The problem was that so little was known about the starfish. To decide whether outbreaks were natural or human-induced, what their long-term consequences might be and what, if anything, we should do to control them, answers were needed to some very basic questions:

1. What is the starfish's reproductive biology – when does it breed, where do the larval stages live, how and how far do they disperse?
2. What is the interaction between the starfish and other species in its environment, especially its natural enemies at all stages of its life cycle?
3. How do human activities, such as those that cause nutrient run-off into the waters near coral reefs, affect starfish populations?
4. How long does the reef take to recover after attack by the starfish?
5. Is there more or less biological diversity on coral reefs after a starfish attack?

It also required the skills of:

1. working in a marine environment and designing and implementing appropriate field survey techniques
2. knowing how to sample without causing excessive damage to the reef or killing organisms unnecessarily
3. identifying marine plants and animals
4. designing effective experiments both in the laboratory and the field to test hypotheses about starfish biology, and
5. mathematical and statistical techniques for developing population models.

These should also be added to the environmental biologist's toolkit.

Back from the brink – the Corrigin grevillea

The Corrigin grevillea is a small, prostrate shrub with pincushion-like inflorescences of white flowers (Figure 1.5).

It is famous as one of the first rare and endangered species in Australia for which new populations have been established in the field. It was first collected in 1954, when it occurred over a region of 100 km^2 centred on the Western Australian wheat-belt town of Corrigin. In this agricultural area 95% of its habitat has been cleared. By 1986 it was believed extinct in the wild, with only one plant in cultivation. Fortunately, an

Figure 1.5 *Grevillea scapigera* (Corrigin grevillea). An integrated conservation project resulted in the establishment of translocated populations of this extremely rare species. (Source: S. Krauss)

extensive search of the area in 1989 found 47 plants, the largest population being on a narrow road verge. In this population no new recruits were surviving, and concern for the survival of the species was well justified, because by 2002 only five plants survived in the wild. It appears that the reduction in the wild populations was caused mainly by clearing for agriculture, but the full impact of other factors such as introduced herbivores and weeds was not known. To conserve this species, off-site conservation and the establishment of one or more new, large populations in secure areas was necessary.

Scientists at the Kings Park Botanic Gardens in Western Australia began work on this species in 1986. They found it impossible to germinate the few seeds available. Seeds of wild species often require mechanical abrasion or particular environmental cues such as fire, smoke, heat, cold or light to break their dormancy. Although many of these factors can be simulated in the laboratory, the precise combination of factors required to germinate many wild species remains unknown.

They had no success growing the grevillea from cuttings, and grafting onto another grevillea (a rootstock of a horticultural hybrid) had limited success. It was clear that the quickest way to get the maximum number of genotypes conserved safely was to use tissue culture, a technique of growing clones (identical copies) using micro-cuttings under sterile conditions in an agar gel containing a defined medium. If clones were available, new, self-sustaining populations might be set up. To rehabilitate this species the following biological and environmental information was needed:

1. Does the species cross-pollinate or self-pollinate to set seed, and what bird or insect pollinators are involved?
2. What is the biology of seed germination? How many seeds set on a plant each year, how do they disperse, how can they be induced to germinate, and do they get eaten in the field before they have a chance to germinate?
3. Can the plant propagate by methods other than seed? Does it resprout after fire? Can it be grown from cuttings?
4. Can we develop a tissue culture medium suitable for the species based on a knowledge of plant nutrition and responses to plant hormones?
5. Is the rarity of the species caused only by habitat destruction, or have other factors such as rabbits and weeds influenced plant survival and lack of seedling recruitment?
6. What is the appropriate soil type and what are the establishment requirements of the plants, if we are to establish a new population?
7. To establish a new population with highest possible genetic diversity and avoid genetic erosion, is it necessary to propagate from all individuals in the relict

populations or will a sample contain enough of the total genetic diversity still existing in the species?

8. Will the translocated population perpetuate itself and maintain the genetic diversity?

It also required the skills of:

1. designing effective experiments both in the laboratory and the field to test hypotheses about seed germination and plant propagation
2. specialist laboratory skills in tissue culture and assessing genetic diversity
3. interacting with the Department of Environment and Conservation, which is responsible for managing and protecting the state's flora, and the local community to get enthusiastic help in the establishment and maintenance of the new populations of plants (Figure 1.6).

These are all further tools for the environmental biologist's kit.

The remainder of this book is designed to provide you with much of the background knowledge and many of the skills highlighted in these examples. In short, it is a toolkit for solving environmental problems. It is organised in four broad themes. All our case studies required sound scientific method, so we discuss this topic first. We cover procedures applicable to all scientific investigations, but with special emphasis on the skills of designing experiments under field conditions. We introduce some limited mathematics and statistics, but not in great detail – you will almost certainly study specialist units in these subjects. We then discuss the two unifying theories of modern biology: the cell theory and the theory of evolution. The study of cells introduces, among other things, the cellular basis of inheritance and genetic diversity shown to be so important in conservation work, as in the case of the Corrigin grevillea. In this context, we explain the skills of tissue culture, microscopy and genetic analysis. An understanding of evolution

Figure 1.6 Establishing wild populations of *G. scapigera* from tissue-cultured clones. Note the area has been cleared of weeds, has irrigation lines in place and is protected by a rabbit-proof fence. Members of the Corrigin Landcare Group are helping Kings Park staff with the planting. (Source: R. Dixon)

reinforces these principles and also allows us to examine why species become extinct. Such knowledge helps biologists anticipate the problems likely to be encountered by Leadbeater's possum or the Corrigin grevillea and take action to avert them.

Applying the scientific method is the foundation for all the case studies, so the second theme of the book explains the basic precepts of the scientific method and the two unifying theories of modern biology: cells and evolution. Basic biological knowledge was fundamental too, so the third and fourth themes of the book apply the scientific method to an understanding of the range of species in the environment, the full spectrum of the world's ecological communities and their associations with the physical environment – in short, biodiversity. In Theme 3 we explain how biological classification works, and describe the diversity of life on Earth and the conservation pressures facing different groups. Theme 4 explores the interactions between different species in biological communities as well as the factors that determine the distribution and abundance of populations. Specific skills for studying these issues are described. These points underpin investigations into the crown-of-thorns outbreaks, as well as the habitat requirements of Leadbeater's possum and the pollination strategies of the Corrigin grevillea. The theme concludes with a discussion of the properties of marine, freshwater and terrestrial environments, because each provides organisms with a unique set of problems and opportunities, as well as requiring specialised techniques for successful ecological studies.

In the final theme we turn our attention to applying the knowledge and skills gained to solving pressing environmental problems using the modern disciplines of conservation biology and restoration ecology. We revisit all three of the case studies to illustrate principles of diagnosis and treatment and to reinforce the relevance and utility of the toolkit of knowledge and skills you gained in studying this book.

Chapter summary

1. Rapid increases in the human population, accompanied by increasing sophistication in tool use and fossil-fuel technologies, are leading to increasing human impacts on the natural environment.
2. Modern humans are characterised by large brain capacity, high learning ability, comprehensive language skills and sophisticated tool use.
3. The collected experience of a human society transmitted from generation to generation is its culture. Science is a special part of culture concerned with understanding the natural world, and technology is the application of scientific understanding.
4. Science and technology, in association with other areas of human endeavour, can both create and solve pressing environmental problems.
5. Biologists can contribute to solving environmental problems through their understanding of:
   - scientific method and the cellular basis of life
   - the evolutionary origins of biodiversity and the extent of Earth's current biodiversity
   - the dynamics of populations and the patterns of interactions of populations of different species with the physical environment in natural communities
   - the new disciplines of conservation biology and restoration ecology, and how these contribute to protecting species and protecting, restoring and even re-creating natural environments.

Key terms

agricultural ecosystem
biosphere
bipedal
clone
ecosystem
environment
pre-agricultural ecosystem
science
technology
tissue culture
urban ecosystem

Test your knowledge

1. List four reasons why people should be concerned about environmental degradation and explain why each is important.
2. What were the main trends in human evolution and what are the unique features of the modern human species?
3. What are the most pressing problems facing (a) Leadbeater's possum and (b) the Corrigin grevillea?
4. Why was there such controversy concerning the possible impacts of the crown-of-thorns starfish on the Great Barrier Reef?

Thinking scientifically

Before they were protected in the 1960s and 1970s, saltwater crocodiles (*Crocodylus porosus*) were almost hunted to extinction in northern Australia. Their numbers have now recovered to the point where some people are arguing that limited hunting should be permitted, with the proceeds going to conservation programs. Which of the following statements could be described as a scientific response to the proposal? Justify your answer.

1. It is wrong to cull wildlife in any circumstances.
2. Hunting should be permitted if there is strong popular support.
3. Before reaching a decision, we should determine whether the proposed level of hunting is likely to cause a decline in crocodile populations.
4. The decision should be left to the indigenous landholders in the area.
5. Crocodiles should be used for meat and skins, not just hunting trophies.
6. Hunting should be permitted only if the crocodiles are a threat to human safety.
7. Hunting is good because it will develop a new tourist industry.

Which of these statements do you think will be most important in the political decision about whether or not to permit hunting?

Debate with friends

Consider the list of features common to all human cultures shown in Box 1.1. Debate the topic: 'Human culture must inevitably transform the environment'.

Further reading

Bolton, G. (1992). *Spoils and spoilers: a history of Australians shaping their environment.* Allen & Unwin, Sydney.

Costanza, R., d'Arge, R., deGroot, R., Farber, S., Grasso, M., Hannon, B., Limburg, K., Naeem, S., O'Neill, R.V., Paruelo, J., Raskin, R.G., Sutton, P. and van den Belt, M. (1997). 'The value of the world's ecosystem services and natural capital' *Nature* 387: 253–60.

Goudie, A. (2006). *The human impact on the natural environment: past, present and future.* Blackwell, Oxford. 6th edn.

Marshall, A. (1966). *The great extermination: a guide to Anglo-Australian cupidity, wickedness and waste.* Heinemann, Melbourne.

Pimm, S. (2001). *The world according to Pimm: a scientist audits the earth.* McGraw-Hill, Boston.

Ponting, C. (2007). *A new green history of the world: the environment and the collapse of great civilizations.* Penguin Books. 2nd edn.

Theme 2

The scientific method and the unifying theories of modern biology

From Theme 1 you understand the central place of the human species in environmental biology today. You also have a general understanding of science and how it might be applied to solving biological problems.

Biologist Hugh Gauch listed five key resources for research in the sciences:

1. equipment to collect data
2. computers and software for data analysis
3. infrastructure, including libraries, colleagues and internet access
4. technical training to use all the above, and
5. a knowledge of the scientific method.

He argued that good research was often most impeded by a poor understanding of the fifth point, the scientific method.

Therefore in this second theme of *Environmental Biology* we explore the nature of the scientific method in detail. We also examine two of its most important contributions to modern biology: the great unifying theories of the cell and of evolution.

2

Science and the environment

Chris Dickman and Mathew Crowther

The mystery of the flying fox populations

In the 1920s the population of the grey-headed flying fox (*Pteropus poliocephalus*) in Australia was estimated to be many millions (Figure 2.1). Their exodus from a camp was an awesome spectacle. By 2000 they had declined dramatically, with a 30% fall in numbers over just the previous 10 years. At the same time, however, flying foxes increased in number in some urban areas. They have long been considered a nuisance in the Royal Botanic Gardens in Sydney, and at the Royal Botanic Gardens in Melbourne their number increased sharply from perhaps a dozen animals in 1986 to over 20 000 in 2006. What has allowed this species to increase in number so dramatically in the urban environment, especially when populations elsewhere are declining? This question can be answered by applying scientific method.

Chapter aims

In this chapter, you will learn the basics of scientific method and how its application identifies and addresses environmental problems.

How does science work?

At one level, science progresses because of new technology. For example, telescopes make it possible for astronomers to discover new stars, cyclotrons allow physicists to probe the structure of matter and gene sequencers enable biologists to identify hereditary codes. Of course, scientists design these tools, guided by the questions they want to answer. The philosophy of science helps to identify these questions and provides the framework for answering them, but this has changed dramatically over the last 200 years. It is instructive to review this briefly to appreciate how science now works.

Figure 2.1 Flying foxes in flight. (Source: Lochman Transparencies)

Early in the 19th century the doctrine of natural theology held sway in the western world. Proponents believed that the natural world was part of God's creation and science was that part of natural theology that studied God's work. Natural theology stimulated great interest in natural history by the clergy of the time. Discoveries of new species, fossils, crystals and elements were taken as further evidence that such a complex world could have been created only by God, foreshadowing by two centuries the arguments for 'intelligent design' used by creationists today. The natural theology world view was held strongly in Australia, as indicated by publications such as the *Australian Quarterly Journal of Theology, Literature and Science* in the late 1820s.

However, other ways of thinking about the natural world were emerging. *Materialists* thought the complexity of nature arose simply from chemical and physical interactions explained by the physical laws; there was no need for God. *Rationalists* wished to improve life through intellectual stimulation and education, while *positivists* asserted that all knowledge of the natural world should be based on empirical observations and experimental data. Publication of Darwin's book, *On the Origin of Species*, in 1859, supported the positivist viewpoint and towards the end of the 19th century this philosophy guided the work of most professional biologists. An important part of the scientific mindset then was that observations could be used to make predictions about how nature would behave under different circumstances, even if no observations could be made to confirm these predictions.

In the 20th century, science added new and exciting dimensions. These included the famous 'thought experiments' of Albert Einstein and, most importantly, the so-called hypothetico-deductive or hypothesis-testing approach advocated by Karl Popper. We explore Popper's ideas in more detail below, but note here that Popper provided a particularly rigorous process for distinguishing the best among several competing explanations for any observation. In this philosophy, there is no certain knowledge, but simply a logical way of improving understanding. More recently, the hypothesis-testing approach has been challenged for being arbitrary and unrealistic, and for failing to appreciate the complexity of nature. The strongest challenge comes from scientists who advocate what has been termed the information-theoretic or strength of evidence approach. This view holds that observations can best be explained by testing their fit against a very wide range of possible models, and this concept is explored in more detail below.

Despite differences in philosophy, most scientists would agree that good science proceeds in a logical and structured manner. Observations are made, evaluated and explained by careful sifting of evidence. The explanations may then be tested by making predictions about the outcome of any new situation, and refined or rejected if found to be wrong. Thus science seeks a mechanistic understanding of nature, and does not inquire into the realms of religious or spiritual belief.

Is science also objective? It is a human enterprise, and as a consequence the observations and discoveries that are made are likely to have a human bias. In 1962, Thomas Kuhn published a critique – *The Structure of Scientific Revolutions* – of how he saw science developing. In Kuhn's view, ideas often become entrenched in science and continue to be supported by the scientists who propose them even when others find them to be wrong. If the scientists supporting a failed idea are senior and have their reputations at stake, change might take place only when they retire or die. In this view, science is not necessarily an ordered and seamless march towards truth, but a more convulsive process characterised by 'paradigm shifts' and mini-revolutions with a strong basis in human interactions.

Applying science – the hypothesis-testing approach

Background

In general, this approach begins by making observations of an interesting pattern, event or puzzle, and then follows a logical sequence of constructing models and hypotheses that can be tested to explain the observations (Figure 2.2). For example, consider a species observed to be declining in population size or in the area over which it ranges. The next step is to identify a plausible explanation, or set of explanations (sometimes also called models), for the decline. They might include loss of habitat, disease, over-harvesting or other factors.

When models have been developed, the next step is to derive hypotheses, or testable predictions, from them. For example, if a model explains the decline of a species as caused by loss of habitat, a reasonable hypothesis would be that the species will be more abundant in areas where its habitat remains intact than in areas where its habitat has been degraded or removed. Alternatively, if a model explains the decline as because of over-harvesting, a plausible hypothesis is that the species will be more abundant in areas where harvesting pressure is light or absent than in areas where harvesting is intense. It is quite possible that several models will exist to account for the decline of a species. In this situation, the models can be distinguished only if different, contrasting predictions are drawn from them.

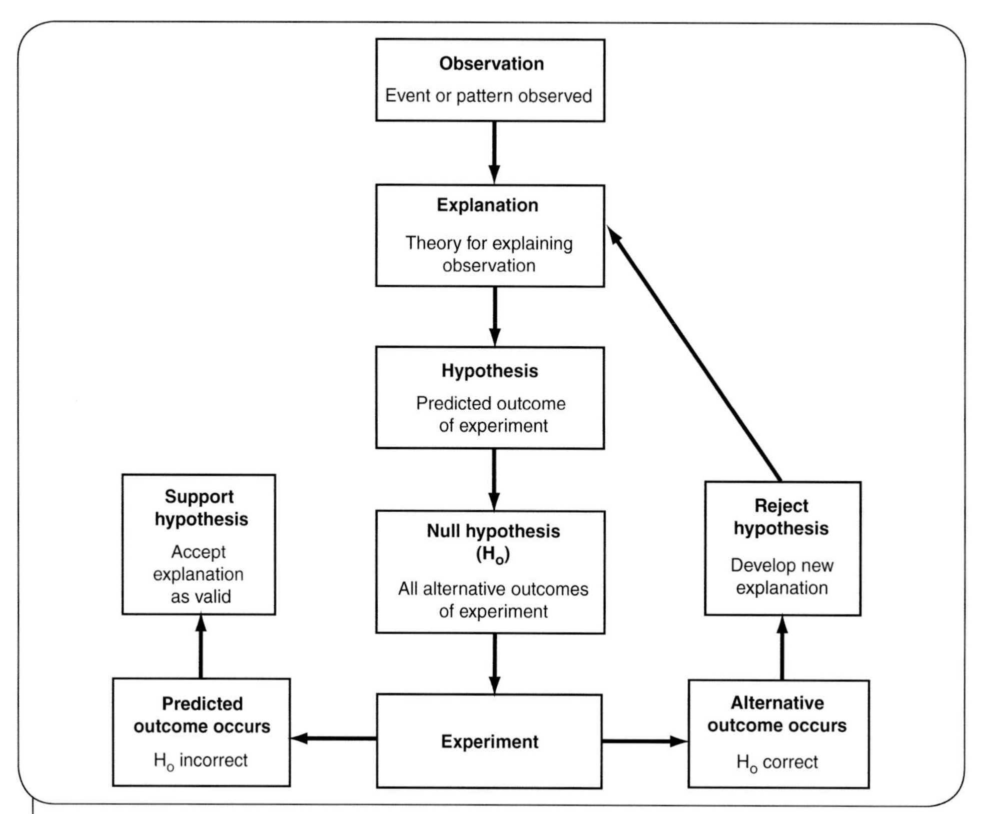

Figure 2.2 The sequence of steps in scientific investigations. Even when the hypothesis is supported, more questions normally arise and are the subject of further investigations. (Source: Copyright Institute of Biology, London, UK)

The hypotheses must then be tested as the final step in the process, and this is achieved most effectively by the use of experiments. For example, if the predictions of the habitat loss model are being tested, two kinds of experiments could be devised. First, mensurative experiments would involve simply measuring the abundance of the target species in intact and degraded habitats. Secondly, manipulative experiments would involve restoring areas of degraded habitat and measuring the abundance of the target species before and after restoration, or in restored and still-degraded habitats. Manipulative experiments could also be designed to deliberately degrade or destroy intact habitat and see if this leads to a decline in the abundance of the target species, but this would not be a good experiment from a conservation perspective (see also Box 2.1).

Statisticians often refer to the variables chosen or manipulated by the experimenter to determine their effectiveness as independent variables. The variables measured to evaluate the response to the independent variables are called dependent variables. In the above example, the condition of the habitat (restored or degraded) is an independent variable and the abundance of the target species in the two habitat types is a dependent variable.

Box 2.1 Animal ethics and environmental biology

Researchers and students working with animals must remember that they are dealing not with machines, but with living creatures able to feel distress and pain. Ethical behaviour is defined through codes of practice proposed and enforced by professional and national bodies and by government legislation. The first Australia-wide *Code of Practice for the Care and Use of Animals for Scientific Purposes* was instigated by the research community and produced by the National Health and Medical Research Council in 1969. It is now in its seventh edition.

Under this code, experiments on animals require approval by an independent Animal Ethics Committee (AEC), which evaluates a detailed written proposal from the investigators outlining the expected value of the research outcomes and the welfare implications for the animals involved. Membership must include a veterinarian, an experienced animal researcher, a person with active involvement in animal welfare and a person independent of the institution and with no experience in animal research. AECs encourage replacement of animals with other techniques when they are available, reducing the number of animals used without compromising scientific rigour and refining methods to prevent or at least minimise pain or distress. They may refuse permission for studies they believe to be unethical.

AECs also approve all teaching exercises involving animals at Australian universities. As well as meeting the same welfare standards required for research projects, teaching exercises must also help students attain learning objectives that could not be reached without using animals. You can download the code at <http://www.nhmrc.gov.au/publications/synopses/eA16syn.htm.>

One point in Figure 2.2 requires more explanation, and this is the step labelled null hypothesis. This step is very important in simplifying the hypothesis-testing procedure. For example, if a hypothesis predicts that a target species does not occur in degraded habitat, it can be proven true only if all areas with degraded habitat are searched and the species is shown to be absent. Even then, confirmation would depend on using very reliable survey methods, and on deploying them at the same time so that the species could not move unnoticed between the areas being surveyed. In practice, such a comprehensive survey would be difficult or impossible to implement. Conversely, the hypothesis would be disproved by just a single observation of the target species in degraded habitat at any time during the survey program. Clearly, it is much easier to refute a hypothesis than to prove it, and this is the real value of the null procedure.

Returning to the example of a species that may be declining because of habitat loss, the null hypothesis would be that there is no difference in the abundance of the target species between intact and degraded habitats, or that it is more abundant in degraded habitat. This null hypothesis would then be refuted if sampling showed the target species to be most abundant in intact habitat. An important point here is that rejection of the null hypothesis provides support for the hypothesis and the model on which it was based. Conversely, if the null hypothesis is accepted and retained, there is no support for either the model or the hypothesis, and it is then necessary to go back to the original observations and come up with some new ideas (Figure 2.2). The sampling data collected during the test procedure often allow the original observations to be interpreted in a new light and thus help to formulate new models. When a new model has been identified, the hypothesis construction and testing process is then repeated. If the study involves animals, an important part of designing tests is to consider the welfare of the animals involved in the tests as well as the value of the results that might be gained (Box 2.1).

Three examples of the hypothesis-testing approach are given below. The first shows its value in diagnosing causes of population changes in species of terrestrial mammals. The second illustrates how development effects can be detected. The final example concerns how to distinguish the best option for management when several possibilities are available.

Example 1: Predation by the red fox on black-footed rock-wallabies

The black-footed rock-wallaby (*Petrogale lateralis*) (Figure 2.3) used to occupy large parts of central and western Australia, favouring rocky outcrops that provided shelter and access to foods including grasses, herbs and low shrubs. By the 1930s populations were in sharp decline, often to the point of local extinction. How can we determine the cause, or causes, of this catastrophic decline? There are five steps:

1. *Observations* – Populations remained reasonably stable until the 1930s. Species often decline because their habitat is altered or destroyed, but the outcrops preferred by rock-wallabies provide an excellent buffer against changes caused by agriculture and other forms of intensive land use. Many native mammals in Western Australia suffered declines in abundance before the turn of the 20th century, apparently due to epidemic disease, but there are no historical records of rock-wallabies being affected. Could competition with other species provide an explanation for the declines? Again, this seems unlikely. The most obvious species of potential competitor are the introduced European rabbit (*Oryctolagus cuniculus*) and the feral goat (*Capra hircus*). However, rabbits occur rarely or not at all in rocky environments and goats are absent from many areas where rock-wallabies have declined.

Figure 2.3 Black-footed rock-wallaby. (Source: Lochman Transparencies)

What about predators? Weighing up to 5 kg, black-footed rock-wallabies are too big to be eaten regularly by feral cats (*Felis catus*). Dingoes (*Canis lupus dingo*) kill wallabies, but because wallabies have co-occurred with this predator for about 4000 years it is difficult to understand why dingoes would cause declines only in recent decades.

Whereas cats and dingoes have alibis, the European red fox (*Vulpes vulpes*) emerges as a strong suspect. Introduced to Victoria in the second half of the 19th century, it spread rapidly through mainland Australia in the early 20th century, arriving in many districts at the same time that rock-wallabies and other native mammals disappeared. Foxes readily kill rock-wallabies and pursue them on the gentler slopes of break-aways and other rocky outcrops. Where foxes are absent, such as on offshore islands, rock-wallabies thrive.

2. *Model* – From the available evidence, foxes are the most likely cause of the decline of the black-footed rock-wallaby. There is little to indicate that other factors or processes may be responsible.
3. *Hypothesis* – If foxes are removed from areas where rock-wallabies have declined, the number of rock-wallabies there will increase.
4. *Null hypothesis* – Removal of foxes will have no effect, or detrimental effects, on the number of rock-wallabies.
5. *Test* – Remove foxes from two or more areas, and compare the number of rock-wallabies there with those in equivalent areas where foxes are left alone. In an actual experiment designed to test the fox-predation hypothesis, Jack Kinnear and colleagues in Western Australia used poison baits to remove foxes from two sites where black-footed rock-wallabies persisted in low numbers. They also studied rock-wallabies in three sites where foxes were not baited. After 8 years the number of rock-wallabies in the fox-removal sites exceeded those in the non-removal sites by four- to fivefold (Figure 2.4), and animals there were venturing away from the rock outcrops to feed. These results reject the null hypothesis and support the hypothesis that fox predation is responsible for the decline.

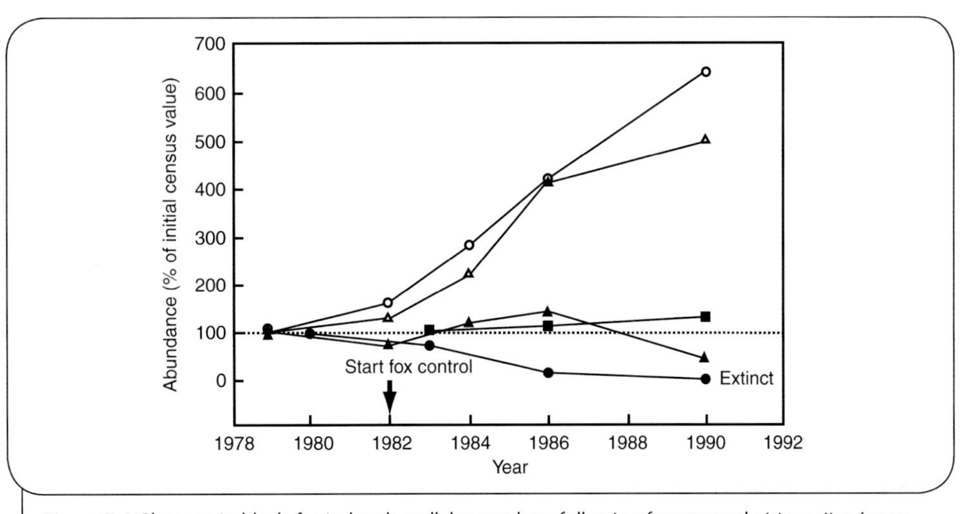

Figure 2.4 Changes in black-footed rock-wallaby numbers following fox removal at two sites (open symbols), and three sites (closed symbols) where foxes were not removed. (Source: Kinnear, J.E, Onus, Onus, M.L. and Sumner, N.R. (1998) 'Fox control and rock-wallaby population dynamics. II An update' *Wildlife Research* 25: 81–88; CSIRO Publishing <http://www.publish.csiro.au/nid/145/issue/253.htm>)

Example 2: Impacts of development on urban wildlife

What are the biological effects of the expansion of Australia's cities? We can expect that any habitat in the path of a new road or construction site will be crushed or removed. Biodiversity at these immediate 'ground zero' sites will be inevitably destroyed. Organisms in the remaining habitat are likely to be disturbed and, as the area of habitat left will have been reduced, we might expect that species requiring large undisturbed areas will gradually disappear. In many parts of the world, attempts are made to predict the impacts of a development and to mitigate them if they seem likely to be severe or to affect threatened species or ecological communities. This evaluation is important in environmental impact assessment. Documents resulting from this process are called environmental impact statements and, if the development is legally contested, they will be crucial evidence.

Let us consider an example. In 2003, a new office building was planned at the top end of Cooper Park, a large bushland park in eastern Sydney. Some of the habitat was targeted for removal for the building itself, but an unspecified area around the construction site was also expected to be disturbed as the building work progressed. How could potential disturbance effects be identified? For simplicity, let us consider a single indicator group, the lizards known as skinks (Figure 2.5).

1. *Observations* – Construction requires access for heavy machinery, employs many people and generates large amounts of rubbish, debris, noise and other disturbance. Two species of garden skinks (*Lampropholis delicata* and *L. guichenoti*) and the water skink (*Eulamprus quoyii*) occur commonly in Cooper Park. As skinks generally avoid disturbance to their habitat, the building process could have negative effects on populations of all species.
2. *Model* – From a range of previous studies and observations, building construction work is likely to have negative effects on populations of skinks living close to the construction site.

Figure 2.5 A species of garden skink. (Source: Lochman Transparencies)

3. *Hypothesis* – If skink populations are sampled before and after construction work has taken place, populations before the development will be larger than after it has taken place.
4. *Null hypothesis* – Construction work will have no effect, or a positive effect, on skink population size.
5. *Test* – Estimate the size of skink populations near the development area before construction disturbance occurs and again afterwards. At the same time, estimate skink population sizes in other, non-disturbed areas at the same times that the populations in the affected area are sampled. Skinks were captured, marked and released near the development site on two occasions before construction started and at a further five sites nearby where no disturbance was expected. All sites were re-sampled three times after construction work finished. On average, skink populations near the construction site fell by 60% following the building work but increased by about 5% at undisturbed sites over the same period (Figure 2.6). These results support the hypothesis that building construction would have negative effects on skink population size, and thus allow rejection of the null hypothesis. Mitigating actions are called for both in this development and for others being planned.

When evaluating the impacts of developments it is common to sample before and after the disturbance has taken place, and to compare changes in the number of species present, their population sizes or other variables measured at these times with observations in undisturbed sites. The approach is a 'before-after, control-impact' (BACI) design. Usually BACI designs are used to investigate the impact of single developments or disturbances so replication is not possible. We explore this issue in greater detail below; here we note

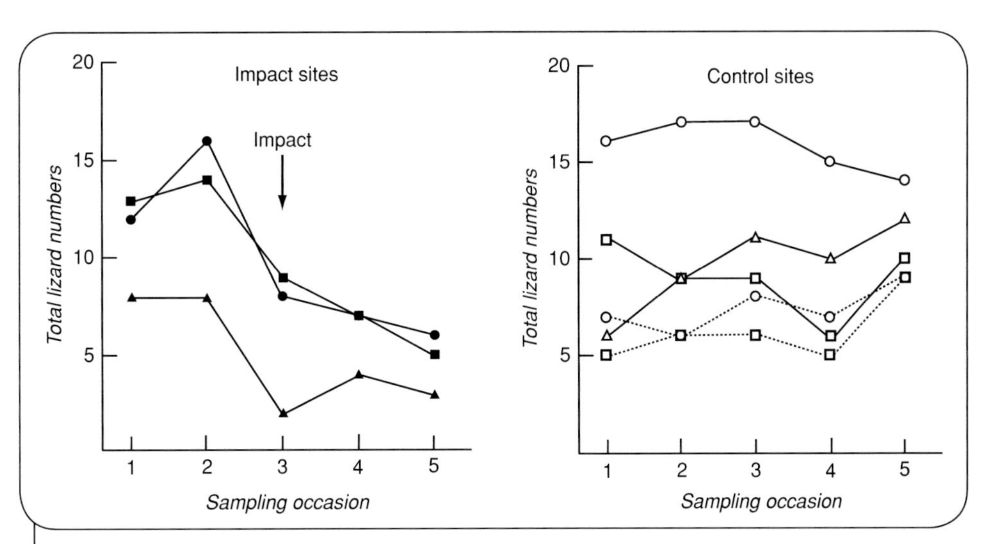

Figure 2.6 Changes in skink numbers in response to construction disturbance. The left-hand figure (closed symbols) shows three disturbed sites, and the right-hand figure (open symbols) shows five undisturbed sites.

only that reliable results can be obtained if there are several sampling occasions before and after the impact has taken place, and if sampling is carried out in multiple control sites that are not disturbed.

Example 3: Reintroduction of threatened species

Many species of plants and animals have disappeared from parts of their original ranges, but persist in small areas. Their prospects of survival are often enhanced if additional populations can be established. Ideally, any reintroduction should take place in an area where the species once occurred and individuals that have been reintroduced should be monitored closely to determine how they fare. Ideally, too, there should not be any threats awaiting reintroduced individuals in their new home. A manager should carry out pilot studies to confirm that all potential threats at a reintroduction site have been identified and controlled. Such studies can utilise small 'probe groups' of individuals before larger breeding groups are introduced, and use the same stepwise process as outlined above to identify the most important threats.

Our example here is the tuan or brush-tailed phascogale (*Phascogale tapoatafa*) (Figure 2.7). This small, insectivorous marsupial is active mostly in trees. It is fast moving and spectacularly agile but, like many other marsupials of similar size, it has declined from large parts of its original range. In Victoria, the species occurs at low density in dry forest and woodland habitats. It disappeared from Gippsland in the eastern part of the state in the late 1960s, making this broad region potentially suitable for a reintroduction. Todd Soderquist, at Monash University, led the program described below and it is instructive to review the stepwise procedure that he used.

1. *Observations* – Phascogales have declined over about 40% of their original range in Victoria, with no evidence of persistence in the Gippsland region. Surveys elsewhere suggest that phascogales are less abundant in fragmented habitats (where habitat is patchy following clearing) and that box–ironbark forests provide more food than the stringybark–silvertop forests predominating over much of Gippsland. Densities of introduced foxes and feral cats are high in Gippsland and both predators eat phascogales. Food availability or predation may have caused the extinction of phascogales in Gippsland and might derail a reintroduction program.

Figure 2.7 Brush-tailed phascogale. (Source: Lochman Transparencies)

2. *Models* – Several equally plausible models can be proposed to explain the phascogale declines: (a) predation from introduced predators; (b) lack of food; and (c) habitat fragmentation. A combination of two factors, or even all three, could be responsible.
3. *Hypotheses* – (a) Survival of phascogales will be higher for animals released into areas where introduced predators have been removed than into areas where they

remain. (b) Phascogales released into stringybark–silvertop forest will have lower foraging success (ability to find food), body weight, growth and condition and ultimately survival than animals released into box–ironbark forest. (c) Phascogales released into fragmented habitat will have lower survival rates than those released into continuous habitat.

4. *Null hypotheses* – (a) Survival of phascogales reintroduced into areas with foxes and feral cats will be the same as, or higher than, that of animals reintroduced into predator-free areas. (b) Foraging success, body weight, growth, condition and survival will be the same for phascogales reintroduced into the two forest types, or higher for animals released into stringybark–silvertop forest. (c) Survival rates of phascogales released into fragmented habitat will be the same as, or higher than, survival rates of animals released into continuous habitat.
5. *Tests* – (a) Remove predators from some areas but not from others, and compare the survival rates of reintroduced groups of phascogales. Feral cats proved difficult to remove, but phascogale survival rates increased where fox numbers were reduced. This allows the first null hypothesis to be rejected. (b) Release groups of phascogales into the two forest types, and monitor their subsequent feeding success, condition and survival. Field sampling showed that the phascogales ate similar foods in the two forest types, but foraging success and body condition were reduced for animals in the stringybark–silvertop forest relative to the box–ironbark. This null hypothesis can be rejected. (c) The appropriate test is to release groups of phascogales into areas of continuous and fragmented forest and monitor survival. Field testing was not undertaken because of the clear support for both the predator and food-based hypotheses. However, several studies have shown that both foxes and feral cats are more active around forest remnants than in intact habitats, suggesting that phascogales would not survive well if released into fragments.

   Taken together, the results of these tests indicate that brush-tailed phascogales are likely to persist if reintroduced into box–ironbark forests where introduced predators, especially foxes, are controlled. Reintroductions into stringybark–silvertop forest with uncontrolled predator populations would fail.

Designs for hypothesis-testing

The above examples provide clear results, but the results of hypothesis-testing are not always clear cut. In general, hypothesis-testing will be most effective if we can design tests appropriately so that we can have confidence in the results. What is needed to ensure sound experimental design? This is a broad topic, but important generalisations can be made.

Replication

Most hypothesis tests require comparisons, whether it is the population size of a target species with and without predators, the same population at two different times, or the cost-effectiveness of alternative management techniques. An important requirement in all such comparisons is that there is replication of the measures being compared. For example, if we suspect foxes of depressing numbers of rock-wallabies, our suspicions are confirmed if the number of rock-wallabies increases at a colony site after foxes are removed. However, if we made censuses of rock-wallabies at just one site, it would be impossible to know whether the number had increased because of fox removal or other reasons such as climatic improvement, a change in food or shelter resources, recovery from disease or pure chance. If rock-wallabies recover in a similar manner at two or more sites after foxes have been removed, the likelihood that other events have driven the increase simultaneously in all sites is reduced and our confidence in the impact of foxes is increased. The use of multiple sites is replication.

Ensuring true replication is often problematic. First, replicates should be independent and not correlated in any way. In the rock-wallaby example, we might be tempted to find a colony site, remove foxes, and then count the rock-wallabies at two or three places at the same colony. This would be an example of pseudoreplication, a situation where multiple measurements are correlated with each other because of their proximity – foxes and wallabies can move between the different places used. It takes a lot more work to find several independent sites and measure the responses of wallabies to fox removal, but this is necessary if we wish to arrive at any general conclusions about fox impacts.

Consider an analogy. In the course of our studies of rock-wallabies, it becomes important to find out the average, or mean, body weight of females and males. We could take the first animal of each sex that we catch and then weigh it 50 times. Or we could attempt to catch 50 different animals of each sex and weigh them individually to get the mean. The first example represents pseudoreplication and is obviously not desirable. The second example provides true replication, and allows us to obtain both the mean weight as well as an estimate of weight variation in the population sample.

Controls

In testing the hypothesis that number of rock-wallabies would increase after the removal of foxes, for example, we saw that we first need replicated sites from which foxes were removed. Removal might require deployment of poison baits, shooting or other methods on a regular basis. If rock-wallabies do increase in number after removal has begun, however, it is difficult to conclude that this was because of the removal of foxes. Perhaps shooting drives away non-target species such as goats, or rock-wallabies eat fox baits with no ill effects and thus use them as extra food. In either case, the number of rock-wallabies could increase because of these artefacts of the experiment rather than in response to the removal of foxes. To overcome this criticism, we need to monitor numbers of rock-wallabies in other sites where no fox control is done. At these sites we should shoot blanks (to copy the noise disturbance effect) and provide baits with no toxin (to allow for the added food effect) at the same times that the real fox removals were taking place at the experimental sites.

These other sites are called controls and they are yardsticks against which experimental treatments are compared. They account for the effects of factors such as climate and other chance changes that might happen even if our experiments are not carried out. Importantly, we should also design controls to cancel out the effects of any artefacts that were introduced by the experimental procedures, so that outcomes of the experiment can be interpreted as being caused only by the factor of interest.

Placement of treatments and controls

The relative placement of control and treatment groups is important, even if all treatment and control replicates are independent. In the rock-wallaby example we saw that foxes were removed from two sites containing wallabies but were left untouched at three other sites. If the two fox-removal sites and the three control sites were a long way apart, any change in rock-wallaby numbers at the removal sites relative to the controls could be caused by the fox-removal treatment or a geographical difference between the two sets of sites. If, for example, the fox-removal sites had received more rainfall than the controls, we could not say whether the increase in rock-wallabies was caused by reduced predation or by increased plant growth and thus food following the rain. The ideal is to have the sites far enough apart to prevent animals moving between them but close enough to minimise differences in weather or other factors.

It is also important that the locations of treatment and control sites should be mixed up. Otherwise, if all the treatment sites are clumped together and all the control sites are clumped together elsewhere, there may be weather or other unknown differences between them. If there are large numbers of replicates, mixing is best achieved by randomising which sites are allocated to experimental treatments and which to controls. True randomisation is best achieved by reference to random number tables or to computer-generated random numbers. Randomisation is particularly appropriate because it removes any subjectivity or even unconscious bias that might occur if the researcher allocated treatments and controls. However, if replication is low, as in the rock-wallaby example, randomisation is less appropriate as it can still lead to the clustering of all experimental treatments together and all controls together. In such cases, researchers can deliberately alternate control and treatment sites. This process is called interspersion.

Control and experimental treatments should be monitored at the same times. For example, if the number of wallabies at the fox-removal sites and the control sites was monitored in different years, any experimental effect on rock-wallaby numbers would be difficult to distinguish from changes in weather and other factors at the treatment and control sites during the two different time periods.

Applying general principles in different situations

Researchers need to keep in mind the species, habitats, processes or other aspects of the system they are studying to ensure that measurements are specific and suitable. For example, if we predict that a species will increase in abundance when a predator is removed, the census technique used to estimate population numbers must be sensitive enough to detect any differences that might occur. The study should also continue for long enough (at least one generation) to allow time for any response to be detected. Measuring and monitoring requirements are likely to vary for each system under study, and for this reason the specific protocol to follow will often depend on the judgment of individual researchers. However, the general principles of replicates, controls and their relative placement must be considered in all studies.

Interpreting the results of hypothesis tests

Now that we have designed a test of a hypothesis and obtained results, how should we interpret the results? Let us say, in the rock-wallaby example, that the number of wallabies increased by an average of 10% in the fox-removal sites compared with the control sites. Is this a real difference, and does it allow rejection of the null hypothesis? A difference of 10% between control and experimental treatments may be taken as enormous by some and trivial by others. Therefore the results of even the best-planned experiments should be evaluated carefully, and there is a very large body of statistical theory available to do this. Statistical evaluation allows us to evaluate the probability of obtaining the observed result by chance and thus get an objective measure of what the difference means.

In much biological and environmental research it is common to find statements of probability (p) along the lines that 'x is greater than y, $p = 0.05$', or 'x and y are no different, but both are less than z, $p < 0.01$'. In these situations x, y and z are mean values for variables that we have measured or estimated as part of a hypothesis test. In the first example, x might be the mean abundance of rock-wallabies at sites from which foxes have been removed, and y the mean abundance at sites where foxes remain. The associated value ($p = 0.05$) tells us that there is a 5% probability of obtaining this result by chance. Because a 5% chance is quite small, the result would usually be interpreted as significant or, biologically, that rock-wallaby numbers increase when foxes are removed. In the second example, x, y and z might be mean efficiency values for three techniques used to detect a rare species, or effectiveness values for methods of pest control. The higher value of z indicates that this method would be preferred, and the low associated probability (< 1%) suggests that the result is most unlikely to be from chance alone. If a null hypothesis had specified x, y and z to be the same, it would clearly be rejected. Statisticians would say that there was a significant difference.

Applying science – the information-theoretic and Bayesian approaches

Researchers still commonly use the hypothesis-testing approach in environmental science, but growing numbers of them are becoming critical of its use. Many statisticians and philosophers of science have also attacked it, arguing that it is too biased by sample size (very large sample sizes can produce trivial but still significant results), that probability values provide no measure of evidence (even very high probability values cannot confirm that a hypothesis is true), and that the null hypothesis is always false (two populations will never be exactly the same). There is concern too that conventionally used levels of significance are arbitrary (there is no logical reason for $p = 0.05$ to be the cut-off point for tests of significance). In practice, environmental managers are often dissatisfied that with the acceptance of a null hypothesis there is nothing upon which to base a decision. In response to these critiques, two other approaches have become increasingly used in environmental science: the information-theoretic approach and the Bayesian approach.

The information-theoretic approach differs from the null hypothesis approach in that it tests the consistency of data against multiple competing hypotheses, not just one. It assesses the 'strength of evidence' for each competing hypothesis rather than comparing 'significance' to 'non-significance' at an arbitrary probability level. Someone using the information-theoretic approach asks the question: 'What is the probability of obtaining the actually observed results under each of the competing hypotheses?' Another tenet of the information-theoretic approach is that when there are multiple hypotheses with similar 'strengths of evidence', the simpler and most parsimonious hypotheses are favoured. This is consistent with the philosophical concept of Occam's razor, which states that the simplest explanation is usually closer to the truth than more complex ones.

Although the statistical and mathematical methods involved in the information-theoretic approach are beyond the scope of this chapter, we will briefly discuss one of the more commonly used measures of evidence for evaluating the merits of competing hypotheses, the Akaike information criterion (AIC). The AIC adjusts the level of support for each competing hypothesis, so that the simplest hypothesis is favoured. Hypotheses with lower AIC scores are more parsimonious and have higher levels of support than hypotheses with higher AIC scores.

Let us examine the information-theoretic approach by using an example of a study of factors that influence the distribution of an endangered native rodent, the broad-toothed rat (*Mastacomys fuscus*). This rat lives in patches of heath surrounding swamps in subalpine areas of south-eastern Australia. Although its diet consists mainly of grass, animals require patches of heath for protection against predators. Three main competing hypotheses can be constructed to explain the distribution of broad-toothed rats in heathland patches: (1) large patches best sustain populations of the rats, (2) patches that are closer together (less isolated) best sustain populations and (3) patches that are both larger and closer together best sustain populations of the rats. Patches of heath were searched for the presence of rats in the subalpine area of Barrington Tops, New South Wales, and the size of each patch and the distance to its nearest patch were measured. Hypothesis 1 had an AIC value of 9.3, hypothesis 2 had a value of 3.1 and hypothesis 3 a value of 3.3. Hence, the most parsimonious hypothesis with the highest level of support was hypothesis 2. We can therefore suggest that less-isolated patches are more likely to support populations of broad-toothed rats. The efforts of environmental managers would be better directed at conserving clumps of neighbouring patches rather than larger areas of heath.

The other method becoming increasingly popular in environmental science is the Bayesian approach. The Bayesian approach is actually older than all the approaches mentioned above, being based on the work of the 18th century English mathematician, the Reverend Thomas Bayes. Its popularity has increased with the advent of more powerful computational methods to cope with its implementation. The Bayesian approach is similar to the information-theoretic approach described above in that it evaluates multiple competing hypotheses. However, it differs from the information-theoretic approach in that it incorporates prior information from previous, related studies to form prior probabilities. The Bayesian method combines these prior probabilities with the probabilities of obtaining the data under the different competing hypotheses to produce updated and more refined levels of support for each of the hypotheses. The probability of one of the competing hypotheses is increased if the data support that hypothesis more than its competitors. Someone using the Bayesian approach asks the question 'What is the probability of each hypothesis being true, given the observed data?' as opposed to the question 'What is the probability of observing the data given that the various hypotheses are true?' The statistical and mathematical methods involved in Bayesian methods are beyond the scope of this chapter (and in fact are very complicated), but you may study them in advanced units. For the moment, it is helpful to understand the approach. The main criticism of the Bayesian method is that the selection of relevant prior information is subjective, and selection of inappropriate 'priors' can lead to misleading conclusions.

Let us consider the Bayesian method by using a study of a carnivorous marsupial, the mulgara (*Dasycercus blythi*). Near Uluṟu in central Australia, the managers of a local resort remove spinifex grass (*Triodia basedowii*) to use as mulch on garden beds. This could have a negative effect on mulgara numbers as they (and their prey) use the spinifex for shelter from predators. Researchers were interested in comparing mulgara numbers in replicated sites where spinifex was removed with sites where it was not removed (control sites). A test of the traditional null hypothesis that 'there will be no difference in mulgara numbers in spinifex removal areas compared to control areas, or increased numbers of mulgaras in the spinifex removal areas' led to a finding of no significant difference and hence acceptance of the null hypothesis.

However, previous studies in the same region had demonstrated that mulgara numbers were higher in areas that had last burnt 11 years ago compared with areas that had burnt within the previous year. The main difference between the two areas was that there was much higher spinifex cover in the areas that had burnt 11 years ago. In this situation, there is considerable prior information to add to the spinifex removal experiment. This allows comparison of three competing hypotheses: (1) spinifex removal will decrease mulgara numbers (as consistent with the previous study), (2) spinifex removal has no effect on mulgara numbers and (3) there is no useful prior information. Incorporating the prior information and using a measure of support called the deviance information criterion (DIC, a Bayesian equivalent of the AIC), hypothesis 1 was found to have the highest level of support. Hence, using a Bayesian approach with prior information we can conclude that the removal of spinifex has a negative impact on mulgara numbers.

Publishing the findings

In Chapter 1 we defined culture as the shared experience of human beings transmitted from generation to generation. Scientific findings can enter culture, but only if they are written down and published. Once published, they are accessible worldwide to other scientists who can read and apply the findings, thus expanding knowledge.

The main method of scientific publication is the peer-reviewed journal article. Once a study is completed, scientists write a report documenting clearly what they did, why they did it, what they found, what the results mean and how they can be applied. These reports are submitted to the editor of a scientific journal in the field. He or she then consults experts in the area. They offer their opinions on the quality of the study design (with careful attention to many of the points covered in this chapter) and the significance of the findings. Their reports may lead to the editor publishing the paper unchanged or, more commonly, returning it to the authors for correction or further work before publication. Sometimes the paper is rejected as having flaws that make the findings unsound and hence unsuitable for publication, or a paper may be redirected to a journal more appropriate to the subject matter. Papers may also be rejected if they have not considered issues of animal ethics (Box 2.1). This process is called peer review and it helps to ensure that published work meets high scientific standards.

Peer review may correct errors, help authors reach sound conclusions from their data before the work becomes widely available, make significant contributions to style and improve readability. Although peer review is not perfect, authors who feel their work was criticised unfairly may submit their work to other journals to seek another evaluation. Work not published in the peer-reviewed literature makes no lasting contribution because it is difficult to access and can be seen as unfinished or even not done. This is why report writing is such an important part of undergraduate education and why your instructors put such emphasis on it!

Chapter summary

1. Good science proceeds in a logical and structured manner. Observations are made, evaluated and explained by careful sifting of evidence. The explanations may then be tested by making predictions about the outcome of any new situation, and refined or rejected if found to be wrong.
2. Human nature means that science cannot be truly objective. However, its claims are always open to test.
3. The hypothesis-testing (hypothetico-deductive) approach begins by making observations of an interesting pattern, event or puzzle, and then proposes explanations (hypotheses) for the observations. We make predictions arising from those explanations and then test them by further observations or by experiments. If the predictions are verified, the explanation is supported, but if they are not, we must seek another explanation.
4. Experiments involving animals cannot proceed without approval of an Animal Ethics Committee.
5. Mensurative experiments take measurements to test predictions, but do not actively change any part of the environment. Manipulative experiments involve actively changing one or more parts of the environment to determine the effects relative to other areas that are left unchanged.
6. Manipulative experiments require:
   - treatment groups, where a single, deliberate change is carried out
   - control groups where the change is not carried out – any differences between control and treatment groups should be caused only by the single factor changed
   - replicates, meaning that there must be more than one treatment group and one control group to allow statistical testing of differences
   - randomisation or interspersion of control and treatment groups to ensure that any difference between them is not caused by changes across the landscape or in time.

   Mensurative experiments have the same requirements, but make comparisons between groups that are already thought to differ in some way.
7. Differences between treatment and control groups are determined by statistical tests, which give the probability that any observed difference is caused by chance. By usual convention, to be accepted as significant a difference must have only a 5% probability or less of having occurred by chance.
8. The information-theoretic approach asks: 'What is the probability of obtaining the actually observed results under each of several competing hypotheses?'
9. The Bayesian information-theoretic approach incorporates prior information from previous, related studies to form 'prior probabilities'. These are combined with the probabilities of obtaining the data under the different competing hypotheses to produce updated and more refined levels of support for each of the hypotheses.
10. Publication is an essential final step if research findings are to make a lasting contribution to human understanding.

Key terms

BACI
control
hypothesis
hypothesis-testing approach
information-theoretic approach
interspersion
manipulative experiment
mensurative experiment
null hypothesis
pseudoreplication
randomisation
replication

Test your knowledge

1. Outline the main steps in the hypothesis-testing approach to science and explain why each is necessary.
2. What is meant by replication and why is it important in experiments?
3. What is meant by controls and why are they important in experiments?
4. How does the information-theoretic approach differ from the null hypothesis-testing approach?
5. How does the Bayesian approach differ from the null hypothesis-testing approach?
6. A firm of biological consultants was engaged to test the effectiveness of a new fertiliser and a new potting mix for growing tomato plants. The consultants prepared 20 pots containing the new potting mix and 20 pots containing clean, washed beach sand. Two tomato seeds were planted in each of the 40 pots. The biologists then placed the 20 pots with the new potting mix in a plant growth cabinet set to 24°C and watered them daily with the recommended solution of the new fertiliser. The 20 pots containing sand were placed in a second growth cabinet, also set to 24°C, and watered daily with distilled water only. After a month the biologists measured the height of the seedlings in each cabinet and found that the seedlings in the new potting mix and watered with the new fertiliser were, on average, 10 cm taller. They concluded that the new potting mix and the new fertiliser were superior to the old potting mix and the old fertiliser. Prepare a critique of this experimental design, paying particular attention to issues of controls, replication, independence and choice of dependent variables.

Thinking scientifically

Consider the case of the changes in populations of flying foxes described at the beginning of the chapter. Using the detailed examples given in this chapter, outline what actions could be taken under the hypothesis-testing approach to investigate this situation. Summarise your points under the headings:

- observations
- models
- hypotheses
- null hypotheses
- tests.

Debate with friends

Assessing the impact of introduced foxes on rock-wallaby numbers required killing of foxes at some sites by shooting and poisoning. Debate the proposition: 'It is ethically wrong to kill wildlife for experimental tests'.

Further reading

Barnard, C., Gilbert, F. and McGregor, P. (2007). *Asking questions in biology: a guide to hypothesis-testing, experimental design and presentation in practical work and research projects*. Pearson Education, Harlow, Essex. 3rd edn.

Gauch, H.G., Jr. (2003). *Scientific method in practice*. Cambridge University Press, Cambridge.

Ruxton, G.D. and Colegrave, N. (2006). *Experimental design for the life sciences*. Oxford University Press, Oxford. 2nd edn.

3

Cell theory I – the cellular basis of life

Carolyn Jones and Mike Calver

Poison the cell, poison the animal

In January 1849, members of Western Australian Surveyor-General John Roe's expedition were forced to halt after crossing the Blackwood River near Kojonup, in the south-west of the state. Their horses were lethargic after browsing some fleshy-leafed vegetation along the route and had to be rested for a day. The horses were lucky to recover. The plants they had eaten contained fluoroacetate, one of the most toxic substances known. Although the poison does not act rapidly, doses of as little as 1 mg per kg of body weight are lethal to a wide range of animals, and some species are killed by even weaker doses. Although Roe's horses fell ill and other introduced European livestock died after browsing these plants, many of the Australian native herbivores in this region are unharmed by eating these plants.

Why is fluoroacetate so toxic? How do some native species overcome this toxicity? Is resistance to fluoroacetate poisoning transmitted from generation to generation? The answers to these questions lie in the properties of cells.

Chapter aims

This chapter covers the common characteristics of all life, the basic structure and functions of cells and the significant differences between the two main groups of organisms, the prokaryotes and the eukaryotes. The fluoroacetate toxicity example illustrates how a knowledge of cell metabolism can be important when dealing with environmental problems.

What is life?

Before we consider the role of cells as the basic unit of life, we should clarify our understanding of life itself. All living things show these properties:

- a cellular structure, meaning that they are composed of one or more basic units of complex molecules enclosed in a membrane
- homeostasis, the ability to maintain a constant internal environment often different from the immediate surroundings
- responsiveness, the ability to detect and respond to changes in their environment
- reproduction, including the inheritance of features from generation to generation
- growth and development over their life span
- a complex metabolism of biochemical reactions sustained by an intake of external energy.

The smallest structural unit showing all these properties is the single cell, composed primarily of four kinds of organic (carbon-containing) compounds (Box 3.1). Organisms may be unicellular (solitary), as occurs in many micro-organisms, or multicellular. In a few cases, such as the marine animals called sponges, cells aggregate with little or no specialisation to form the multicellular organism. Cells of a multicellular organism usually differentiate into different tissues, each with a unique appearance and function. In turn, different tissues aggregate to form organs with specific structures and functions, while sequences of organs with related functions form organ systems. Regardless of the structural and functional complexities of these formations, they are all underpinned by cellular structure.

Box 3.1 The main molecules of life

At the chemical level, the complexity and diversity of living things are built on four broad classes of organic compounds (compounds that contain atoms of the element carbon):

1. *Carbohydrates:* These are molecules composed of carbon, hydrogen and oxygen atoms in the ratio of 1 carbon: 2 hydrogen: 1 oxygen. They include sugars, starches and cellulose. The many C–H bonds in carbohydrate molecules store energy, which can be released to power cellular functions. Simple carbohydrates such as sugars can be transported within multicellular organisms to concentrate energy stores where needed, whereas complex carbohydrates such as starch in plants and glycogen in animals are used for longer-term energy storage. Cellulose in plants is an important structural carbohydrate, giving rigidity to cell walls.
2. *Lipids:* These are the fats and oils. At the molecular level they contain many hydrogen and carbon atoms, but few oxygen atoms. The energy-storing C–H bonds make lipids very effective stores of chemical energy, ideal for long-term reserves of energy. They are also good insulators, maintaining body warmth for a range of animals in cold environments. Lipids are also insoluble in water, so both animals and plants use lipid

layers in the form of wax to reduce or prevent evaporative water loss. Furthermore, when placed in water the lipid molecules aggregate, with the most strongly water-repellent (hydrophobic) ends of the molecule facing inwards, away from the water. This property of lipids makes them a key component in plasma membranes.

3. *Proteins:* Protein molecules contain nitrogen atoms as well as carbon, hydrogen, oxygen and sometimes sulfur. They are built from smaller structural units called amino acids. Proteins perform an extraordinary range of functions for organisms including:
   - enhancing cellular activity (enzymes, which are biological catalysts that increase the rate of key reactions, are proteins)
   - defence (many toxins produced by animals and plants are proteins, and proteins on the cell surface help the organism's defences recognise invading micro-organisms)
   - regulation of transport into and out of cells
   - support
   - movement
   - binding and storage of some important minerals (e.g. ferritin is a protein that binds iron)
   - regulation (important hormones, which are chemicals produced in specific organs and released into the circulation to control activities elsewhere in the organism, are proteins).
4. *Nucleic acids:* These are large, complex molecules encoding for the production of all other cellular components. DNA (deoxyribonucleic acid) contains the specific instructions for cell functioning, whereas two types of RNA (ribonucleic acid) transcribe these instructions and transport them to ribosomes where they are translated into the production of proteins.

Development and precepts of the cell theory

The development of the modern understanding of the cell theory is a fascinating example of scientific method at work. Much of the early work was observational and tied to the invention and constant improvement of the microscope. Some early microscopists made important observations using microscopes with only a single lens. The first compound microscopes (those using two lenses) were developed probably in Holland in the late 16th century. In 1665, the Englishman Robert Hooke used one of these early instruments to examine thinly sliced cork, and published the first observations of cells. By 1809 the Dutchman Anton van Leeuwenhoek had proposed that living organisms were collections of cells, and in 1831 the Scotsman Robert Brown observed a subcellular structure, the nucleus, in plant cells. In 1838 the Germans Matthias Schleiden and Theodor Schwann observed that animal and plant cells shared many similarities and, in 1839, Schwann published his idea that all life is composed of basically similar cells. Later that century another German, Rudolf Virchow, stated that new cells arise by the division of existing cells and not as outgrowths. The basic tenets of the cell theory were then complete:

- All living things are composed of one or more basic units called cells, which are essentially the same in composition.
- All the biochemical processes of life occur within cells.
- All new cells arise from pre-existing cells by cell division.

Improvements in microscopy in the late 19th century, coupled with sophisticated staining techniques to highlight cell structures, led to the discovery of thread-like chromosomes within the nucleus and observations of their behaviour during cell division. Later work during the 20th century added the final important precept that cells contain hereditary information coded in DNA (deoxyribonucleic acid) molecules and transmitted on the chromosomes during cell division. The basic unit of inheritance is the gene, which is a length of DNA. Each gene influences the organism's form and function by coding for the synthesis of a protein important in cell structure or function, or for an RNA (ribonucleic acid) molecule that may activate other genes or form part of cell ribosomes (see below).

Thus the development of the cell theory spanned centuries and involved the work of scientists of several nationalities, not all of whom were mentioned in this brief history. It was only possible because findings were recorded and published. This illustrates the vital importance of publication in scientific advances and the power of human culture to pass ideas from generation to generation.

Cell size

The existence of cells was confirmed following the development of the compound light microscope, which passes light through a thin section of an organism and then uses glass lenses to magnify and project the image. The simplest compound microscope uses a single objective lens to collect light from the sample and a single ocular or eyepiece to focus the image for the eye. Modern compound microscopes can magnify about 1000 times. Although they often have two eyepieces (binocular) to reduce operator fatigue, the image transmitted to each eye is the same (Figure 3.1). The stereo or dissecting microscope uses a light source shone onto the specimen from above. There are two separate optical paths with two objectives and two eyepieces that provide different images to each eye, so the specimen is viewed in three dimensions. This is especially useful for examining small, three-dimensional structures. Magnifications up to 200 000 times are possible with the electron microscopes developed in the 20th century. These produce images either by transmitting a beam of electrons (subatomic particles) through a thin section and using electromagnets to produce an image (transmission electron microscope, TEM), or by directing an electron beam onto the surface of a prepared specimen and creating an image from the electrons emitted from the specimen when struck by the beam (scanning electron microscope, SEM) (Figure 3.2).

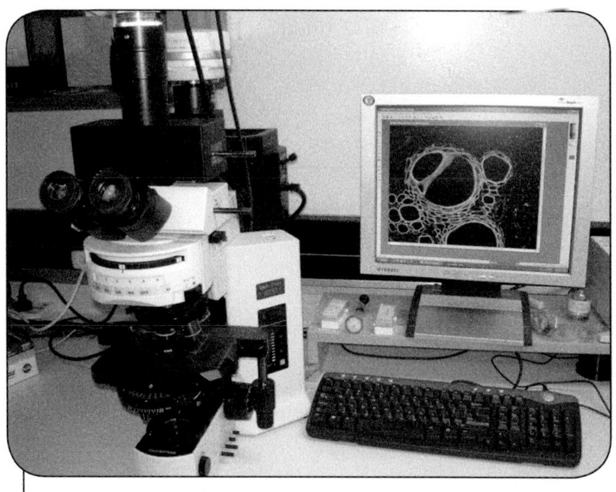

Figure 3.1 A modern compound microscope. The microscope in this example projects its image to a computer screen. (Source: Gordon Thompson)

The development of microscopy was important for our understanding of cells because most cells would need to be enlarged at least tenfold to be seen with the unaided human eye. There are, for example, between 1000 and 100 000 bacterial cells on each tooth in a clean human mouth. An unclean mouth might have up to one billion bacteria on each tooth, so in one mouth there might be more bacteria than there are people in the world. As a further example, a teachers' manual from the 1970s observed that if a 9 L bucket was filled with human ova and a sewing thimble was filled with human sperm, there would be enough sperm and ova to produce every person in the world. Of course, in the early

Figure 3.2 A modern scanning electron microscope. The image appears on a computer screen. (Source: Gordon Thompson)

Table 3.1 Scales used to measure the size of organisms, cells and subcellular organelles.

Scale	Example	Can be seen by
Metre (m)	Human	Unaided eye
Centimetre (cm) = 10^{-2} m	Large insect, very large cell such as an emu egg	Unaided eye
Millimetre (mm) = 10^{-3} m	Large cell such as some fish eggs	Unaided eye
Micrometre (μm) = 10^{-3} mm = 10^{-6} m	Most plant and animal cells measure 10–100 μm Most bacteria and eukaryotic cell organelles measure 1–10 μm	Light microscope Electron microscope
Nanometre (nm) = 10^{-3} μm = 10^{-9} m	The smallest bacteria are just over 100 nm in size whereas viruses are 50–100 nm in size	Electron microscope

21st century one would need a little more of each. Scientists use less imaginative scales to describe the size of cells (Table 3.1), but the essential point is that most cells are very small.

The upper and lower limits of cell size are set by simple logistics. A cell must have sufficient volume to contain all the internal structures it needs to survive and reproduce, setting a lower size limit. It must also have sufficient surface area for the volume of its contents to obtain essential nutrients from the environment and dispose of wastes. As a cell increases in volume, its surface area also increases, but at a slower rate. Therefore large cells have smaller surface area for their volume than small cells, setting an upper limit for cell size. The concept is familiar to any cook. A large carrot will cook faster if cut into smaller sections because the surface area exposed to the heat is greater, although the volume stays the same.

Despite their small size, cells are surprisingly complex. We now describe different types of cells and their internal structures.

Prokaryotic and eukaryotic cells

All living cells are characterised by (1) a surrounding plasma membrane separating the cell's contents from the external environment, (2) cell content known as cytoplasm within the plasma membrane, (3) hereditary material coded in DNA molecules and passed to daughter cells during cell division, and (4) ribosomes, which are subcellular structures that synthesise protein molecules following the directions coded in DNA. Despite these common features, two very different types of cells, called prokaryotic and eukaryotic cells, can be distinguished.

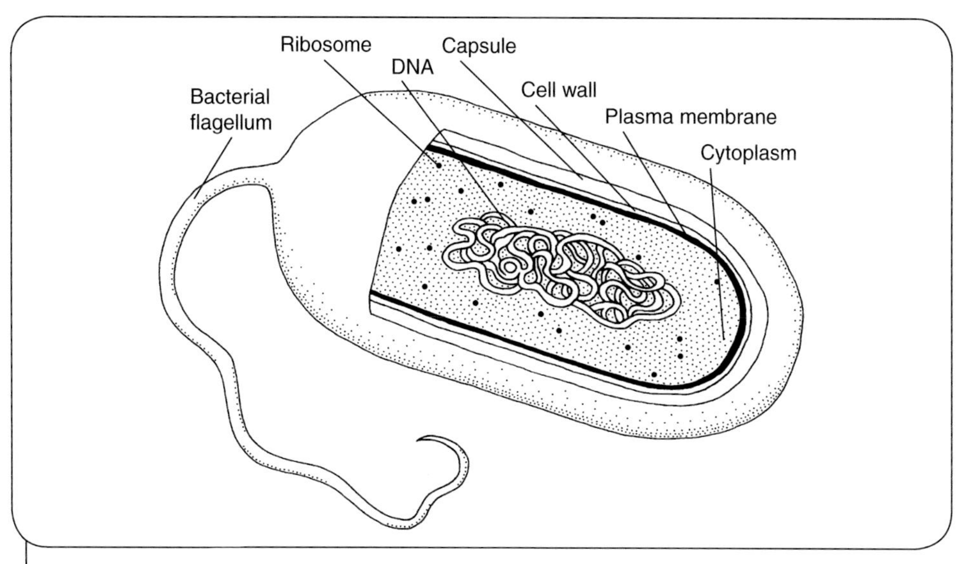

Figure 3.3 The main structures of a prokaryotic cell. (Source: Adapted from a public domain image, Wikimedia Commons. This file is licensed and freely available for use under the Creative Commons Attribution ShareAlike 2.5 License)

Prokaryotic cells

Prokaryotic cells (Figure 3.3) are small, rarely exceeding 10 µm in length. Their plasma membrane is encased within a protective and supportive cell wall. This is sometimes surrounded by a capsule, or sticky outer coat, which provides further protection and may attach the cell to external surfaces. Long, whip-like flagellae may be present, allowing movement in liquid surroundings. The DNA exists as a circular strand coiled within the cell, so a great deal of DNA is packed into a confined space. If stretched out, prokaryote DNA may be hundreds of times longer than the cell itself. Ribosomes present in the cytoplasm produce protein molecules from the instructions coded in the DNA. Despite their simple structure, prokaryotic cells are the basis of an extraordinary diversity of life from the domains Bacteria and Archaea. They are important in many life-sustaining processes on Earth and are covered in more detail in Chapter 9.

Eukaryotic cells

The typical eukaryotic cell is about 10 times larger than a prokaryotic cell. It is much more complex internally and contains up to 10 000 times as much DNA. There are some unicellular eukaryotes, and eukaryotic cells are the basis of the multicellular plants, algae, animals and fungi. The larger size means that eukaryotic cells have a lower surface area to volume ratio than prokaryotic cells, but they compensate by having specialist membrane-bound organelles within the cytoplasm. Many of the chemical processes of eukaryotic cells, called their cellular metabolism, occur within these organelles, which have their own distinct internal conditions optimised for the processes they perform. The experimental study of the structure and function of organelles is an important field in cell biology, often using radioactive isotopes (Box 3.2).

Box 3.2 Radioactive isotopes and stable isotopes in biological research

Just as organisms consist of cells as their fundamental units, matter is also made up of basic particles. Each element consists of unique, tiny atoms, the smallest unit of matter that still keeps all the properties of the element. Each atom in turn comprises more than a hundred subatomic particles, of which only three need concern us here: the proton, the neutron and the electron.

Protons and neutrons are located in the core nucleus of the atom. Protons are positively charged and neutrons have no charge. Negatively charged electrons orbit the nucleus. The number of protons in an atom is a unique number for each element, called the atomic number, whereas the sum of the protons and the neutrons for an atom is its mass number. Unlike the atomic number, the mass number need not be the same for all atoms of an element because some have different numbers of neutrons. These variants in mass number all behave chemically as though they were identical, despite their differences in the number of neutrons in their nuclei, and are called isotopes of the element. Different isotopes are written in chemical shorthand, with the mass number in a superscript followed by the symbol for the element. For example, the most common isotope of carbon has a mass number of 12 and is written ^{12}C. Less common, but of great interest biologically, is another isotope, ^{14}C, which has two extra neutrons in its nucleus. Some isotopes are stable, meaning that their nuclei remain intact, whereas radioisotopes decay spontaneously, releasing particles and energy.

Radioactive isotopes are useful for studying cellular processes because they behave identically to stable isotopes of the same element but their presence is easily detectable. Therefore tracking the path of radioactive isotopes through chemical reactions allows biologists to determine where in a cell particular reactions take place and the pathway of particular atoms through those reactions.

Stable isotopes are very useful in studying animal diets because of two important properties: (1) the isotopic ratios in an animal's tissues reflect the isotopic ratios of its foods and (2) isotopic ratios vary in different locations. By combining information on isotopic ratios in animals and in their environments, biologists can determine what the animals are eating and sometimes also evidence of what caused extinctions (Box 7.3).

While all eukaryotic cells are similar, some differences occur between those of plants, fungi and animals. Plant (Figure 3.4) and fungal cells are enclosed within a thick, slightly elastic cell wall that provides strength and support. Plant cells may also contain chloroplasts, organelles that absorb solar energy and use it to manufacture organic compounds in the process of photosynthesis (Chapter 4). They often also contain a large central vacuole, which stores water, mineral nutrients and other chemicals. This compresses the living cell contents into a narrow layer at the sides with overall a large surface area, so plant cells can reach larger sizes than animal cells with small vacuoles. Centrioles are small structures

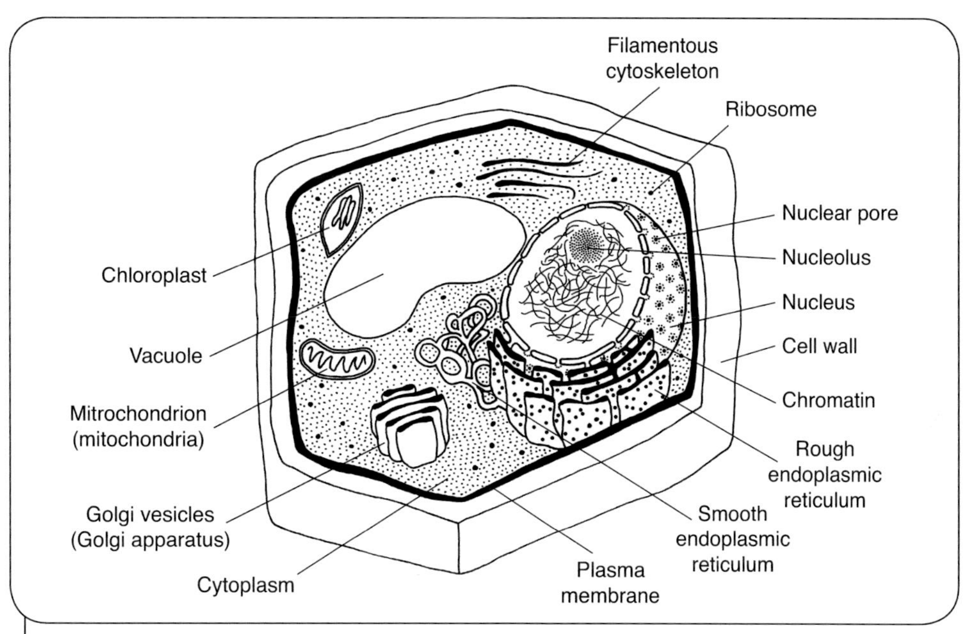

Figure 3.4 The main structures of a plant cell. (Source: Adapted from a public domain image, Wikimedia Commons. This file is licensed and freely available for use under the Creative Commons Attribution ShareAlike 2.5 License)

made of fine protein tubes found in animal and fungal cells and a small number of plant cells. Animal cells (Figure 3.5) may also have whip-like flagellae for movement, but in plants these occur only in the male reproductive cells of some species. We now consider some of the more important organelles in detail.

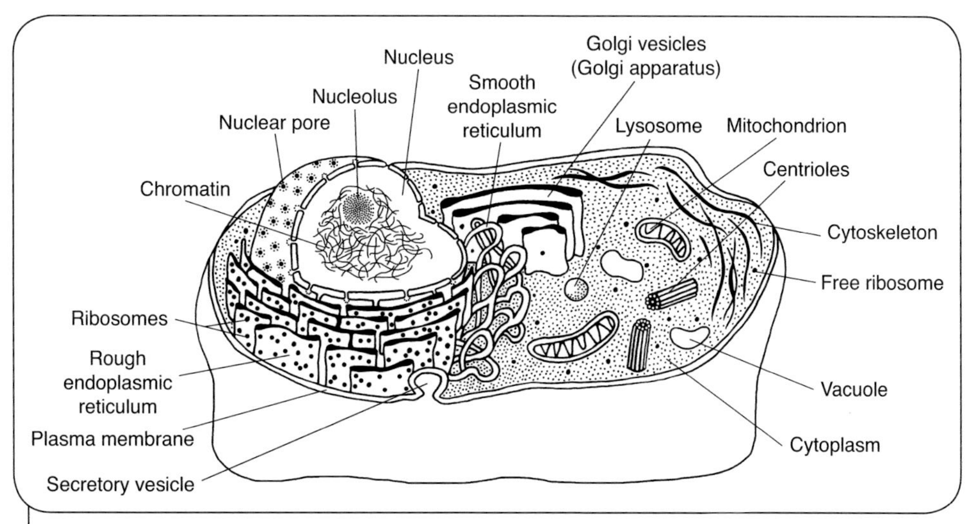

Figure 3.5 The main structures of an animal cell. (Source: Adapted from a public domain image, Wikimedia Commons. This file is licensed and freely available for use under the Creative Commons Attribution ShareAlike 2.5 License)

Functions of eukaryotic cell organelles

Properties of membranes and the role of the plasma membrane

Eukaryotic cells and many of their organelles are enclosed by plasma membranes. These regulate the movement of substances in and out of the cell and its organelles and thereby create the optimum conditions in particular organelles for the functions they perform. Therefore we will begin by considering the structure of these membranes and how substances move across them.

The fundamental structure of a membrane is based on phospholipid molecules, which consist of twin strands of oil attached to a phosphate head. These molecules are organised into two layers, with the phosphate groups that are water-attractant oriented to the outside and the oil strands, which are water-repellent, to the inside of the cell or organelle (Figure 3.6). Interspersed with the phospholipid molecules are different proteins, often unique to particular organelles within a cell and performing specialist functions. Some are enzymes or biological catalysts for biochemical reactions, accelerating the reactions on the membrane without being consumed themselves. Others may receive chemical messages from other cells, or assist in moving molecules of dissolved substances across the membrane. Many pollutants or poisons such as lead actively interfere with either enzymes or messengers (Box 3.3).

Plasma membranes are supple and fluid with a consistency like cooking oil. This means that the phospholipids and the proteins are always moving, although at any one time the proteins may form patterns or mosaics. This gives the modern name 'the fluid mosaic model' to our understanding of membrane structure.

Water and small dissolved molecules may cross the plasma membrane either by passive transport with no energy cost to the cell, or by active transport that requires energy. Larger molecules are moved into cells by endocytosis or out of cells by exocytosis. An understanding of all types of transport is important to explain how cells interact with their environment.

Passive transport

If you add a few drops of a dye to a glass of water, the dye will ultimately diffuse (spread out) evenly throughout the glass. At the molecular level, the particles of dye are moving

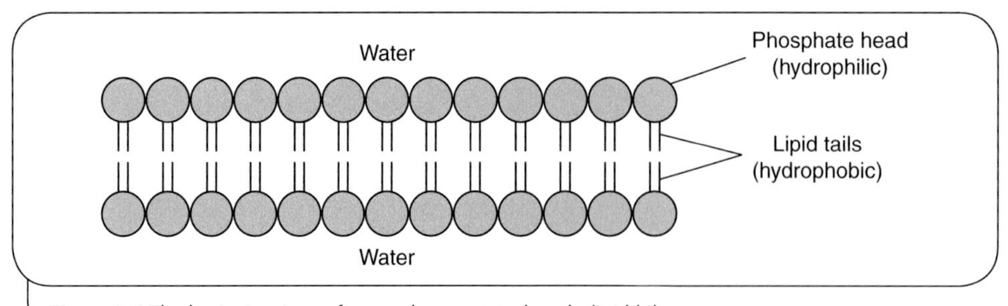

Figure 3.6 The basic structure of a membrane as a phospholipid bilayer.

Box 3.3 Lead poisoning in the Californian condor

With a weight of nearly 9 kg and a wingspan of almost 3 m, the Californian condor is one of the largest birds on the North American continent. Individuals mate for life and may live for over 50 years. At the time of European settlement of North America, the range of Californian condors extended along the mountains of the Pacific coast, but their numbers and range decreased so markedly that the species was listed as endangered under US federal legislation in 1967 and under Californian legislation in 1973. The last wild individuals were trapped for a captive breeding program in the 1980s and releases of captive bred animals into the wild began in 1992.

Habitat destruction and shooting by landholders played their part in the decline of the Californian condor but, unexpectedly, surviving birds are at risk from lead poisoning. They are carrion feeders and ingest fragments of lead shot when feeding on the carcasses of game animals killed by hunters. Splintered bones are especially attractive to the birds as a calcium source and they mistake lead fragments for bones.

Lead has many harmful effects on animals. It interferes primarily with proteins that normally bind zinc or calcium as part of cellular metabolism, displacing the zinc or calcium and preventing normal functioning. One well-known example is anaemia caused by the inhibition of the zinc-binding enzyme ALAD (δ-aminolevulinic acid dehydratase), which is important in the production of red blood cells. Other zinc-binding proteins that may be inhibited by lead are involved in transcription (the production of cellular proteins from the instructions coded in the nucleus of the cell); this could explain developmental abnormalities in animals with lead poisoning. Lead also interferes with the role of calcium in neurotransmission, the sending of messages between nerve cells. Poisoned animals become disoriented.

Solutions to protect the Californian condor include establishing reserves where hunting is prohibited and the promotion of alternatives to lead shot to reduce the chances of poisoning from carcasses. Bullets made from copper or a combination of tin, tungsten and bismuth (TTB) are now available and may become cheaper following the US Army's decision to switch to lead-free bullets. These measures, in association with protection of habitat and reintroductions from the captive breeding program, are essential for the long-term survival of the species. At present, released Californian condors are recaptured regularly to monitor their blood lead levels and provide treatment if required.

randomly (Brownian motion), without the addition of an external source of energy, until the even distribution is achieved. The particles do not remain in one place from then on. They continue to move at random, but the overall result is that even mixing is maintained. Now imagine a glass separated into two halves by a porous membrane. If dye is added to one side of the membrane it will still spread evenly throughout the glass, with particles moving through the pores of the membrane. A range of dissolved substances can cross the plasma membranes of cells in the same way, assuming that there are transport

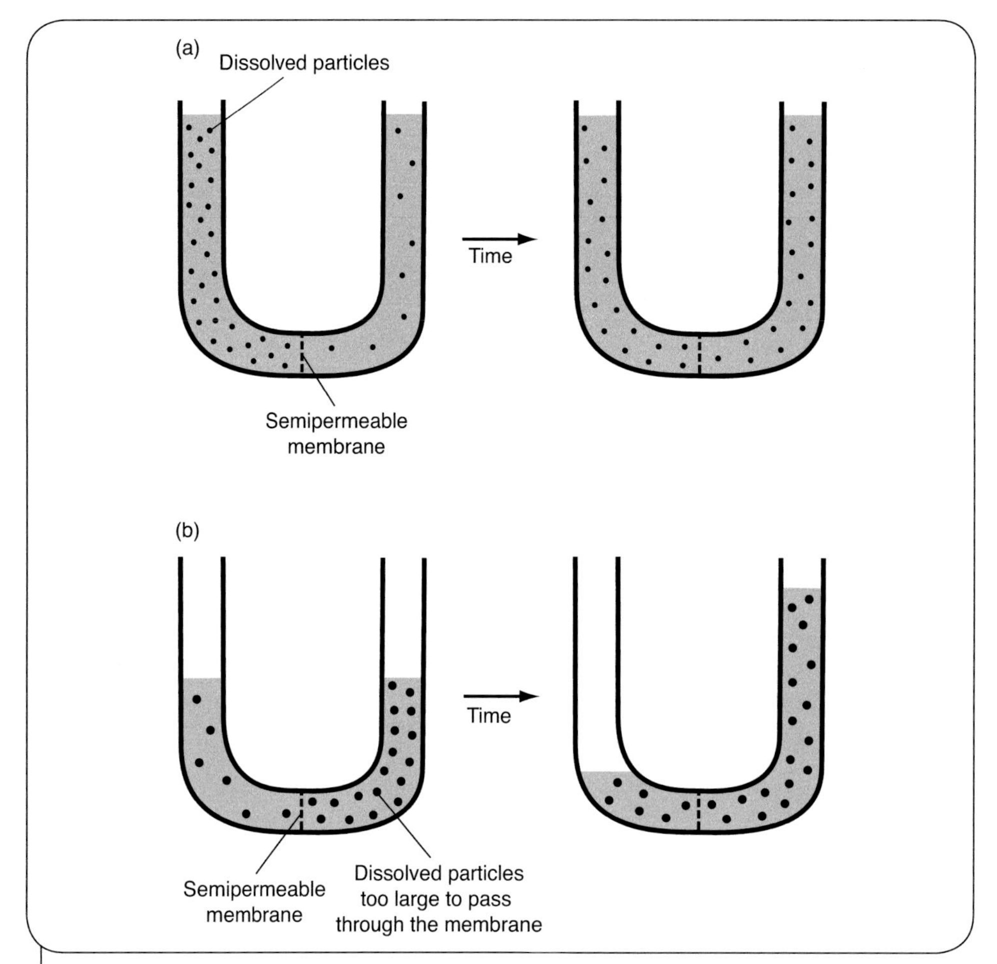

Figure 3.7 (a) Molecular movement during passive transport. Dissolved substances diffuse along their concentration gradients across the membrane. (b) Molecular movement during osmosis. Large dissolved molecules cannot cross a semipermeable membrane, but smaller water molecules can. The overall movement of water molecules across the membrane equals the concentrations on both sides of the membrane. (Source: B. Cale)

proteins in the plasma membrane to provide a route. The cell does not use energy, so this simple diffusion process is called passive transport (Figure 3.7a).

Osmosis, or the diffusion of water across a membrane, is a special case of passive transport. It occurs when a membrane is permeable to water (the solvent) but not to substances dissolved in water (the solutes). If the solute concentrations (osmolarities) are unequal on the two sides of the membrane, solute particles cannot diffuse across the membrane to equalise them. However, water molecules can cross the membrane and will do so until the solute concentrations on each side of the membrane are equal (Figure 3.7b).

Osmosis may cause problems for organisms (Figure 3.8). In situations where solute concentrations are equal inside and outside cells (isotonic), there is no overall tendency

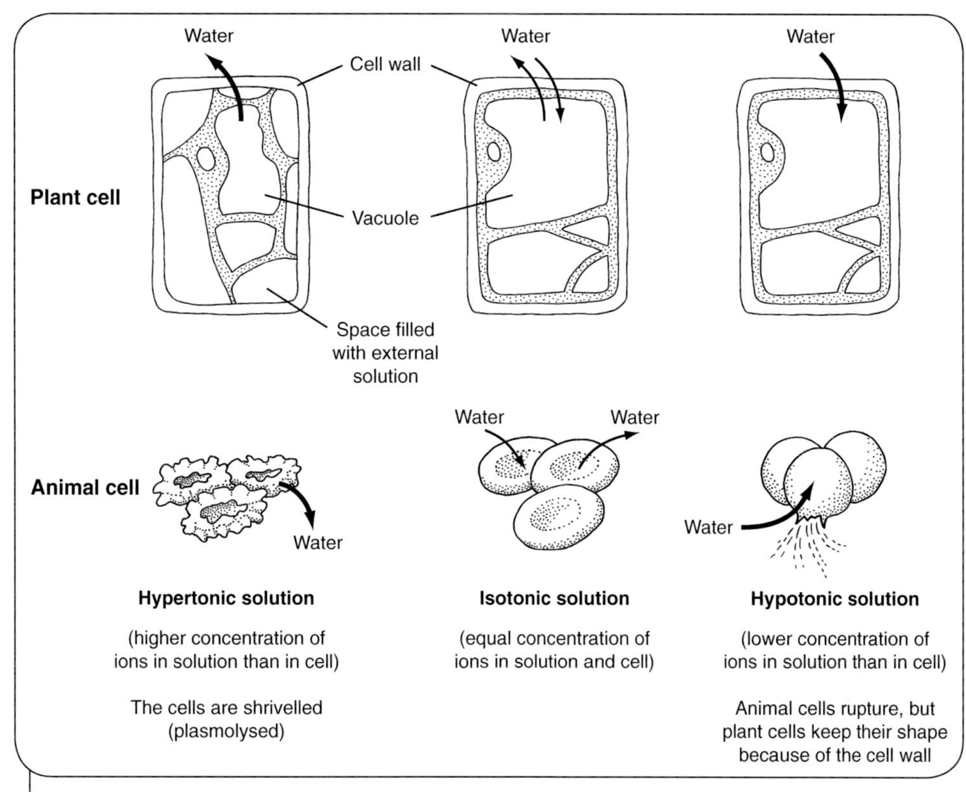

Figure 3.8 The responses of plant and animal cells to isotonic, hypertonic and hypotonic conditions depend on the presence or absence of a cell wall. The arrows indicate water movement under the different conditions. In hypertonic conditions salt concentrations are higher outside the cell than in it. The animal cells shrivel as water leaves the cell by osmosis. Plant cells do not shrivel because of the cell wall, but the plasma membrane contracts within the cell wall. In isotonic conditions salt concentrations outside and inside the cell are equal. Plant and animal cells neither gain nor lose water overall. In hypotonic conditions salt concentrations are lower outside the cells than inside and water enters the cell. An animal cell eventually bursts from the build-up of pressure within. A plant cell is prevented from bursting by its cell wall, but becomes swollen or turgid. (Source: B. Cale)

for the cells to gain or lose water. Water molecules are continually crossing the plasma membrane, but the rates of movement into and out of the cell are the same. Should the concentration of solutes be greater inside the cell (hypotonic), it will gain water by osmosis, and if it lacks a firm outer cell wall it may burst if the internal pressure becomes too great. Conversely, if the solute concentrations outside the cell are greater than those inside (hypertonic), the cell will lose water by osmosis and shrivel. Animals living in hypertonic or hypotonic environments must osmoregulate, or control their water balance to survive. Plants are in a slightly different situation because of their cell walls. They thrive best in hypotonic conditions where there is an input of water that keeps the cell turgid, while the cell wall prevents it from bursting. However, in hypertonic conditions plant cells lose water, the plasma membrane contracts from the cell wall and plant shoots wilt.

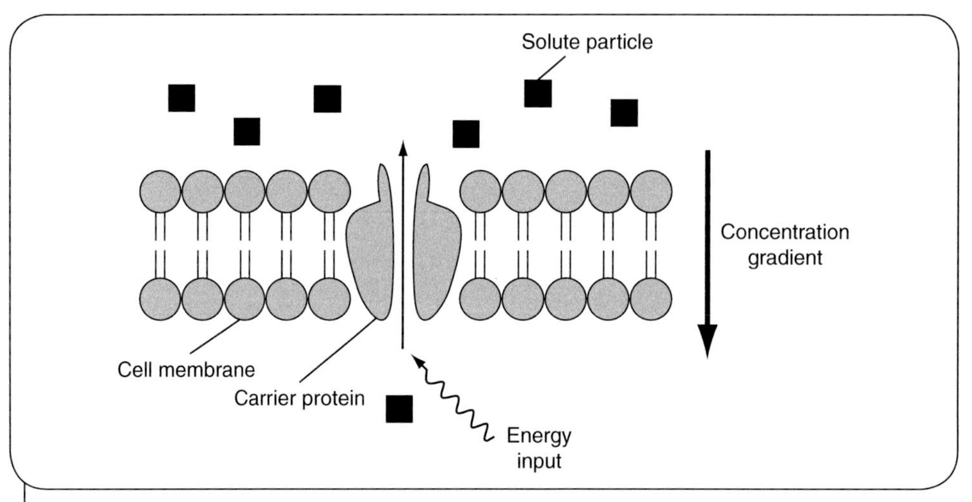

Figure 3.9 Molecular movement during active transport. A carrier protein in the cell membrane uses energy to pump in ions against a concentration gradient. (Source: B. Cale)

Active transport

Cells often maintain internal concentrations of small molecules (such as mineral nutrients) higher than those of their surroundings, against the direction of diffusion. This requires energy to distort the shape of a transport protein in the membrane, causing it to attract a solute particle and release it on the other side of the membrane (Figure 3.9).

Exocytosis and endocytosis

Large particles do not pass through the plasma membrane when exiting or entering a cell. In the process of exocytosis, a cell removes large particles by first enclosing them in a membrane-bound bag called a vesicle. The vesicle fuses with the plasma membrane and spills its contents outside the cell. Endocytosis is the reverse process for uptake of large particles. The plasma membrane folds inward and then closes the top of the fold, creating a vesicle within the cell with the large particles trapped within.

Organelles of the endomembrane system

Many structures inside eukaryote cells are also based on membranes. Some of these form an interconnected endomembrane system within the cell, joined either physically or by the transfer of vesicles. These are the organelles we consider next.

The nucleus

The nucleus is enclosed within its own double membrane and contains the DNA molecules and attached proteins in strands called chromatin. The DNA codes for the production of the cell's other proteins, which have key roles in all cellular activities (Box 3.4). Proteins have roles in plasma membranes and as enzymes regulating cellular processes, so it is through protein synthesis that the DNA regulates the cell. Transcribing the DNA

Box 3.4 Is the hereditary material DNA or protein?

Although good observations based on microscopy described cell structure, careful experimentation was critical in understanding the function of different cell components. One important example was the key experiment conducted in 1952 by the biologists Martha Chase and Alfred Hershey to determine whether DNA or protein was the substance in the nucleus that controlled cell function. To do this, they experimented with a particular type of virus called a bacteriophage.

Viruses meet some of, but not all, the conditions of life. They consist of a nucleic acid in a protein coat, but cannot reproduce unaided. Instead, part of the virus must penetrate the DNA of a living cell and direct the cell to produce more virus. Viruses that infect bacterial cells are called bacteriophages.

Chase and Hershey reasoned that if the protein component of a bacteriophage contained the hereditary material, then that protein should penetrate a bacterial cell, leaving the nucleic acid outside. However, if the nucleic acid was the hereditary material, then the nucleic acid should penetrate the bacterial cell, leaving the protein outside. In one experiment the bacteriophage DNA was labelled with an isotope of phosphorus, ^{32}P, while in a second the protein was labelled with an isotope of sulphur, ^{35}S. In the first experiment most ^{32}P was detected inside the infected bacteria, whereas in the second experiment most ^{35}S was found outside the infected bacteria. Furthermore, in both experiments the bacterial cells went on to produce more virus. It was the DNA in the bacteriophage, not the protein, which entered the bacterial cells and caused them to produce more virus. This is but one example of how experimentation, as well as observation, has developed our understanding of cellular processes.

code in the nucleus produces messenger RNA (mRNA), which passes through the pores in the nuclear membrane and proceeds to ribosomes in the cytoplasm or attached to endoplasmic reticulum. Using transfer RNA (tRNA), the ribosomes translate the DNA code into a linear sequence of amino acids, which collectively form proteins (Chapter 5). Components of the ribosomes are also produced in a special region of the nucleus, the nucleolus.

The DNA of eukaryotic cells is not in a circular strand as it is in prokaryotes. Instead, the chromatin is in several extended threadlike strands that, before cell division, coil and condense to form visible chromosomes. Different eukaryotic species have different numbers of chromosomes in the nuclei of their cells – for example, a human being has 46 chromosomes, a platypus has 26 and a silky oak has 20. They usually exist in matched pairs called homologous chromosomes, one of which came from the organism's mother (a maternal chromosome) and one from the father (a paternal chromosome). Each member of a pair carries genes controlling the same inherited characteristics. Matching genes relating to the same characteristic on homologous chromosomes are called alleles and their position on the chromosomes is called a locus.

Endoplasmic reticulum

The endoplasmic reticulum, often abbreviated to ER, is a network of flattened tubes providing a large surface area of membrane within the cell. Rough endoplasmic reticulum is studded with ribosomes that have two main functions: producing more membrane and modifying proteins for transport to other organelles or secretion by the cell. Membrane produced by rough endoplasmic reticulum may increase the size of the organelle, or be moved to other organelles. Proteins for secretion from the cell are first packaged in transport vesicles, which move first to the Golgi apparatus (or Golgi body) (described below) for further processing before secretion. Smooth endoplasmic reticulum lacks ribosomes and produces lipids for use by the cell or for secretion.

Golgi apparatus

The Golgi apparatus is the only cell organelle named after its discoverer, Camillo Golgi. It consists of a layer of flattened sacs and in function can be thought of as a series of docks, factories and warehouses. Substances synthesised in the rough and smooth endoplasmic reticulum are received in a 'dock' on one side of the Golgi apparatus, modified in 'factories' and sorted in 'warehouses' for transport to various destinations. Final products leave the Golgi apparatus via a 'shipping dock', which is different from the part of the organelle that received the substance originally. Substances are moved within the Golgi apparatus in transport vesicles, which also transport the substances to the plasma membrane for secretion or to become part of other organelles. Cells actively involved in secretion have large numbers of Golgi bodies.

Lysosomes

These organelles arise from the rough endoplasmic reticulum and the Golgi apparatus and contain enzymes that break down food particles and recycle damaged organelles. They also play a role in controlled cell death, or apoptosis. This process is important throughout development, where some cells die in a planned sequence. For example, death of the cells between the fingers in human embryos gives rise to distinct digits rather than flaps or paddles.

However, the importance of lysosomes is shown most clearly by the health problems arising when they malfunction. For example, in the hereditary human condition Tay-Sachs disease, sufferers' lysosomes lack an enzyme for the breakdown of a group of lipids called gangliosides. These build up in the nerve cells of the brain causing physical and mental problems and death within the first 4 years of life.

Vacuoles

Vacuoles are membranous sacs with a variety of functions. For example, in some unicellular eukaryotes contractile vacuoles are important in osmoregulation. Those living in fresh water are hypotonic to their environment and take up water constantly by osmosis. The contractive vacuoles collect excess water from the cytoplasm and expel it through a pore to the outside, preventing the cell from swelling and bursting. More commonly, vacuoles store chemicals, waste products or nutrients for later release. Plant cells often have a large central vacuole that may occupy up to 80% of the cell volume. Plant vacuoles may also work as lysosomes. Animal cells have smaller vacuoles.

The vacuoles are the last component of the interconnected membranous organelles forming the endomembrane system of cells. However, there are very important organelles that are not part of this system and we consider these next. Two of these, the chloroplast and the mitochondrion, are energy-converting powerhouses. Others are involved in protein synthesis or in cell support and movement.

Energy conversion: chloroplasts and mitochondria

Energy is the capacity to perform work. The energy of motion is called kinetic energy. Heat (based on the movement of particles) and light (electromagnetic radiation visible to the eye) are two familiar examples. By contrast, potential energy is stored energy that an object has because of its position or structure. Thus a book on a shelf has potential energy that is released if it falls and molecules have chemical energy in their chemical structure. Energy can be neither created nor destroyed (the first law of thermodynamics), although it can be changed from one form to another. However, energy transfers are not 100% efficient (randomness increases, the second law of thermodynamics). For example, in burning wood to heat water not all of the energy released is heat (some will be light, for instance) and not all the heat is transferred to the water. All life depends on conversion of energy sources to forms usable by organisms, and the two cellular organelles that do this are the chloroplast and the mitochondrion.

The chloroplast

Solar, geothermal, wind, tidal and many other forms of energy are abundant on the Earth, but not in a form readily usable by organisms to power their cellular reactions. The most common solution is that provided by the chloroplast, the organelle in some eukaryotic cells that converts solar energy (light energy from the sun) into chemical energy in sugar molecules in the process of photosynthesis (Chapter 4). Animals and fungi lack chloroplasts and do not photosynthesise, but most still depend indirectly on photosynthesis because their food sources can almost all be traced back to photosynthetic organisms (the bacteria and animals living around hydrothermal vents are an exception; see Box 4.1).

The mitochondrion

Once the chloroplast converts solar energy into the chemical energy within sugar molecules, cells are still presented with the problem of releasing that chemical energy in a form that can power cellular reactions. The mitochondrion (plural: mitochondria) does this by converting the chemical energy of organic molecules such as sugars into the chemical energy of the molecule ATP (adenosine triphosphate) that then powers work within the cell. Oxygen is required for this process, called aerobic respiration. Limited amounts of ATP are also produced in the cytoplasm in the absence of oxygen (anaerobic respiration) as a precursor to aerobic respiration in the mitochondria. We consider ATP and respiration in more detail in Chapter 4.

Cells use ATP continuously and rapidly, consuming and regenerating ATP molecules millions of times a second. Clearly, any disruption to production will be lethal to the cells and to the organism they comprise. The fluoroacetate that sickened Roe's horses as described at the beginning of the chapter does exactly that, although the precise mechanism of toxicity is complex and not yet fully understood. One possibility is inhibition of two enzymes critical to the production of ATP so the cells cannot produce the energy they require for normal functioning. Another is that transport of critical molecules into and out of the mitochondria is disrupted, leading to toxic effects. Resistant native reptiles may protect themselves by detoxifying fluoroacetate, but resistant native mammals cannot do this and use a different mechanism not yet understood fully.

DNA in chloroplasts and mitochondria

Chloroplasts and mitochondria are unusual in having their own loop of DNA, coding for proteins specific to their special functions. However, they are not autonomous and also rely on proteins coded in the nuclear DNA. The presence of DNA suggests that these organelles may have evolved from prokaryotes that became integrated in larger eukaryotic cells. In support of this idea are the facts that they are similar in size to prokaryotes, their DNA is organised in a ring and divides similarly to DNA in prokaryote reproduction. This widely accepted theory of the origin of chloroplasts and mitochondria is called the endosymbiont theory (endo: inside; symbiosis: two or more species living in direct contact), and is discussed further in Chapters 7 and 10.

Ribosomes

The instructions to produce proteins are encoded in the DNA in the nucleus, but assembly occurs outside the nucleus at the ribosomes. These are complexes of a particular form of RNA known as ribosomal RNA (rRNA) together with several key proteins. Under the electron microscope ribosomes appear as small, round bodies. Those associated with rough endoplasmic reticulum produce proteins that function in plasma membranes or are exported from the cell. Ribosomes free in the cytoplasm produce proteins that function in the cytoplasm.

Cytoskeleton and movement

Networks of protein fibres called the cytoskeleton extend throughout the cytoplasm of eukaryotic cells for support and movement. They anchor organelles, provide tracks for organelle movement and are at the core of cilia and flagellae, which are protuberances from the cell enclosed in extensions of the plasma membrane that provide locomotion. Cilia are short, hairlike in appearance and numerous, whereas flagellae are longer, whip-like and fewer in number.

Cell division

Body cells (somatic cells) of an organism reproduce by dividing in the process of mitosis, with one cell giving rise to two daughter cells. The process is complex because the cell's DNA, either the circular DNA of prokaryotes or the chromosomes of eukaryotes, must also be copied, packaged and allocated to the daughter cells. In addition, in almost all cell divisions the cytoplasmic organelles are equally distributed between the daughter cells. As the daughter cells grow, new organelles are synthesised. The details of cell division and heredity, together with some of their environmental consequences, are given in more detail in Chapters 5, 6 and 9.

Chapter summary

1. Living things are cellular, homeostatic and responsive, with complex metabolisms. They reproduce, grow and develop over their lives.
2. Living things are made of four classes of organic compound: carbohydrates, lipids, proteins and nucleic acids.
3. The smallest unit displaying all the properties of life is the individual cell.
4. The development of the cell theory shows the scientific method in action, especially in relation to observation, experimentation and publication.
5. Cells are the fundamental units of life and all organisms are made up of one or more cells.
6. Cells arise only by the division of other cells under current conditions on the Earth.
7. Most cells are small, typically rounded and about 10 μm in diameter (they would need to be enlarged 10 times to be visible to the human eye). Cell size is limited by diffusion, which is the process by which substances enter and exit the cell and move within it. Small cells have a large surface area for their volume, and efficient diffusion.
8. Prokaryotic cells from domains Bacteria and Archaea possess ribosomes for protein synthesis and DNA to transmit hereditary information, but do not have their DNA in membrane-bound nuclei or other membrane-bound organelles.
9. Eukaryotic cells are the basis of many unicellular and most multicellular organisms. They possess ribosomes for protein synthesis, a distinct nucleus surrounded by a nuclear membrane enclosing paired chromosomes that carry the DNA, and other specialised organelles bounded by membranes.
10. Important structures in all eukaryotic cells include the:
    - nucleus, which isolates and organises DNA
    - ribosomes, which synthesise proteins
    - endoplasmic reticulum, which modifies new proteins and makes lipids
    - Golgi bodies, in which proteins and lipids are packaged for secretion or for use within the cell
    - cytoskeleton, which aids in movement, shape and organisation
    - plasma membrane, which regulates movements of substances into and out of the cell.
11. In addition to the features common to all eukaryotic cells, plant cells have chloroplasts for converting light energy to chemical energy, commonly a large storage vacuole and a cell wall.
12. In addition to the features common to all eukaryotic cells, animal cells have centrioles visible during cell division.
13. Mitosis is a form of cell division that produces two cells, each with the same amount of genetic material as the parent. It occurs in normal growth and wound healing.

Key terms

active transport
Brownian motion
cell wall
centrioles
chloroplast
chromatin
cilia
cytoskeleton
deoxyribonucleic acid (DNA)
diffusion
endocytosis
endoplasmic reticulum

endosymbiont theory
eukaryote
exocytosis
flagellae
germ cell
Golgi apparatus
homeostasis
hypertonic
hypotonic
isotonic
life
locus
lysosomes
meiosis
metabolism
mitochondrion
mitosis
nucleus
organ
osmolarity
osmosis
plasma membrane
prokaryote
ribonucleic acid (RNA)
ribosome
semipermeable membrane
somatic cell
symbiosis
tissue
vacuole

Test your knowledge

1. Produce a summary table showing the features common to all eukaryotic cells and indicate whether they occur in both animal and plant cells.
2. Produce a summary table showing the similarities and differences between prokaryotic and eukaryotic cells.
3. Produce a summary table showing the main organelles of eukaryotic cells and their functions.
4. What are the three central precepts of the cell theory?

Thinking scientifically

Giardia duodenalis is a single-celled, eukaryotic organism living in the guts of mammals and birds. This organism is unusual because it lacks several cell components present in most other eukaryotic cells.

Cellular component	Present in many eukaryotic cells	Present in *Giardia*
Nucleus	Yes	Yes
Vacuole	Yes	Yes
Ribosomes	Yes	Yes
Mitochondria	Yes	No
Endoplasmic reticulum	Yes	No
Golgi bodies	Yes	No
Centrioles	Yes	Yes
Cilia	Yes	No

What are the functions of the four cell components missing from *Giardia* and how might *Giardia* survive without them?

(Source: This question is adapted from Curriculum Council of Western Australia, *Western Australian Tertiary Entrance Examination 2005*.)

Debate with friends

Which (if any) of these suggestions should be enforced to help the conservation of the Californian condor?

1. Ban hunting in California condor habitat.
2. Replace all lead ammunition sold with TTB ammunition.
3. Impose a 'conservation tax' on ammunition to fund condor conservation.

Further reading

Becker, W.M., Kleinsmith, L.J. and Hardin, J. (2006). *The world of the cell.* Pearson Benjamin Cummings, San Francisco.

Harris, H. (2000). *The birth of the cell.* Yale University Press, New Haven.

Nielsen, J. (2006). *Condor: to the brink and back: the life and times of one giant bird.* Harper Collins, New York.

4

Cell theory II – cellular processes and the environment

Carolyn Jones and Mike Calver

Power to the people

Wentworth shire in rural Victoria may become the site of one of Australia's most ambitious engineering projects. A private company plans to build a 1 km high solar tower surrounded by a 5 km wide greenhouse to generate electricity to power up to 200 000 homes. If it proceeds, the large-scale venture will be a commercial version of a 200 m tall, 50 kW prototype built near Manzanares, south-eastern Spain, in 1982. It ran with minimal maintenance for 7 years, delivering power night and day to the local grid.

The principles of solar tower technology are simple. Hot air is generated from solar energy in a glass or polycarbonate greenhouse surrounding the tower. Within the tower, temperatures fall by 1°C per 100 m of altitude, so the air at the top of a 1 km-tall tower will be about 10°C cooler than at the base. Heated air entering the tower from the surrounding greenhouse increases the temperature differential, so the hot air rises by convection, just like the updraught in a chimney. The rising air spins wind turbines mounted in the tower to generate electricity. At night, heat stored in solar cells during the day is released to continue heating air and powering the turbines. Although the capital cost of building a solar tower is high, the final product is non-polluting and very cheap to run and maintain.

Solar towers show how human ingenuity can transform solar energy into forms suitable for our use. But using solar energy is hardly original. Green plants and some prokaryotic organisms have done that for millennia, converting solar energy into organic compounds in the process of photosynthesis. In turn, the energy stored in those organic compounds can be released in cellular respiration to power cellular functions.

Chapter aims

In this chapter the important processes of photosynthesis and respiration, which make solar energy available to all organisms, are described. This requires an appreciation of how energy can be trapped, stored and released from chemical bonds so a knowledge of the key molecules and metabolic pathways is important.

Uses and sources of energy for organisms

Cells use energy for a wide range of functions including:

- movement of the whole cell
- movement of cell organelles
- building complex molecules
- maintenance and repair
- active transport
- cell division (mitosis and meiosis)
- transmission of electrical impulses (nerve cells).

Given that the laws of thermodynamics tell us that energy can be neither created nor destroyed and that transfers of energy from one form to another are not 100% efficient, cells must have an ongoing external energy supply to survive.

The sun is the direct or indirect external energy source for most organisms on land and in the surface water of oceans and freshwater bodies (chemosynthesis is a much rarer alternative occurring around hydrothermal vents in the deep ocean; see Box 4.1).

In the process of photosynthesis, solar energy is trapped in the chloroplasts and converted to chemical energy in the form of the bonds in ATP (adenosine triphosphate) molecules. The energy in ATP is then used to produce organic molecules such as carbohydrates, using CO_2 as the sole carbon source. Photosynthetic organisms are called

Box 4.1 The chemosynthesis alternative

In cold, dark waters deep beneath the Pacific and Atlantic oceans, hydrothermal vents arise near the junctions of giant plates of the Earth's crust. As these plates move apart, sea water rushing into the gaps created in the ocean floor is heated as high as 400°C by the molten magma beneath. The superheated water rises back into the ocean, but cannot boil because of the great pressure. Dissolved minerals including iron, sulfides, barium, calcium and silicon are carried out with the heated water, precipitate on contact with the cold surroundings and sometimes form chimneys. Surprisingly, life thrives in these hostile environments, drawing its energy for producing organic compounds not from the sun but from the chemicals arising from the vents. This is the process of chemosynthesis.

Energy for chemosynthesis is generated by reactions between O_2 and inorganic chemicals such as hydrogen sulfide. A simple word equation for the reaction is:

carbon dioxide + water + oxygen + hydrogen sulfide organic compounds + sulfuric acid

Chemosynthesis appears independent of sunlight, but it still requires oxygen that is produced elsewhere as a by-product of photosynthesis. Chemoautotrophs, which use chemosynthesis, are the basis of deep-ocean communities clustered around the hydrothermal vents. They are the exception to the rule that all ecological communities on the planet rely on photosynthesis to fix solar energy in a form useable by organisms.

autotrophs (literally 'self-feeding') and, with the addition of nitrogen from inorganic sources such as nitrates, they biosynthesise all the organic molecules they need. Autotrophs make up about 20% of all known multicellular species and there are also many unicellular examples. The other 80% of organisms that cannot photosynthesise are heterotrophs. They obtain energy by ingesting autotrophs or other heterotrophs and accessing the energy stored in the bonds of their organic molecules. These organic molecules are also their source of carbon. Although only autotrophs photosynthesise, both autotrophs and heterotrophs use cellular respiration. This breaks down organic molecules to carbon dioxide and water and converts the chemical energy released into the high-energy chemical bonds of ATP molecules. ATP molecules function as energy currency in cells, powering activities that require energy input. The interrelationships of photosynthesis and respiration are shown in Figure 4.1.

ATP molecules consist of the nitrogenous base adenine, the attached sugar ribose and three linked phosphate groups (Figure 4.2). These phosphate groups are negatively

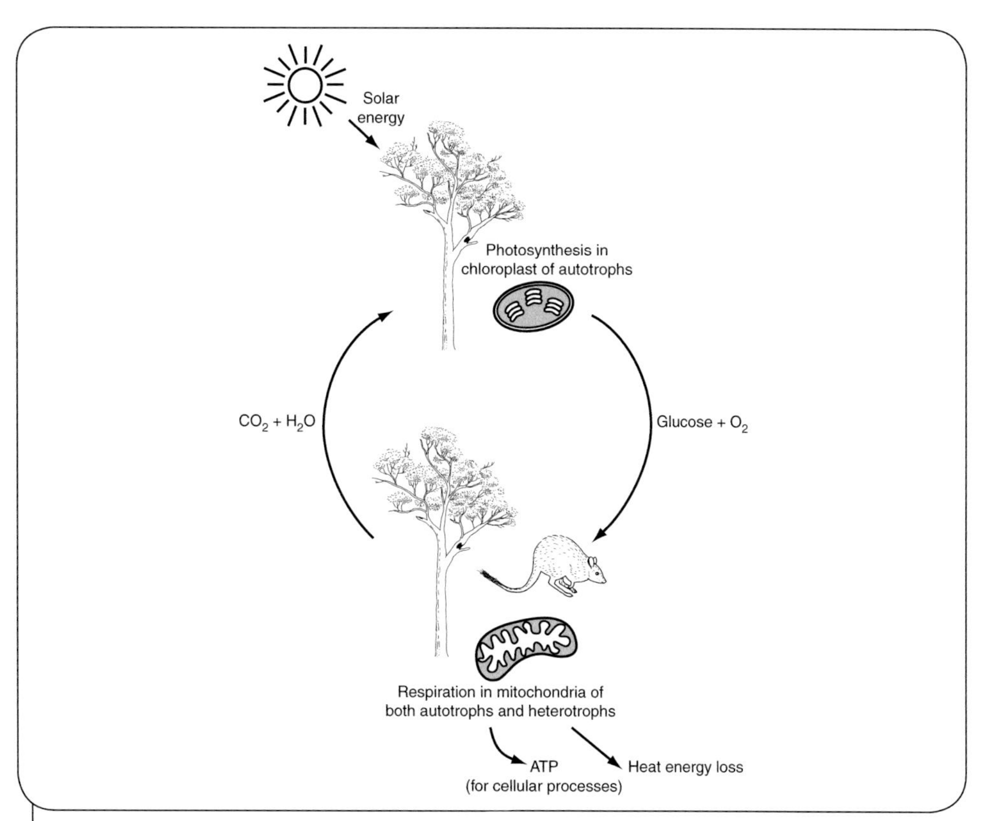

Figure 4.1 Photosynthesis and respiration are interlinked because of their inputs and products. Both autotrophs and heterotrophs can generate energy for cellular processes by breaking down organic molecules in respiration in their mitochondria in the presence of oxygen. This releases carbon dioxide and water, and some energy is lost as heat. Only autotrophs can produce organic molecules from carbon dioxide and water, using solar energy converted to chemical energy by their chloroplasts. (Source: B. Cale)

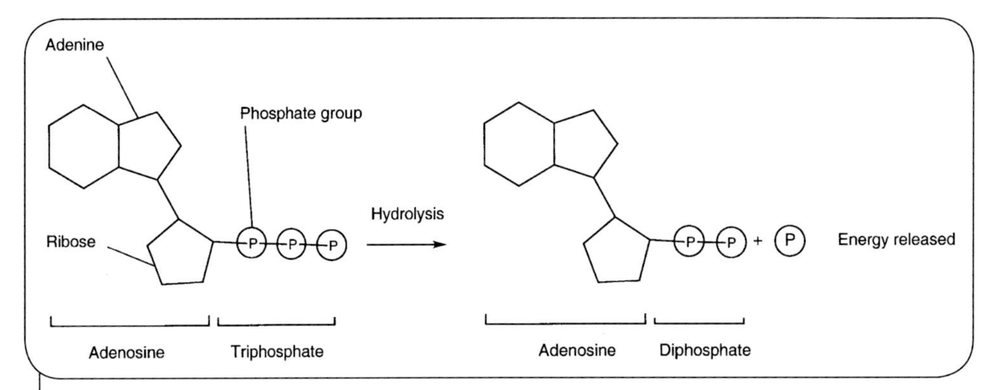

Figure 4.2 ATP molecule. Breaking the bond to the third phosphate group in a chemical reaction involving water (hydrolysis) produces a free phosphate group and a remaining molecule called ADP (adenosine diphosphate), with a release of energy from the broken chemical bond. (Source: B. Cale)

charged and hence repel each other, so the chemical bond that links them is very strong. When it is broken much energy is released. Breaking the bond to the third phosphate group in a chemical reaction involving water (hydrolysis) produces a free phosphate group and a remaining molecule called ADP (adenosine diphosphate), with a release of energy from the broken chemical bond. The transfer of a phosphate group from ATP to another molecule with release of energy is called phosphorylation and is the basis of most energy-consuming work in cells (Figure 4.2). Cells of both autotrophs and heterotrophs convert ADP to ATP continuously, using the energy released from organic compounds such as sugar. In eukaryotic cells, this process occurs mainly in the mitochondria.

The rest of this chapter is concerned mainly with the energy flow involved in the production of ATP. First, we examine how photosynthesis transfers solar energy into the bonds of ATP molecules and then uses the ATP to produce glucose. Secondly, we turn to how cellular respiration releases the energy stored in the bonds of organic molecules and packages it in ATP. Lastly, we take a 'big picture view' of how photosynthesis and respiration are involved in the flow of energy through the biosphere.

Photosynthesis

Landmarks in the study of photosynthesis

Our understanding of photosynthesis developed over several centuries through a combination of careful observation and experimentation, aided by the production of more powerful microscopes and the development of new techniques such as the use of radioisotopes to trace the fate of atoms and molecules in chemical reactions. It involved scientists from many countries and the sharing of findings and technical innovations through publication.

By the 17th century scientists realised that plants could live and grow in containers with minimal changes in the volume of the soil, so plants clearly did not convert the soil into plant tissue. In the 18th century British scientists established that air, light and water were needed for plant growth and that green plants gave off oxygen. Swiss scientists then discovered that carbon dioxide had to be present if plants were to produce oxygen, and also that plant growth required uptake of water as well as carbon from carbon dioxide. In the late 19th century Theodor Engelmann made the next major advance. In a series of elegant observations he used careful microscopy and the habits of oxygen-seeking bacteria to demonstrate that oxygen was released from the chloroplasts of green plants exposed only to particular wavelengths of light (Box 4.2). Thus by the dawn of the 20th century carbon dioxide and water were known as inputs of photosynthesis whereas sugars, oxygen and water were known as products. The German Otto Warburg made the first major 20th century advance in understanding photosynthesis by deducing from the experimental findings of British scientists that there were two distinct classes of photosynthetic reactions: those that required light (the light-dependent reactions) and

Box 4.2 What wavelengths of light power photosynthesis?

Light moves as a wave of oscillating electric and magnetic fields with the wavelength measured in units called nanometres (nm). The shorter the wavelength, the greater is the energy. The light visible to humans is only the small part of the spectrum between 400 and 740 nm. Some other organisms can see in other parts of the spectrum (insects, for example, can see in the short wavelength – ultraviolet – part of the spectrum).

When light strikes a molecule, the energy is either lost as heat or it is absorbed by the molecule. Pigments are molecules that absorb light well in the visible spectrum, and chlorophylls and carotenoids are the most important of the plant pigments for photosynthesis. In combination, they absorb light in the range of about 400–525 nm (the violet region of the spectrum) and 625–700 nm (the red region of the spectrum). Theodor Engelmann first demonstrated this phenomenon in an elegant experiment he conducted in the late 19th century.

Engelmann reasoned that if photosynthesis occurred equally in all parts of the visible spectrum, then sections of a green plant illuminated by different parts of the spectrum should all produce similar amounts of oxygen. To test this he used a light microscope with a prism mounted above the light source to split the light into a spectrum of colours. He placed a strand of the freshwater alga *Spirogyra* on a slide, arranging it across the colours of the spectrum. Finally, he added oxygen-seeking bacteria to the slide because they would concentrate in areas where the oxygen production was highest. The bacteria did not spread out evenly across the filament but concentrated in the regions of red and violet light, indicating that those wavelengths of light were mainly absorbed during photosynthesis (Figure B4.2.1).

continued ›

Box 4.2 continued ›

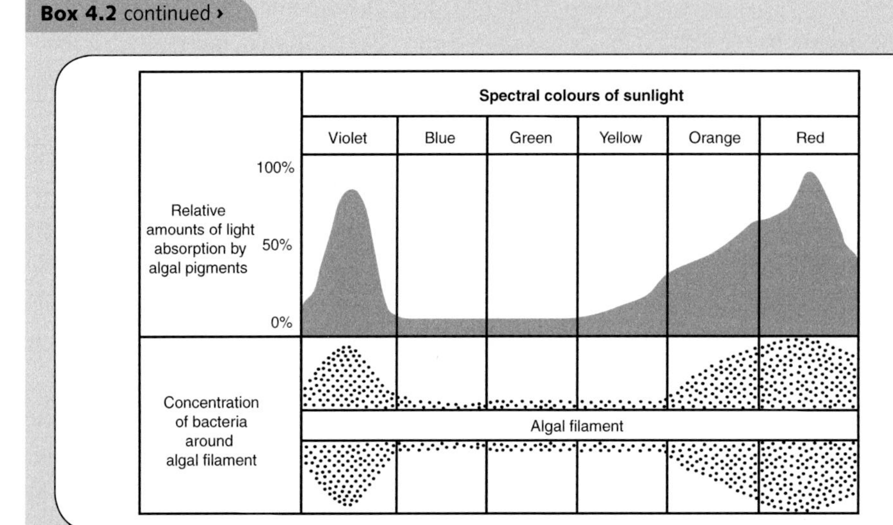

Figure B4.2.1 In an early photosynthesis experiment, a strand of a freshwater alga was placed on a slide, arranging it across the colours of the spectrum. Oxygen-seeking bacteria added to the slide concentrated in areas of high oxygen production, indicating where photosynthesis was occurring. The bacteria concentrated in the regions of red and violet light, indicating that those wavelengths of light were mainly absorbed during photosynthesis (Source: Adapted from Brown, D.W. and Sewell, J.J. (1984) *Australian biology test item bank*. Australian Council for Educational Research, Melbourne. Reproduced by permission of the Australia Council for Educational Research.)

those that did not (the light-independent reactions). Later, the development of radioisotope techniques showed that the oxygen produced during photosynthesis originated from the input of water molecules, the water produced contained oxygen atoms from the carbon dioxide input and hydrogen atoms from the water input, and the sugar molecules produced contained carbon and oxygen atoms from the carbon dioxide input and hydrogen atoms from the water input.

In combination, these scientific studies give us this picture of the broad nature of photosynthesis:

- Using solar energy, autotrophs make sugar molecules from water and CO_2, releasing oxygen as a by-product. The equation is:

$$6CO_2 + 12H_2O + \text{solar energy} \rightarrow C_6H_{12}O_6 + 6H_2O + 6O_2$$

- Pigments for trapping solar energy occur in the chloroplasts and photosynthesis takes place there.
- Photosynthesis occurs in two distinct stages: energy-trapping reactions (also known as the light-dependent reactions) and the synthesis of sugar from CO_2 (also known at the light-independent reactions).

We will now consider the structure of the chloroplast and these two processes of photosynthesis in more detail.

The structures and organelles for photosynthesis

The leaf

Although all green parts of a plant contain chloroplasts and can photosynthesise, chloroplasts are concentrated in the mesophyll tissue within leaves where numbers can reach 500 000 mm^{-2} of the surface area. Carbon dioxide reaches the mesophyll tissue through tiny pores on the leaf's surface called stomates, which connect to air spaces within the leaf. Oxygen exits in the reverse direction. Water is supplied by transport tissue called xylem, whereas phloem, another transport tissue, removes the sugars produced in photosynthesis (Figure 4.3a).

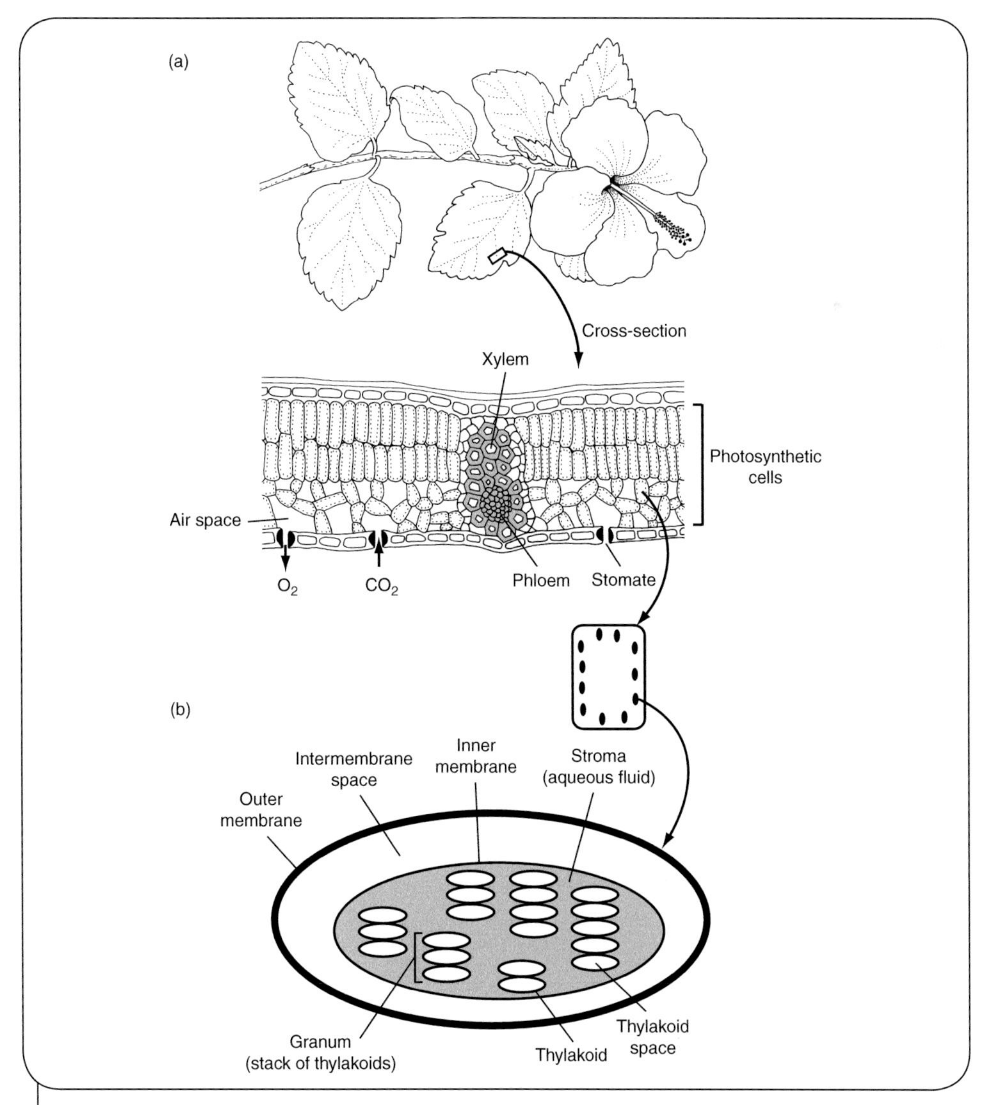

Figure 4.3 Photosynthetic structures. (a) Leaf cross-section showing air spaces, transport tissue (xylem for water and phloem for sugars) and photosynthetic cells. (b) Chloroplast detail. (Source: B. Cale)

The chloroplast

Chloroplasts are flat and disc-like. Internal membranes separate them into three major partitions (Figure 4.3b). Externally, each chloroplast has an outer and an inner membrane, with a small space between them making the 'first partition'. The inner membrane encloses the 'second partition', filled with a fluid called stroma. This is where sugars are made from carbon dioxide and water during photosynthesis. Within the stroma are stacks of interconnected hollow discs called thylakoids, which may be organised in stacks called grana (singular: granum). The contents of these discs, the thylakoid spaces, are the 'third partition' of the chloroplast. Their membranes contain the chlorophyll molecules that trap solar energy in the light-dependent reactions of photosynthesis and power the sugar synthesis occurring in the stroma.

There are slight variants in the structure of the chlorophyll molecule known as chlorophyll a, chlorophyll b, chlorophyll c1, chlorophyll c2 and chlorophyll d. Many photosynthetic organisms also have accessory pigments, which absorb specific wavelengths of light and transfer the light energy to chlorophyll molecules. They include the red, orange and yellow carotenoids and the phycobilins. The types of accessory pigment and chlorophyll present vary in different groups of autotrophs.

Energy-trapping or light-dependent reactions

Energy-trapping reactions occur in two different light-collecting units, photosystem I and photosystem II, in different areas of the thylakoid membrane. They are named in order of their discovery, although photosystem II actually functions first. The steps can be traced in Figure 4.4. The process begins in photosystem II where solar energy is used

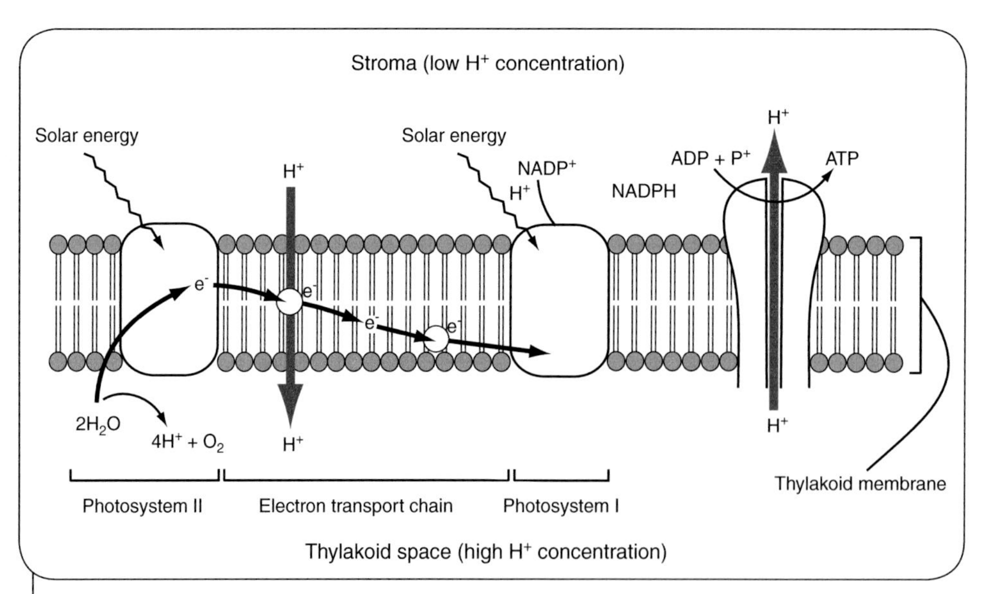

Figure 4.4 The light-dependent reactions of photosynthesis produce NADPH and ATP for use in glucose synthesis in the Calvin cycle. (Source: B. Cale)

to split a water molecule to produce oxygen, hydrogen ions (H^+) and energised electrons. Oxygen diffuses out of the leaf. The energised electrons are shuttled down a series of carrier molecules to photosystem I via a series of oxidation-reduction (redox) reactions, in which one substance loses electrons (oxidation) and another gains them (reduction). Potential energy is lost in each redox reaction and the carrier molecules use the electrons' energy for the active transport of H^+ ions from the stroma into the thylakoid space. In photosystem I, solar energy re-energises the energy-depleted electrons, which are then used to combine $NADP^+$ and H^+ to produce NADPH (nicotinamide adenine dinucleotide phosphate) molecules. The electrons are thus maintained in a high state of potential energy and are used later to produce glucose.

There is now a high concentration of H^+ ions in the thylakoid space, creating a concentration gradient across the thylakoid membrane. This is a store of potential energy. A useful analogy is that of water in a dam that, if released, can be used to turn turbines to produce hydroelectric power. Thus when the H^+ ions move across the membrane to the stroma via ATP synthase (an enzyme that functions as a molecular motor), energy is converted and used to add a phosphate group to ADP, producing ATP. The ATP is then available for use in the light-independent reactions to make glucose from CO_2.

Light-independent reactions: synthesis of sugar in the Calvin cycle

The ATP and NADPH produced in the light reactions of photosynthesis provide the energy and high-energy electrons respectively needed for the production of glucose from CO_2 in the light-independent reactions that make up the Calvin cycle (named after its discoverer, Melvin Calvin). The fate of different molecules can be traced in Figure 4.5. In the first stage of this cycle, the enzyme rubisco combines three molecules of CO_2 with three molecules of the five-carbon sugar RuBP to produce

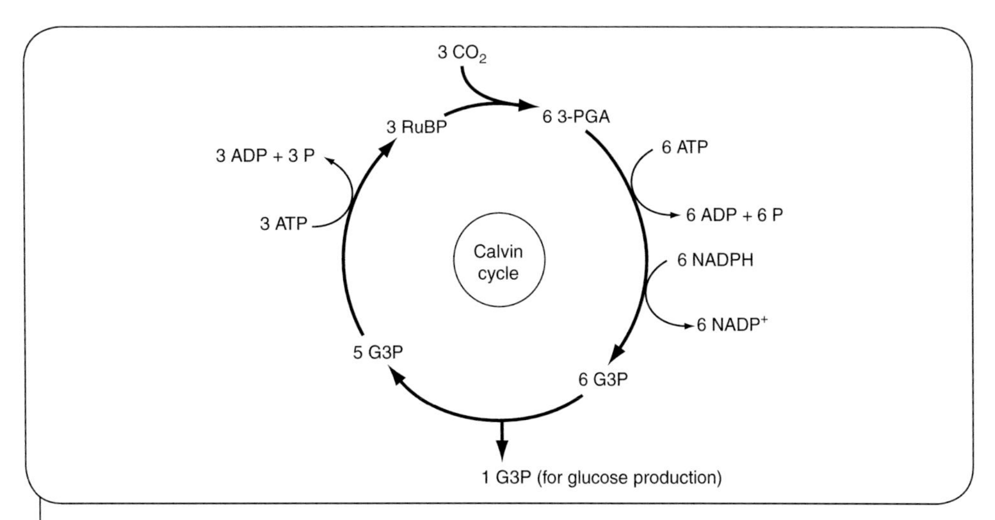

Figure 4.5 In the Calvin cycle the NADPH and ATP produced in the light reactions are used to produce glucose. (Source: B. Cale)

six molecules of the three-carbon sugar PGA. Energy from six ATP molecules and high-energy electrons from six NADPH molecules are then used to convert the six molecules of PGA to six molecules of G3P. One molecule of G3P leaves the cycle for conversion into glucose, whereas the other five are converted to three molecules of RuBP, using a further three ATP molecules produced in the light reactions. This completes the cycle. Figure 4.6 shows the overall integration of the energy capture and sugar synthesis reactions.

An interesting twist on the Calvin cycle and the light reactions is that photorespiration reduces the efficiency of photosynthesis (Box 4.3). Some species called C_4 and CAM plants modify the basic photosynthetic pathway to reduce this inefficiency and photosynthesise when water is restricted (Box 4.4).

Fate of the glucose

Photosynthesis occurs in the leaves, but autotrophs need energy in all the cells of their bodies. They achieve this by transporting the glucose produced in the Calvin cycle to all cells of the plant. Some is used to generate ATP in cellular respiration to power cellular functions, some is combined into structural molecules such as the cellulose in cell walls

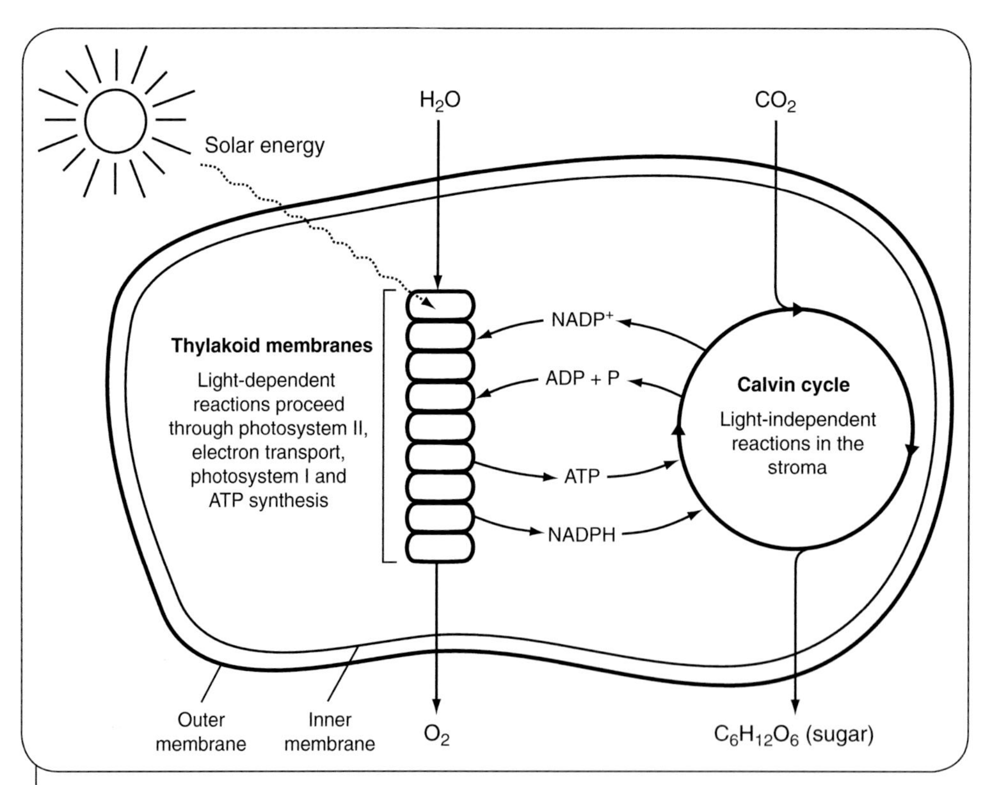

Figure 4.6 An overview of photosynthesis. The light reactions occur on the thylakoid membranes within the chloroplast and produce NADPH and ATP. The Calvin cycle in the stroma of the chloroplast uses the NADPH and ATP from the light reactions to produce glucose. (Source: Adapted from Figure 10.21, p. 198 of Campbell, N.A., Reece, J.B. and Meyers, N. (2006). *Biology*. Pearson Education Australia. 7th edn. Copyright © 2005 by Pearson Education Inc. Reproduced by permission.)

Box 4.3 Photorespiration

At the beginning of the Calvin cycle the enzyme rubisco attaches CO_2 to RuBP in carbon fixation. However, rubisco can also attach O_2 to RuBP, producing CO_2 and consuming ATP in the process. This is called photorespiration because it can only occur in the light when RuBP is available from the Calvin cycle.

Whether carbon fixation or photorespiration occurs depends on the concentrations of O_2 and CO_2 in the immediate environment. Although rubisco has a higher affinity for CO_2 than for O_2, photorespiration will occur if O_2 concentrations are high. It reduces significantly the efficiency of photosynthesis by releasing, not binding, CO_2. It also uses ATP. Many plants lose over 25% of the carbon that could be fixed during photosynthesis to photorespiration.

How could such an inefficiency arise? In the early history of the Earth when photosynthesis first evolved, O_2 was probably absent or at very low concentrations in the atmosphere (Chapter 7) so photorespiration was negligible. However, the success of photosynthesis produced large atmospheric concentrations of O_2, leading to the possibility of photorespiration. It is one of many examples of imperfect structures or processes that persist in organisms because evolution can work only by modifying existing structures and biochemical processes (Chapter 6).

Box 4.4 Increasing the efficiency of photosynthesis: C_4 and CAM plants

In the standard example of photosynthesis described in this chapter, the first organic compound produced by photosynthesising plants contains three carbon atoms (it is a three-carbon compound). Plants using this method are called C_3 plants. They are common, and many, such as wheat and rice, are important as crops. However, these plants cope badly with hot, dry weather because under these conditions they close their stomates to reduce water loss. Therefore CO_2 cannot enter the leaf and O_2 levels build up, leading to wasteful photorespiration, which reduces overall production of sugar and ATP (Box 4.3).

C_4 plants, so named because the first organic compound produced by photosynthesis contains four carbon atoms, have a solution. The enzyme responsible for synthesising this four-carbon compound has a very high affinity for CO_2 and continues to fix it even at low concentrations when the stomates are closed. The four-carbon compound is then moved to specialised bundle-sheath cells surrounding the veins of the leaf and broken down to release CO_2. The bundle sheath cells are largely impermeable to CO_2. Therefore the CO_2 concentration remains high enough for the Calvin cycle to begin, avoiding the inefficiencies of photorespiration. Sugar cane is an example of a C_4 commercial crop plant.

continued ›

Box 4.4 continued ›

A further adaptation enabling photosynthesis in dry conditions is found in crassulacean acid metabolism (CAM) plants, named after the plant family Crassulaceae in which the process was first found. CAM plants include cacti, pineapples and many other succulents (water-storing plants). They open their stomates only at night and fix CO_2 into a four-carbon compound that is stored and then released to the Calvin cycle only during the day when the stomates are closed. CO_2 produced from the four-carbon compound is then fixed in the chloroplasts as usual.

C_4 plants and CAM plants are well suited to dry conditions. Despite their special adaptations, they still ultimately use the Calvin cycle to complete photosynthesis.

and some is converted to starch for long-term storage. Heterotrophs ingesting starch and other plant carbohydrates break these down into glucose, which they use for their own respiration to generate ATP. We now turn to this process of respiration, used by both autotrophs and heterotrophs.

Respiration

Landmarks in the study of respiration

Our understanding of respiration, like that of photosynthesis, is based on centuries of observation and experimentation. In the 18th century the Frenchman Lavoisier showed that burning requires oxygen and releases carbon dioxide. By extension, the function of breathing was to draw oxygen into the body to make internal combustion possible, because carbon dioxide is breathed out. As microscopy developed and techniques for staining cell structures improved, mitochondria were described and named in the second half of the 19th century. They were found in almost all eukaryotic cells. The German Otto Warburg, who also contributed to our understanding of photosynthesis, introduced the next major steps in the early 20th century. Using manometry to monitor the volumes of oxygen produced by tissues under different conditions, Warburg discovered the membrane-bound proteins called cytochromes that either carry out or catalyse the important series of redox reactions leading to the formation of ATP during respiration. Ultracentrifugation techniques were developed shortly afterwards, enabling separation of cell components for study in isolation. Mitochondria were found to be the site of the oxidative reactions that produce energy for cellular functions after the initial stages were completed in the cytoplasm. Electron microscopy further advanced our understanding of the internal structure of mitochondria.

In combination, these scientific studies give us this picture of the broad nature of cellular respiration:

- During respiration, both autotrophs and heterotrophs break the carbon–hydrogen bonds of glucose molecules to produce water and carbon dioxide and release energy, about half of which is stored in ATP. The rest is lost in heat. The equation is:

$$C_6H_{12}O_6 + 6O_2 \rightarrow 6CO_2 + 6H_2O + \text{energy}$$

- Respiration occurs in three broad stages: glycolysis, which is anaerobic (occurring without oxygen) and occurs in the cytoplasm, and the Krebs cycle and oxidative phosphorylation, which both occur in the mitochondria.

The organelle for respiration: the mitochondrion

Mitochondria (Figure 4.7) have a double external membrane and an inner component called the matrix where the main energy-producing reactions occur. Numerous folds called cristae increase the surface area of the inner membrane within the matrix, and enzymes embedded in the membrane produce ATP. Whereas the outer membrane is structurally similar to the plasma membrane of cells, the inner membrane is far richer in protein. This reflects its richness in enzymes for the process of oxidative phosphorylation (see below) and in ATP synthase for producing ATP. The inner membrane is also highly impermeable and most molecules and ions need special transporters to cross it.

Almost all eukaryotic cells have mitochondria and their number and complexity vary with the functions of the cell. Cells producing energy for muscular movement or using a great deal of active transport have many mitochondria with complex internal cristae, whereas cells with smaller energy needs have fewer, simpler mitochondria. They are often

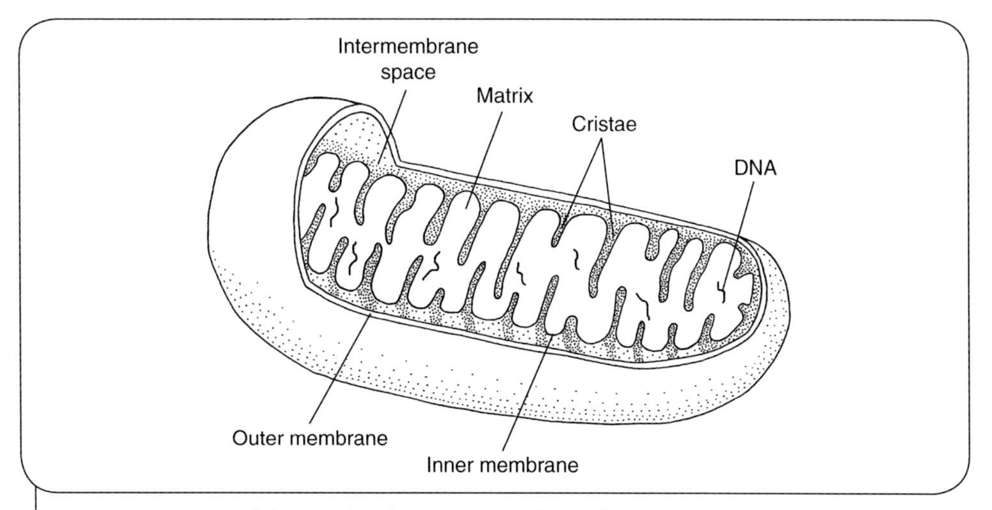

Figure 4.7 Structure of the mitochondrion. (Source: Adapted from a public domain image, Wikimedia Commons. This file is licensed and freely available for use under the Creative Commons Attribution ShareAlike 2.5 License)

randomly distributed within cells, although sometimes they are associated closely with other organelles requiring large amounts of ATP.

Energy from glucose

Glycolysis

Glycolysis occurs in the cell cytoplasm and involves the conversion of the six-carbon glucose molecule to two three-carbon pyruvate molecules. First, two ATP molecules are used to convert glucose to two molecules of glyceraldehyde 3-phosphate (G3P). The energy pay-off happens in the next stage, where the G3P molecules are converted into two molecules of pyruvate. The energy produced is stored in four ATP molecules and the high potential energy of electrons in two NADH (nicotinamide adenine dinucleotide) molecules (Figure 4.8). Overall, there is a gain of two ATP and two NADH. About 2% of the energy in the glucose molecule is released in glycolysis, and the remainder is bound in pyruvate.

The Krebs cycle

In the next stage of respiration, pyruvate is transported into the mitochondrion across the double membrane and into the matrix. Here it is converted into a compound called acetyl CoA, with the formation of NADH. It then enters the Krebs cycle (sometimes also called the citric acid cycle or the tricarboxylic acid (TCA) cycle, depending on whether you prefer to honour the biochemist who first unravelled its mysteries or name it after some of the important reactants). You can trace the main events of the cycle in Figure 4.9.

The Krebs cycle turns twice to break down the acetyl CoA from one glucose molecule, forming four CO_2, two ATP, six NADH and two $FADH_2$ (another high-energy electron carrier, flavin adenine dinucleotide) molecules. It is called a cycle because it begins with the formation of a six-carbon compound (citrate) and finishes with a four-carbon compound, oxaloacetate, ready to accept two carbon atoms from acetyl CoA and begin the process anew. Converting the energy in the high-energy-electron carriers NADH and $FADH_2$ into ATP requires the final step in cellular respiration, oxidative phosphorylation in the electron transport chain.

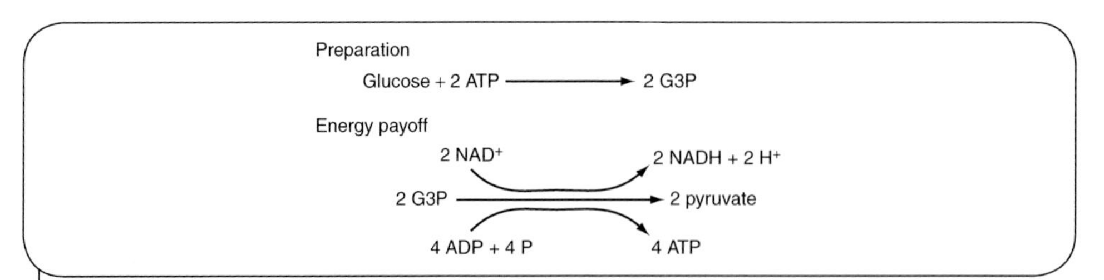

Figure 4.8 Glycolysis, a metabolic pathway beginning with glucose and ending with pyruvate and the production of two ATP and two NADH molecules. (Source: B. Cale)

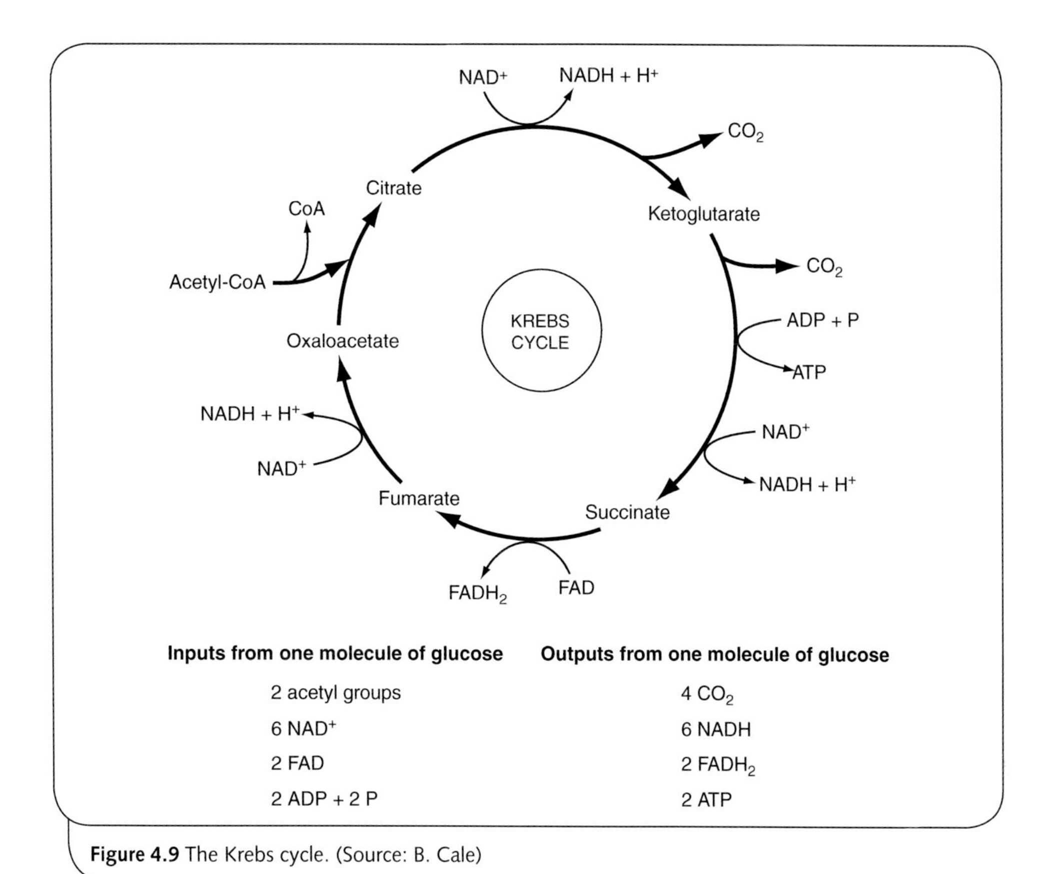

Figure 4.9 The Krebs cycle. (Source: B. Cale)

Oxidative phosphorylation

Oxidative phosphorylation occurs on the inner membrane (the cristae) of the mitochondrion and converts the potential energy of the high-energy electrons in NADH and $FADH_2$, which were produced in glycolysis and the Krebs cycle, into ATP. It takes its name from the series of redox reactions (similar to those we discussed in the light reactions of photosynthesis) with which it begins and its ultimate goal, the addition of a phosphate group to ADP (phosphorylation) to form ATP. You can trace the process in Figure 4.10.

The process begins when electrons are released from their carriers, NADH and $FADH_2$, into an electron transport chain of oxidation/reduction (redox) reactions. As they are passed between cytochromes (carrier molecules), the energy released is used to pump H^+ ions across the inner mitochondrial membrane into the intermembrane space and create a concentration gradient. Ultimately, electrons are accepted by oxygen, which combines with free H^+ ions to form water molecules. The concentration gradient created across the inner mitochondrial membrane leads to movement of H^+ ions back across the membrane via the enzyme ATP synthase, powering the production of ATP from ADP.

Oxidative phosphorylation can add a maximum of 34 ATP molecules to the two ATP molecules produced in glycolysis and the two ATP molecules produced in the Krebs

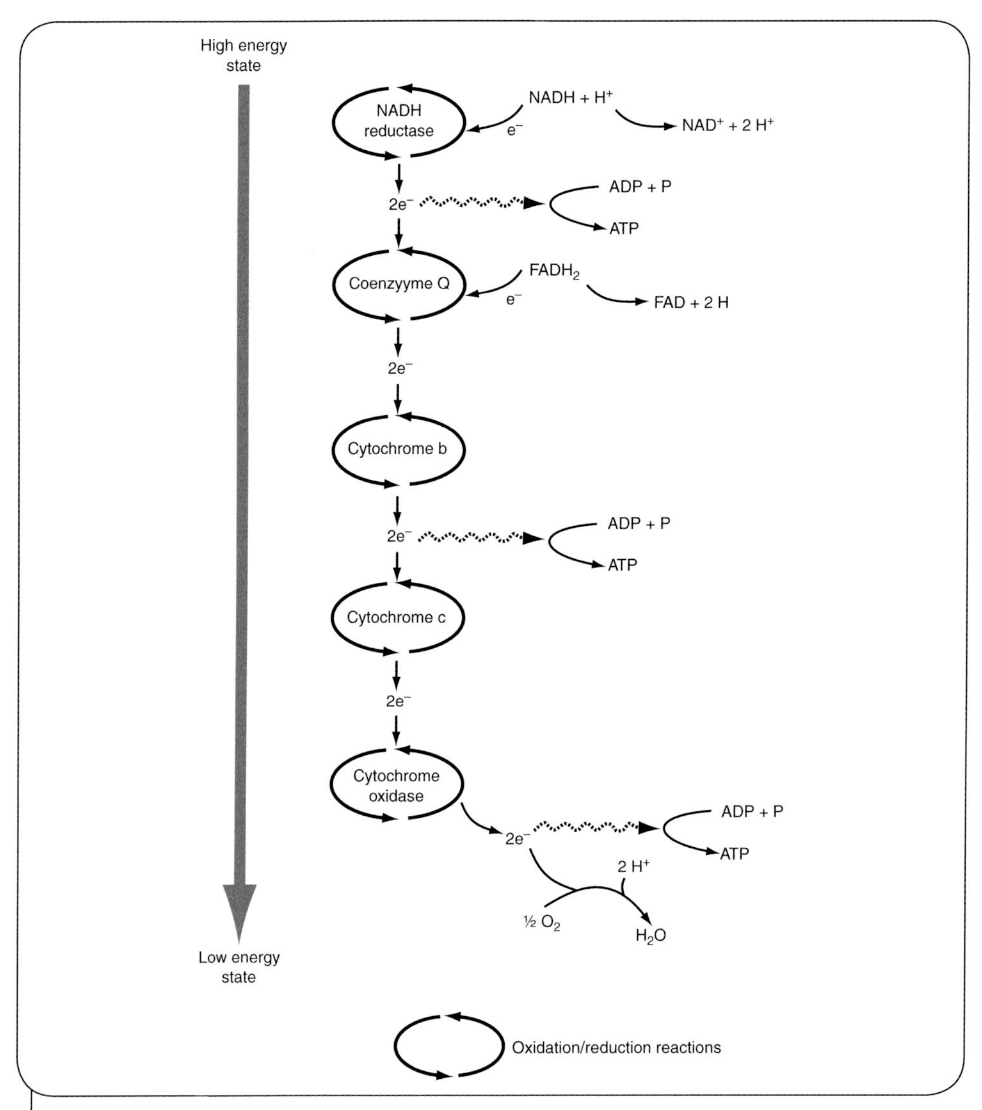

Figure 4.10 Oxidative phosphorylation in the electron transport chain. (Source: B. Cale)

cycle. Overall, this means that up to 40% of the potential energy in a glucose molecule is transferred into ATP through respiration, while much of the remainder is lost in heat. Figure 4.11 overviews all stages of aerobic respiration.

Energy from fat and protein

Although the respiration of glucose is a common source of energy, lipids and proteins can also be respired. Lipids are first hydrolysed to form glycerol and fatty acids. The glycerol is then converted into G3P, a compound in the glycolysis pathway. The fatty acids are converted into acetyl CoA, which is then passed into the Krebs cycle. Proteins are first decomposed into their constituent amino acids, which may be converted into pyruvate, acetyl CoA or intermediates of the Krebs cycle.

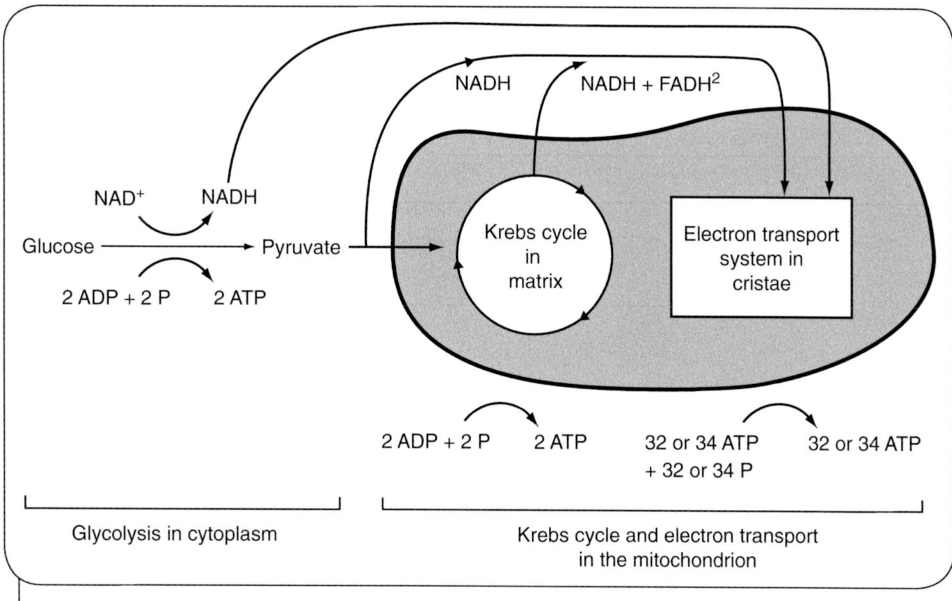

Figure 4.11 Overview of respiration. (Source: B. Cale)

Anaerobic respiration

Although glycolysis is the first stage in aerobic respiration, it requires no oxygen. Some bacteria and yeasts make do with that as their sole energy source, converting the pyruvate produced in glycolysis into alcohol, NAD^+, for reuse in glycolysis and CO_2 (alcoholic fermentation). The equation is:

$$\text{pyruvate} + \text{NADH} \rightarrow \text{ethanol} + \text{NAD}^+ + \text{CO}_2$$

Controlled alcoholic fermentation is important in many industrial and food production processes such as wine making. Animal cells convert the pyruvate to lactic acid rather than alcohol (lactic acid fermentation) as follows:

$$\text{pyruvate} + \text{NADH} \rightarrow \text{lactic acid} + \text{NADH}^+$$

Muscle cells do this when oxygen supplies run short during strenuous exercise. Ultimately, the lactic acid is transferred via the bloodstream to the liver for conversion back into pyruvate.

Energy flow in the biosphere

The earliest life on Earth was prokaryotic and generated energy for cellular processes via fermentation of organic molecules in an anaerobic atmosphere. The earliest photosynthetic prokaryotes appeared about 3.5 billion years ago and sunlight became the main source of energy for all life through the generation of glucose. Fermentation remained as a source of energy for cellular processes. However, the oxygen released as a waste product of photosynthesis transformed the early Earth. Over the next 1.7 billion years it combined with iron in the rocks and oceans and, when that option was saturated, it accumulated in the atmosphere.

Given the dependence of our lives upon atmospheric oxygen, it can be difficult to comprehend that the atmospheric oxygen accumulation was toxic to early life. Proteins, nucleic acids and other organic compounds in cells were oxidised (combined with oxygen and changed in structure, while producing carbon dioxide and heat). This was not a fierce process as in a fire, but a more gradual one such as the browning of cut fruit. Nevertheless, many organisms were killed. Some survived in places where oxygen could not reach and continued to live anaerobically, while others adapted by oxidising the waste products of fermentation in the process we call respiration. Not only did this protect other cellular components from oxidation, but it also produced a great deal more energy from a single molecule of glucose and made possible the evolution of larger organisms.

In the biosphere today, photosynthesis and cellular respiration remain linked as key processes making solar energy available to all organisms. Autotrophs convert solar energy into the energy of chemical bonds, whereas heterotrophs convert the energy in chemical bonds into other chemical forms. Each energy conversion is not 100% efficient because energy is lost as heat, so solar energy must be added continually to the biosphere for life to continue. Both these processes play critical roles in two major environmental issues: the enhanced greenhouse effect (Box 4.5) and efficiencies in food production (Box 4.6).

Box 4.5 Photosynthesis and global warming

CO_2 supports life on Earth not only by its role as a raw material in photosynthesis, but also by maintaining the temperature of the planet's surface within a range suitable for life. It does this by trapping some of the heat energy radiated from the Earth after it is warmed by solar radiation.

Solar energy reaches the Earth as both ultraviolet (UV) radiation and visible light. Much of the UV is filtered by the ozone layer in the upper atmosphere, but the visible light reaches and warms the Earth's surface. The warmed planet then radiates heat. Some of this heat escapes back into space whereas gases such as CO_2 in the atmosphere absorb some and radiate it back to Earth. This process, critical for keeping the surface of the Earth warm enough for life, is known as the greenhouse effect (Figure B4.5.1). The concentration of CO_2 in the atmosphere is kept in balance through the carbon cycle in which the twin processes of photosynthesis and respiration keep carbon cycling between organic and inorganic forms on a planetary scale (Chapter 17). Photosynthesis fixes the carbon in organic compounds, whereas respiration and fire release it as CO_2 into the atmosphere. At any one time, carbon is present in the atmosphere as CO_2 and also locked in the bodies of living organisms and in the residues of long-dead organisms stored as fossil fuels beneath the Earth's surface.

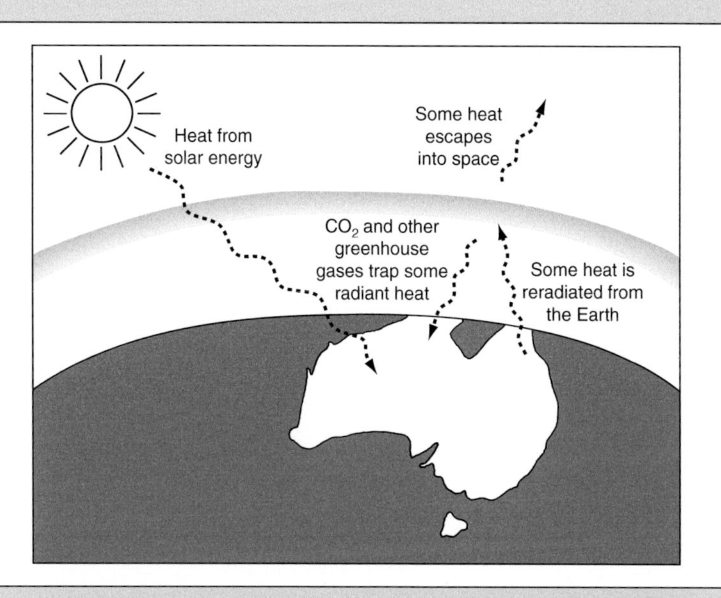

Figure B4.5.1 The mechanisms of the enhanced greenhouse effect. Greenhouse gases trap some radiant heat from the Earth's surface, preventing it escaping into space. (Source: B. Cale)

The great Industrial Revolution of the late 18th and early 19th centuries inadvertently began a global experiment into the influence of atmospheric CO_2 concentration on surface temperatures. The widespread burning of fossil fuels such as coal and oil released large amounts of CO_2, whereas agricultural clearing and deforestation reduced the abundance of photosynthetic organisms to convert the CO_2 to organic compounds. The increased atmospheric CO_2 concentrations are believed to be creating an enhanced greenhouse effect, warming the Earth's surface further. Although this may appear a pleasant thought in mid-winter, changes of just a few degrees in the surface temperature are believed to be already causing major environmental changes including melting of polar ice caps and associated rises in sea levels, increased severity of storms and climate change. There will be serious consequences for human agriculture as well as habitat loss for many organisms.

Photosynthesis fights against the enhanced greenhouse effect. Plantations of rapidly growing trees sequester atmospheric CO_2 in plant tissue, while providing timber for construction and furniture. Furthermore, humans can learn from the example of photosynthesis to explore non-polluting solar power to generate electricity.

Box 4.6 The luxury diet

In nutrient-rich waters off the Antarctic continent, unicellular floating autotrophs called phytoplankton are the main photosynthetic organisms. They are eaten by small animals called krill, which in turn are eaten by squid, fish and even whales. Still larger fish and squid feed on the smaller ones sustained by the krill. Organisms connected in this way by feeding relationships form a food chain and the exact position of each species within the food chain is called a trophic level. The autotrophic phytoplankton are called producers, the heterotrophic krill feeding on them are primary consumers, the fish eating the krill are secondary consumers and so on (these ideas are developed further in Chapter 17). Human fisheries normally concentrate on the larger fish and squid, making people tertiary or even quaternary consumers. However, during the late 20th century some Russian fisheries experimented with harvesting krill. There is a sound biological rationale for the experiment based on inefficiencies in the transfer of energy.

To begin with, phytoplankton convert only about 1% of the solar energy available to them into organic compounds in their bodies. When the krill feed on the phytoplankton, only a fraction of the energy (about 10%) in the phytoplankton tissues is converted into krill biomass available for its predators to eat. The rest is lost because of inefficiencies in feeding and digestion and because some of the energy is used in the krill's own respiration. A similar loss occurs when fish feed on the krill. Thus only about 1% of the energy in the bodies of phytoplankton comes to be stored in the bodies of the fish feeding as secondary consumers. If humans ate the krill rather than the fish living as secondary consumers, about 10 times more energy would be available to human consumers. In this case, although the mathematics is simple the logistics are not. Harvesting krill efficiently and preventing spoilage is difficult, as is the preparation of krill as a human food.

A similar logic applies to grain-fed farm animals. Feeding grain to cattle and then eating the meat provides 10 times less energy to human populations than eating the grain directly. In this case, storing, distributing and processing the grain is not difficult, so large quantities of grain-fed beef provide a luxury diet.

Chapter summary

1. Cells use energy for movement of the whole cell, movement of cell organelles, building complex molecules, growth, maintenance and repair, active transport and cell division (mitosis and meiosis).
2. The sun is the primary energy source for almost all living things, regardless of whether they are autotrophic or heterotrophic. The primary energy currencies within cells are the energy in chemical bonds in ATP molecules and high-energy electrons in molecules of NADH, NAP and $FADH_2$.
3. Photosynthesis occurs in organelles called chloroplasts, which are located in the green parts of plants and algae, mainly in the leaves. In photosynthesis, autotrophs use solar energy to make sugar molecules from water and CO_2, releasing oxygen as a by-product. The equation is:

$$6CO_2 + 6H_2O + \text{solar energy} \rightarrow C_6H_{12}O_6 + 6O_2$$

   Photosynthesis occurs in a series of linked 'light-dependent reactions', which fix solar energy, and 'light-independent reactions', which produce glucose.
4. The energy trapping or light-dependent reactions occur on the thylakoid membrane. Solar energy is used to split water molecules and shuttle electrons down an electron transport chain to form NADPH molecules with high-energy electrons. Meanwhile, a H^+ gradient is created across the thylakoid membrane that is used to energise the production of ATP from ADP.
5. In the light-independent reactions (the Calvin cycle) glucose is synthesised from CO_2 with the energy provided by NADPH and ATP produced in the light reactions. Glucose is transported throughout the plant for use in structural compounds, food stores and as a basis for cell respiration.
6. Both autotrophs and heterotrophs use cellular respiration to power the energy needs of cells. In the presence of oxygen, it breaks the carbon-hydrogen bonds of glucose molecules to produce water and carbon dioxide and release energy, about half of which is stored in ATP and the rest is lost in heat. The equation is:

$$C_6H_{12}O_6 + 6O_2 \rightarrow 6CO_2 + 6H_2O + \text{energy}$$

7. Cellular respiration in the presence of oxygen (aerobic respiration) occurs in three broad stages: glycolysis (which occurs in the cytoplasm), and the Krebs cycle and oxidative phosphorylation (which occur in the mitochondria).
8. Glycolysis yields a gain of two ATP molecules and two NADH molecules. Its end product, pyruvate, is moved across the mitochondrial membrane, converted into acetyl CoA and taken into the Krebs cycle.
9. The Krebs cycle breaks down acetyl CoA to form carbon dioxide and two ATP, six NADH and two $FADH_2$ molecules.
10. Oxidative phosphorylation occurs on the inner membrane of the mitochondrion and converts the potential energy of the high-energy electrons in NADH and $FADH_2$ into ATP. It adds a maximum of 34 ATP to the two ATP molecules produced in glycolysis and the two ATP molecules produced in the Krebs cycle. Overall, this means that up to 40% of the potential energy in a glucose molecule is transferred into ATP through respiration, whereas much of the remainder is lost in heat.
11. Proteins and lipids can also be respired by first breaking them down into simpler compounds that are intermediates in either glycolysis or the Krebs cycle.

12. Anaerobic respiration leading to the production of ethanol (plants, bacteria and fungi) or lactic acid (animals) occurs in the absence of oxygen. It is the sole form of energy production for some organisms.
13. In the biosphere today, photosynthesis and respiration remain linked as key processes making solar energy available to all organisms. Autotrophs convert solar energy into the energy of chemical bonds, and both autotrophs and heterotrophs convert the energy in chemical bonds into other chemical forms. Each energy conversion is less than 100% efficient because energy is lost as heat, so solar energy must be added continually to the biosphere for life to continue.

Key terms

aerobic respiration
anaerobic respiration
ATP
autotroph
Calvin cycle
carotenoid
chlorophyll
cristae
cytochromes
fermentation
greenhouse effect
glycolysis
grana
heterotroph
hydrolysis
Krebs cycle
light-dependent reactions
light-independent reactions
mitochondrion
NADH
NADPH
oxidative phosphorylation
photosynthesis
reactions
redox (oxidation-reduction)
respiration
stroma
thylakoid

Test your knowledge

1. Describe what occurs in the light-dependent and the light-independent reactions in photosynthesis.
2. Explain why C_4 plants and CAM plants are more water-efficient than C_3 plants.
3. How does the double-membrane structure of the mitochondrion assist the reactions of respiration?

Thinking scientifically

Over the next few decades, several missions by spacecraft will look for evidence of life elsewhere in the solar system. Two places that will be explored by spacecraft landers are the planet Mars and Europa, a satellite of the planet Jupiter.

Mars is known to have a thin atmosphere of carbon dioxide and nitrogen. Although water vapour is present in the atmosphere and water ice is present in polar regions, there are no oceans, rivers or lakes. Temperatures on Mars vary from −128°C at the poles in winter to 37°C at the equator in summer. Scientists believe liquid water may be found somewhere beneath the surface of the planet.

Europa is much further than Mars from the sun. It has no atmosphere and its surface is covered mostly by ice several kilometres thick. There is good evidence that oceans of liquid water lie beneath the ice. The average surface temperature of Europa is –180°C, but it is thought that the oceans underneath are warmed by tidal forces and may have temperatures similar to those on Earth.

1 What particular difficulties would be faced by Earth-like organisms living on:
   (a) Mars?
   (b) Europa?

2 In terms of the requirements of living organisms, explain why the conditions on Mars and Europa may be too hostile for life.

(Source: This question is adapted from Curriculum Council of Western Australia, *Western Australian Tertiary Entrance Examination* 2002.)

Debate with friends

Biofuels – liquid fuels including ethanol and biodiesel derived from organic matter – will be important energy sources as oil supplies decrease and costs increase. The use of land for producing biofuels may reduce food supplies, causing higher food prices. Debate the topic: 'Biofuels are a solution to declining oil supplies'.

Further reading

Bechtel, W. (2006). *Discovering cell mechanisms: the creation of modern cell biology.* Cambridge University Press, Cambridge.

Becker, W.M., Kleinsmith, L.J. and Hardin, J. (2006). *The world of the cell.* Pearson Benjamin Cummings, San Francisco.

Harris, H. (2000). *The birth of the cell.* Yale University Press, New Haven.

5

Cell theory III – the cell cycle

Annette Koenders

Is extinction really forever?

People are fascinated with extinct organisms. How exciting would it be to see a trilobite on the ocean bed, or observe a *Tyrannosaurus rex*? Past books and films about this fantasy involved time travel, but more recent stories use ancient DNA to clone and resurrect extinct organisms. It is a marvellous example of art imitating and leading science. The past 50 years have seen the discovery of the structure of DNA and the development of techniques for manipulating DNA, spawning the biotechnology industry. We now have genetically modified foods, Dolly the cloned sheep (now deceased) and other cloned animals. We have also extracted ancient DNA from fossils such as insects embedded in amber and even the leg bone of a *Tyrannosaurus rex*. A natural progression of the imagination is to use ancient DNA to re-create extinct organisms. Recently the Australian Museum attempted unsuccessfully to clone the extinct thylacine (Tasmanian tiger) (Plate 5.1) using DNA from a preserved specimen, but they were able to show that parts of some genes could function when inserted into mouse cells.

Chapter aims

In this chapter, we examine the molecular processes of cells, giving you the background to understand these techniques and to evaluate their environmental significance. We examine the hereditary material and how it codes messages, controls cells and is copied and distributed over generations through cell division; and we apply this knowledge to conservation.

What is the hereditary material and how does it code messages?

The hereditary material is DNA

As explained in Chapter 3, deoxyribonucleic acid (DNA) contains hereditary information and transmits it between generations. These are its only roles. The products of DNA – the proteins and ribonucleic acids (RNA) (see Chapter 3) – perform the work. DNA contains information for making all of the cell's proteins and RNA molecules and, largely, when to make them. Each piece or sequence of DNA coding for a protein or RNA molecule is called a gene, so genes are DNA sequences coding for functional products. Examples are proteins biosynthesising lipids, carbohydrates or cellular structures, extracting energy from fuel molecules and assisting in transport or communication. Alternatively, the product can be RNA, which activates genes (gene expression).

Genes occur along strands of DNA molecules called chromosomes. A gene's position on a chromosome is called its locus (see Box 5.1 for an explanation of basic genetic

Box 5.1 Genetic terminology

1. The phenotype of an organism is all of its observable properties. A trait or character is a specific phenotypic attribute. The genotype of an organism is its genetic constitution. The phenotype is determined by the genotype and by all the environmental factors that act on the organism.
2. A gene is a specific sequence of DNA that determines a trait by coding for a functional product: either ribonucleic acid (RNA) or protein. A locus (plural: loci) is the position of a gene on a chromosome. Alleles are different forms (DNA sequences) of a gene that occupy the same locus. Different alleles are usually written as upper case or lower case letters (e.g. A or a), or as a letter with different superscripts (e.g. A^r or A^w).
3. Haploid cells have only one copy of each gene whereas diploid organisms have two. Diploid cells can therefore be homozygous at a locus, meaning that the alleles are the same (e.g. AA), or heterozygous, meaning that the alleles are different (e.g. Aa).
4. The gene action of an allele is the way in which that allele interacts with the alternative allele at a locus to produce a phenotype in a diploid organism. The gene action of an allele can be dominant, recessive or co-dominant. Dominant alleles are expressed to the same degree in the phenotype whether in heterozygous or homozygous form. Recessive alleles are expressed in the phenotype only when homozygous. By convention, if we are using the upper case/lower case notation, we usually use upper case for the dominant allele and lower case for the recessive allele. Co-dominant alleles are expressed when heterozygous, but not to the same extent as when homozygous.

terminology). Chromosomes also contain non-coding regions. Some of these are important in gene expression. In eukaryotic cells there is also much DNA with no known function at all. Some of this non-coding 'junk' DNA is used in DNA fingerprinting, a molecular mechanism used in criminal forensic applications for identifying individuals, and also in tracing breeding history in endangered organisms.

The hereditary information is stored on chromosomes a bit like written language. In the same way that we use individual letters to form words, DNA uses individual chemical subunits, called nucleotides, to form genetic words. The genetic words combine to make a gene (or a sentence). In many ways, the language of genetics is simpler than human languages: there are only four genetic letters, all genetic words are three letters long and all organisms use the same genetic language.

The structure of the DNA molecule was discovered in Nobel Prize–winning research by Watson and Crick (Box 5.2). Each DNA nucleotide contains three parts, only one of which holds information, the other parts having structural functions. The part holding information is the nitrogenous base. There are four different nucleotides, each with a different base in DNA: thymine (T), cytosine (C), adenine (A) and guanine (G) (Figure 5.1). T and C belong to a chemical group called pyrimidines, while A and G belong to a chemical group called purines. When in solution together, a pyrimidine is attracted to a purine, forming hydrogen bonds between them (Figure 5.1). A pairs with T by forming two hydrogen bonds, and C pairs with G by forming three hydrogen bonds. This is termed complementary base pairing and, in essence, provides the entire basis for DNA's functioning.

Box 5.2 Watson, Crick and the structure of DNA

Possibly the most famous scientific paper in all of biology was published by James Watson and Francis Crick in the journal *Nature* on 25 April 1953. It begins with what, in hindsight, is a masterly understatement: 'We wish to suggest a structure for the salt of deoxyribose nucleic acid (DNA). This structure has novel features which are of considerable biological interest'.

In their paper, Watson and Crick proposed that DNA has the form of a double helix, with the two strands of the helix wound round each other like a twisted ladder. The rails of the ladder are made of alternating units of sugar and phosphate and the rungs consisted of a pair of bonded nucleotides. The structure they proposed was essentially correct, and in 1962 they were awarded the Nobel Prize for their discovery.

Watson and Crick arrived at their proposed structure through a theoretical approach, by constructing three-dimensional models. To do this, they relied on data from other people, particularly X-ray crystallographic analyses by Rosalind Franklin, which showed that DNA had a helical structure and that the phosphate backbone was on the outside of the molecule.

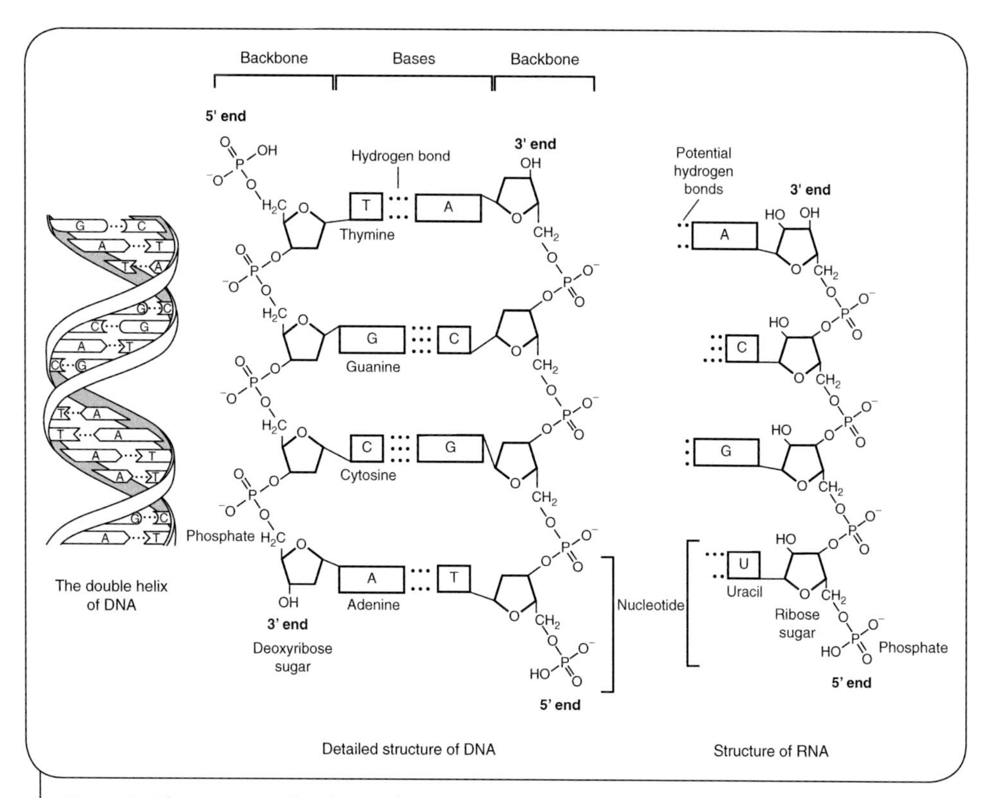

Figure 5.1 The structure of nucleic acids. There are four different nucleotides or bases in DNA: thymine (T), cytosine (C), adenine (A) and guanine (G). RNA substitutes uracil (U) for thymine. Pairing between complementary bases is the key to the genetic code that underlies the functioning of DNA. (Source: Adapted from Figure 16.7, p. 297, of Campbell, N.A., Reece, J.B. and Meyers, N. (2006). *Biology*. Pearson Education Australia. 7th edn. Copyright © 2005 by Pearson Education Inc. Reproduced by permission.)

The other two parts of DNA nucleotides, a deoxyribose sugar and a phosphate group, form the structural backbone. By connecting the phosphate of one nucleotide to the sugar of the next, individual nucleotides are connected via strong chemical bonds, leaving the informational parts, the bases, free and protruding (Figure 5.1). Although this arrangement keeps the sequence of the bases intact, it does expose them. A single strand of DNA will, in the presence of free DNA nucleotides, finish with each of its own nucleotides bonded via hydrogen bonds to a free complementary nucleotide. Within living organisms, DNA occurs as a double-stranded molecule with the sugar–phosphate backbones on the outside and the bases paired in the centre. This protects the bases from chemical damage and orders them in sequence, like letters on a long page. Although each individual hydrogen bond between the bases is weak, several hundred thousand of them along a chromosome are a strong connection! This gives DNA enormous stability while providing a mechanism for DNA expression, replication and repair.

The DNA of prokaryotes consists of one circular chromosome. Many prokaryotes also possess additional, smaller, circular pieces of DNA, called plasmids. These are important in transferring genetic information between organisms and also are important tools in scientific research and biotechnology. Eukaryotic organisms generally have much more DNA than do prokaryotes, held in one or more linear chromosomes within the nucleus and separated from the cell's cytoplasm by a double membrane. Eukaryotic chromosomes consist of DNA and a similar amount of closely associated, specialised proteins called histones. Histones keep the chromosomes organised and regulate gene expression. Collectively, the DNA and associated histones are termed chromatin.

Unravelling the genetic code

For genes the product of which is a functional RNA molecule, gene expression is relatively straightforward, because DNA and RNA share similar chemical structures (Figure 5.1). Most final gene products, however, are not RNA, but are proteins made of a series of different subunits: amino acids. Cells have mechanisms for translating the DNA code into amino acid sequences. Each three-letter DNA word (a codon) codes for a unique amino acid, or for starting or stopping protein synthesis (Table 5.1). The code has redundancy because in some cases different codons specify the same amino acid, but no ambiguity because each codon specifies only one amino acid. The DNA code is universal among all living organisms and is further evidence that all life on Earth shares a common ancestor. It also makes gene transfer possible between unrelated organisms – a gene from one organism expressed in another produces the same protein or RNA as in the donor organism.

Table 5.1 The genetic code as designated by the three bases of (a) a DNA codon, and (b) an mRNA codon reading in the 5′ → 3′ direction. Each codon codes for a specific amino acid. The code for the amino acid methionine is also a start signal for translation. Three codons code for 'stop' signals in transcription or translation.

(a) DNA genetic code					
1st base	**2nd base**				**3rd base**
	A	G	T	C	
A	AAA Phenylalanine	AGA Serine	ATA Tyrosine	ACA Cysteine	A
	AAG Phenylalanine	AGG Serine	ATG Tyrosine	ACG Cysteine	G
	AAT Leucine	AGT Serine	ATT *Stop*	ACT *Stop*	T
	AAC Leucine	AGC Serine	ATC *Stop*	ACC Tryptophan	C
G	GAA Leucine	GGA Proline	GTA Histidine	GCA Arginine	A
	GAG Leucine	GGG Proline	GTG Histidine	GCG Arginine	G
	GAT Leucine	GGT Proline	GTT Glutamine	GCT Arginine	T
	GAC Leucine	GGC Proline	GTC Glutamine	GCC Arginine	C
T	TAA Isoleucine	TGA Threonine	TTA Asparagine	TCA Serine	A

(a) DNA genetic code					
1st base	**2nd base**				**3rd base**
	A	**G**	**T**	**C**	
	TAG Isoleucine	TGG Threonine	TTG Asparagine	TCG Serine	G
	TAT Isoleucine	TGT Threonine	TTT Lysine	TCT Arginine	T
	TAC Methionine *Start*	TGC Threonine	TTC Lysine	TCC Arginine	C
C	CAA Valine	CGA Alanine	CTA Aspartic acid	CCA Glycine	A
	CAG Valine	CGG Alanine	CTG Aspartic acid	CCG Glycine	G
	CAT Valine	CGT Alanine	CTT Glutamic acid	CCT Glycine	T
	CAC Valine	CGC Alanine	CTC Glutamic acid	CCC Glycine	C

(b) mRNA genetic code					
1st base	**2nd base**				**3rd base**
	U	**C**	**A**	**G**	
U	UUU Phenylalanine	UCU Serine	UAU Tyrosine	UGU Cysteine	U
	UUC Phenylalanine	UCC Serine	UAC Tyrosine	UGC Cysteine	C
	UUA Leucine	UCA Serine	UAA *Stop*	UGA *Stop*	A
	UUG Leucine	UCG Serine	UAG *Stop*	UGG Tryptophan	G
C	CUU Leucine	CCU Proline	CAU Histidine	CGU Arginine	U
	CUC Leucine	CCC Proline	CAC Histidine	CGC Arginine	C
	CUA Leucine	CCA Proline	CAA Glutamine	CGA Arginine	A
	CUG Leucine	CCG Proline	CAG Glutamine	CGG Arginine	G
A	AUU Isoleucine	ACU Threonine	AAU Asparagine	AGU Serine	U
	AUC Isoleucine	ACC Threonine	AAC Asparagine	AGC Serine	C
	AUA Isoleucine	ACA Threonine	AAA Lysine	AGA Arginine	A
	AUG Methionine, *Start*	ACG Threonine	AAG Lysine	AGG Arginine	G
G	GUU Valine	GCU Alanine	GAU Aspartic acid	GGU Glycine	U
	GUC Valine	GCC Alanine	GAC Aspartic acid	GGC Glycine	C
	GUA Valine	GCA Alanine	GAA Glutamic acid	GGA Glycine	A
	GUG Valine	GCG Alanine	GAG Glutamic acid	GGG Glycine	G

To 'read' the hereditary information, it is vital to know not only the genetic code, but also which DNA strand contains the genetic information and in which direction to read. The particular strand of the DNA containing a gene, the template strand, is identified by promoter sequences where transcription initiation complexes form.

In the language of heredity, the chemical structure of the bases indicates the direction in which to read. The backbones with alternating sugar and phosphate groups provide direction, because a sugar molecule will sit at one end, the 3′ end, of the nucleic acid, and a phosphate group at the other end, the 5′ end. In fact, the two strands in double-stranded DNA are anti-parallel; one strand runs in the 3′ to 5′ direction, and the other runs 5′ to 3′ (Figure 5.2). The hereditary language is read in the 5′–3′ direction. This is the only direction in which DNA is replicated and transcribed into messenger RNA (mRNA) to be carried out of the nucleus, and in which transfer (tRNA) translates these instructions into proteins in the ribosomes.

How does the hereditary material control cell function?

The complete genetic make-up of organisms written in their DNA is called the genotype. The forms that gene products take are the phenotype and vary throughout an organism's life. The phenotype includes invariant characteristics, such as molecular machinery and the number of limbs or eyes. It also includes characteristics varying with age or season, such as hair colour or hormone production. One way by which phenotype varies is through regulation of gene expression.

Many genes cannot be expressed, particularly in eukaryotic organisms, until the appropriate initiation complex is assembled on their promoter sequence on the DNA. Initiation complexes consist of several to many proteins and RNA molecules inhibiting or promoting gene expression. Once all systems are in place, DNA expression proceeds in a two-step process: transcription (of the DNA code into RNA) and then translation (of the RNA code into an amino acid sequence).

Transcription

Transcription produces a strand of RNA complementary to the coding strand of the DNA. RNA is a similar molecule to DNA with some important differences: (1) RNA has a ribose sugar which has an additional hydroxyl (–OH) group on the 2′ carbon, (2) RNA also has four nitrogenous bases, but has uracyl instead of thymine and (3) RNA is usually single-stranded (Figure 5.1). Transcription starts with the enzyme RNA polymerase unwinding the double helix and exposing free DNA bases (Figure 5.2). During transcription, free RNA nucleotides present in the nucleus pair with their complementary DNA bases by hydrogen bonding, similar to the way that DNA nucleotides pair up in the two complementary strands of chromosomes. RNA polymerase attaches these aligned RNA nucleotides to each other by connecting their phosphate and sugar groups in a way similar to the DNA backbones. RNA polymerase 'runs' along the DNA sequence in a 5′ to 3′ direction until it reaches a termination sequence. The RNA strand that results from this process is the primary transcript (Figure 5.2).

In prokaryotic cells the primary transcript produces proteins immediately, providing rapid responses to environmental changes, such as encountering a different food source. In eukaryotic cells the process is slower, because transcription occurs within the nucleus before translation (the formation of a protein from RNA) in the cytoplasm.

Much eukaryotic DNA does not code for RNA or protein sequences. Most of this non-coding DNA lies between genes and appears functionless, although some of it is regulatory. There are also stretches of non-coding DNA (introns) within genes. They are transcribed into the primary transcript, but then removed (spliced out) before translation of the coding regions (exons). Splicing occurs within the nucleus and varies for many genes, so it is possible that some genes may produce more than one type of RNA molecule (with the potential for more than one gene product: RNA or protein). Forming two, three or even more proteins with different functions from a single DNA sequence reduces the amount of DNA required and simplifies the regulation of gene expression. This allows multicellular organisms to attain complexity without cluttering their cells with DNA.

How does the cell's machinery distinguish introns, the non-coding sequences, from exons, the coding sequences? Introns have end regions that tend to base pair with one another, producing a hairpin loop along the strand of RNA. Spliceosomes, complexes of specialised RNA and protein molecules, recognise these hairpin structures, remove those sections of RNA and splice the exons together. This splicing apparently does not proceed in the same way each time for a particular primary transcript; sometimes exons are skipped and removed with their adjacent introns, causing a different amino acid sequence in the protein.

The primary transcript of eukaryotic cells undergoes two more processes before leaving the nucleus. A modified guanine (G) cap is placed at the 5′ end, and a tail of many adenine subunits (poly-A tail) is added to the 3′ end, protecting the RNA from being broken down too quickly and facilitating its transport from the nucleus into the cytoplasm for translation.

In summary, the primary transcript is an RNA molecule that is a complementary copy of the template strand on the DNA. This is the messenger RNA (mRNA) in prokaryotic cells, but is further processed in eukaryotic cells by splicing out introns, adding the G-cap on the 5′ end and adding a poly-A tail at the 3′ end to form mRNA.

Translation

The next step in gene expression is translation of the mRNA into the amino acids combined in a protein. The sequence of codons on mRNA determines the sequence of amino acids in the protein and ultimately the protein's functional properties. Translation from codons, three-letter nucleotide words, to amino acids is effected by transfer RNA (tRNA). There is a different tRNA molecule for each codon (Table 5.1), resulting in

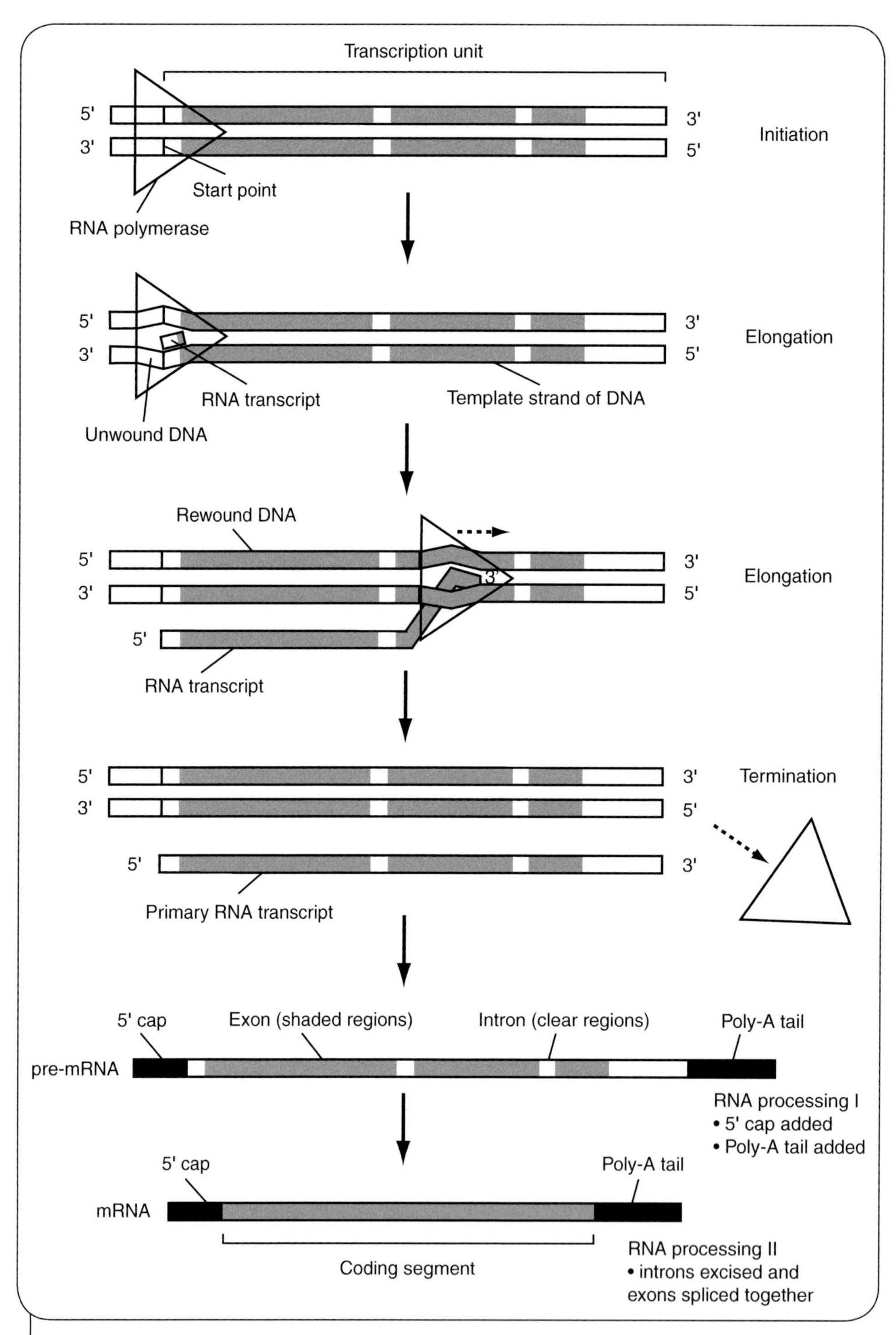

Figure 5.2 Transcription of DNA to mRNA in eukaryotes. Both exons (coding sequences) and introns (non-coding sequences) are transcribed and a cap and a poly-A tail are added. Introns are then excised and the exons are spliced together before the mRNA molecule leaves the nucleus. (Source: Adapted from Figure 17.7, p. 315 and Figure 17.8, p. 316 of Campbell, N.A., Reece, J.B. and Meyers, N. (2006). *Biology*. Pearson Education Australia. 7th edn. Copyright © 2005 Pearson Education Inc. Reproduced by permission).

several different tRNA molecules for each of the 20 amino acids. Each tRNA has a site for recognising a specific codon on mRNA, termed the anti-codon, and another site where the amino acid corresponding to that codon is bound. Transfer RNA molecules have a specific shape, facilitating complementary base pairing of the anti-codon with a codon on mRNA. The anti-codon consists of a three-nucleotide sequence complementary to one of the codons, providing a direct chemical link between the genetic code and the amino acid sequence.

Once the mRNA molecule is exported from the nucleus, a ribosome clamps onto the mRNA, enabling the first tRNA to pair with the start codon. There are three tRNA binding sites within a ribosome: an amino acid (A) site, a peptide (P) site and an exit (E) site. They provide a structural framework for the operation of the tRNA molecules (Figure 5.3). The first tRNA molecule attaches to the P site. The next tRNA molecule attaches at the A site, where the anti-codon of the tRNA is base-paired with the codon on the mRNA. The amino acid from the tRNA in the P site separates and becomes covalently attached to the amino acid in the P site, beginning a chain that is now two amino acids long. The ribosome then moves one codon along the mRNA, so that both tRNA molecules have shifted one ribosome binding site. The tRNA that has lost its amino acid is now in the E site and leaves the ribosome. The tRNA that was in the A site is now in the P site,

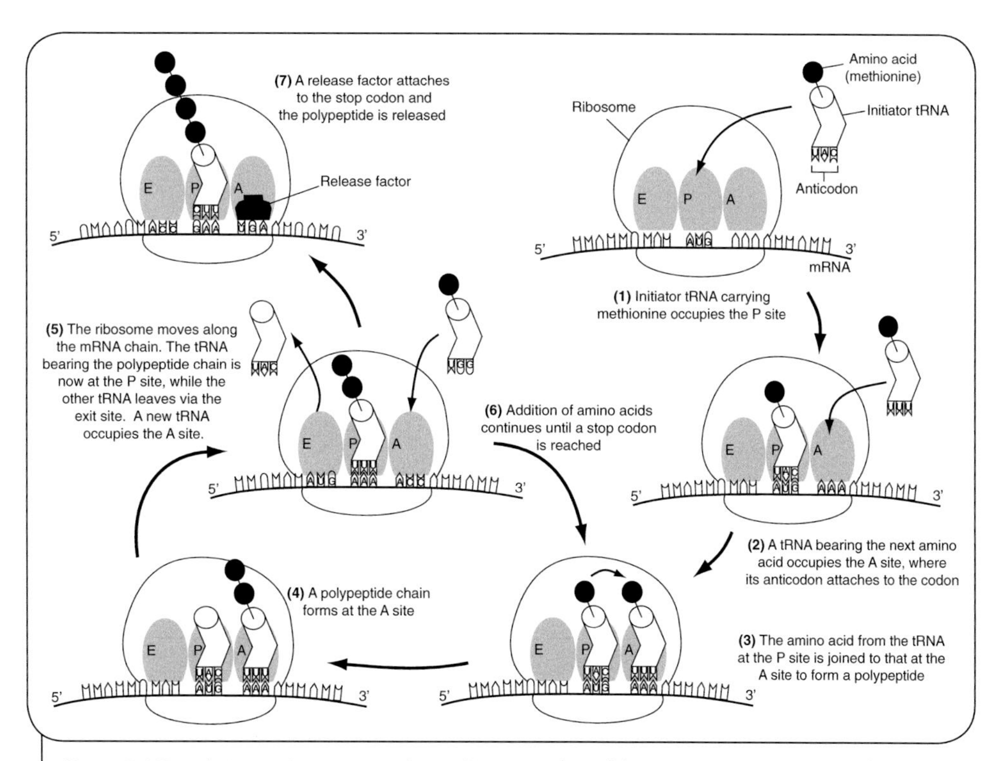

Figure 5.3 Translation and protein synthesis. (Source: Adapted from Figure 17.26, p. 331 of Campbell, N.A., Reece, J.B. and Meyers, N. (2006). *Biology*. Pearson Education Australia. 7th edn. Copyright © 2005 Pearson Education Inc. Reproduced by permission.)

leaving the A site open for the next tRNA molecule. This process continues until a stop codon is reached. A special protein termed a release factor pairs with the stop codon and releases the polypeptide chain from the last tRNA and translation ends. Transfer RNA molecules are continually reused by attaching a new amino acid once the tRNA molecule has left a ribosome. In addition, a single mRNA molecule is generally translated many times over before it is broken down into nucleotide subunits. Finally, many polypeptide chains undergo post-translational modifications and are folded into the correct three-dimensional shape before they are functional proteins.

The sequences of bases (termed genes) in DNA therefore manage the processes of cells by specifying the properties of the functional molecules, proteins and RNA. The first step in the expression of a gene is the formation of an initiation complex on the promoter region of the gene in response to environmental and/or internal cues. Next, transcription results in a complementary RNA molecule that, with or without modification, either carries out cellular functions itself or is translated into a protein. Figure 5.4 summarises the process.

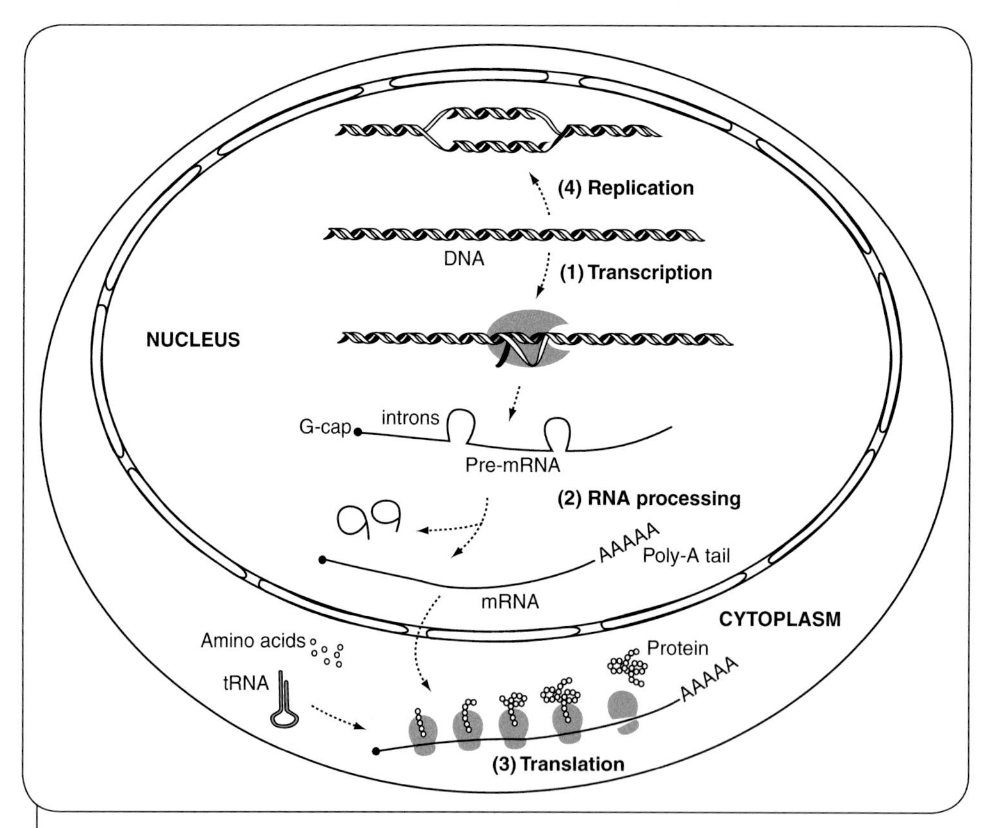

Figure 5.4 Overview of (1) transcription, (2) RNA processing, (3) translation and (4) DNA replication (Source: Adapted from Figure 17.26, p. 333 of Campbell, N.A., Reece, J.B. and Meyers, N. (2006). *Biology*. Pearson Education Australia. 7th edn. Copyright © 2005 by Pearson Education Inc. Reproduced by permission.)

Now that we have examined the way in which the hereditary information is stored in DNA and put into action by cells, we look at the effect of mutations on these factors and the mechanisms cells use to transmit the hereditary information between generations.

Mutations in the genetic code and their effects

Mutations, which are uncorrected changes or errors in the DNA sequence, are the ultimate source of genetic variation. Mutations do not necessarily change gene products: there is redundancy in the genetic code, with the third base frequently interchangeable (Table 5.1). In addition, amino acid substitutions in proteins do not always affect protein function. In this way, changes in DNA and amino acid sequence can accumulate between species in a process termed neutral evolution. Neutral mutations can be used to unravel the evolutionary histories of groups of organisms and can even be calibrated to provide times of divergence. On the other hand, mutations that do change the function of gene products are mostly deleterious, although occasionally they are beneficial. These rare, beneficial mutations are the raw material on which natural selection acts and are the basis for adaptive evolutionary change (see Chapter 6).

It is a basic characteristic of life that genes ultimately specify what organisms can do and their appearance (see Box 5.3). However, there is much scope for variation in gene expression, in response to environmental cues and to mutations in regulatory

Box 5.3 Genome mapping

On 14 April 2003, the International Human Genome Sequencing Consortium announced the successful conclusion of the Human Genome Project. The end result of this project was the complete sequencing of the approximately three billion base pairs that make up human DNA. This international effort has been hailed as one of the most ambitious scientific undertakings of all time, but the human genome is not the only one that has been studied. The genomes of many other organisms have also been completely sequenced or at least mapped (which involves identifying and sequencing a series of landmarks throughout the genome) in detail.

Genome maps help scientists find genes, particularly those involved in diseases or in commercially important characteristics of crop plants or livestock. This may have many practical applications, such as helping to prevent or treat inherited diseases, and selectively breeding more productive animals and plants. However, genome maps or even complete genome sequences are not ends in themselves. They simply provide a starting point in trying to unravel how the genome influences the phenotypic characteristics of human and other organisms. As one of the project leaders of the Human Genome Project said when the entire human genome had been sequenced: 'It's like being given the best book in the world, but it's in Russian, and it's incredibly boring to read'.

genes. Many of the differences between species are not caused by the evolution of novel genes, but by subtle changes in the regulation of gene expression, particularly during embryonic development. The picture that is emerging is complex, with modulation of gene expression a vital component of evolution.

How is the hereditary material copied and distributed through cell division?

Cells reproduce by accumulating resources (growing) and then dividing. The overall process is called the cell cycle and the specific process of division is called mitosis. DNA is duplicated before division and identical copies passed to each progeny cell. Mitosis occurs in normal growth and healing. Meiosis, the cell division that produces gametes (eggs and sperm), differs in that the gametes contain only half the DNA information of the parent cells. The next sections describe how cells copy DNA and distribute it to progeny cells in both mitosis and meiosis.

DNA replication prior to cell division

DNA replication starts with unwinding the double helix, exposing individual nucleotide bases. Single-strand-binding proteins keep the two strands apart for free bases to pair with their complementary bases on each strand and DNA polymerase connects the sugar–phosphate backbones. This method of replication is called semiconservative, because at its completion two identical double-stranded pieces of DNA have been formed, each containing one old and one new strand. It also provides some options for DNA repair. DNA polymerase actually 'proofreads' during DNA replication. If one of the complementary bases is degraded, the cell can remove and replace it using the base on the other strand as a template.

Cell division in prokaryotes

Cell division in prokaryotes is known as binary fission, literally 'splitting in two'. It begins with duplication of the single, circular prokaryote DNA molecule at one particular place. Copying proceeds in both directions until complete. Meanwhile, the cell elongates and the two DNA molecules are moved towards the opposite ends of the growing cell. A septum or division occurs at the midpoint, dividing the original cell in two.

Cell division in eukaryotes

The cell cycle

A eukaryotic cell has from four to more than 100 chromosomes to copy, sort and distribute to progeny cells. Imagine having a very small sewing box (nucleus) packed with 40 very thin and long sewing threads (chromosomes) loosely wound on small bobbins (histones). Now you have to make an exact replica of each thread without taking any of the threads out of the box. Then you must split the box, ensuring that each new box ends up with exactly the same number and types of thread. This happens in cell division.

Cell division in eukaryotes is part of a cell cycle of alternating phases (Figure 5.5). A non-dividing cell is in interphase, which has three stages: growth (G1), DNA duplication (S) and a second period of growth (G2). Cell division itself is called the mitotic phase (Figure 5.5). The mitotic phase is further divided into the overlapping stages of mitosis, involving the division of the nucleus and its contents into two distinct progeny nuclei, and cytokinesis, in which the cytoplasm divides and two progeny cells form. There are

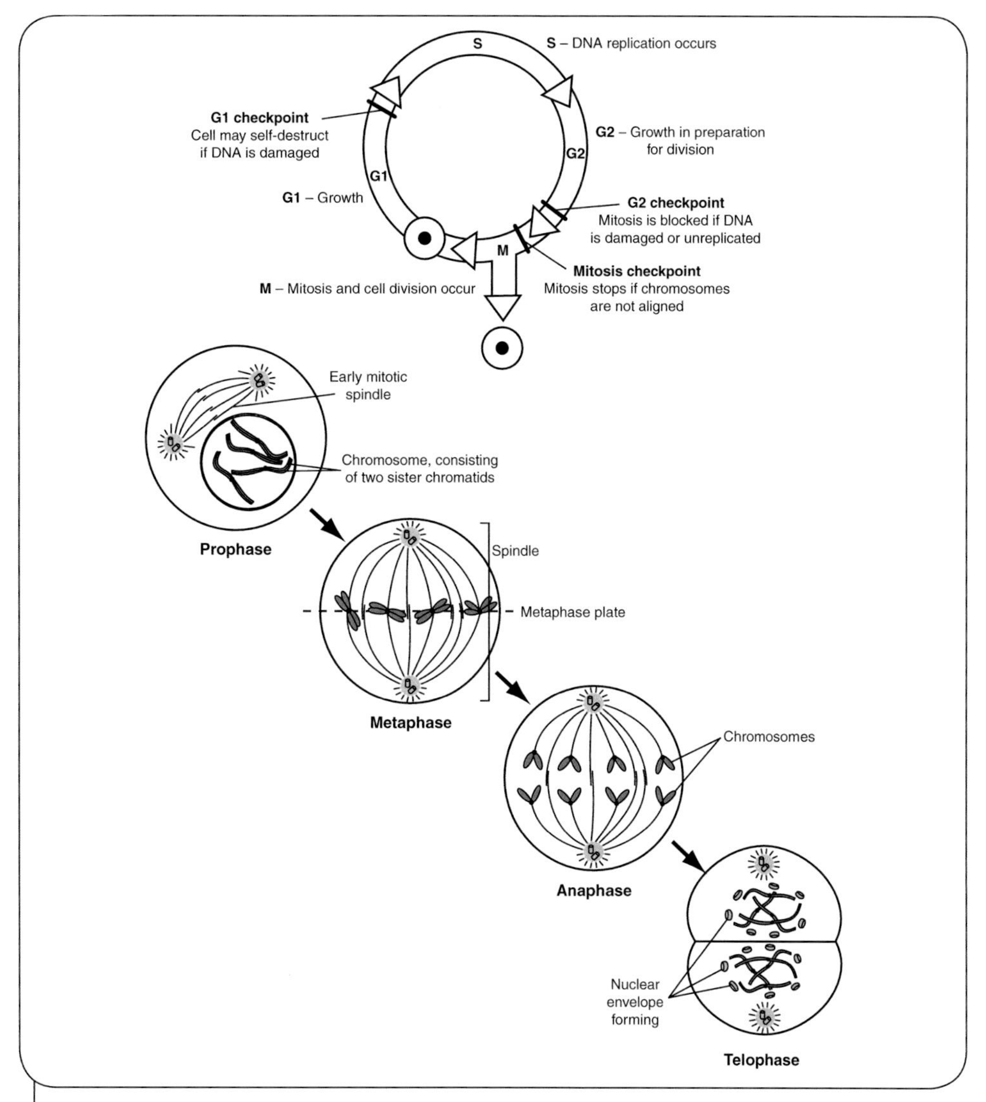

Figure 5.5 The cell cycle and mitosis. (Source: Adapted from Sherman, I.W. and Sherman, V.G. (1983). *Biology: a human approach*. Oxford University Press, New York. 3rd edn)

many checkpoints in the cell cycle to prevent proceeding to the next phase until the current one is completed correctly. This is vital, because a cell that proceeds to mitosis before completing DNA replication, for example, will produce mutant offspring. There are also repair mechanisms if some processes go wrong. If the error is not recoverable, the cell self-destructs.

Mitosis and cytokinesis

DNA duplication has already occurred before the cell nucleus divides. As a result, each chromosome consists of two identical copies called sister chromatids, joined at a centromere. When they separate each is called a chromosome again and each daughter cell will have one of each pair, giving each daughter cell a complete and identical set of chromosomes.

Mitosis, the division of the nucleus, is a continuous process like a moving picture. But just as 'stills' can be selected from a film to illustrate moments of action, snapshots of mitosis can be taken to illustrate critical stages (Figure 5.5). In summary, mitosis begins with the chromatin condensing into distinct chromosomes and the nucleolus disappearing (prophase). The nuclear membrane then breaks down and spindle fibres develop, radiate from the two poles of the cell, and attach to the centromeres of the chromosomes (prometaphase). Next, the chromosomes align across the middle of the cell equidistant from the poles (metaphase), and the sister chromatids separate (and are thereafter called chromosomes) before moving towards the poles of the elongating cell (anaphase). Lastly, the chromosomes uncoil at each pole and a new nuclear envelope membrane forms around each, accompanied by the reappearance of nucleoli (telophase).

Cytokinesis, the division of the cell into two daughter cells, overlaps telophase. In animal cells a cleavage furrow develops at the cell surface, deepens and finally divides the cell. In plant cells, which have a cell wall, a cell plate forms in the middle of the parent cell and grows outwards to divide it.

Mitosis underlies growth and cell replacement. It is also the basis of asexual reproduction in unicellular eukaryotes, or multicellular eukaryotes where new individuals arise from outgrowths from a parent. Sexual reproduction, uniting special sex cells from a male and a female parent, is made possible by a different cell division called meiosis.

Meiosis

In sexual reproduction, sex cells called gametes from each parent (ova from the female, sperm from the male) unite at fertilisation to form a new individual. If gametes had the same chromosome number as the body cells of the species, the new individual would have twice as many chromosomes as each of its parents. However, this does not occur because the gametes are formed by meiosis, a form of cell division ending with four daughter cells, each with half the normal chromosome number. Cells with this reduced number of chromosomes are haploid (n), while normal body cells (also called somatic cells) are diploid ($2n$). Cells undergoing meiosis are called germ cells.

In a diploid eukaryotic cell, therefore, one set of chromosomes is inherited from each parent and the pairs of chromosomes are called homologues. If one member of each homologous pair is assigned to a gamete, the gamete will have genes for the control of all the characteristics of the organism. Homologous chromosomes have the same genes at the same positions (loci), but the actual DNA sequences frequently differ. These different forms of a gene are called alleles (see Box 5.1). Most animals and many plants are diploid; they have two sets of homologous chromosomes in their somatic cells, with their gametes being haploid.

Meiosis reduces the chromosome number by relying on the properties of homologous chromosomes. Like mitosis, meiosis is preceded by DNA replication so that at the beginning of cell division each chromosome consists of two sister chromatids. There are two cell divisions: first, a reduction division called meiosis I in which the homologous chromosomes are separated into two haploid daughter cells; and then a second division called meiosis II in which the sister chromatids in each of the two daughter cells divide, producing a total of four haploid daughter cells (Figure 5.6). Each of the two cell divisions is divided into stages describing chromosome behaviour.

Meiosis I begins with prophase I, when the chromosomes condense and chromosomes from each homologous pair align so that each of the two chromatids in one chromosome pairs with one of the two chromatids of the homologous partner chromosome. Paired

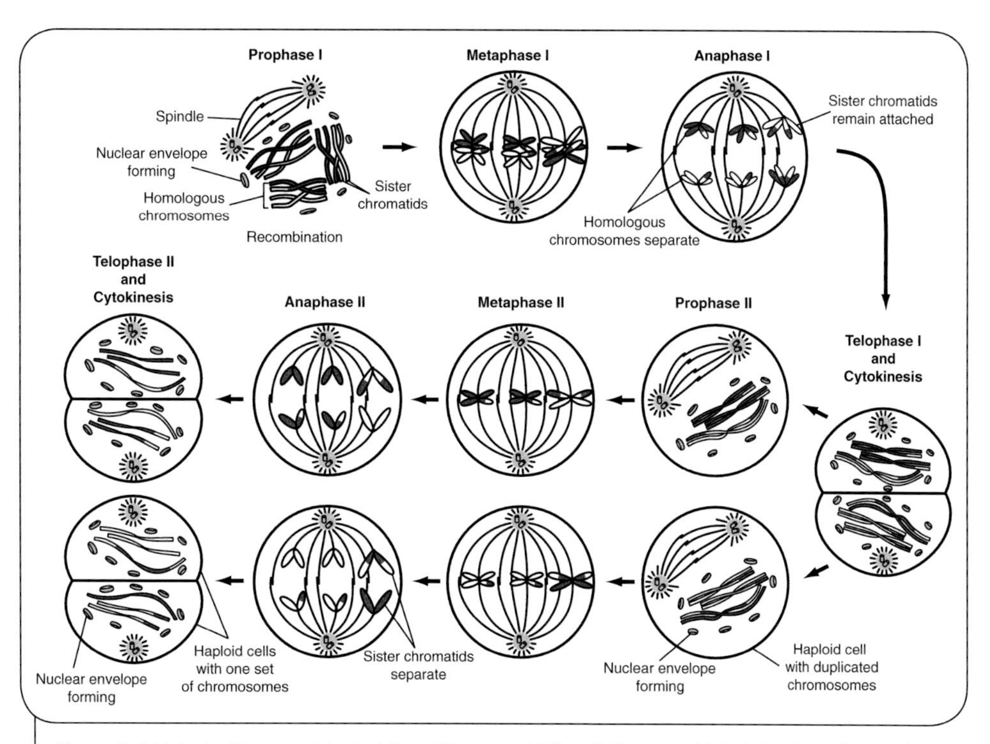

Figure 5.6 Meiosis. (Source: Adapted from Sherman, I.W. and Sherman, V.G. (1983). *Biology: a human approach*. Oxford University Press, New York. 3rd edn)

strands of chromatids from homologous chromosomes may break where they meet and swap parts of their DNA when the broken ends join, not with the original end, but with that from the homologous chromosome. This is recombination, and it increases genetic variation in the gametes (Figure 5.7). Meanwhile the nuclear membrane disappears and a spindle forms. In metaphase I the homologous pairs align across the middle of the cell and spindle formation finishes. During anaphase I, homologous

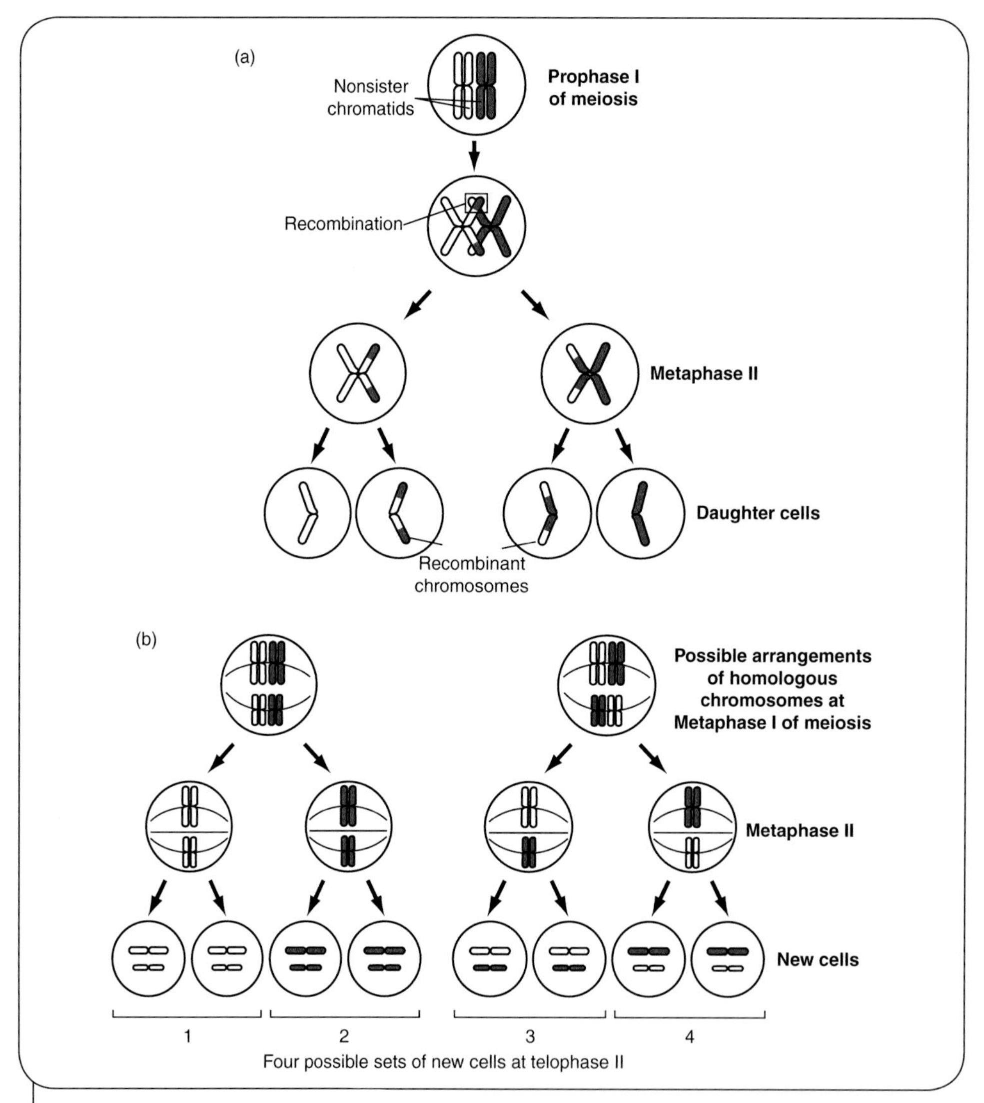

Figure 5.7 Sources of genetic variation in gametes. (a) Recombination exchanges genes between homologous chromosomes during prophase I of meiosis. (b) Independent assortment of homologous chromosomes during the first division of meiosis means that chromosomes from each homologous pair are allocated randomly to each daughter cell. (Source: Adapted from Figure 13.10, p. 248 of Campbell, N.A., Reece, J.B. and Meyers, N. (2006). *Biology*. Pearson Education Australia. Copyright © 2005 Pearson Education Inc. Reproduced by permission.)

pairs of chromosomes separate, with the partners moving towards opposite poles. Maternal and paternal chromosomes are allocated randomly to each pole (independent assortment), which increases the range of genetic diversity in the gametes. Telophase I follows, with the chromosomes arriving at opposite poles of the cell. Each pole has only the haploid number of chromosomes, one member from each homologous pair, with each chromosome consisting of two chromatids. Cytokinesis normally follows. In some species there is an interphase before meiosis II begins, while in others it follows immediately. There is no further chromosome duplication because each chromosome still has two chromatids.

Meiosis II proceeds similarly to mitosis in each of the two daughter cells arising from meiosis I. In prophase II, the spindle forms and the chromosomes move to the middle of the cell. In metaphase II they align across the middle of the cell and in anaphase II the centromeres separate and the sister chromatids, each now again called a chromosome, move towards the opposite poles. In telophase II, nuclei form at each of the poles and cytokinesis occurs. Meiosis is complete and there are four haploid daughter cells. Table 5.2 summarises the similarities and differences between mitosis and meiosis, while Box 5.4 compares the benefits and drawbacks of asexual and sexual reproduction.

Table 5.2 A summary of the similarities and differences between the stages of mitosis and meiosis.

Stage	Mitosis	Meiosis I	Meiosis II
Preparation	DNA replication	DNA replication	
Prophase	Spindle forms	Spindle forms. Recombination between homologous chromosomes	Spindle forms
Metaphase	Sister chromatids line up at the metaphase plate	Homologous chromosomes line up at the metaphase plate	Sister chromatids line up at metaphase plate
Anaphase	Chromosomes divide at the centromere. Sister chromatids, now called chromosomes, move to opposite ends of the cell	Homologous chromosomes move to opposite sides of cell, sister chromatids remain attached	Sister chromatids move to opposite ends
End result	Two progeny cells, genetically identical to each other and the parent cell	Two haploid progeny cells, genetically distinct from the parent cell and each other	Four haploid progeny cells, genetically distinct from the parent cell and each other

Box 5.4 Why do we have sex?

The late wildlife biologist Dennis King reputedly refused to sign his name to a paper with the opening sentence: 'Sexual reproduction is a formidable problem for biologists'. Regardless of jokes and sensitivities, the core issue of that statement remains unresolved. Some unicellular organisms reproduce asexually by simply dividing. Many multicellular organisms reproduce asexually by regeneration of a whole organism from a body part. For example, in misguided early attempts to control the crown-of-thorns starfish (*Acanthaster planci*) on the Great Barrier Reef it was common practice to collect the starfish, cut them up and throw back the pieces! This made the situation much worse as the starfish reproduced rapidly asexually via regeneration. Other multicellular organisms may grow a new individual as a bud arising from the parent and then separating, or by development of an unfertilised egg into a new individual (parthenogenesis). All these methods allow a rapid build-up of numbers from a small starting population, but the progeny are genetically identical to the parents.

Few multicellular organisms rely on asexual reproduction entirely. Most can reproduce sexually, even if they do so infrequently, and sexual reproduction is dominant in most higher animals and plants. Sexual reproduction has some important drawbacks, however. Most importantly, it is inefficient compared to asexual reproduction, because each offspring is only a half genetic copy of each parent, rather than a full genetic copy. Meiosis is also a much more complicated process than mitosis, takes longer and is more prone to error. Production of gametes is also costly as many are not fertilised – a waste of resources. In addition, sexual reproduction requires finding a suitable mate. This can be difficult if the organism is unable to move over distances (such as plants or corals), or occurs only sporadically within a landscape, and may be dangerous if the organism is predatory (for example, female spiders and praying mantises often eat their mates) or has to advertise itself to potential mates and inadvertently does so to predators as well (for example, frogs calling or birds displaying).

Given these disadvantages, why did sexual reproduction evolve and why is it so widespread? One benefit of sexual reproduction is that diploid cells with homologous pairs of chromosomes can repair chromosome damage during meiosis. When homologous chromosomes align during meiosis, an undamaged partner can be used as a template to repair another damaged by radiation, chemicals or heat. This may be why some unicellular eukaryotic cells, which are normally haploid, fuse to form a diploid cell under environmental stress.

A second explanation lies in the long-term advantages of genetic diversity. Genetic diversity is favoured in a changing environment, because characteristics that are advantageous in one set of circumstances may be disadvantageous when things change. Sexual reproduction increases diversity among the offspring through recombination, independent assortment and random fertilisation. Recombination of homologous chromosomes during meiosis produces novel combinations of alleles. Independent assortment of maternal and paternal chromosomes during meiosis also produces novel combinations of alleles by producing a variety of chromosome combinations in the gametes. For example, following meiosis an organism with four chromosomes (two homologous pairs) can produce four different combinations of maternal and paternal chromosomes in the gametes (Figure 5.7). Mathematically, this can be represented as 2^n, where n is the haploid number. As the number of chromosomes increases, the number of possible combinations rises rapidly. For example, the silky oak with a haploid number of 10 has 2^{10} or 1024 possible chromosome combinations in its gametes.

The biology of mating further increases genetic diversity, because which gametes fuse is generally thought to be random. As each gamete is likely to be genetically distinct from other gametes, even offspring from two individuals are generally genetically distinct. In the case of our silky oak above, the number of possible random gametes combinations is 1024×1024 or 1,048,576. An additional source of heritable variation is mate selection, because in a single population there are generally many potential mates for each organism. The choice of mates also introduces genetic variability among the individuals within the population.

Overall, sexual reproduction does not produce new genes or mutations, but it produces individuals with different combinations of alleles, thus increasing the probability that some individuals will survive to reproduce in a variable environment. Sexual reproduction also comes at a cost: a suitable mate must be found, potentially exposing individuals to danger, and some new combinations of alleles may be less successful than those of the parents.

How is an understanding of genetic diversity used in environmental biology?

Molecular techniques for reading the sequences of amino acids in proteins and the sequence of nucleotides in DNA and RNA in organisms have revealed much genetic variation not apparent from body form and structure (morphology). It is now common to study genetic characteristics (usually called genetic markers) together with the phenotypic characteristics of organisms, such as morphology, behaviour and habitat requirements. Together, they are powerful tools for elucidating the evolutionary history of populations and determining how they should be managed and conserved.

Tracing breeding patterns of individuals and genetic exchange between populations

Much can be learned about the breeding behaviour of individuals and populations using short, repeated DNA sequences called microsatellite markers. Generally, these sequences are about four or five bases long and are repeated a varying number of times. They do not code for a gene product as far as we know and mutate rapidly, so individuals have detectable differences in the lengths of their microsatellite fragments. Microsatellite studies are used widely to determine the parentage of individuals and to infer genetic exchange within and between populations.

For example, in winter, humpback whales migrate north from Antarctic waters along the west and east coasts of Australia to calve in warm tropical waters. This migration was once exploited for whaling in Australia, but now it supports whale-watching ecotourism. Molecular studies revealed little interbreeding (or gene flow) between the west- and east-coast whales. Furthermore, there may even be subgroups among the west-coast whales, with little gene flow between them. This knowledge is important for the management of humpback whales, because managers must conserve all the subgroups to maximise overall genetic variability.

A second case concerns control strategies for feral pigs in Australia. Pigs escaping from farms or released for hunting established feral populations across wide areas of eastern Australia, parts of the tropical north and in the south-west. They cause significant environmental and agricultural damage and may spread diseases of people and livestock. In the south-west, microsatellite studies identified eight largely self-contained feral pig populations with little genetic exchange between them. These populations would be difficult to distinguish using conventional ecological methods. A major implication from this work is that outbreaks of wildlife diseases would spread quickly within one of these populations, but more slowly between them. This suggests that controlling pigs at the boundaries between populations might be effective as a barrier to disease spread. A further disquieting discovery was that wild pigs from distant areas had been introduced into wilderness areas by irresponsible people wanting to establish populations for hunting.

Microsatellite techniques may also reveal secrets of a species' breeding biology. For example, the tiny nectar-feeding honey possum (weighing between 5 and 12 g) (Figure 5.8) from south-western Australia has had much of its original habitat destroyed through land clearing. Microsatellite studies of honey possum litters revealed that over 85% of them were sired by multiple males. Females accepting multiple mates increase the genetic diversity of their offspring, a behaviour which may be important in coping with the habitat variability created in the honey possum's fire- and drought-prone environment.

Tracing the tree of life

Other widely used genetic markers for evolutionary studies include ribosomal RNA (rRNA), and mitochondrial and chloroplast DNA. Their value lies in determining the relationships between larger groups of organisms in the tree of life (see Chapter 8).

Ribosomal RNA is the central component of the ribosome, performing essential functions in translation for all organisms. Translation is so vital that changes in rRNA over time are small, so that even distantly related organisms show similarities in their rRNA nucleotide sequences. Nevertheless, sufficient differences accumulate so that organisms may be grouped according to similarities in their rRNA. Those with the most similar rRNA are likely to have shared a recent common ancestor and are therefore most closely related.

Figure 5.8 Microsatellite techniques illuminate the breeding biology of the tiny honey possum. (Source: Lochmann Transparencies)

Mitochondrial and chloroplast DNA are also useful markers, because they are not recombined during meiosis, and are inherited mainly through the maternal line because small male gametes generally do not contribute organelles to the zygote. This property has enabled researchers to trace human origins to the 'mitochondrial Eve', the most recent common mother to modern humans as determined by mitochondrial DNA.

The bar code of life

Scarce information about which species are present in an area is a common barrier to effective management. Currently, expert knowledge is required to identify many plant and animal groups. This is expensive and time-consuming, so an international consortium was formed to establish a rapid and reliable method for identifying individual organisms to species level. This is done using a database of the same short DNA sequence for all known species. The sequence selected is the cytochrome oxidase I (COI) gene of mitochondrial DNA. Almost all eukaryotic organisms have mitochondria and possess this gene. The mutation rate of the sequence is fast enough to discriminate individual species, but not so fast that individuals from a single species are likely to have different sequences. This database assists with the management of species, populations and communities of organisms by providing a method for identification independent of the organismal group.

Chapter summary

1. Deoxyribonucleic acid (DNA) is the hereditary material. DNA occurs in cells as long molecules called chromosomes.
2. DNA nucleotides are the letters of the genetic language. They have a ribose sugar with a phosphate group and one of four nitrogenous bases: adenine, guanine, thymine or cytosine. A nucleotide sequence is formed by a covalent bond between the phosphate of one nucleotide and the 3′ carbon of the ribose sugar of the next.
3. DNA occurs as a double-stranded molecule, with the sugar–phosphate backbones on the outside and the nitrogenous bases on the inside. The strands are held together by hydrogen bonds between complementary nitrogenous bases on opposite strands; adenine pairs with thymine and cytosine pairs with guanine. Complementary base pairing provides stability for DNA and a mechanism for expressing genes and replicating DNA.
4. Genes are DNA sequences coding for a functional product, either ribonucleic acid (RNA) or protein. They occur at specific sites (loci) on chromosomes.
5. Gene expression proceeds once an initiation complex of proteins and RNA is assembled at a special promoter region on the DNA. RNA polymerase then transcribes the template strand into a complementary RNA strand, termed the primary transcript. In eukaryotes the primary transcript is modified into messenger RNA (mRNA) by removal of introns and addition of a G-cap at the 5′ end and a poly(A)-tail at the 3′ end. The mRNA is transported into the cytoplasm, where it is translated via tRNA on ribosomes into a polypeptide chain.
6. Cells undergo cycles of growth and division. Before cell division, or mitosis, DNA is replicated. During mitosis, the chromosomes, each consisting of two identical sister chromatids, are lined up in the midline of the cell, with one sister chromatid on each side. The sister chromatids separate and move to opposite poles of the cell. The cell cytoplasm then undergoes division to form two genetically identical progeny cells.
7. Mitosis is responsible for growth and cell replacement, and is also the basis of asexual reproduction. Asexual reproduction results in clonal offspring that are genetically identical to the parent and each other. This is a useful strategy in stable environments.
8. Sexual reproduction introduces genetic variation among offspring. This is a useful strategy in variable conditions. In sexual reproduction, diploid organisms, with two copies of every chromosome (one from each parent), produce haploid gametes, with only one copy of every chromosome, through meiosis. A new diploid organism is produced when the gametes of two individuals fuse during fertilisation.
9. Meiosis involves two divisions. Before these divisions, DNA is replicated, so that each chromosome consists of two sister chromatids. In the first division, homologous chromosomes pair in the midline of the cell prior to being pulled to opposite poles of the cell. The cell then divides to produce two daughter cells, each with half the original number of chromosomes. In the second division, the sister chromatids are separated and move to opposite poles in each daughter cell. Cell division then produces a total

of four haploid daughter cells. During meiosis, new combinations of alleles are produced in the gametes by recombination of homologous chromosomes and independent assortment of non-homologous chromosomes.

10. Molecular genetics play an important role in conservation by revealing genetic diversity and genetic exchange within and between populations and by elucidating evolutionary relationships and relatedness between organisms.

Key terms

allele
asexual reproduction
binary fission
chromatid
chromosome
co-dominant alleles
codon
cytokinesis
diploid
DNA fingerprinting
DNA sequencing
dominant allele
exon
gamete
gene
genetic code
genotype
haploid
heterozygous
homologous chromosome
homozygous
independent assortment
intron
locus
meiosis
microsatellite
mitosis
mutation
parthenogenesis
phenotype
recessive allele
recombination
sexual reproduction
transcription
translation
zygote

Test your knowledge

1. Explain the function of the genetic material using its chemical structure.
2. Describe the genetic code using language as a metaphor.
3. Describe the processes involved in gene expression in a eukaryotic cell.
4. Describe modification of the primary transcript in eukaryotic cells.
5. Compare the modes of cell division in prokaryotes and eukaryotes.
6. What is meiosis and what are its major stages?
7. Describe the sources of genetic variation in the formation of gametes through meiosis.
8. Describe how molecular genetics can be applied in environmental biology, using a real example.

Thinking scientifically

The team attempting to clone the thylacine had just one individual from which to extract DNA. They were planning to insert the cloned thylacine DNA into cells from a related marsupial, the numbat, from which the nuclear material had been extracted. Assuming this had successfully produced a viable animal, describe the ways in which you predict this animal would be similar to original thylacines, and in which ways it might be expected to be different.

Debate with friends

Fittingly, in 2007 James Watson, one of the discoverers of the DNA structure, was the first person to be given information on the sequence of his entire genome. He said he wanted to know it all – except for the region known to control whether or not he might be expected to suffer from Alzheimer's disease. Would you like to know about your genome? Given the information, who would you wish to (or who should you have to) tell about any deleterious genes: your relatives? your employer? your insurance company?

Further reading

Dawkins, R. (1991). *The blind watchmaker.* Longman Scientific and Technical, Harlow.

Dawkins, R. (2004). *The ancestor's tale: a pilgrimage to the dawn of evolution.* Houghton Mifflin, Boston.

Gould, S.J., (ed.) (1993). *The book of life.* Random House, Milsons Point, New South Wales.

Watson, J.D. and Crick, F.C.H. (1953). 'A structure for deoxyribose nucleic acid'. *Nature* 171: 271–8.

6

Evolutionary theory – the origin and fate of genetic variation

Alan Lymbery and David Ayre

The genetic challenges of captive breeding

Australia has a unique freshwater fish fauna, with over 75% of species found nowhere else in the world. Unfortunately, more than 10% of these species are under serious threat of extinction, mostly from loss and degradation of their freshwater habitat. One option which is now being considered to help conserve native fish species is breeding them in captivity and then releasing captive-bred fish to boost population sizes in the wild. This raises many challenging questions.

First, where should we collect the fish for the captive-bred population? Can we assume that all fish of the same species are the same, or is there important genetic variation between populations in different rivers?

Secondly, how should we go about a successful breeding program? How many fish do we need to collect to found the breeding population? Should we put all the breeding animals together and let nature take its course, or establish male and female pairs and let them breed separately?

Thirdly, how should we release the captive-bred fish? Will captive-bred animals survive and reproduce in the wild? Should they be released only into rivers from which that species has disappeared?

These are all questions involving evolution and to answer them requires an understanding of evolutionary theory.

Chapter aims

In this chapter we study the genetic basis of variation among organisms, the way this genetic variation is inherited from one generation to the next, the forces that shape the distribution of genetic variation within and between populations, and what it means to say that populations of organisms belong to the same or different species.

Evolution

All known human cultures have had explanations for the origin of the Earth and its biota. Some of these explanations regarded different species as fixed and unchanging over time, whereas others argued that they changed gradually in a process of evolution and that living

Box 6.1 Evidence for evolution

Evolution is supported by so many diverse lines of evidence (Figure B6.1.1) that it is now accepted as a fact by almost all biologists.

Evidence from fossils

Fossils are the preserved remains or traces of once-living organisms. Hard parts of organisms, such as teeth, bones, shells, leaves, spores and pollen, are most often preserved, but soft parts, footprints and tracks may also be fossilised. Fossils are found most commonly in rocks, but may also be preserved in amber or in permafrost. Rock fossils are created when organisms are buried in sediment, the calcium in their bones or other hard tissues mineralises and the surrounding sediments harden to form a rock. Fossilisation is rare, so only a few of the species that have ever existed are known from fossils.

Eighteenth-century naturalists studying fossils realised that the Earth once supported very different species of organisms to those with which they were familiar, although clear similarities suggested that existing species might be derived from extinct ones. Prior to the end of the 19th century, rocks and the fossils they contained were dated by their position relative to each other (naturalists assumed the rocks in deeper strata were older). With the discovery of radioactivity, rocks and fossils were dated absolutely using radioactive elements that decay at a known rate into non-radioactive isotopes. Once a rock is formed, no additional radioactive elements are added, so the ratio of radioactive to non-radioactive isotopes in the rock estimates its age.

Dated fossils provide a history of life extending back at least 3.5 billion years on an Earth formed about 4.5 billion years ago (see Chapter 7). The oldest fossils are traces of prokaryotes, followed by simple eukaryote forms and then the sequential appearance of more-complex forms. For example, fish are the oldest fossil vertebrates, followed by amphibians, then reptiles, then birds and lastly mammals. At finer levels of detail, fossils show transitions between forms, such as the gradual change from large land-dwelling mammals to aquatic whales about 55 million years ago.

Evidence from biogeography

Naturalists travelling on the great European voyages of discovery in the 19th century encountered unique animals and plants on different islands and continents. This suggested that the living species were descended from ancestors that had been isolated in these

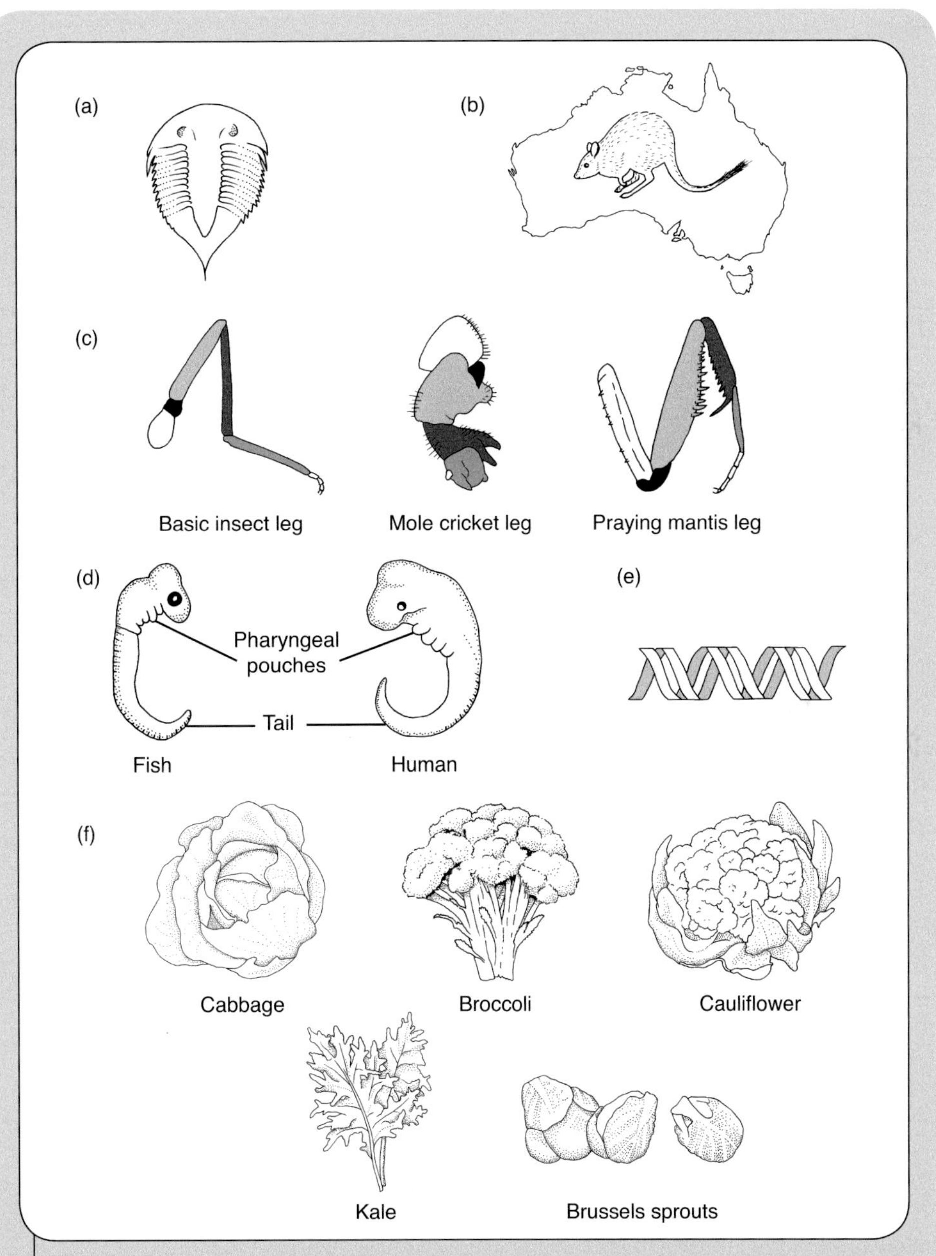

Figure B6.1.1 Evidence for evolution comes from studies of (a) past life forms revealed in the fossil record, (b) unique collections of species in isolated continents and islands (biogeography), (c) homologous structures such as insect limbs, (d) homologous structures in embryos, such as these vertebrates, (e) the common molecular basis of life, (f) artificial selection in domestic species, such as these five vegetables all derived from wild mustard. (Source: B. Cale)

continued ›

Box 6.1 continued ›

areas. Australia's marsupials are an example of a diverse fauna developing in isolation. Similarly, in New Zealand a great diversity of birds, including many flightless species, arose in the absence of ground-dwelling mammals.

Evidence from comparative anatomy

Comparative anatomy is the study of the comparison of body structures in different species. In some cases, such as the legs of different insects in Figure B6.1.1, the structures are variations on the same plan, but modified for different functions. Such structures are called homologous (see Chapter 8), and could logically arise by modification of an ancestral form.

Evidence from comparative embryology

The early embryological stages of many organisms are strikingly similar. For example, early embryos of all vertebrates (fish, amphibians, reptiles, birds and mammals) have pharyngeal pouches (slits at the side of the throat) and a tail. Although they develop into different structures or are lost in the adult animal, they do indicate common ancestry.

Evidence from molecular biology

The DNA sequence of organisms documents not only the hereditary history of individuals but also of species. All species share a universal genetic code based on DNA and RNA (see Chapter 5), suggesting a common origin for all life. Just as siblings have DNA more similar to each other than to unrelated members of their species, species with similar genetic sequences share a recent common ancestor (see Chapter 8).

Evidence from domesticated plants and animals

People have long been interested in pedigrees, both of themselves and also of their domestic plants and animals. By deliberately breeding from stock with desirable characteristics, breeds with very different appearance or behaviour were developed. For example, the vegetables broccoli, kale, cabbage and cauliflower look very different, but they were all derived by selective breeding from a single species of wild mustard. Similarly, different breeds of domestic dog arose from a common ancestor through selective breeding. This process, called artificial selection, illustrates the capacity of species to diverge in form within a few generations of selective breeding.

species are descendants of earlier ones. From the 1700s onwards, a steady accumulation of evidence has supported the evolutionary view of life on Earth (Box 6.1).

The famous naturalist Charles Darwin defined evolution as 'descent with modification'. Darwin was not the first to propose that organisms changed over generations, but he made an unrivalled contribution to biology by identifying a process, which he called natural selection, which could both generate evolutionary change and explain why organisms seemed well suited to their environment. Natural selection is an inevitable consequence of four conditions: (1) there is inherited variation within species; (2) some inherited variations are favourable relative to others; (3) not all young produced in each generation survive; and (4) individuals with inherited variations best suited to their environment are more likely to survive and breed. Over time, individuals with characteristics best suited to the environment increase in the population.

Natural selection cannot produce perfect organisms, because it works by modifying existing structures. It cannot scrap the entire body plan or biochemical processes of an organism and build a new design from scratch (hence the problem of photorespiration – see Chapter 4, Box 4.3). Furthermore, the characteristics of an organism are often compromises between competing functions. A waterbird's feet, for example, must both swim effectively in water and support the bird on land. Nevertheless, natural selection constantly refines the match between a species' characteristics and the environment.

Darwin's theory of evolution through natural selection is one of the great unifying themes of biology. The geneticist Theodosius Dobzhansky famously said 'nothing in biology makes sense except in the light of evolution'. Although Darwin raised many of the key questions that continue to fascinate modern biologists, our understanding of evolution has advanced considerably since his time. The key to this advancement has been a much more detailed understanding of how genetic variation originates and how it is inherited from generation to generation.

The origin of genetic variation

Phenotypic characters

The phenotype of an organism consists of all its observed structural and functional features (see Chapter 5). Different phenotypic features are called characters or traits. Characters are, to a certain extent, artificial constructs, because what we call a character depends on how we wish to divide up the phenotype of an organism into its component parts. In general, we recognise something as a character if it varies, either among different individuals of the same species, or among different species.

For example, consider a flock of budgerigars drinking at a water hole in central Australia. The sex of the birds is a character: some are male and some are female. Body weight is also a character: even birds of the same sex and age may differ in weight. Plumage colour is a character: although most birds are predominantly green, occasional blue or yellow birds are seen. Frequency of preening is a character: like all flying birds, budgerigars usually clean their feathers after any activity, but some birds preen more than others. If we made a list of all the characters we could think of, there would be two fairly distinct types. First, there are those characters such as sex and plumage colour, the values of which fall into discrete categories: male or female for sex; green, blue or yellow for plumage colour. These are discrete or categorical characters. Secondly, there are characters such as body weight and frequency of preening, the values of which may fall anywhere along a continuum: usually between 25 and 35 grams for adult body weight; and between 60 and 180 minutes per day for time spent preening. These are continuous or quantitative characters.

The effect of genetic variation on phenotypic characters

Phenotypic characters vary among individuals of the same species for two reasons – genetic differences and environmental differences. The phenotype of an organism is determined by its underlying genetic constitution (the genotype), and by its physical, chemical and biological surroundings (the environment). Differences among individuals in genotype or environment will lead to differences in phenotypic characters.

Consider genetic variation first. The ultimate source of genetic variation is mutation, which is a change in the sequence or arrangement of DNA. Mutations occur randomly throughout the genome, either arising spontaneously or being induced. Spontaneous mutations are mistakes occurring during DNA replication in mitosis or meiosis. The molecular 'proofreading' systems to correct DNA replication errors are not perfect, and mistakes occur at the rate of about one nucleotide per billion replicated. Induced mutations arise from damage to DNA by environmental mutagens, such as radiation and certain chemicals. With both spontaneous and induced mutations, we can make a distinction between point mutations, which are changes to one or a small number of nucleotides in the DNA sequence, and chromosome mutations, which are changes in the number or structure of chromosomes and affect many nucleotides.

The effect of a mutation depends on what it does to the organism's phenotype and where in the body it occurs. Because of redundancy in the genetic code (Chapter 5, Table 5.1), some mutations change the DNA sequence of a gene but have no effect on the amino acid sequence of the corresponding protein, and therefore no effect on the phenotype. Even when mutations change the protein sequence and the phenotype, the change may not affect survival or reproduction. Such mutations are neutral. Mutations altering survival or reproduction are generally harmful, because they are unlikely to improve the function of the phenotypic character that the protein influences. Very occasionally, beneficial mutations arise and improve survival or reproduction. Mutations occurring in the somatic (non-reproductive) cells of a sexually reproducing organism may change the phenotype, but are not passed on to future generations. To be inherited by the next generation, mutations must occur in the reproductive or germ-line cells.

The inheritance of genetic variation

Although the chances of a mutation occurring at any particular nucleotide in a genome are quite low, the average mutation rate across the entire genome is surprisingly high (Table 6.1). Consequently, many genes in an organism exist in different forms or alleles (see Chapter 5, Box 5.1), which have arisen from mutations over evolutionary time. The concept of alleles, and how they are inherited from generation to generation, was first deduced by Gregor Mendel in the 1860s. This pattern of inheritance is now called Mendelian inheritance.

To understand how Mendelian inheritance works, consider a phenotypic character determined by a single gene, such as plumage colour in budgerigars. Wild budgerigars have predominantly green feathers (Plate 6.1a). Other colours, such as blue, yellow and white, do occasionally arise (Plate 6.1b), but probably make their bearers easy targets

Table 6.1 Estimated mutation rate per genome per generation for different organisms (Source: Data from Drake, J.W. (1999). *Proceedings of the National Academy of Sciences, USA* 88: 7160–4; Drake, J.W., Charlesworth, B., Charlesworth, D. and Crow, J.F. (1998). *Genetics* 148: 1667–86; Grogan, D.W., Carver, G.T. and Drake, J.W. (2001). *Proceedings of the National Academy of Sciences, USA* 98: 7928–33)

Species	Kingdom	Number of mutations per genome per generation
Escherichia coli	Bacteria	0.0025
Sulfolobus acidocaldarius	Archaea	0.0018
Neurospora crassa	Fungi	0.003
Drosophila melanogaster	Animalia	0.14
Homo sapiens	Animalia	1.6

for predators. These colour varieties are, however, interesting to breeders of domestic budgerigars and the genetic control of many of them has been studied in detail.

Whether the plumage is predominantly green or blue is determined by a single gene at the *f* (for feather colour) locus. There are two alleles at this locus: F, which gives green plumage; and *f*, which gives blue plumage. F is dominant to f. Therefore, three genotypes are possible (FF, Ff and ff), but only two phenotypes (FF and Ff are both green, whereas ff is blue).

To understand how plumage colour is inherited, recall (from Chapter 5) that when gametes are produced in meiosis, homologous chromosomes (and the alleles they contain) segregate. FF birds will produce only gametes with the F allele, ff birds will produce only gametes with the f allele, but Ff birds will produce two types of gametes, half carrying F and the other half f at the *f* locus. When two birds mate, the union of their haploid gametes to form a diploid zygote is random (uninfluenced by which alleles they contain). Therefore, we can use the laws of probability to predict the ratio of different types of genotypes in the offspring when we know the genotypes of the parents. This is called the segregation ratio, often visualised in a simple diagram called a Punnett square (Figure 6.1).

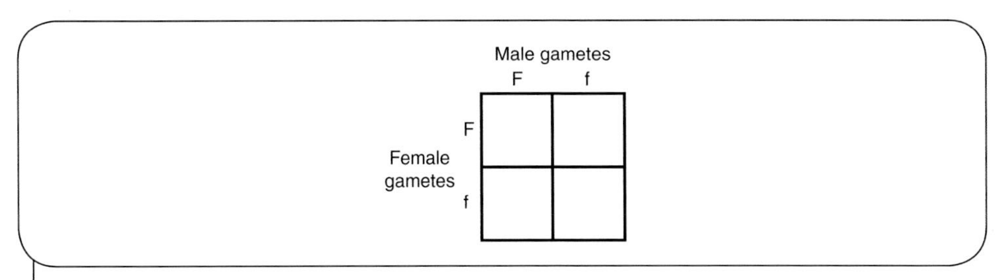

Figure 6.1 Procedure for using a Punnett square. Place the different possible types of male gametes along the top of the square, and the different possible types of female gametes along the side of the square. Each potential zygote is then represented in the boxes inside the square, by the union of male and female gametes. (Source: B. Cale)

Using a Punnett square, we predict that a green bird of genotype FF crossed to a blue bird of genotype ff will produce offspring that are all green, with the genotype Ff. We also predict that when two heterozygous green birds (Ff) cross, their offspring will contain three different genotypes (FF:Ff:ff) in the ratio 1:2:1, and two different phenotypes (green: blue) in the ratio 3:1 (Figure 6.2).

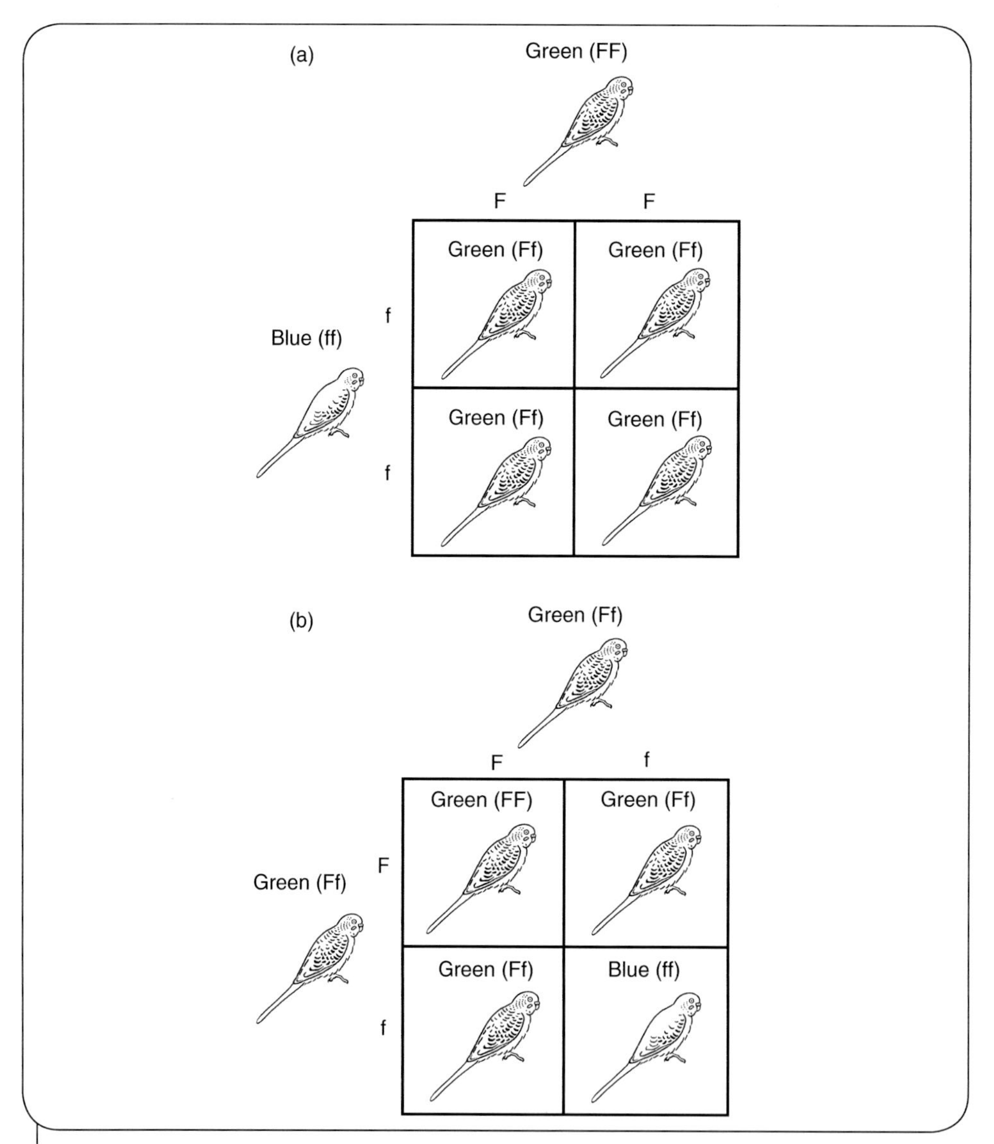

Figure 6.2 The inheritance of plumage colour in budgerigars. (a) The offspring of a green FF bird and a blue ff bird will all have the genotype Ff and a green phenotype. (b) The offspring of two Ff birds can be of three genotypes (FF, Ff, ff; in the ratio 1:2:1), but only two phenotypes (green or blue; in the ratio 3:1). (Source: B. Cale)

The inheritance of continuous characters

Discrete characters are determined by single genes and, because their inheritance follows a Mendelian pattern, they are sometimes referred to as Mendelian characters. However, most characters of importance to organisms, such as body size, speed, endurance, intelligence, fertility and reproductive age, are not Mendelian. They are continuous characters and, at first glance, their inheritance does not seem to follow Mendelian patterns. When individuals with different values for a continuous character are crossed, their offspring do not segregate for the parental values, although they usually resemble the parents to some degree. Nevertheless, we can explain this pattern of inheritance in Mendelian terms.

Continuous characters are determined not by a single gene, but by many genes, each with a small effect on the phenotype. These genes are often called quantitative trait loci (QTLs), and at each QTL influencing a continuous character, different alleles will be segregating from one generation to the next in a Mendelian fashion. To see why this leads to a continuous distribution of phenotypic values, first consider a character that is determined not by many loci, but by only two.

Grain colour in wheat is determined by two different genes (g_1 and g_2), each with two alleles: R_1 and r_1; and R_2 and r_2. R_1 and R_2 produce red pigment, whereas r_1 and r_2 do not. There are nine possible genotypes in a population of wheat plants with both alleles segregating at both loci (Figure 6.3). These nine genotypes give five different phenotypic values for grain colour, depending on how many R alleles are in each genotype. Note that the phenotype is determined by adding together the number of

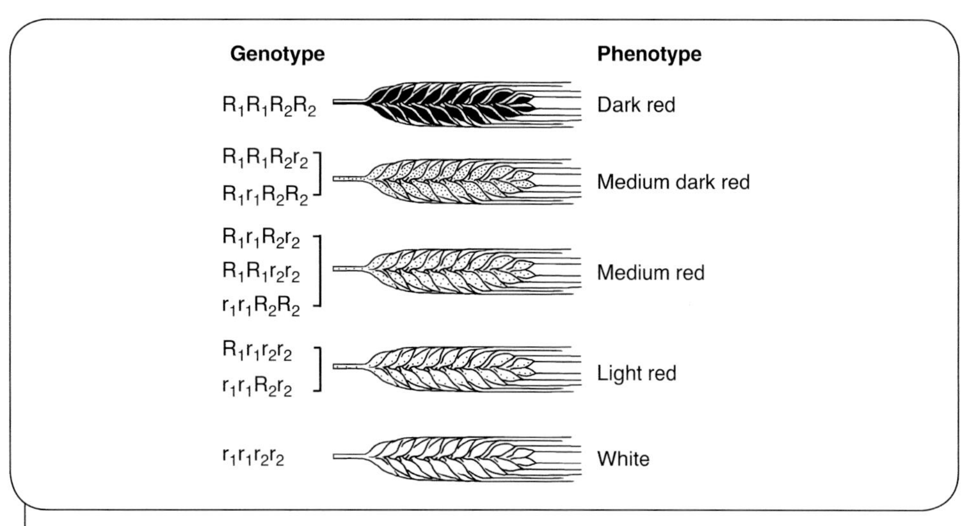

Figure 6.3 The genotypes and phenotypes of seed colour in wheat. The amount of red pigment is determined by two loci, each with two alleles (R_1 and r_1; R_2 and r_2). R_1 and R_2 produce red pigment, whereas r_1 and r_2 do not. In a population of wheat where both alleles are present at both loci, there are nine possible genotypes, giving five possible phenotypes. (Source: Adapted from Knox, B., Ladiges, P., Evans, B. and Saint, R. (2005). *Biology*, McGraw-Hill Australia. 3rd edn)

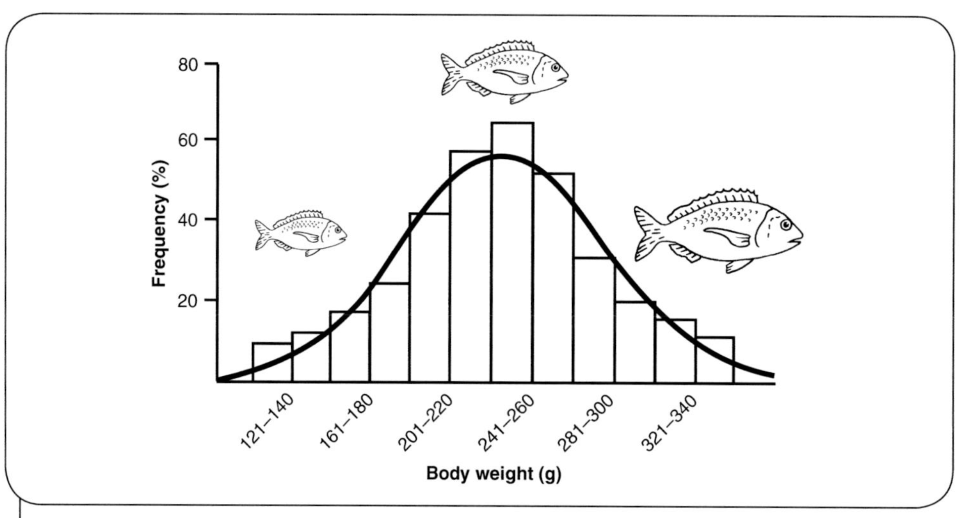

Figure 6.4 Frequency distribution of body weight in a population of black bream (*Acanthopagrus butcheri*). These fish were all spawned in a hatchery at the same time, so they are all the same age. They are destined to be released into the wild to boost declining stocks of this popular recreational angling fish. Body weight is continuously distributed, and a normal curve has been superimposed on the frequency distribution. (Source: B. Cale)

R alleles, regardless of which locus they come from; the two loci are said to act additively. Note also that the extreme values (dark red and white) are determined by only one genotype each ($R_1R_1R_2R_2$ and $r_1r_1r_2r_2$ respectively), whereas the more intermediate values are determined by progressively more genotypes. Therefore, in populations where all alleles are segregating, most plants will have intermediate grain colours and few will have extreme colours.

Now imagine a character determined by the additive effects of alleles at many loci. The range of phenotypic values possible is very large and, when they are plotted, they will tend to follow a normal distribution (Figure 6.4). Because most continuous characters are influenced by many QTLs, they are often called polygenic characters, to distinguish them from discrete characters that are determined by a single gene.

Genetic variation and environmental variation

Variation in a phenotypic character may be caused not only by effects of different alleles at the genes determining the character, but also by effects of different environments. For discrete characters determined by a single gene, environmental variation is usually minor; differences in phenotypic values among individuals almost always mean that they have different genotypes. This is not necessarily true for continuous characters, where variation is always influenced by a combination of genetic and environmental effects. Body weight, for example, is a continuous character determined genetically to some extent; heavier individuals tend to have heavier offspring. However, body weight is also influenced by environment, especially food; genetically identical individuals fed different diets usually have different body weights. The extent to which phenotypic variation

Box 6.2 The meaning of heritability

Although heritability is often used to mean the capacity of a character to be passed from one generation to the next, in genetic terms it has a much more precise meaning. Heritability is the extent to which phenotypic differences between individuals in a population for a particular continuous character can be explained by genetic differences.

There are two different senses in which the word 'heritability' is used. Heritability in the broad sense, sometimes called the degree of genetic determination, describes the extent to which differences between individuals in their phenotypes are determined by differences in their genotypes. Heritability in the narrow sense describes the extent to which differences between individuals in their phenotypes are determined by differences in the genes transmitted from their parents.

The narrow-sense heritability of a continuous character is one of its most important properties, because it determines the rate at which that character will change from one generation to the next in response to natural or artificial selection.

in a continuous character is caused by genetic or environmental variation is called the heritability of the character (Box 6.2).

Microevolution

Genes in populations

Evolution occurs when some individuals are more successful than others in passing their genes to succeeding generations. Evolution is therefore a property of populations. In genetic terms, a population is a group of organisms of the same species, in a defined geographic area, capable of interbreeding or exchanging genes. For many loci, there will be no genetic variation within a population. All birds in most wild budgerigar populations will have genotype FF at the *f* locus. Such loci are called monomorphic and the absence of genetic variation is described as monomorphism. At other loci, however, individuals may possess different alleles and therefore have different genotypes. For example, coat colour in Shorthorn cattle is determined by a single gene with two alleles and all three genotypes (RR, giving red; rr, giving white; and Rr, giving roan, a mixture of white and red, coat colour) are common in Shorthorn herds. Loci where different alleles are segregating within a population are called polymorphic and the presence of genetic variation is polymorphism.

Where loci are polymorphic, the basic description of the genetic constitution of a population is given by genotype frequencies, or the proportion of each genotype at each locus. For example, a herd of 100 Shorthorn cattle may contain 50 red, 25 roan and 25 white animals. The genotype frequencies in this herd are 0.5 RR, 0.25 Rr and 0.25 rr.

In genetic terms, a population is a breeding group and it is alleles, not genotypes, that are transmitted from generation to generation. Therefore, to describe completely the genetic constitution of a population we need to calculate allele frequencies as well as genotype frequencies. Allele frequencies are calculated by counting the number of alleles in each genotype. For the 100 Shorthorn cattle above, there are 2 × 100 = 200 alleles in the population (because every individual has two alleles at a locus). The total number of R alleles is 2 × 50 (every red cow has two R alleles) + 1 × 25 (every roan cow has one R allele) = 125. The frequency of the R allele is therefore 125 ÷ 200 = 0.62. The frequency of the r allele can be similarly calculated as 75 ÷ 200 = 0.38. Note that allele frequencies, like genotype frequencies, always sum to 1.

Hardy-Weinberg equilibrium

Changes in genotype and allele frequencies within a population over time are referred to as microevolution. Microevolutionary change can be produced by several evolutionary forces. The Hardy-Weinberg equilibrium law provides a means of identifying whether or not evolutionary forces are causing microevolutionary change at any locus.

In essence, the Hardy-Weinberg law describes what the genotypic composition of a population should be in the absence of forces promoting evolutionary change. In a large, closed population where no forces are acting to change the genetic constitution, the Hardy-Weinberg law states that (1) allele and genotype frequencies are constant from generation to generation, and (2) there is a simple relationship between allele frequencies in one generation and genotype frequencies in the next. A population satisfying both conditions for any locus is in Hardy-Weinberg equilibrium for that locus. The manner in which genotype frequencies differ from these equilibrium expectations allows us to predict the dominant evolutionary forces acting on a population. To understand this, consider how random mating will generate new individuals in a large, idealised population.

Truly random mating in which all individuals are equally likely to mate with one another may be approximated in the spawning of corals or pelagic fish, where clouds of eggs and sperm are shed into the water. If, for a particular locus, there are two alleles, A and a, which occur with frequencies p and q respectively (such that $p + q = 1$) then we expect p of the sperm and p of the eggs will each have the A allele. Similarly, q of both sperm and eggs will have the a allele. The proportion of the resulting population that should be AA or aa homozygotes is p^2 and q^2 respectively, and the proportion of Aa heterozygotes is $2pq$ (Figure 6.5). The genotype frequencies in the offspring can be expressed in terms of the allele frequencies of the parental generation: p^2 (AA); $2pq$ (Aa); q^2 (aa). Although we will not give the proof here, it can be shown that, in the absence of any other forces causing deviation from random mating, these allele and genotype frequencies remain constant from one generation to the next.

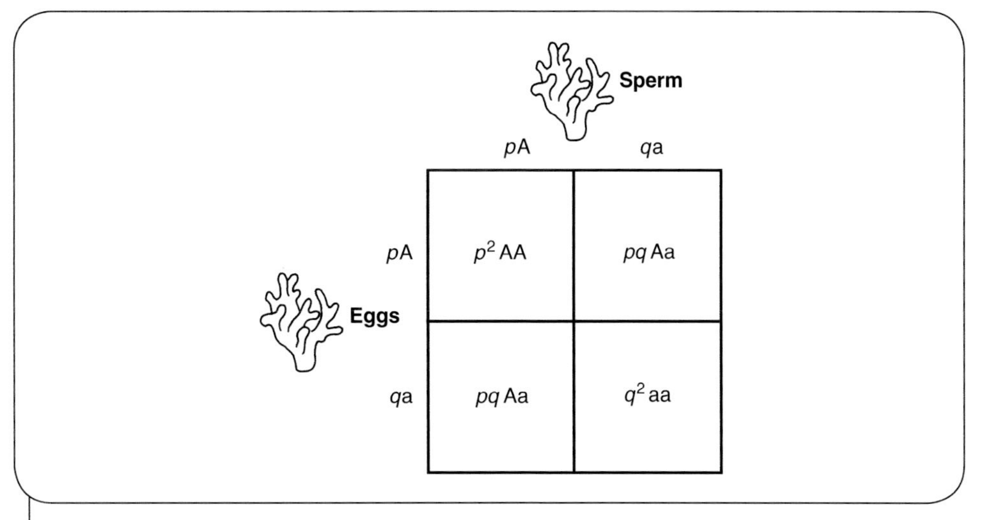

Figure 6.5 The relationship between allele and genotype frequencies in a population at Hardy-Weinberg equilibrium. In a population of spawning coral, sperm and eggs are released into the water. At any given locus, the frequency of each type of sperm or egg will reflect the allele frequencies in the parents; p sperm and eggs carrying the A allele and q sperm and eggs carrying the a allele. The chance that they will combine to form offspring of a particular genotype is given by the product of their frequencies: p^2 for AA, $2pq$ for Aa and q^2 for aa. (Source: B. Cale)

Evolutionary forces causing deviation from Hardy-Weinberg equilibrium

The Hardy-Weinberg law states that allele and genotype frequencies will remain in equilibrium from generation to generation, as long as the population is very large, mating is random, and there is no migration, mutation or selection. Microevolutionary change, therefore, implies that one or more of these assumptions is violated. We will now consider how removing these assumptions causes allele and genotype frequencies to change.

Small population size

The Hardy-Weinberg law assumes that populations are infinite in size. Of course, they are not. In small populations, allele and genotype frequencies may change from generation to generation simply because by chance the gametes from some individuals contribute more to the next generation than the gametes from other individuals. This causes random changes in allele and genotype frequencies between generations, a process called genetic drift. Genetic drift occurs in all finite populations, but is greater in small ones. Founder and bottleneck effects are special cases of genetic drift.

The founder effect occurs when a small number of individuals start a new population. These individuals are likely to carry only a proportion of the genetic variation present in the original population, and therefore the allele frequencies in the new population will differ from the original. Founder effects are particularly important in the colonisation of islands. For example, silvereyes (*Zosterops lateralis*) are small songbirds native to Australia. In 1830, birds from Australia colonised New Zealand's South Island. From there, they colonised Chatham Island, off the east coast of New Zealand, and Norfolk Island, to the northwest. A recent molecular genetic study of these populations found that the average number of alleles at several loci declined along the migratory route, leading to different allele frequencies in Australian, mainland New Zealand and island populations.

The bottleneck effect occurs when normally large populations are drastically reduced in size, for example by catastrophic environmental changes. The survivors are likely, as in the founder effect, to carry only some of the original genetic variation. The Corrigin grevillea (*Grevillea scapigera*) (Chapter 1) went through a severe bottleneck. Extensive habitat destruction in the Western Australian wheat belt reduced the total population size to 47 plants. Genetic diversity among these plants was low, because they represented only a fraction of the original population, raising concerns for the long-term survival of the Corrigin grevillea (see also Box 6.3).

Non-random mating

The Hardy-Weinberg law assumes that individuals within a population mate at random, with respect to the particular locus being considered. Non-random or assortative mating

Box 6.3 Genetic problems with captive breeding

In a captive breeding program, animals or plants in danger of extinction are held in captivity, bred to increase their numbers and, if possible, offspring are returned to the wild. Captive breeding programs were once a collection of haphazard, individualised efforts undertaken by zoos or conservation organisations around the world. These days, however, they are highly organised and coordinated international programs, designed and implemented to conserve genetic diversity. This is commonly expressed as the 90%/200 year rule; the aim being for the captive breeding population to retain 90% of its original genetic diversity for 200 years.

There are two common problems when attempting to retain genetic diversity in captive breeding populations. The first is that, because they are of necessity established from small and endangered wild populations, they often have few founders. This small founding population is unlikely to contain all the genetic variation present in the original wild populations. Rare alleles obviously have low chances of being included.

The second main problem is inbreeding depression. There is usually only a small number of animals in the program and, until recently, pedigree records were often not kept. Therefore matings with relatives can be frequent, reducing fertility and increasing juvenile mortality as a result of inbreeding depression. Detailed pedigrees, exchanging animals between zoos, and storing and transferring sperm can overcome the problems.

occurs in two ways. In positive assortative mating, individuals mate preferentially with others of similar phenotype, whereas in negative assortative mating, individuals mate preferentially with others of dissimilar phenotype. A special case of positive assortative mating is inbreeding, where individuals mate preferentially with relatives. Positive assortative mating and inbreeding do not change allele frequencies in a population, but they do change genotype frequencies, leading to more homozygotes than the Hardy-Weinberg law predicts. Inbreeding occurs naturally in some species, especially plants, but in other species is often associated with reduced survival, growth or fertility. This is called inbreeding depression and is caused by increased homozygosity of harmful, recessive alleles (Box 6.3).

Migration

Migration is the movement of individuals between populations. The genetic consequence of migration is gene flow, or movement of genes between populations. The Hardy-Weinberg law assumes no migration or gene flow. Although this assumption is rarely completely satisfied, the extent to which gene flow changes allele and genotype frequencies depends upon its rate and whether allele frequencies differ between migrants and residents.

Mutation

While mutation is the ultimate source of all genetic variation, mutation rates are normally so low that they do not affect Hardy-Weinberg equilibrium. There is very little chance that mutation will lead to a change in allele or genotype frequencies, unless other evolutionary forces (such as selection) are also acting.

Selection

Selection, the change in allele and therefore genotype frequencies from one generation to the next caused by the different survival or reproductive ability of different genotypes, is often the strongest evolutionary force. It occurs whenever some genotypes contribute more offspring to the next generation than others. Selection is a relative term; it acts in favour of (for) certain genotypes at the expense of (against) other genotypes. The fitness of a genotype is the contribution of offspring it makes to future generations, relative to other genotypes, and the phenotypic characters that enhance the reproductive success of that genotype are called adaptations.

In artificial selection, people decide which genotypes are most fit by allowing some individuals, but not others, to breed. Domesticated plants and animals, such as wheat, corn, sheep, cattle, dogs and cats, are the result of many generations of artificial selection, enhancing characters people find desirable. More recently, the techniques of molecular biology have been used to transfer genetic material between organisms. This is a new, very rapid form of artificial selection.

In natural selection, environmental conditions determine which genotypes in a population are most fit. Natural selection is the main process by which evolution occurs.

Box 6.4 Evolving in response to cane toads

The cane toad (*Bufo marinus*) was introduced into Australia in 1935 to control insect pests in sugarcane fields. Although the toads did little to control insect pests, they proved highly successful invaders, spreading from their original introduction sites in Queensland across much of the north of Australia. Cane toads eat native animals and out-compete native frogs, and are also highly toxic to native species, such as snakes, preying upon them. About 50 species of Australian snakes appear to have been adversely affected by cane toads.

A snake's risk of poisoning by eating cane toads depends on its head size relative to its body mass. The greater the snake's head size, the bigger the toad that can be eaten and therefore the greater the amount of toxin ingested. However, heavier snakes can tolerate more toxin. Ben Phillips and Richard Shine from the University of Sydney found that two of the snakes most vulnerable to poisoning by cane toads, red-bellied black snakes (*Pseudechis porphyriacus*) and green tree snakes (*Dendrelaphis punctulatus*), have evolved smaller head sizes and greater body sizes since they were first exposed to cane toads. This provides strong evidence that these predators have adapted to the invasion of a toxic prey species.

Natural selection can produce very rapid changes in allele frequencies, and therefore in genotype and phenotype frequencies, and there are many contemporary examples of microevolutionary change brought about by natural selection (Box 6.4). Over long periods of time, natural selection can produce big changes in the genetic composition of populations, leading eventually to the formation of new species.

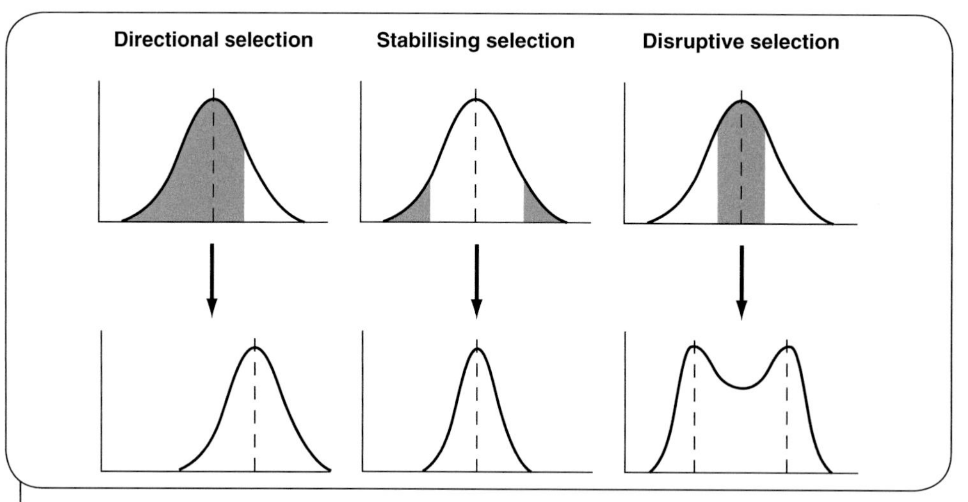

Figure 6.6 Modes of selection on a continuous character. The top panels show the distribution of a continuous character, such as body weight or fecundity, in a population before selection. The phenotypes that will be favoured by selection are white and the phenotypes that will be selected against are shaded black. The bottom panels show the distribution of the character after selection has occurred. (Source: B. Cale)

Although we define natural selection in terms of changes in allele frequencies, caused by the differential survival and reproduction of genotypes, the selecting environment actually exerts its effects on the phenotype of the organism. The phenotypic characters of most importance to survival and reproduction are usually continuous characters, which are normally distributed in a population. Natural selection may act in one of three different ways on a continuous character (Figure 6.6). In directional selection, phenotypes towards one extreme of the normal curve are most fit, and selection shifts the mean or peak of the curve. In stabilising selection, phenotypes near the mean are most fit, and selection reduces spread around the peak of the curve. In disruptive selection, phenotypes at both extremes are most fit, and selection forms two (or more) peaks.

Speciation

The most influential of all books in biology is Charles Darwin's *On the Origin of Species*. Despite the title, however, Darwin did not discuss how species form. Instead, he was concerned with how organisms of the same species change over generations and adapt to their environment – what we now refer to as microevolution. What is the relationship between microevolution and the formation of new species? Before we can answer that question we need to understand what we mean by the term 'species'.

Species concepts

A very general definition of species is that they are the smallest evolutionarily independent units; organisms of the same species are on the same evolutionary pathway, but organisms of different species are on different pathways. Although this is an intuitively appealing idea, it is not always easy to decide whether different populations are on the same or different pathways, and this is why it has been so difficult for biologists to formalise the concept of what constitutes a species.

The most widely used species definition is the biological species concept, proposed by Ernst Mayr in 1942. It defines a species as a group of actually or potentially interbreeding populations that is reproductively isolated from other such populations. There are two complementary parts to this definition. First, members of the same species must be able to mate and produce fertile offspring when in contact. Secondly, members of different species cannot mate and produce fertile offspring when in contact. Supposedly, populations freely exchanging genes will not accumulate genetic differences, and will remain on the same evolutionary pathway.

Although the biological species concept has proven to be a very useful way of describing and understanding the existence of different species in nature, it has not been universally successful, for a number of reasons. The concept has no real meaning for many bacteria, plants and protists, which do not reproduce sexually or regularly exchange genes between individuals. Even for sexually reproducing organisms, the biological species concept is difficult to apply if populations do not normally come into contact in nature. Not only is it impractical to test all these populations for reproductive compatibility, but such tests may be misleading. In addition, it seems that for many plants, and a surprising number of animals, different species coexist for long periods, following different evolutionary pathways despite substantial interbreeding (hybridisation). For example, brown stringybark (*Eucalyptus baxteri*) is quite distinct from messmate (*Eucalyptus obliqua*), yet in coastal Victoria their ranges overlap and they hybridise freely.

For these and other reasons, many alternative species concepts have been proposed. Most are similar to the biological species concept in recognising species as evolutionarily independent units, but they use criteria other than (or additional to) reproductive isolation to determine independence.

Speciation

Speciation is the process by which new species arise. This requires genetic divergence between populations that were once part of the same species. According to the biological species concept and most other species definitions, restriction of gene flow between populations is crucial for genetic divergence.

Barriers to gene flow arise in several ways. Those responsible for most speciation events are geographic barriers such as oceans, deserts or mountain ranges. Populations isolated by geographic barriers are said to be allopatric, and speciation occurring because of geographic isolation is called allopatric speciation (Figure 6.7). Although we rarely witness speciation events, allopatric speciation has been convincingly inferred from the current geographic distributions of many closely related species. For example, in the south-east and south-west of Australia we see spectacular, independent species radiations within the plant family Proteaceae, including banksias, grevilleas and persoonias. The persoonias or geebungs are shrubs with bright green leaves and masses of tubular yellow flowers. There are 74 and 40 species with eastern and western Australian distributions respectively, and only one occurring in both areas. These disjunct distributions provide strong evidence that speciation occurred because the eastern and western populations are isolated by unsuitable dry habitat. Sometimes we can see examples of allopatric speciation in action. The spectacular Papuan kingfisher (*Tanysiptera hydrocharis*) has bright plumage, a strong bill and exceptionally long tail feathers. It is classified as a single species throughout its wide range in New Guinea, but differences between allopatric populations hint strongly at geographic barriers causing genetic differentiation. On mainland New Guinea, populations differ very little, despite variations in topography and climate. Isolated island populations, however, differ from each other and from mainland birds in plumage colour, bill size and tail feathers.

Although allopatric speciation appears to be the most important means by which new species arise, speciation may sometimes occur without geographic isolation. Populations that are not geographically separated are said to be sympatric, and speciation that

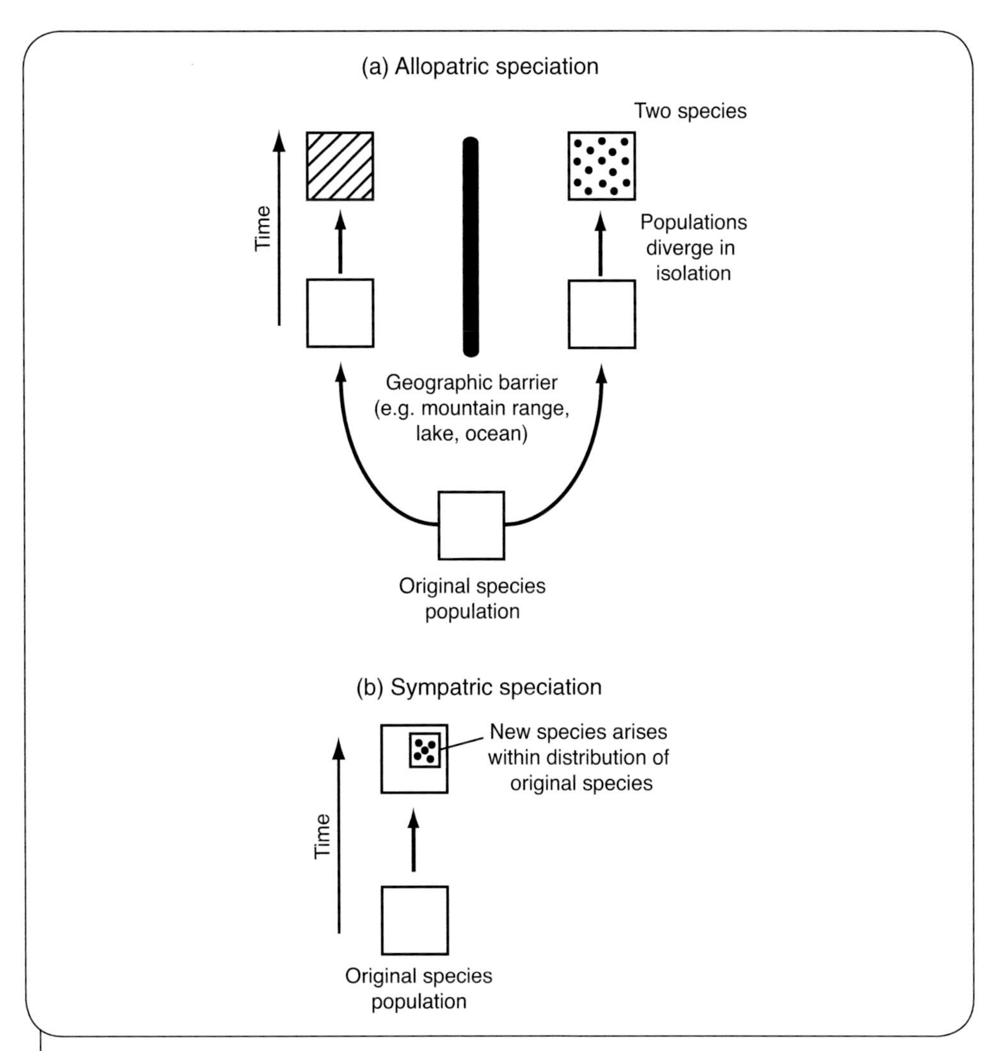

Figure 6.7 Outline of allopatric and sympatric speciation. In the allopatric model (a), two populations diverge in isolation, whereas in the sympatric model (b), they diverge within each other's geographic range (Source: Adapted from Knox, B., Ladiges, P., Evans, B. and Saint, R. (2005). *Biology*. McGraw-Hill Australia, 3rd edn).

occurs without geographic isolation is called sympatric speciation (Figure 6.7). The only type of sympatric speciation that has been convincingly demonstrated occurs through polyploidy, where an individual has more than two sets of chromosomes. Polyploidy often arises when two closely related species hybridise, and from the hybrid a line with double the usual number of chromosomes arises. These polyploid hybrids are usually reproductively isolated from their parental species, because the different chromosome numbers inhibit successful mating. If, however, the polyploids can successfully mate among themselves or are able to reproduce asexually, then a new species has instantaneously arisen. Polyploidy is common in plants and appears to have been important in the evolution of new species. For example, several recent allopolyploid speciation events (hybridisation events that have combined the diploid genotypes of two species) are believed to have occurred within the Australian genus *Glycine*. These plants are the wild relatives of cultivated soybean. Although less common in animals, speciation through polyploidy has been inferred in some grasshopper species and in the gecko *Heteronotia* in Australia.

Mechanisms of divergence

Geographic barriers and chromosomal duplication encourage speciation by restricting gene flow between populations. For speciation to occur, however, the populations must then diverge genetically until they are incapable of successful interbreeding should they come back into contact. Genetic divergence may occur passively through genetic drift, especially if one or both of the populations are small. Natural selection is also likely to play a part in this divergence, particularly if the populations are subjected to different environmental conditions.

Natural selection may also play a much more direct role in enhancing reproductive isolation between populations. If two populations have partially diverged in isolation, to the point where they are still capable of interbreeding but with reduced efficiency (for example, the hybrid offspring have lower fertility), and then come back into contact, selection will favour any genetic changes reinforcing reproductive isolation. Male frogs, for example, attract mates by calling, and different species have distinctive calling patterns to attract females of the appropriate species. Two closely related species in south-eastern Australia, *Litoria ewingi* and *L. verreauxi*, have overlapping ranges. Where both species occur, the calls of each species are distinct, but where only one species occurs, their calls are less distinct. The differentiation of the calls in the area of overlap is an adaptation to prevent males attracting the wrong species of female.

From microevolution to macroevolution

As far as we know, life on Earth evolved only once, so all the species that have ever existed on Earth trace back to a common ancestor (Figure 6.8). Microevolution and speciation produced from this humble beginning the staggering diversity of life that we see around us: an estimated 6–10 million species of organisms and an endless variety of forms at the population level. Our understanding of how this diversity has arisen over evolutionary time has increased enormously in recent years, particularly with new developments in molecular genetics (Box 6.5).

The history of life has not, of course, been one cumulative increase in biodiversity. Populations and species may disappear as well as appear, a process called extinction. The disappearance of a population is called local extinction, whereas global extinction refers to the disappearance of a whole species. The appearance and extinction of species over time is called macroevolution, and the macroevolutionary history of the Earth is the topic of Chapter 7.

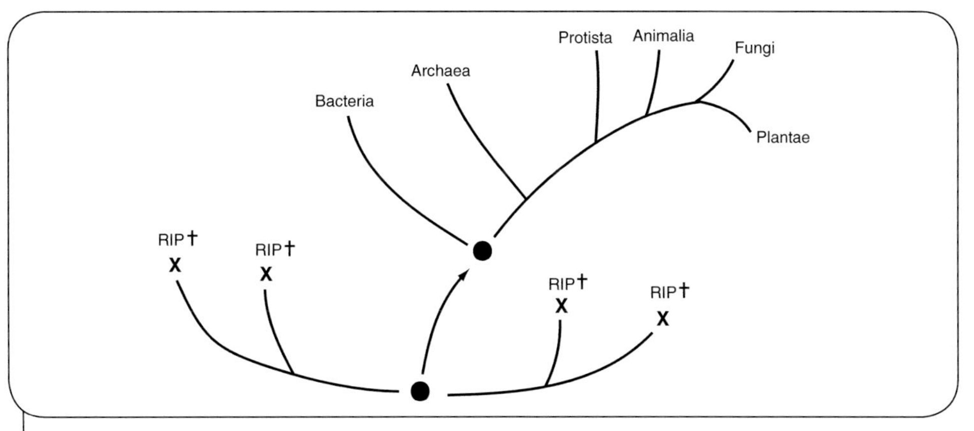

Figure 6.8 A very simplified version of the tree of life. The first living organism presumably had several descendant lines, all but one of which died out. The surviving line gave rise to all the organisms that we see today, represented on the tree by the six major kingdoms of life. (Source: B. Cale)

Box 6.5 A new evolutionary synthesis?

The framework of modern evolutionary theory, as briefly outlined in this chapter, was laid down in the first half of the 20th century, first when Darwin's theory of gradual evolutionary change was reconciled with the newly rediscovered Mendelian genetics, and then when the theoretical population genetic studies of Ronald Fisher, J.B.S. Haldane and Sewell Wright were applied to natural populations by Theodosius Dobzhansky, Ernst Mayr and George Gaylord Simpson. The resulting 'Modern Synthesis' has ever since been the prevailing paradigm of evolutionary biology. In recent years, it has been suggested that this paradigm should be expanded to incorporate findings from a number of new fields of research. Some of these are outlined below.

First, the Modern Synthesis did not really incorporate the study of the ontological development of individual organisms (embryology and developmental biology) into its theoretical framework. In the past 10–15 years molecular genetic techniques have been used to identify many of the genes that control development and investigate how these genes differ among different species. One of the most startling discoveries has been the phylogenetic conservation of genes that regulate morphological development by controlling the expression of other genes. The *Hox* genes, for example, have been found in all animal phyla. They are region-specific selector genes that ensure that the right genes are switched on to cause the correct structures to develop in each region of the body, such as the head at the top of the body. Although *Hox* genes are found in all animals, it seems that expansion in the number of these genes has been associated with some major evolutionary transitions, from animals with two cell layers to those with three cell layers (see Chapter 14) and from invertebrates to vertebrates.

continued ›

Box 6.5 continued ›

Secondly, the Modern Synthesis is essentially a theory of genes, which considers that genetic information flows only in one direction; from DNA to RNA to proteins (sometimes called the central dogma of molecular biology). It is now known that things are actually a little more complicated than this, and that sometimes the regulation of gene activity which occurs during an organism's lifetime can be inherited by the next generation (this is called epigenetic inheritance). For example, in many organisms the expression of genes is controlled by modification of the structure of DNA through the addition of a methyl group ($-CH_3$); genes can effectively be switched off through methylation. In some cases it has been discovered that genes are transmitted to offspring in this new, methylated state. Although the role of epigenetic inheritance in evolution is a topic of much debate, it does provide the potential for an enormous increase in the complexity of inherited genetic information.

Thirdly, most of the important features of organisms, as we have seen, are continuous characters rather than Mendelian characters. The Modern Synthesis, however, was developed from Mendelian genetics and it is only in quite recent times that quantitative genetic concepts have been incorporated into the study of evolution. This has led to an increased appreciation of the role of environmental modification of the phenotype (phenotypic plasticity) in evolutionary changes. The extent of phenotypic plasticity of a character, for example, can itself be selected for and become an adaptation to variable environments; this may complicate in many ways the role that natural selection plays during evolution.

Finally, the application of a relatively new class of mathematical techniques (often called complexity theory) to biology has suggested that living organisms, like other complex systems, show a tendency for self-organisation. This has led to the theory, still very controversial, that the increase in complexity of form during evolution may in part be due to the spontaneous behaviour of biological systems, as well as the influence of external evolutionary forces such as natural selection.

Chapter summary

1. Evolution is the gradual process by which the present diversity of life arose from a common beginning about 3.5 billion years ago. The theory of evolution is strongly supported by evidence from the fossil record, biogeography, comparative anatomy, comparative embryology, molecular biology and artificial selection.
2. Natural selection, a mechanism of evolution proposed by Charles Darwin, says that individuals with hereditary variations best suited to their environment are more likely to survive and breed. Over time, their characteristics predominated in the population and can accumulate to create new species.
3. The phenotype (the observable characteristics) of an organism is determined by the genotype (genetic constitution) of the organism and the environment. Different phenotypic features of an organism are called characters. Discrete characters are those (such as sex or coat colour) the values of which fall into a small number of separate categories. Continuous characters are those (such as body weight and length) the values of which are normally distributed along a continuum.
4. Phenotypic characters differ among organisms because of genetic and environmental variation. Genetic variation in characters arises through changes in the sequence or arrangement of DNA (mutations). If mutations occur in germ cells, they are inherited. As a result of mutations, genes often exist in a number of different alleles. During meiosis, alleles segregate and the ratio of offspring genotypes from a given mating is the segregation ratio. Discrete characters are determined by single genes with little, if any, environmental influence. Their inheritance can be described entirely by the segregation ratio of alleles. Continuous characters are determined by many genes, each of small effect, and the environment in which those genes are expressed. Their inheritance is more difficult to predict.
5. The genetic constitution of a population can be described by genotype frequencies (the proportion of each type of genotype at a locus) and allele frequencies (the proportion of each allele at a locus). In a large, randomly mating population with no migration, mutation or selection, genotype and allele frequencies do not change, and bear a constant relationship to each other (the Hardy-Weinberg equilibrium).
6. Microevolution is a change in allele frequencies in a population over generations, occurring when the Hardy-Weinberg equilibrium assumptions are not met because of small population size, non-random mating or selection. Small population size leads to random changes in allele frequencies through genetic drift. Founder and bottleneck effects are special cases of genetic drift, often reducing genetic diversity in threatened species. Non-random mating may also change allele frequencies. Inbreeding is a case of non-random mating in which related individuals mate, causing decreased survival or fertility (inbreeding depression). Selection is the most important factor changing allele frequencies. Natural selection occurs because of differences in fitness (reproductive contribution) of genotypes due to the actions of the natural environment, whereas artificial selection occurs because people regulate mating in domesticated species.

7. There is no single, universally agreed definition of a species. The most widely used definition is the biological species concept, which says that a species is a group of actually or potentially interbreeding populations, reproductively isolated from other such groups. New species arise when populations diverge because of restricted gene flow between them. The most important barriers to gene flow, leading to the evolution of new species, are geographic barriers such as oceans, deserts or mountain ranges.
8. The appearance and extinction of species over evolutionary time is macroevolution.

Key terms

adaptation
allele frequency
allopatric speciation
artificial selection
biological species concept
directional selection
disruptive selection
evolution
extinction
fitness
founder effect
gene flow
genetic drift
genotype
genotype frequency
Hardy-Weinberg equilibrium
heritability
heterozygote
homozygote
inbreeding
inbreeding depression
macroevolution
microevolution
mutation
natural selection
phenotype
polymorphism
polyploid
population
recessive allele
selection
speciation
stabilising selection
sympatric speciation
trait

Test your knowledge

1. How do discrete and continuous characters differ?
2. In a population of 95 captive budgerigars, we know that there are 75 birds with the genotype FF, 15 with the genotype Ff and 5 with the genotype ff. What are the genotype and allele frequencies at this locus in the population?
3. Explain the differences between artificial selection and natural selection.
4. What is the biological species concept? What is the main problem in its practical application?
5. Define the terms 'microevolution' and 'macroevolution'.
6. Summarise the evidence that biological evolution occurs.

Thinking scientifically

In a captive breeding program, organisms under threat of extinction are bred in captivity and their offspring returned to the wild to boost declining populations. Captive breeding programs have helped to bring organisms such as the numbat (*Myrmecobius fasciatus*) and the Corrigin grevillea back from the brink of extinction. List the factors to consider when establishing a captive breeding population of an endangered species. In arriving at your answer, think about what a captive breeding program should be trying to achieve.

Debate with friends

Debate the proposition: 'Evolution tells us nothing about conservation'.

Further reading

Darwin, C. (1859). *On the origin of species by means of natural selection, or, the preservation of favoured races in the struggle for life.* John Murray, London.

Ridley, M. (2004). *Evolution.* Blackwell, Malden, MA. 3rd edn.

Stearns, S.C. and Hoekstra, R. (2005). *Evolution, an introduction.* Oxford University Press, Oxford. 2nd edn.

Zimmer, C. (2001) *Evolution: the triumph of an idea.* HarperCollins, New York.

7

The history of life on Earth

Stephen Hopper

A very distinctive biota

When Charles Darwin invited the British botanical artist Marianne North to visit him in 1879, she was delighted. Her intrepid adventures, late in life, travelling the world to paint plants in landscapes had clearly impressed Darwin, and he conveyed to Miss North an important message. She related: 'He seemed to have the power of bringing out other people's best points by mere contact with his superiority. I was much flattered at his wishing to see me, and when he said he thought I ought not to attempt any representation of the vegetation of the world until I had seen and painted the Australian, which was so unlike that of any other country, I determined to take it as a royal command and go at once'.

The biodiversity of Australia and (to a lesser extent) South America stands out as the most distinctive among the Earth's continents. This has been evident since biologists explored them during the age of European global colonisation and imperialism. Understanding why these biotas are so distinct has occupied many pages of text and produced some very vigorous debates. New approaches and insights now allow us to apply unprecedented scientific rigour to investigating historical biogeography (the distributions of organisms and the factors determining them) and there is growing consensus about how life on Earth has emerged.

This is not to say that all is known. Far from it. There remain extraordinary opportunities still for exploration and major discovery, even in well-settled corners of the planet such as Australia.

Chapter aims

This chapter provides a broad overview of the evolutionary changes that have occurred since life began on Earth, and of the geological and climatic forces that drove them. It starts with a description of the geological timescale and then proceeds to an account of the major evolutionary transitions in the history of life. The recent evolutionary history of the Australian biota is studied in detail.

A deep-time perspective

Life on Earth today is the culmination of geological and evolutionary events extending back possibly 3.8 billion years. To better understand and conserve the biodiversity around us on the Earth today we need to understand the history of these evolutionary events and the mechanisms by which they unfolded.

Geological eras

The geological timescale is divided into four eras (in order from oldest to most recent: Pre-Cambrian, Palaeozoic, Mesozoic and Cenozoic) (Figure 7.1). They were originally delimited because the boundaries between them are marked by abrupt changes in the fossil record. Eras are subdivided into periods and, for the Cenozoic era, periods are further subdivided into epochs (Figure 7.1). Estimation of the ages of materials from different periods is based on radiometric dating, calculated from the rates of decay of radioactive elements (see Chapter 6, Box 6.1).

A deep-time perspective shows how today's interactions between organisms and their environments were shaped in the past. Organisms function with an array of adaptations, many of which evolved in different times and places and often for different purposes than seen today. We rarely see and should not expect a perfect match between organisms and their environments at any specific place and time. Populations thrive, decline or become extinct, depending upon the match of their biological characters with changing environmental circumstances.

This is most evident when considering the impact of humans on environments since the advent of agriculture. Few organisms have responded positively to the high level of disturbance, selective cultivation and domestication accompanying agricultural practice. The prime beneficiaries have been crops and domesticated animals, as well as a small number of weeds and feral animals with fortuitously attuned life histories. By contrast, the vast majority of plants and animals in agricultural regions have simply been swept aside, their habitats destroyed. Their adaptations for survival and reproduction evolved in environments dissimilar to those brought about by massive and rapid agricultural land clearance.

Biogeography

Biogeography is the study of the geographical distributions of organisms and of the historical forces responsible for these distributions. Today, biogeographers recognise six zoological regions and five botanical regions on land, each characterised by vertebrates and seed plants restricted to those regions (endemics) (Figure 7.2). At present, there is no

Era	Period	Epoch	Millions of years ago	Biological events
Cenozoic	Neogene	Recent	0.1–0	Expansion of human civilisations. First hominids. 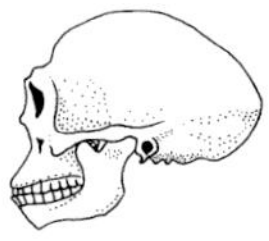
		Pleistocene	1.8–0.1	
		Pliocene	5–1.8	
		Miocene	24–5	
	Palaeogene	Oligocene	34–24	Radiation of flowering plants, insects, birds and mammals.
		Eocene	55–34	
		Palaeocene	65–55	
Mesozoic	Cretaceous		144–65	Extinction of dinosaurs, flying reptiles and many marine animals. First flowering plants. 
	Jurassic		206–144	First birds. Dinosaurs dominant. Radiation of seed plants – conifers, cycads and ginkgos.
	Triassic		248–206	First dinosaurs and mammals.
Paleaozoic	Permian		290–248	Mass extinction of marine and land life. Radiation of reptiles.
	Carboniferous		354–290	First reptiles. First seed plants. Diversification of land plants – ferns and fern allies. 
	Devonian		417–354	Extinction of many marine species. First land vertebrates (amphibians). Radiation of land plants. 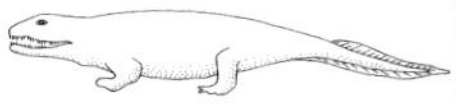
	Silurian		443–417	Colonisation of land by plants, fungi and invertebrates. First jawed fishes. 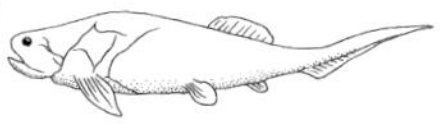
	Ordovician		490–443	Mass extinction of marine life. First jawless fishes.
	Cambrian		543–490	Cambrian explosion of marine life. Earliest chordates. 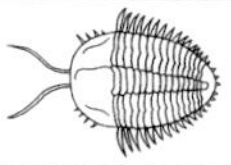
Pre-Cambrian			4600–543	Radiation of Ediacaran fauna. First land organisms. First multicellular organisms. Origin of eukaryotes. Origin of prokaryotic life. 

Figure 7.1 Geological timescale, with major biological events. In some schemes, the Cenozoic era is divided into Tertiary (Palaeocene to Pliocene) and Quaternary (Pleistocene to recent) periods rather than Palaeogene and Neogene periods. (Source: B. Cale)

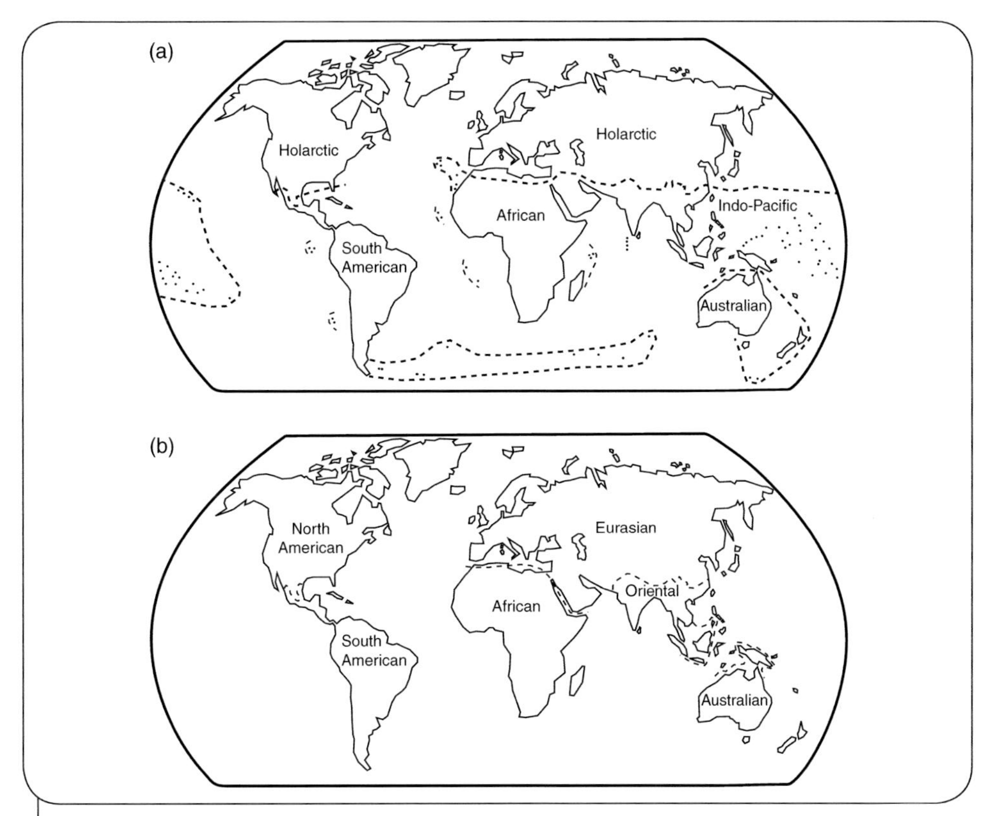

Figure 7.2 Major (a) botanical and (b) zoological regions on land. (Source: Redrawn from Cox, B. (2001). 'The biogeographic regions reconsidered' *Journal of Biogeography* 28: 511–23)

universally agreed classification of bioregions for the oceans. Terrestrial biogeographers have recently begun considering strategic approaches for conserving global biodiversity. In 2000, 25 global biodiversity hotspots were mapped, highlighting places that are richest in unique, threatened species of plants and vertebrates (Figure 7.3). These hotspots deserve priority for conservation action if limited resources are to be deployed effectively to minimise extinctions.

Biologists have puzzled over the reasons for such global distributions of organisms since the 1700s. Our understanding of the events that have led to these distributions has been immeasurably enriched in recent times with new theoretical developments in geology, and with the application of DNA sequencing and new computer-based methods of revealing evolutionary relationships among organisms (see Chapters 5 and 8). Together with the discovery of new fossils, these developments are placing increasingly accurate time frames around major evolutionary events in the history of life and patterns of relationships in organisms.

One of the most important of these revolutionary new approaches has been plate tectonics theory, which recognises that the Earth's crust is divided into a number of plates moving relative to each other, carrying biota with them (Box 7.1). Some movements

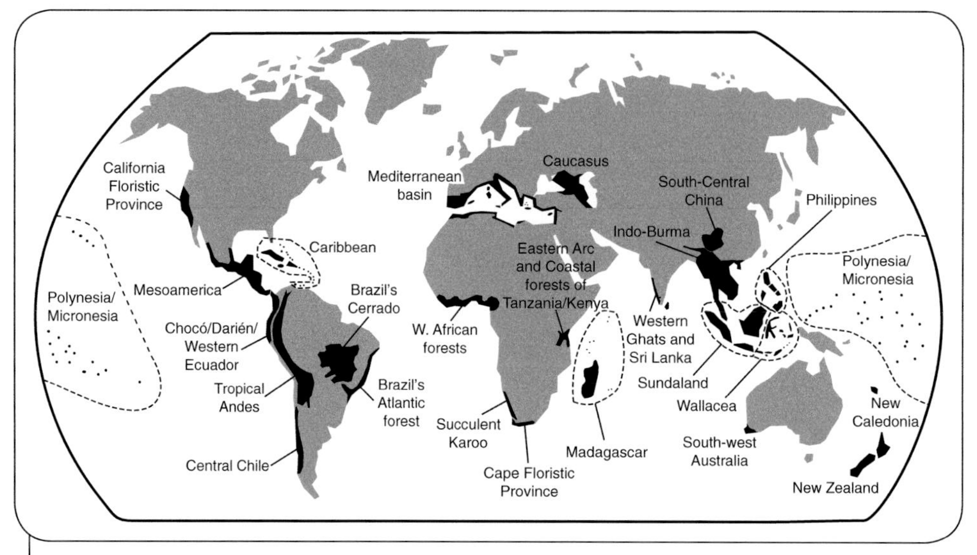

Figure 7.3 Global biodiversity hotspots (dark shading), rich in unique, threatened species of plants and vertebrate animals. (Source: Redrawn from Myers, N., Mittermeier, R.A., Mittermeier, C.G., da Fonesca, G.A.B. and Kent, J. (2000). 'Biodiversity hotspots for conservation priorities.' *Nature* 403: 853–8)

have isolated biotas on different continents, each evolving independently in response to local conditions. Others have brought land masses together, combining their biotas and sometimes creating major environmental features such as mountain chains.

Box 7.1 The moving continents

The belief that the continents of the Earth have not always been fixed in their current positions was first suggested by the Dutch mapmaker Abraham Ortelius in 1596, but it was not until the early part of the 20th century that the idea was developed in detail. In 1912, Alfred Wegener, a German meteorologist, published his theory of continental drift, proposing that the continents we see today split from two huge land masses (Laurasia in the Northern Hemisphere and Gondwana in the Southern Hemisphere) and these were themselves derived from a single supercontinent (Pangaea).

Wegener's theory was controversial and initially rejected by scientists, and it is only in recent years that new geological evidence has resurrected the concept of continental drift in the new theory of plate tectonics. The theory of plate tectonics recognises that the Earth's crust is fragmented into a number of plates, and that these plates move relative to one another. Plate tectonics theory has revolutionised our understanding of the history of the Earth and provided explanations for questions as diverse as why earthquakes and volcanic eruptions are localised to very specific areas, how mountain ranges such as the Himalayas were formed and why identical fossil organisms are found on widely separated land masses.

Through careful geological studies, scientists have been able to reconstruct the position of the Earth's continents through geological time. This reconstruction becomes more difficult as we move back in time and, although there is widespread agreement on continental movements since the Permian (about 270 million years ago), movements prior to that period are less certain. Figure B7.1.1 indicates current views on the position of the Earth's land masses over the last 650 million years.

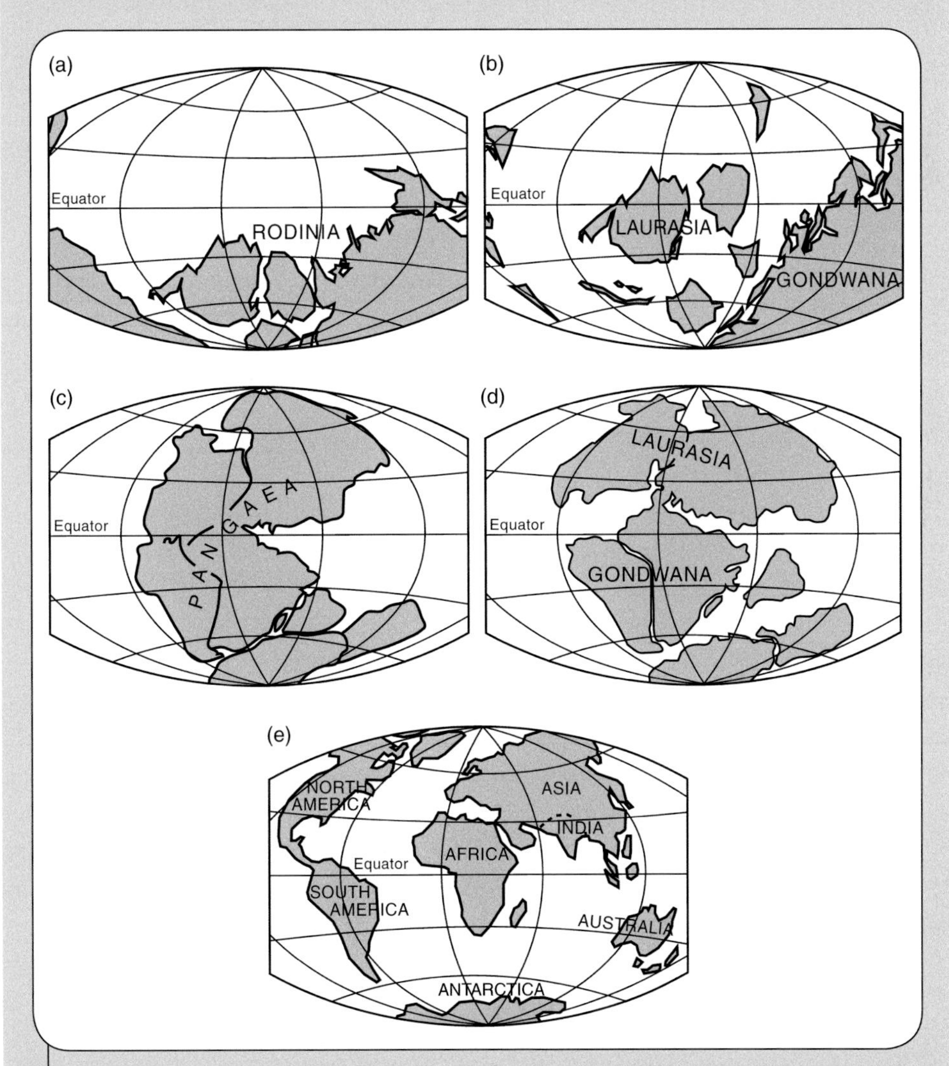

Figure B7.1.1 Global reconstructions of land masses through time: (a) Pre-Cambrian, about 650 million years ago; the supercontinent Rodinia is beginning to break apart. (b) Cambrian, about 500 million years ago; Rodinia has fragmented into a number of separate land masses, of which the largest are Gondwana and Laurasia. (c) Permian, about 270 million years ago; Gondwana and Laurasia have reunited to form the supercontinent Pangaea. (d) Jurassic, about 150 million years ago; Pangaea has fragmented into Laurasia and Gondwana, and these supercontinents are themselves beginning to break apart. (e) Pliocene, about 5 million years ago; the continents are in roughly their present day positions. (Source: B. Cale)

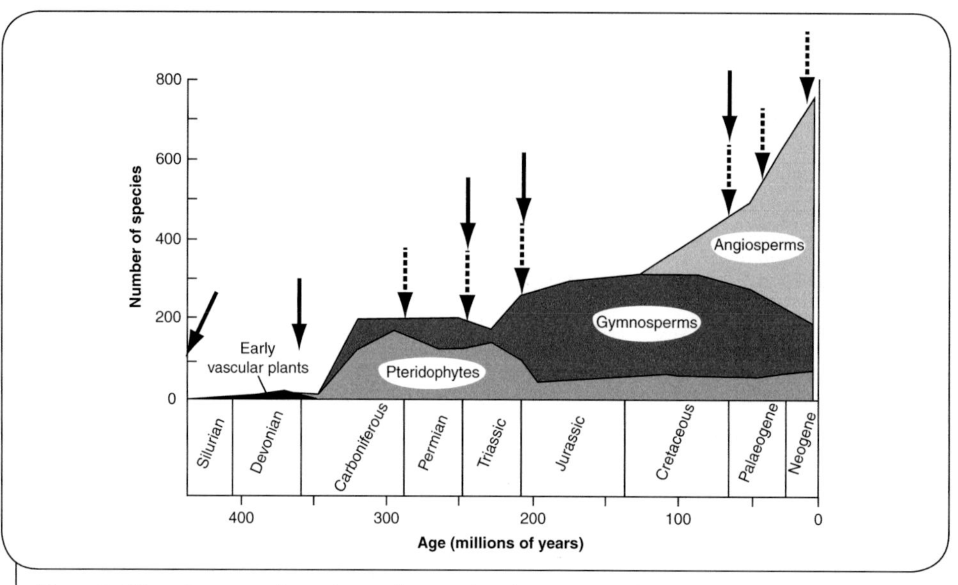

Figure 7.4 Diversity curves through time for vascular plant species, showing the lack of correlation with mass extinction events for marine invertebrate families (solid arrows) and tetrapod mammals (dashed arrows). (Source: Redrawn from Willis, K.J. and McElwain, J.C. (2002). *Evolution of plants*. Oxford University Press)

It is also becoming clear that the current global distribution of organisms has been influenced by major environmental changes, causing mass extinctions and radiations of organisms during evolutionary history (Figure 7.4). Marine and terrestrial animals have endured several mass extinctions. Causes remain controversial, with theories ranging from the effects of large asteroid impacts, to massive vulcanism, basalt flows and ejection of subterranean CO_2 and SO_2 into the atmosphere, causing acid rain and anoxic marine conditions. Interestingly, plants do not display mass extinctions to the same extent as animals, and plant extinctions have been more gradual. However, the large advances in plant evolutionary forms from green algae to land plants, mosses to ferns, and gymnosperms (cycads, ginkgos and conifers) to flowering plants are all associated with high global CO_2 levels at different periods in the Earth's history.

Beginnings

The earliest evidence of life is carbon of biological origin in Greenland's sedimentary rocks dated at 3.8 billion years ago. The likely source was marine prokaryotes such as anaerobic (their metabolism does not require oxygen) archaea and bacteria (see Chapter 9) living in extreme habitats such as undersea volcanic vents.

Photosynthesising prokaryotes called cyanobacteria originated within the next few hundred million years. Today, cyanobacteria form slippery black biofilms on rock outcrops,

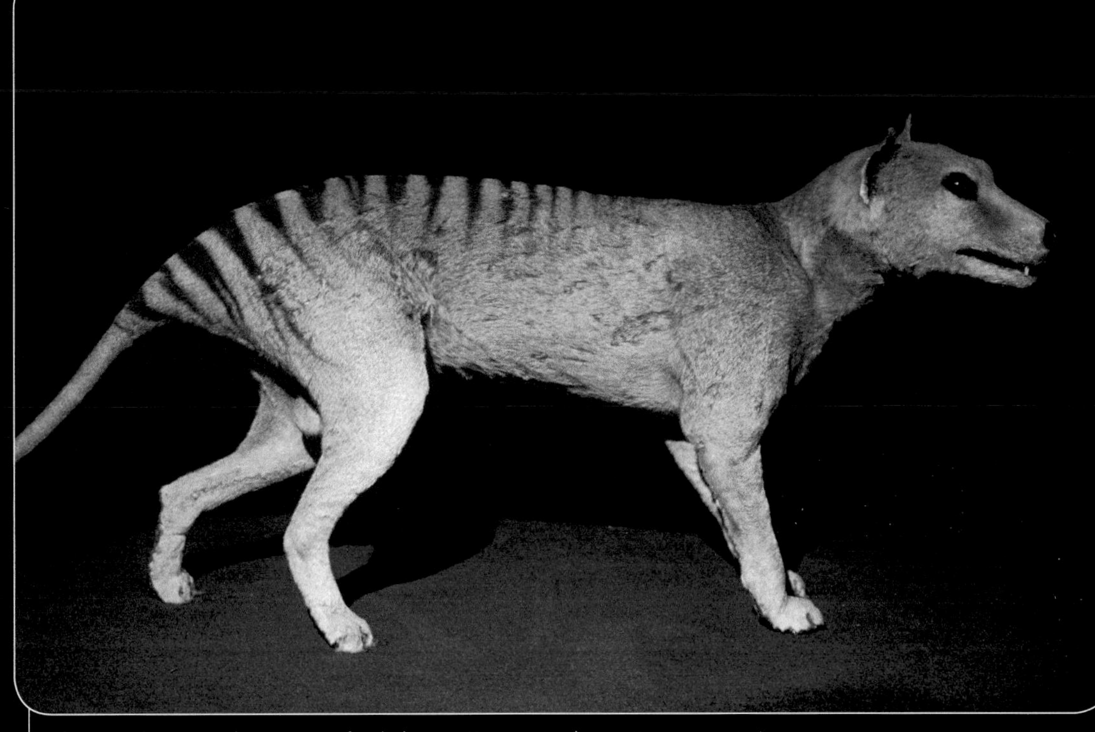

Plate 5.1 A preserved specimen of a thylacine. (Source: Lochman Transparencies)

Plate 6.1 (a) Wild budgerigars are predominantly green in colour. (Source: Lochman Transparencies)

Plate 6.1 (b) Blue and white budgerigars are rare in the wild, because they are conspicuous to predators. However, they can be bred in captivity. (Photograph: John Martin)

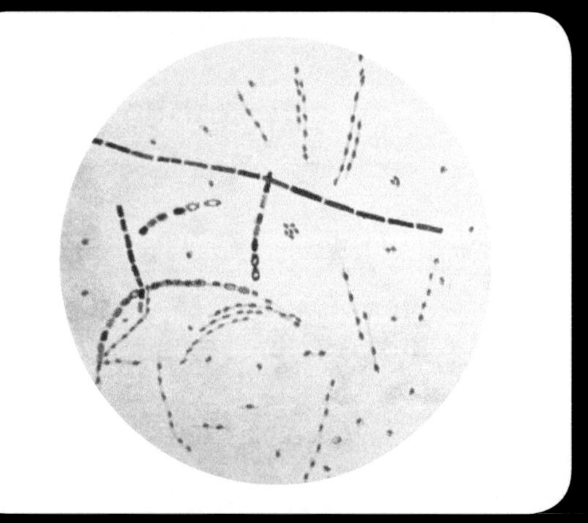

Plate 9.1 Examples of prokaryote shapes. (a) Rod-shaped (bacillus). This example is Bacillus anthracis. (Source: Centers for Disease Control and Prevention, USA)

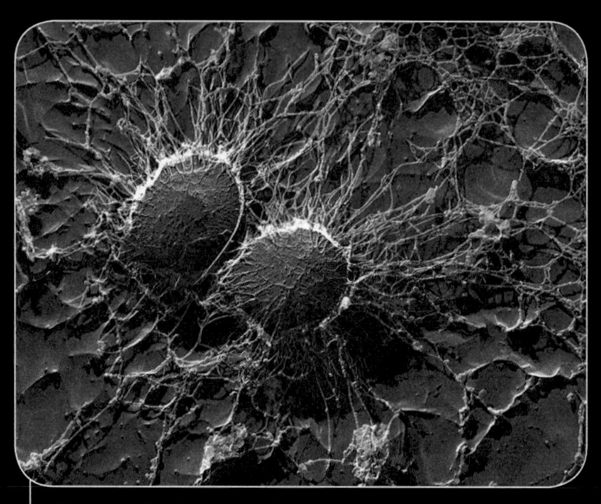

Plate 9.1 (b) Spherical (coccus) greatly magnified in an electron micrograph. This example is Staphylococcus aureus. (Source: Agricultural Research Service, USA)

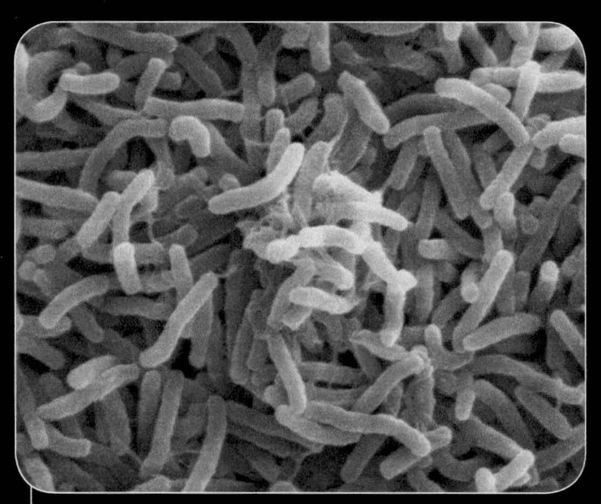

Plate 9.1 (c) Curved rod (vibrio) greatly magnified in an electron micrograph. This example is *Vibrio cholerae*. (Source: Image in public domain)

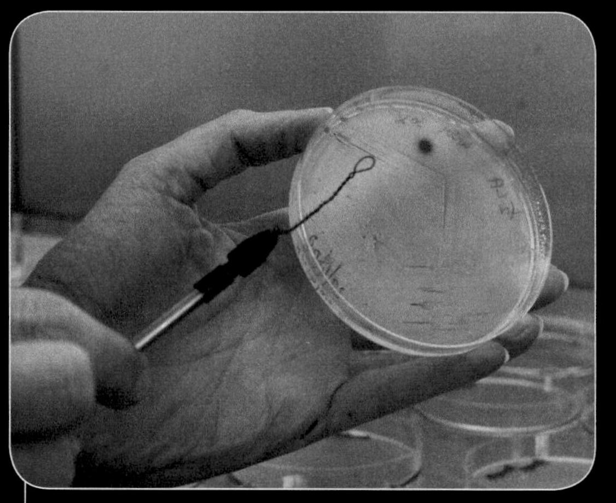

Plate 9.2 Colonies of pink-pigmented, nitrogen-fixing bacteria isolated from a South African legume. (Source: M. Waters)

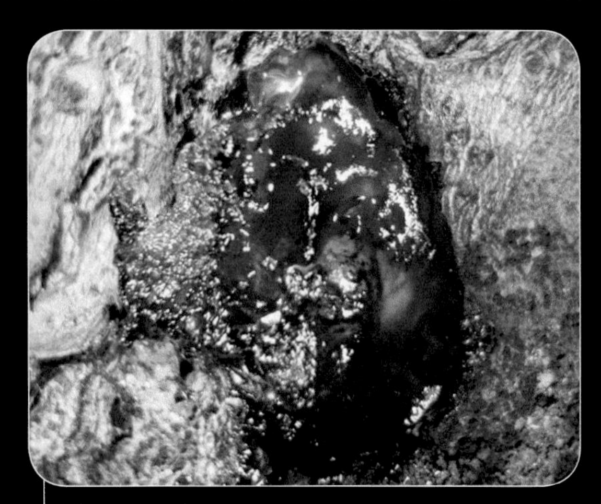

Plate 9.3 Pathogenic protists cause disease. (a) Gum oozing from the stem of apricot infested with bacterial disease (*Pseudomonas syringae*). (b) Symptom of bacterial wilt disease in a eucalypt stem. Droplets of bacteria can be seen oozing from freshly cut wood. (Source: B. Dell)

(b) Symptom of bacterial wilt disease in a eucalypt stem. (Source: B. Dell)

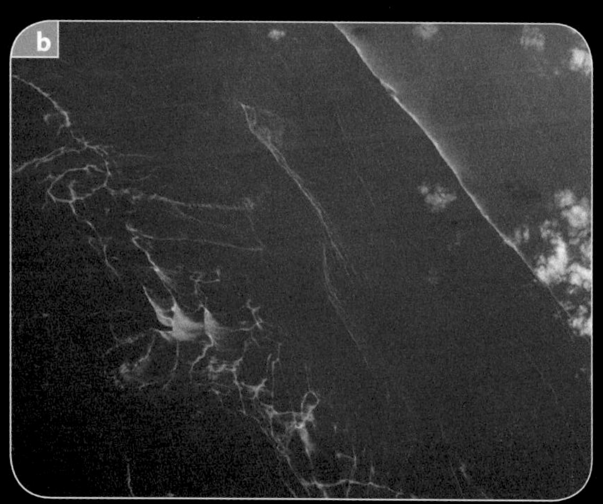

Plate 9.4 The filamentous cyanobacterium *Trichodesmium* (a) under the microscope and (b) an aerial view of the blooms it forms – they can cover ocean areas greater than 10 000 km^2 (Source: a S. Kranz/Alfred Wegener Institute, Germany; b NASA)

Plate 9.5 A rabbit with the viral disease myxomatosis (Source: Agriculture Protection Board of Western Australia).

Plate 9.6 Prokaryotes have been isolated from extreme environments such as (a) hot springs, (b) acidic sites and (c) deep-ocean hydrothermal vents (Source: a <http://commons.wikimedia.org/wiki/Image:Emerald_spring.jpg>; b G. Wardell-Johnson; c <http://commons.wikimedia.org/wiki/Image:Nur04506.jpg>)

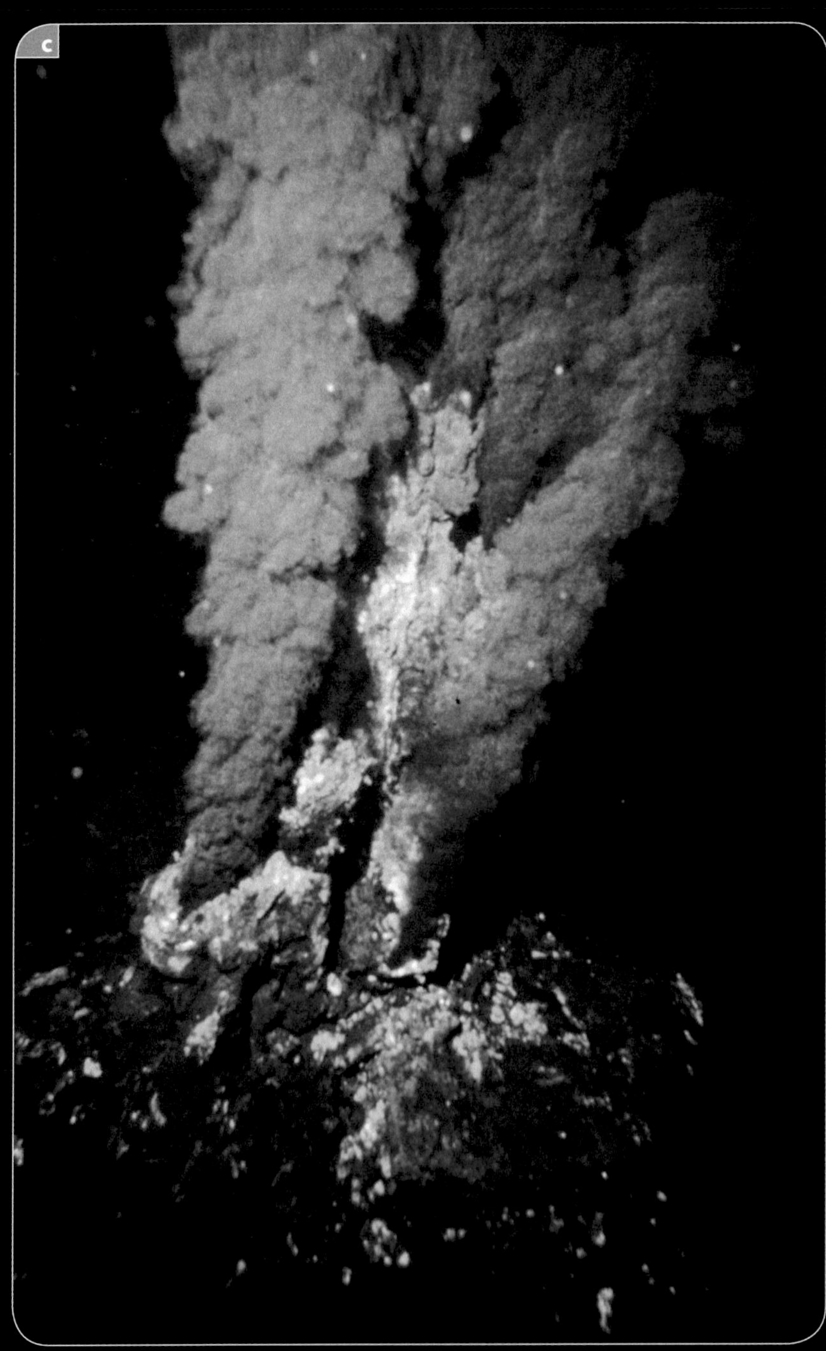

Plate 10.1 (a) Healthy jarrah (*Eucalyptus marginata*) forest. (b) Jarrah forest infested with *Phytophthora cinnamomi*. The death of trees and understorey has left 'graveyard areas'. The surviving trees are marri (*Corymbia calophylla*), which is resistant to the *P. cinnamomi*. (Source: a, b N. Marchant)

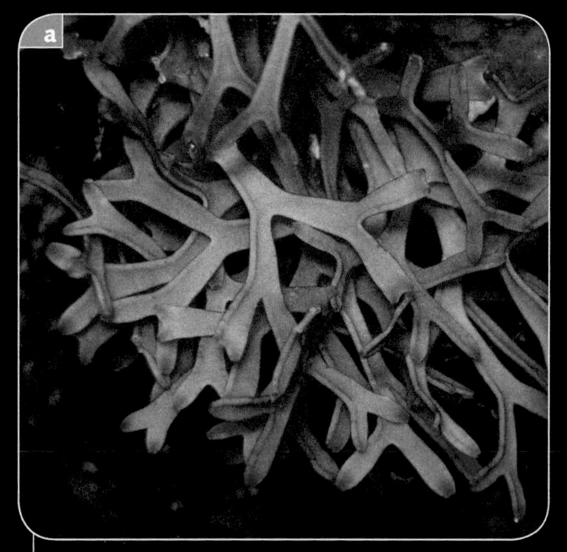

Plate 10.2 Examples of red algae. (a) *Dichotomaria* sp. (b) *Schizmenia dubyi* (Source: a, b J. Huisman)

Plate 10.3 Examples of green algae. (a) *Caulerpa sertularioides*. (b) *Udotea glaucescens*. (c) *Microdictyon* sp. (d) *Ulva fasciata*. (Source: a–d J. Huisman)

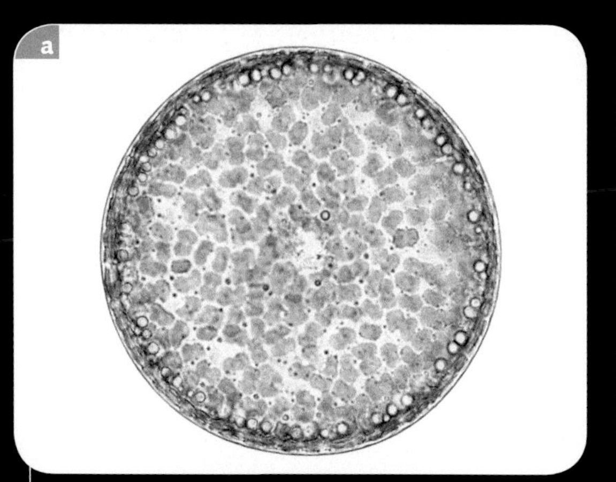

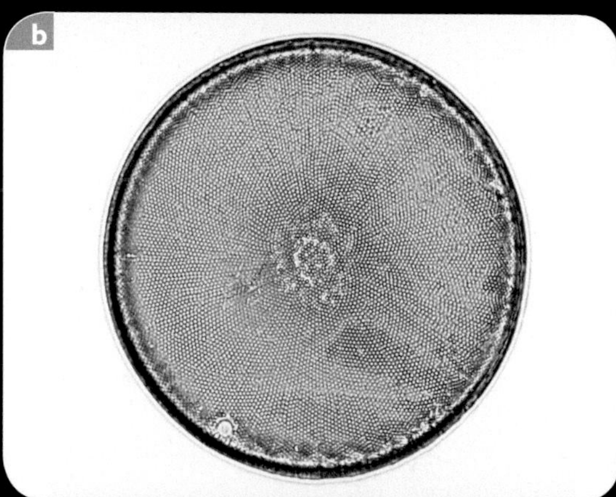

Plate 10.4 (a) A living cell of the diatom *Actinocyclus*. (b) The cleaned silica shell of this diatom. (Source: a–b J. Huisman)

Plate 10.5 Examples of brown algae. (a) *Turbinaria ornata*. (b) *Padina boryana*. (c) *Cystophora racemosa*. (Source: a–c J. Huisman)

Plate 10.6 A scientist in full protective gear injects the chemical metham sodium into soil to control an infestation of the plant pathogen *Phytophthora cinnamomi*. (Source: W. Dunston)

Plate 10.7 People cultivate many types of fungi for food. (Source: B. Dell)

Plate 10.8 The fruiting bodies of some fungi reach impressive sizes. (a) The giant bolete (*Phaeogyroporus portentosus*) can grow to 1 m in diameter. (b) The sturdy fruiting bodies of shelf or bracket fungi often occur in layered horizontal rows. This orange bracket fungus is pathogenic on many trees. (Source: a, b B. Dell)

and the dome-shaped stromatolites seen, for example, in shallow saline waters of Western Australia's Shark Bay (Figure 7.5). Fossil stromatolites 3.5 billion years old are found at the whimsically named North Pole mine site in the Pilbararegion of Western Australia. These are among the oldest fossils known. Stromatolites reached their peak diversity between 1.3 and 1.0 billion years ago. At this time, major marine animal groups may have evolved, with their grazing possibly causing stromatolite species numbers to subsequently decline by more than 75%.

Figure 7.5 Stromatolites, built by cyanobacteria in Shark Bay, Western Australia. (Source: Lochman Transparencies)

Photosynthesising marine cyanobacteria generated oxygen as waste. This turned the seas red for more than a billion years as iron precipitated, forming the massive iron-rich sediments and red rocks mined today from places such as Western Australia's arid Pilbara. As oxygen slowly saturated the oceans, the atmosphere also became oxygenated so that, about 2 billion years ago, oxygen accounted for 1% of the atmosphere, which was dominated by carbon dioxide and methane from volcanic degassing.

The land at this time received high doses of ultraviolet radiation from the sun, because the atmosphere had no mechanism to reduce it. Intense ultraviolet light can kill organisms, so life on land was inhibited. However, as oxygen increased in the atmosphere it interacted with ultraviolet light to form ozone. A layer of ozone in the upper atmosphere reduced the amount of ultraviolet light reaching the Earth's surface, making the land more hospitable.

Sometime between 2.8 and 2.4 billion years ago, oxygen levels reached the point where some organisms evolved aerobic metabolism (reliant on oxygen). Aerobic respiration produces far more ATP from organic compounds than does anaerobic respiration (Chapter 4), so these organisms had an energetic advantage over their anaerobic competitors.

The eukaryotes (with complex nucleated cells) may have evolved from the archaea as early as 3.5 billion years ago. Convincing microfossil evidence of single-celled eukaryotes, however, first appears much later at 2.1 billion years ago in rocks from Michigan, USA, containing the giant unicellular alga *Grypania*.

Multicellular life

Multicellular marine organisms evolved about 1.2 billion years ago. Plant diversification intensified after eukaryote cells incorporated photosynthesising cyanobacteria as endosymbionts (internal shared life), which eventually evolved into chloroplasts. This is called the primary endosymbiotic event (Chapter 10). Marine red algae are the first multicellular organisms in the fossil record. Green plants evolved later, between 1 billion and 825 million years ago. In Spitsbergen, 800–700-million-year-old shales contain 1 cm-tall fossils of branching filamentous green algae remarkably similar to today's green alga, *Cladophora*. Marine fungi appear as fossils at 1 billion years ago. Fossil and DNA evidence suggest that multicellular animals evolved later, following massive global glaciation some 800–700 million years ago.

Atmospheric oxygen levels increased about 1 billion years ago, a time of enhanced continental drifting and mountain building. This may have accelerated the diversification of eukaryotic organisms, because of the potential for aerobic respiration provided by mitochondria. Mitochondria are also endosymbionts, formed from the inclusion in eukaryote cells of purple, non-sulfur proteobacteria.

First life on land

Exactly when life colonised land is not known. Non-photosynthetic bacteria, archaea and eukaryotes may have avoided the harmful solar radiation penetrating an oxygen-poor atmosphere by living underground or in spaces in rocks. Life on the surface of the land was likely to have been the province of cyanobacteria. They coped with prolonged exposure to dry air as they do today, by closing down photosynthesis and other metabolic functions when desiccated, and reactivating them when wet. Perhaps it was not until the burst of oxygenation of the atmosphere around 1 billion years ago that photosynthetic terrestrial organisms could survive beneath a protective ozone layer.

Terrestrial microbe crusts resulted in the breakdown of rock and soil formation, releasing iron and phosphorus and fixing nitrogen from the atmosphere. This paved the way for future life on land in the form of grazers, detrital feeders and vascular plants. Fossil single-celled eukaryotes, for example photosynthetic chlorophytes and animals such as amoebae, have been found in rocks 750 million years old. This was the time when the first land plants diverged from green algae, estimated from DNA sequence data. A single global continent, Rodinia, existed at this time, undoubtedly with continental deserts present. However, the earliest land plants could not have grown far from permanent water, as their reproductive cycles relied on the transmission of free-swimming male gametes to receptive female eggs.

At this stage, the Rodinian terrestrial deserts were perhaps the sole preserve of desiccation-resistant cyanobacteria, mat-forming green algae and possibly lichens, although unequivocal fossil lichens do not appear until about 400 million years ago, so their occurrence on Rodinia is speculative. Lichens are symbiotic organisms combining fungi with either cyanobacteria or green algae. The fungi provide structural support, water and protection from dehydration, while the algae or cyanobacteria provide food through photosynthesis. Such partnerships, if they existed, would have formed biocrusts and mats on Rodinian soils and rocks, resisting erosion by wind and water. Many landscapes and soils today still depend for their stability and nutrient cycling on these humble crusts and mats of symbiotic organisms.

Marine animal life radiates

Fossils of soft-bodied marine animals, first recognised in South Australia's Ediacaran Hills, are now known from rocks 620–550 million years old from several countries, including Russia, Canada and Namibia. This Ediacaran fauna contains many strange forms whose way of life remains speculative, as they are unrelated to today's animals.

Superbly preserved fossils of corals and sponge embryos parasitised by nematode worms can be found in marine phosphatic rocks of the Doushantuo Formation of China, dated at 580 million years. By 540 million years ago, animals intermediate between invertebrates (animals without backbones) and chordates (including the backboned animals and their close relatives) are evident as fossils alongside obvious chordates, invertebrates and other animal groups.

During the early Cambrian, the most common marine multicellular animals were crab-like trilobites, now an extinct group. The diversity of fossil animals dramatically increases in rocks aged from 540 to 500 million years ago, an event called the 'Cambrian explosion'. Continental plates were reorganised through this time as the single global continent, Rodinia, fragmented into Laurasia and Gondwana, with the further break-up of Laurasia into many islands. Melting of ice sheets by 590 million years ago created large, shallow continental shelves and the Cambrian explosion occurred in these intricate and extensive environments.

The first fish-like animals appear in the Ordovician; there are examples in rocks in Australia's Northern Territory dated at 470 million years ago. By the late Silurian, larger armoured fishes had evolved, and modern groups such as sharks and bony fishes appeared in the Devonian.

A major glaciation occurred about 445 million years ago, with glaciers extending from the South Pole as far as 60°S latitude. This affected shallow-water marine animals in particular, and is known as the late Ordovician mass extinction. Although shallow-water shelf environments largely disappeared, new opportunities for moving onto land were created as the sea level dropped. The atmosphere in the Cambrian/Ordovician has been estimated to have had 18 times more CO_2 than at present, enhancing soil formation and favouring the growth of microbial mats. These provided food for animals evolving from an aquatic to a terrestrial existence.

Conquest of the land

Moving onto land posed several challenges to animals and plants, including abandoning the structural support of water, coping with increased exposure and dehydration, and reproducing without using water to transmit gametes. Various solutions appeared by about 400 million years ago. Trackways of giant, scorpion-like eurypterids, for example, are to be seen in 420 million-year-old Murchison River sandstones from Kalbarri on Australia's west coast (Figure 7.6). Their hard external skeleton, evolved in marine conditions for protection from predators, now gave structural support for movement on land.

At this stage, known as the Silurian period, the west coasts of the Antarctic–Australian peninsula of east Gondwana were in mid-latitudes and part of a substantial Gondwanan desert. Deserts still remained the province of lichens, algal mats and cyanobacteria, with localised patches of mosses, liverworts and hornworts close to fresh water. Today, mosses, liverworts and ferns still require free water for sexual reproduction and are confined to moist habitats. The evolution of pollen, resistant to desiccation because of a hard coat, enabled transport of the male gametes to receptive female gametes by wind rather than water. The aqueous links with plant sexual reproduction were finally broken.

The evolution of desiccation-resistant spores and seeds blown by the wind was the other major step enabling plants to colonise land beyond wetland habitats or their margins. Resistant spores and seeds protect the embryo from desiccation, predators and pathogens, and provide nutrients for germination and early growth. Green algae in seasonal ponds were probably the earliest plants to evolve desiccation-resistant spores. Land plants evolved protective cuticles for the plant body and specialised conducting tissue for water and nutrients, recorded in fossil form 430 million years ago. Gas exchange was hampered by thickened cuticles, and stomates (pores in the leaf epidermis that rectify this problem) are first seen in fossils 408 million years old. Evolving cylindrical stems anchored by roots enhanced structural support and improved access to light, developments seen in fossils from 408 million years ago. Terrestrial fungi are first evident as fossils about 440 million years ago, enabling mycorrhizal (fungal–root) associations with vascular plants to evolve early in the colonisation of land. In a mycorrhizal association, the plant absorbs mineral nutrients from the fungus, which can extract them from soil more efficiently than the plant, while the fungus gains carbon compounds from the plant roots. With these relationships, plants were able to occupy many terrestrial landscapes (see Chapter 10, Box 10.6).

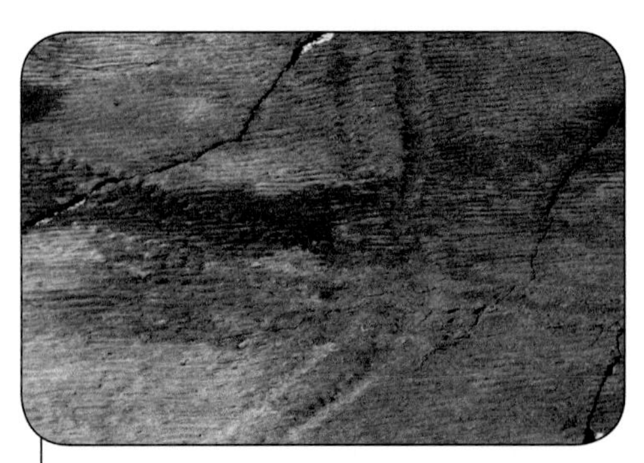

Figure 7.6 Tracks of eurypterids, preserved in sandstone at Kalbarri, Western Australia. (Source: ©Jiri Lochman, Lochman Transparencies)

The essential precursor to seeds was the transition from plants producing spores of the same size (homospory) to producing two sizes – megaspores and microspores (heterospory). This occurred around 400 million years ago in groups such as the horsetails, tassel ferns, spike mosses, quillworts and ferns. Megaspores were precursors to ovules and microspores were precursors to pollen. By 370–354 million years ago, fossil megaspores developed an outer protective coating, becoming the first ovules or unfertilised seeds. Simultaneously, pollen evolved from microspores.

At the same time as these reproductive innovations evolved, terrestrial plants increased their size and stature, attaining heights to 35 m in wet, fertile environments. This was a time of remarkably high atmospheric CO_2 concentrations and global temperatures. As plants evolved ovules and seeds, and increased their biomass as trees, they could survive and reproduce over all terrestrial landscapes.

Invertebrates colonised land over 400 million years ago and the first land vertebrates, amphibians, appeared in the late Devonian, about 370 million years ago. In the Carboniferous, as land plants diversified and expanded their range, so did vertebrates, and the first reptiles appeared on land during this time. Over the next 50 million years, reptiles increased in abundance as amphibians, which depend on water, declined.

Deep freeze and the world's greatest extinction event

This global greening of the land during the Carboniferous caused atmospheric CO_2 to plummet, resulting in global climate cooling. Massive glaciation occurred on the southern part of the supercontinent Pangaea from 320 to 270 million years ago. Deserts were likely to the north, in areas that were to become North America, western Europe and the northern parts of South America and Africa. The future continents of Australia and Antarctica were too far south to contain deserts, and were mainly under ice. Ferns and fern allies dominated global vegetation, but seed plants were also present.

The earliest seed plants were pteridosperms (seed ferns) and cordaites (conifers with long, strap-like leaves). Seed ferns had fern-like foliage (but were not closely related to ferns) produced on plant forms ranging from vines to trees. They survived for more than 200 million years, overlapping in time with the earliest flowering plants. Although cordaites were prominent in Carboniferous and Permian forests together with seed ferns and giant horsetails, they were extinct by 240 million years ago.

The world's greatest mass extinction of animals occurred about 250 million years ago, in the Permian period. As many as 95% of marine animals, including trilobites, went extinct and, on land, mammal-like reptiles (our ancestors) almost disappeared. Plant life was also reduced to simpler communities, although extinction of major lineages was not as pronounced as for animals. Causes of the Permian extinction remain speculative, with hypotheses ranging from the impact of a large asteroid off the north-west Australian coast to volcanic events, including massive gaseous eruptions creating acidic conditions in the seas and atmosphere. The evolutionary stage was reset, with flowering plants, dinosaurs, birds and mammals first appearing in the fossil record after the Permian extinction.

Emerging modern lineages

Global warming through the Triassic period melted the glaciers and there was widespread aridity following the formation of Pangaea, which was centred on equatorial latitudes. Massive mountain ranges were pushed up following the collision of Gondwana with its Northern Hemisphere counterpart Laurasia and probably blocked moisture-laden oceanic winds from penetrating inland, further increasing aridity. The Earth was inherited by plants capable of surviving and reproducing in arid environments. The age of seed plants had arrived.

Seeds evolved in the Devonian and true seed plants, including gymnosperms such as cycads, ginkgos and conifers, first appeared about 280 million years ago. They survived the Permian mass extinction, but did not dominate global vegetation until the world became a hothouse 200 million years ago. Flowering plants first appeared in the late Cretaceous and quickly became abundant worldwide.

The dominant land vertebrates through this period of a drying and warming climate were reptiles: pterosaurs, crocodiles, dinosaurs and one lineage that gave rise to mammals in the late Triassic.

By the early Jurassic (195 million years ago), a band of desert stretched across southern Pangaea, in the lands that were to become South America and contiguous parts of Africa to central Antarctica, while in the north it was the future western North America east of the Rockies that was arid. Most of the future continents of Eurasia and Australia were in high latitudes and removed from desert-like conditions. As Pangaea began to fragment into Laurasia and Gondwana, and these two supercontinents further fragmented, the world's climate changed again. Although some desert areas still persisted between 100 and 50 million years ago, much of the world was moist, with temperatures and CO_2 levels higher than those today and no polar glaciers. Indeed, the highest sea levels known (250 m above modern shorelines) occurred 90 million years ago, reducing land area considerably, creating many islands and favouring an explosion of new species of seed plants, dinosaurs and insects.

Towards the middle and end of the Cretaceous, coniferous rainforests were gradually replaced by flowering plants. Dinosaurs reigned on land until their ultimate demise, believed to be caused by an asteroid impact in the Caribbean Sea off the Yucatan Peninsula of Mexico 65 million years ago. Many shallow-water marine animal species also suffered mass extinction at this time, but relatively few plants species were lost. Perhaps the ability to resprout from underground structures or germinate from seed was their salvation. Similarly, burrowing animals on land avoided global demise. Of the vertebrates, one surviving lineage related to the dinosaurs – the birds – and mammals inherited the Earth.

A drier, colder and more familiar world

Final separation of Australia from Antarctica 42 million years ago heralded the onset of modern patterns of global climate, enhanced when Drake's Passage opened between

South America and Antarctica 40 million years ago. The southern circumpolar current then started up, Antarctica froze and aridity worsened in Australia, Africa and South America. Essentially, cooler and drier modern climates were established by the Pliocene about 5 million years ago. Radiation of many modern animal and plant lineages was well advanced, including the emergence of bipedal hominids in Africa about 6 million years ago and the spread of C_4 grasses adapted to arid environments (see Chapter 4). Fire was a big feature of many landscapes, especially on southern continents.

Two events around 3–4 million years ago altered global oceanic circulation profoundly – the constriction of the Indonesian throughway from the Pacific to the Indian Ocean as the Australasian plate drifted north, and the closure of the Atlantic–Pacific Ocean connection with the joining of South America to North America. The latter created the Gulf Stream, forcing warm tropical water up into the North Atlantic. The greater precipitation in high northern latitudes formed Arctic glaciers from 3 million years ago, beginning major glacial cycles further south across Eurasia and North America; these have continued over the past 2 million years. Climatic fluctuations intensified and aridity increased during these glacial cycles. Extreme aridity occurred on southern continents during the last glacial maximum around 18 000 years ago.

The history of the Australian biota since the demise of the dinosaurs

Australia provides a compelling story of insular evolution on a continental scale during the dramatic cooling and drying that the world has experienced over the past 40 million years. The nascent island-continent started with a Gondwanan inheritance of animals and plants that had evolved while Australasia remained connected to South America and Africa through Antarctica. For example, many genera present in the now widely separated countries of Australia and South Africa belong to the family Proteaceae (for example, *Banksia* in Australia and *Protea* in South Africa). Fossil eucalypts have recently been confirmed from both South America and New Zealand, where they have long been extinct. Marsupials are common both in Australia and South America, reflecting their land-based connection through Antarctica until 40 million years ago.

About 65 million years ago, as the age of the dinosaurs drew to a close at the Cretaceous–Palaeogene boundary, Antarctica and Australia separated, beginning in the west, and pivoting on Tasmania. Despite being close to the South Pole at this time, Australian landscapes were dominated by rainforests. Fossil deposits 30–60 million years old as far inland as the presently arid Coolgardie area of Western Australia contain material from rainforest cycads, podocarps and conifers, as well as angiosperm rainforest and mangrove plants.

Sclerophyllous (hard-leaved) plants in the families Myrtaceae and Proteaceae, so dominant today, also formed a significant component of the vegetation, presumably on deeply weathered, sandy, infertile soils and in communities marginal to the luxuriant subtropical rainforests. Adaptations of this flora to nutrient-poor soils included sclerophyllous leaves, underground food storage organs (lignotubers, rhizomes, corms), and bulky wooden fruits concentrating scarce nutrients. These adaptations later became critical for other purposes; sclerophylly helps resist drought-induced wilting, and lignotubers aid survival from drought, grazing and fire.

The fossil record in Australia contains much evidence over the last 40 million years of emerging arid biomes such as mid-latitude deserts, tropical and subtropical savanna, temperate grasslands and steppes, and Mediterranean-type winter-wet communities. DNA sequence studies of many animal and plant groups reinforce this perspective on the age of Australian deserts. Most radiations started more than 40 million years ago, intensified at 25–15 million years ago, and showed explosive speciation over the past 5–10 million years. This pattern is seen in diverse plants (eucalypts, wattles, peas and she-oaks), marsupials, and bird, frog and reptile lineages.

Marsupial groups common today in arid habitats, for example, have estimated ages of origin of 45 million years ago for marsupial moles, and 25 million years ago for macropods such as kangaroos and wallabies, and for smaller carnivorous dunnarts. Blind, wingless water beetles living beneath these deserts today in calcrete caverns and aquifers had several independent origins from terrestrial ancestors about 15–10 million years ago, when Australia's aridity intensified. All this suggests that the onset of aridity in Australia, first felt on the Nullarbor, and moving north and westwards by about 30–20 million years ago, prompted radiation of today's arid-adapted flora and fauna throughout the desert fringe (Box 7.2).

The onset of aridity in the Palaeogene also imposed climatic barriers in the centre and the north of the continent, isolating the south-west biota. Desert dominated much of the Australian landscape at this time, and explosive speciation occurred in the flora along the arid zone's south-west margin. Climatic fluctuations during the Neogene period had the greatest impact in semiarid transitional regions and less pronounced impacts in the peripheral, wetter margins of the continent, as well as in the arid centre.

In summary, the combination of an ancient landscape remaining unglaciated and above sea level for well over 200 million years provided unparalleled opportunities for the persistence of ancient terrestrial plants in Australia. Many taxa of the ancient rainforests persist today in the east and north, but were driven to extinction in the west and centre as arid conditions developed over the past 40 million years. Other groups, specialising on deeply weathered infertile soils, were pre-adapted (already suited) to new arid environments. They speciated explosively, especially in southern Western Australia, providing the characteristic modern vegetation of eucalypt, wattle, spinifex and shrubs.

Arrival of Aboriginal people

The expansion of modern humans from Africa followed several innovations associated with coastal living, as seen in cave deposits in South Africa – fishing, ochre-based artwork, and more sophisticated stone-tool technologies, all of which are evident around 80 000 years ago. A huge volcanic eruption in Sumatra, 72 000 years ago, probably rendered coastal living unsuitable for humans across much of south-east Asia and the Indian subcontinent, stimulating extensive migrations. Hence the ancestors of Australian Aboriginal people probably migrated through Indonesia and ultimately across the Timor

Box 7.2 Aridity and fire in the Australian landscape

In the 1940s and 1950s, there was a widely promoted view that most of Australia's deserts are less than 10 000 years old, having been created from widespread tropical rainforests by Aboriginal land management practices involving fire (so-called 'firestick farming'). This hypothesis needs to be considered in the context of fossil evidence of fire regimes arising since Australia became an island-continent 40 million years ago. Well before human occupation there were frequent fires, as evidenced by charcoal in fossil deposits in eastern New South Wales over long periods. Moreover, charcoal levels increased dramatically about 10 million years ago, when aridity intensified.

The presence of significant charcoal seems anomalous during the presumed dominance of rainforest vegetation from 45 to 15 million years ago. Rainforest as the common vegetation is inferred from the abundant pollen of southern beech (*Nothofagus*) and other rainforest trees in fossil deposits. However, this may reflect the bias of the fossil record to wet depositional sites. Although rainforests must have been common, other seasonally dry woodlands and forests also occurred, as shown by the trace of Myrtaceae (eucalypts and related species) pollen in some deposits. There is no reason to expect vegetation from 45 to 15 million years ago to have been any less complex and geographically variable than it is today, with rainforests on wetter sites and woodlands, shrublands and grasslands on drier sites.

Further evidence that there was not a continuous blanket of rainforests across Australia at this time comes from the rich deposits of animal fossils in the 25 million-year-old limestone rocks of Queensland's Riversleigh, south of the Gulf of Carpentaria. The Riversleigh fauna represents the heyday of Australian marsupial diversity. At the time, Riversleigh contained twice the number of families that are alive today in north-east Queensland's wet tropics region.

Palaeozoologists (zoologists studying extinct animals) originally interpreted the Riversleigh environment as rainforest. Typical reconstructions of life 25 million years ago at Riversleigh have animals immersed in a riot of plant forms matching those seen today in the wet tropics of coastal north Queensland. A recent examination of large plant fossils, however, demonstrated otherwise. She-oaks were common, as were light-loving shrubs and vines typical of monsoonal woodland or 'dry jungle', as seen today in northern Australia. Studies of other fossil sites and faunas, including those containing the remarkable giant thunder birds or mihirungs (flightless, goose-like browsing and fruit-eating animals), support the view that much of Australia's contemporary fauna and many extinct megafaunal lineages originated in semiarid environments rather than green rainforests. The 'Green Cradle' hypothesis needs to be replaced by a 'Semiarid Cradle' hypothesis.

Figure 7.7 Two extinct members of the Australian fauna in the Pleistocene. The giant goanna *Megalania prisca* was 7 m long and is depicted here ambushing the flightless bird *Genyornis newtoni*. The artist (Peter Trusler) reconstructed the animals by placing the expected musculature and skin over the fossilised bones. They shared the Pleistocene landscape of Australia with giant wombats, carnivorous kangaroos and marsupial lions. (Source: Reproduced from Vickers-Rich, P. and Rich, T.H. (1999). *Wildlife of Gondwana*, Indiana University Press, with permission of the artist)

Sea after this eruption, occupying Australia 45 000–42 000 years ago. There are reliably dated sites of this age occurring as far apart as at Devil's Lair in the extreme south-west and at Lake Mungo in the south-east desert fringe. Speculation on earlier dates of occupation as far back as 72 000 years ago is supported by molecular DNA studies, but there is no supporting archaeological evidence.

Some scientists hypothesised that the extinctions of about 23 of 24 genera of vertebrates larger than 45 kg, the Australian fossil megafauna (Figure 7.7), were caused by rapid overkill by humans about 45 000 years ago. However, at least one reliably dated site, at Cuddie Springs in north-west New South Wales, has six megafaunal species persisting at 36 000–29 000 years ago, at least 10 000 years after Aboriginal people arrived. Moreover, evidence of butchered megafaunal bones was found there, so humans were hunting megafauna some 10 000 years after most genera had gone extinct. Rapid overkill seems unlikely in light of these studies. It is more likely that climate change towards severe aridity reduced the food species, and thus the survival, of large browsers and associated carnivores. Some scientists suggest that Aboriginal burning of vegetation, rather than overhunting, reduced forage and therefore contributed significantly to the extinctions (Box 7.3).

The age of technological expansion

Today, we are in the midst of what many contemporary biologists term the sixth global extinction event (Box 7.4). Expanding agriculture followed by the Industrial Revolution

Box 7.3 What killed the megafauna?

Gifford Miller and his colleagues used stable isotopes (see Chapter 3, Box 3.2) in fossil egg shells of emus and the extinct mihirung *Genyornis* (see Figure 7.7) to determine their diets before and after the arrival of Aboriginal people. They found that emus endured a significant change and narrowing of diet towards more arid-adapted shrubby plants at 45 000 years ago. Before this they ate nutritious grasses in wet years and arid-adapted shrubs in dry years. *Genyornis* favoured the nutritious grasses, and went extinct about 45 000 years ago at three widely separated localities in arid central and south-east Australia, presumably because it could not adapt its diet to changes in the vegetation. As an independent test, stable isotopes from wombat teeth were examined at the same sites. These showed similar changes to those seen in emu eggshells. Wombats feed on grasses and reeds, and narrowed their diet to more arid-adapted food at 45 000 years ago.

An abrupt ecological shift in semiarid Australian vegetation at 50 000–45 000 years ago, when Aboriginal people arrived, is inferred from these isotope studies of megafaunal diet. Miller and his colleagues concluded: 'We speculate that systematic burning practised by the earliest human colonisers may have converted a drought-adapted mosaic of trees and shrubs intermixed with palatable nutrient-rich grasslands to the modern fire-adapted grasslands and chenopod/desert scrub. Nutrient-poor soils may have facilitated the replacement of nutritious C_4 grasses by spinifex, a fire-promoting C_4 grass that is well adapted to low soil-nutrient concentrations'. It seems, therefore, that browsing megafauna unable to adapt their diets became extinct, followed by their animal predators.

irrevocably altered the habitat of many species. Human activity is visibly changing global climate now, and consumption of resources has reached or will soon approach critical thresholds.

Biodiversity conservation is intimately intertwined with human decisions. Moreover, approaches to conservation need to consider the age of the landscape. Studies of plant and animal lineages on the world's oldest landscapes such as in south-west Australia, the Greater Cape Floristic Region of South Africa, and Venezuela's tepui mountaintops, suggest that evolution produces extraordinary diversity under difficult environmental circumstances, given long periods of time. This richness is perhaps the most vulnerable of all when humans settle and manage ancient landscapes in the same way as was successful for much younger, more fertile and regularly disturbed places, such as Western Europe.

Darwin perceived the essential divergence of Australia's plant and animal life from all other continental biotas when he urged Marianne North to go there to complete her paintings of the world's botanical highlights. Subsequent studies of the history of life on Earth now give us an exquisite insight into why such patterns are evident to those who care to look at the remarkable biological heritage with which we are entrusted.

Box 7.4 Is there an extinction crisis?

Our most reliable information on extinction rates in recent times comes from birds and mammals, because these organisms are conspicuous and well studied. Estimates based on the best available evidence suggest that about 85 species of mammals (about 2.1% of known species) and 113 species of birds (about 1.3% of known species) have become extinct since 1600. Most of these extinctions occurred in the last 150 years and the rate of extinctions rose steadily over this time; between 1986 and 1990 an average of four species became extinct every year. Australia has one of the highest recent extinction rates, especially of mammals, where 27 species have become extinct since European settlement of the country.

Mass extinctions have occurred at other times in the Earth's history, but this most recent extinction crisis is notable in two respects. First, it is almost totally attributable to the increasing population size and spread of a single species, humans, rather than to a catastrophic abiotic event. Secondly, the extinctions are associated with a fundamental loss of ecosystem resources, which may prevent any rebound in species numbers.

We are a long way from understanding how best to chart a sustainable path into tomorrow's new world of changed environments of our own making. Hopefully, we will have the wisdom to understand the fundamental messages from evolution, geology, plate tectonics and biogeography, and use them to conserve both biodiversity and the processes that produce it.

Chapter summary

1. Life on Earth is the culmination of geological and evolutionary events beginning 3.8 billion years ago. The geological timescale is divided into four major eras (Pre-Cambrian, Palaeozoic, Mesozoic and Cenozoic), subdivided into periods and epochs. Understanding this history provides valuable insights into present and future management of biodiversity.
2. Biogeographers recognise six different zoological and five botanical regions on land, each characterised by vertebrates and seed plants exhibiting high levels of endemism at the family level or above.
3. Our understanding of the evolution of this diverse biota is enhanced by plate tectonics theory, radiometric techniques for accurately dating rocks and applying DNA sequencing to trace evolutionary pathways.
4. The earliest evidence of life is carbon of biological origin from Pre-Cambrian rocks 3.8 billion years old. The oldest known fossils are stromatolites, formed from the activities of photosynthesising cyanobacteria 3.5 billion years ago. The first eukaryotes, single-celled algae, appeared 2.1 billion years ago.
5. The first fossil multicellular organisms are marine red algae, about 2.1 billion years old. Green plants arose between 1 billion and 825 million years ago, while multicellular animals first evolved 800–700 million years ago.
6. Marine animal life radiated enormously 540–500 million years ago, at the beginning of the Palaeozoic. This was driven by the fragmentation of the Earth's land masses and melting of ice sheets, creating large areas of shallow continental shelf. A major glaciation at 445 million years ago destroyed these environments and led to a major extinction of marine invertebrate animals.
7. Land was probably first colonised by life around 1 billion years ago, with fossil single-celled eukaryotes recorded from rocks 750 million years old. Plants and animals appeared on land just over 400 million years ago during the Silurian period. The evolution of desiccation-resistant spores and seeds, and structural adaptations for increasing size and capturing more light, enabled plants to expand their range over all landscapes during the Carboniferous period, 354–290 million years ago. The first land vertebrates, amphibians, appeared about 370 million years ago. Reptiles also first appeared in the Carboniferous. A mass extinction event around 250 million years ago, at the end of the Palaeozoic, dramatically reduced the diversity of terrestrial and marine animal life.
8. Global warming at the beginning of the Mesozoic increased aridity on land and led to the expansion of seed plants, including gymnosperms, which could survive and reproduce in arid environments. During the Cretaceous period, flowering plants gradually replaced gymnosperms over much of the land. Reptiles, especially dinosaurs, were the dominant land vertebrates throughout the Mesozoic. Their extinction, 65 million years ago, marked the end of the Mesozoic.
9. The Cenozoic saw the onset of modern patterns of global climate and extensive radiation of flowering plants and mammals. Hominids appeared in Africa about 6 million years ago and modern humans began dispersing around the world about 80 000 years ago. Australia was probably occupied around 45 000–42 000 years ago.
10. The fossil history of the Australian biota suggests widespread aridity on the continent for 20–30 million years. The extraordinary diversity and uniqueness of much of Australia's biota arose in arid or semiarid environments. This rich biota is, however, extremely vulnerable

to extinction unless we recognise the differences between the ancient Australian landscape and the younger, more fertile and regularly disturbed landscapes dominating most of the Earth's land surface.

Key terms

biodiversity hotspots
biogeography
Cambrian
Cenozoic
endosymbiont
epoch
era
heterospory
homospory
Gondwana
Laurasia
mass extinction
Mesozoic
Palaeozoic
Pangaea
period
plate tectonics
Pre-Cambrian
Rodinia
stromatolites

Test your knowledge

1. List the four major geological eras, in order from oldest to most recent.
2. What was the Cambrian explosion and what was its likely cause?
3. Explain why global warming during the Triassic period led to an expanded distribution of seed plants.
4. Briefly describe the evidence that aridity has been widespread on the Australian continent for at least the last 20 million years.

Thinking scientifically

The impact of humans on other species of organisms, on ecosystems and even on the global climate is now universally accepted. There is, in most parts of the world, an urgent need for conservation action to protect threatened plants and animals, and the environment in which they live. Explain why conservation strategies that are informed by an understanding of evolutionary history are more likely to be effective than strategies that are based solely on contemporary ecological studies of the organisms we wish to conserve.

Debate with friends

Extinction is a natural part of evolution. There have been a number of mass extinction events throughout the history of life on Earth. After each of these mass extinctions, biological diversity has rebounded, with new species evolving to utilise the resources previously used by the species that disappeared. Why, then, should our current extinction crisis, caused by human activities, be a cause for concern?

Further reading

Cox, B.C. and Moore, P.D. (2000). *Biogeography*. Blackwell, Boston. 6th edn.
Fortey, R. (2002). *Fossils: the key to the past*. Natural History Museum, London. 3rd edn.
Willis, K.J. and McElwain, J.C. (2002). *The evolution of plants*. Oxford University Press, Oxford.
Zimmer, C. (1998). *At the water's edge*. Free Press, New York.

Theme 3

Applying scientific method – understanding biodiversity

Early this century biologists recognised that some regions of the Earth were unusually rich in the number of unique species living there but also threatened by human activity. They called these regions 'biodiversity hotspots'. Thirty-four are currently recognised, including the South-west Botanical Province in Western Australia. This area of about 360000 km^2 is characterised by low soil fertility and a Mediterranean climate, with most rainfall during winter and dry summers. It is bounded to the south and west by ocean and by desert to the north and east. The climate, low soil fertility and isolation are the driving forces in the evolution of a diverse and unique biota. The province contains nearly 6000 described plant species (including about 4500 unique to the region) and 456 vertebrates (including 100 unique to the region). Although the invertebrate diversity is poorly described, it is likely to be extremely rich as well.

In this third theme of *Environmental Biology* we apply your understanding of the scientific method to the problems of classifying the diversity of life in biodiversity hotspots and elsewhere on Earth. You will also learn how the abundance and distribution of species is determined by interactions within and between species and between species and their environment.

8

Coping with cornucopia – classifying and naming biodiversity

Alan Lymbery and Howard Gill

Conserving the unknown

Since the mid-20th century, our ability to exploit the world's oceans has increased tremendously, largely using technology originally designed for sea warfare. Fishing boats now reach the furthest oceans, schools of fish are tracked underwater with pinpoint precision, and huge nets and lines harvest many fish rapidly.

Despite this vastly increased efficiency, the total world harvest of fish has hardly changed in the last 30 years. Most of the world's important fisheries, including Australia's, are either fully exploited or overexploited; many have collapsed completely, with no fish left to catch. It is not just the targeted fish species that have suffered. Overfishing is often compounded by collateral damage to the marine environment and other inhabitants of the ocean.

In a CSIRO study of damage caused by trawling off deepwater sea mounts south of Tasmania, video cameras revealed that chains and nets dragged by boats fishing for orange roughy removed dense communities of benthic organisms. The tragedy is not only the number of species lost from these communities, but also that the number could not be estimated because we know little about the inhabitants of the ocean floor. Most of the diversity found in marine ecosystems consists of invertebrates living in or on bottom sediments, with estimates of the number of marine benthic species varying from 0.5 to 5 million. Very few are formally described and named. For example, in a recent study of deep-sea polychaete worms, 64% of the species found were previously unknown to science. How can we protect something that we do not know exists?

Chapter aims

In this chapter we explore the way in which biological diversity is ordered and named. We describe the advantages of a classification based on evolutionary relationships among organisms, investigate the ways in which evolutionary histories are reconstructed, and outline the principles for naming and identifying the groups in a classification.

Classifying biodiversity

There are many definitions of biodiversity (Table 8.1). Most have a consistent theme: biodiversity refers to all biological variety existing at multiple biological levels, from populations of a single species to entire ecosystems. This biodiversity is under unprecedented threat from human activities, and in Chapters 24 and 25 we explore different strategies for conserving it. We cannot determine how successful these strategies are, however, unless biodiversity has been documented and described. This is a daunting task. We do not know how many species of living organisms there are, but a conservative estimate is 6–10 million. Of these, only about 1.5 million are named. Species are currently becoming extinct at a rate estimated to be 50–100 times faster than they are being described. For many species, we will never know that they existed, much less know what we need to do to save them.

The taxonomic hierarchy

Humans make sense of the world by ordering it, by placing like with like. Taxonomy is the science of ordering or classifying living organisms and the scientists doing this

Table 8.1 Definitions of biodiversity.

Definition	Source
The variety and relative abundance of species.	Magurran, A.E. (1988). *Ecological diversity and its measurement*. Princeton University Press.
The variety of the world's organisms, including their genetic diversity and the assemblages they form.	Reid, W.V. and Miller, K.R. (1989). *Keeping options alive: the scientific basis for conserving biodiversity*. World Resources Institute.
The genetic, taxonomic and ecosystem variety in living organisms of a given area, environment, ecosystem, or the whole planet.	McAllister, D.E. (1991). *Canadian Biodiversity* 1: 4–6.
The variety of organisms considered at all levels, from genetic variants belonging to the same species through arrays of species to arrays of genera, families and still higher taxonomic levels; includes the variety of ecosystems, which comprise both the communities of organisms within particular habitats and the physical conditions under which they live.	Wilson, E.O. (1992). *The diversity of life*. Belknap Press.
The diversity of life in all its forms, and at all levels of organisation.	Hunter, M.L. Jr (1996). *Fundamentals of conservation biology*. Blackwell Science.

Box 8.1 The terminology of systematics

Systematics: A branch of comparative biology that aims to infer the evolutionary relationships among organisms, classify them into a hierarchical series of groups and name the groups. Systematic studies may include biogeographical, genetic, biochemical, physiological, ecological and behavioural features of the organisms.
Taxonomy: The theory and practice of describing, classifying and naming organisms. Various protocols, rules and regulations must be adhered to during these procedures. Although systematics and taxonomy are not strictly the same, many workers use the terms interchangeably.
Classification: A hierarchy of groups and subgroups of organisms based on their phylogenetic relationships, or the process of arranging organisms into such groups. In a biological classification the hierarchical series of groups are:

- kingdom
- phylum
- class
- order
- family
- genus
- species.

Nomenclature: The naming of taxonomic groups. Nomenclature is a formal process, governed by various rules, which aim to minimise confusion while maximising the information content of a classification.

work are called taxonomists. 'Systematics' is sometimes used as a synonym of taxonomy, but is in fact a broader field, encompassing the study of the diversity of living organisms and their natural relationships (Box 8.1).

The earliest classifications came from the ancient Greeks, who categorised living things as either plants or animals. They then divided animals into those of the sea, the land and the air, and plants into herbs, shrubs and trees. The modern system of classification dates from the great Swedish naturalist Carolus Linnaeus in the mid-18th century. Linnaeus created the standard binomial (two-part) naming system for species (e.g. *Homo sapiens* for humans, *Canis familiaris* for dogs). Linnaeus, and the taxonomists who followed him, also grouped species into larger, more inclusive categories, and this hierarchical system of classification is the one we use today.

In the hierarchical system, species with similar characteristics are placed in the same genus, similar genera in the same family, similar families in the same order, orders in the same class, classes in the same phylum and phyla with similar properties in the same kingdom (Table 8.2). (Note that the term 'phylum' is always used for animals but some authorities use 'division' for plants and fungi.) A group of organisms at any particular level or rank in the hierarchy is called a taxon (plural: taxa). All living organisms are

Table 8.2 The taxonomic hierarchy. Taxon names for five common Australian organisms.

Rank	Red kangaroo	Boobook owl	Wood white butterfly	Lemon-scented gum	Kangaroo paw
Kingdom	Animalia	Animalia	Animalia	Plantae	Plantae
Phylum	Chordata	Chordata	Arthropoda	Anthophyta	Anthophyta
Class	Mammalia	Aves	Insecta	Magnoliopsida	Liliopsida
Order	Marsupiala	Strigiformes	Lepidoptera	Myrtales	Liliales
Family	Macropodidae	Strigidae	Pieridae	Myrtaceae	Haemodoraceae
Genus	*Macropus*	*Ninox*	*Delias*	*Eucalyptus*	*Anigozanthos*
Species	*M. rufus*	*N. novaeseelandiae*	*D. aganippe*	*E. citriodora*	*A. manglesii*

classified into ever more inclusive taxa at the level of species, genus, family, order, class, phylum and kingdom. All organisms must be formally recognised as belonging to each of these categories when they are named. That is, a species must be assigned to a genus, a genus to a family, a family to an order, an order to a class, a class to a phylum and a phylum to a kingdom. Sometimes, intermediate ranks (e.g. subclass, infraclass, superorder, suborder, tribe) are also used in a classification, although their use is not compulsory.

Evolutionary classification

The main concern of the science of systematics is recognising groups as taxa at each level of the taxonomic hierarchy. We could group organisms by size, colour, whether they are edible or inedible, or whether they are dangerous or harmless. However, since the acceptance of the theory of evolution by natural selection (see Chapter 6), systematists have attempted to classify organisms according to their evolutionary relationships or phylogeny.

The phylogeny of a group of organisms is a hypothesis of their evolutionary relationships. It may be depicted verbally in a classification or pictorially in the branching pattern of a phylogenetic (or evolutionary) tree. As in any other family tree, a phylogenetic tree groups organisms according to how recently they shared a common ancestor and has an implicit time axis, with descendants located higher up the tree than ancestors.

The phylogenetic tree in Figure 8.1, for example, depicts evolutionary relationships between three common Australian animals: a flying fox, a kangaroo and a kookaburra. Despite the superficial similarity between the flying fox and the kookaburra (both have wings), the phylogenetic tree indicates that the flying fox and the kangaroo are more closely related to each other than either is to the kookaburra. In an evolutionary classification, flying foxes and kangaroos are placed in a different taxon (class Mammalia) to kookaburras (class Aves). Of course, at a lower taxonomic level flying foxes and kangaroos also belong to different taxa (order Chiroptera and order Marsupialia, respectively), while at a higher taxonomic level all three animals are placed in the same taxon (phylum Chordata).

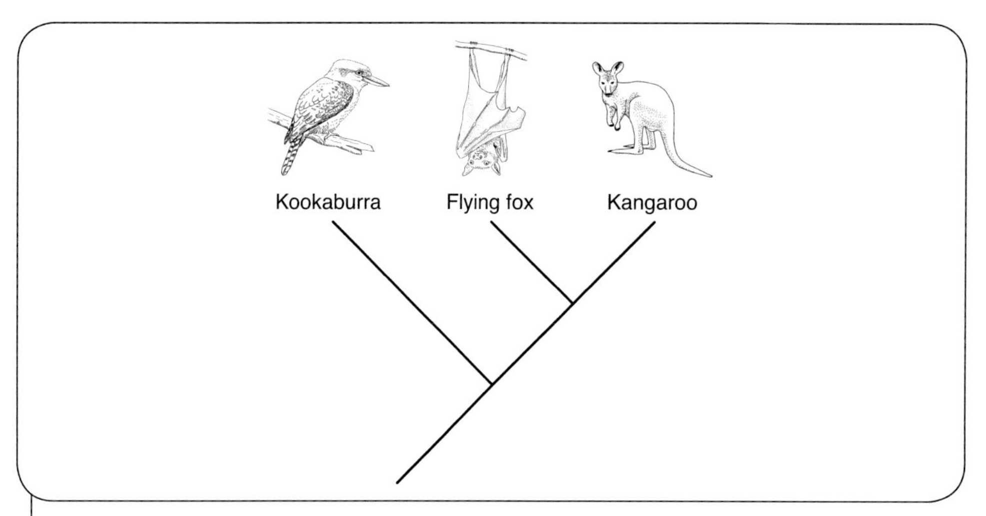

Figure 8.1 A phylogenetic tree, showing the evolutionary relationships between a flying fox, a kangaroo and a kookaburra. (Source: B. Cale)

Evolutionary classifications have several important features. First, the basal taxonomic unit, the lowest rank into which all organisms are classified, is the species. Secondly, a higher taxon is a group of species that all descended from a single ancestral species. For example, the genus *Macropus* is a group of about 14 extant (currently living) species of kangaroos and wallabies. These are all classified in the same genus because we believe that they are descended from a single ancestor, represented as point A in Figure 8.2. There are 10 other genera of kangaroos and wallabies (*Wallabia*, *Onychogalea*, *Lagochestes*, *Petrogale*, *Thylogale*, *Setonix*, *Dendrolagos*, *Dorcopsulus*, *Dorcopsis* and *Lagostrophus*), which, along with *Macropus*, are placed in the family Macropodidae. These are all believed to be descended from another, older ancestral species (point C in Figure 8.2).

Thirdly, although most taxonomists regard species as natural entities (see Chapter 6), the exact limits of higher taxa are less certain. For example, should only the species descended from ancestral species A in Figure 8.2 be placed in the genus *Macropus*, or should some other species (perhaps those descending from ancestral species B, now classified as *Wallabia*) also be included in *Macropus*? Should the family Macropodidae be restricted to the descendents of ancestral species C, or should the family be broadened to include the descendents of ancestral species D (to also include the genera *Bettongia*, *Aepyprymnus*, *Caloprymnus* and *Potorus*, currently placed in the family Potoroidae or rat kangaroos), or of ancestral species E in Figure 8.2 (to include, as well as the rat kangaroos, the genus *Hypsiprymnodon*, currently placed in the family Hypsiprymnodontidae)? There are no definitive answers to these questions from the phylogenetic tree. It tells us the pattern of evolutionary relationships among species and therefore which species can legitimately be grouped to form higher taxa, but the decision as to which of these groups should actually constitute higher taxa is subjective.

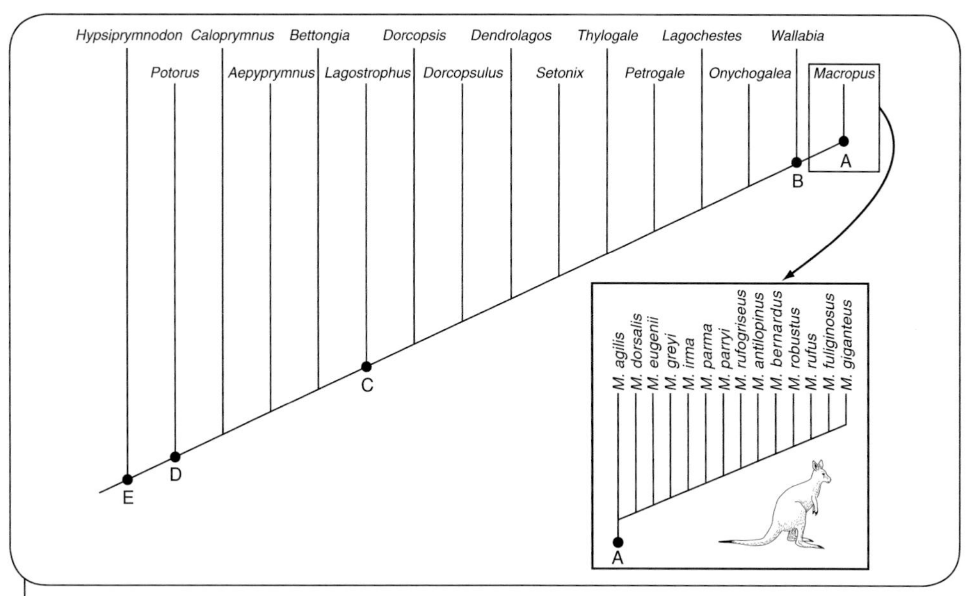

Figure 8.2 A phylogenetic tree, showing presumed evolutionary relationships between kangaroos and wallabies. In the main figure, currently recognised genera are shown at the tips of the branches, while the inset shows the 14 species classified in the genus *Macropus*. The branching points or nodes (labelled A, B, C, D and E) represent hypothetical ancestral species which gave rise to modern-day descendants. (Source: B. Cale)

The classification of organisms into species and higher taxa is of great practical significance, because taxonomic rank is often important in assessing the conservation priority of an endangered group of organisms. In general, the higher the taxonomic rank, the more powerful the arguments in favour of conservation. There is therefore much interest in systematic studies of rare and endangered organisms (Box 8.2).

Box 8.2 The political ramifications of classification

Conservation action is usually directed at species. This makes biological sense, because most biologists would agree that species are real entities in nature, representing independently evolving lineages (see Chapter 6). The species rank also appeals to the public and to policy makers. Lists of rare and endangered taxa are almost always species lists. These lists, such as the Red List of the International Union for the Conservation of Nature, form the basis of national and international conservation planning and legislation. Great political significance can, therefore, be attached to the classification of a group of organisms as a species.

Marine turtles are a good example of the political significance of species status. There are currently seven or eight recognised species of marine turtles, and most are threatened by pollution, harvesting and disruption of nesting habitat. The most endangered marine turtle species is Kemp's ridley (*Lepidochelys kempi*), found in the Gulf of

continued ›

Box 8.2 continued ›

Mexico and western Atlantic. Kemp's ridley is very similar morphologically to the olive ridley (*Lepidochelys olivacea*), which has a circumglobal distribution in tropical and warm-temperate oceans. The species status of Kemp's ridley was therefore disputed until molecular genetic studies in the 1990s confirmed that it was distinct from the olive ridley. The confirmation that Kemp's ridley is a separate species bolstered conservation efforts for this turtle.

Taxonomic progress can also work in the other direction. The black turtle (*Chelonia agassizii*), found in the eastern Pacific, has long been classed as a separate species from green turtle (*Chelonia mydas*), which has a much wider distribution in tropical and subtropical seas around the world. Separate species status has been based on differences in colouration, size and carapace shape. Recent molecular genetic studies failed to find an evolutionary distinction between green and black turtles and strongly suggest that the black turtle should not be regarded as a separate species. These findings sparked a major controversy, with fears that downgrading the taxonomic status will compromise the protection of black turtle populations.

The purpose of an evolutionary classification

All biological classifications are essential information storage and retrieval systems. By using an internationally recognised and agreed name for a group of organisms, such as *Eucalyptus camaldulensis* for the ubiquitous river red gum of inland Australia, taxonomists indicate the group without having to list all the properties occurring in each group member. It is easier and more efficient to say that a tree is in the species *Eucalyptus camaldulensis* than it is to say that it is one of a group of trees found naturally along river courses in arid and semiarid areas of Australia, growing to a height of 30 m, with smooth white bark, elongate leaves and small, conical fruits with protruding, triangular valves.

An evolutionary classification also predicts information that we do not yet have. For example, if we find a new species of furred animal that suckles its young, then we can predict that it will also have other traits of the class Mammalia, including endothermy (maintaining a high internal body temperature), heterodont dentition (different, specialised types of teeth), a four-chambered heart with a double circulation system, internal fertilisation and (unless it is a monotreme) the ability to bear live young, with the embryo nourished by a specialised placenta.

Predictive ability is an inherently useful feature of an evolutionary classification. For example, the Moreton Bay chestnut (*Castanospermum australe*), found along river banks in east-coast Queensland and New South Wales rainforests, produces large seeds, rich in starch. They are also poisonous, but Aborigines removed the toxins by grating, soaking and baking before eating the seeds. One of the toxins is the alkaloid castanospermine, which has potential in combating cancer and the AIDS virus. To find other sources of castanospermine, scientists turned to plants related to *C. australe*. Although this is the only species in the genus *Castanospermum*, another genus within the same family (*Alexa*, native to tropical South America) is very similar. Castanospermine has now been found in plants of the genus *Alexa*.

Phylogenetic reconstruction

Although there are different thoughts on the best way to determine (or reconstruct) evolutionary relationships, most evolutionary biologists now accept that the most objective and logical approach is to use the techniques of cladistics (or phylogenetic systematics), first developed by the German entomologist Willi Hennig.

Cladistic analysis starts with the characters shared by a group of organisms. In systematics, a character is any feature of an organism that can be described or measured and thus be potentially useful in determining relationships. They can be DNA sequences, biochemical products, morphological features or behaviours. Some systematists use the term 'character' to describe a feature and its appearance whereas others use the terms 'character' to describe the feature and 'character state' to describe the character's appearance. For example, one worker may say that a character of adult sailfish is an enlarged dorsal fin and a character of marlin is a relatively small dorsal fin, whereas another worker may say that the character is dorsal fin and that in adult sailfish the state of that character is enlarged, while in marlin the state is not enlarged. We will use the latter approach, but neither is incorrect.

The logic of cladistic analysis is deceptively simple; common ancestry is indicated by shared character states, and more recent common ancestry is indicated by shared, derived character states, rather than shared, ancestral character states. Let us explain this in more detail. Figure 8.3 shows the phylogeny of three species of land plants: a conifer,

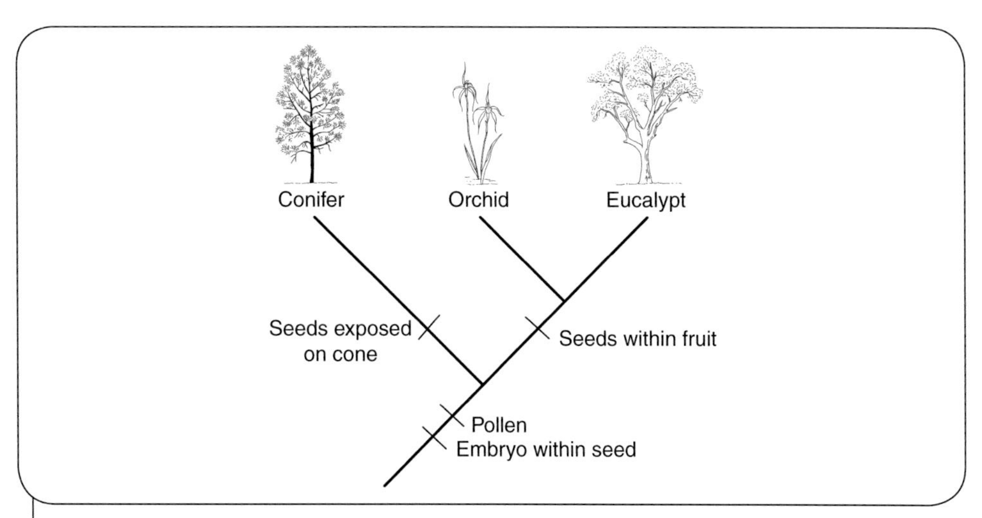

Figure 8.3 A phylogenetic tree, showing the evolutionary relationships among a conifer, an orchid and a eucalypt, based on reproductive characters. (Source: B. Cale)

an orchid and a eucalypt. All of these plants have certain character states in common; for example, they all produce pollen and in all of them the embryo is contained within a seed. These character states are ancestral within this group of plants. They tell us that conifers, orchids and eucalypts share a common ancestor, but give no indication of phylogenetic relationships within the group. For this, we have to look to the states of other reproductive characters, in particular the presence or absence of a specialised structure called a fruit. In both orchids and eucalypts the seeds are contained within a fruit, whereas in conifers the seeds lie exposed on the female cone. The presence of a fruit is therefore a derived character state, indicating that orchids and eucalypts share a more recent common ancestor with each other than either of them did with conifers.

Put like this, cladistic analysis sounds obvious. In practice, however, there are many complications that introduce uncertainty into the process of reconstructing phylogenies. Two of the main areas of uncertainty involve determining what characters to use in cladistic analysis and distinguishing 'derived' from 'ancestral' character states.

Homologous and analogous characters

For the cladistic analysis of a group of organisms, the evidence consists of several characters, each with different character states. For example, for a group of plant species, one character might be 'structure of stem', with the states 'woody' and 'non-woody', while another character might be 'form of seed', with the states 'enclosed in fruit' and 'not enclosed in fruit'. If all of these characters pointed to the same phylogeny, there would be no problem. In practice, however, different characters point to different phylogenies. In the case of the conifer, the orchid and the eucalypt referred to above, the character 'form of seed' indicates that orchids and eucalypts are more closely related to each other than either is to conifers. If we chose the character 'structure of stem' to produce our phylogeny, we would probably group conifers and eucalypts together as sharing a more recent common ancestor because they both have woody stems, whereas the stem of an orchid is not woody.

When different characters indicate different phylogenies, one of them must be misleading. A set of species has only one phylogeny because it can have only one ancestral history. We need to distinguish reliable characters, called homologous, from unreliable characters, usually called analogous (or homoplasious). Only homologous characters should be used to reconstruct phylogenies.

Homologous characters always have a similar structure, although not necessarily a similar function. Most importantly from a systematic viewpoint, a homologous character is found in two or more taxa and in their common ancestor. For example, all birds have wings supported by an elongated second digit of the forelimb. Although wings may function differently in a flightless emu and a wedge-tailed eagle, their structural similarities and their ubiquity indicate that wings are a homologous character for the class Aves (birds). An analogous character, by contrast, is found in two or more taxa, but not in their common ancestor. Analogous characters usually perform similar functions, but are structurally different. For example, the wings of insects, birds and bats are analogous.

They are all adaptations for flight, but they have evolved independently. Insect wings are composed of sheets of chitin and protein and, although the wings of birds and bats superficially look similar, the wings of bats are supported by all five digits of the forelimb rather than just one as in birds.

There are several ways to distinguish homologous from analogous characters. Occasionally, fossil evidence shows that a character found in two or more taxa is not present in their common ancestor and is therefore analogous. More often, we undertake careful anatomical and developmental studies to ensure that homologous characters have a fundamentally similar structure and embryonic origin. Sometimes, we use the weight of phylogenetic evidence to infer which characters are homologous and which are analogous. If the majority of characters indicate a particular phylogeny, while a few characters suggest a different phylogeny, we might assume that these few characters are providing misleading information and are therefore analogous.

Distinguishing derived and ancestral character states

Once we have determined which characters are homologous and therefore reliable indicators of phylogeny, the next stage in a cladistic analysis is to distinguish ancestral and derived character states. Only shared derived character states, not shared ancestral character states, indicate most recent common ancestry.

Sometimes we can distinguish the derived and ancestral states of a character from the fossil record. If a group of organisms has a fairly complete fossil record, then the character states appearing first in that record are ancestral. Where the fossil record is imperfect we need other approaches. The method most commonly used is called out-group comparison. This involves examining the character not only in the group of organisms for which we are trying to determine phylogeny (the in-group), but also in one or more closely related organisms (the out-group). The character state found in the out-group is presumed to be ancestral. For example, consider the character of reproductive biology among three different species of mammals – water rat, grey kangaroo and echidna (Figure 8.4). The rat and the kangaroo produce live young (they are viviparous), while the echidna lays eggs (it is oviparous). Which is likely to be the ancestral and which the derived state? Using the method of out-group comparison we would look at the reproductive biology of closely related species of vertebrates that we know are not mammals, such as birds, reptiles, amphibians or fish. Almost all of these breed oviparously, so viviparity is likely to be a derived character state, shared by rats and kangaroos, indicating that they share a more recent common ancestor with each other than either does with the echidna.

Cladistic analysis of molecular characters

Although cladistic analysis was first formulated for morphological characters, molecular characters, particularly the sequences of amino acids in proteins and nucleotides in

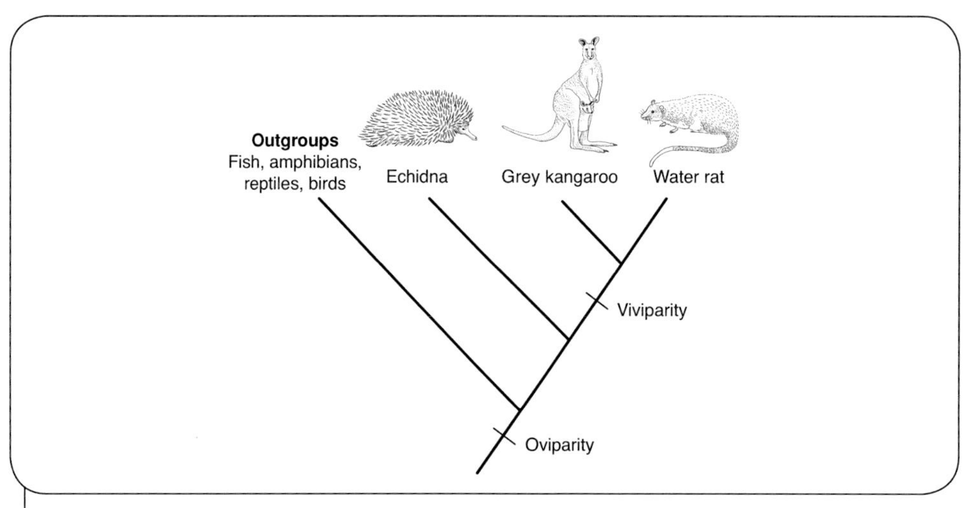

Figure 8.4 A phylogenetic tree, showing the evolutionary relationships among an echidna, a kangaroo and a water rat. Based on out-group analysis, oviparity is presumed to be the ancestral state and viviparity the derived state for this group of mammals. (Source: B. Cale)

DNA, are now used increasingly in systematic studies because they have several advantages. First, they provide almost unlimited amounts of information. Every amino acid in a protein molecule and every nucleotide in a DNA sequence represents a separate character, whereas morphological studies are typically based on only 20–30 different characters. Secondly, evolutionary change in molecular characters is more easily comparable than evolutionary change in morphological characters. A change at one amino acid or nucleotide site is equivalent to a change at another site, whereas a change in one morphological character (e.g. limb bone structure) does not necessarily represent the same amount of evolution as a change in another character (e.g. viviparity.) Finally, molecular characters reveal relationships among organisms with no apparent morphological homologies. It is very difficult, for example, to compare the morphology of a seaweed, a fungus and a mammal, but they all have nucleic acids so we can compare their DNA sequences.

Although the logic of cladistic analysis is identical for morphological and molecular characters, the practice often differs. For morphological characters, distinguishing homologous and analogous characters is vital. When characters provide conflicting information, the standard cladistic solution is to examine them in more detail to determine which are homologous. This is rarely possible with molecular characters; we cannot tell whether identical nucleotides or identical amino acids were derived from a common ancestor (homologous) or arose independently (analogous). However, given the large number of molecular characters available in a phylogenetic analysis it is possible to use statistical techniques to determine the most likely phylogeny from all the molecular sequence data. Three main types of statistical techniques – distance methods, parsimony methods and maximum likelihood methods – can be implemented using readily available software. This active research area has radically altered some of our long-held views on evolutionary relationships (see Chapter 13).

From phylogenetic reconstruction to classification

Cladistic analysis can reconstruct a phylogeny, but these species still have to be grouped into a hierarchy of taxa for classification. There are two different schools of thought as to how this should occur: the traditional (or evolutionary) approach and the phylogenetic (or cladistic) approach. To see how these approaches differ, we need to distinguish monophyletic, paraphyletic and polyphyletic groups. A monophyletic group contains one ancestor and all of its descendant species (Figure 8.5a), a paraphyletic group includes an ancestor and some, but not all, of its descendent species (Figure 8.5b), while a polyphyletic group includes species that do not have a most recent common ancestor (Figure 8.5c).

Polyphyletic groups usually arise when we are misled by analogous characters. For example, grouping birds, bats, grasshoppers and some butterflies together because they have wings forms a polyphyletic group, because winged grasshoppers and butterflies are more closely related to other insects than they are to bats and birds. Flight has apparently evolved independently in these groups. Both traditional and phylogenetic approaches agree that polyphyletic groups should not be included in a classification.

Paraphyletic groups arise when we use ancestral and derived character states to unite species. This usually occurs because differential rates of evolution in different lineages lead to some descendants of a common ancestor retaining ancestral character states, whereas other descendants evolve new, derived character states. A phylogenetic approach would always reject paraphyletic groups in a classification, whereas a traditional approach would accept them in certain circumstances.

A good example of these different approaches is the classification of reptiles. Figure 8.6 shows the phylogeny of birds, crocodiles, snakes and lizards. In a traditional classification, crocodiles are grouped with snakes and lizards in the class Reptilia, of equal taxonomic rank to the class Aves (birds). Reptilia, however, is a paraphyletic group, because crocodiles are more closely related to birds than they are to snakes and lizards. Birds have evolved rapidly while crocodiles, snakes and lizards have not, and so have been left looking more like each other than like birds. In a traditional classification the rapid evolutionary change of birds is recognised by putting birds in a separate class, whereas a phylogenetic approach groups crocodiles with birds so that only monophyletic groups are included in the classification.

Most of our classifications are traditional rather than strictly phylogenetic, sometimes including paraphyletic groups. A purely phylogenetic classification would produce many unfamiliar groupings. Whether this would be a good or a bad thing is a hotly debated topic among systematists.

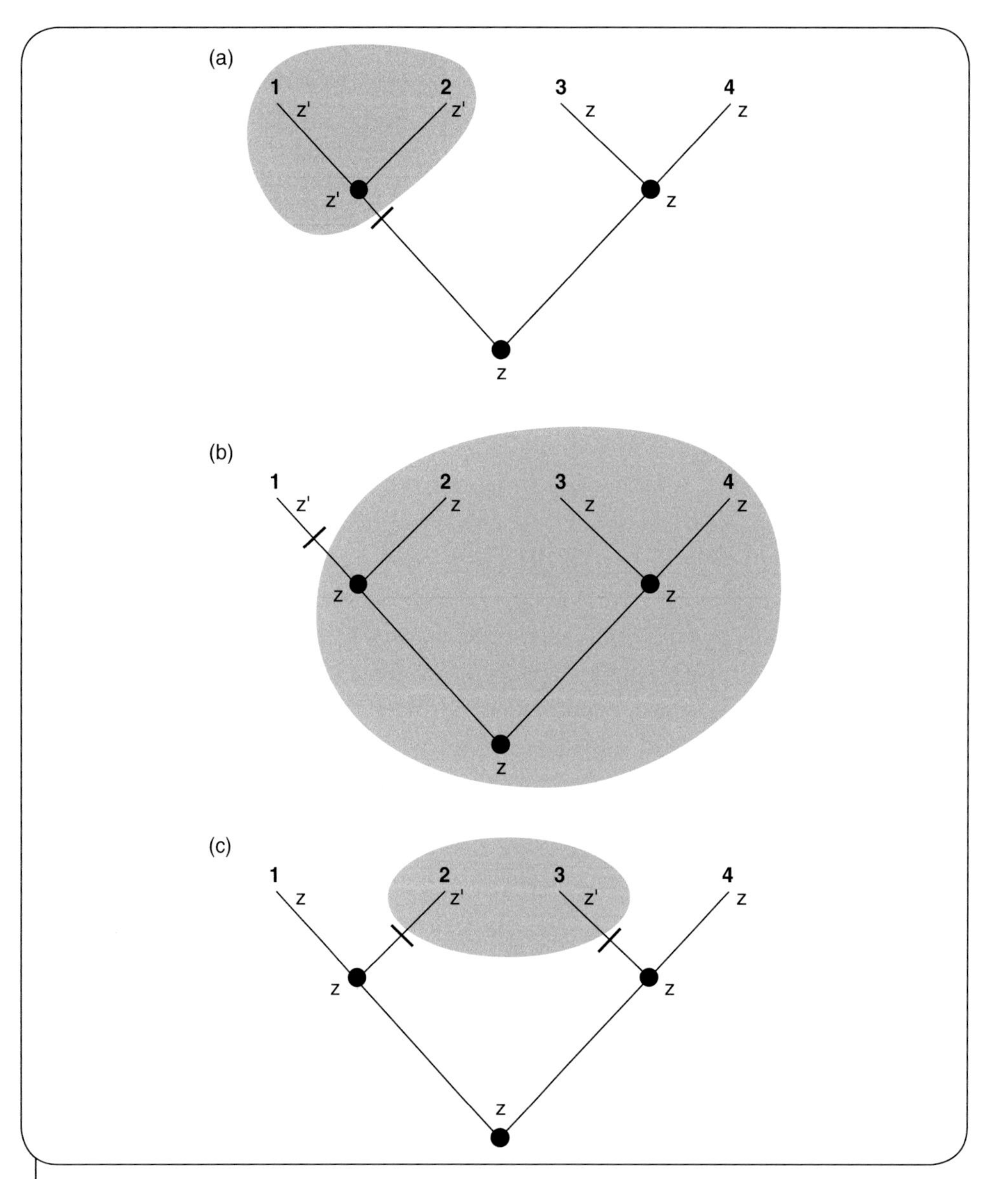

Figure 8.5 An example of monophyletic, paraphyletic and polyphyletic groups formed from four different species (1–4) using the character Z, which has two states: z (ancestral) and z′ (derived). The tree shows the true phylogeny of the species. (a) In a monophyletic group, shared derived character states for Z are used to unite an ancestor and all of its descendent species. (b) In a paraphyletic group, shared, ancestral character states are used to unite an ancestor and some of its descendent species. (c) In a polyphyletic group, analogous character states are used to unite species that do not share a most recent common ancestor. (Source: Redrawn from Ridley, M. (2004) *Evolution*. Blackwell Publishing, 3rd edn)

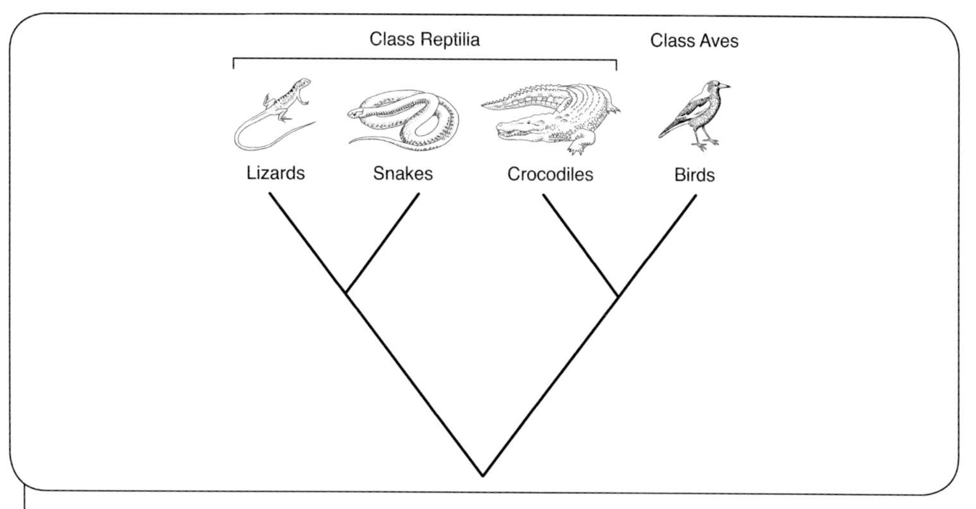

Figure 8.6 Phylogeny of birds, crocodiles, snakes and lizards. In a traditional classification, snakes, lizards and crocodiles are grouped into the class Reptilia and birds are grouped into the class Aves. Reptilia is a paraphyletic group. (Source: B. Cale)

Naming biodiversity

Classification puts organisms into groups, whereas nomenclature names these groups. For effective communication it is essential that each group of living organisms at each level in the taxonomic hierarchy has a single, universally accepted name. Common names for organisms, beautiful and evocative as they often are (think of the mountain ash, the blue-eyed smokebush, the whistling water duck and the feathertail glider), fail this criterion. Apart from the multiplicity of languages, even within one language the same group of organisms is often known by more than one name, or the same name is used for different groups. For example, the rare bilby (*Macrotis lagotus*) is also known as the rabbit-eared bandicoot, rabbit bandicoot, pinkie and dalgyte. River red gum, or simply red gum, is the accepted common name for trees of the species *Eucalyptus camaldulensis* throughout Australia, but red gum also refers to the species *Corymbia calophylla* in south-western Western Australia, and *Eucalyptus tereticornis* along the eastern coast of Victoria and New South Wales.

Scientific nomenclature eliminates this confusion, and ensures one common language in biological classification. There are strict rules, called codes of nomenclature, for assigning names to groups. Similar, but separate, codes exist for animals (the International Code of Zoological Nomenclature), plants and fungi (the International Code of Botanical Nomenclature) and micro-organisms (the International Code of Nomenclature of Bacteria).

By tradition and convention, taxon names are given in Latin form. Names of taxa above the rank of genus consist of only one word, a plural noun written with an initial capital letter. The codes often stipulate a standardised ending for the names of all taxa of

Table 8.3 Standardised endings for the names of taxa, as recommended by the international codes of zoological, botanical and bacteriological nomenclature.

Category	Zoological	Botanical	Bacteriological
Phylum		-phyta (plants) -mycota (fungi)	
Subphylum/subdivision		-phytina (plants) -mycotina (fungi)	
Class		-opsida (higher plants) -phyceae (algae) -mycetes (fungi)	
Subclass		-idae (higher plants) -phycidae (algae) -mycetidae (fungi)	
Order	-iformes*	-ales	-ales
Suborder		-ineae	-ineae
Superfamily	-oidea		
Family	-idae	-aceae	-aceae
Subfamily	-inae	-oideae	
Tribe	-ini	-eae	-eae
Subtribe	-ina	-inae	

* the ending '-iformes' is not specified in the zoological code, but is extensively used.

a given rank (Table 8.3). For example, names of animal families always end in '-idae', such as Canidae for the dog family, while names of plant families always end in '-aceae', such as Myrtaceae for the family of eucalypts and their relatives. The names of genera also consist of one word with an initial capital letter, but they are singular nouns and always written in italics, e.g. *Canis*, *Eucalyptus*, *Bacillus*.

The names of species are different to the names of other taxa, because they always consist of two words, first the name of the genus into which the species is classified, followed by a second word (the specific epithet) unique to that species within the genus. Specific epithets, like genus names, are always written in italics, but the initial letter is not capitalised. The specific epithet is never used alone, because it may be the same for species in different genera. New Holland, for example, is an old Dutch term for Australia; it is used as the specific epithet for the red-necked avocet (*Recurvirostra novaehollandiae*), the Australian little grebe (*Podiceps novaehollandiae*) and the masked owl (*Tyto novaehollandiae*). When the same species name is used repeatedly in one document, the generic name is commonly abbreviated to the initial capital (e.g. *R. novaehollandiae*).

Although stability of nomenclature is important for scientific communication, it is not always possible within evolutionary classifications based on our knowledge of phylogeny. Each time a new organism is discovered or our understanding of the structure and function of described organisms deepens, deficiencies in our existing classifications

Box 8.3 What's in a name? Raising money for conservation

Funds are always in short supply for conservation programs. Several non-profit conservation groups now offer naming rights to new species in exchange for donations to conservation science. The German group BIOPAT (Patrons for Biodiversity), for example, has raised more than US$450 000 in this way, with the cost of naming a species ranging from US$3000 to US$13 000. The money is typically split between the institution of the species's discoverer and conservation research projects in the country of origin of the species.

This practice is not without its critics. The International Commission on Zoological Nomenclature is worried that selling species names may lead to fraudulent species descriptions and undermine the scientific system of nomenclature. Nevertheless, with the huge backlog of undescribed species, and an ever-declining public investment in systematic research, selling naming rights is likely to become even more prevalent.

become apparent. Systematics is not static – classifications change as our phylogenetic knowledge increases, causing nomenclatural instability.

Naming taxa has practical implications beyond communication (see Box 8.3). To cope with nomenclatural instability, rules in the codes of nomenclature must be followed when giving or changing names. First, the name, and a description of the taxon to which the name is being given, must be published, usually in a printed scientific journal. Secondly, the name must be associated with a type specimen; for a species name this is a representative individual on which the description is based, while for higher level taxa it is a representative taxon of the immediately preceding taxonomic rank (a species for a genus name, a genus for a family name, and so on). Type specimens are usually held in museums or herbaria, in the city in which the taxonomist who first described the taxon lived. Therefore, many of the types of Australian organisms are held in London and other European cities. Thirdly, when two or more names have been given to the same taxon, the correct name is the oldest, by the principle of priority. For example, *Brontosaurus* was the name given to a genus of giant herbivorous dinosaurs in 1879. It was recently discovered that the fossilised *Brontosaurus* bones were actually from the same group of organisms that were first described from fossils in 1877 and given the name *Apatosaurus*. By the principle of priority, *Apatosaurus* is the correct genus name.

Identifying taxa

Identification is the process of finding the taxon to which an organism belongs. Identification requires that the organism has been previously described, named and placed within the taxonomic hierarchy. Identification is sometimes straightforward; a dog can be easily recognised as the species *Canis familiaris*, within the family Canidae, order Carnivora, class Mammalia, phylum Chordata, kingdom Animalia. Less familiar organisms cause problems. For example, we might recognise the tall and elegant salmon gum of inland Western Australia as a member of the genus *Eucalyptus*, but not know that its species name is *Eucalyptus salmonophloia*. Often we cannot identify an organism to a class or order level. Web spinners occur in silk-lined tunnels under rocks, bark or leaf litter. Although they produce silk like spiders, a careful examination shows that they are insects. Most of us, however, could not identify them as belonging to the order Embioptera, let alone recognise the different families, genera or species.

Correct identification precedes serious biological study, because unless we know what organism we are dealing with we cannot generalise our observations to other organisms. It is also of immense practical importance (see Box 8.4). To know how widespread a species is, or whether it is common or rare, we must identify it. For example, the central rock rat (*Zyzomys woodwardi*) is one of the rarest Australian rodents and was recently placed on the Critically Endangered list. It is restricted to rocky ranges around Alice Springs in the Northern Territory. The central rock rat is very similar to the common rock rat (*Zyzomys argurus*), wide-ranging across Queensland, the Northern

Box 8.4 Using macroinvertebrate keys to monitor water quality

Australia is the driest permanently inhabited continent in the world. Fresh water is therefore a critical resource in Australian ecosystems, but unfortunately it is under increasing threat from climate change, pollution, salinity, river regulation and water harvesting. Waterwatch Australia is a national, community-based monitoring network for water quality, established by the Australian government. Waterwatch encourages landholders, community organisations and school groups to monitor local waterways.

One of the water-monitoring techniques used by Waterwatch is a macroinvertebrate grade score. Aquatic macroinvertebrates are those invertebrates that are visible to the naked eye (usually insects, crustaceans, arachnids, molluscs and annelids; see Chapter 14). Different macroinvertebrates have different sensitivities to pollutants and it is possible to grade each type from 1 to 10, where 1 is most tolerant and 10 is most sensitive to the pollutant. These grade numbers can be assigned at any taxonomic level, but in practice are usually applied to different orders or families of macroinvertebrates.

The procedure is to sample macroinvertebrates from the water body being monitored, identify them to the appropriate taxonomic level (order or family), assign individual grade numbers (obtained from Waterwatch manuals) to each group and calculate the average grade score. The greater the grade score, the less polluted is the water body. The critical part is accurate identification of the macroinvertebrates using several region-specific identification keys available in the Waterwatch manuals.

Territory and the Kimberley in Western Australia. Determining the endangered status of the central rock rat requires identifying specimens as belonging to *Z. woodwardi* and not *Z. argurus*.

Keys

An identification key is a means of deducing the taxon to which an organism belongs. Keys can be designed for use at any level of the taxonomic hierarchy. For example, in trying to identify an insect found singing in our garden at night, we might first use a key to the orders of insects, which enables us to place it within the order Orthoptera (grasshoppers and crickets), then a key to the families within this order, which places it within the family Tettigoniidae (bush crickets or katydids) and finally a key to the species within this family, which identifies our specimen as *Mygalopsis marki*, a rarely seen but quite common native Australian species.

The simplest keys use a series of photographs or diagrams against which to compare the organism to be identified. Most popular field guides are of this form. Picture keys are easy to use, but limited (they are effective only when the number of possible taxa is small) and often of doubtful accuracy (many taxa, especially at the species level, look similar in a photograph or diagram). Most serious identification uses written keys, providing a series of choices relating (usually) to the external features of the organism to be identified. Often, two choices are offered at each step, and the key is then called a dichotomous key. Selection of one character choice that matches the organism in one question leads to another character choice and so on until the organism is identified (Figure 8.7).

In conventional dichotomous keys, the correct choice must be made at each stage to reach the true identification. This is not a problem if the key is carefully constructed

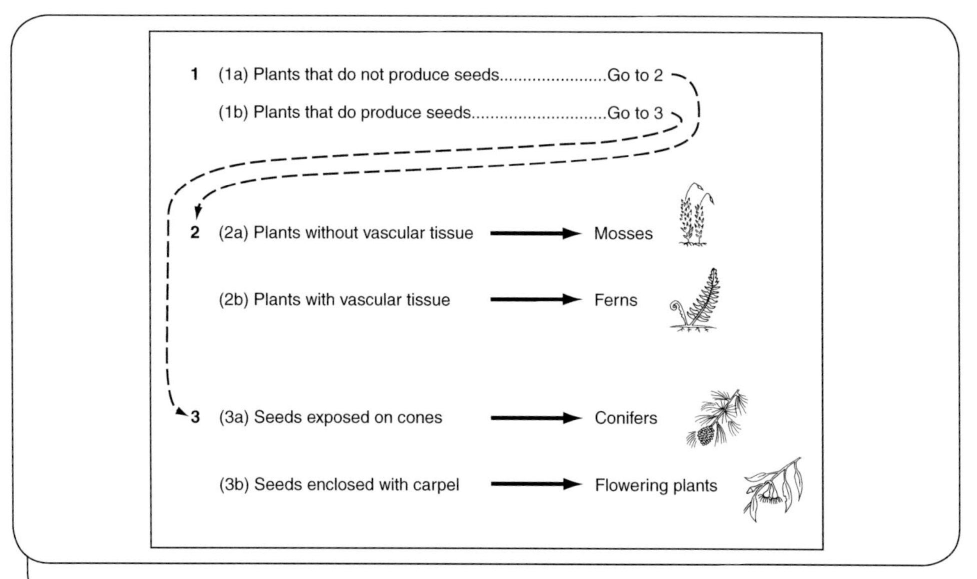

Figure 8.7 A simple dichotomous key to distinguish the major groups of land plants. (Source: B. Cale)

and the users are careful, but potentially an error can be made at each dichotomous step in the key. There is, therefore, increasing interest in interactive computer keys, in which users enter the character states of the organism to be identified. The program then eliminates taxa the attributes of which do not match those of the organism and the process is continued until only one taxon is left. Interactive keys are difficult to construct, but have desirable features absent in conventional dichotomous keys. In particular, users can enter characters in any order and still arrive at a correct identification, even if errors are made in one or more characters.

Kingdoms of life

The earliest classification systems recognised only two kingdoms of living organisms: plants and animals. However, first with the discovery of micro-organisms, and more recently with the use of biochemistry and molecular biology to investigate evolutionary relationships, our knowledge of the kinds of organisms on Earth has increased dramatically. Most biologists now use a six-kingdom classification system, first proposed by the microbiologist Carl Woese in the 1970s. These six kingdoms are often further grouped into three superkingdoms or domains (Figure 8.8).

Two of the domains, Bacteria and Archaea (in older books sometimes called Eubacteria and Archaebacteria), contain only one kingdom each. Both consist of microscopic, single-celled prokaryotes lacking nuclei and membrane-bound organelles. Bacteria are the most abundant organisms on Earth, playing key roles in the functioning of all ecosystems. Like bacteria, archaea are found in many environments and are crucial in the flow of nutrients and energy through ecosystems. They diverged from bacteria very early in evolutionary

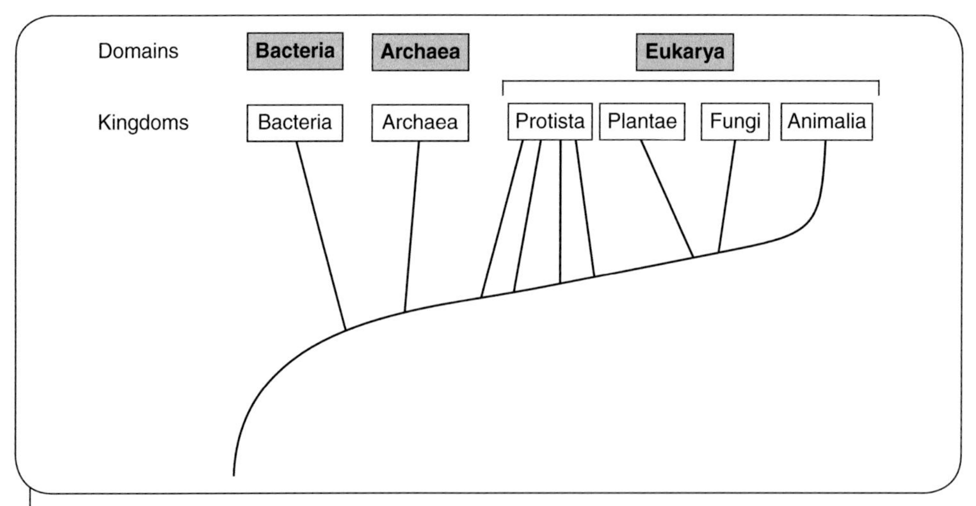

Figure 8.8 Simplified version of a tree of life, showing division into domains and kingdoms. (Source: B. Cale)

history, and, despite the structural and functional similarities between these two groups, DNA sequencing studies suggest that archaea are more closely related to the third domain of life, the Eukarya. Eukarya includes all eukaryotic organisms, classified into four different kingdoms: Protista, Plantae, Fungi and Animalia (although some biologists classify the fungi within the Protista – see Chapter 10). Although the first eukaryotes were unicellular, their complex cellular organisation with functional organelles enabled the evolution of larger cells and, eventually, multicellular life.

In the next seven chapters we explore the diversity of structure and function within these six kingdoms of life. Some problems and contradictions will become apparent as we do. The four-kingdom classification of the Eukarya, for example, is not entirely satisfactory, because it does not completely match evolutionary relationships. In particular, the protists are not a monophyletic kingdom because some (such as some groups of green algae) are more closely related to plants than they are to other protists. Also there is one notable absentee from the tree of life shown in Figure 8.8. Viruses are not usually placed within any of the kingdoms, because they possess only some of the properties of living organisms; they are not cellular and cannot reproduce on their own. Nevertheless, they are important components of ecosystems and we will briefly consider them in Chapter 9.

Chapter summary

1. Systematics (or taxonomy) is the science of classifying organisms into groups (called taxa) and naming the taxa. Biological classification is hierarchical, with living organisms classified into ever more inclusive taxa from species (least inclusive), genus, family, order and class through to phylum and kingdom.
2. Classification into taxa is based on the phylogeny or evolutionary history of organisms, depicted visually in a phylogenetic tree. An evolutionary classification serves as an essential information storage and retrieval system, and also has predictive capability because it is based on evolutionary relationships.
3. Phylogenies are determined or reconstructed by cladistic analysis of the characters or features of organisms. In cladistic analysis, recent common ancestry is determined by the presence of shared, derived character states between organisms. The two essential steps in cladistic analysis are (a) distinguishing characters with common ancestry (homologous characters) from those that are structurally or functionally similar but are derived independently (analogous characters); and (b) for homologous characters, distinguishing derived character states from ancestral character states.
4. Naming of taxa is governed by international codes of nomenclature. Species names differ from the names of other taxa because they consist of two words: first the name of the genus into which the species is classified, followed by the specific epithet, which is peculiar to that species. Genus names and specific epithets are always written in italics, with the initial letter of the genus name capitalised.
5. An identification key is a means of deducing the taxon to which an organism belongs. Most are dichotomous keys, which have a series of steps with two choices, usually relating to external morphological features, at each step. A specimen is compared to the choices at each step until it is identified to a taxon.
6. Most biological classifications of living organisms now recognise six kingdoms: Bacteria, Archaea, Protista, Plantae, Fungi and Animalia. Protista, Plantae, Fungi and Animalia are usually grouped into a superkingdom or domain, Eukarya, while the Bacteria and Archaea are each placed in a separate domain.

Key terms

analogous
ancestral
binomial system
biodiversity
character
character state
cladistics
class
classification
codes of nomenclature
derived
dichotomous key
division
domain
family
genus
homologous character
kingdom
monophyletic
nomenclature
order
paraphyletic
phylogeny
phylum
polyphyletic
rank
species
systematics
taxon (taxa)
taxonomy

Test your knowledge

1. Why is it useful to classify organisms?
2. Explain the advantages of a classification based on evolutionary relationships among organisms.
3. What type of character states indicate that two species shared a more recent common ancestor with each other than either did with a third species?
4. Explain the difference between homologous and analogous characters.
5. What are monophyletic, polyphyletic and paraphyletic groups?
6. What is the difference between a classification and a key?
7. Name the three domains and six kingdoms of life.

Thinking scientifically

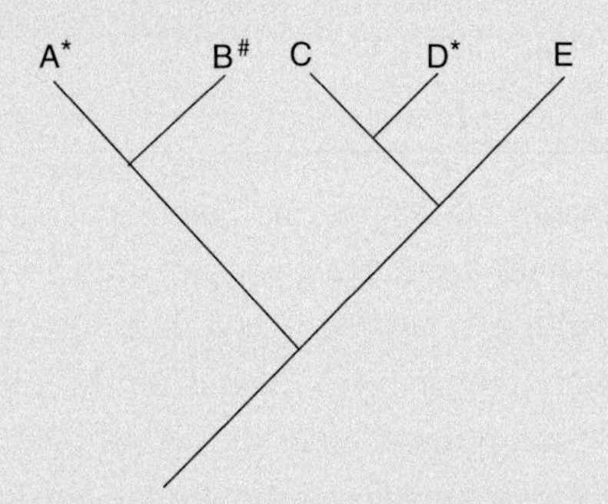

Cladistic analysis of five species of terrestrial plants indicated the phylogenetic relationships shown above. Species A and D (marked *) are both rare and endangered, while species B (marked #) is extinct and known only from herbarium specimens. Conservation of species A and D will require protection of their habitat and an expensive breeding program. There is only enough money available for a breeding program for one of the species. Which should receive the highest priority and why?

Debate with friends

Ernest Rutherford, the famous physicist, is quoted as saying: 'In science there is only physics; all the rest is stamp collecting'. This jibe has often been directed at systematics, which has become in many ways a poor relation among biological disciplines. Not only is funding for taxonomic studies scarce, but fewer and fewer taxonomists are now being trained in universities or other institutions. Debate the topic: 'Systematics should have a high priority in university teaching in biology'.

Further reading

Hillis, D.M., Moritz, C. and Mable, B.K. (eds) (1996). *Molecular systematics.* Sinauer, Sunderland, MA. 2nd edn.

Ridley, M. (2004). *Evolution.* Blackwell, Malden, MA. 3rd edn.

Wilson, E.O. (1992). *The diversity of life.* Belknap Press, Cambridge, MA.

9

Microscopic diversity – the prokaryotes and viruses

Graham O'Hara

Some like it hot

The hot springs and geysers of Yellowstone National Park in the USA have fascinated tourists and scientists for many years. But it was not until 1964 that microbiologist Thomas Brock tested for microbial life in waters as hot as 82°C. He found what are now called thermophilic bacteria, astounding micro-organisms thriving in high temperatures that would denature the enzymes of most other life forms. This inspired others to search for microbes in inhospitable environments. It also facilitated a vital technique in molecular biology, the polymerase chain reaction, or PCR, for copying small amounts of DNA.

The PCR reaction mixture contains DNA nucleotides, DNA polymerases and some other ingredients. It is heated to 90°C, causing the double-stranded DNA to separate into two single strands. Then it is cooled slightly and the DNA polymerase makes complementary strands of DNA. The mixture is reheated and the process continued, with the amount of DNA doubling each cycle. The DNA polymerase of most organisms is denatured at 90°C, so new DNA polymerase needs to be added for each cycle. However, DNA polymerase from thermophilic bacteria is made to order and, because DNA structure is identical in all living beings, it applies to any DNA. Kary Mullis perfected PCR in 1983, later receiving a Nobel Prize. The first DNA polymerase used widely in PCR (commercially known as Taq-polymerase) came from *Thermus aquaticus*, a thermophilic bacterium isolated by Brock.

PCR is valuable in forensic and medical science. Environmental biologists use PCR to study DNA from remains of ancient organisms, trace relationships in endangered species and unravel breeding behaviour in populations. All of this is possible because of the properties of unique prokaryotes, organisms often ignored in biological surveys.

Chapter aims

In this chapter we will describe the broad diversity of prokaryotes from the domains Archaea and Bacteria and also consider the viruses, microscopic infectious agents that hover on the boundaries of life. As well as describing the distinctive characteristics of these micro-organisms, we will also describe their important roles in different ecosystems.

Identification, structural features and classification of prokaryotes

What makes a prokaryote?

All living cells share a plasma membrane separating their contents (cytoplasm) from the external environment, hereditary material that is coded in DNA molecules and passed to daughter cells during cell division, and subcellular ribosomes that synthesise protein molecules following directions coded in DNA. However, prokaryotic and eukaryotic cells have several important differences (Table 9.1; see also Chapter 3).

Table 9.1 Principal differences between prokaryotic and eukaryotic cells.

Property	Eukaryotes	Prokaryotes
Size	Larger, mainly > 10 μm in length	Smaller, mainly < 10 μm in length
Membrane-enclosed nucleus with nucleolus	Yes	No
Complex internal organelles	Yes	No
Cell division	Mitosis and meiosis	Cell fission, budding or fragmentation
Ribosomes	Larger 4 RNAs present and 78 proteins	Smaller 3 RNAs present and 55 proteins
Cell wall	If present, no muramic acid	Muramic acid the building block of the cell wall

Prokaryotic cells are small, rarely exceeding 10 μm in length. Their plasma membrane is encased within a protective and supportive cell wall that may be surrounded by a capsule, or sticky outer coat. The cell wall is not based on cellulose as in plant cells or chitin as in fungi, but on other compounds. The DNA exists as a coiled, circular strand and is not enclosed in its own membrane. Indeed, prokaryotic cells contain no membrane-bounded organelles at all. Ribosomes present in the cytoplasm produce protein molecules from the instructions coded in the DNA. Most prokaryotes are unicellular, but there are colonial and multicellular forms.

Despite their simple structure, prokaryotic cells are the basis of an extraordinary diversity of microbial life extending back to the origins of life on Earth. Surprisingly, they can fossilise, mainly through replacement of cellular structures with minerals that can be detected through examination of ancient rocks. The first records are estimated at 3.5 billion years old, and prokaryotes had the Earth to themselves for the next two billion years (see Chapter 7). Today, prokaryotes are best known for those species that cause disease (pathogens). Tuberculosis, bubonic plague and cholera, for example, are all caused by prokaryotes. However, prokaryotes also enable the existence of all eukaryotic life through their critical roles in decomposing the wastes and dead bodies of larger organisms. This returns key chemical elements to the biosphere, ready to be absorbed by other organisms. Other prokaryotes live in close association with multicellular eukaryotes, such as the bacteria in the human digestive tract that synthesise essential vitamins.

Recognising different species of prokaryotes is difficult, because organisms with very similar physical appearance can have very different genomes revealed by molecular analysis. Therefore microbiologists regard two prokaryotes as being of different species if their genomes overlap by less than 70%. This is a much less stringent application than the definitions of species applied to eukaryotes. For example, if it was used for eukaryotes, all furred animals (the mammals, including mice, elephants and humans) would be regarded as a single species!

Using this approach, microbiologists have identified about 5000 prokaryote species. However, the true diversity could range between 400 000 and 4 million species and searches in new environments continually discover new species. The structural features of prokaryotes help to explain their success and their diversity.

Structural features

Early microscopists recognised that prokaryotes come in a range of shapes (Plate 9.1):

- *cocci* – spherical cells that may occur in long strands like beads on a necklace (streptococci), or in clusters like bunches of grapes (staphylococci)
- *bacilli* – rod-shaped prokaryotes can occur singly or in chains
- *curved or spiral* – a range of shapes including 'comma shaped' prokaryotes called vibrios, short corkscrew shapes (spirilla) and longer, more flexible corkscrew shapes (spirochetes).

These shapes are all bounded and maintained by a protective cell wall. This may be covered by a sticky capsule for adhering to substrates and, in the case of pathogens, protecting the cell from the host's immune system. If present, hairlike pili on the surface also help in attachment. Many prokaryotes move using whip-like flagellae, seeking out nutrients and avoiding toxins. They are structurally different from the flagellae of eukaryotic cells.

Internally, prokaryotic cells lack the membrane-bound organelles that perform critical functions such as respiration and photosynthesis in eukaryotic cells. However, some have infoldings of the cell membrane or other specialised membranes as substitutes. Ribosomes are present for protein synthesis, but they are smaller than those in eukaryotic cells and show differences in their proteins and RNA content. These differences are sufficiently great for some antibiotics to disrupt protein synthesis in prokaryotic pathogens but not in their eukaryotic hosts.

Archaea and Bacteria: two main branches of prokaryotic evolution

The substantial differences between prokaryotic and eukaryotic cells led some biologists to propose that all life was organised into two principal domains, based on these fundamental differences in cellular structure. However, in a highly influential paper published in 1990 microbiologist Carl Woese and his colleagues proposed a revolutionary new classification. Based on molecular biology studies using some of the techniques described in Chapter 5, they concluded that prokaryotic life actually consisted of two broad and very different groupings that they called the Archaea and the Bacteria. They proposed a classification of three domains of life: the prokaryotic Archaea and Bacteria, and the eukaryotic Eukarya (see Chapter 8).

Under the microscope, archaeans look similar to the bacteria. Some cells are circular, some are rod-shaped and there is a range of other forms. They have cell walls and in some cases flagellae, and their DNA is organised in a single, circular chromosome. However, detailed studies of their ultrastructure and their nucleic acids reveal fundamental differences supporting separation of the two groups (Table 9.2).

Table 9.2 Principal differences between cells of archaea and cells of bacteria.

Property	Archaea	Bacteria
Introns (non-coding sections of DNA)	Present in some genes	Mostly absent
tRNA sequences	Some unique tRNA sequences, some match eukaryotic sequences	Some unique tRNA sequences
Flagellae	Powered by ATP Flagellar filaments rotate as one unit	Powered by flow of H^+ ions Many flagellar filaments rotate independently
Peptidoglycan in cell wall	Mostly absent	Present
Membrane lipids	Side chains of phospholipids are not fatty acids, but isoprene	Side chains of phospholipids are fatty acids
Ribosomes	Not sensitive to the antibiotics chloramphenicol, streptomycin and kanamycin	Sensitive to the antibiotics chloramphenicol, streptomycin and kanamycin

Archaean tRNA molecules contain unique introns, as well as some that match those of eukaryotic cells. Bacterial tRNA mainly lacks introns. Protein synthesis in archaean ribosomes is not inhibited by the antibiotics chloramphenicol, streptomycin and kanamycin, which are effective against bacterial infections. Thus the two groups show fundamental differences not only in nucleic acids but also in their key protein-assembling processes.

Differences also occur in the structure of the cell walls and plasma membranes of archaeans and bacteria. Most archaean cell walls lack the compound peptidoglycan, which is a fundamental constituent of bacterial cell walls. Furthermore, the phospholipids in archaean plasma membranes do not have fatty acids, but a different compound called isoprene, in their side chains.

Archaeans are classified into four main groups based on their rRNA: the Euryarchaeota, the Cronarchaeota, the Korachaeota and Nanoarchaeota. Many of the early species studied were adapted to extreme environments (Box 9.1), although archaeans are now known to be much more widespread. The bacteria are separated into 24 groups, based largely on characteristics of their nutrition. They are the common prokaryotes we encounter every day as our normal microflora, and inhabit the environments and ecosystems around us. The following discussion applies broadly to both the archaea and the bacteria, with unique features noted as appropriate.

Box 9.1 Extremophiles – at the limits of life

The first archaeans discovered came from environments so harsh that they were regarded as unsuited for life (Plate 9.6). They were called extremophiles, literally 'lovers of extreme conditions'. They survive, or in some cases require, conditions that would destroy other life forms by denaturing their proteins or nucleic acids or interfering with the function of lipids in the plasma membrane. Not all extremophiles are archaeans and there are both bacterial and eukaryotic examples. However, most are prokaryotic. Types of extremophiles include:

- thermophiles – thrive in hot environments up to 90°C
- halophiles – either tolerate or require environments several times saltier than sea water
- methanogens – obtain energy by using CO_2 to oxidise H_2, releasing methane as a waste; oxygen is toxic to them
- psychrophiles – true psychrophilic organisms are killed by room temperatures and grow well under cold conditions
- barophiles – display optimum growth under high-pressure conditions, such as those at great depth in the ocean
- acidophiles – grow under very acidic conditions (low pH)
- alkaphiles – grow under very alkaline conditions (high pH)
- radiation-resistant – can survive high radiation exposure, although high radiation is not a requirement for their growth and survival.

Studies of extremophiles are difficult because of the demanding culture conditions required, but the payoff is high. Biotechnologists have been quick to exploit the properties of extremophiles for a wide range of applications.

One application is the use of whole organisms, sometimes in groups of different species, to extract metals such as copper, gold and uranium from ores. More common applications are based on molecules extracted from extremophiles. These are mainly enzymes, but can include other proteins such as cryoprotectants (biological 'antifreezes') and lipids stable under extreme conditions. Only a small number of the total of prokaryotic species have been identified, and extremophiles have been studied intensively only for about the last 25 years. Many useful compounds and critical applications remain to be discovered.

Prokaryote metabolism

Nutrition

In Chapter 4 we introduced the concept of autotrophic organisms that utilised CO_2 and inorganic elements as their sole building blocks for biosynthesising all the organic compounds they require, supporting biosynthesis with energy from sunlight. Heterotrophic organisms, by contrast, ingest organic molecules as their source of carbon and energy. These definitions work well when considering animals and plants, but they do not cover the full range of options for prokaryotes, as shown in Table 9.3. This is because some autotrophic prokaryotes do not use sunlight as their energy source and some heterotrophic prokaryotes do not obtain energy from the organic compounds they ingest.

Table 9.3 Diversity of metabolism in prokaryotes and eukaryotes.

Nutritional type	Energy source	Carbon source	Eukaryotes	Prokaryotes
Autotrophs				
Photoautotroph	Sunlight	CO_2	Plants	Yes
Chemoautotroph	Inorganic chemicals	CO_2	No	Yes
Heterotrophs				
Photoheterotroph	Sunlight	Organic compounds	No	Yes
Chemoheterotroph	Organic compounds	Organic compounds	Animals	Yes

This diversity of metabolism among prokaryotes is directly related to their diverse environments, where energy, carbon and other essential nutrients are available in many forms. The options are:

- *photoautotrophs* – like green plants, these prokaryotes capture light energy to drive the biosynthesis of their organic compounds, using only CO_2 and inorganic elements
- *chemoautotrophs* – these are a unique nutritional class of prokaryotes. CO_2 provides their sole carbon source, but their energy comes from the oxidation of inorganic compounds such as hydrogen sulfide or ammonia. They are the ultimate primary producers, and in some environments are the very first organisms in food chains. (See Box 4.1 for a detailed example).
- *photoheterotrophs* – this mode of nutrition is also unique to prokaryotes, occurring in some marine species. Sunlight provides their energy source, but they obtain their carbon from organic compounds.
- *chemoheterotroph* – this category covers many prokaryotes as well as animals, fungi and many protists. Organic compounds are ingested as a source of both energy and carbon. Individual species of chemoheterotrophic bacteria can be very nutritionally diverse in the range of organic compounds they can use for energy and carbon. For example, *Burkholderia cepacia* can metabolise organic compounds such as hydrocarbons, pesticides, herbicides, detergents and antibiotics.

Nitrogen metabolism

Nitrogen is a key component of nucleic acids, as well as the amino acids of proteins in all organisms. Although it is the largest component of the atmosphere, eukaryotes obtain their nitrogen either in inorganic forms from the soil or in ingested organic compounds. Prokaryotes are far more versatile. In the process of nitrogen fixation, some species transform atmospheric nitrogen to ammonia, which is then used in biosynthesis (Plate 9.2). Nitrogen fixation by prokaryotes is critical in making nitrogen available to eukaryotic organisms (see Chapter 17).

Prokaryote reproduction

Binary fission and population growth

An individual prokaryotic cell grows by increasing in size, duplicating its single-stranded 'circular' chromosome (following the process described in Chapter 5) and then dividing by binary fission (Figure 9.1). In addition to binary fission, prokaryotes may also reproduce asexually by fragmentation – the parent cell breaks into pieces, each of which may form a new cell. Alternatively, a bulge or bud may form at one end of a cell, eventually splitting away to form a new cell. Finally, in response to adverse environmental conditions, some prokaryotes undergo sporulation, producing a resistant stage called an endospore. Sporulation begins with DNA replication. The copy is then isolated together with some cytoplasm by an ingrowth of the plasma membrane, forming an isolated endospore. The endospore then dehydrates and is encased in a thick, protective protein coat. The original cell then dies and its cell wall dissolves, leaving only the dormant, highly resistant endospore that can withstand extremes of temperature, dehydration and toxins. When suitable conditions return enzymes in the endospore destroy the protein coat, the cell rehydrates and normal activity resumes. Strictly, although DNA replication is involved, endospore formation is not reproduction because one cell does not give rise to two or more viable cells.

Asexual reproduction can be extremely rapid, with many prokaryotes dividing again after an interval of only a few hours in a process known as exponential growth. This can produce very large numbers in a short time (Figure 9.2a, b). However, a population of prokaryotes does not grow exponentially forever because eventually one or more environmental factors will limit growth. When a population of prokaryotes is introduced to a new environment, it often grows with a standard growth curve consisting of four phases (Figure 9.2c). When initially introduced into the new environment, the prokaryotes show a lag phase – the population is not increasing while the prokaryotes adjust to the new environment. For example, they may need to adapt to changes in carbon source, temperature, pH or aeration. After adjusting to the new conditions comes the exponential phase, when the prokaryotes grow and divide at a constant rate expressed as the mean generation time (MGT), which is

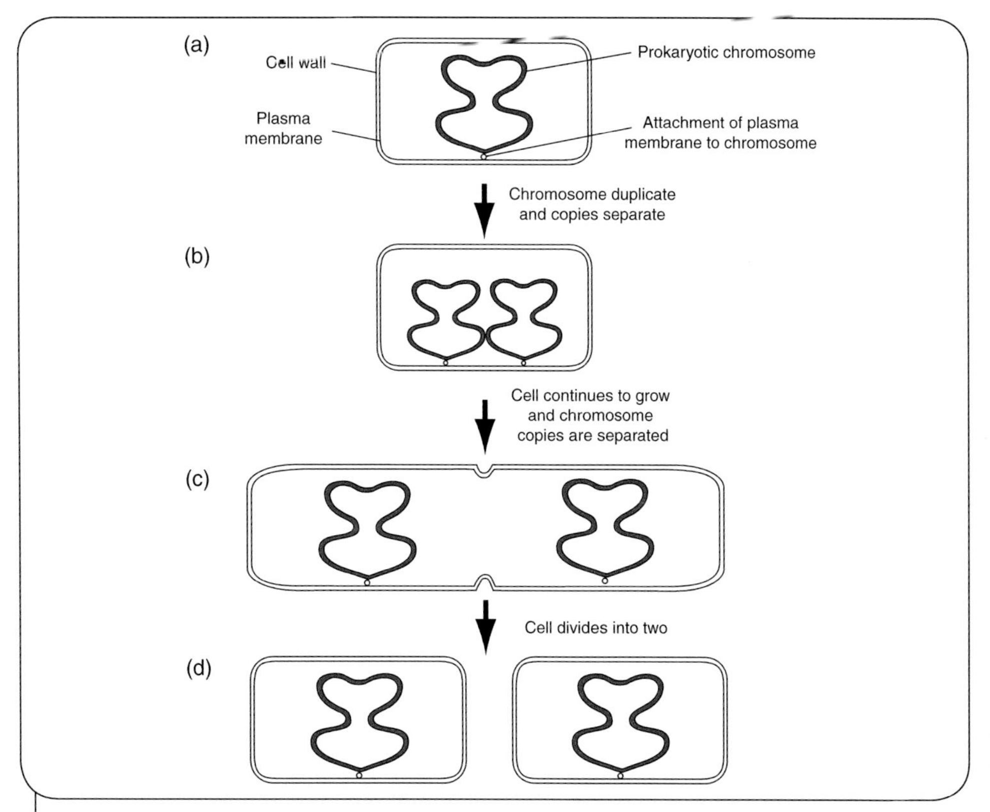

Figure 9.1 Binary fission in prokaryotes. (a) The prokaryotic cell showing basic structures, including a simplified representation of the single-stranded chromosome attached to the plasma membrane. (b) The chromosome duplicates to give two chromosomes, each attached to a separate point on the plasma membrane. (c) Cell growth increases the distance between the two chromosomes and the cell begins to divide. (d) Division is complete, forming two cells. (Source: B. Cale)

the time taken for the population to double. This depends on factors such as the species of prokaryote, the type of growth medium and the environmental conditions. Ultimately, cell growth and reproduction stop (the stationary phase) as a consequence of changes in the environment such as nutrient depletion, pH change, exhaustion of oxygen and the accumulation of toxic wastes. A population of prokaryotes in the stationary phase will eventually enter the death or decline phase because something in the environment kills them. This is often the factor that caused them to enter the stationary phase.

DNA transfer in prokaryotes

Binary fission allows rapid increases in prokaryote numbers, but the cost is that the progeny of binary fission are genetically identical because they arise from a single parent cell. This does not provide the constant shuffling of genetic material that increases variation in populations of sexually reproducing organisms. In prokaryotes three different

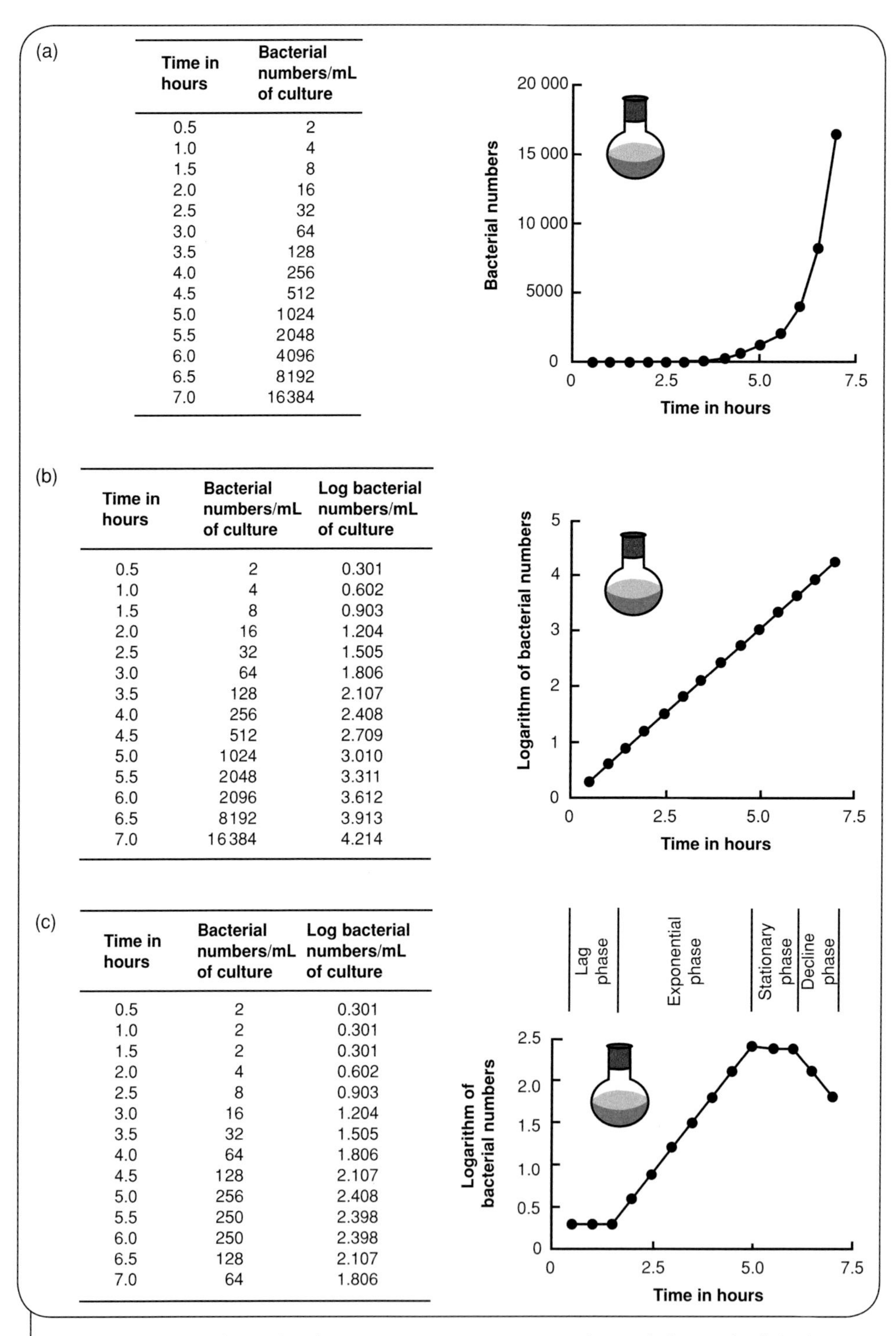

(a)

Time in hours	Bacterial numbers/mL of culture
0.5	2
1.0	4
1.5	8
2.0	16
2.5	32
3.0	64
3.5	128
4.0	256
4.5	512
5.0	1024
5.5	2048
6.0	4096
6.5	8192
7.0	16384

(b)

Time in hours	Bacterial numbers/mL of culture	Log bacterial numbers/mL of culture
0.5	2	0.301
1.0	4	0.602
1.5	8	0.903
2.0	16	1.204
2.5	32	1.505
3.0	64	1.806
3.5	128	2.107
4.0	256	2.408
4.5	512	2.709
5.0	1024	3.010
5.5	2048	3.311
6.0	2096	3.612
6.5	8192	3.913
7.0	16384	4.214

(c)

Time in hours	Bacterial numbers/mL of culture	Log bacterial numbers/mL of culture
0.5	2	0.301
1.0	2	0.301
1.5	2	0.301
2.0	4	0.602
2.5	8	0.903
3.0	16	1.204
3.5	32	1.505
4.0	64	1.806
4.5	128	2.107
5.0	256	2.408
5.5	250	2.398
6.0	250	2.398
6.5	128	2.107
7.0	64	1.806

Figure 9.2 Exponential growth in bacteria. (a) Growing unrestrained, a single bacterial cell dividing every hour could give rise to over 16000 cells in 7 hours. (b) Plotting logarithms of the population numbers represents the curvilinear exponential growth as a straight line. (c) Normally, growth is not exponential for a long period and the population passes through a lag phase, an exponential phase, a stationary phase and finally a decline phase.

techniques evolved to exchange genetic material between cells of the same species but different parentage:

- *transformation* – in this method, cells take up fragments of DNA from the surrounding environment. British medical officer Frederick Griffith first demonstrated this in 1928 – he found that when he mixed a culture of dead pathogenic bacteria with a live culture of non-pathogenic bacteria, some of the non-pathogenic bacteria became pathogenic and the pathogenicity was inherited. At the time, the role of DNA in inheritance was not established, but we now know that uptake of DNA fragments caused the change.
- *transduction* – this method uses a bacteriophage, a subcellular infectious particle that attacks prokaryotes, to transfer prokaryote DNA. A fragment of DNA from a prokaryote cell is accidentally packaged with a phage particle when it reproduces inside a prokaryote host and is carried with it when it infects a new prokaryote cell.
- *conjugation* – this involves two prokaryote cells, a donor or 'male' and a recipient or 'female'. The two cells attach using structures called sex pili and their cytoplasm joins. The donor cell replicates a portion of its DNA while passing it to the recipient cell. During conjugation cells may also transfer plasmids, which are small, circular DNA molecules separate from the prokaryote chromosome (Box 9.2).

New DNA introduced into a prokaryote cell may be incorporated into the cell's chromosome in a process similar to recombination in eukaryotic chromosomes (Chapter 5). This increases genetic diversity within prokaryote populations.

Box 9.2 Gene transfer and antibiotic resistance

In addition to their chromosome, prokaryotes carry DNA in separate small DNA molecules called plasmids, which have many functions. Some, known as F-factor plasmids, carry genes coding for the production of pili and other structures or substances needed for conjugation. Others carry genes for specific enzymes. When cells conjugate, plasmids replicate and pass from donor to recipient cells.

Problems arise when individuals of some pathogens carry plasmids coding for enzymes that inactivate antibiotics. These plasmids are coded R for resistant. When prokaryotes are exposed to antibiotics, small numbers of resistant cells survive while susceptible ones are killed. The survivors multiply, passing the resistant plasmids to their progeny and to other cells through conjugation. Plasmids may even be transmitted between prokaryote species. You will recognise this as an example of natural selection, proceeding rapidly because of the short generation time of bacteria.

Humans contribute to antibiotic resistance by adding antibiotics to livestock feed as a growth promoter, by prescribing antibiotics unnecessarily (for viral infections that do not respond to antibiotics) and by patients not completing courses of antibiotics so some of the prokaryotes exposed to the antibiotic survive. New antibiotics are continually developed in an attempt to stay a step ahead of evolving resistance.

Environmental influences on prokaryote growth

Water, temperature, pH and oxygen are four important environmental factors affecting both the capacity of bacteria to grow and their rate of growth.

Water

Liquid water is essential for prokaryotes to grow because it is required for cell metabolism. The amounts and types of dissolved solutes determine the availability of water. These solutes include inorganic ions (Na^+ , Cl^-, K^+, Mg^{++}) and organic compounds (sugars, amino acids, organic acids).

High concentrations of solutes decrease the availability of water to prokaryotes, because water is withdrawn from the cells by osmosis. For example, many saline habitats such as the ocean, salt lakes and even human skin have abundant liquid water, but because of high salt concentrations the water is not available for all prokaryotes. Special types of halophilic bacteria, specifically adapted to growing in saline environments, occur in these habitats.

Temperature

Environmental temperature has dramatic effects on the growth and survival of prokaryotes, which lack mechanisms of insulation or internal regulation of temperature. If it is too hot then prokaryotes die because of coagulation and denaturation of proteins, degradation of nucleic acids, and disruption and melting of cell membranes. If it is too cold they survive but cannot grow. At intermediate temperatures they can grow at a rate directly affected by temperature.

pH

The pH of their external environment affects prokaryotes because they can grow only in aqueous environments. The harmful effects of adverse pH on prokaryotes are a result of disruption of cell membranes, inhibition of enzymes, inhibition of transport and uptake systems, or effects on the availability of nutrients.

Oxygen

Oxygen is a substantial component of atmosphere (about 20%) and also a very reactive element. It is essential for aerobic respiration. There are five groups of prokaryotes in terms of their interaction with oxygen. Aerobes use oxygen for respiration, and cannot grow anaerobically using fermentation. Microaerophils require low levels of oxygen (5–10%) in their environment. Facultative anaerobes may grow aerobically using respiration or anaerobically using fermentation. Obligate anaerobes can grow only without oxygen being present. They always use fermentation and are killed by oxygen. Aerotolerant anaerobes also always use fermentation, but they can tolerate the presence of oxygen.

Viruses – at the boundaries of life

Viruses were not discovered until the late 19th century and the story again illustrates the power of the scientific method through observation and controlled experiments to increase our understanding of the biological world. By this time prokaryotes were known causes of disease in humans, other animals and plants, so biologists sought to isolate them in their quest for treatments. While working on a disease of tobacco plants now known as tobacco mosaic virus (TMV) at an agricultural station in the Netherlands in 1879, German biologist Adolf Mayer showed that sap extracted from diseased tobacco plants infected other tobacco plants. He could not isolate a prokaryote or a fungus from the extracts, so he concluded that a new form of 'soluble bacterium' was the cause. However, in 1892 when Russian Dmitri Ivanovsky filtered extracts of diseased plants, they remained infectious even after passing through pores small enough to trap all known bacteria. Furthermore, after treating the extracts with alcohol or formalin, which would kill any bacteria, they remained infectious. Ivanovsky concluded that a toxin was responsible, or that his filters were defective.

Six years later Dutch biologist Martinus Biejerinck tackled the same question. He had not read Ivanovsky's publications (one instance of the failure of scientific communication!), but he repeated and extended Ivanovsky's experiments. Biejerinck confirmed that extracts from infected tobacco plants infected others, even after filtration, treatment with alcohol or formalin or time delays of up to 3 months. However, heating the extracts to 90°C did prevent infection. He concluded that a contagious living liquid caused the disease. This view was overturned convincingly in 1935 when American biochemist Wendell Stanley crystallised the infectious agent for study by electron microscopy, showing clearly that it was a particle. He also showed that it remained infective even after crystallisation. These particles are now known as viruses and they infect eukaryotes and other prokaryotes.

What is a virus?

Simply, a virus is a subcellular, small infectious particle that can replicate itself only inside the cell of a host organism. Outside their living host cells viruses are inert and incapable of metabolic activity or replication. As such, viruses are not living organisms, and they challenge our understanding of what 'being alive' means.

Virus structure

A virus particle is composed of genetic material that can be either DNA or RNA, but very rarely both. This is surrounded by a protein coat or capsid, sometimes surrounded by a lipid-based envelope derived from the membrane of its host cell.

Viruses are very small, ranging in size from about 25 nm to about 400 nm, and they vary markedly in the size of their genomes. Some, such as the poliovirus, have only enough genes to make three or four different proteins, whereas others such as the poxviruses contain enough genes to make more than 100 proteins. Viruses come in a wide variety of different shapes such as cylinders, spheres and even icosahedrons (polygons with 20 faces). The phages, viruses that infect bacteria, are some of the most interestingly shaped viruses, looking like space vehicles designed for a moon or Mars landing. These particles consist of an icosahedron head containing the viral genetic material, on a protein cylinder with tail fibres that attach the particle to the host bacterium.

Virus replication

Viruses are very specific in the hosts they can infect. At the broader level there are animal, plant, fungal and bacterial viruses, but some are species-specific or even operate at the subspecies level. This host specificity has important implications for epidemic diseases – for example, the concerns about particular influenza-causing viruses that may jump species from birds or mammals to humans.

The initial step in virus reproduction is adsorption of the virus particle onto the specific host cell. The cell itself may engulf the virus through endocytosis, or the virus may fuse with the plasma membrane. Having entered the cell, the viral nucleic acid (either DNA or RNA) uses the host's metabolic processes to replicate itself many times before each replicated nucleic acid is packaged into a protein coat. These may be released by lysis (rupturing) of the host cell. If the virus particles are packaged into lipid envelopes, release of the virus particles and envelopes occurs simultaneously after modification of the infected host cell plasma membrane. The infected cell is not lysed.

Viruses harm host cells in many ways during infection and replication. These include effects on the host cell plasma membrane, inhibition of host DNA, RNA and protein synthesis, and the formation of aggregations of viral capsid proteins called inclusion bodies.

Virus evolution

Viruses leave no fossil traces, so studies of their origins use molecular techniques to investigate viral nucleic acids, which are subject to natural selection in the same ways as the genomes of cellular organisms. One theory of viral origins is that they are degenerate parasites, originally with a cellular structure, that have lost all features other than those required for replicating themselves with the assistance of a host. Some modern bacteria from the genera *Rickettsia* and *Chlamydia* can reproduce only within host cells, possibly illustrating an intermediate step in the development of viruses. This may be the origin of some of the larger viruses such as the poxes. Alternatively, smaller viruses may be parts of the genomes of other organisms. For example, many prokaryotic and some eukaryotic cells have plasmids, which are DNA molecules separate from chromosomal DNA and capable of replicating themselves (Box 9.2). Plasmids separated from cells might persist as viruses. A third explanation is that viruses are primitive forms of self-replicating DNA that are precursors to cellular life.

Viruses also have a significant role in the evolution of other organisms. It is becoming evident that virus DNA has been incorporated into the DNA of host organisms. For example, up to 8% of the human genome is DNA from viruses (Box 9.3).

Box 9.3 Viruses and evolution

In 1970 American scientists Howard Temin and David Baltimore, working independently, discovered an enzyme called reverse transcriptase, which makes DNA from RNA in a process called reverse transcription. The discovery unlocked the mystery of the reproduction of a particular family of viruses called the retroviruses, and also gave insights into the role of viruses in evolution.

A retrovirus consists of an RNA genome and molecules of reverse transcriptase within a capsid. When the retrovirus infects the cytoplasm of a host cell, it synthesises a DNA strand from its RNA by reverse transcription and then adds a second, complementary DNA strand. This double-stranded DNA enters the host cell's nucleus and penetrates the chromosomal DNA, where it is called a provirus. The provirus in turn is transcribed into RNA by normal cellular processes and in turn translated into viral proteins, which are assembled into new viruses that leave the cell and infect others. Reverse transcription lacks some of the checks involved in normal DNA synthesis (see Chapter 5) and, as a result, retroviruses often mutate. This makes them difficult to prevent with vaccination or treat with antiviral drugs, but particularly interesting subjects for evolutionary study. However, their significance for evolution extends well beyond that.

In the rare event that an infected cell is in the germ line (the cells that produce gametes), it may give rise to an infected gamete. If that gamete survives and forms a zygote and ultimately an embryo that also survives, the retrovirus becomes part of the host species's genome and is inherited as an endogenous retrovirus (as distinct from exogenous retroviruses, which are not incorporated into the genome and are not inherited). Endogenous retroviruses rarely persist beyond a few generations before being disabled by mutation, but parts of their DNA persist in the host's genome. Although such a sequence of events is uncommon, over long periods of time retroviruses contribute surprising amounts of DNA to host species – for example, up to 8% of the human genome consists of sections of retrovirus DNA. These fragments of DNA serve as useful markers for determining relationships between species. If the same fragments of retrovirus DNA occur in the same places on the same chromosomes in different species, these species probably shared a common ancestor. The alternative is the highly unlikely outcome that the different species were infected by the same retroviruses inserting DNA into the same places on the same chromosomes in the germ line.

Even more interestingly, some endogenous retroviruses may be activated by their hosts to perform specific functions. For example, during pregnancy in viviparous (live-bearing) mammals (see Chapter 15), endogenous retroviruses are activated when the embryo implants. They are immunodepressors and may help prevent the rejection of the embryo by the mother's immune system. Other viral proteins limit the migration of cells between the mother and the developing embryo. Thus biologists are actively considering the hypothesis that endogenous retroviruses may have been critical in the evolution of the mammalian placenta.

continued ›

Box 9.3 continued ›

Until early this century, the most recent known case of a retrovirus becoming endogenous was in pigs, estimated at about 5000 years ago. However, it now appears that an endogenous retrovirus is developing in Australian koala (*Phascolarctos cinereus*) populations.

Koalas were nearly exterminated on the Australian mainland by hunting for fur during the 19th and early 20th centuries, but some animals were translocated to offshore islands where they have remained isolated since around 1920. They thrived, and koalas were later translocated from these islands to restock the mainland. Many of these newly established mainland populations are now infected with a retrovirus (koala retrovirus, or KoRV) that has become endogenous. Infected animals are more susceptible to a range of diseases. The incidence of KoRV is extremely low in the offshore island populations, indicating that KoRV was not present when these populations were founded and has only recently infected the mainland populations (Figure B9.3.1). Researchers therefore have a unique opportunity to study the incorporation of an endogenous retrovirus into a species's genome.

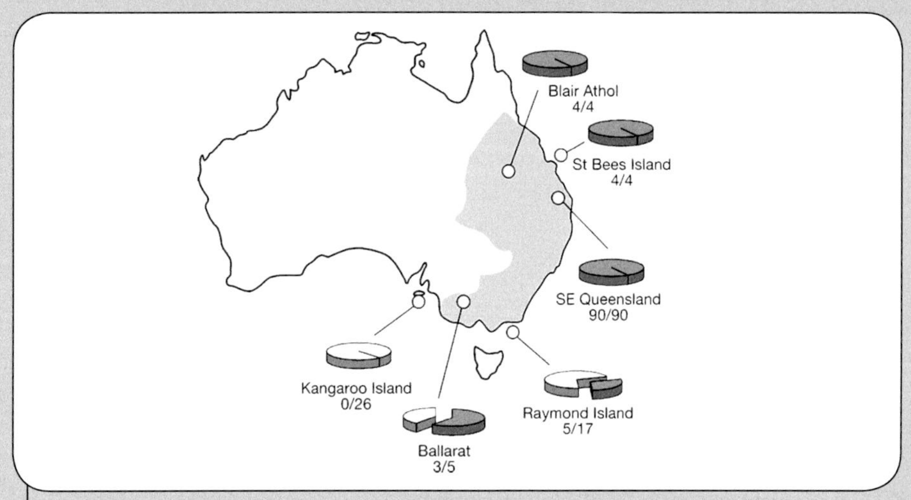

Figure B9.3.1 The grey shading shows the distribution of koala populations in Australia. Prevalence rates of koala retrovirus are represented as pie graphs with infected animals (grey) and uninfected (white). Values indicate the number of infected animals/number of animals tested. (Source: Adapted from Tarlinton, R.E., Meers, J. and Young, P.R. (2006). 'Retroviral invasion of the koala genome' *Nature* 442: 79–81)

Virus classification

Virus classification is complicated by the 'pseudo-living' nature of viruses. Originally, it was based mainly on phenotypic characteristics including morphology, the type of nucleic acid (RNA or DNA), the host organism, the type of disease caused and the means of replication. Modern molecular biology permits a more detailed consideration of the viral genome. Several different classification systems have been proposed, but the International Committee on the Taxonomy of Viruses developed an internationally accepted classification, including agreed names. It classifies viruses within a range of orders, each of which contains one or more families, genera and species.

Subviral particles

Recent work revealed that viruses are not the smallest disease-causing agents and that three subviral particles exist: prions, viroids and satellites. Prions are infectious agents consisting entirely of protein. They convert normal protein molecules into the abnormally structured prion form, causing disease by altering the structure of the brain and neural tissue. The restructured proteins resist denaturation by chemical physical agents and hence accumulate in infected tissue. They are very difficult to treat. All prion diseases have long incubation periods, but progress rapidly after initial symptoms and are invariably fatal. Examples include bovine spongiform encephalopathy (BSE) in cattle, which can be transmitted to humans where it causes Creutzfeldt-Jakob disease (CJD). Other human diseases believed to be caused by prions are Gerstmann-Straussler-Scheinker syndrome and fatal familial insomnia. Viroids are very short stretches of circular, single-stranded RNA without a protein coat. They are unique to plants and cause several crop diseases. Satellites are very similar to viroids, but occur within the capsid of other viruses on which they depend for replication.

The environmental importance of prokaryotes and viruses

Prokaryotes and viruses play many critical roles in the environment. Some cause infectious disease and some are toxic, while others are vital links in nutrient cycling within ecosystems or offer novel methods for pollution remediation or biological control of pest species.

Infectious disease

Infectious diseases are major killers of humans and other wildlife, including plants (Plate 9.3). In a recent survey of threats to endangered mammals throughout the world, biologists from the University of Georgia (USA) and the University of Sheffield (UK) identified 13 viruses and eight bacteria that posed significant risks to one or more of these species. One well-known example concerns epidemics of fatal morbillivirus infections in pinnipeds (seals and their relatives) and cetaceans (whales, dolphins and their relatives) over the last 20 years.

General precautions against the establishment and spread of infectious disease in wild populations include restricting contact between humans and their domestic animals and wildlife, especially interactions such as unauthorised feeding. In cases of gorilla tourism, for example, tourists may transmit a range of ailments including measles and respiratory diseases to wild gorilla populations. Although vaccination and antibiotic treatments are sometimes used to protect gorilla populations visited by tourists, this approach is not applicable in most wildlife situations. Care must also be taken if organisms are moved between locations, and captive-bred organisms should be screened for disease before reintroducing them to wild populations.

Prokaryote toxins

Cyanobacteria, a phylum of photosynthetic bacteria, may reach high concentrations in lakes, rivers, dams or other water bodies, especially after pollution with fertilisers or other nutrients (Plate 9.4). They may produce toxins fatal to aquatic animals and stock or wildlife. The link between animal deaths and cyanobacteria was first discovered in Australia in 1878, when sheep died after drinking from Lake Alexandrina (near the mouth of the Murray) when it had a large 'bloom' of *Nodularia spumigena*.

Prokaryotes in ecosystems

According to the proverb, the squeaky wheel gets the grease. In the same way, our natural concern with disease overshadows the enormous contribution micro-organisms make to ecosystems.

One important contribution is to the continuing cycling of nutrients between organisms and the non-living environment. Many prokaryotes are decomposers, reducing the complex organic molecules in animal wastes and dead bodies into inorganic chemicals that can be absorbed by plants and cycled again through ecosystems. Without this, vital nutrients would remain locked in wastes and corpses. Marine viruses make an important contribution here. They infect cells in surface waters and, when these cells lyse, they release carbon compounds that would otherwise sink with the dead cell. These carbon compounds remain available for use by cells in the surface waters.

A second important contribution is to energy flow. Whereas nutrients are finite and are recycled through ecosystems, energy is ultimately lost in heat and must continually be added. Producer organisms are critical because they convert energy and inorganic compounds into organic molecules, which then become a source of energy and nutrients for consumers. The most common energy source is the sun, and cyanobacteria in the ocean are often important producers at the base of many oceanic food chains. Through photosynthesis they also release oxygen to the atmosphere.

The final important contribution of prokaryotes to ecosystems is nitrogen fixation. Bacteria called *Rhizobium* living in nodules on the roots of legumes (beans, acacias and their relatives) convert atmospheric nitrogen into forms accessible to plants (Plate 17.1d), as do cyanobacteria living mutualistically with cycads and the water fern *Azolla*.

Pollution remediation

Modern industries produce new chemicals at a great pace. Release of these chemicals or their by-products into the environment can poison organisms directly. In other cases, chemicals may accumulate within organisms without being degraded or excreted. Predators of these organisms may accumulate still higher quantities of the pollutant in a process of biomagnification, ultimately leading to toxic effects in organisms near the tops of food chains.

Prokaryotes may provide a solution because many degrade different toxic compounds. For example, hydrocarbons such as those found in oil spills are present in the environment from natural seeps. Their high-energy bonds are rewarding to those prokaryotes able to degrade them, and there is a range of species able to degrade oil molecules.

Bioremediation uses living organisms to degrade toxic compounds in the environment. Prokaryotes can be used to rectify pollution problems, either by introducing them to polluted areas or encouraging the activity of naturally occurring ones by the use of fertilisers or other means. One significant example occurred in 1989 when the oil carrier *Exxon Valdez* ran aground in the Gulf of Alaska, spilling 20% of its cargo. Chemical clean-ups were used at all beaches, but scientists seized the opportunity to also use bioremediation by fertilising some areas of contaminated coast to encourage the growth of oil-degrading bacteria. Control areas were not fertilised. Comparing the fertilised plots and the controls showed that adding fertiliser significantly enhanced the effectiveness of the chemical clean-up.

Prokaryotes and biological control

Biologist E.O. Wilson referred to the common human tendency to respond to pest organisms with KCK – 'kick, cuss and kill'. This often involves the application of pesticides or other toxins, which can have harmful consequences for non-target species. Biological control, which introduces a natural predator, competitor or pathogen specific to the pest to control its numbers, presents an environmentally friendly alternative.

In Australia, introduced mice, rabbits and foxes cause enormous agricultural and environmental damage. Control is largely by poisoning, shooting and trapping, although the myxoma virus and the calici virus also provide some regulation of rabbit numbers (Plate 9.5). Poisoning, shooting and trapping are all expensive, sometimes dangerous to non-target species, and they require ongoing use. There may also be welfare concerns about animals, even though they are pests, dying in these ways.

One promising alternative line of research is virally vectored immunocontraception. It works on the principle that animals can be immunised against some of their own key reproductive proteins, leaving them healthy but sterile. Species-specific viruses could be genetically engineered to spread these contraceptive agents. These viruses include canine herpes virus I for foxes, myxoma virus for rabbits and murine cytomegalovirus or ectromelia virus for mice. Concerns regarding the development and use of genetically modified viruses must be met as part of the research, but the promises of significant savings and increased environmental protection are a strong stimulus for the research.

A promising bacterial biological control is based on the soil bacterium *Bacillus thuringiensis*, some strains of which kill insects. *B. thuringiensis* is essentially non-toxic to vertebrates and, because it must be ingested before it kills the insect, generally safe to non-target insects. It can therefore be sprayed onto crops to control insects. Its chief disadvantage is that it degrades quickly in sunlight and usually persists for less than a week.

It is not always necessary to use bacteria or viruses to take advantage of their unique properties. For example, the gene for synthesis of the *B. thuringiensis* toxin has been isolated and transferred to plant species (notably cotton) through genetic engineering. This protects the plant from insect attack by producing the toxin in the plant cells.

Chapter summary

1. Prokaryotic cells are small (less than 10 μm in length) with a cell wall, a single circular chromosome and ribosomes, but no membrane-bound organelles.
2. Prokaryotes are the oldest life forms on Earth and only a small fraction of prokaryote species have been described.
3. Prokaryotic cells can be spherical, rod-like or corkscrew-shaped and may exist in chains or clumps.
4. Prokaryotes are classified into two domains, the Archaea and the Bacteria. These differ in characteristics of their nucleic acids, ribosomes, cell walls and plasma membranes.
5. There are four broad types of prokaryote nutrition:
   - Photoautotrophs use light energy to drive biosynthesis and CO_2 as their carbon source.
   - Chemoautotrophs obtain both energy and carbon from inorganic sources.
   - Photoheterotrophs obtain energy from sunlight and carbon from organic molecules.
   - Chemoheterotrophs obtain both energy and carbon from organic sources.
6. Many prokaryotes are able to fix atmospheric nitrogen into forms accessible by eukaryotic cells.
7. An individual prokaryotic cell grows by increasing in size, duplicating its circular chromosome and then dividing by binary fission. This results in exponential growth, which can be limited by factors such as temperature, oxygen, pH and water.
8. Genetic diversity in prokaryotes is increased by absorbing DNA from the environment (transformation), transfer of DNA by a bacteriophage (transduction) and transfer of DNA from one cell to another (conjugation).
9. A virus is a non-living, subcellular, small infectious particle that can replicate itself only inside the cell of a host organism. Outside their living host cells, viruses are inert and incapable of metabolic activity or replication.
10. Subviral particles that cause disease include prions, viroids and satellites.
11. Micro-organisms are important as agents of infectious disease and nutrient cycling, and are valuable tools in pollution remediation and biological control.

Key terms

bacillus
bacteriophage
binary fission
bioaccumulation
chemoautotroph
chemoheterotroph
coccus
conjugation
eukaryote
extremophile
halophile
lysis
pathogens
peptidoglycan
photoautotroph
photoheterotroph
prion
prokaryote
satellites
thermophiles
transduction
transformation
viroids
virus

Test your knowledge

1. Summarise the key differences between (a) prokaryotes and eukaryotes, and (b) archaea and bacteria.
2. Define, with examples, the four major nutritional strategies of organisms. Which ones are unique to prokaryotes?
3. Explain the factors that can limit the population growth of prokaryotes.
4. In the absence of sexual reproduction, how do prokaryotes increase genetic diversity in their populations?
5. What is a virus and how does it cause disease?
6. Describe four examples of the environmental importance of micro-organisms.

Thinking scientifically

1. Fertilising areas contaminated with oil spills is sometimes recommended to encourage the growth of soil prokaryotes that can degrade the oil. Using your knowledge from this chapter and the principles of experimental design in Chapter 2, design an experiment to test the hypothesis that fertilisation increases the rate of decay of oil spills.
2. The infectious proteins called prions produce more prions by converting normal protein molecules into the prion form. This does not use the genetic code of RNA and DNA. Is it correct to say that prions reproduce? Are prions 'alive'?

Debate with friends

Smallpox, a deadly viral disease, was an important lethal and disfiguring infection of humans for most of the last 3000 to 6000 years. In 1967 the World Health Organization launched an extensive campaign to eradicate smallpox, based on mass vaccination campaigns and a surveillance system. In 1980, smallpox was pronounced eradicated. However, stocks remain in research laboratories in Russia and the United States. Researchers claim that these stocks should be maintained to assist viral research and the development of new vaccines, while others are concerned that they might escape or be used in biological warfare. Debate the topic: 'The world's remaining stocks of smallpox virus should be destroyed'.

Further reading

Dimmock, N., Easton, A. and Leppard, K. (2008). *Introduction to modern virology*. Blackwell Publishing, Oxford. 6th edn.

Gross, M. (1996). *Life on the edge: amazing creatures thriving in extreme environments*. Perseus Books, Cambridge, MA.

Tortora, G.J., Funke, B.R. and Case, C.L. (2006). *Microbiology: an introduction*. Benjamin Cummings, San Francisco. 9th edn.

10

Mysterious diversity – the protists (including the fungi)

Treena Burgess and Luke Twomey

The biological bulldozer

Phytophthora dieback, colloquially known as the 'biological bulldozer' because of its clearing of areas of susceptible native vegetation, is now widespread in native forests, woodlands and heathlands of south-western and south-eastern Australia, including Tasmania. It is caused by the oomycete *Phytophthora cinnamomi*, which was introduced in soil attached to trees imported for horticulture following European settlement. The oomycete infects plants through the roots and kills by blocking the uptake and circulation of water and nutrients. The disease spreads naturally through ground water, but much more rapidly by human transfer of infected plants and soil. Deaths of infected plants can reduce vegetation cover by up to 50% and alter species composition, facilitating invasion by introduced weeds. These in turn can increase fuel loads and change the frequency of fire in infected areas. Such radical changes in plant communities reduce the range of food and shelter available for local animals, so animal communities are affected too.

In the botanically rich areas of south-western Australia, eucalypts in the subgenus monocalyptus (especially *Eucalyptus marginata* or jarrah in Western Australia) and at least 40% of the understorey plants are susceptible. Large areas of the jarrah forest are known as 'black gravel' or 'graveyard' sites as they have been so severely impacted by *P. cinnamomi* over the last 70–100 years: they are essentially denuded of many of the plants and animals they supported (Plate 10.1). The cost to the community is substantial through reduced nature tourism opportunities, devastation of the wildflower-picking industry, loss of valuable timber trees and infection of horticultural stocks. Currently, there is no eradication treatment possible for phytophthora dieback and management centres on hygiene and quarantine to contain the spread of the disease. This destructive oomycete is just one example of the impacts of protists on the environment and human wellbeing.

The kingdom Protista presented in this chapter has enormous diversity and its classification is far from settled. Many, such as the marine 'algae', are keystone species in their ecosystems and others have an impact on the biota through the diseases they cause. Details of some minor groups are included as they offer clues as to the evolution of major groups such as the plants and animals. Along with the diverse morphology there is great variation in modes of nutrition, and a number of complex life cycles, one of which – the 'alternation of haploid and diploid generations' – is also found in plants. Perhaps the greatest sign of the turmoil in classification is that some authorities consider the fungi members of the Protista (as in this text), whereas others place the fungi in a kingdom of their own.

Chapter aims

In this chapter we will explore protistan and fungal diversity in detail, including many more examples of the impacts of protists on the environment and human wellbeing.

Is there a kingdom Protista?

The protists were the first organisms to evolve eukaryotic cells with membrane-bound organelles, including a nucleus. Beyond that they are so diverse that it is easiest to define them as eukaryotes that are not fungi, animals or plants (but to complicate matters, some biologists now include the fungi within the protists). They are of great interest because of their diversity of form and lifestyle and because, as the first eukaryotes and therefore the precursors of animals and plants, they have much to tell us about the early evolution of eukaryote life. As you will see, they have significant environmental roles as well.

Diversity of form and lifestyle

The protists have a bewildering diversity of forms and lifestyles. Some are unicellular, whereas others form colonies (such as the diatom didymo, *Didymosphenia geminata*, a serious pest in New Zealand lakes; see Figure 10.1), or large multicellular organisms (such as the giant brown seaweeds). Most are aquatic and cannot survive desiccation, but many live in soil, on rocks, on the bark of trees or in the atmosphere. Others are parasites, living in the circulatory systems and organs of animals and drawing nourishment from their hosts. Some live their adult lives attached to a substrate, and others move using flagellae or cilia. The algae are autotrophic protists containing plastids (cytoplasmic organelles that photosynthesise). Other protists are heterotrophic, feeding on organic matter.

Protists also show great diversity in reproduction. Most unicellular forms reproduce asexually via fission, essentially a mitotic division forming two genetically equal cells. Sometimes genetic diversity can be enhanced when two different cells align and exchange genetic material prior to cell division. Other unicellular species undergo meiosis and

Figure 10.1 In New Zealand, the boating public is encouraged to follow the 'check, clean and dry' routine to prevent spread of the diatom pest didymo. (Source: J. McComb)

produce male and female haploid gametes, which fuse to form a diploid individual. Multicellular protists often have complex reproductive cycles, characterised by some form of alternation of sporophyte (spore producing) and gametophyte (gamete producing) generations (Box 10.1).

Box 10.1 Alternation of generations

Understanding the life cycle of protists is important when they are pathogens, are cultivated for food, or are introduced pests such as the alga *Undaria pinnatifida* (wakame), which was accidentally introduced to Tasmania and is spreading along Tasmanian and Victorian coasts. Many protists and all land plants alternate haploid and diploid stages in their life cycles. In some species these stages look similar (e.g. *Ulva*, Figure B10.1.1a) but in others they differ in shape and size (e.g. *Ecklonia*, Figure B10.1.1b). It is often necessary to grow the spores of protists in culture to find out what the various stages of the life cycle look like, as there is little hope of finding or identifying the small stages, particularly in the ocean.

The haploid stage has a single set of unpaired chromosomes (the 'n' or haploid number for the species). This stage produces, through mitosis, the gametes, and is thus called the gametophyte. In some species male and female gametes are similar in shape and size, but in most the female gamete is larger and immobile (the egg), while the male gamete is small and motile (the sperm). Gametes always have the haploid number of chromosomes.

Gametes fuse to form a diploid (2n) cell, the zygote. In single-celled organisms this may be the whole organism, but in multicellular ones the zygote divides by mitosis to form the organism's body or thallus. This is called the sporophyte because at some stage

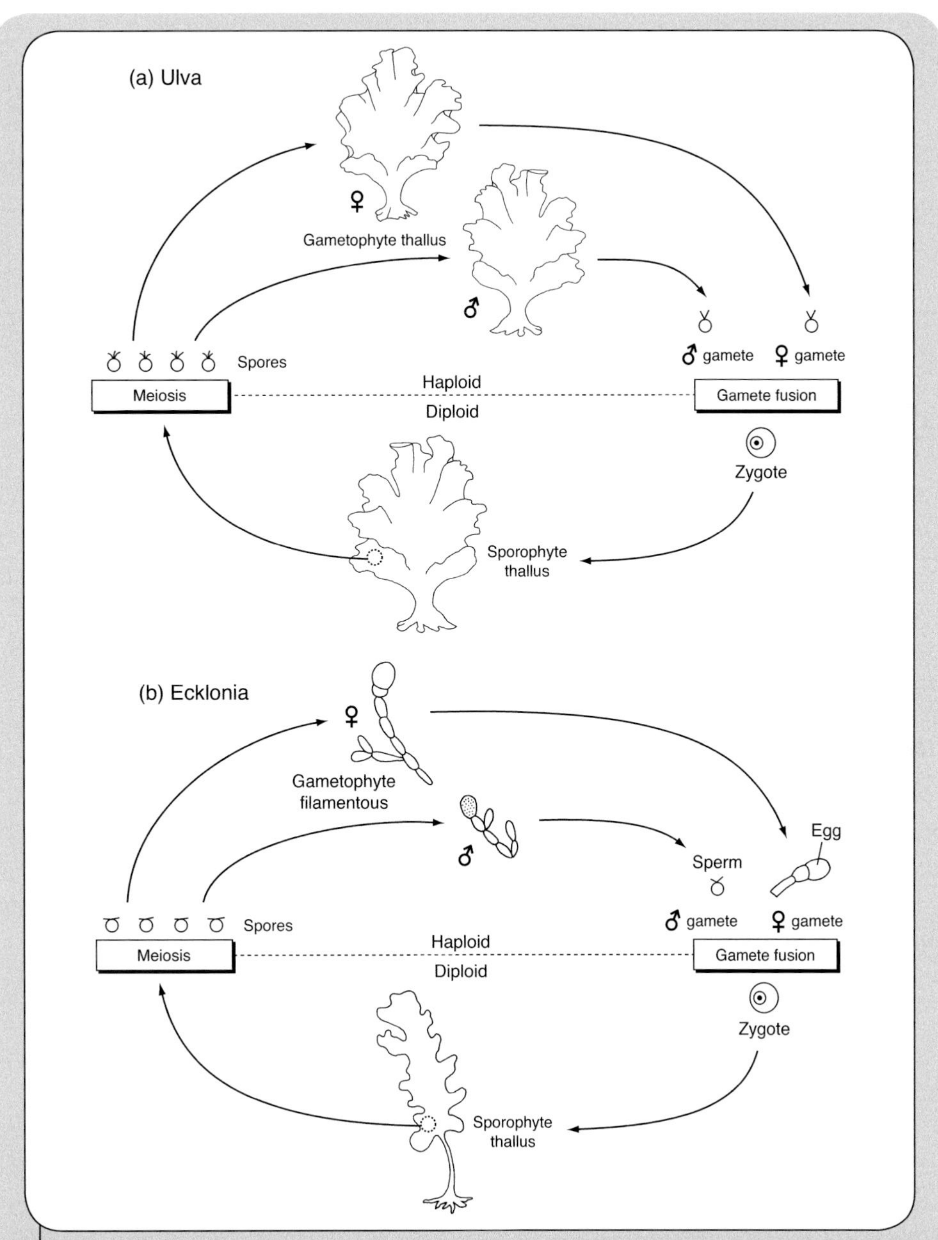

Figure B10.1.1 The alternation of a haploid gametophyte generation and a diploid sporophyte generation is characteristic of many protists and all land plants. The two stages can live separately, or one stage can live attached to the other. (a) An example of a cycle in which the sporophyte and gametophyte generations look similar is shown by *Ulva*, a green alga, (b) In *Ecklonia*, a brown alga, the generations are very different in appearance. (Source: B. Cale)

continued ›

Box 10.1 continued ›

certain cells undergo meiosis and form haploid spores. Each haploid spore may develop into an organism, or divide to form a multicellular haploid thallus. The cycle can be complicated by the presence of other types of spores for producing new copies of either the gametophyte or the sporophyte.

The basic principle of alternation of generations is shown in Figure B10.1.1 and life cycles for specific plants are described in Chapters 11 and 12. In each life cycle the key stages are the haploid gametophyte (producing gametes by mitosis), the diploid zygote (from the union of two gametes) and the diploid sporophyte (producing spores by meiosis).

Classification and evolution

From the above descriptions you can appreciate that there are few unique characteristics to describe the protists. Furthermore, there are obviously fewer observable characteristics for the classification of small unicellular organisms than for larger multicellular ones. Thus features such as the placement and surface or internal structure of the flagellum may be very important. Often these details can be seen only using an electron microscope. However, using molecular biology as well as classical and modern morphology and cell ultrastructure, biologists classify protists into six clusters (supergroups).

Rather alarmingly for students who have just come to grips with the hierarchical system of biological classification (Chapter 8), protist experts say it is impossible to put formal categories against the clusters and that each cluster might be a kingdom! Consequently, in Figure 10.2 the protists are presented in six supergroups and below this are 'first rank' groupings. The names of these groupings do not have the traditional endings that show if they are ranked as an order, class or family.

It is now accepted that the animals arose from within the protists similar to the Choanomonada and the plants from the Charophytes. We should not be surprised at the diversity among the protists, nor that the phylogenetic lines are sometimes hard to work out – after all, they have had 1.5 billion years of evolutionary experiments.

One intriguing concept underlying protist evolution is endosymbiosis. An organism initially lives inside another, but over time the symbiont integrates into the host cell machinery. Traces of the symbiont's history are seen in the extra membrane that bounds it, and sometimes a different type of mitochondrion with remnants of its own DNA. In some groups of protists, information from ultrastructure and DNA analysis shows that colourless, heterotrophic organisms are related to autotrophs with chloroplasts, so the colourless organisms lost their plastids during evolution. To make things even more interesting, there is evidence that some organisms have indulged in a further cycle of secondary endosymbiosis (Box 10.3).

Environmentally important protists

We describe several groups of protists in detail to show the fascination of these organisms. Some, such as the green, red and brown algae, underpin life in the ocean, some cause

Super groups	First rank groups	Examples
Amoebozoa	Tubulinea Flabellinea Stereomyxida Acanthamoebae Entamoebida Mastigamoebidae Eumycetozoa	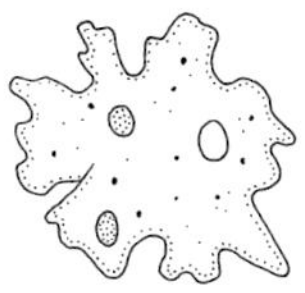
Rhizaria	Cercozoa Haplosporidia Foraminifera Radiolaria	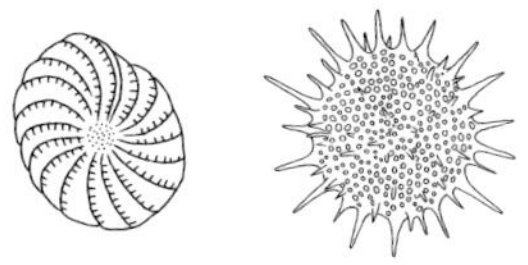
Archaeplastida	Glaucophyta Rhodophyceae Chloroplastida	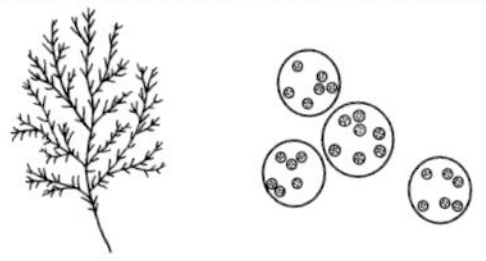
Chromalveolata	Stramenopila Alveolata	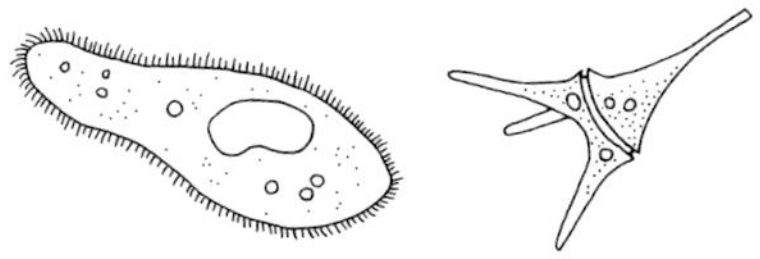
Excavata	Excavata	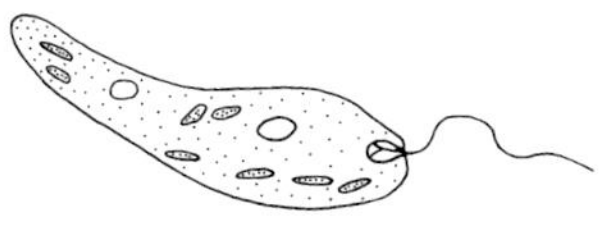
Opisthokonta	Fungi Choanomonada	

Figure 10.2 The supergroups of protists. (Source: B. Cale)

major diseases such as potato blight or phytophthora dieback, and others catch our attention when we are looking at pond water under the microscope. Finally, some smaller groups are included as clues to the evolution of other eukaryotes. Where possible, we describe the diversity of the group in Australia. Many of the groups of protists that make up phytoplankton – photosynthetic protists that drift freely in the water column of aquatic ecosystems – are not included despite their important roles as food for filter feeders and in carbon fixation (Box 10.2).

Box 10.2 Fertilising the ocean

Phytoplankton are photosynthetic protists that drift freely in the water column of aquatic ecosystems. They play an important role as primary producers, and are generally considered the first link in aquatic food webs. In the ocean their growth is regulated by the supply of light and the important plant nutrients nitrogen (N), phosphorus (P) and silicon (Si). Generally, when the supply of these factors is met, phytoplankton primary productivity is high. However, about 40% of the world's ocean surface waters are replete with light and the main nutrients, yet phytoplankton biomass remains low. These high-nitrate, low-chlorophyll (HNLC) regions occur in the equatorial Pacific Ocean, the Southern Ocean near Antarctica and the subarctic north-east Pacific.

In the 1980s scientists hypothesised that there were insufficient micronutrients, in particular iron, in these waters. Iron is essential for synthesising chlorophyll and several key photosynthetic proteins. In coastal continental margins, iron from terrestrial sources is blown into the ocean as aeolian dust, particularly in the dry and sandy equatorial regions. Current estimates suggest that dust from the Sahara Desert supplies around 50% of the iron to the world's oceans. Dust flux at the poles is significantly lower than at the equator because of the absence of land masses and the reduction of erosion on the ice-covered terrestrial margins, so the supply of iron is limited in neighbouring oceans.

In a unique experiment called IRONEX (iron experiment), biological oceanographer John Martin and co-workers fertilised a 64 km^2 area of the Pacific Ocean with iron. Within 10 days of fertilising, they observed a large bloom of phytoplankton in the experimental area. In fact the impact of the iron fertiliser was so large that satellite scanners resolved the bloom from space. Shipboard measurements indicated a twofold increase in phytoplankton biomass, a threefold increase in chlorophyll concentration and a fourfold increase in primary production. The results from the IRONEX experiments led to several further investigations in HNLC regions. The research gained momentum in the 1990s through the hypothesis that enhanced phytoplankton production in HNLC regions could draw down atmospheric carbon dioxide levels, possibly reducing global warming.

Supergroup Amoebozoa

Amoebozoans are unicellular and surrounded only by their cell membrane, lacking external shells of any kind. They move using outward projections of the cytoplasm called pseudopodia (literally 'false feet') created by directional flowing of the cytoplasm. They are heterotrophic and trap bacteria or other protists with their pseudopodia in a process called phagocytosis. Pseudopodia are advanced on either side of the prey and ultimately close around it, bringing it into the cell.

This distinctive body form and lifestyle is sometimes called amoeboid. It is not unique to this supergroup, also occurring in some rhizarians, algae and fungi shown by molecular data to be evolutionary lines distinct from the amoebozoans. Therefore the amoeboid body form evolved independently several times.

Amoebozoans live at densities of thousands of individuals per cubic centimetre in moist soil, fresh water and the ocean. Environmentally, they are important consumers of soil bacteria, recycling bacterial nutrients back into the ecosystem. Medically, some cause serious human diseases including amoebic meningitis (a potentially fatal inflammation of the brain membranes caused by *Naegleria fowleri*) and amoebic dysentery (a severe gastrointestinal disease caused by *Entamoeba histolytica*).

Supergroup Rhizaria

Rhizarians are also heterotrophic with unicellular amoeboid bodies, but their pseudopodia are finer than those of amoebozoans and are strengthened internally with microtubules. In one ecologically significant group, the foraminiferans, the fine pseudopodia project through small pores in an external shell (test) of calcium carbonate to form a branching network used for movement and prey capture.

The total number of living species of foraminiferans is uncertain, because forms indistinguishable in appearance may be genetically distinct. They are mainly marine, with few freshwater species. Reproduction involves alternation of generations. The tests fossilise readily so about 30 000 extinct species can be recognised, dating back to the early Cambrian period. Some fossils reach 15 cm in diameter, an extraordinary size when compared to the largest living species of up to 9 mm in diameter (few exceed 1 mm in diameter). The fossil remains are important ingredients in products ranging from cement to blackboard chalk and also indicate potential oil deposits.

Supergroup Archaeplastida (glaucophytes, red algae and green algae)

These autotrophic protists all have photosynthetic plastids with chlorophyll a from an ancient endosymbiotic cyanobacterium (Box 10.3), cellulose cell walls and starch as a storage product from photosynthesis. They include unicellular and multicellular forms.

Box 10.3 Endosymbiosis theory on the origin of chloroplasts

Studies of the photosynthetic organelles (plastids) in protists provided many clues to the ancestry of the algae (autotrophic protists) and modern-day plants. The most accepted theory is that all algae acquired their plastids from a single ancestor which later branched into three main lineages within the supergroup Archaeplastida: the Glaucophyta, the Rhodophyceae and the Chloroplastida (Figure B10.3.1). The monophyletic

continued ›

Box 10.3 continued ›

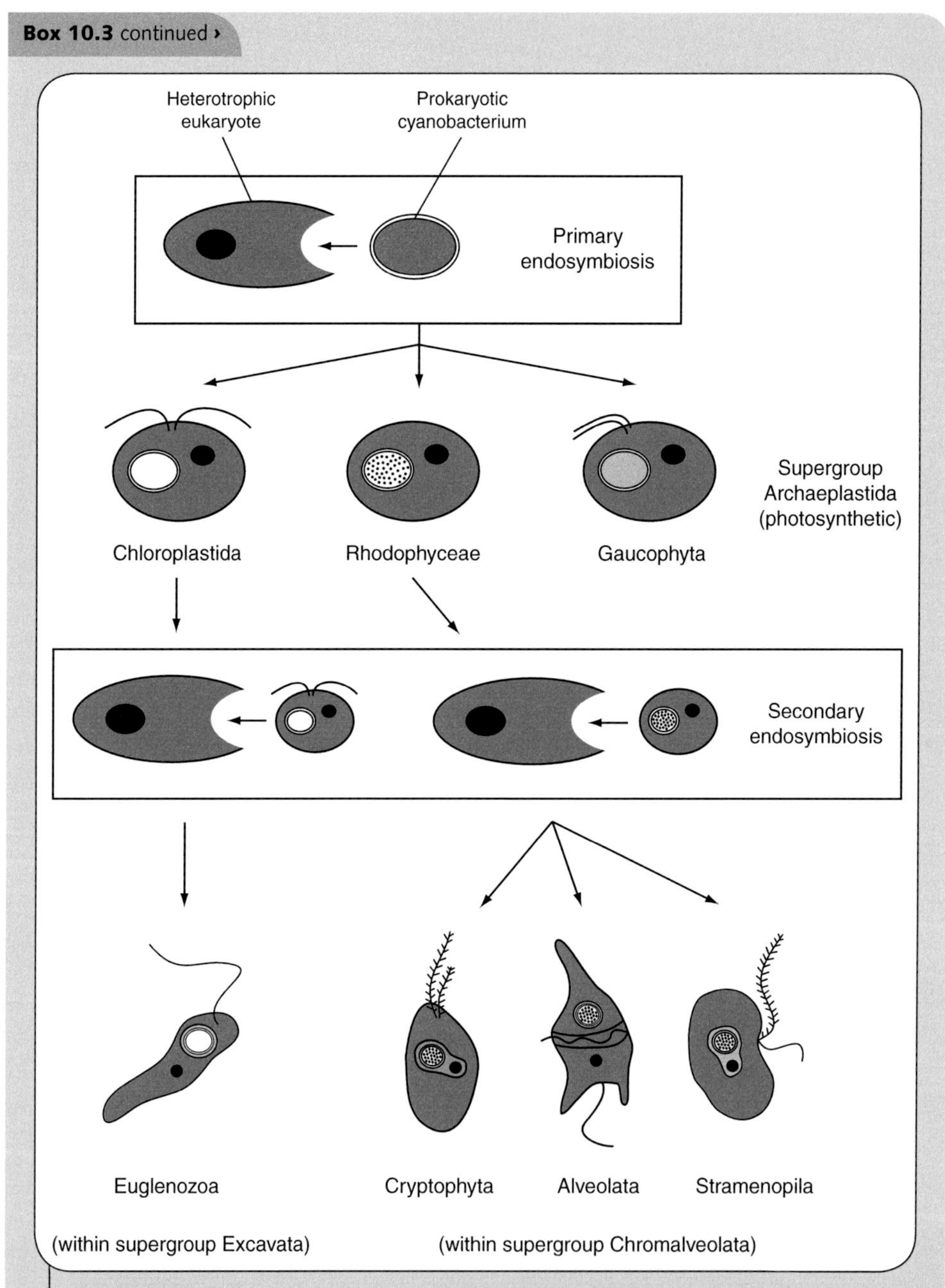

Figure B10.3.1 Conceptual diagram of plastid evolution through endosymbiosis. In the process of primary endosymbiosis, a prokaryotic cyanobacterium was engulfed by a heterotrophic eukaryote to give rise to the photosynthetic protists in supergroup Archaeplastida. Secondary endosymbiosis of a photosynthetic protist from first rank group Chloroplastida (within the Archaeplastida) and a heterotrophic host led to the formation of the Euglenozoa (within supergroup Excavata). Secondary endosymbiosis of another photosynthetic protist from supergroup Archaeplastida and a heterotrophic host led to the formation of the supergroup Chromalveolata. (Source: B. Cale)

(single ancestor) origin of plastids theory gathered momentum from studies of the glaucophytes showing that the plastid (termed a cyanelle) shares several molecular and ultrastructural characteristics with photosynthetic cyanobacteria. These characteristics suggest that the plastid was once a free-living bacterium that was captured and incorporated into the host through a single symbiotic event, called primary endosymbiosis. In time the single ancestor diverged into the three modern-day lineages. In these organisms the plastids are bound by two membranes, a result of the endosymbiosis process.

However, many present-day algae have three or four membranes surrounding their plastids, a fact that initially complicated their taxonomy and studies of evolution. Another group of singled-celled algae, the cryptophytes, provided strong evidence of additional endosymbioses, this time involving a eukaryote as the endosymbiont. Biologists discovered that three membranes surrounded the plastid of cryptophytes and that there was a tiny area that contained DNA, called the nucleomorph. Recent DNA sequencing of the nucleomorph indicated that the plastid DNA shared similarities to the DNA within the nucleus of red algae. Consequently it is considered that cryptophytes evolved from secondary endosymbiosis, the process whereby a heterotrophic eukaryote engulfs a photosynthetic eukaryote. Other groups also appear to have evolved through secondary endosymbiosis (Figure B10.3.1).

Glaucophyta

Glaucophyta is a small, little-studied group of microscopic freshwater protists. Their main significance is evolutionary, because the common ancestor of the glaucophytes, red algae and green algae is believed to have arisen from an endosymbiotic relationship with a prokaryote (Box 10.3).

Rhodophyceae (the red algae)

Rhodophytes are common on coastal rocks, with some species occurring to 250 m deep. Others grow in fresh water. Some are unicellular, but most are macroscopic, multicellular organisms recognised as 'seaweeds' (Plate 10.2). There are about 5500 species worldwide with 1300 known from Australia. In the ocean the red seaweeds are the most diverse in Australian waters, but they are not as conspicuous as the larger brown and green seaweeds.

In multicellular red algae, the algal body (thallus) comprises entwined filaments within a gelatinous matrix of polysaccharides (complex carbohydrates) such as agar and carrageenan. Some reds have cell walls impregnated with calcium carbonate, which contributes to beach sand when the organisms die and decompose. Some have chloropyll d as well as chlorophyll a, and all have accessory pigments called phycobilins, a characteristic inherited from endosymbiosis with a cyanobacterium. The phycobilins broaden the range of light wavelengths their cells can absorb for photosynthesis.

Red algae have complex life cycles. Some have alternation of gametophyte and sporophyte stages that may look very different (the gametophytes can be small, filamentous organisms and the sporophyte a large red seaweed), but many have three phases (triphasic). There is alternation of a gametophyte and sporophyte generation, but in the sporophyte phase there are two types of plants – the sporophyte arising from the zygote stays as a small parasitic plant on the gametophyte before producing spores that grow into a free-living sporophyte stage. Red algae do not have any flagellate stages and therefore their spores rely on ocean currents. Production of thousands of spores increases the probability of fertilisation between male and female gametes.

Red algae are of great economic and cultural importance. Some species produce the commercial gelling agents agar or carrageenan. In Japan, surrounding regions of the Pacific Ocean basin and south-east Asia, the dried thallus of *Porphyra* (nori) is eaten extensively. Seaweed culture, made possible from research into the life cycles and cultivation techniques, is more efficient than collection from the wild.

Chloroplastida (green algae)

The diverse green algae range from simple unicells to large, complex multicellular organisms with cellular differentiation. They occur in nearly all moist habitats with light for photosynthesis (Plate 10.3). Some 8000 have been described, including 2000 from Australia.

The large green algae have a thallus of parenchyma, a solid tissue construction in which neighbouring cells share a common cell wall. They have life cycles of alternating generations of haploid and diploid stages in which the different generations may or may not look similar (Figure B10.1.1).

Green algae have two or more flagellae in at least one part of their life cycle. Flagellum ultrastructure and molecular biology separate them into two monophyletic branches: the Chlorophyta and the Charophyta. All chlorophytes have plastids with similar structure and photosynthetic pigments to land plants, including chlorophylls a and b, b-carotene and several other carotenoids. The substances produced during photosynthesis, most notably starch, are identical to those produced in land plants. However, molecular evidence suggests that, rather than the chlorophytes being the precursor to land plants it was the ancient charophytes, because they have all the above features and, in addition, their reproductive structures, mitosis and the ultrastructure of their motile sperm closely resemble land plants.

Supergroup Chromalveolata (diatoms, brown algae and water moulds; dinoflagellates, ciliates and apicomplexans)

Chromalveolatans arose from an ancestor that acquired a plastid containing chlorophylls a and c through secondary endosymbiosis (see Box 10.3) of a red alga. This is supported by phylogenetic analyses of ribosomal RNA sequences and cytological evidence of four membranes surrounding the plastid. At some stage in their life cycle they have motile cells with two flagellae. These are very distinctive as the anterior (forward) pointing one is hairy (tinsellated) while the trailing one is smooth. Within this group the diatoms, brown algae and water moulds are placed in the subgroup Stramenopila because of their flagellum structure. Some groups (such as the water moulds) lack chlorophyll, but because they have the unique flagella structure they are considered to belong to this group and to have lost their chloroplasts during evolution.

Other important organisms in this group are in a subgroup called Alveolata because they all have sacs of endoplasmic reticulum (called alveoli) located beneath the plasma membrane of the cell. They include the dinoflagellates, ciliates and apicomplexans.

Bacillariophyta (diatoms)

We admire the autotrophic unicellular Bacillariophya, or diatoms, for their ornate and elaborate silica cell walls (Plate 10.4), but they have greater significance as primary producers in the marine food chain. Where light and nutrients are available, diatoms grow and divide up to four times daily. During photosynthesis they assimilate nutrients and convert them into high-energy fatty acids, desired by marine heterotrophs including other protists, zooplankton, filter-feeding molluscs, and fish. Hence, in regions where nutrients and light are abundant, diatoms are productive and the energy is transferred to heterotrophs. Where the supply of nutrients is low, diatom productivity is reduced, supporting fewer heterotrophs.

Phaeophyceae (brown algae)

The 2000 or so species of brown algae are autotrophic, multicellular and mainly marine (Plate 10.5). Although only 350 species occur in Australia, they include our largest seaweeds, and kelps such as *Macrocystis* can form dense 'forests' 20 m tall. Their chloroplasts contain chlorophylls a and c, but they are masked by a brown accessory pigment, the carotenoid fucoxanthin. They store products of photosynthesis as laminarin or mannitol, not the starches used in land plants. The large multicellular ones show some differentiation of the thallus morphology (into a holdfast to attach the alga to the substrate, a stem-like stipe and a photosynthetic, leaf-like blade). Internally, there are special food-conducting cells similar to those seen in land plants, but which evolved independently. Brown algae have several different life cycles based on the alternation of gametophyte and sporophyte generations. People from several cultures eat brown algae, but there is little evidence of Aboriginal people eating seaweeds except the bullkelp, *Durvillaea potatorum*.

Oomycetes (water moulds)

This group of filamentous, unicellular protists was classified as fungi before molecular analysis placed them with the brown algae and diatoms. Their cell walls are strengthened with cellulose, whereas fungal cell walls are strengthened with chitin. Oomycetes prefer high humidity and running surface water, hence the name water mould. They absorb their food from surrounding water or soil or invade a host to feed, thus causing disease (pathogens).

Some oomycetes cause diseases of fish and invertebrates, but others are particularly significant because they are aggressive plant pathogens. They fall into three main groups:

1. *pythium* – this is the most prevalent group. Most species have a large host range and diseases such as pythium dampening off are common in glasshouses
2. *phytophthora* – phytophthora dieback kills trees in natural ecosystems and in orchards, and numerous collar rots are phytophthora diseases of crops, especially in the tropics (Plate 10.1). Potato blight, caused by *Phytophthora infestans*, reached Europe from South America in the 1830s. It thrives in warm, moist conditions and is spread by wind and rain. Potato plants are infected through the leaves and occasionally through the tubers (potatoes), leading rapidly to death. Failure of the potato crop following a potato blight outbreak in the 1840s in Ireland resulted in the death or emigration of over 3 million people. Irish emigration to Australia was significant and it is commemorated in a memorial in the Hyde Park Barracks, Sydney. Modern horticulturalists control potato blight with a combination of techniques, including growing resistant strains, watering from driplines rather than overhead sprinklers, and using chemical sprays and drenches.
3. *downy mildews* – downy mildews (not to be confused with powdery mildew, caused by a fungus) cause a white mould or film on leaf surfaces. The best known of these is *Plasmopara viticola*, the downy mildew of grapes. It is native to North America but was introduced to Europe on contaminated stock in the 1870s and nearly wiped out the French wine industry. *Plasmopara viticola*, first reported in Australia in Victoria in 1917, is the most important pathogen in the Australian wine industry. In wet years (when infection is hardest to control with chemicals), losses can be huge.

Dinoflagellates

These mainly marine unicellular protists have two flagellae. One is directed backwards for propulsion, while the other spins the cell. They may be naked, scaled or covered with thick cellulose plates (Figure 10.3). Some dinoflagellates are autotrophic, containing chloroplasts with chlorophylls a and c and the carotenoid peridinin, and others are colourless heterotrophs. Some dinoflagellates switch between autotrophy and heterotrophy, responding to the availability of light and foods. They exist in the water column, in sediments, or as parasites or endosymbionts in animals. For example, the body tissue of corals, sea anemones and some molluscs contain dinoflagellate endosymbionts called zooxanthellae, which supply the host with nutrition derived from photosynthesis and in return receive protection and excretory nitrogen from the host.

There are about 2100 species worldwide and some produce highly toxic compounds. 'Blooms' of planktonic dinoflagellates under conditions of high nutrients and high water temperatures may turn the water red. When the dinoflagellate involved is toxic, these 'red tides' kill marine animals and poison people eating fish, oysters or mussels from contaminated waters.

Figure 10.3 Scanning electron micrograph of the dinoflagellate *Protoperidinium* spp. The thick interlocking plates called theca provide protective armour, and the equatorial groove called cingulum houses a transverse flagellum (not seen here). (Source: H. Patterson)

Ciliates

Ciliates are unicellular with cilia on the outer cell surface for swimming and, in some species, capturing prey. They are a highly diverse group containing about 8000 species. Ciliates are heterotrophic, ingesting bacteria and smaller protists through a specialised buccal cavity. Some are parasitic.

Ciliates typically have one or more macronuclei and a single micronucleus. The macronucleus controls metabolism and development, and the micronucleus is involved in sexual reproduction. The most commonly observed methods of sexual reproduction are conjugation, whereby genetic material is swapped between two individuals, and autogamy, whereby the nucleus is rearranged within an individual. However, the dominant form of reproduction is asexual through binary fission.

Apicomplexans

Nearly all the 5000 species of apicomplexans are unicellular and parasites of animals. Their life cycles are complex. They derive their name from the unique organelles forming the apical complex, an apparatus used to gain entry to their host by secreting enzymes and proteins that erode extracellular material. Important parasites include the species of *Plasmodium* that cause malaria in humans. The infective cell (sporozoite) passes from mosquitoes to the human bloodstream, collects in the liver and divides to produce merozoites. These enter the circulatory system and attack the red blood cells, releasing toxins causing the characteristic malarial fever and chills. After many replications within the circulatory system, the merozoites produce gametocytes, which are ingested by a blood-sucking mosquito. There they release gametes that fuse and form a zygote. This divides through meiosis and releases new sporozoites, which move into the mosquito's salivary glands to complete the life cycle.

In the mid-1990s research by Geoff McFadden at the University of Melbourne showed that the apicomplexans have a remnant plastid, which seemed functionless and was certainly not photosynthetic. The plastid is of red algal origin and can be traced back to the same endosymbiotic event that gave rise to plastids in the other Stramenopila. These findings support the classification of the apicomplexans with the other Alveolates and stimulated research into new medicines to combat apicomplexan parasitises in humans.

Humans can alter the incidence and geographic distribution of diseases caused by apicomplexans. For example, the mosquitoes that transmit malaria require warm, swampy areas to breed, and climate change may create suitable habitats in parts of the world now free of the disease. *Toxoplasma gondii*, an apicomplexan that causes the disease toxoplasmosis in many mammals and birds, completes the sexual phase of its life cycle in cats. By introducing cats to new environments, humans also spread *T. gondii* (Box 10.4).

Box 10.4 Cats and wildlife – the protistan connection

Cats are loved and valued pets in up to a third of Australian homes, and many responsible owners neuter their pets and confine them at night to prevent unwanted breeding and to reduce attacks on wildlife. Unfortunately, hunting by pet cats or dumped, unwanted kittens is not the only way cats may harm wildlife. They may also spread the protist *Toxoplasma gondii*, which can cause potentially fatal toxoplasmosis in wildlife and humans.

T. gondii belongs to a group of parasites called the Apicomplexa, classified within the protist supergroup Chromalveolata. Its complex life cycle involves sexual and asexual stages, but the sexual stage occurs only within felids (the cat family, including the domestic cat) so the cat is called the definitive host (Figure B10.4.1). Sexual reproduction occurs in cells lining the cat's small intestine, producing infectious oocysts shed with the cat's faeces. Oocysts resist desiccation and temperature extremes and persist in soil for long periods. When ingested with plant food or contaminated water they infect many mammals and some birds as intermediate hosts. The parasite enters the bloodstream across the intestinal lining and ultimately encysts within muscles or the brain, developing into bradyzoites, which slowly reproduce asexually. Mature cysts burst, releasing blood-borne trachyzoites, which reproduce asexually. They may encyst in further cells or, if the host is pregnant, cross the placenta to infect the embryo. If this occurs in humans, miscarriage may result. If a predator other than a cat eats an infected animal, it too becomes infected with cysts. However, if a cat eats an infected animal the parasite establishes in the cat's intestinal lining, reproduces sexually and completes the life cycle.

Healthy intermediate hosts often suffer no serious symptoms following infection, although there is evidence that infected mice and rats lose their fear of cats. However, stress or other diseases that lower the effectiveness of the host's immune system encourage rapid proliferation of the parasite. This causes inflammation and severe tissue damage when infected cells burst, often accompanied by behavioural changes

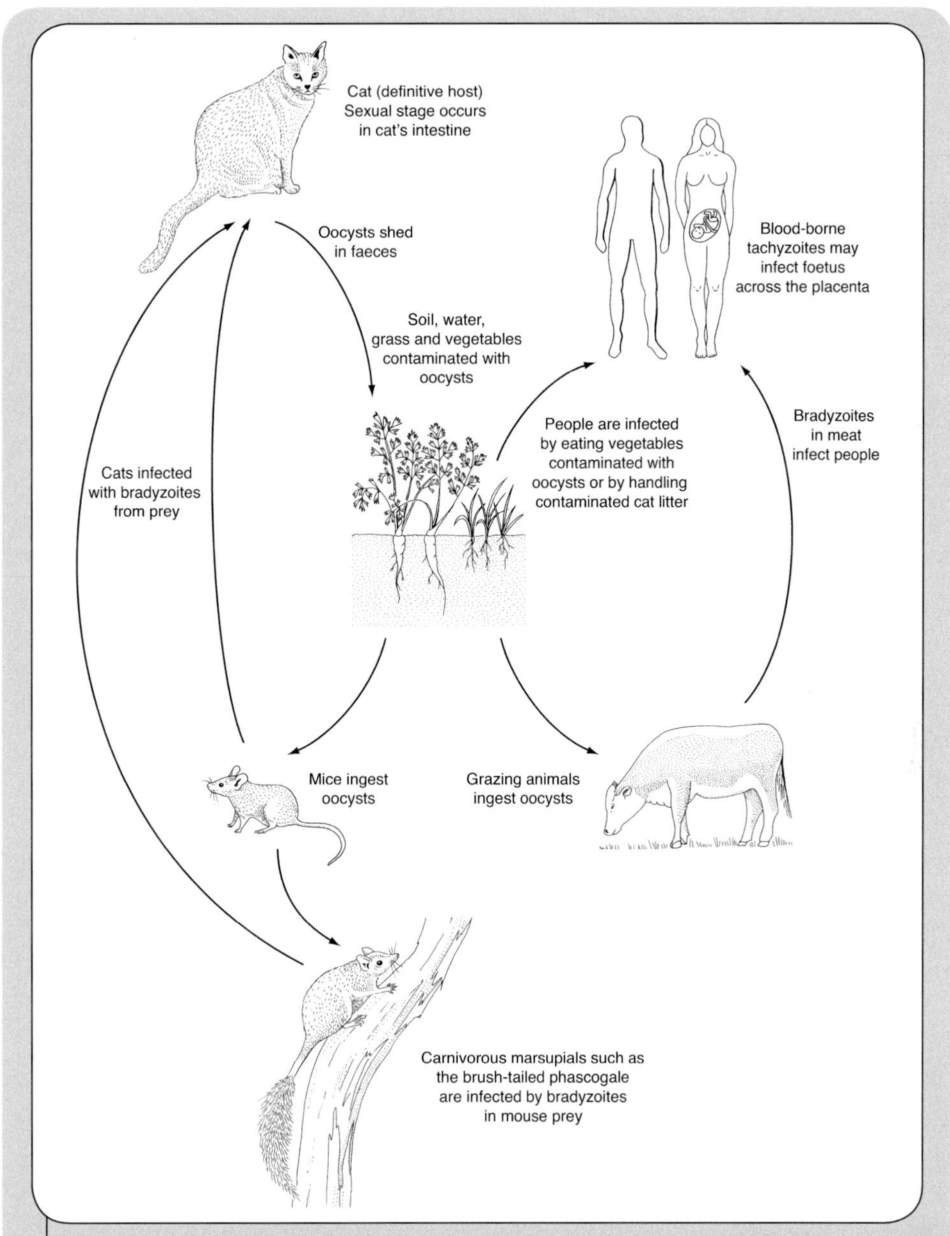

Figure B10.4.1 The life cycle of *Toxoplasma gondii*. (Source: B. Cale)

and loss of coordination because of damage to muscles or the nervous system. This clinical condition is called toxoplasmosis and it kills wild and captive Australian mammals including wombats, bandicoots, koalas, possums and kangaroos. A recent study of brushtail possums in Sydney revealed that about 6% had been exposed to *T. gondii*. Confinement of pet cats and sanitary disposal of droppings can contain infection in urban areas.

Supergroup Excavata (euglenids and trypanosomes)

The Excavata are named for their feeding groove (gullet). Our examples come from the Euglenozoa, which includes the euglenans and the trypanosomes.

Euglenida

Euglena are free-living and unicellular, with about 100 species worldwide. They are mostly in fresh water, although marine species do occur. They display both plant and animal characteristics and have several nutritional modes: a species can be heterotrophic, autotrophic or able to switch between heterotrophy and autotrophy (mixotrophic). Photosynthetic euglena have chloroplasts containing chlorophylls a and b and b-carotene. These pigments also occur in green algae and land plants, but euglena can be differentiated by the presence of three chloroplast membranes, rather than two. The third membrane indicates that euglena obtained their chloroplasts through secondary endosymbiosis (see Box 10.3). The product of photosynthesis is called paramylon, which forms solid aggregates in the cytoplasm. Euglena have a flexible cell shape because of proteinaceous strips under the plasma membrane, and move using two flagellae arising from the gullet. A small, light-sensitive organ called the stigma, located at the anterior (front) of the cell, enables autotrophic euglena to seek well-lit areas for photosynthesis.

Trypanosomes

Trypanosomes are single-celled parasites from the bloodstream of humans and other mammals, causing debilitating illnesses such as sleeping sickness or Chagas disease. They are never photosynthetic and have only one flagellum.

Supergroup Opisthokonta (choanomonads)

These organisms are all heterotrophic and are grouped together based on the ultrastructure of the cells (particularly the centrioles) and molecular evidence.

Choanomonada

Choanomonada has only about 150 species, but it is important to describe them here because they are considered the closest living protist relatives of the animals. Their structure is nearly identical to the choanocytes of sponges (Chapter 14). They are characterised by a collar of tightly packed hairlike extensions called microvilli, surrounding a single flagellum. The beating flagellum draws water through the collar, which filters organic particles and bacteria for food. In free-swimming choanomonads the flagellum propels the cell. Most choanomonads, however, are sessile, attaching by a thin stalk opposite the flagellum. Colonial forms are typically a cluster of cells on a single stalk, although some form aggregates with collared cells on the outside and non-motile cells within. Reproduction is by simple division and no sexual cycles have been observed. They inhabit both freshwater and marine systems.

Is there a kingdom Fungi?

Fungi are a distinct and recognisable group of heterotrophic organisms. Some authorities consider them among the protists and group them with the Opisthokonta (as shown in Figure 10.2), while others consider them distinctive enough to be placed in their own kingdom (as is shown in Chapter 8). There are about 77 000 described species, but thousands remain undescribed. They have many interactions with humans (Box 10.5, Plate 10.7).

Diversity of form and lifestyle

Fungi are mostly terrestrial, with some from marine and freshwater habitats. They generally grow as filamentous tubular hyphae, collectively known as the mycelium. The hyphal walls contain a complex carbohydrate called chitin, and may or may not have cross walls (septa), dividing the cytoplasm into chains of cells. Hyphal diameter varies between species and cells may have one or several nuclei. At certain times and under certain conditions the mycelium aggregates to form either fruiting bodies for sexual reproduction (the familiar

Box 10.5 Fungi and humans

Humans and fungi have a long and complicated relationship, with beneficial and detrimental encounters. The beneficial encounters are obvious. Fungi are eaten (mushrooms and true truffles), used in baking and brewing (yeasts), and flavour many cheeses and soy products. The first antibiotic came from a fungus (*Penicillium*) and its discovery dramatically altered medicine. It could also be argued that natural LSD, a hallucinogen produced by ergot (*Claviceps purpurea*), and other hallucinogens are also beneficial as their use had spiritual importance in many societies. One downside of ergot was that the hysterical frenzies produced by consumption of infected grain may have been the cause of medieval witch hunts and trials. Another downside of ergot is that another toxin caused limbs to drop off!

Detrimental encounters with fungi occur whenever a plant pathogen devastates a crop, especially if this is the only food source. Catastrophic tree diseases have also been caused by fungal pathogens (for example, chestnut blight caused by *Cryphonectria parasitica* and Dutch elm disease caused by *Ophiostoma ulmi*). In all cases, the epidemic followed the introduction of the pathogen from another country, highlighting the importance of quarantine and biosecurity. In Australia, there is particular concern about eucalypt diseases emerging on eucalypts introduced for use in plantations in Asia, South America and Africa. These diseases are absent in Australia and their impact on our native eucalypts is unknown.

mushroom is one example) or resting structures (Plate 10.8). Fungi grown in laboratories on nutrient agar form dense colonies of hyphae, but in the environment where nutrition is limited the hyphae are sparse until they find a new food source. Usually after the food is depleted reproduction occurs. Because of this growth habit, a single fungal colony can cover a large area. One of the largest living organisms is considered to be *Armillaria gallica*, a pathogen of oaks in Michigan, USA. A single individual covers over 600 ha of forest. The rapid and extensive growth of the hyphae compensates for immobility by extending towards food sources without moving the whole body.

Most fungi reproduce both sexually and asexually, resulting in complex life cycles. In a generalised example, sexual reproduction involves two haploid mycelia that grow together and fuse. The separate haploid nuclei need not fuse promptly and the mycelia may persist with two genetically distinct haploid nuclei for long periods. When they eventually do fuse, the diploid zygote is short-lived. It divides by meiosis to produce haploid spores, which are dispersed by specialised reproductive structures in fruiting bodies. In asexual reproduction, spores are commonly produced at the ends of specialised hyphae. Some unicellular fungi reproduce asexually by cell division or by budding, in which small buds grow out from a parent cell.

Most life cycles involve releasing many haploid spores. These are wonderfully varied in size, colour and ornamentation (Figure 10.4). Some spores are actively expelled

Figure 10.4 Scanning electron micrographs showing the diversity of fungal spores. (Source: B. Dell)

from the fruiting structure, whereas others have sticky surfaces and are dispersed by insects. Fungal spores can travel far by wind, even crossing continents and oceans. More localised dispersal can occur by rain splash or flowing water. Animals such as bettongs and quokkas eat the above- and below-ground fruiting bodies of fungi, and their diet can be deduced from the spores in their faeces.

Environmental roles of fungi

There are essentially three groups of fungi: (1) the saprophytes, which decompose dead organic matter, (2) the parasites, which feed on living hosts, and (3) the mutualists, which form close associations with other living organisms in which both parties benefit (see also Chapter 17). Saprophytes are absolutely essential because decomposition releases nutrients in dead bodies back into the environment for use by other organisms. For example, only fungi can decompose lignin, a structural compound in plant cell walls, producing simpler compounds usable by other organisms. Parasites cause important diseases of crops, livestock and people. In plants, parasitic fungi called biotrophs can live within the plant for some time without causing any symptoms. Other parasitic fungi are nectrotrophs, killing plant cells and forming a lesion. Chytridiomycosis, caused by *Batrachochytrium dendrobatidis*, is a fungal disease deadly to amphibians (frogs and related animals), causing population declines and possibly even extinctions in eastern Australia and North and South America. Mycorrhiza (see Box 10.6) is a mutualistic association between fungi and plant roots that, instead of causing disease, benefits both the fungus and its host.

Box 10.6 Mycorrhiza

(Contributed by Bernie Dell)

Careful removal of the litter (fallen leaves and twigs) on the floor of a eucalypt forest reveals extensive mats of fungal hyphae colonising the fine feeder roots of the eucalypt trees to form structures known as ectomycorrhizae (ecto: outside, because these fungal hyphae are outside the root cells) (Figure B10.6.1a). The hyphae form a mantle around the root and penetrate between the epidermal cells, creating the Hartig net (Figure B10.6.1b). Further hyphae ramify through the soil, extending the surface area of the eucalypt roots by more than a factor of 100. The plant benefits by obtaining some of the water and minerals, such as nitrogen, phosphorus and copper, absorbed by the fungi and passed to the eucalypt host via the Hartig net, while the fungi receive sugars and other organic nutrients in exchange.

Many ectomycorrhizal fungi associated with woody plants in forests and woodlands in Australia are still unnamed. Fruiting bodies may form above or below ground (Figure B10.6.1c, d) but there are more species that fruit below ground, mostly producing small pebble-like underground fruiting bodies 1–2 cm in diameter. They

continued ›

Box 10.6 continued ›

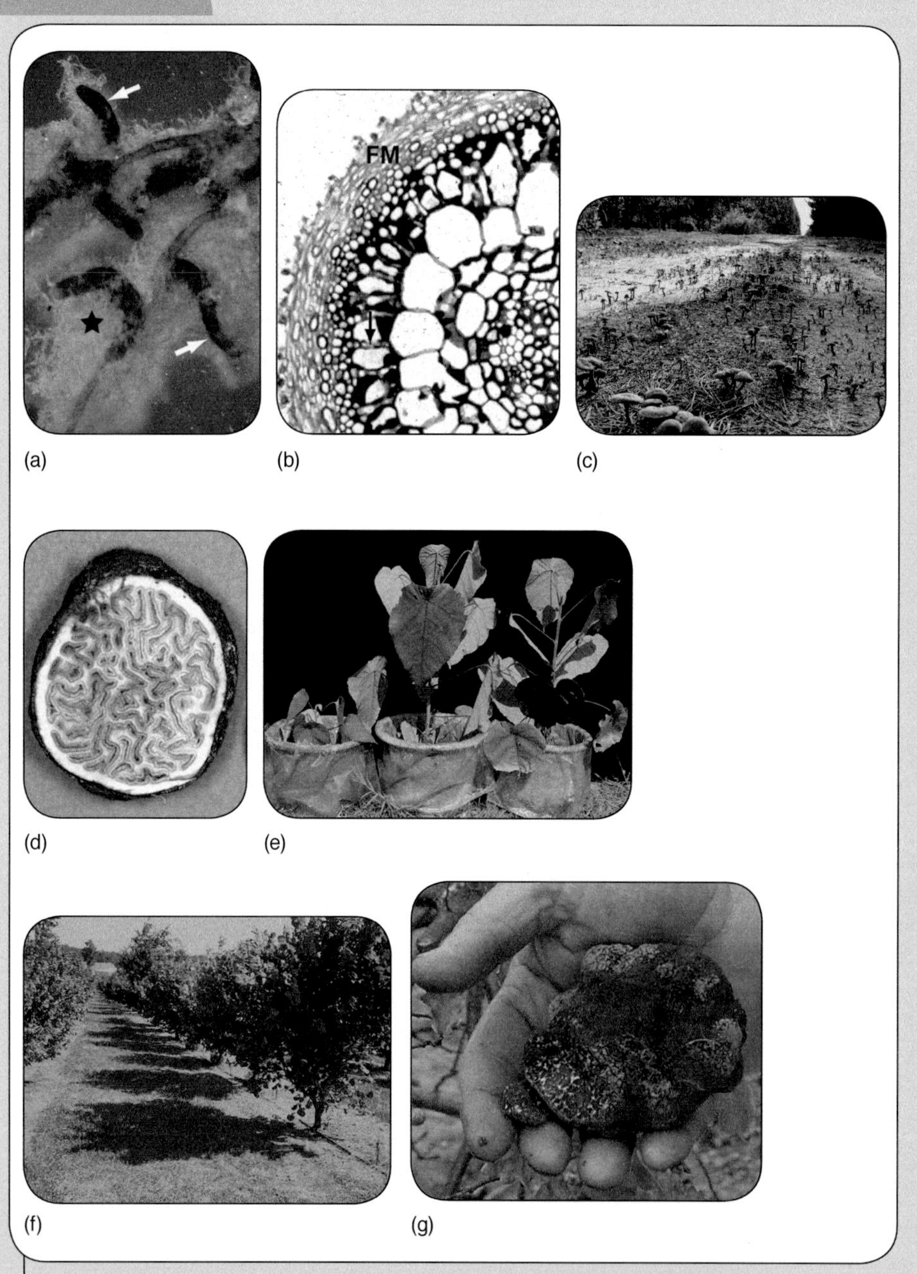

Figure B10.6.1 Eucalypt ectomycorrhizae. (a) The swollen black ectomycorrhizae (arrowhead) surrounded by fungal mycelia (star) extending into the soil. (b) Section through an ectomycorrhiza showing the sheathing fungal mantle (FM) and Hartig net (arrowed) extending between the epidermal cells. (c) *Laccaria* fruiting at the edge of a pine plantation. (d) A hypogeous stone-like truffle in the genus *Labyrinthomyces*. (e) Response of *Macaranga denticulata* to inoculation with spores of arbuscular mycorrhizal fungi (centre) compared to the uninoculated control (left) and the addition of superphosphate (right). (f) A young truffle farm in Western Australia. (g) Black gold! A fruiting body of the French black truffle. (Source: B. Dell)

contain nutritious tissue eaten by fungus-eating (mycophagous) marsupials, which locate hypogeous fruiting bodies by smell. Fungal spores pass undamaged through the digestive tract and are dispersed in faeces. This tripartite relationship is important in Australian ecosystems: the fungus provides the host tree with access to scarce soil nutrients, and the marsupial benefits from the food it receives, and disperses spores as well as turning over the soil, burying litter and increasing rain infiltration.

Ectomycorrhizal fungi are collected for food in many parts of the world. Most of these cannot be grown in cultivation, but one exception is the French black truffle that is cultivated on hazelnut tree roots in truffle farms in temperate Australia (Figure B10.6.1f, g). Little information is available on fungi eaten by indigenous peoples. Be cautious when eating fungi collected from the wild because poisonous species are common and correct identification requires specialist advice.

All eucalypts are ectomycorrhizal in the wild, as are acacias, casuarinas, myrtle beech, some legumes in the pea family and some daisies. Most of the other plants form mycorrhizal associations with microscopic soil-inhabiting fungi that do not produce macroscopic fruiting bodies. Unlike the ectomycorrhizae, the fungal hyphae pass through the cell wall in these plants to form branching structures called arbuscules. These associations are called endomycorrhizae (endo: inside, because these fungal hyphae extend into the root cells).

As well as increasing the plant's access to limiting soil nutrients (Figure B10.6.1e), mycorrhizal fungi also aid plants by enhancing drought tolerance, increasing tolerance to some pests and diseases, and increasing the rate of photosynthesis. When revegetating disturbed landscapes or planting Australian trees in foreign landscapes, the absence of compatible mycorrhizal fungi may limit success and inoculation is recommended.

Classification of fungi

Modern molecular biology techniques, together with detailed examination of reproductive structures, have identified five major types of fungi: the Chytridiomycetes, Zygomycetes, Glomeromycetes, Ascomycetes and Basidiomycetes. Some ascomycete and basidiomycete fungi live in mutualistic associations with cyanobacteria or green algae, forming organisms known as lichens (Plate 17.1).

Chytridiomycetes

Chytrids provide the oldest fungal fossils and are placed ancestrally to other fungal groups in phylogenies. There are some 1000 species, mostly aquatic, suggesting that fungi originated in water, as did plants and vertebrates. Unlike all other fungi, they have flagellated spores, suggesting that other fungi lost flagella during evolution. Chytrids also differ from all other fungi in that they are generally unicellular or form short chains of cells, which are often multinucleate (coenocytic).

Chytrids are important saprophytes and parasites in both aquatic and terrestrial habitats and have been collected from the arctic to the tropics. They degrade tough biological materials such as chitin, keratin and cellulose, and hence are important in nutrient recycling. As noted above, they may be parasites of aquatic organisms.

Zygomycetes

Zygomycetes include saprophytes (a familiar species is *Rhizopus*, the black bread mould), soil fungi, insect and fungal parasites, and symbionts of arthropods. There are about 800 species that are characterised by large, thick-walled, coenocytic spores and thin-walled hyphae. They reproduce asexually using stalked sporangiophores, each bearing one sporangium containing hundreds of spores. Sexual reproduction may also occur when hyphae of two different mating types grow towards each other. The hyphae produce gametangia, which fuse to form a zygosporangium. An indefinite time later, the gametes within the zygosporangium unite, forming a multinucleate zygospore protected by a thick cell wall. It may remain dormant for long periods before the nuclei undergo meiosis, producing haploid spores. A sporangiophore then arises to disperse the spores.

Spores from both sexual and asexual reproduction germinate into haploid mycelia, with the hyphae functioning as gametangia (spore-producing organs) during the sexual stage. The hyphae are long and tubular, with a cytoplasmic lining and a large vacuole in the centre. Like chytrids, the zygomycota have coenocytic cells except to seal off the reproductive structures or sporangiospores.

Glomeromycetes

This small group of coenocytic, strictly terrestrial fungi only reproduces asexually. Fewer than 200 species have been described worldwide. Molecular biology studies only recently revealed them as distinct from Zygomycetes. Their significance lies in the fact that they all form endomycorrhizae with plant roots, assisting the plants in absorbing nutrients and receiving glucose in return (Box 10.6).

Ascomycetes

This is the largest fungal group, including unicellular yeasts, green moulds, powdery mildews, morels, cup fungi and true truffles. In addition, most lichens have fungi from this group. Hyphae produce sexual ascospores in a sac-like ascus. Asexual spores are produced directly from the hyphae, but may be contained within an asexual fruiting structure. These may be spherical with an apical opening (pycnidia) or saucer-shaped (acervuli). Many plant diseases are caused by fungi belonging to this group. They generate diversity by reproducing sexually and then propagate rapidly asexually, so inoculum increases rapidly, causing an epidemic within a single growing season.

Tuber melanosporum, the French black truffle, is considered the finest of the edible fungi and is certainly the most expensive. It occurs naturally as an ectomycorrhizal associate of forest trees in France. In the past 50 years there has been a significant reduction in production because of clearing and decline of forests containing hazelnut and oak. Efforts to domesticate the farming of the truffle have been successful, and the black truffle is now produced in Australia and New Zealand from artificially inoculated trees. The truffles are harvested using trained pigs and dogs, which locate them by scent (Figure B.10.6.1f, g).

Basidiomycetes

This group, which also features septate hyphae, is characterised by reproductive basidiospores borne upon a club-like structure called a basidium. Species in this group include all the large fungi with obvious fruiting structures such as mushrooms, polypores, crusts, corals, clubs, basidiolichens and jellies. Smaller members of the group include rusts and smuts, which are parasites of insects or plants. The sturdy fruiting bodies of the characteristic shelf or bracket fungi resemble protruding shelves in layered horizontal rows (Plate 10.8b).

The lichens

In lichens, the fungus and its partner, a cyanobacterium or a green alga, are so closely entwined that they are regarded as a single species. The photosynthetic partner provides carbon compounds, while the fungus provides a suitable substrate and helps in absorbing water and mineral nutrients.

Lichens can survive with little or no soil and so are important colonisers of rock faces and other exposed sites. Some tolerate extreme cold and provide vital animal food in northern climates, whereas others are drought-tolerant. However, they are very susceptible to air pollution, so deaths of lichens indicate deteriorating air quality.

Chapter summary

1. Protists are eukaryotic organisms and mainly unicellular. In some classifications they include all the organisms not classified in the fungi, plant or animal kingdoms, while other classifications regard the fungi as a group within the protists.
2. Six supergroups of protists are recognised on the basis of morphology, ultrastructure and molecular genetics. These groups are not yet arranged into a taxonomic hierarchy within the kingdom. The charophytes share a common lineage with land plants, and the choanomonads with animals.
3. Many groups of protists have plastids derived form primary endosymbiotic relationships with cyanobacteria, and secondary endosymbiosis with various single-celled eukaryotic organisms with red or green plastids.
4. Supergroup Amoebozoa includes heterotrophic, single-celled organisms that move by pseudopodia and capture food by phagocytosis.
5. Supergroup Rhizaria includes heterotrophic protists with amoeboid bodies with fine pseudopodia strengthened internally with tubules.
6. Supergroup Archaeplastida includes protists with photosynthetic plastids. Important groups are the mainly marine Rhodophyceae (red algae) and the marine and freshwater Chloroplastida (green algae). Charophyta are similar to green algae but have features of morphology and ultrastructure close to plants.
7. At some stage in their lives all members of supergroup Alveolata have motile cells with two distinctive flagellae. The diatoms and brown algae are photosynthetic, whereas the water moulds, ciliates and apicomplexans are heterotrophic and include important pathogens.
8. Supergroup Excavata all have a feeding groove or gullet. The euglenids are unicellular and can switch between autotrophy and heterotrophy. Trypanosomes are similar in shape and size to euglenids, but are parasitic in the blood of mammals and other hosts.
9. Supergroup Opisthokonta includes a range of heterotrophic protists. The Choanomonads are considered to have an ancestor common to the line that gave rise to animals.
10. Fungi are classified here in supergroup Opisthokonta within the kingdom Protista, but are sometimes treated as a separate kingdom. They are heterotrophic, obtaining nutrition from decomposition of organic matter. They are grouped depending on features of their reproduction and nutrition.

Key terms

alternation of generations
amoeboid
autogamy
basidium
blade
coenocytic
endosymbiosis
fission
fruiting body
fucoxanthin
gametophyte
holdfast
hyphae
macronucleus
micronucleus
mixotrophic
mycelium
mycorrhiza
paramylon
parenchyma
pathogen
phagocytosis
phytoplankton
plastid

pseudopodia
spore
sporophyte
stipe
thallus
zooxanthellae

Test your knowledge

1. Which supergroups of protists are (a) exclusively heterotrophic, (b) exclusively photoautotrophic and (c) contain both autotrophic and heterotrophic examples? Name an example of a protist using each mode of nutrition.
2. Which supergroups of protists include parasitic species? Describe for each supergroup an example of a parasitic life cycle and name a parasitic species.
3. What is meant by endosymbiosis, and what is its role in the evolution of the protists?
4. Fungi can be divided into three broad nutritional groups. Name these, give an example of each and explain the importance of each to the functioning of ecosystems.

Thinking scientifically

Debate the topic 'Some species of Australian eucalypts are grown in plantations overseas as sources of fuel and timber. However, not all plantations grow well. Design an experiment to test the hypothesis that the growth of eucalypt trees outside Australia could be enhanced by inoculating trees with mycorrhizae'.

Debate with friends

Debate the topic: 'To reduce the risk of spread of infection, public access should not be allowed in areas of National Parks infested with *Phytophthora cinnamomi*'.

Further reading

Hausmann, K., Hülsmann, N. and Radek, R. (2004). *Protistology*. Schweizerbartsche, Verlagsbuchhandlung, Stuttgart, Germany. 3rd edn.

Money, N.P. (2007). *The triumph of the fungi: a rotten history.* Oxford University Press, Oxford.

Tudge, C. (2000). *The variety of life: a survey and a celebration of all the creatures that have ever lived.* Oxford University Press, Oxford.

11

Plant diversity I – the greening of the land

Barbara Bowen, Mike van Keulen and Jen McComb

The plant dinosaur

In 1994 David Noble, a New South Wales National Parks and Wildlife officer, abseiled into a gorge in the Great Dividing Range west of Sydney. He found himself in a small stand of distinctive trees, with bark like bubbling chocolate, strappy, leathery leaves and crowns emerging above the rainforest. Experts examined the specimens and recognised the plant as belonging to the family Araucariaceae, which includes the hoop, kauri, Norfolk Island and the bunya pines. Excitingly, it was close to fossil specimens of pollen and leaves dating from the Cretaceous (Plate 11.1a). The species has thus existed for possibly 150 million years, once occurring over much of Gondwana. Its discovery was equivalent to finding 'a small living dinosaur'.

The species was named *Wollemia nobilis*, a nice tribute to its discoverer, and is commonly known as the Wollemi pine, from the national park where it occurs. There are about 100 trees in two groves, and molecular techniques detected no genetic diversity between individuals although seeds are produced and germinate in nature.

The future of the remnant population is precarious because fire, pest or disease introduced into its small canyon could wipe out the species with its limited genetic diversity. The best measure to protect it is keeping the exact location secret. Unauthorised visitors may already have introduced the root-rot pathogen *Phytophthora cinnamomi* to the population. The species has been multiplied through collection of seed (Plate 11.1b), cuttings and tissue culture and it was released commercially in 2005. This is a species that was close to extinction before even being discovered! That such a large, distinctive new species could be found 150 km from the biggest Australian city is astonishing, but indicative of the discoveries still to be made in the diverse Australian land flora.

Chapter aims

In this chapter we introduce you to the characteristics shared by all land plants and the phyla that include mosses and liverworts, ferns, and gymnosperms (cycads and conifers). We describe how the morphology and life cycle of each group determines its interactions with the environment, as well as the diversity of each group in Australia and its conservation status.

Plants on the land

Plants are multicellular eukaryotes that evolved from the green algae. They produce organic molecules by photosynthesis. Their evolution was linked strongly to obtaining water, carbon dioxide and mineral nutrients on land, supporting their bodies without the buoyancy of water, reducing evaporative water loss, and reproducing in air rather than water.

Transition from ocean to land habitats

The ocean and land habitats are very different, yet the essentials for protists (algae) and plants are the same so far as light, water, mineral nutrients and gases are concerned. The difference is that the aquatic ancestors of land plants obtained all these requirements (excepting sunlight) from the surrounding water, whereas land plants draw their requirements from soil and air.

All green land plants trap solar energy in chloroplasts. Sunshine is unfiltered by passage through water so they photosynthesise efficiently and have abundant sugars available for synthesising thick cell walls. Therefore land plants can grow much larger than plants or algae in the ocean. However, land plants lack the support of surrounding water to keep them upright and maximise exposure to light for photosynthesis. Simple air bladders or oil droplets are enough to support plants in water, but on land plants of any size require internal strengthening, often provided by the specialised vascular tissue used to transport water. Shoots must be strong enough to overcome gravity, and the plant must be firmly embedded in the ground to resist wind. The embedding structures are called roots if they contain vascular tissue and rhizoids if they do not.

As in marine algae, water makes up 90% of the living cells of land plants. Mosses and liverworts absorb water all over their surface, but in the larger plants only the roots absorb water. Thus most land plants obtain water from the soil, but the availability and quality of soil water vary considerably. The water may be fresh or brackish, and rain may come in torrential tropical downpours or rare desert showers. Further, for water to be transported to the tops of tall plants it must be partly dragged up by forces generated by evaporation from the stems and leaves, so that the roots must constantly access water from the soil. Roots, stems and leaves must also be linked for water transport, and most land plants have specialised vascular tissue to do this.

Land plants have their shoots surrounded by air, which has 21% oxygen and 0.03% CO_2. In the soil small air pockets supply these gases to the roots or rhizoids. The concentration of CO_2 is much the same in air as in water, but water contains only

0.636 oxygen. Furthermore, gases diffuse more slowly in water than in air. Thus land plants are adapted to higher oxygen levels, and most die in flooded soils because their roots cannot function in low oxygen concentrations.

Finally, the ocean provides a fairly uniform habitat. Admittedly, phytoplankton may drift into areas of higher or lower nutrients or temperature, and benthic algae may be in deep, still water or battered by waves. However, these differences are trivial when compared with the diversity of habitats on land. Plants grow from the icy tundra to tropical forests or sandy deserts. Some land plants have even re-invaded aquatic habitats. The more varied the habitats, the more varied the types of plants. Thus there is a uniform planktonic flora, a more diverse benthic algal flora, and a very rich and diverse land flora.

Features common to all land plants

Land plants include the bryophytes (mosses, hornworts and liverworts), ferns and their relatives, cone-bearing plants and the familiar flowering plants. They all share these features, many of which are adaptations to life on land:

- growth arising from specialised regions called apical meristems at the tips of roots and shoots
- alternation of sporophyte and gametophyte generations. Sporophytes and gametophytes may be separate distinct individuals of various sizes, or one generation may be greatly reduced and live within the other. Their requirements may be very different.
- an embryo arising from the union of male and female gametes that is protected and nourished by the female gametophyte (and the endosperm in flowering plants – Chapter 12)
- spores produced by the sporophytes and dispersed by air, not water
- multicellular sex organs with cellular walls that do not develop into gametes
- photosynthetic pigments (chlorophylls a and b) and accessory photosynthetic pigments (carotenoids). Thylakoids in the chloroplasts are arranged in stacks (grana). Energy reserves are in the form of starch.
- distinctive, specialised holes in the epidermis (outer layer) called stomates, which control gaseous exchange between the air and the plant. The epidermis is commonly covered with a cuticle to reduce water loss. (Cuticle and stomates are absent only in some of the bryophytes.)
- in meiosis and mitosis, the breakdown of nuclear membrane and nucleoli in prophase, and centrioles only in cells that develop into motile sperm.
- cell walls of cellulose.

Trends in the evolution of land plants

Which group of algae gave rise to the land plants? The list of distinctive features of land plants supports the Chloroplastida, which share with land plants features of cell division, cell-wall composition, chlorophyll and plastid structure, and storage of starch. Although some authorities claim a marine green algal ancestry, compelling molecular evidence points to the charophytes, a green algal group found today in fresh or brackish water, as likely progenitors.

The primitive bryophytes are still almost entirely dependent on moist environments. They lack external cuticles and a vascular system for internal water transport or support and, like the green algae, their flagellated sperm need a film of water to swim to the egg. Ferns and their allies evolved vascular tissue for transport and support and well-developed cuticles to reduce water loss. However, they retained the primitive flagellated sperm, restricting them to moist habitats. The more advanced gymnosperms (cone-bearing plants) and angiosperms (flowering plants) do not require water for reproduction. The sperm cells contained in the pollen are carried to the egg via wind or animals and their embryos are protected by seed coats. Along with protective cuticles and well-developed vascular tissues for support and for water and food transport, these adaptations enabled seed plants to colonise and dominate almost every corner of our planet.

The non-vascular bryophytes and the vascular plants (ferns, conifers and flowering plants) possibly arose independently from the algae. The first fossil plant acknowledged to be a land plant had short, leafless stems, which divided dichotomously and had globular sporangia at their tips. It is named *Cooksonia* after Isobel Cookson, a Melbourne University palaeobotanist who devoted her life to examining fossil remains that she extracted

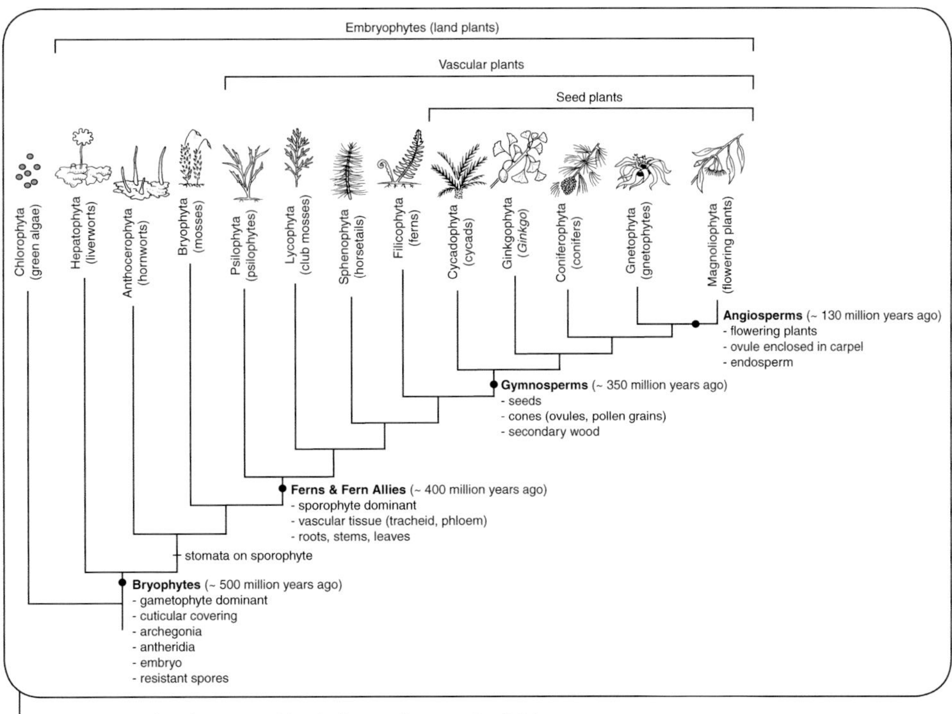

Figure 11.1 The diversity of land plants. (Source: B. Cale)

from rocks using various potentially explosive solutions, much to the alarm of her colleagues. *Cooksonia* occurred 450 million years ago in the Silurian and was followed by further vascular plants in the Devonian. These were probably confined to swamps as their dichotomously branching stems, with small, spiny appendages, had rhizoids at their base rather than roots. From these primitive ancestors diverse plants arose that were sufficiently distinctive to be placed in separate phyla of the plant kingdom. Most groups are now totally extinct or have a few species surviving to the present day. The main groupings in the living plants are those of non-vascular and vascular plants and of seedless plants and seed plants. We will now explore the important characteristics of the plants in these groups (Figure 11.1).

Seedless plants (non-vascular)

The bryophytes

'Bryophytes' is a collective name for three phyla of non-vascular plants: the Hepatophyta (liverworts), Anthocerophyta (hornworts), and Bryophyta (mosses) (Figure 11.2). In this text we use 'bryophytes' to collectively name all three phyla of seedless, non-vascular plants and 'mosses' when we mean that phylum specifically.

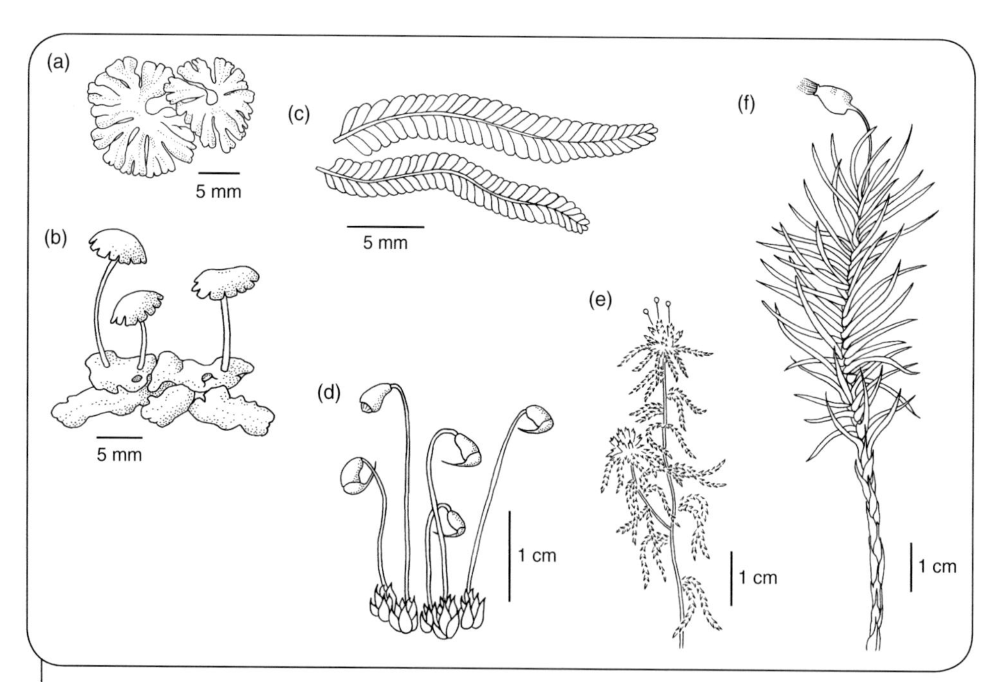

Figure 11.2 A range of Australian bryophytes. (a)–(c) Liverworts. (a) *Hymenophyton* (a thalloid gametophyte). (b) *Marchantia* (thalloid gametophyte with archegoniophores). (c) *Lophocolea* (a leafy liverwort). (d)–(f) Mosses, all gametophytes with attached sporophytes. (d) *Funaria*. (e) *Sphagnum* (note that sporophytes rarely form on the Australian mainland). (f) *Dawsonia* (one of the tallest Australian mosses). (Source: B. Cale)

As with all land plants, bryophytes undergo alternation of generations in their life cycles, but they are unique in that the dominant generation is the gametophyte. We now consider the main characteristics and traits of the liverworts and mosses, which have enabled these early plants to persist for millions of years, albeit now in restricted land habitats. The remaining group, the hornworts, are a very small group, in some ways intermediate between mosses and liverworts.

Phylum Hepatophyta – the liverworts

Liverworts are the most primitive bryophytes. Their earliest fossils are found in Devonian rocks about 400 million years old, but it is very likely that they arose earlier than this. There are about 6500 species, found throughout the world from the Arctic to the tropics. Most are well adapted to moist habitats, although some are found in drier areas and there are some aquatic examples. The conspicuous liverwort plants are the gametophyte stage of the life cycle and are generally flat and prostrate (Plate 11.2a). Some have simple branching lobed structures (leafy liverworts), but the most recognised form consists of a flattened 'liver-shaped' thallus from which its name is derived. The upper surface comprises photosynthetic cells. As with their green algae ancestors, liverworts lack stomates, but these upper cells are exposed to the air through small pores that facilitate gas exchange. Uptake of water and mineral salts is via osmosis directly through the surface cells of the plant body. Simple, single-celled rhizoids extend from the lower epidermal cells. These anchor the thallus to the substrate and act as a simple wick, absorbing water. There are no specialised internal tissues for transporting food and water, so the liverwort relies on diffusion for nutrient transport within the plant body. The flat, thin thallus is in contact with surface moisture, but it loses water quickly in dry conditions because it lacks a cuticle and has only rhizoids.

The life cycle of a liverwort starts with a haploid spore that germinates to produce the green thallus or gametophyte (Figure 11.3). This produces special umbrella-like structures bearing the male and female gametes. Some liverworts produce these on the same gametophyte (they are monoecious) whereas others have separate male and female gametophytes (they are dioecious). Archegoniophores bear the female gametes and eggs are produced in archegonia (singular: archegonium) on their lower surface. The male structure is the antheridiophore and sperm are produced on the upper surface in antheridia (singular: antheridium). Liverwort sperm have two flagella and swim to the egg, attracted by chemicals released from the archegonium. Without water, fertilisation cannot occur. Raindrops on the upper surface of the antheridiophore can also splash sperm onto the archegonium. Fertilisation of the egg results in a diploid zygote, which undergoes cell division (mitosis) to form the small sporophyte. This is not photosynthetic and it remains attached to the female gametophyte, on which it is entirely dependent for food and water. In the sporophyte capsule 'spore mother cells' undergo meiosis to produce haploid spores (n). These are released into the air and, if they land in a suitable place, germinate to form a new gametophyte.

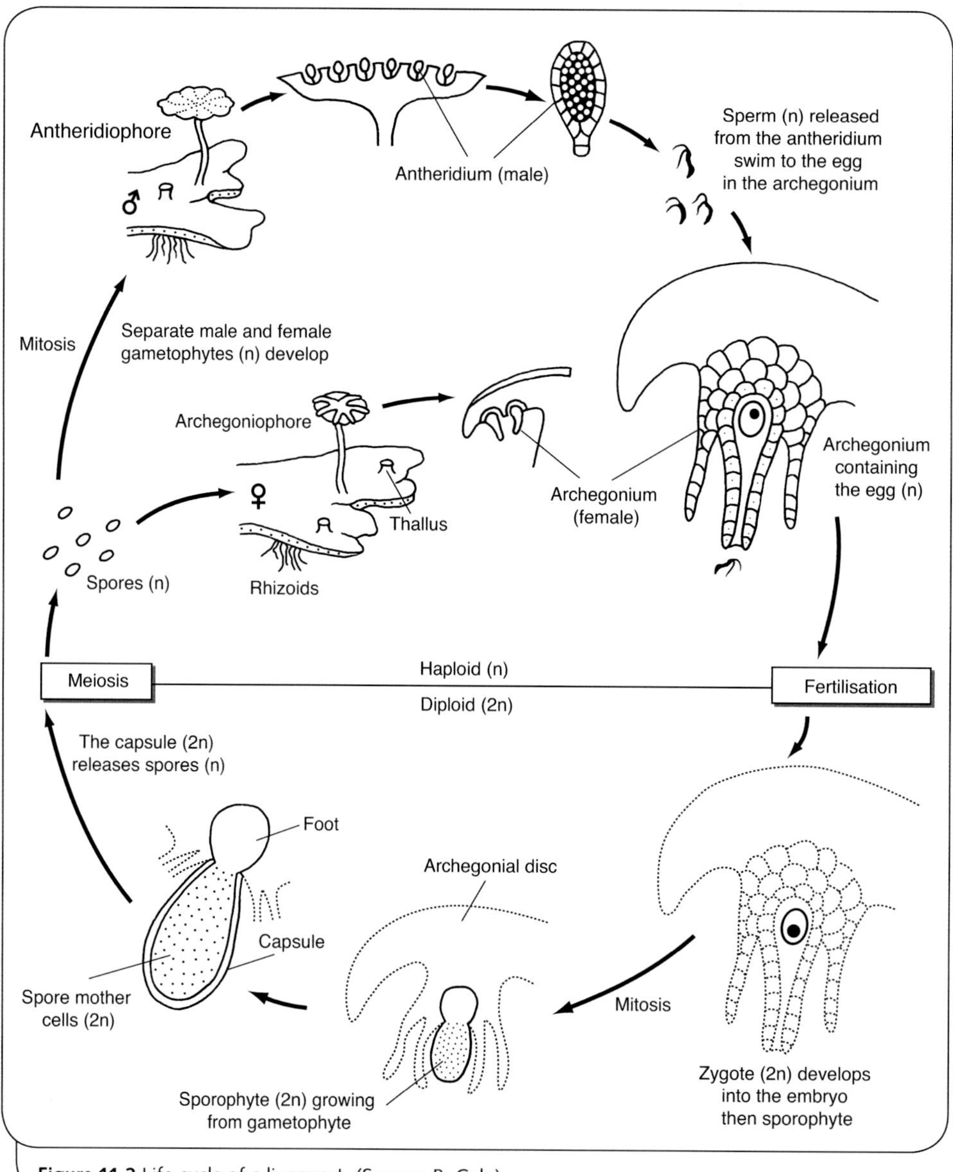

Figure 11.3 Life cycle of a liverwort. (Source: B. Cale)

Liverworts also reproduce asexually either by separation of parts of the thallus or from structures on the thallus called gemma cups. In these, numerous gemmae (small buds of liverwort) are formed. They are released when the cup fills with water and grow into new plants.

Phylum Bryophyta – the mosses

The Bryophyta or mosses comprises about 10 000 species, which have been extremely successful in growing in most parts of the world alongside the more advanced vascular plants (Plate 11.2b). Moss morphology ranges from small compact clumps to large plants almost fern-like in appearance. The most conspicuous plant body in mosses is the leafy gametophyte stage. The leaves generally comprise a single layer of photosynthetic cells except in the centre, where most have a thickened region that helps supports the leaf. These leaves thus are different in structure to the true leaves of vascular plants (Chapter 12). Moss plants are anchored to the substrate by branched, multicellular rhizoids.

A significant advance in mosses is the gametophyte stem, which has specialised elongated cells for transporting water and sugars. Although these cells have a similar function and resemble the vascular tissue of the ferns, they lack the specialised walls that are thickened with lignin (stiffening material) seen in ferns. They are therefore called hydroids (water-conducting cells) and leptoids (food-conducting cells) to indicate their structural differences from the conducting tissues of true vascular plants.

Most mosses have thin cuticles, absorb water and nutrients directly through the epidermis and generally rely on diffusion for internal movement of water and sugars. The turgor pressure of water in their cells supports the small erect plants, so when moss plants dehydrate they shrink and become brown and brittle. They also achieve support by growing in dense colonies and entangling the rhizoids and leafy stems.

The moss life cycle is similar to that of the liverwort (Figure 11.4). Following germination of the haploid spore, the gametophyte starts off as a branched, photosynthetic

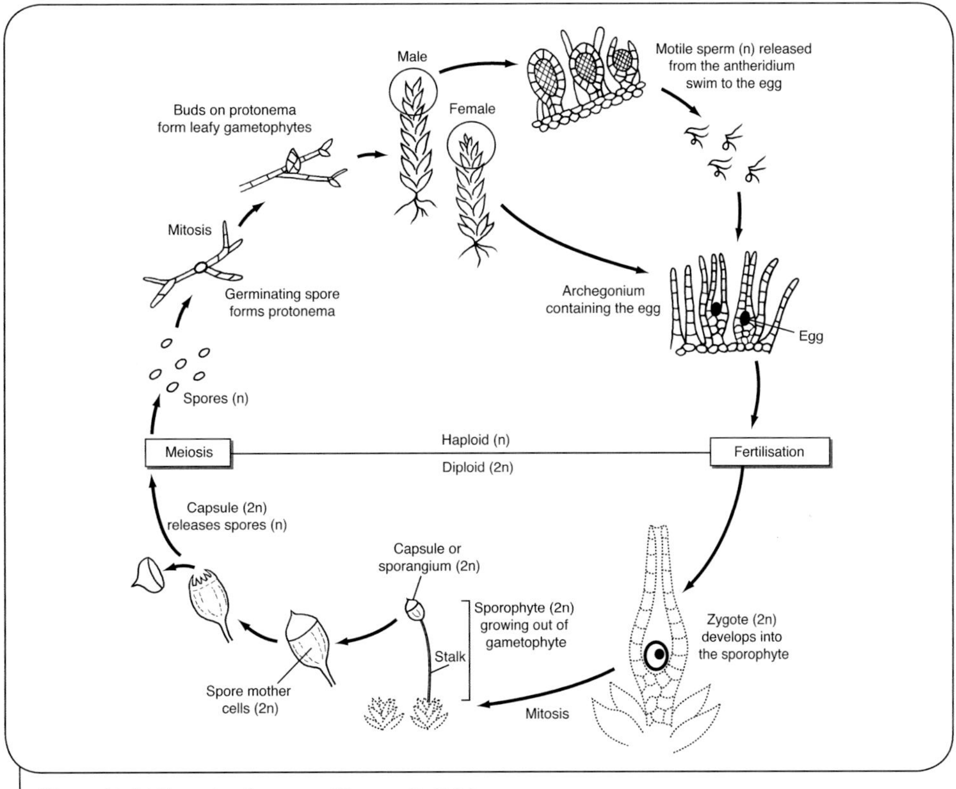

Figure 11.4 Life cycle of a moss. (Source: B. Cale)

filament. This soon develops one or more buds that grow into a sward of closely packed leafy moss gametophytes. As with liverworts, moss gametophytes can be monoecious or dioecious. The antheridia (containing numerous bi-flagellated sperm) and the archegonia (housing a single egg) are normally located in terminal clusters at the end of the shoots. The antheridia shed sperm that swim or are dispersed via raindrops. Attracted by secretions from the archegonium, the sperm swim to the egg where fertilisation takes place and the zygote is formed. The sporophyte develops as a stalked capsule with its base embedded in the gametophyte on which it remains dependent. It may take as long as 18 months for the capsule to mature. Capsules have a lid or operculum covered by a protective cap of cells. When the capsule is mature these drop off and as many as 50 million haploid spores are released.

Habitat of the liverworts and mosses

Liverworts and mosses occur worldwide, and with knowledge of their structure and life cycles it is easy to predict their habitat. Absorption of water and minerals from the surface of the plant body, no specialised conducting and support tissue, and the requirement of free water for moving sperm explain why liverworts and mosses generally grow in moist environments and remain small. In Australia and New Zealand they are common in the tropical and temperate rainforests, where they grow on wet rocks and logs, in gullies, and around creeks and soaks. Despite being mostly very small, some are conspicuous and much of the distinctiveness of particular habitats depends on them. G.M. Scott wrote: 'Without them the rainforests would be merely wet, not lush, bogs, if they existed at all, would be merely quagmires, the mallee would be a sandy desert, and mountain rocks would be barren. Much of the attractiveness of these habitats would be gone'.

However, some liverworts and mosses grow in areas where moisture is not always available. If the gametophyte dries out it does not die, but remains shrivelled and dormant. When water is available the dried gametophyte absorbs moisture and becomes green and photosynthetic again. Plants with this capacity are known colloquially as 'resurrection plants', and there is interest in transferring, by genetic engineering, some of the attributes of these plants to other species such as pasture grasses.

Liverworts and mosses also occur in the Arctic and Antarctic. Antarctica is a polar desert and plants grow only where there is free water, usually water that melts in summer from snow or glaciers, or where steam and gases escaping from volcanoes warm the ground. They must survive repeated freezing and thawing, and they depend on a covering of snow to prevent desiccation by strong winds and to insulate against extreme temperatures. They survive as tightly packed shoots, frequently with orange pigment that is thought to prevent photosynthetic damage. Reproduction is asexual from deciduous shoots and gemmae. Growth is extremely slow (0.3–1.0 mm/year) and clumps may be hundreds of years old.

The environmental value of liverworts and mosses may be overlooked because of their small size. However, they are important pioneer plants, establishing on bare rock surfaces or disturbed sites. They stabilise soil by building up humus and capturing soil particles, thereby creating soil conditions suitable for more advanced plants. They create microhabitats for other organisms, providing bacteria, algae and small animals with food, protection and a humid microclimate. Mosses are also useful bioindicators of air quality and pollution. They absorb pollutants as well as moisture and other particles through their leaves, so accumulation of pollutants such as heavy metals can be measured in their cells. Different species have different levels of sensitivity and changes in moss communities can be one of the first indications of an environmental problem.

Liverworts and mosses – conservation status

The number of liverworts and mosses in Australia is uncertain as they have not been studied in as much detail as the higher plants, but about 10% of the 400–500 liverwort and 600–700 moss species are listed as endangered or rare. This is probably an underestimate because few people collect them and they are easy to miss in surveys. For most liverworts and mosses it is almost impossible to count the number of individual plants in a population, let alone determine how many genotypes are present, so that concepts of 'rarity' are based more on the number of locations in which swards of particular species are found, rather than the actual number of plants.

Certainly some are very restricted. *Sphagnum leucobryoides* grows with shoots mostly buried in wet quartzite sand and is known only from one location in south-western Tasmania – the exact position of which is secret. Another moss, *Aecidium filiform,* is found only on the top of a single rock stack in a Queensland National Park. At present Australia has a limited commercial harvest of *Sphagnum* for horticultural potting mixes and for wrapping bare-rooted roses and fruit trees for transport. There are very few places where this harvest is sustainable, because Australia lacks large areas of *Sphagnum* bogs. *Sphagnum cristatum*, the main species harvested, occurs in high-rainfall areas of south-eastern Australia at altitudes of 300–1500 m. Its maximum growth is 5 cm a year at 900 m and less at higher altitudes.

The major factors endangering liverworts and mosses are habitat destruction, fire, drought, flooding and salinisation. Conservation is important not only because of their aesthetic value, but also because these ancient plants may have valuable compounds such as antibiotics not found in more highly evolved plants.

Seedless plants (vascular)

The ferns and fern allies

Early in the history of the invasion of the land, plants developed internal transport tissues for water and mineral nutrients (xylem) and sugars and other metabolites (phloem). The water-conducting cells were thickened and strengthened with lignin, enabling tall, branching growth. Thus, the problem of water and food transport throughout the body was solved and plants could invade drier habitats. Another major change in the seedless vascular plants is that the sporophyte now dominates the life cycle. In ferns and fern allies the gametophyte is always smaller, sometimes microscopic, and independent of the sporophyte.

Vascular plants first appeared in the fossil record about 410 million years ago. Rhyniophyta is the earliest known phylum of these plants, represented by several genera including the ancestral *Cooksonia* and *Rhynia*. Fossils of *Rhynia* found in Scotland are so well preserved that it is possible to recognise different cellular structures and tissues. They lacked leaves and true roots and generally consisted of branched or unbranched photosynthetic stems (usually less than 50 cm tall) and terminated with spore capsules. Aerial stems arose from horizontal stems (known as rhizomes) connected to the soil by rhizoids.

There was a massive radiation of seedless vascular plants during the Carboniferous period (354–290 million years ago) and many reached tree size, forming forests so vigorous they left vast amounts of fossilised fuel in the form of coal. In fact, the term 'Carboniferous' refers to the rich deposits of coal (carbon fuel) occurring throughout northern Europe, Asia and North America from this time. Australia also has rich coal deposits dating from the late Carboniferous and earliest Permian times (300–250 million years ago). This period followed an ice age and coincided with a warming climate and rapid evolution of a diverse southern flora. Australian coal deposits are the product of cold swampy bogs in which seedless vascular plants grew in immense profusion. They are exploited today and are an important part of the Australian economy.

Today, there are four phyla of living seedless vascular plants. From simplest to most complex they are Psilophyta (whisk ferns), Lycophyta (club mosses), Sphenophyta (horsetails) and Filicophyta (ferns). The first three phyla are commonly grouped as fern allies and we do not consider them here because they are rare in the present-day flora. The main changes from simple to complex are the development of leaves and roots. About 90% of all living seedless vascular plants belong to the Filicophyta. Members of this group show great diversity in both form and habit and form a major component of many ecosystems around the world.

Phylum Filicophyta (Pteridophyta) – the ferns

There are about 11 000 species in the phylum Filicophyta (called Pteridophyta in many texts) worldwide. The prominent sporophyte ranges in size from small epiphytic ferns that are a single cell thick and look similar to liverworts, to tall tree ferns growing to 10 m (Figure 11.5, Plate 11.2c).

The basic morphology comprises a rhizome with leaves and roots arising at the nodes. They have true leaves or megaphylls with vascular tissue running through the midrib and branching out into the photosynthetic tissue. These are called fronds and are generally large in relation to the total plant size. In nearly all ferns the young fronds are tightly coiled, protecting the delicate young leaf during early development. This is called circinate vernation and, as the leaf matures, it uncoils and spreads out its large photosynthetic surface.

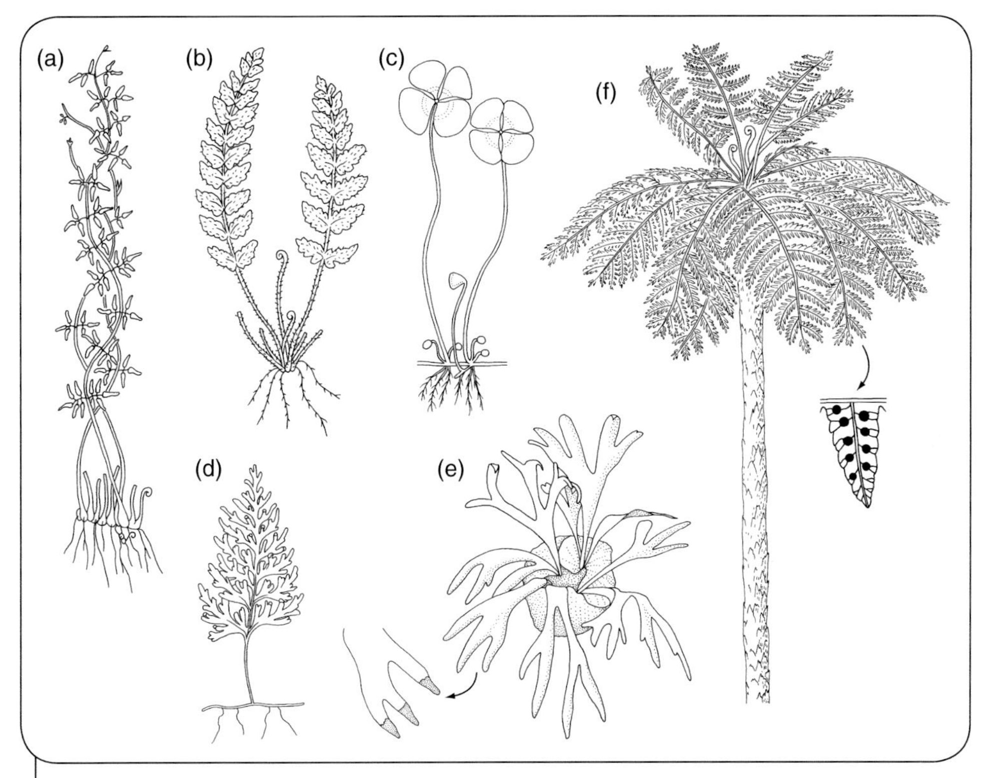

Figure 11.5 A range of Australian ferns. (a) *Lygodium*, a climber. (b) *Cheilanthes lasiophylla*, a drought-resistant 'resurrection' species. (c) *Marsilea* (nardoo) grows in shallow water or wetland margins. (d) *Hymenophyllum* (filmy fern) from cool temperate rain forests. (e) *Platycerium* (elkhorn), an epiphyte. (f) *Cyathea*, a tree fern (the inset shows the enlarged underside of a frond, showing sori). (Source: B. Cale)

Leaves give an advantage to ferns over non-vascular plants by increasing their photosynthetic capacity. A waxy, waterproof outer cuticle overcomes problems of excessive water loss from the leaf surface. Well-developed stomatal pores on the leaves use specialised guard cells to control the opening under different conditions, an important adaptation to life on land. The tissue below the epidermis consists mainly of photosynthetic cells and branching vascular bundles. The vascular bundle is the most complex tissue of the frond, consisting of xylem and phloem. The sieve cells of the phloem (Chapter 12) transport the primary product of photosynthesis, glucose, from the frond to the rest of the plant. They are elongate, and the sugars flow through perforations in their end walls. The xylem of ferns consists of elongated, hollow cells that are dead at maturity (tracheids, Plate 11.4e), interconnected by pores or pits. Tracheids transport minerals and water from the roots to the photosynthetic tissue. They were the first type of water-conducting cell to evolve in vascular plants and are generally the only type of water-conducting cells found in non-flowering vascular plants.

The principal functions of the underground stem or rhizome are anchorage, storage and transport. Also, the outer epidermis of the rhizome has a thick waxy cuticle to prevent water loss. The packing cells (parenchyma) of the rhizome store starch and water. The vascular bundles of the rhizome are similar to the fronds, but are also surrounded by a single modified layer of cells called the endodermis. This has a waterproof fatty band, the Casparian strip, around the cells. Water and mineral salts cannot simply diffuse into the vascular bundle through the permeable cell walls, but must pass through the permeable cell membrane of the living endodermal cells, which can selectively uptake different ions.

The roots absorb water and minerals from the soil. They are long and thin and emerge along the entire length of the rhizome. The outer epidermal cells near the root tips have root hairs that absorb water and dissolved nutrients. Each root has only a single central vascular bundle with surrounding endodermis, linking the roots directly with the vascular tissue of the rhizome and the rest of the fern body.

The major difference between the life cycle of ferns and that of bryophytes is that the sporophyte generation is dominant (Figure 11.6). In addition, fern sporophytes lack capsules for spore production. Instead, cells in sporangia on the underside of fronds undergo meiosis to form spores. Many sporangia are grouped together in a sorus (plural: sori), seen as brown patches along the edge or on the underside of fertile fronds

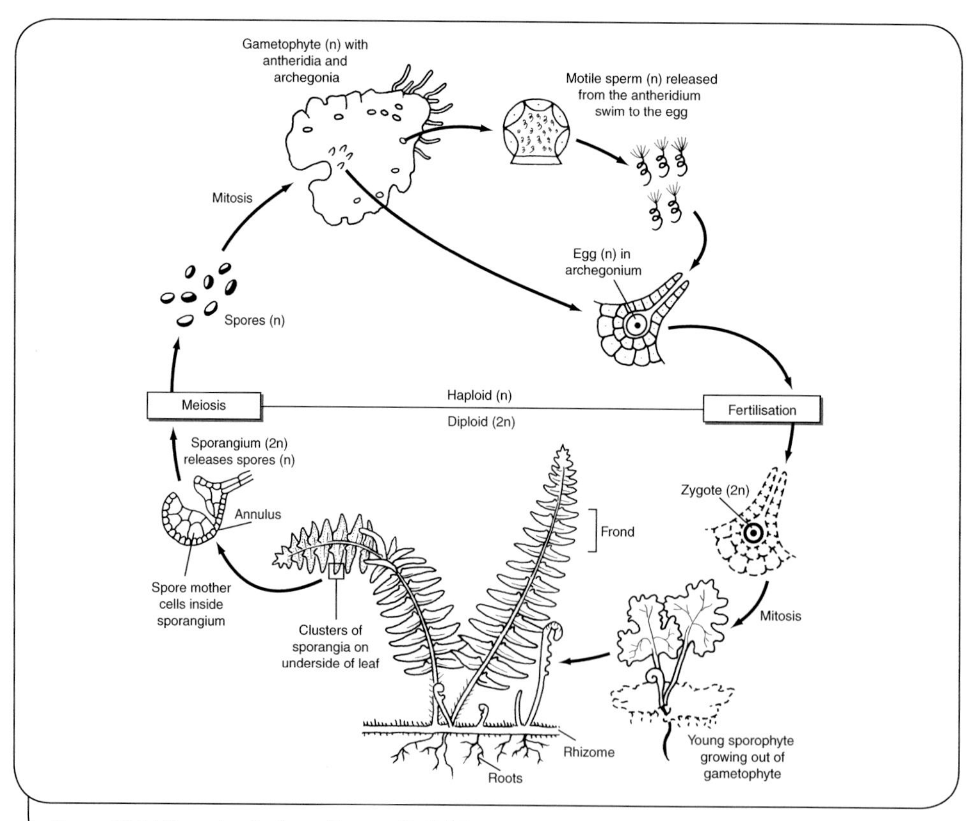

Figure 11.6 Life cycle of a fern. (Source: B. Cale)

(Figure 11.5f). Each sorus may be protected by a thin membrane or enfolded in the edge of the leaf. When the haploid spores mature, the sporangium is ripped open by the drying and distortion of the annulus, a ring of specially thickened cells. The elasticity of the walls of the annulus restores the original shape, flinging out the spores.

Spores landing on a suitable substrate germinate and grow into a small, thalloid gametophyte anchored by simple rhizoids. In ferns the gametophyte is usually bisexual, producing both antheridia and archegonia on the underside. The antheridia sit on the surface near the edge of the gametophyte; the archegonia are sunken into tissue near the centre with only the neck slightly protruding.

The antheridia and archegonia usually mature at different times so there is cross-fertilisation between different gametophytes. Motile sperm produced in the antheridia swim through a layer of water to the neck of the archegonia, attracted to chemicals released by neck canal cells. They ultimately reach and fertilise the egg. Following fertilisation, the zygote divides to form the young, leafy sporophyte that soon becomes independent of the gametophyte. Ferns can also reproduce asexually, either from growth and fragmentation of the rhizome or from buds on the tips of the leaves that develop into a new plant when they break off and fall to the ground.

Habitat of ferns

With their water-conducting tissue, strengthened stems, stomates, well-developed cuticle and true roots, ferns grow into much larger plants than the bryophytes. Tree ferns have a stout upright stem or trunk supported by vascular tissue and surrounding leaf bases. Despite this, most ferns are small and confined to moist places because they still depend on water for the motile sperm to reach the egg and effect fertilisation, and the sporophyte must grow where it forms on the gametophyte. Some ferns such as bracken (*Pteridium* sp.) growing throughout Australia and New Zealand do well in dry environments by having very efficient asexual reproduction; others may be deciduous and dormant in the dry season or confined to water-collecting crevices in rocks in arid areas.

Filicophyta – conservation status

Ferns are common in the Australian fossil record, but at present there are only 456 species in 112 genera. Only five genera are endemic to Australia, meaning that they occur nowhere else in the world. Ferns occur widely from tropical rainforests to cracks in rocks in the inland desert and alpine regions, but are most common in moist tropical environments in north-east Queensland where the species have evolutionary links to those in south-east Asia and the Pacific Islands. There is also a distinct southern group in south-east Australia with evolutionary links to species in New Zealand and the Antarctic Islands. Some 21 species are considered endangered or vulnerable, mostly from north-east Queensland.

Tree ferns are highly desirable for horticulture and are used for fencing and buildings in New Zealand. They dominate the understorey vegetation in many wet forests and support many epiphytes (plants living on the fern, but not drawing nourishment from it). Trunks of the tree fern *Dicksonia antarctica* in eastern Tasmania may support up to 97 species of mosses and other ferns as epiphytes. However, some species grow slowly and large specimens can be 500 (possibly up to 1000) years old.

In 1975, the tree fern families Cyatheaceae and Dicksoniaceae were listed in CITES (the Convention on International Trade in Endangered Species). This obliges Australia to report any trade involving tree ferns or tree-fern products, even though the species may not necessarily be endangered in Australia. Harvesting of *D. antarctica* has removed it from many parts of its former range in South Australia. It is now protected on the mainland, but in Tasmania tree ferns are harvested commercially, mainly salvaged from land to be cleared for roads, agriculture or logging. There is a system of permits and tagging for plants harvested legally, to discourage poaching of these ecologically and aesthetically important plants.

Seed plants (vascular, non-flowering)

The evolution of seeds was one of the most important events in the evolution of vascular plants, further reducing their dependency on water. A seed forms from an ovule, which is essentially a megasporangium (a sporagium that produces a female gametophyte), wrapped in two outer coats in which there is a small hole, the micropyle. Seeds are the products of fertilisation of ovules, so they are actually embryonic plants within a hard coat protecting the fragile embryo against physical damage, extremes of heat and cold, and desiccation. Inside seeds are tissues that nourish the embryo, assisting germination and establishment of young seedlings. Furthermore, seeds can be dispersed over long distances and remain viable for long periods until conditions are suitable for germination. Concurrent with seed evolution was the evolution of pollen grains to house and transport the male sperm cells.

The gymnosperms

The first gymnosperms evolved about 350 million years ago. The term 'gymnosperm' means 'naked seed' and gymnosperms differ from flowering plants (Chapter 12) in that the seed is not protected by an ovary, but grows on the surface of a modified leaf in a structure called a strobilus or cone. Pollen grains are produced in male cones and are carried by wind or by animals to the female cones, removing the dependency of the male sperm on water.

Gymnosperms also have a well-developed vascular system of xylem and phloem and have true roots, stems and leaves. Their vascular system is more efficient than that of ferns, transporting water to great heights. Most gymnosperms are woody and can grow into trees of great stature, some giants in the plant world.

Modern gymnosperms are classified into four phyla: the Cycadophyta (cycads), the Ginkgophyta (*Ginkgo* sp.), the Coniferophyta (conifers) and the Gnetophyta (gnetophytes).

One of these, the Ginkgophyta, is represented today by only a single species, *Ginkgo biloba* or maidenhair tree, which has remained basically unchanged for over 100 million years. The fourth group, the gnetophytes, shares some characteristics of their vascular system and reproduction with the flowering plants. However, they are considered to have evolved in parallel with the angiosperms, rather than being ancestral to them. They are not discussed further, as there are no Australian examples. Here we will consider the main characteristics of the remaining two gymnosperm groups, the cycads and the conifers.

Phylum Cycadophyta – the cycads

Cycads are an ancient group sometimes known as 'dinosaur plants' because they were prominent in the Jurassic landscape when dinosaurs dominated the animal world. In fact, the Jurassic is often referred to as the 'Age of the Cycads'. Living members share many features with their ancient relatives, but comprise only 11 genera and 140 species.

Cycads resemble palms. They can be quite large, some close to 20 m high, and have compound leaves in a cluster at the top of a stout trunk covered by the bases of old leaves. The central part of the trunk consists of soft, sponge-like pith with surrounding vascular tissue. In addition to normal roots, cycads form specialised coralloid roots inhabited by nitrogen fixing cyanobacteria of the genus *Nostoc* (Plate 17.1). The cycad provides energy and nutrients for the cyanobacteria and in turn receives nitrogen for growth.

The life cycle of the cycads (Figure 11.7) shows the important changes associated with the development of seed. Alternation of generations is still present, but now the gametophyte generation is reduced to a few cells. The male gametophyte is the pollen grain and the female gametophyte is in the ovule. These are produced in large cones at the apex of the plant (Plate 11.3a–d). Pollen cones have many sporangia on each microsporophyll whereas each megasporophyll of ovulate cones bears only two to eight ovules. The female gametophyte develops inside the ovules, each from a haploid cell from meiosis. On the gametophytes, archegonia differentiate. The female gametophyte is packed with starch, providing the food source for the embryo when it forms. In contrast to seedless plants, the female gametophyte remains in the ovule on the cone attached to the parent sporophyte.

In cycads male and female cones are borne on different plants. The minute pollen is transported to the female cone by wind or insects. Some insects are closely dependent on cycads, suggesting a long evolutionary relationship. Once a pollen grain reaches the micropyle of an ovule (pollination) it germinates, producing a tube that grows towards the archegonium containing the egg. Sperm shed from the pollen tube swim the last small distance to fertilise the egg. Pollination occurs when the female cone is soft and has immature ovules. Fertilisation occurs as long as 4 months later when the cone is woody and the ovule has matured. Following fertilisation, the ovule wall develops into the seed coat and the fertilised egg within develops into the embryo. Only one embryo survives in each seed, even though the eggs of several archegonia may be fertilised.

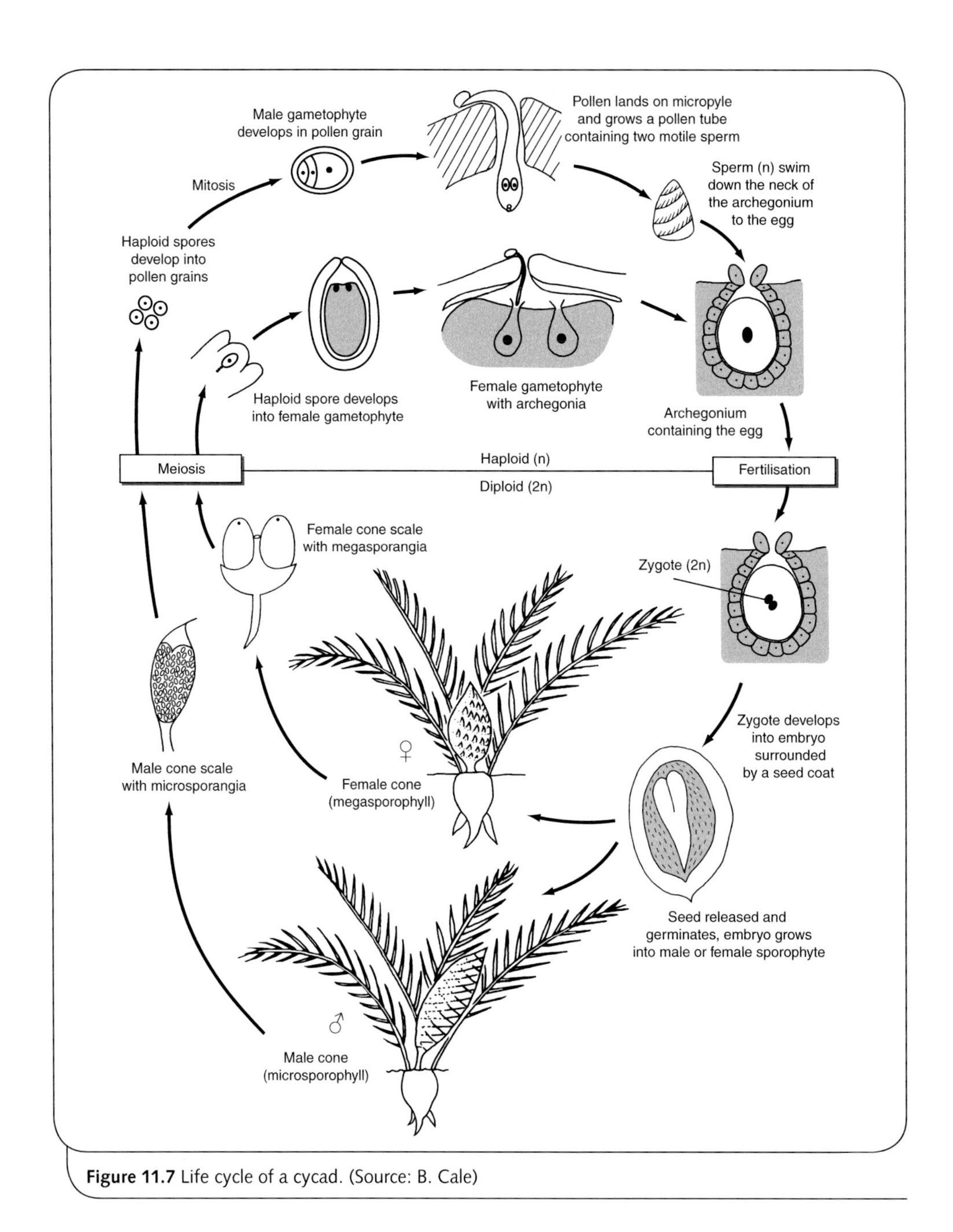

Figure 11.7 Life cycle of a cycad. (Source: B. Cale)

Cycad seeds are large, with a fleshy outer coat overlying a hard protective layer (Plate 11.3d). The seed contains very strong neurotoxins (see Box 11.1), probably as a defence against herbivory, so dispersal is restricted to where the seed falls near the parent. However, these seed plants are the first group freed from the constraint of having the sporophyte grow only where the gametophyte can survive, facilitating the invasion of drier areas.

Box 11.1 Toxic cycads

The zamia 'palms' or cycads were a useful carbohydrate for indigenous Australians and early settlers. Both the fleshy seed coat and the large starchy gametophyte are valuable, particularly as seeds are produced in large numbers after fire and can be stored (Plate 11.3d). An arrowroot substitute is derived from the central pith. Today, 'bush food' shops carry Burrawang flour derived from cycads.

Careful preparation is needed, though, because all parts of the plants contain a deadly toxin, macrozamin or cycasin, to resist insect attack. Aboriginal children were forbidden to touch the attractive red seeds and women processing cycads coated their hands with clay or washed them frequently to avoid poisoning. Various techniques for removing the toxin were developed, including leaching in water, storing and baking. It is doubtful that all the toxin is removed. In South America toxins from cycads were used in magic and medicine, and in Africa the fermented starch made an alcoholic drink.

Seeing Aboriginals eating cycads, many early Australian explorers ate them without proper preparation and were poisoned. Vlaming's men in 1696 ate cycad seeds and 'began to vomit so violently that there was hardly any difference between us and death' and 'crawled all over the earth and made ungovernable movements'. Joseph Banks obviously hadn't read Vlaming's journal, recording that after eating the seeds Cook's men were violently affected 'both upwards and downwards'.

Livestock can be poisoned by eating cycad fronds, which are often the first to sprout after a fire because of the massive food store in the stem and root tuber. Affected livestock suffer from rickets, neurological damage and tumours.

On the island of Guam in the Pacific the Chamorros people used to eat large quantities of *Cycas circinalis* flour, particularly during times of famine as during the Japanese occupation of World War 2. A high number of people suffer from a motor neurone disease called amyotrophic lateral sclerosis. This is caused by long-term ingestion of small quantities of a toxin, not the cycasin, but an amino acid, beta-N-methylamino-L-alanine (BMAA), which causes neurodegeneration in the brain and spinal cord. This amino acid is not found in proteins and its mode of action may be the effective binding of so much copper and zinc that there is permanent damage to nerves. The disease may take 20–40 years manifest, but since widespread concern about eating *Cycas* flour was raised in the 1980s its incidence has declined.

Other examples of the female gametophytes of gymnosperms being used for human food include bunya nuts, pine seeds and ginkgo seeds. They do not have problematic toxins as do the cycads.

Habitat of cycads

Cycads are distributed throughout the world, but are concentrated near the equator in tropical and subtropical habitats. With their reduced dependency on free water, they occupy habitats ranging from rainforests, to savannas and even semidesert scrublands.

In the moist tropical north of Australia, cycads are an integral part of the vegetation. They also occur in suitable areas of central Australia and in the understorey of coastal eucalypt forests.

Cycadophyta: conservation status

There are four genera (*Macrozamia*, *Bowenia*, *Lepidozamia* and *Cycas*) and 68 species of cycads growing in Australia. The first three genera have extensive Australian fossil records dating from the Cenozoic. Climate change on the geological scale, rather than from anthropogenic causes, restricts the distribution of some relict species. For example, *Macrozamia macdonnellii*, which has seeds weighing up to 50 g (the largest of any cycad), grows only on the bare sandstone sides of valleys of ancient river systems in central Australia (Plate 11.3e). Some 13 other species, mostly from Queensland, are endangered or vulnerable.

The attractive form of cycads and the mystique of their ancient lineage draws many admirers and there are active cycad societies cultivating species from Australia and elsewhere, aiding in their conservation. Interestingly, Australian cycads were protected only recently. Before this they were considered noxious weeds because they poisoned livestock.

Phylum Coniferophyta – the conifers

The conifers are the most numerous and widespread of the gymnosperms, with 50 genera and 550 species worldwide (most from the Northern Hemisphere) (see Australian examples in Figure 11.8). Their basic habit is that of a tree with an extensive root system and a trunk with many branches and leaves. Their wood is made up of tracheids and termed 'soft wood', whereas the wood of angiosperms has both tracheids and vessels and is termed 'hard wood'. The leaves have thick cuticles with sunken stomates and may be long and narrow (the typical 'pine needles') or flattened and scale-like as in many Australian conifers. These features reduce evaporative water loss, enabling conifers to grow where water is scarce. The roots consist of a main taproot and branching laterals. As with cycads, conifer roots also form special associations with other organisms. However, instead of forming coralloid roots, many conifer roots are infected with a beneficial fungus forming mycorrhiza (literally 'fungus root'). The fungus enhances uptake of nutrients to the plant and in return receives sugars from photosynthesis (see Chapter 10, Box 10.6).

The life cycle of the conifers is similar to that of the cycads, although there are important differences (Figure 11.9). The cones of conifers are borne along the branches, and male and female cones may be produced on the same tree. The male cones are usually only a few centimetres long and produce many pollen grains. These are carried by wind to the vicinity of the female gametophytes in the ovulate cones. Ovulate cones have a whorl of megasporophylls (scales), each of which bears two ovules. These ovules may exude a liquid from the micropyle that traps floating pollen grains, increasing

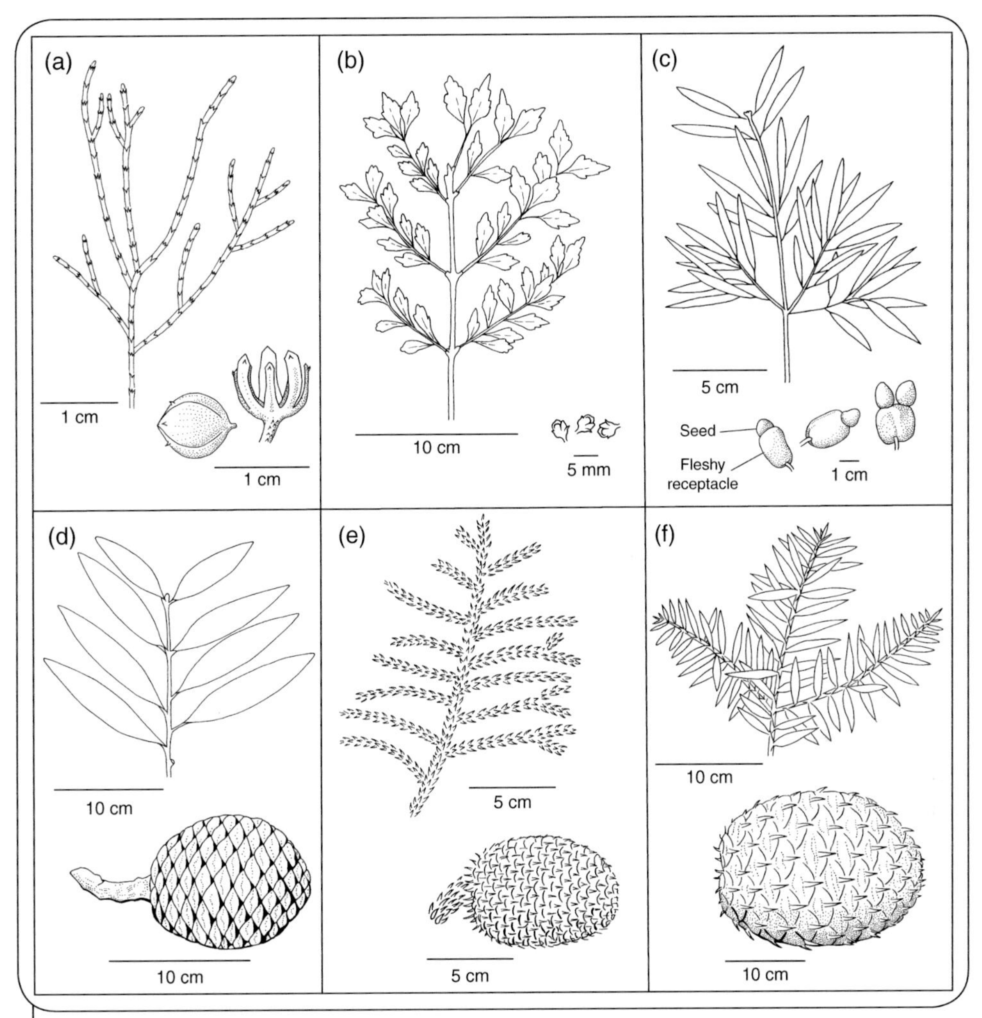

Figure 11.8 Shoots and female cones of a range of Australian conifers. (a) *Callitris glauca* (white cypress pine). (b) *Phyllocladus aspleniifolius* (celery top pine). (c) *Podocarpus elatus* (brown pine). (d) *Agathis robusta* (kauri pine). (e) *Araucaria cunninghamii* (hoop pine). (f) *Araucaria bidwillii* (bunya pine). (Source: B. Cale)

the chance of pollination. As in cycads, pollination occurs when the female cone is soft and immature, and fertilisation may not occur until months later. When a pollen grain contacts the ovule, it produces a tube that carries the sperm cell to the egg in the archegonium embedded in the female gametophyte. However, unlike the cycads, the sperm of conifers are non-motile so the pollen tube must reach the egg cell before it ruptures and releases the sperm cells. The zygote grows into an embryo sporophyte nourished by the female gametophyte, and the seeds remain in the cones until they are mature. Many conifers have winged seeds that float in the wind for dispersal from the parent.

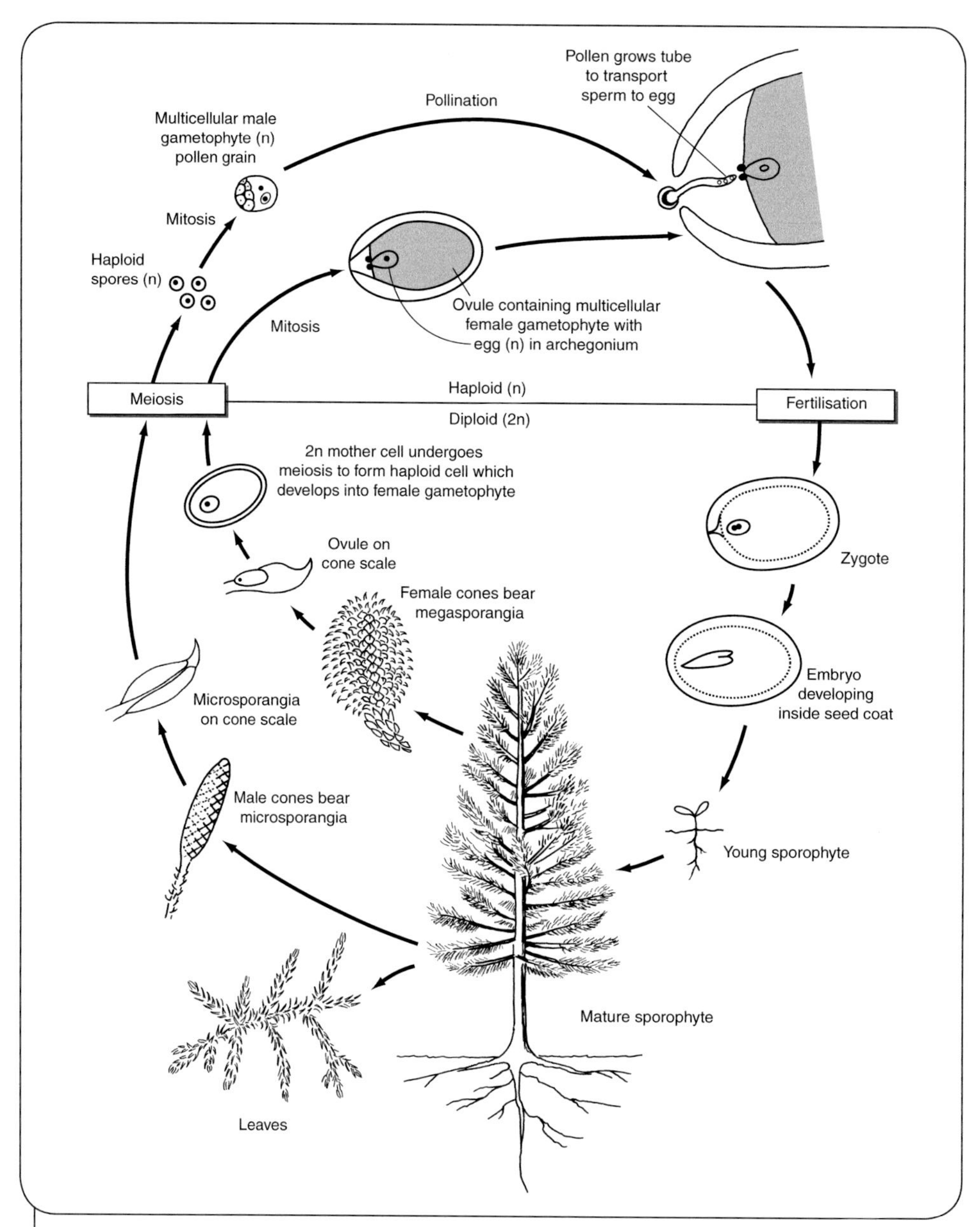

Figure 11.9 Life cycle of a conifer: hoop pine (*Araucaria cunninghamii*). (Source: B. Cale)

Habitat of conifers

Conifers have adaptations for growth in stressful environments and many survive drought, heat, wind and frost. As a result, conifers occur in habitats ranging from the subarctic to the subtropics. They are particularly well adapted to high altitudes and latitudes, where they tolerate prolonged cold weather.

Coniferophyta: conservation status

Three families of Coniferophyta occur in Australia: Araucariaceae (Plate 11.4a), Cupressaceae (Plate 11.4b) and Podocarpaceae (Plate 11.4c, d). The genus *Athrotaxus* that includes the King Billy pine is sometimes placed in the family Cupressaceae, while other authorities recognise a separate family, Taxodiaceae. In terms of the number of species, the Coniferophyta are poorly represented in Australia. The trees occur from rainforests to arid inland areas in mixed stands, usually with emergent crowns taller than the broadleaf species. Three species from New South Wales are endangered or vulnerable, including the monospecific endemic *Wollemia*. There are no true pines native to Australia, though several species are grown for softwood production (Box 11.2).

Box 11.2 Introduced pines in Australia

Many Australian gymnosperms colloquially known as pines (such as Huon pine, Wollemi pine) are not true pines as they do not belong to the family Pinaceae. However, species of many genera of Pinaceae are familiar in the Australian landscape as *Pinus* species are grown in plantations, while *Abies* (fir), *Larix* (larch) and *Cedrus* (cedar) are grown in parks and gardens in the southern states. *Pinus* has a fine-grained, soft wood ideal for sawn timber, ply and paper making. Plantations of pine species, mainly from America and the Mediterranean, began in Australia around 1880 and there are now some 1 million ha of mainly *P. radiata* (Monterey pine) in the south, and *P. elliottii* (slash pine) and *P. carabaea* (carabean pine) in Queensland. These introduced species have ectomycorrhizal fungi on their roots, and it was not until the appropriate fungi were also introduced that plantations thrived. In some areas pines have become environmental weeds invading heathlands and open eucalypt forest. In contrast *P. radiata* is becoming rare in its native habitat and its conservation is of concern. Pine plantations have a rather gloomy, sterile appearance, but may support more animal diversity than the farmland that they replace, though less than native forest.

The key characteristics of pines that differ from the Australian conifers are that pines have needle-like leaves in clusters; each cluster is actually an extremely short side branch (Figure B11.2.1). Pine also has distinctive pollen with two air-filled bulges, which assist in wind dispersal. As in other gymnosperms, pine pollen has to reach the ovule while the cone scales are soft enough to open out to allow pollen access to the ovule, and at this stage the female gametophyte is very immature. At fertilisation more than one archegonium can be fertilised (this is called polyembryony). This results in a competition between young zygotes, but one eventually outgrows the others. To add to the oddities, the surviving embryo then splits into four, with again only one surviving to maturity. Seeds of pine have a flat wing, which aids in wind dispersal. Mature seeds are shed from the cone 2 years after the cones first appear (Figure B11.2.1).

continued ›

Box 11.2 continued ›

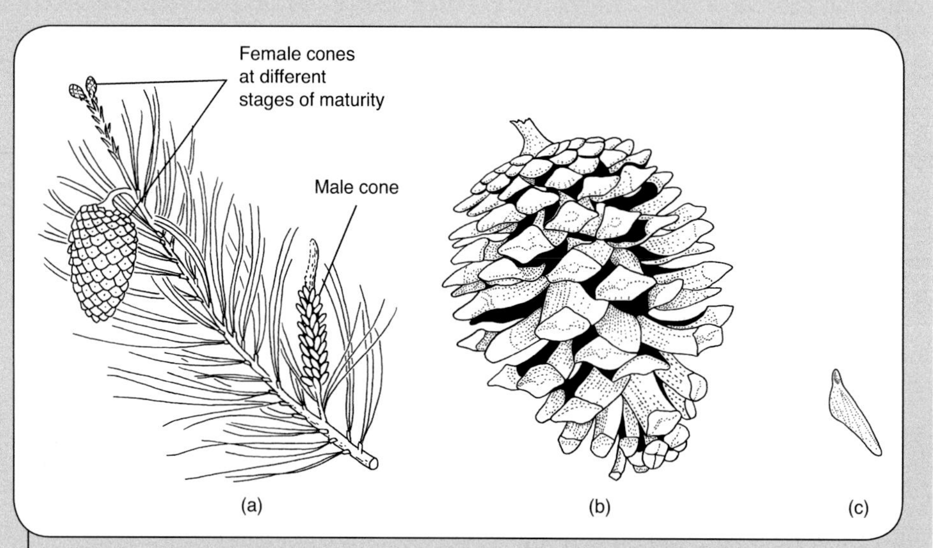

Figure B11.2.1 (a) Shoot of *Pinus* with needle-like leaves. Female cones of two stages of maturity are shown, and mature male cones. (b) A mature female cone at the seed-shed stage. (c) Winged seed. (Source: B. Cale)

Chapter summary

1. The ocean and the land offer very different habitats for plants in terms of support, light intensity, supply of water and nutrients, and availability of oxygen.
2. All land plants have an embryo, a haploid gametophyte and a diploid sporophyte generation, stomates in an epidermis covered with cuticle and sex organs covered with a layer of sterile cells. Subcellular characteristics include features of the mitotic and meiotic divisions, the presence of chlorophylls a and b, and cell walls of cellulose.
3. Mosses and liverworts have a dominant (larger) gametophyte generation with the sporophyte never totally independent of the gametophyte. Both generations lack roots and fully developed vascular tissue, so plants are vulnerable to desiccation. On the gametophyte, eggs are held in the archegonia and motile sperm are shed from the antheridia. Spores produced from meiosis in the capsule of the sporophyte are dispersed by air.
4. The dominant generation in ferns is the sporophyte, which has roots, stems and leaves with well-developed phloem and xylem. It thus tolerates dry conditions. The gametophyte is a small thallus, susceptible to drying, which produces archegonia containing the egg and antheridia from which motile sperm are dispersed.
5. In gymnosperms, ovules are formed by the enclosure of the female gametophyte in sporophyte tissues. After fertilisation and the formation of a zygote, the ovule becomes a seed. The male gametophyte is released from the plant as a pollen grain. Reproduction is slow as the ovuleis pollinated when immature and fertilisation follows sometimes many months later; subsequent development of the zygote to a fully formed embryo is also slow. The dominant sporophyte in gymnosperms ranges from vines and shrubs (rare), to palm-like cycads and tall trees.
6. In cycads, the pollen is produced from large male cones. After transfer to the ovule it grows a pollen tube to transmit the sperm to the region of the archegonia. The motile sperm swim the short distance to the egg cell. The developing embryo feeds on food stored in the female gametophyte.
7. The conifers are large trees producing small male and larger, woody female cones. Pollen (the male gametophyte) is blown from the male cones onto the ovules and grows a pollen tube to deliver sperm cells directly to the egg cell. There is no motile stage. The developing embryo in the seed depends on food in the female gametophyte.

Key terms

alternation of generations
angiosperms
antheridia
antheridiophore
archegonia
archegoniophore
bioindicator
Casparian strip
circinate venation
cone
cone scale
coralloid root
cuticle
dioecious
diploid
endemic
endodermis
epidermis
fertilisation
frond
gamete
gametophyte
gymnosperm
haploid

hydroids
lamina
leptoids
lignin
megaphyll
meiosis
microphyll
midrib
mitosis
monoecious
mycorrhiza
ovary
ovule
parenchyma
petiole
phloem
pinnae
pits
pollen
pollination
rachis
rhizoids
rhizome
root hairs
sieve cells
sorus
sporangia
spores
sporophyll
sporophyte
stomate
suberin
thallus
tracheids
vascular system
xylem
zygote

Test your knowledge

1. Which features of life cycle, structure and physiology are common to the protist green algae and land plants, and which are unique to land plants?
2. In which major phyla of the land plants is the sporophyte larger than the gametophyte, and in which is it smaller?
3. Explain why bryophytes and ferns are found mainly in moist areas.
4. Describe which parts of the cycad life cycle are diploid and which are haploid.
5. Compare the mode of sperm distribution and egg fertilisation in a pine and in a free-living fern gametophyte.

Thinking scientifically

Some species of mosses growing near a new industrial area are turning yellow and dying. What experiments would you carry out to determine whether this is caused by emissions from the factories, or some other factor?

Debate with friends

Debate the topic: 'Given limited resources, it is not important to include liverworts and mosses in botanical surveys of remnant areas proposed for development'.

Further reading

Farjon, A. (2008). *A natural history of conifers.* Timber Press, Portland, Oregon.
Hill, K.D. and Osborne, R.O. (2001). *Cycads of Australia.* Kangaroo Press, East Roseville NSW.
Raven, P.H., Evert, R.F. and Eichhorn, S.E. (1999). *Biology of plants.* Freeman Worth Publishers, USA.
Sacks, O. (1998). *The island of the colorblind and Cycad Island.* Random House, New York.
Tudge, C. (2000). *The variety of life: a survey and a celebration of all the creatures that have ever lived.* Oxford University Press, Oxford.
White, M.E. (1998). *The greening of Gondwana.* Kangaroo Press, East Roseville NSW.

12

Plant diversity II – the greening of the land

Bernie Dell, Mike van Keulen, Jen McComb and Barbara Bowen

The end of Eden – extinctions in the Australian flora

In April 1770, Lieutenant James Cook's ship *Endeavour* sailed into a natural harbour on the east coast of Australia, south of where Sydney now stands. Cook, originally struck by the number of stingrays, named it 'Stingray Bay'. Later, impressed by the many plants collected by the ship's naturalists, Joseph Banks, Daniel Solander and Herman Spöring, he renamed it 'Botany Bay'.

Cook was rightly impressed. We know now that about 90% of the Australian flora is endemic, occurring nowhere else in the world. Wildflower tourism is important in the domestic economy and the international market in Australian cut flowers is valued at more than $440 million annually. Ironically, only about 10% of that revenue is generated locally. The rest comes from countries with extensive horticultural enterprises, such as Israel. Australia also has industries based on products such as timber and eucalyptus oil, but they too are challenged by overseas growers of Australian plants.

Today, Botany Bay, where the first great collection of Australian plants was made, is the site of Sydney's Kingsford Smith airport, but the headlands are protected in the Botany Bay National Park. The park's management plan stresses preserving sites of historical significance, and endangered species and ecological communities. The issue of conservation in the face of development occurring at Botany Bay is reflected throughout Australia. Since European settlement 60 Australian plant species have become extinct and 1180 are threatened or rare, from a total of about 22 000. The vast majority of these are flowering plants. The extinctions are happening at a far higher rate than expected from natural processes over geological time, and are driven by habitat reduction, pests and pathogens, and fragmentation associated with agriculture, urbanisation and mining.

Chapter aims

This chapter covers the phylum of plants that dominates the world's land surface, phylum Anthophyta (the angiosperms). The main features of their morphology, internal structure, nutrition and reproduction are described to help explain the basis of their success. We also consider some of their special adaptations to the Australian environment.

Phylum Anthophyta – flowering plants dominate the land

We live in a world surrounded by plants from phylum Anthophyta (or Magnoliophyta), the flowering plants or angiosperms. There are at least 250 000 different species from over 450 families. From their beginnings in the early Cretaceous, some 130 million years ago they evolved to become the dominant plant group on land and even recolonised the waters. Flowering plants occur in dry, scorching deserts, tropical rainforests, frozen tundra and the sea. They display many forms, sizes and lifestyles, including some of the tallest trees in the world (Box 12.1). They have complex physiology and some of

Box 12.1 Tallest trees

The tallest Australian trees (and the world's tallest hardwoods) are mountain ash (*Eucalyptus regnans*), so called because the pale wood superficially resembles the pale wood of the unrelated European ash. These trees grow in regions of 1000–1500 mm rainfall in south-eastern Victoria and Tasmania. Trees over 100 m have been recorded in Victoria, but logging and fires removed these old giants and the tallest specimen is now from the Styx valley in Tasmania, measuring 96 m in 2001. There is a report from 1880 by a licensed surveyor of a felled *E. regnans* at Thorpdale, Victoria being 115 m tall. A less reliable 1872 record is of a felled tree with a broken top from Watts River in Victoria measuring 133 m. The estimated height with the top restored was 152 m. If these 19th-century reports were true, *E. regnans* in the past was even taller than the tallest conifers.

An account of the effort required to fell one of these huge trees in the 1930s tells that two axemen took two and a half days to do the job, and ended with the statement that one tree made '6,770 cubic feet of wood and was pulped into 75 tons of newsprint'. These trees appear to live for 400–450 years and most of the remaining giants are declining in height as they shed dying branches. Other exceptionally tall species include *E. viminalis* at 91 m and an *E. delegatensis* at 90 m. The karri (*E. diversicolor*) from Western Australia has a recorded maximum height of 85 m. Eucalypts do not usually have spectacularly thick trunks. However, in the 1930s one tree near Melbourne, called the King Edward tree, was 24 m in girth, and another, known as the Bulga stump, 31 m. *Eucalyptus jacksoni* in the south of Western Australia does not reach great heights, but may be up to 20 m in girth.

The world's tallest living trees are conifers, not angiosperms. The coast redwood *Sequoia sempervirens* occur along a narrow strip of the Pacific coast of California and many specimens over 107 m tall are known, with the tallest measuring 115 m. The tallest conifer ever recorded was 127 m, a Douglas fir *Pseudotsuga menziesii* from Lynn Valley in British Columbia.

Measuring height of extremely tall trees is difficult. Often they are growing on sloping ground, the crown may lean or be off-centre, and there are centuries of accumulated compost and debris at the base, leading to uncertainty as to where the 'ground' really is. Furthermore, the easiest techniques of measurement need a clear sighting to the top of the tree from at least a tree-length away. This is a problem in dense forest, and in national parks it is inappropriate to hack out a clear line of sight. Height can be measured using a laser range finder, with a theodolite, or by climbing the tree and dropping a tape to the base – all of which are better than chopping the tree down! Measurements of record-breaking trees are most reliable when a licensed surveyor uses a theodolite, or at least two different types of measurement have verified the height.

Tall trees are iconic and may be pivotal to raising public interest in plant conservation. However, they often occur in small over-mature stands that will naturally deteriorate over the next century (Figure B12.1.1). Reservations of tall trees are justified on the basis of their grandeur and age, but no management techniques can prevent their deterioration with age. Biodiversity values do not necessarily coincide with the occurrence of these ancient trees. For future generations to be able to marvel at tall trees, we need to conserve areas of fast-growing younger trees and to protect such areas from fire.

(a)

(b)

Figure B12.1.1 A giant *Eucalyptus regnans* from southern Tasmania. (a) In 1962 the tree was 98 m tall and had a healthy crown. (b) By 1998 the crown had deteriorated and it was only 91 m tall. (Source: a T. Mount; b J. Hickey) the chemicals they produce, especially those to protect them against browsing animals, provide important drugs. Most crop plants are angiosperms and interactions with animals profoundly influenced angiosperm evolution.

Key characteristics of angiosperms

Angiosperms are distinguished by several key features, mostly associated with reproduction:

- *flowers* – special shoots with a limited life containing stamens that produce the male gametophyte (pollen) in two pairs of pollen sacs, and an ovary where the female gametophyte (embryo sac) develops
- *double fertilisation* – each pollen grain gives rise to two sperm, one of which fertilises the egg to form a zygote, while the other fuses with the large, diploid cell of the embryo sac to form an endosperm (a nutritive tissue within the seed that feeds the plant embryo)
- *enclosed seed* – the fertilised ovules (seeds) are enclosed within the ovary, which forms the fruit that protects them
- *uniquely structured phloem tissue* – sieve tube cells with sieve plates between them facilitate transport, and companion cells regulate the function of the sieve tubes.

Major groups of angiosperms

Molecular analysis based on chloroplast DNA and a gene coding for ribosomes separates the angiosperms into several evolutionary lines. However, about 97% are placed in just two groups, based on the number of seedling leaves (cotyledons) present on the embryo. The monocotyledons (monocots, including grasses, lilies, orchids and palms) have only a single cotyledon and the eudicotyledons (eudicots, including most other familiar plants) have two. The number of cotyledons is linked with significant differences in anatomy and morphology (Table 12.1). Although these two groups do not represent evolutionary lines, the division is still useful and used widely.

The Australian flora is rich in eudicot species (Plates 12.1, 12.2) and nearly half of these come from these families: the legume group including Mimosaceae (wattles), Fabaceae (peas) and Caesalpiniaceae (sennas and their relatives); Myrtaceae (including the eucalypts or gum trees); Proteaceae (banksias, grevilleas and their relatives; Asteraceae (the daisies); Epacridaceae (leucopogons and their relatives); Rutaceae (boronias and correas); and Euphorbiaceae (including the wedding bushes, *Ricinocarpos* spp.). There are also many monocot species in the families Poaceae (spinifex and other grasses), Cyperaceae (sedges) and Orchidaceae (orchids) (Plate 12.3).

Identification of angiosperms

The huge diversity of the flowering plants gives rise to a key problem – how do we know which species do what? We need to name species before we can discuss their ecological functions and their conservation.

Table 12.1 Morphological and anatomical traits of monocotyledons and eudicotyledons.

Trait	Monocotyledon	Eudicotyledon
Vegetative morphology		
Seedling	One cotyledon	Two cotyledons
Root	Many roots arising from the stem base	Dominant taproot
Leaf	Major veins parallel, in grasses the leaf base forms a sheath around the stem	Veins form a network
Reproductive morphology		
Flowers	Parts tend to occur in multiples of 3	Parts often in multiples of 4 or 5
Pollen	One pore or furrow	Often 3 pores or furrows
Anatomy		
Root	Vascular core with numerous groups of xylem and phloem; central pith present	Vascular core with 2 to 5 groups of xylem and phloem; central pith absent
Stem	Vascular bundles scattered and surrounded by fibres	Vascular bundles arranged in a ring and with 1 or 2 fibre caps
Leaf	Guard cells elongated in grasses	Guard cells kidney-shaped

Traditionally, plants were identified using printed regional floras or descriptive keys, or by comparing them with named specimens in herbaria. Recent electronic keys and floras are:

- *Flora of Australia Online* – www.environment.gov.au/biodiversity/abrs/online-resources/flora/main/index.html
- *Acacias of South Australia* – http://www.flora.sa.gov.au/id_tool/acacia.html
- *Flora of New South Wales* – http://plantnet.rbgsyd.nsw.gov.au/floraonline.htm
- for identification of eucalypts – www.chah.gov.au/cpbr/cd-keys/euclid3/index.html.

Australia's Virtual Herbarium (http://www.anbg.gov.au/avh/) captures over 200 years' data on the flora of Australia. It links descriptions of the flora to nationwide data and information. Check with state and regional herbaria for availability and updates to e-floras for your region.

Angiosperm architecture – the winning plan for land plants

Angiosperm bodies consist of roots and stems for anchorage and support, leaves for photosynthesis, vascular tissue for both support and transport of water, mineral nutrients and sugars, and flowers for reproduction. When studying plant anatomy, the three-dimensional plant body is usually sliced in two dimensions for microscopic analysis (Box 12.2).

Box 12.2 Sectional planes of plants

Although plants are three-dimensional structures, sections of tissue are cut in two dimensions for microscopic analysis. For general purposes, the most useful types of section for plants are:

- transverse section (TS)/cross-section (CS) – cut at right angles to the long axis of a structure, for example, across a stem
- longitudinal section (LS) – cut along the long axis. If a longitudinal section passes exactly through the middle of a structure, it is called a radial longitudinal section (RLS) while, if it is off-centre, it is termed a tangential longitudinal section (TLS).
- oblique sections (OS) – sections that are cut at other angles not parallel or at right angles with one of the main axes. These are often very hard to interpret under the microscope.

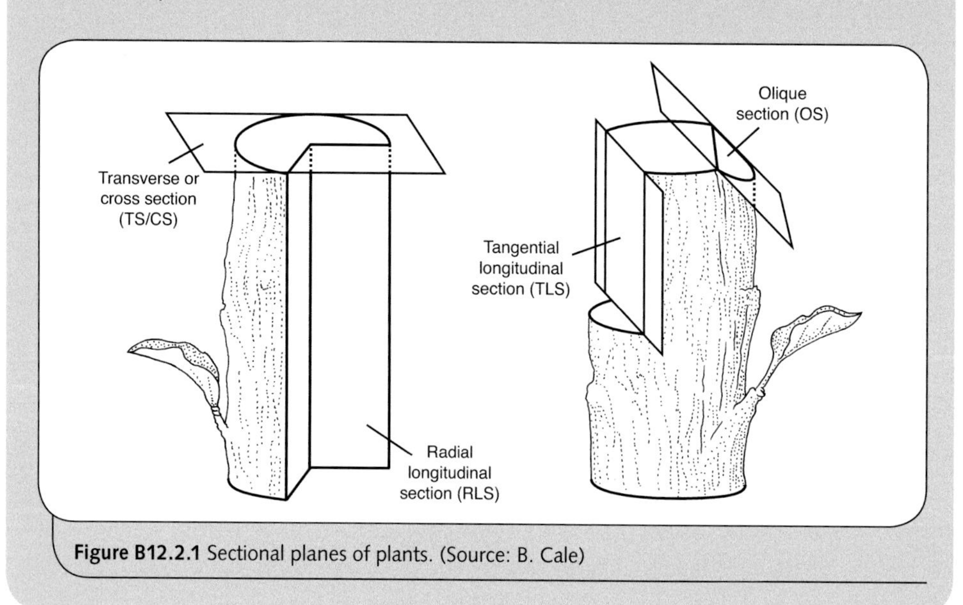

Figure B12.2.1 Sectional planes of plants. (Source: B. Cale)

Basic structure of seedlings and plant body form

Seedlings consist of root and shoot systems (Figure 12.1). The root system has an orderly branched structure, with secondary roots (also called lateral roots) arising from the primary root formed on germination. Small, water-seeking root hairs occur close to the root tip in most species (Plates 12.4a, 23.4, 23.5). The shoot consists of a stem bearing leaves and nodes, and in the axil of each leaf an axillary bud. Axillary buds develop into side shoots, and the apical meristem or the axillary buds may develop into flowers. In the embryo in the seed, the first node has leaves called cotyledons.

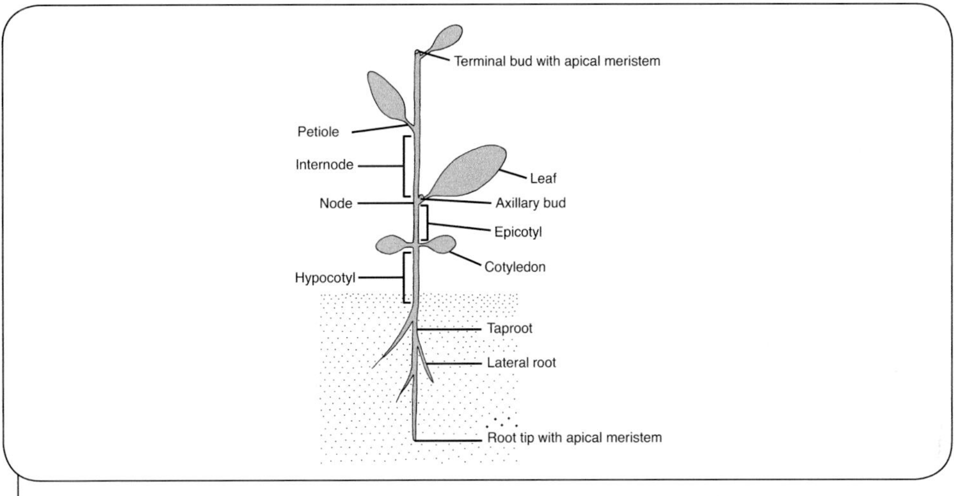

Figure 12.1 The seedling shoot and root systems of a eudicot. (Source: B. Cale)

In eudicots the primary root from the embryonic root apex usually persists as the taproot, growing directly downward. Taproot systems are extensive, accessing water deep underground. Secondary roots produce a much-branched root system. In monocots, the primary root is generally short-lived. Most roots arise from the stem, forming a shallow, fibrous root system (Figure B12.3.1) ideal for unstable substrates. Monocots are frequently used in coastal restoration, where the dense, shallow root system stabilises sandy soils. Monocots also include many wetland plants and seagrasses, adapted to waterlogged and submerged habitats.

In both eudicots and monocots, the shoot develops into the stem supporting the leaves, flowers and fruits, and connecting the leaf veins with the vascular system of other parts. In woody eudicots, the stem increases in girth through secondary growth. Although monocots do not produce wood, some form tree-like structures by enclosing the stem with large leaf sheaths (as in bananas), by increasing the number of vascular bundles in the stem (as in palms, bamboo and pandanus) or by cementing leaf bases with resin (as in *Xanthorrhoea*, the grass trees, Box 12.3). Grasses are the most common monocots. Some clump (e.g. wheat) and others run (e.g. couch). In the former, new upright stems develop from tillers produced at the base of the main stem. In the latter, rhizomes (underground stems) extend the plant and upright stems (culms) arise at intervals. In arid Australia, species of porcupine grass or spinifex (*Triodia* spp.) form large clumps known as hummock grasslands (Plate 22.5d,g). The coastal spinifex (*Spinifex* spp.) colonises dunes using rhizomes.

Apical meristems and tissue formation

All of the root and shoot parts of seedlings develop from apical meristems near the tips of roots and stems. They give rise to the protoderm, ground meristem and procambium,

Box 12.3 Grasstrees – a distinctive monocot

Some of the most distinctive plants in the Australian flora are the grasstrees, belonging to the genera *Xanthorrhoea, Dasypogon* and *Kingia* (Figure B12.3.1). Early last century a paper claimed that tall grass trees were 6000 years old, and this concept caught the public imagination. The blackened trunks have a crown of long, thin leaves and a skirt of persistent dead leaves providing important nesting sites for small marsupials. The apex of the plant is broad and flat, and, at 1600 μm wide, the broadest of flowering plants. From the apex the whorl of leaves is produced at a rate of about three leaves per day in spring, slowing but not stopping in winter and autumn. The young leaf bases and stem apex are tender and nutty, and were enjoyed by Aboriginals. Parrots also seek out these parts, sometimes damaging the apex and destroying the plant.

Grasstrees are monocots, yet they have a tree-like form. Leaf bases and rings of secondary vascular bundles strengthen the stem. An additional cambium, termed a desmium, produces a layer of cortical cells in the stems, storing starch and nutrients withdrawn from drying leaves. The roots of *Xanthorrhoea* arise from the base of the stem, but in *Kingia* they originate near the top of the stem and burrow down to the soil through the leaf bases. A further peculiarity of *Xanthorrhoea* roots is that they are contractile, and the apex of young plants can be 10 cm below the ground surface. The plant grows as a basal rosette for up to 30 years before the above-ground trunk becomes visible. Another distinctive feature of *Xanthorrhoea* is their copious production of resin, filling gaps between the leaf bases and forming a tough sheath.

The plants are highly flammable, burning quickly at high heat (up to 1000°C). Tightly packed moist young leaves that do not burn protect the apex. Plants are rarely killed by fire and are fertilised by nutrients in the ash from the burned skirt. When the outer burned surface of the stem is ground off, the cut surfaces of the leaf bases are seen to be in alternating bands of cream (from growth in spring and summer) and brown (from growth in autumn and winter) (Figure B12.3.1). This allows the age of the plant to be determined. Most old plants are around 250 years old with an exceptional 6 m individual (Figure B12.3.1) estimated to be 450 years old. They are slow growing, but far from the 6000 years old suggested earlier. Furthermore, some of the layers are blackened by fire. The frequency of fires can be determined from the positions of these burned layers (Figure B12.3.1c, d), although if burning of an area is patchy the results may be difficult to interpret. This has shown that the frequency of fires in south-western Australia is much lower under the current management strategy of prescribed burning than it was in pre-European times. The optimal period between fires has yet to be determined, and it appears that plants left too long unburned are susceptible to stems rotting and breaking and attack by termites.

Plants flower in the spring after fire, and the apex is transformed into a spike of flowers that can elongate at up to 7–10 cm per day – fast for an otherwise slow-growing species. Although a few plants may flower in years when there has been no fire, fewer seeds survive to maturity than in the synchronous post-fire flowering because of damage by moth larvae.

Figure B12.3.1 Grasstrees. (a) Exceptionally tall (6 m) grasstree (*Xanthorrhoea preissii*) in wandoo woodland, south-western Australia. (b) Fibrous root system of a monocot, the grasstree (*Xanthorrhoea preissii*). (c) Mantle of leaf bases of *Xanthorrhoea acanthostachya* undergoing removal to reveal colour bands (d) Banding pattern on surface of a *X. preissii* plant in jarrah forest. Black bands represent the effect of fires, whereas the pairs of alternating pale and dark bands represent annual growth increments. ((Source: a R. Armistead; b B. Dell; c, d B. Lamont)

continued ›

Box 12.3 continued ›

The species are attractive for horticulture and are transplanted successfully to parks and gardens from areas being cleared for development. Aboriginals used the resin for waterproofing and attaching stone blades to wooden handles. Contemporary Australians used resin, mainly from Kangaroo Island, to manufacture chemicals, varnishes, metal lacquer and wood stain. This export industry slowly declined and ceased in 1997 because it was not environmentally sustainable.

which in turn generate cells for the epidermis protecting the seedling, the cortex and pith comprising the bulk of the plant body, and the vascular bundles transporting water, minerals and organic compounds (Figure 12.2). As cells mature away from the apical meristems, their anatomy differentiates, depending on final position and function (Plate 12.4e). The main support, food storage and protective tissues contain one or a few types of cells:

- *epidermis* – the external cell layer that protects the plant, produces root hairs below ground and excretes a cuticle to reduce water loss above ground. It may also produce protective or glandular hairs, and contain pores (stomata) for gas exchange.
- *parenchyma* – the most abundant tissue in seedlings. It may contain starch (as in potato) or chloroplasts in leaves and green stems. It is the most abundant cell type in the cortex and pith of stems and roots. In leaves it is called mesophyll.
- *collenchyma* – a support tissue in young stems located close to the epidermis and comprising elongated, narrow living cells

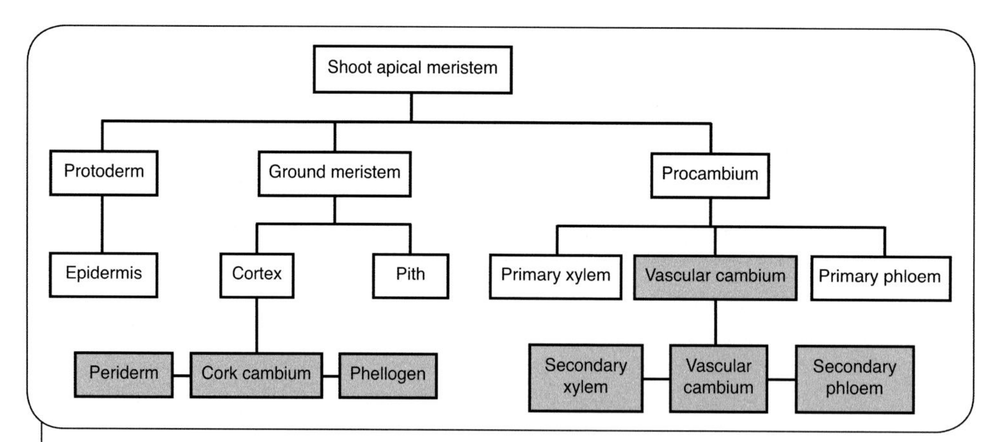

Figure 12.2 Tissues formed from the shoot apical meristem. All flowering plants have primary growth, but only woody plants have secondary growth (dark boxes) and form bark and wood. (Source: B. Cale)

- *sclerenchyma* – non-living support cells strengthened with the chemical lignin. One type, fibres, have a length to width ratio of about 100:1 (Figure 12.3) and are harvested commercially for products such as industrial hemp. A second type, sclereids, are irregularly shaped and occur, for example, in the brittle leaves of *Banksia* (Plate 12.4c) and waterlily (Plate 12.4d).
- *xylem* – water transport cells (tracheids and vessels), with fibres for support and parenchyma for support and food storage. Tracheids (Plate 11.4e) are elongated, narrow cells from which the more efficient vessels evolved. Angiosperms have both tracheids and vessels in their xylem. In advanced angiosperms, vessels have a large bore relative to their length and the end walls disappear completely during development. Thus, once water enters it flows unimpeded through interconnected tubes. Protoxylem are the first xylem vessels to arise from the procambial strand and can stretch as the tissue in which they are placed elongates. Metaxylem are more robust cells formed after completion of cell elongation.

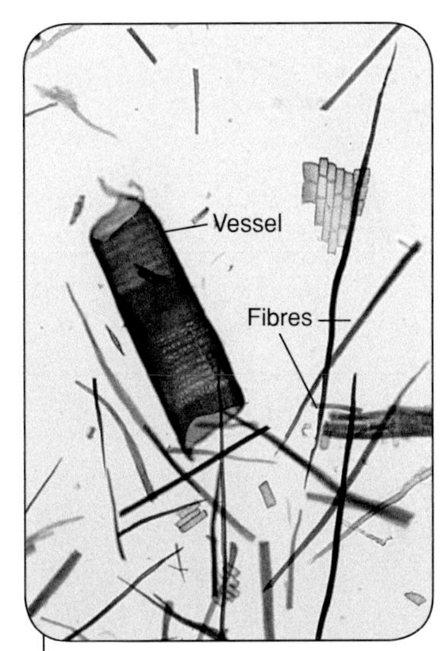

Figure 12.3 Macerated wood of Tasmanian blue gum (*Eucalyptus globulus*) showing a large vessel and many fibres. (Source: G. Thompson)

- *phloem* – carbohydrate transport cells (sieve tubes) are living cells without a nucleus at maturity. They have perforated end-walls (sieve plates), companion cells for loading and unloading the sieve tubes, fibres for support and parenchyma for food storage.

Xylem and phloem develop within a few millimetres of the apical meristems from the procambial strands, and are collectively called vascular tissue. The arrangement of the vascular tissue differs in roots and stems and between monocots and eudicots.

In roots of eudicots (Figures 12.4, 12.6), the procambial strands combine to form a large, complex central vascular bundle (stele) at the core of the root. It contains several protoxylem units or poles (commonly 2–5) alternating with the same number of phloem units. The metaxylem differentiates inside the protoxylem, forming a central core of xylem. By contrast, the stele in monocot roots contains numerous protoxylem and phloem units and the metaxylem does not extend far, so the centre of the stele is occupied by parenchyma (Figure 12.5). External to the vascular tissues are two layers of tissue separating the stele from the cortex. The outermost layer, the endodermis, consists of tightly packed cells with cell thickenings that force water and dissolved ions to pass only through the membranes of living cells, en route to the xylem. The inner layer or pericycle initiates roots in the region of the protoxylem poles (Plate 12.4b).

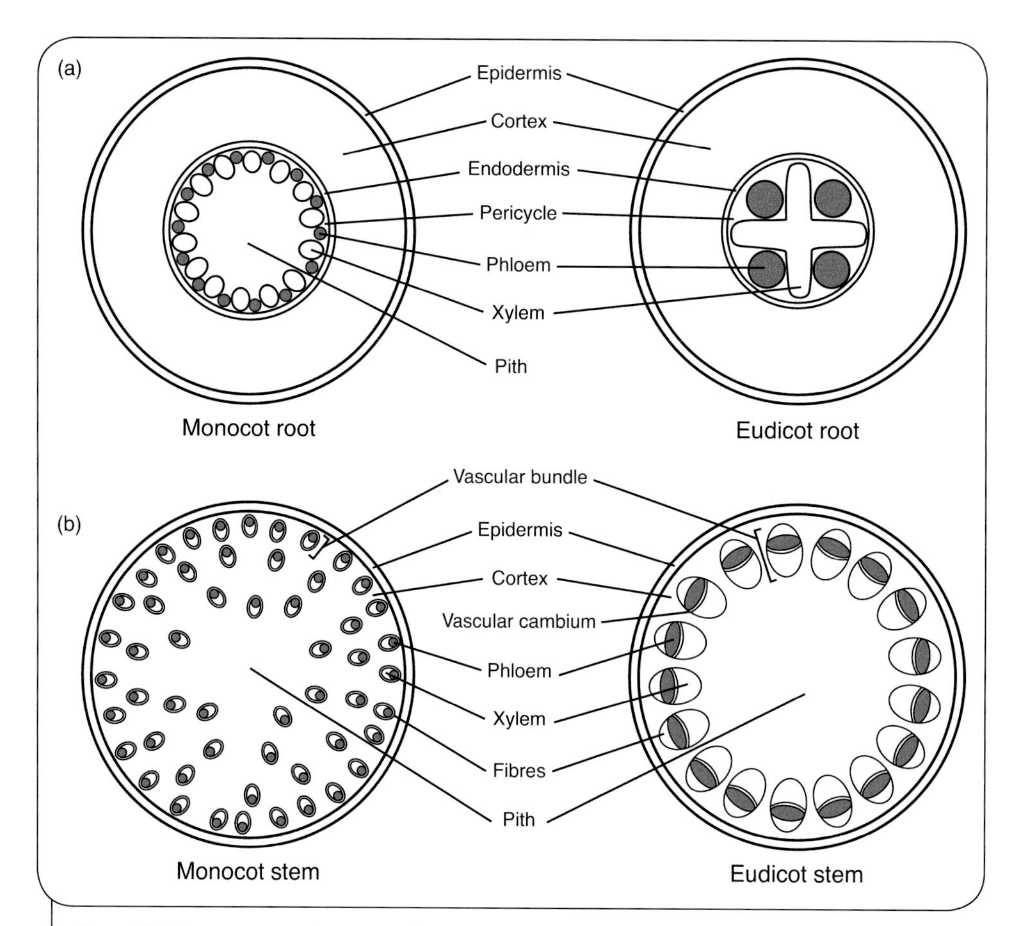

Figure 12.4 Arrangement of tissues in the (a) root and (b) stem of monocot and eudicot seedlings. (Source: B. Cale)

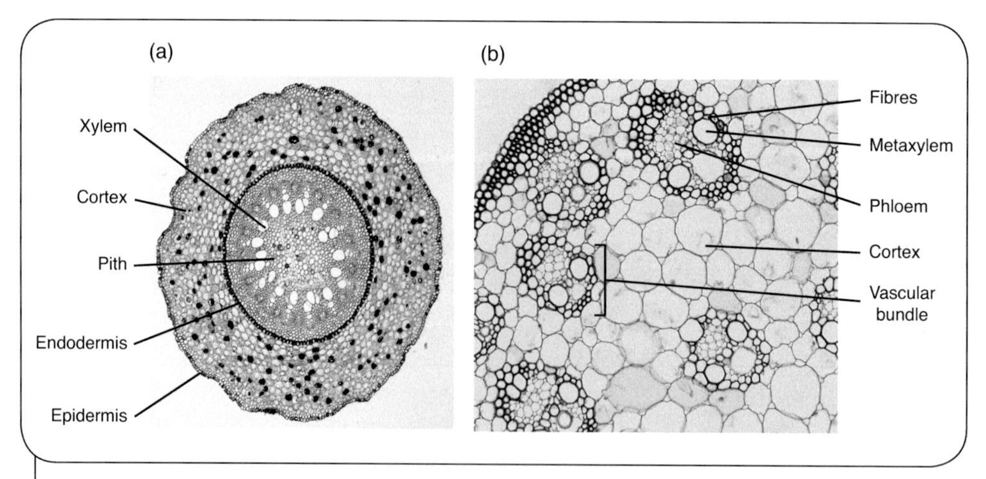

Figure 12.5 Root and stem anatomy of monocots. (a) Root of *Smilax*. (b) Stem of maize (*Zea*). (Source: B. Dell)

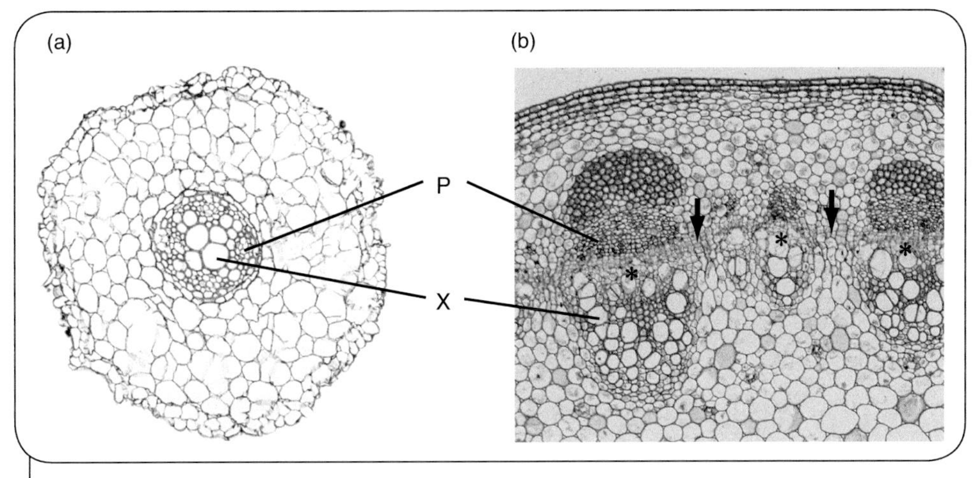

Figure 12.6 Root and stem anatomy of eudicots. (a) TS root of *Ranunculus* showing phloem (P) and xylem (X). (Source: G. Thompson), (b) TS of stem of *Helianthus* (a eudicot) showing the ring of vascular bundles in the cortex being split apart by secondary growth from the vascular cambium. The vascular cambium (*) has formed between the primary phloem (P) and the primary xylem (X) and extends through the cortex to form a continuous sheet of cells (arrows). (Source: B. Dell)

Vascular tissue in monocot and eudicot stems also arises from the procambium. In monocots, the procambial strands remain diffuse throughout the stem, whereas in eudicots they form a ring of vascular bundles (Figures 12.4–12.6).

The herbaceous habit

Cell division in the apical meristems leads to longitudinal growth, resulting in the herbaceous habit – thin plants less than a metre high. They include annual species such as the everlastings that carpet south-western Australia after autumn rains, and complete their life cycle from seed germination to seed production within a year. These plants are drought avoiders. Biennial herbs live for 2 years, producing a rosette of leaves in the first year and flowering after winter in the second year. This habit is rare in Australia. Herbaceous perennials live many years, although they may die back to avoid climatic extremes. Australian examples include kangaroo grass (*Themeda australis*), spear grass (*Stipa* spp.), and kangaroo paw (*Anigozanthos* spp.). Some spread vegetatively from horizontal stems. Many introduced herbaceous perennials and weeds in Australian gardens and natural environments come from cold, temperate climates and die back each year to underground parts with food reserves, regrowing each spring. (Table 12.2).

Underground plant parts were very important in Aboriginal diets. De Grey, an early European explorer, found it difficult to walk for kilometres around Western Australia's Hutt River because the ground was pitted with holes from harvesting yams (swollen root tubers of *Dioscorea*). The onion-shaped tubers of spike-rush (*Eleocharis dulcis*) have long been excavated from water bodies in north-eastern Australia and the tuberous roots of yam daisy (*Microseris lanceolata*) were once a staple of people in south-eastern Australia.

Table 12.2 Underground organs used by some herbaceous perennial angiosperms to survive unfavourable seasons.

Underground organ	Part modified	Example
Bulb	Leaf	Onion, *Crinum, Haemodorum spicatum*
Corm	Stem	*Gladiolus, Stylidium petiolare*
Rhizome	Stem	Ginger, many rushes and sedges
Tuber	Stem	Potato, yams (*Dioscorea*), sundew (*Drosera*), ground orchids
Tuber	Root	Yams, *Burchardia, Platysace, Clematis*

The woody habit

In woody eudicots, an additional lateral meristem, known as the vascular cambium (Figure 12.6b), increases girth and supports the greater height and expanding canopy. The vascular cambium is comprised of a single, thin sheet of dividing cells at the junction of the wood and bark. Daughter cells to the inside differentiate into wood (secondary xylem) and those to the outside form the inner bark (secondary phloem). Many trees are dated using annual growth rings in the wood (Figure 12.7).

The architecture of wood

Wood delivers support as well as the plumbing moving water and mineral nutrients from the roots to the leaves (Figures 12.7, 12.8). In the wood of some trees, the vessels are large enough to be seen with the naked eye and in many eucalypts are often larger in roots than stems. The strength of fibre and water transport cells depends on the chemical composition and structure of the cell wall. Although all the wood supports the tree,

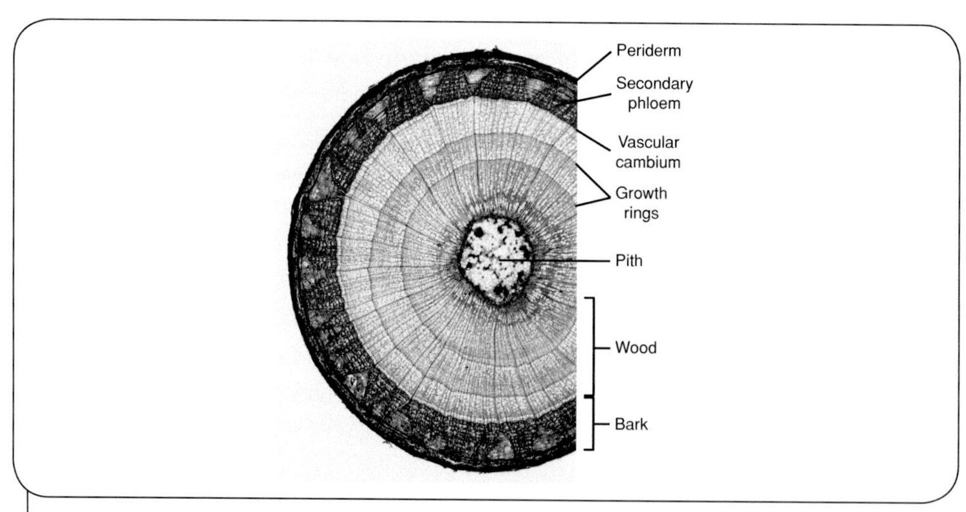

Figure 12.7 Cross-section of woody stem of *Tilia* in the spring of its fourth year of growth. (Source: B. Dell)

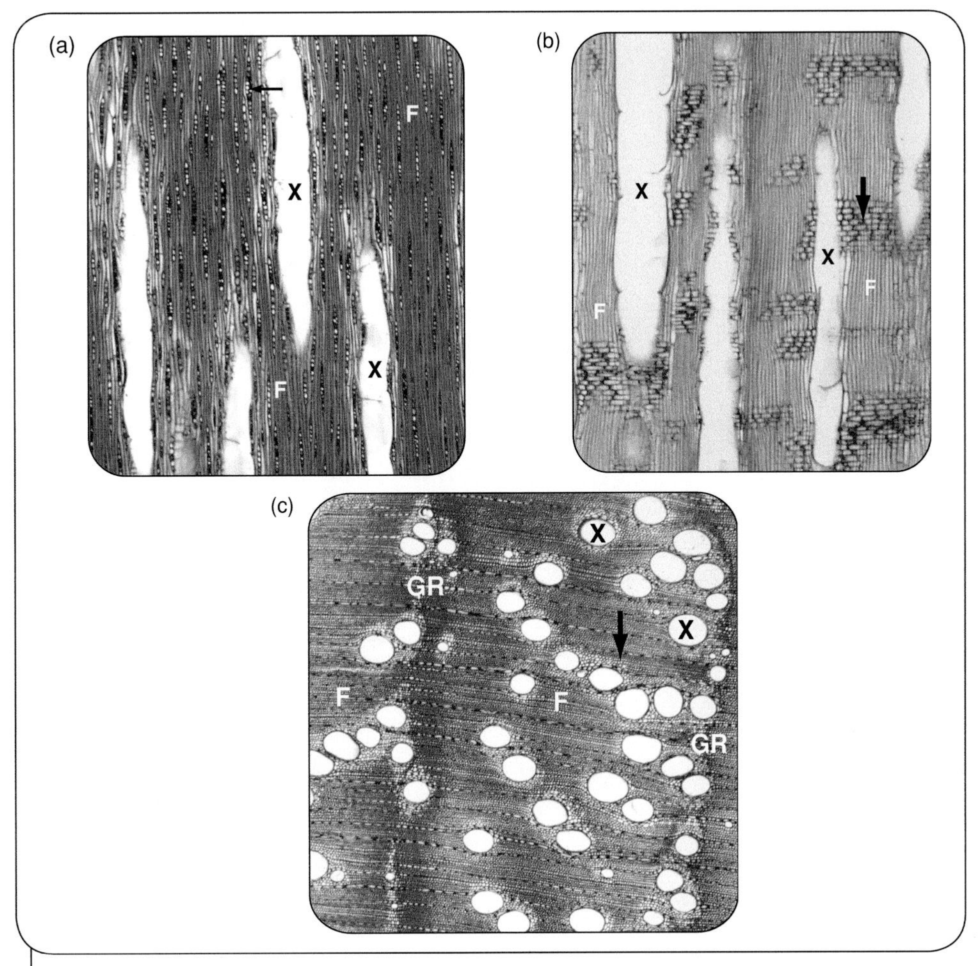

Figure 12.8 Wood anatomy in *Eucalyptus globulus*. (a) Tangential longitudinal section. (b) Radial longitudinal section. (c) Transverse section. (X) xylem vessel, (F) fibres, (GR) growth ring, arrows indicate rays. (Source: B. Dell)

usually the sapwood (the youngest layers closest to the bark) does most of the water transport. As the wood darkens with age it becomes non-functional heartwood. The inner bark (Figure 12.7) is largely living food transport and storage tissue (secondary phloem), transporting carbohydrates such as sucrose, amino acids and some mineral nutrients from source areas such as leaves to sinks such as the meristems and young fruits. Roots depend on transport of organic compounds for respiration by the phloem. The slow death of trees after ringbarking results from the gradual starvation of the roots, although reserves held in the roots may last for months. The functional link for water delivery is not broken by ringbarking.

Corky tissue (periderm), arising from a second lateral meristem, the cork cambium (Figure 12.2), protects the inner bark and the vascular cambium from weather, grazing animals, pests and diseases. In most woody plants this is renewed each new growing season.

Figure 12.9 Epicormic growth in crown of *Eucalyptus diversicolor* after a wildfire. (Source: B. Dell)

Successive years of activity of the cork cambium build up discrete layers of cork separated by remnants of the old food-transport tissue to form the outer bark or periderm (Plate 12.5).

The periderm insulates the cork cambium and the vascular cambium from the intense heat of wildfires. It also protects buds and meristematic tissue buried deep within the bark of many Australian trees. For example, in most eucalypts there are epicormic bud strands in the inner bark and wood. After fire, the strands produce large numbers of vegetative buds that mature into condensed shoots forcing their way out of the scorched bark to form epicormic shoots (Figure 12.9). Even where the crown is completely lost, new epicormic shoots appear from the blackened trunks. However, some eucalypts are killed by fire.

Tall monocots such as palms, grass trees and pandanus lack lateral meristems. When juvenile, these plants have a broad apical meristem producing many vascular bundles, providing support. There are no vegetative axillary buds in the trunk, so decapitation kills them. The persistent leaf bases also provide support. The ability for stems to resprout after fire is not limited to woody plants, as the grass trees regenerate leaves after fire and are stimulated to flower by the ethylene in the smoke (Box 12.3).

Structures for photosynthesis

In most angiosperms photosynthesis occurs in leaves (Figure 12.10), although in many Australian species leaf blades are reduced or lost and photosynthesis occurs in the leaf petioles or the stems. Leaves are generally flattened to maximise exposure to light and have specialised structures to obtain water and CO_2. Specific structures are usually a compromise between maximising photosynthesis and minimising water loss and overheating.

Basic structure of leaves

There is great diversity in form (from simple leaves having one blade to divided compound leaves with many leaflets) and size (a few millimetres to several metres), which has much to do with optimising interception of light and the availability of water in the environment. Leaves are attached by the petiole to the stem at the nodes and arranged in patterns.

Internally, the leaf veins or vascular bundles support the leaf blade and connect it with vascular tissue elsewhere in the plant for transporting water, mineral nutrients

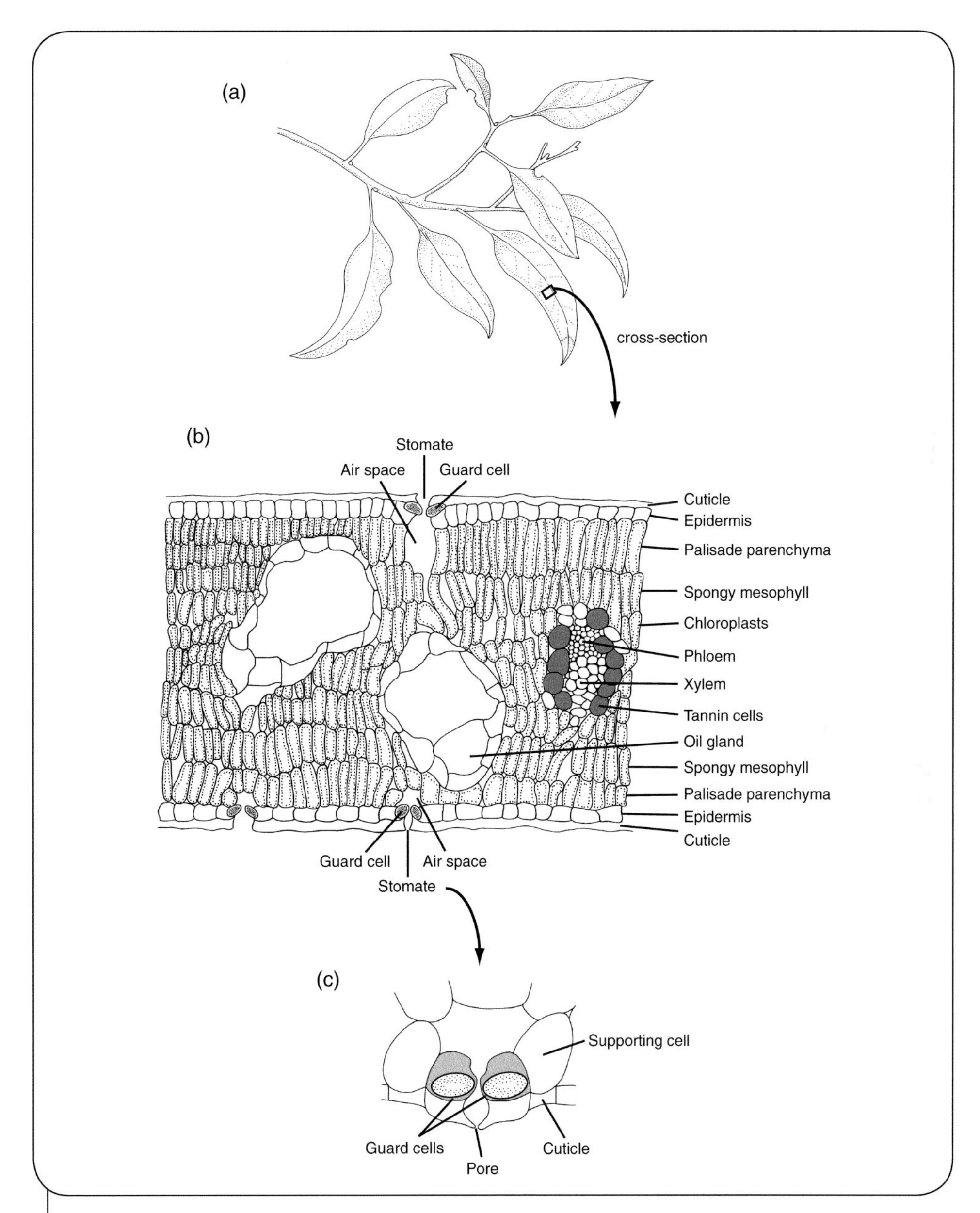

Figure 12.10 Structure of an angiosperm leaf (a eucalypt) and pathways of gas exchange. (a) Leaf morphology. (b) Structure of a typical leaf. (c) A pair of guard cells forming a stomate. (Source: B. Cale)

and photosynthetic products. The external surfaces typically have a cuticle to prevent evaporative water loss other than through special pores (stomates, Figure 12.10). A layer of epidermal cells lies beneath the cuticle. In plants from moist, temperate regions the main photosynthetic cells are packed into a palisade parenchyma located beneath the epidermis and exposed to sunlight. However, leaf anatomy is different in plants adapted to harsher environments (Plate 23.3). In eucalypts, the palisade parenchyma occurs on both sides of the leaf in species with vertical (pendulous) leaves (Figure 12.10). In plants with needle-like leaves, the palisade may extend in a circle beneath the epidermis.

Optimising light

Leaf shape may change throughout a plant's life to optimise light exposure under different conditions. For example, in *Eucalyptus globulus* the juvenile leaves occur in pairs at each node (they are 'opposite'), and the adult leaves are alternate (one leaf at each node). The arrangement of the juvenile leaves maximises light interception on the forest floor. The adult leaves hang vertically, reducing heat load. A less common arrangement is whorled, with three or more leaves at each node. Large leaves are common in rainforests where competition for light is strong and water availability is not a problem, whereas small leaves are common in arid landscapes.

The water dilemma

Water for photosynthesis is delivered in xylem vessels in a process driven largely by evaporative water loss from the stomates in the leaves. Plants are thus confronted with a dilemma. With broad leaf areas and plenty of sunlight, they photosynthesise as long as the water supply continues. But if the soil dries and water supply is interrupted, the leaves wilt and, if rehydration is not restored, the plant dies. To regulate water loss under dry conditions, paired guard cells (Figure 12.10) surrounding the stomates in the leaf surface change shape to open or close the pore.

Evaporative water loss from the stomates is enhanced by airflow across the leaf and low humidity over the stomates. Therefore many plants adapted to dry conditions reduce water loss by recessing their stomates into the leaf surface, surrounding them with hairs or curling their leaves (Plate 23.3b). These measures reduce exposure of the stomates to drying winds and trap a layer of moist air immediately over the stomate, slowing evaporation.

Gas exchange

The stomates also provide a route for gas exchange between the plant and the external environment. CO_2 diffuses into the leaf air spaces through the stomates and the O_2 produced during photosynthesis exits by the same route (Figure 12.10).

Internal transport

Water acquisition and water use

All land plants require access to water while they are growing. Not surprisingly therefore, the roots of angiosperms are highly effective in exploring soils for water. Not all of the soil water can be accessed by roots, however. The amount of water that is held by a soil after it has finished draining after rainfall or watering is called the field capacity, and up to half of this water is so tightly bound to soil constituents that it is unavailable for uptake. More water is stored in a loam than in a sandy soil.

Water enters fine roots passively except where there are impervious barriers such as corky tissue. It passes freely through the walls of root cells (as through blotting paper), but when it reaches the endodermis, the radial wall thickening of these cells means that it must pass into the cytoplasm of the living cells via osmosis. From there it passes into the non-living water transport cells, the vessels or tracheids. In moving across the root, water follows an osmotic gradient because the concentration of solutes in root cells is higher than that in the soil solution. Dissolved minerals are absorbed actively with expenditure of energy. The difference in solute concentration outside and inside the root creates a root pressure sufficient to move water to the leaves of seedlings, but not to the canopy of shrubs and trees.

The main driving force in larger plants results from transpiration of water from the stomates in leaves. Most of the water that is delivered into the leaf undergoes a phase shift, from liquid to gas, at the surface of cells close to the stomates. When the stomates open, water vapour diffuses into the surrounding atmosphere. It is the difference between the humidity inside the leaf (close to 100%) and that of the surrounding air that drives the ascent of sap in trees. The force provided by transpiration pulls water into the leaf to replace water lost to the atmosphere. Like drinking through a straw, water lost at the top is replaced by pulling water into the xylem in the roots. The column of water in the xylem from root to leaf is kept intact by cohesive forces between water molecules, and adhesive forces between the water molecules and the walls of the xylem vessels.

Sensors in the leaf monitoring light, CO_2 and signal compounds from the roots regulate the opening and closing of stomates. Generally, stomates open in the morning and close in the evening. Prior to opening, an osmotic gradient is established by actively pumping potassium ions from neighbouring cells into the guard cells. The resultant influx of water causes the guard cells to elongate and bow outwards so an aperture forms between them. The reverse occurs when the stomates close. Under stressful conditions, such as high temperature and strong winds, stomates may close during the day. On hot days, water loss can temporarily exceed supply in soft, large-leaved plants, and leaves may wilt even when the soil is moist.

Compared to introduced crop plants, the Australian flora is very efficient in its water use. Even when grown in plantations, eucalypts are many times more efficient in their water use than cotton and rice, producing more harvestable biomass per unit of water transpired. During drought, eucalypts' extensive root systems access deep stores of soil water. Although it is unknown how deeply eucalypt roots penetrate, Western Australian jarrah roots were measured at 41 m and tree roots are found at over 50 m depth in caves.

Nutrient uptake

Plants obtain the minerals essential for their growth and reproduction from soil (Table 12.3). The micronutrients such as copper and zinc are required in small amounts (mg/kg dry weight, µM to mM), whereas the macronutrients such as calcium and nitrogen are required in larger amounts (g/kg dry weight, mM to M).

Many Australian soils contain low levels of mineral nutrients because of a long history of weathering. Without the addition of fertiliser containing macro- and micronutrients (Plate 12.6), growth of introduced species, especially crops, is greatly impaired and symptoms of nutrient deficiency are expressed in the foliage. Yet, natural ecosystems with trees have evolved on many lateritic and sandy soils across the continent and these often support high species diversity. How is it that large amounts of standing biomass can accumulate without the addition of fertiliser? There are several explanations for this. First, plant growth may proceed at a pace that is dependent on the nutrient supply and hence symptoms of nutrient deficiency are rarely encountered in natural vegetation, unlike those that occur once the land is cleared for exotic crops that have high nutrient demands. Secondly, native plants

Table 12.3 Mineral nutrients essential for plant growth and examples of their functions.

Element	Function
Macronutrients	
Ca	Structural component of cell walls
S	Component of amino acids cysteine and methione, sulfolipids
K	Maintaining turgor, stomatal function, activator of many enzymes
Mg	Component of chlorophyll, activates many enzymes
N	Component of nucleic acids, amino acids and protein
P	Storage and transport of energy (ATP), photosynthesis and regulation of some enzymes
Micronutrients	
Fe	Redox reactions in photosynthesis and respiration
Cl	Photosynthesis
Co	Required by symbiotic bacteria for vitamin production in nitrogen-fixing nodules of legumes
Cu	Photosynthesis, structural component of some enzymes
B	Structural component of cell walls, development of sex organs
Mn	Splitting of water in photosynthesis, structural component of some enzymes
Mo	Reduction of nitrate
Na	Photosynthesis in some warm-climate plants
Ni	Urea metabolism
Zn	Required for many enzymes and photosynthesis

are very efficient in their use of scarce nutrient resources to grow and produce seed. Thirdly, Australian plants possess many root adaptations that increase their efficiency (Plates 23.4c, 23.5).

Moving sugars

Sugar produced in the leaves by photosynthesis moves via the phloem to other parts of the plant for respiration or conversion to starch for storage. Stored starch may be converted to sugar and also moved through the plant in the phloem, so sugar movement occurs both upwards and downwards. The area where the sugar originates is called a sugar source, and its destination is a sugar sink.

Sugars move faster in the phloem than can be accounted for by diffusion alone. According to the pressure flow hypothesis, sugar at a source is actively concentrated into the phloem. This draws in water by osmosis, increasing pressure in the source end of the phloem. At the sink, sugar leaves the phloem, drawing water after it by osmosis. Pressure reduces in the phloem at the sink end, creating a gradient of pressure from the source to the sink. Water flows down the pressure gradient, carrying sugar with it.

Reproduction and dispersal

Life cycle

The angiosperm life cycle (Figure 12.11) has two phases. The diploid sporophyte is the vegetative body (roots, stem and leaves) and the flowers, fruits and outer parts of the seed. Cells dividing meiotically are localised in the male and female structures of the flower. The gametophytes that form after meiosis are the pollen grains (male gametophyte) and the embryo sac (female gametophyte).

Flower structure

Flowers, the site of sexual reproduction, form at the end of a short stem (peduncle). They may occur singly, but often aggregate into inflorescences. For example, in *Banksia* species large numbers of flowers cover the woody cones (Plate 12.1d), and in daisies the flowers compress into a head (Plate 12.2c). In addition to male and female sex organs, the flower has vegetative parts for protection and attracting pollinators.

The flower consists of two sets of modified leaves at the base surrounding the sex organs (Figure 12.12). The lowermost set is the calyx. Above this lie the petals, forming the corolla. When calyx and corolla parts look similar, they are termed the perianth. Pollen-producing stamens (the male reproductive organs) occur inside the corolla. Each stamen has a terminally located two-lobed anther supported by a filament. At maturity, each stamen has four pollen sacs (Figure 12.13). The female reproductive organ, the carpel, is located centrally in the flower. It has three parts: lowermost is the ovary, above

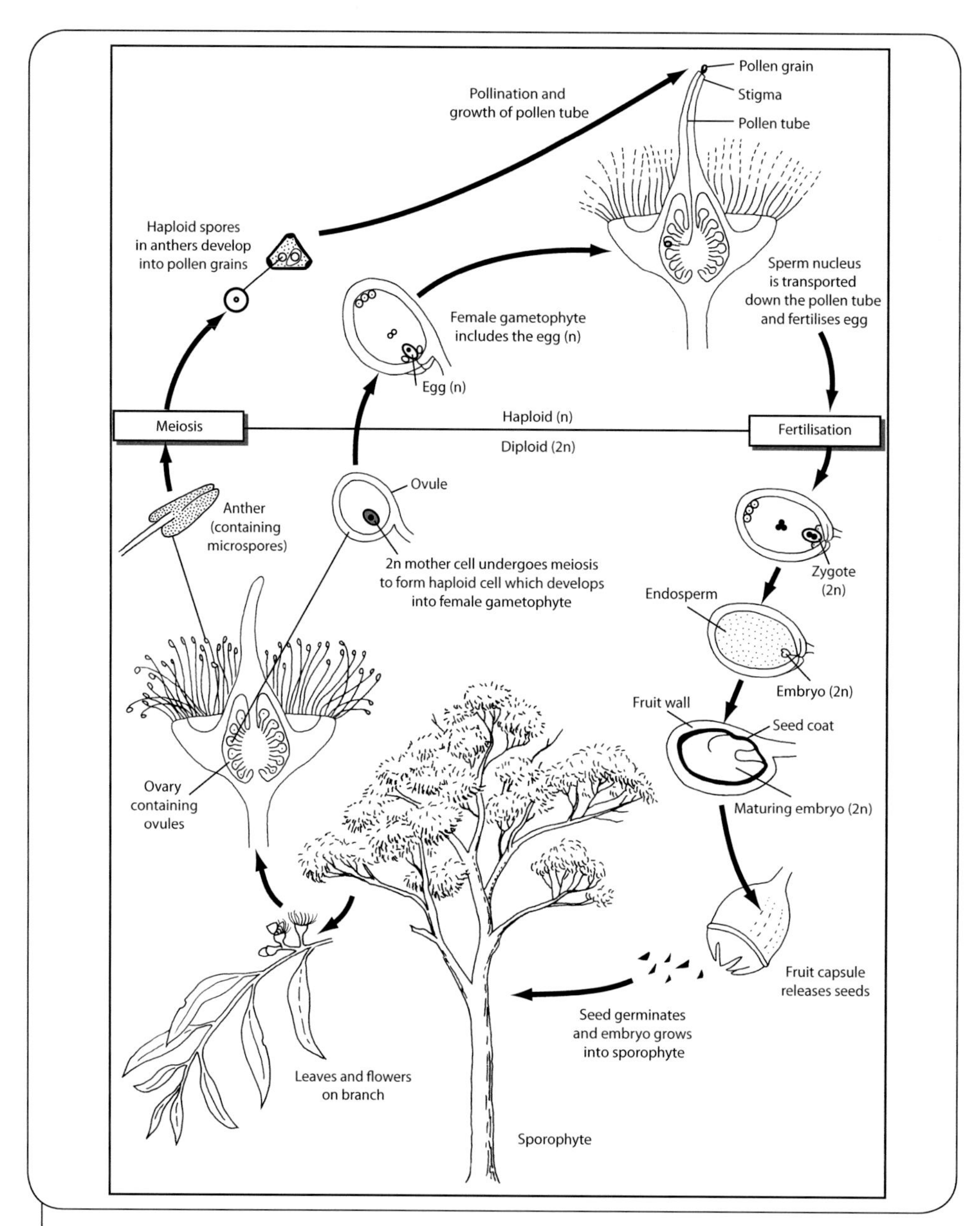

Figure 12.11 Angiosperm life cycle: a eucalypt. (Source: B. Cale)

this is the style and uppermost is the stigma. The stigma captures and provides the site for pollen to germinate. Pollen tubes grow through the style into the ovary to fertilise the ovules. Only flowering plants protect their seeds inside ovaries. The ovary wall becomes the fruit. Most flowering plants have bisexual flowers as described above, but some have unisexual flowers borne on the same plant or separate plants.

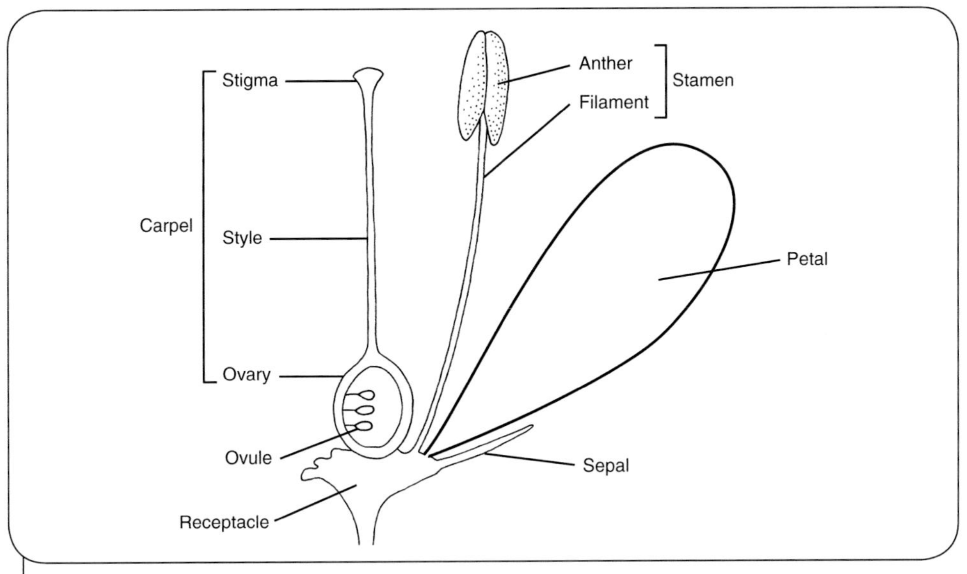

Figure 12.12 Diagram of a stylised flower. (Source: B. Cale)

Sexual reproduction

When the anther is young, a central group of cells in each lobe undergoes synchronous meiosis to form tetrads of haploid cells, which separate into single-celled pollen grains (the male gametophyte) (Figure 12.13). The pollen is surrounded by the tapetum, which provides nutrients. As the anther expands during flower bud growth, each pollen grain

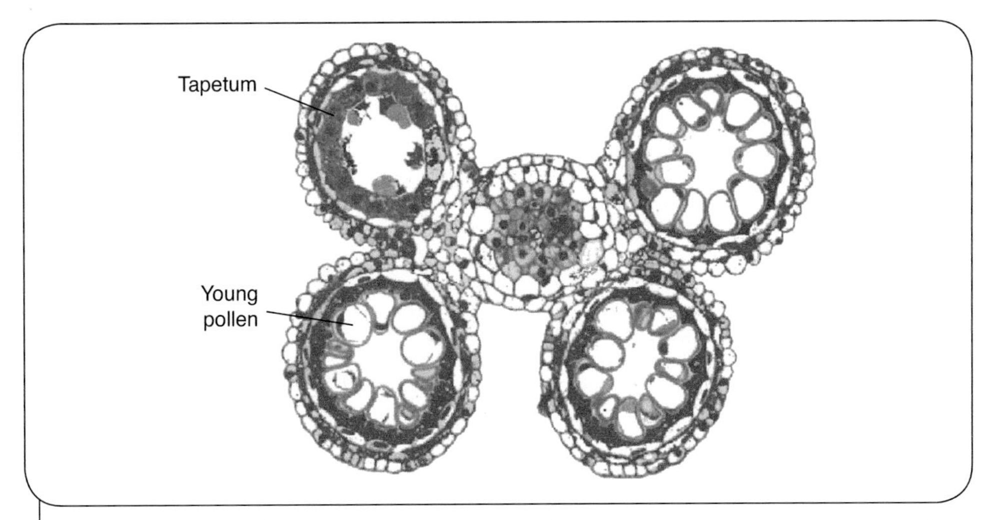

Figure 12.13 TS through the four-lobed anther of wheat showing young pollen grains surrounded by the tapetum. (Source: B. Dell)

undergoes mitosis, forming a two-celled pollen grain with a large vegetative cell and a small generative cell. The generative cell undergoes a second mitotic division before or after shedding, depending on family. At anthesis (when the flower opens), the anther splits and sheds the pollen.

Ovules develop within the ovary (Plate 12.4f, g). Each consists of a nucellus surrounded by two protective layers, leaving a small opening called the micropyle at one end. A single large cell inside the nucellus undergoes meiosis to produce four haploid cells. Three of these usually degenerate, but one expands and divides mitotically (Figure 12.14). Two further mitotic divisions form an eight-nucleate cell. Rearrangement of the nuclei and partial wall development results in a seven-celled embryo sac, the female gametophyte. At the micropylar end are located the egg cell and flanking synergid cells, centrally lie the polar nuclei and at the far end of the sac lie the antipodal cells.

Following transfer of pollen to the stigma (pollination), the pollen grains take up water and germinate. The pollen tubes grow through the style, into the ovary and enter the ovule through the micropyle. The pollen tube then releases the two sperm cells. One sperm cell unites with the egg cell and forms a diploid zygote that develops into an embryo. The second sperm nucleus unites with the polar nuclei to form the endosperm, a nutrient-rich tissue protecting and nourishing the embryo. This 'double fertilisation' is unique to angiosperms.

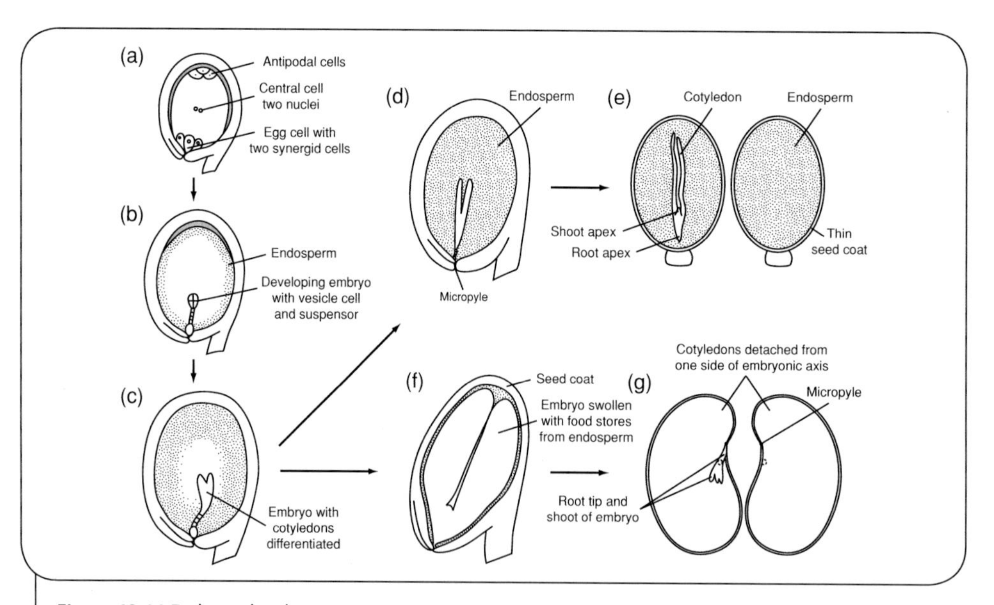

Figure 12.14 Embryo development in angiosperms. (a) Embryo sac at time of fertilisation. (b), (c) Early embryo growth. (d), (e) Endospermic seed with a small embryo embedded in the nutritive endosperm. (f), (g) Non-endospermic seed – nutrients are transferred into the cotyledons, which occupy most of the seed volume. (Source: B. Cale)

Pollination

Most plants evolved strategies that enhance cross-pollination and prevent self-fertilisation, thereby increasing genetic diversity in the offspring. Cross-pollination is aided by abiotic agents such as wind and water or by biotic agents such as insects and birds (Plate 12.7). Grass flowers are wind-pollinated, are green or dull-coloured and produce no nectar or lack scent. The filaments elongate to push the anthers outwards and the released pollen disperses on air currents. Flowers depending on animal pollinators attract them with food rewards such as nectar and pollen and may have brightly coloured petals and scent.

The tiny honey possum feeds exclusively on nectar, including the large inflorescences of banksias (Plate 12.7b). Its pointed snout and long, brush-tipped tongue gather nectar and pollen. Pollen is transferred as feeding animals move between flowers. In the arid-zone shrub *Eremophila*, evolution for insect or bird pollination has resulted in flowers with distinct morphologies (Figure 12.15). Some flowers open at night and are pollinated by moths or bats. These flowers are generally white and scented. Bird-pollinated flowers are often red or yellow, contain copious amounts of nectar and are unscented.

Seed development

The first tissue to develop after fertilisation of the ovule is not the zygote, but the endosperm. This forms a nutritive tissue around the zygote. This is a major advance for angiosperms compared with other plants. Little food is supplied for the growth of the embryo until fertilisation has occurred, in contrast to plants with large food stores in the female gametophyte that are wasted if fertilisation fails. When the zygote divides, it differentiates into the embryo proper and some structures that hold it in the right orientation in the

Figure 12.15 Flower morphology in *Eremophila* has evolved for bird (left; note the exerted anthers and stigma, arrowed) and insect pollination (right; note that the anthers and stigma are inside the tube of the flower, the corolla forms a broad landing platform for insects and there are markings inside the corolla tube to guide the insect to the nectary at the base). (Source: B. Dell)

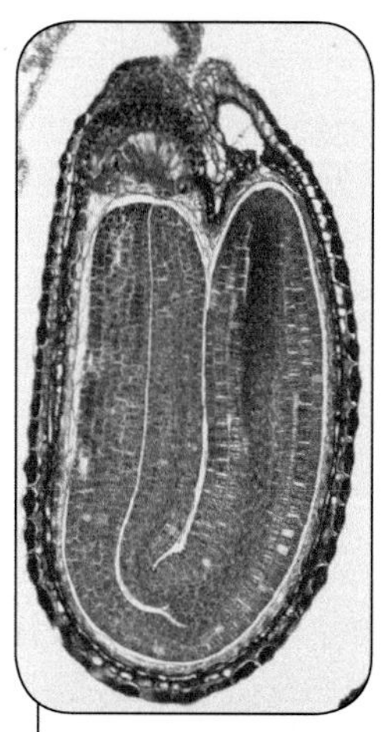
Figure 12.16 Non-endospermic seed of *Capsella* in LS. See Figure 12.14 for labelling. (Source: B. Dell)

mass of endosperm. Thus it first divides into a small apical cell and a large basal cell. The basal cell forms the suspensor and the vesicle that anchors the embryo to the micropyle. Division of the apical cell leads to a proembryo. The protoderm, ground meristem and procambium then develop and eventually a root apex, shoot apex and one or two cotyledons form (Figure 12.14). In some seeds (ambiguously called non-endospermic seeds), the endosperm is gradually absorbed as the embryo enlarges, so at maturity the embryo is tightly packed with nutrients stored in the cotyledons, to be used during germination (Figure 12.16). Examples are beans and acacia seeds. In other species such as wheat, the mature embryo is surrounded by endosperm (they are called endospermic seeds) (Figure 12.17). When seeds germinate the cotyledons withdraw food from this tissue and transfer it to the seedling.

Fruit development

The ovary wall develops into the fruit. Most fruits contain three layers: the exocarp (skin), the endocarp closest to the seed, and the mesocarp between the exocarp and endocarp. Fruits can be conveniently divided into fleshy and dry fruits (Plate 12.8). In simple fleshy fruits such as peaches and plums, the mesocarp is fleshy. In almonds, the mesocarp is shed and the seed is protected by a woody endocarp. The entire pericarp is soft in true berries such as tomatoes and grapes. Dry fruits can be further subdivided into those that split at maturity (dehiscent fruits) and those that do not (indehiscent fruit).

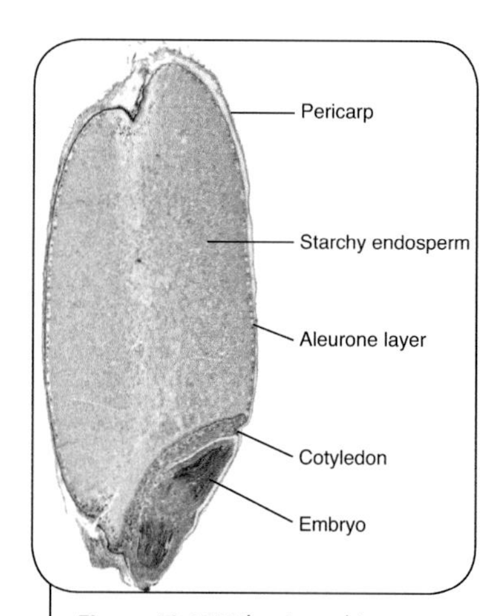

Figure 12.17 Wheat seed in LS. The fruit wall and the seed coat are fused together as the pericarp and surround the starchy endosperm. Beneath the pericarp lies the aleurone layer, the tissue that provides the amylase enzymes for the breakdown of starch. The scutellum or cotyledon is sandwiched between the embryo and the endosperm. (Source: B. Dell)

Fruit and seed dispersal

The attraction of vertebrates to fleshy fruits is obvious. Fruits may be carried short distances and the seeds dropped, or ingested seed may pass unharmed through the digestive tract to be deposited further away. The brush-tailed bettong or woylie caches fruits of sandalwood trees. Not all the cache sites are dug up, hence some seed germinates.

Ants disperse the seed of many Australian legumes. Ant-dispersed seed has a fleshy, nutrient-rich elaisome attached to one end. The seed may be carried underground or discarded above ground around ant nests (Plate 12.9). Hard seed stored below ground by ants can germinate after wildfires heat the soil to sufficient depth.

Fruit and seed are also dispersed by wind and water. Examples of dispersal by wind include the parachute-like seeds of the Asteraceae (daisies), the winged fruits of casuarinas and the minute seed of orchids. The coconut palm is dispersed long distances by ocean currents because the fruit is buoyant and tolerates sea water. The seed of many marsh and intertidal plants is dispersed by water.

Seed germination

Seeds contain only 5–20% of their weight as water, so they must imbibe water before the embryo resumes growth and germinates. On imbibition, cell enlargement and cell divisions begin. The primary root or taproot emerges from the seed coat through the micropyle, anchors the seed to the soil, and absorbs water and mineral nutrients. The shoot then emerges.

Seeds may germinate as soon as they fall, but most enter dormancy. This allows time for seed dispersal, and accumulates seed banks in soil. Dormancy can be caused by the seedcoat or controlled by the endosperm or embryo. Many acacias and other legumes prevent water uptake by impermeable layers in the seedcoat (hard seededness). Scarification (mechanical abrasion) or hot water treatments break this type of dormancy. In the wild, heat from fires cracks the seedcoat and imbibition occurs. An excess of growth-inhibiting hormones (regulatory chemicals that are produced in one part of the plant and control activities elsewhere) in the embryo prevents some seed from germinating. The balance of hormones is altered in favour of germination by treatment with low temperature and growth-stimulating hormones. Smoke enhances germination in many Australian plants (see Box 12.4).

Box 12.4 Where there's smoke, there's germination!

Fire is a major force shaping Australian ecosystems (Chapter 22). Some Australian plants survive fire and resprout from trunks or roots, while others are killed and rely on seed to re-establish. These seeds are often dormant, and habitat burning is the key to germination, making it extremely difficult to propagate seed conventionally. What triggers germination after fire? Is it heat, the age of the seed, the creation of a nutrient-rich ash bed, the death of shading overstorey plants or another factor? When the conservation of threatened species is involved, the problem is significant.

Experiments showed that heating some species of dormant seeds either by dropping them briefly into boiling water or heating them under lamps after planting stimulated germination. But there still remained species that germinated after fire, but did not respond to heat.

continued ›

Box 12.4 continued ›

A breakthrough occurred in South Africa, where biologists studying the rare and difficult to cultivate *Audouinia capitata* succeeded in breaking dormancy by applying smoke. Further studies revealed that gentle fires generating temperatures of 160°C–200°C released water-soluble volatile compounds that stimulated germination, root and seedling growth and even flowering in many species. The approach was adapted successfully to Australian plants in the 1990s by researchers at Kings Park and Botanic Gardens in Perth, Western Australia. They found smoke promoted earlier and more uniform germination and broke seed dormancy in species previously found to be difficult or impossible to germinate. Species with large, woody fruits seem less likely to respond to smoke, which is most successful with small-seeded species.

Smoke was tested initially by drawing smoke from burning foliage into tents put over seed trays or placed directly over soil in unburned areas of bushland. Smoke could be more conveniently applied when bubbled through water and the resultant brownish liquid 'smoke water' diluted and used to soak seeds or water seedling trays. Smoke is effective on many Australian, African and American native species previously difficult or impossible to germinate, and on horticultural and agricultural species including weeds. In some species such as lettuce that need light for germination, smoke water causes high germination even when seeds are kept in the dark. Although this was very exciting, smoke water was inconvenient to use as its composition varies between batches according to the material burned, and smoke may also contain inhibitory compounds that reduce the effect of the active ingredient. Nevertheless, smoke treatment is now important in propagating seed of endangered species and in restoring mine spoils and disturbed land.

Chemists and biologists found that the critical substance burned was cellulose, so a purer form of smoke derived from cellulose filter papers was analysed in a search for the active ingredient in the smoke. Ether extractions of smoke water and fractionation using the techniques such as gas chromotography (GC), mass spectrometry (MS) or high performance liquid chromatography (HPLC) resulted in over 4000 peaks or fractions (possible unique compounds). They were further analysed with a bioassay, testing the ability of each fraction to germinate lettuce seeds kept in the dark, and two Australian species known to be stimulated by smoke. One of the smaller peaks contained the active ingredient, identified as a butenolide (3-methyl-2H-furo[2,3-c]pyran-2-one) (Figure B12.4.1). Identification was confirmed by synthesis of the compound from pyromeconic acid.

It is water-soluble, heat-stable to 99–120°C and active at extremely low concentrations – less than 1 part in a billion (ppb, 10^{-9} M). This compound stimulates germination of many species (Table B12.4.1), opening exciting possibilities. For example, it may be possible to spray fields to induce all weed seeds to germinate so they can then be killed with herbicide before planting a crop. It will also be possible to enhance germination of native grasses for perennial pastures and when revegetating bushland. A spray of 1–20 g per ha replicates the effect of a wildfire.

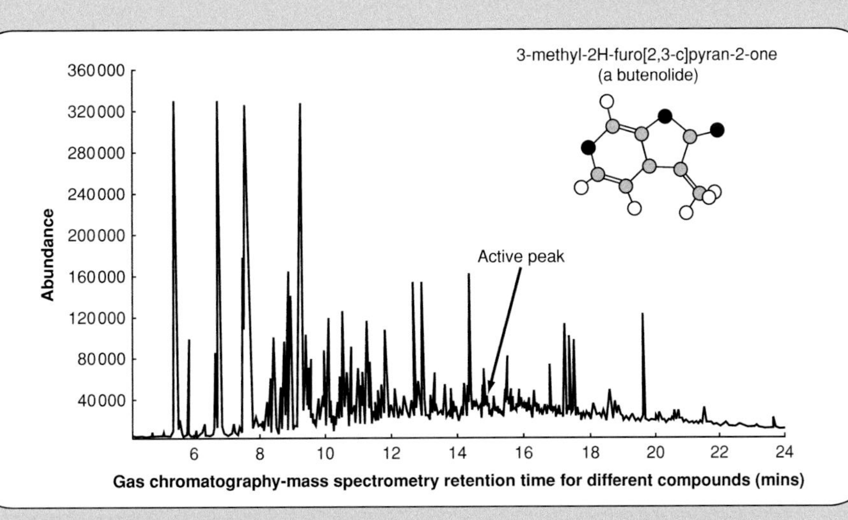

Figure B12.4.1 Analysis of the compounds in smoke using gas chromatography and mass spectrometry techniques. Different compounds are retained in the apparatus for varying times (retention times) depending on their characteristics, resulting in peaks of abundance on the curve when specific compounds are released. The arrow shows the peak active in seed germination. The inset shows the chemical structure of the butenolide responsible for germination. (Source: K. Dixon).

Table B12.4.1 Percentage germination of Australian native species in response to diluted smoke water as 10 ppb butenolide extracted from smoke (data from Flematti et al. (2005). 'A compound from smoke that promotes seed germination.' 305(5686) *Science* 977.

Species	Water control	1/10 dilution smoke water	10 ppb compound
Conostylis aculeata	8	45	50
Stylidium affine	1	82	80
Brunonia australis	8.0	Not tested	25.0
Conostylis candicans	1.0	Not tested	36.0
Lawrencella davenportii	16.7	Not tested	56.3
Ozothamnus cordifolius	0.0	Not tested	24.0
Scaevola thesioides	2.0	Not tested	17.0
Verticordia densiflora	2.0	Not tested	11.1

Chapter summary

1. Angiosperms are vascular plants characterised by enclosed seeds, flowers as reproductive organs, phloem tissue with sieve plates and companion cells and a double fertilisation. About 90% of Australian angiosperms are endemic.
2. For convenience, angiosperms are often divided into monocotyledons (monocots) with only a single seedling leaf present on the embryo and eudicotyledons (eudicots) with two.
3. Seedlings consist of root and shoot systems. In eudicots, the embryonic root persists as the taproot whereas in monocots most roots arise from the stem, forming a shallow, fibrous root system. In both eudicots and monocots, the shoot develops into the stem supporting the leaves, flowers and fruits, and connecting the leaf veins with the vascular system of other parts. In woody eudicots, the stem increases in girth through secondary growth.
4. All of the root and shoot parts of seedlings develop from apical meristems near the tips of roots and stems. They give rise to the protoderm, ground meristem and procambium, which in turn generate cells for the epidermis that protects the seedling plant, the cortex and pith which together make up the bulk of the plant body, and the vascular bundles that transport water, minerals and organic compounds.
5. Vascular tissue consists of xylem (tracheids and vessels) for water transport and support and phloem (sieve tubes with companion cells) for sugar transport.
6. In roots of eudicots, there is a large, central vascular bundle (stele) In monocot roots phloem and xylem are in a ring with pith in the middle. Vascular tissue in eudicot stems forms a ring of vascular bundles, whereas in eudicot stems the bundles are diffused throughout the stem.
7. Lateral meristems (the vascular cambium and the cork cambium) increase the girth in woody eudicots and support large plant bodies. Tall monocots such as palms, grass trees and pandanus lack lateral meristems. When juvenile, these plants have a broad apical meristem producing many vascular bundles, providing support.
8. Photosynthesis primarily occurs in leaves. Inputs are provided via the xylem (water) and stomates in the leaf surface (CO_2). Water loss from the stomates is regulated by the opening and closing of guard cells. Adaptations such as recessing the stomates, surrounding them with hairs and curling the leaf also reduce water loss.
9. The main force moving water from the roots to the leaves is a transpiration stream. Evaporation of water from the stomates pulls up a continuous chain of cohering water molecules, assisted by adhesion between water molecules and the walls of xylem vessels. Root pressure, caused by the osmotic flow of water into the xylem in the roots, is a less important force.
10. Sugars move in the phloem along an osmotic pressure gradient caused by a high concentration of sugar at the source and a low concentration at the sink (destination).
11. The diploid sporophyte is the vegetative body (roots, stem and leaves) and the flowers, fruits and outer parts of the seed. Cells dividing meiotically are localised in the male and female structures of the flower. The haploid gametophytes that form after meiosis are the pollen grains (male gametophytes) and the embryo sacs (female gametophytes).
12. 'Double fertilisation' occurs in the ovule within the flower's ovary. On entering the ovule the pollen tube releases the two sperm cells. One unites with the egg cell, forming a diploid zygote that develops

into an embryo. The second sperm nucleus unites with the ovule's polar nuclei to form the endosperm, a nutrient-rich tissue protecting and nourishing the embryo.

13. Pollination and seed dispersal may occur by wind, water or animals.

Key terms

angiosperm
annual
anther
apical meristem
bark
calyx
cambium
carpel
collenchyma
companion cell
cork cambium
corolla
cortex
cotyledon
culm
double fertilisation
embryo sac
endosperm
epicormic shoots
epicotyl
epidermis
growth rings
guard cells
hormones
hypocotyl
lateral roots
meristem
metaxylem
parenchyma
peduncle
perennial
perianth
pericarp
pericycle
phloem
pith
pollen
pollination
pressure flow hypothesis
primary growth
primary root
procambium
protoderm
protoxylem
rhizome
root
sclerenchyma
secondary growth
secondary root
seed
stamen
stele
stem
stigma
stolon
stomate
style
synergid cells
taproot
vascular bundle
vascular cambium
vessel
wood
xylem

Test your knowledge

1. List (a) the distinguishing characteristics of angiosperms and (b) the distinguishing features of monocots and eudicots.
2. What is the function of apical meristems and where are they located?
3. Describe the structure and function of the two types of vascular tissue: (a) xylem and (b) phloem.
4. Describe the process of wood formation in eudicots. Given that monocots do not form wood, how do some monocots reach a large size?
5. Explain how (a) water and CO_2 enter leaves for photosynthesis and (b) how sugars produced by photosynthesis are transported throughout the plant body.
6. List the male and female parts of a flower and describe the function of each. Where are the male and the female gametophytes located?
7. Describe the ways in which flowering plants achieve dispersal of (a) pollen and (b) fruits and seeds.

Thinking scientifically

The diagram below shows an apparatus prepared to demonstrate how water moves in a plant. It can be built in different sizes and the largest working models are up to 1.7 m tall. Water loss is shown by changes in the readings on the balance.

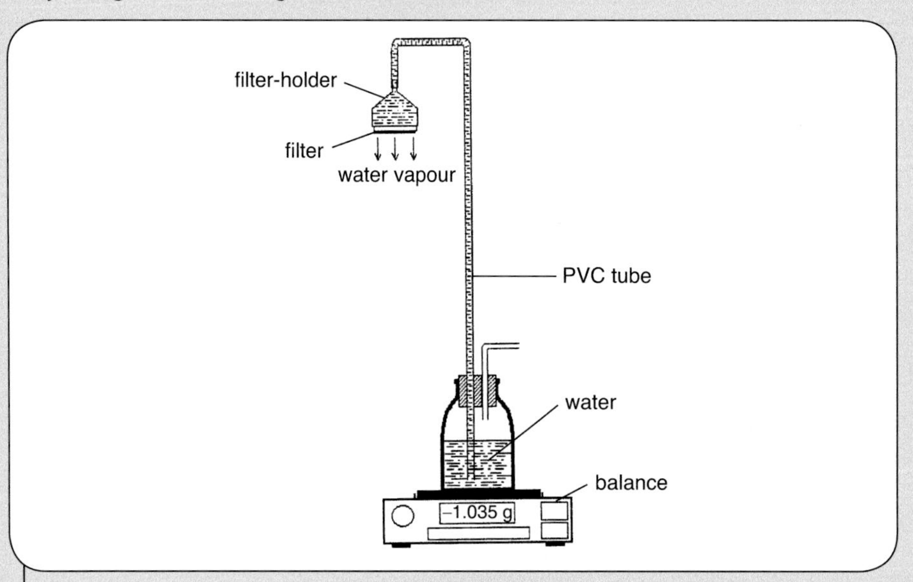

Figure 12.18 Model of a transpiring tree. (Source: Vilalta, J.M., Sauret, M., Duró, A. and Piñol, J. (2003). 'Make your own transpiring tree.' *Journal of Biological Education* 38: 32–35. Copyright the Institute of Biology, London, UK)

1. Explain how the model works. Include in your explanation what parts of a real plant are equivalent to the parts of the model.
2. Design an experiment using the apparatus to test the effects of wind on water movement.
3. Design an experiment using the apparatus to test the effects of humidity on water movement.

Debate with friends

Debate the topic: 'The best way to conserve a plant species is to find a use for it'.

Further reading

Lamont, B.B., Wittkuhn, R. and Korczynskyj, D. (2004). 'Ecology and ecophysiology of grasstrees' *Australian Journal of Botany* 52: 561–82.

Leigh, J., Boden, R. and Briggs, J. (1984). *Extinct and endangered plants of Australia*. Macmillan, South Melbourne.

Hewson, H. (1999). *Australia: 300 years of botanical illustration*. CSIRO Publishing, Collingwood, Victoria.

Thiele, K.R. and Adams, L.G. (eds) (2002). *Families of flowering plants of Australia: an interactive identification guide*. CSIRO Publishing/Australian Biological Resources Study, Collingwood, Victoria.

Tudge, C. (2000). *The variety of life: a survey and a celebration of all the creatures that have ever lived*. Oxford University Press, Oxford.

13

Life on the move I – introducing animal diversity

Mike Calver and Alan Lymbery

All creatures great and small

The animal kingdom includes many large and beautiful creatures. It is easy to engage the public's sympathy for the plight of endangered animals such as tigers and pandas, or, closer to home, Tasmanian devils and numbats, and these animals are often the focus of intense conservation efforts. Parasites, on the other hand, live in or on other organisms. They are typically small, hard to see and often harm their hosts, so they are usually either ignored or regarded as a nuisance in practical conservation programs. This is a simplistic view. Parasites have vital roles to play in the functioning of natural ecosystems. In recent years it has been realised that many parasites are valuable indicator species, and their disappearance from an ecosystem is often a symptom of a deeper, underlying problem resulting from pollutants or other human impacts.

Particularly important as indicators of environmental quality are parasites with complex life cycles, that travel through a range of different host species on their way from egg to adult. Figure 13.1, for example, shows the life cycle of a species of flatworm that lives in freshwater ecosystems in Australia. A parasite such as this may be a very sensitive indicator of environmental quality. The parasite and its various hosts are bound in a complex web of feeding interactions. They all have very different body structures and very different lifestyles and the environment in which they live must support all of these. If changes to the environment adversely affect any of its hosts, the parasite will be unable to complete its life cycle and will disappear.

Despite their wide range of body structures and lifestyles, all the organisms in this complex web are animals. Unlike plants, most animals are motile and feed in many different ways and as a consequence have a much greater diversity of structure and function than do plants.

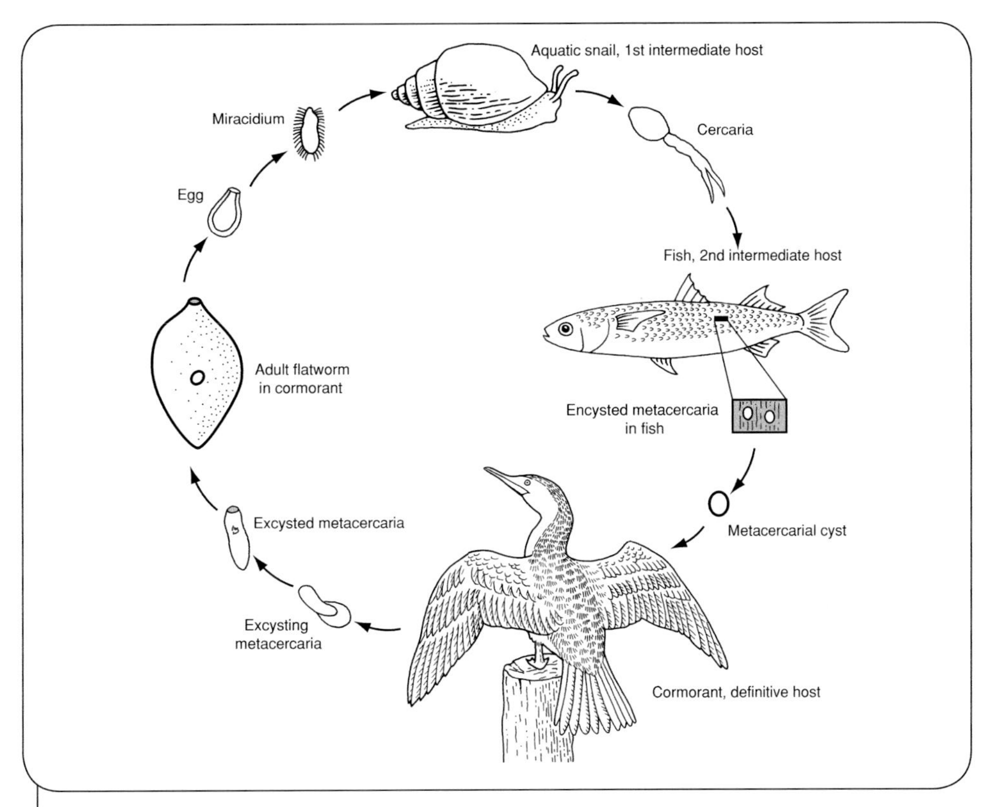

Figure 13.1 The life cycle of an unnamed Australian freshwater fluke. The adult parasite lives in the intestine of fish-eating birds. Eggs passed with the faeces hatch into free-swimming larvae that find and penetrate a freshwater snail. Within the snail, more larval forms are produced. They eventually leave the snail and invade the flesh of a fish. There they wait until the fish is eaten by a bird, so completing the life cycle. (Source: B. Cale)

Chapter aims

This chapter explains what all animals have in common, describes the influence of environment, lifestyle and size on how different animals solve the problems of life, and outlines how the main types of body plan found in the animal kingdom can be used to group and classify animals.

What is an animal?

In common usage, an animal is a furred organism. These, however, make up only about 4600 species among about 1.3 million animal species. Sponges, snails, sea stars, spiders, fishes, birds and many other organisms lack fur, but are all animals. Their internal structures may be described in terms of sectional planes (Box 13.1).

Box 13.1 Sectional planes in animals

Many animals have distinct left and right sides, a definite 'front end' (anterior) and 'rear end' (posterior), and a 'back' (dorsal surface) distinct from a 'belly' (ventral surface). The system for naming the sections of animals is therefore a little more complicated than that for the sections of plants (Figure B13.1.1):

1. transverse section (TS) – any section across an animal or tissue at right angles to the long axis (see also 'frontal section' below)
2. cross-section (CS) – often used interchangeably with transverse sections, but is less defined and most useful when cutting tissue where the terms 'ventral', 'dorsal', 'anterior' and 'posterior' have no meaning (e.g. when cutting a section through a sponge)
3. longitudinal section (LS) – not a very useful term for animals for, unlike plants, it is difficult to know exactly how a section has been cut. The qualifying terms 'tangential' and 'radial' tend not to be applicable for animals and are therefore restricted to use in plant sections. All that can be said of a longitudinal section of an animal is that it is one in parallel with the longitudinal axis of the body. However, the next two terms ('sagittal' and 'frontal') are more explicit and generally preferred.
4. sagittal section (SS) – if a line of a section is vertical and lengthwise, from head to tail, the plane is a sagittal one. If the section passes exactly through the middle of the animal, dividing it into equal left and right halves, then it is called a midsagittal section (MSS), whereas all other sections to one side or the other are termed parasagittal sections (PSS).

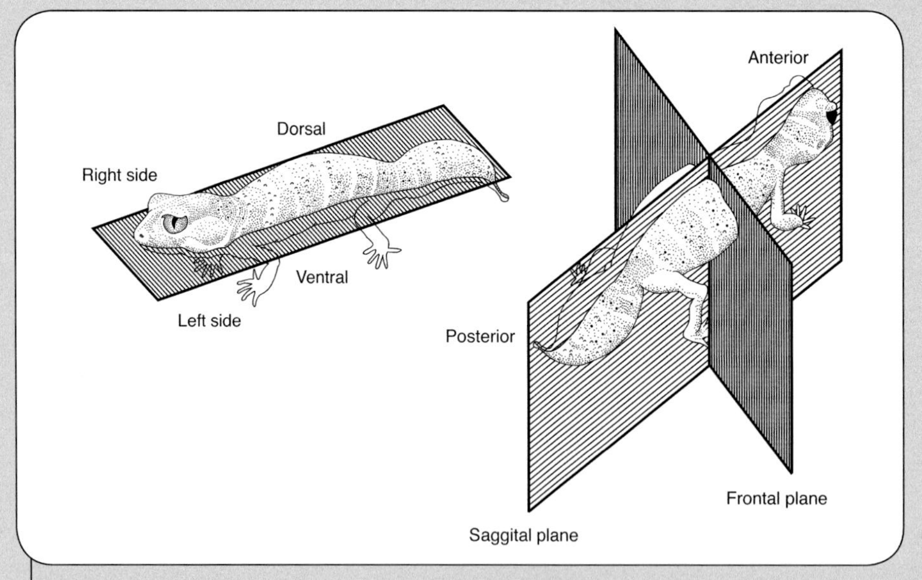

Figure B13.1.1 Sectional planes in animals. (Source: B. Cale)

Plate 11.1 Wollemi pine. (a) Fossil showing leaves of Wollemi pine (*Wollemia nobilis*) dating from the Cretaceous, along with specimens from a living tree. (b) Collecting seeds from tall trees in remote canyons presents special problems: a seed collector suspended from a helicopter to obtain seeds of Wollemi pine (*Wollemia nobilis*) from a gorge in the Wollemi National Park. (Source: a–b J. Plaza)

Plate 11.2 The seedless plants. (a) Conspicuous gametophyte phase of a thallose liverwort. (b) Sward of leafy moss plants growing on a moist rock surface. (c) Australian tree ferns growing in moist subtropical forests can grow to 10 m in height. (Source: a–c Lochman Transparencies)

Plate 11.3 The cycads. (a) Large male cones produced in the apex of the cycad (*Zamia* sp.). (b) Pollen cones produce many sacs containing pollen grains on their microsporophylls. (c) Large female cones exposing brightly coloured, large seeds produced on the cycad known as the Burrawang (*Macrozamia communis*). (d) Close-up of megasporophyll with two large seeds. (e) The MacDonnell Ranges cycad (*Macrozamia macdonnellii*) growing near Palm Valley, Northern Territory. This species has the largest seeds of any cycad, and its survival is threatened by climate change. (Source: a J. Lochman; b–d B. Dell; e J. McComb)

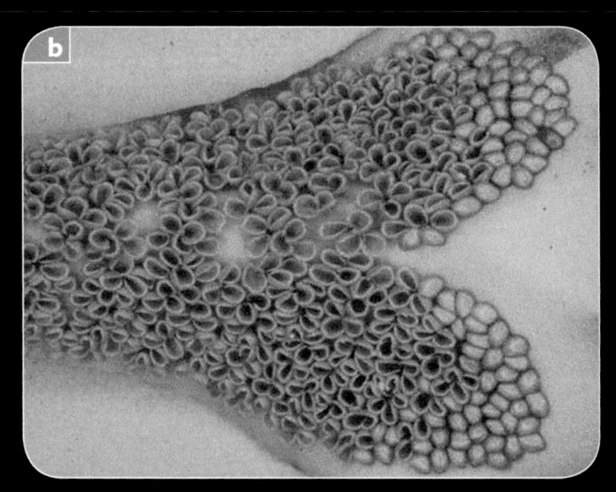

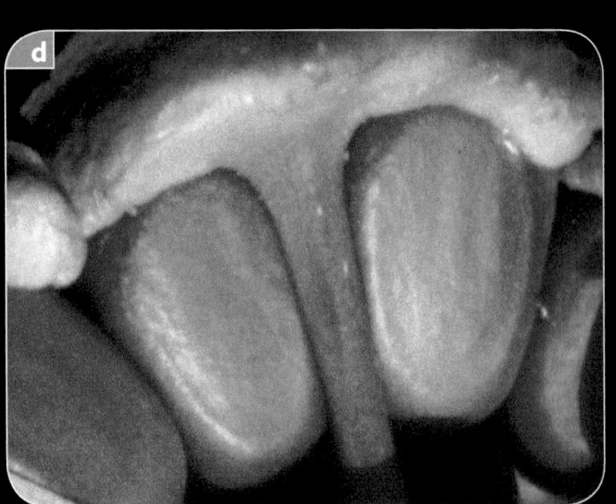

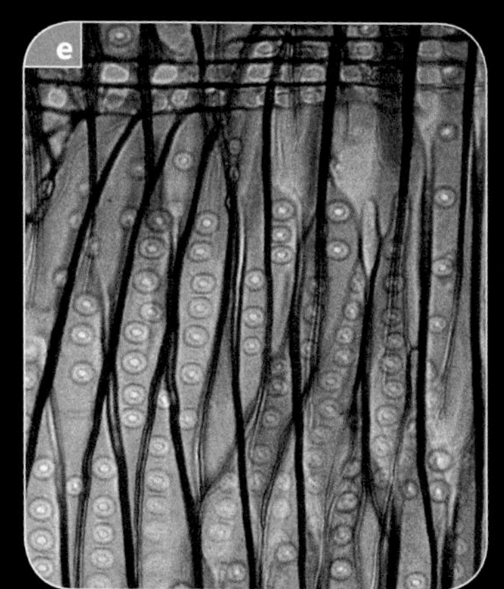

Plate 11.4 The conifers. (a) The Australian hoop pine (*Araucaria cunninghamii*) can grow up to 60 m tall and is generally found growing on drier sites in rainforests, in places that are rocky or have soils with relatively low fertility. (b) The Australian white cypress pine (*Callitris columellaris*) grows in a range of soil types in arid and semiarid areas. (c) Male cones of the Australian conifer wild plum (*Podocarpus drouynianus*). (d) Female cones each with two separate ovules and a fleshy receptacle. (e) Water-conducting tracheid cells in the wood of *Pinus* sp. (Source: a R. Whyte; b Creative Commons; c, d A.S. George; e B. Dell)

Plate 12.1 Three common families of Australian woody plants (a) Proteaceae (foreground), Mimosaceae (left) and Myrtaceae (rear). (b) *Eucalyptus pyriformis* (Myrtaceae). (c) Tasmanian waratah (*Telopea truncata*) (Proteaceae). (d) *Banksia coccinea* (Proteaceae). (e) *Acacia aphylla* (Mimosaceae). (Source: a Lochman Transparencies; b, e B. Dell; c A.S. George; d © Lochman Transparencies)

Plate 12.2 Further examples of common Australian woody plant families. (a) *Kennedia prorepens* (Papilionaceae). (b) *Petalostyles cassioides* (Caesalpiniaceae). (c) Daisy flowers (Asteraceae) *Podolepis gracilis*. (d) *Styphelia hainesii*, family Epacridaceae. (e) Euphorbiaceae (*Phyllanthus calycinus*) male inflorescence (left) and female (right). (f) *Boronia* (Rutaceae). (Source: a–d, f A.S. George; e B. Dell)

Plate 12.3 Examples of Australian monocots. (a) Foxtail mulga grass (*Neurachne alopecuroidea*) (Poaceae). (b) Christmas bells (*Blanfordia grandiflora*) (Blandfordiaceae). (c) White spider orchid (*Caladenia patersonii*) (Orchidaceae). (Source: a–c A.S. George)

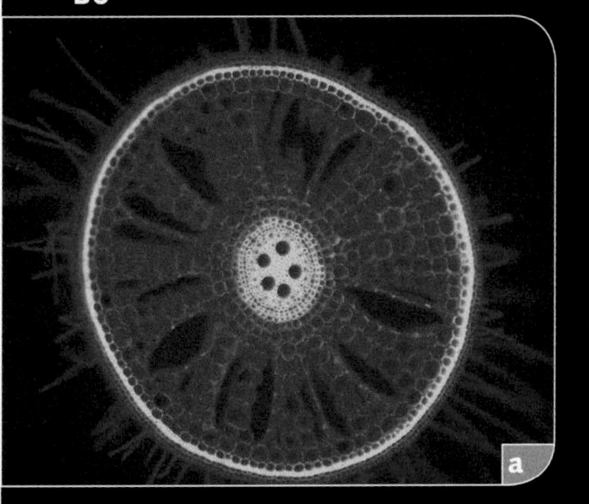

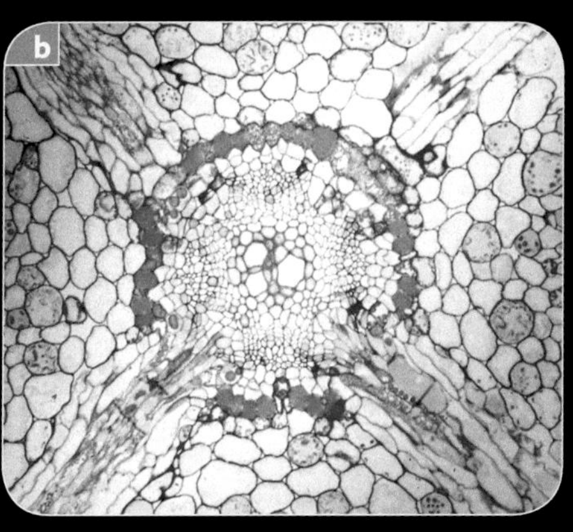

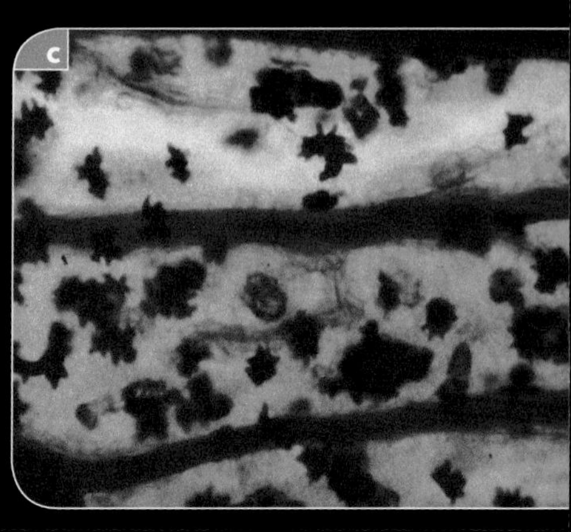

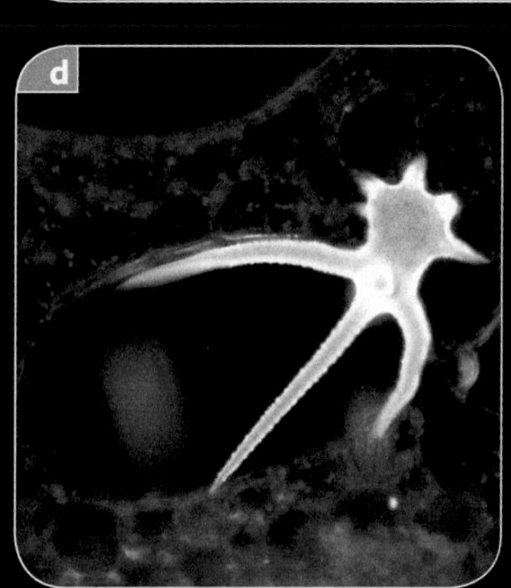

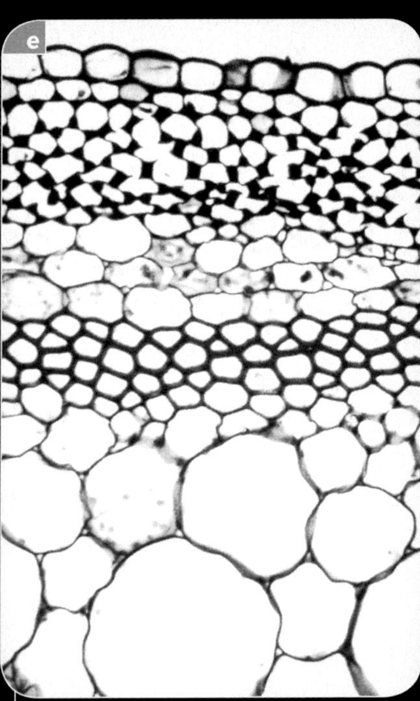

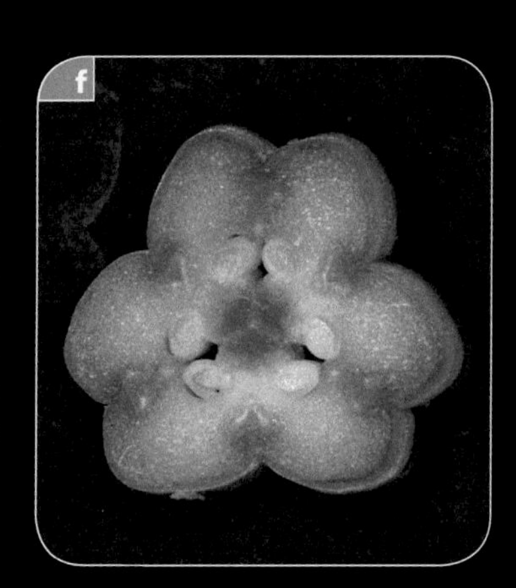

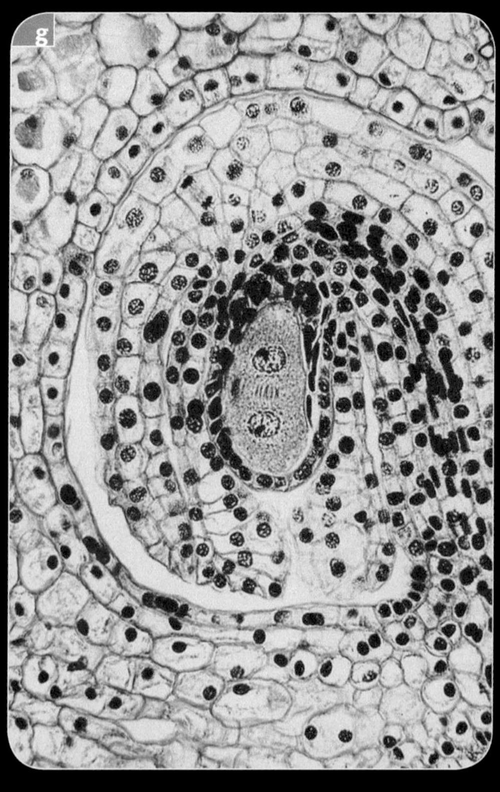

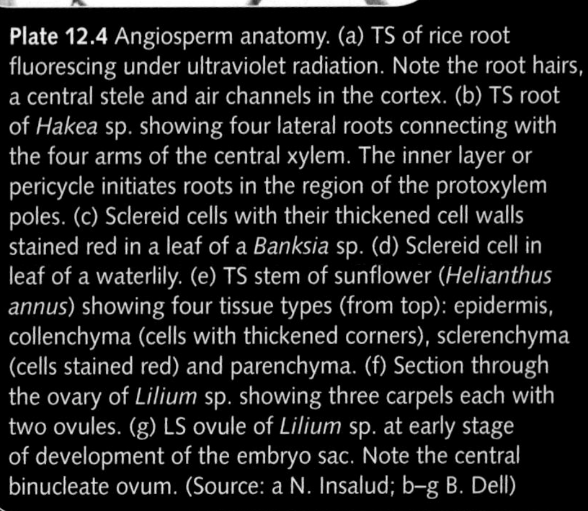

Plate 12.4 Angiosperm anatomy. (a) TS of rice root fluorescing under ultraviolet radiation. Note the root hairs, a central stele and air channels in the cortex. (b) TS root of *Hakea* sp. showing four lateral roots connecting with the four arms of the central xylem. The inner layer or pericycle initiates roots in the region of the protoxylem poles. (c) Sclereid cells with their thickened cell walls stained red in a leaf of a *Banksia* sp. (d) Sclereid cell in leaf of a waterlily. (e) TS stem of sunflower (*Helianthus annus*) showing four tissue types (from top): epidermis, collenchyma (cells with thickened corners), sclerenchyma (cells stained red) and parenchyma. (f) Section through the ovary of *Lilium* sp. showing three carpels each with two ovules. (g) LS ovule of *Lilium* sp. at early stage of development of the embryo sac. Note the central binucleate ovum. (Source: a N. Insalud; b–g B. Dell)

Plate 12.5 Examples of bark types. (a) Exfoliating bark of *Eucalyptus caesia*. (b) Persistent bark of *Corymbia calophylla*. (c) Annual sheet-like layers in a paperbark tree (*Melaleuca* sp.). (Source: a–c B. Dell)

Plate 12.6 Response of Geraldton wax (*Chamelaucium uncinatum*) to the addition of increasing doses of phosphorus fertiliser. (Source: B. Dell).

Plate 12.7 Examples of pollinator vectors for the Australian flora. (a) Singing honeyeater and *Grevillea*. (b) Honey possum and *Banksia*. (c) Jewel beetle and *Chamelaucium*. (d) Bee fly (Note the pollen grains on its back). (Source: a–d Lochman Transparencies)

Plate 12.8 Some fruit types. (a) Fleshy fruit of *Syzgium* (Myrtaceae). (b) Fleshy pericarp in oil palm (*Elaeis*, Arecaceae). (c) Woody follicle of *Hakea* (Proteaceae). (d) Open follicles of *Banksia* (Proteaceae). (e) Legumes of *Cassia* (Fabaceae). (f) Woody fruit capsules of *Eucalyptus macrocarpa* (Myrtaceae). (Source: a–f B. Dell)

Plate 12.9 Acacia and other seeds discarded next to an underground ant gallery. (Source: B. Dell)

Plate 14.1 Sponges are mainly marine animals with a simple body plan not organised into tissues or organs. (Source: Lochman Transparencies)

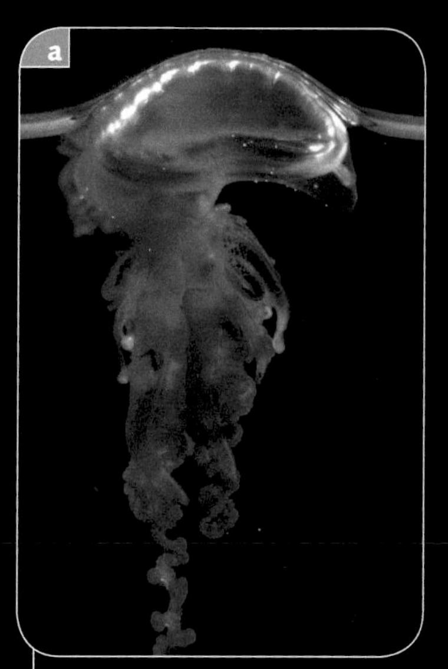

Plate 14.2 The basic body forms of the cnidarians are either (a) a free-floating medusa like the Portuguese man-of-war or (b) an attached polyp like the colonial anemone. (Source: Lochman Transparencies)

Plate 14.3 Soft corals show a great diversity of colour and form. (Source: Lochman Transparencies)

Plate 14.4 The massive calcium carbonate exoskeletons secreted by hard corals form coral reefs, which are hotspots of marine biodiversity. (Source: J. Huisman)

Plate 14.5 The diversity of segmented worms or annelids.
(a) Marine fireworm. (b) Fan worm. (c) Earthworm.
(d) Christmas tree worm. (e) Christmas tree worm. (f) Leech.
(Source: a, c, d, f Lochman Transparencies; b, e J. Huisman)

Plate 14.6 The diverse arthropods occur in marine, freshwater and terrestrial environments. (a) Freshwater crayfish. (b) Mantis shrimp. (c) Ghost crab. (d) Scorpion. (e) Katydid (bush cricket). (f) Centipede. (g) Golden orb spider. (h) Millipede. (Source: a, c, e–h, Lochman Transparencies; b J. Huisman; d R. Armistead)

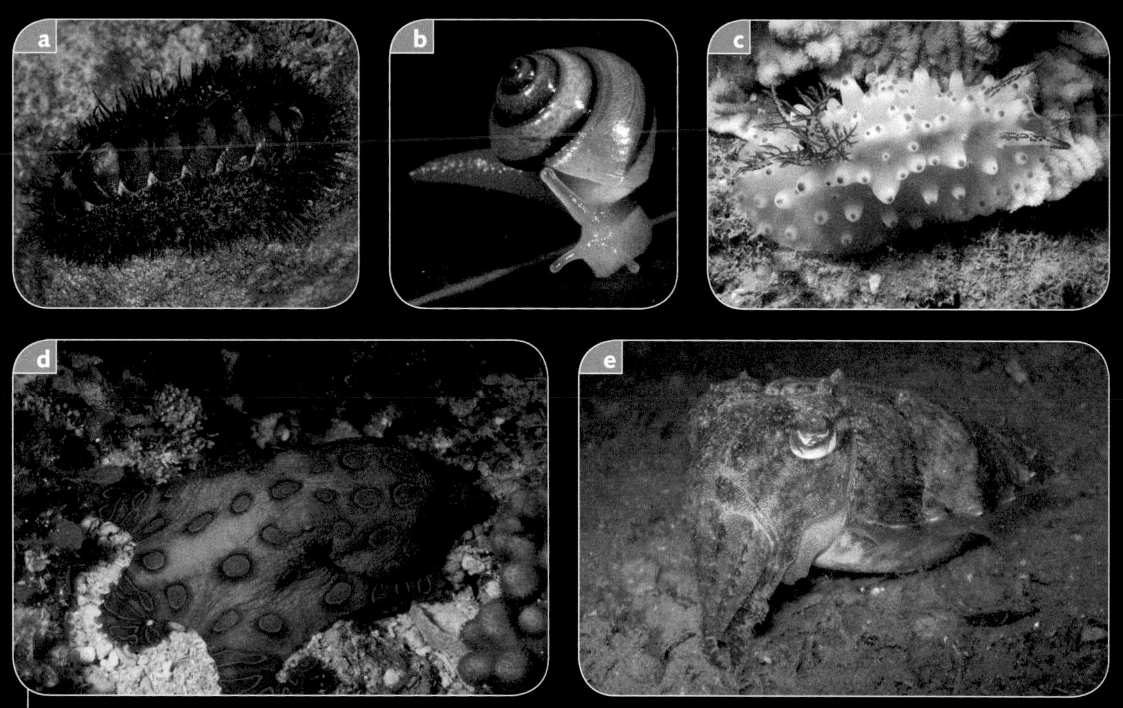

Plate 14.7 The basic mollusc body form of a soft body with a muscular foot and surmounted by a hard shell has been successful in aquatic and terrestrial environments. (a) Spiny chiton. (b) Rainforest snail. (c) Nudibranch. (d) Blue-ringed octopus. In the octopus, squid and their relatives the shell has been internalised or lost and the foot modified into tentacles. (e) Cuttlefish. (Source: a–d Lochman Transparencies; e J. Huisman)

Plate 14.8 Adult echinoderms are radially symmetrical, although the larvae are bilaterally symmetrical. (a) Sea star. (b) Brittle star. (c) Feather star. (d) Sea urchin. (e) Sea cucumber. (Source: a–e Lochman Transparencies)

Plate 15.1 The pouched lamprey is one of the few surviving agnathan fish and occurs in south-western Western Australia, South Australia, Victoria and Tasmania as well as New Zealand, Chile and Argentina. Adults are marine predators of larger fish, but migrate into freshwater streams to breed. The larvae (ammocoetes) are filter feeders that bury themselves in mud for up to 4 years. (a) Larvae are characteristically mud-brown in colour (the scale units are millimetres). (b) Immature adults (just metamorphosed) are silver at the sides with a blue back (the scale units are millimetres). (c) Adult male ready for spawning after upstream migration, with characteristic baggy gular pouch. (d) Toothed, suction-cup jawless mouth and rasping tongue of the adult. (e) Upstream breeding migration of adults. (f) Collecting adults using electrofishing. (Source: a–f David Macey)

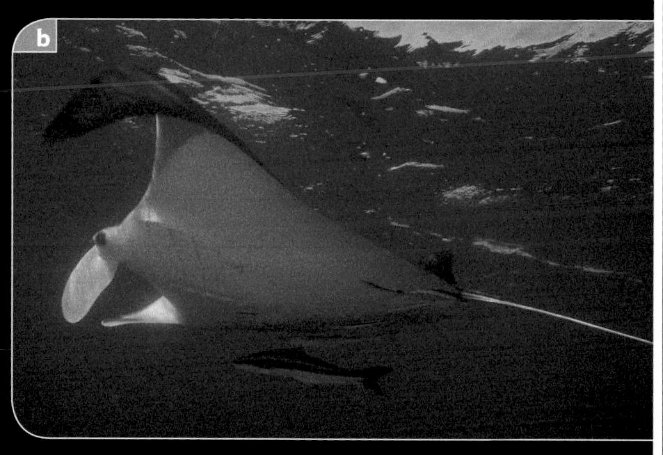

Plate 15.2 The Chondrichthyes, or cartilaginous fish, have skeletons of cartilage and a tough, leathery skin incorporating small, tooth-like scales. All chondrichthyes are carnivorous, for example (a) the grey nurse shark, but some such as (b) the manta ray are specialised for filter feeding. (Source: a, b Lochman Transparencies,)

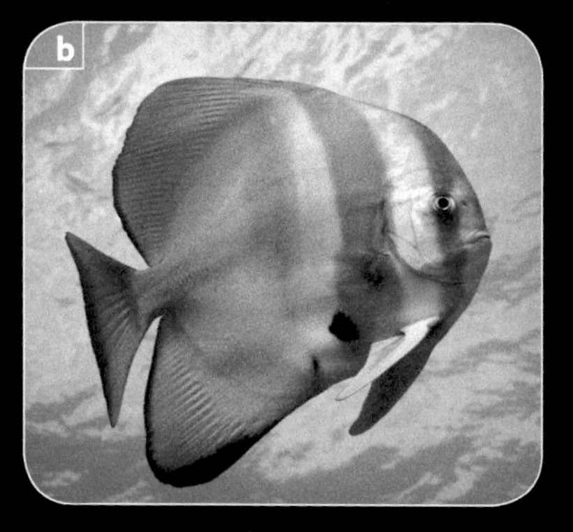

Plate 15.3 The Osteichthyes, or bony fish, have bony skeletons and a body covering of scales. They vary greatly in size, shape and reproductive strategies. (a) Trevally. (b) Batfish. (c) Moray eel. (d) Lion fish. (Source: a–d J. Huisman)

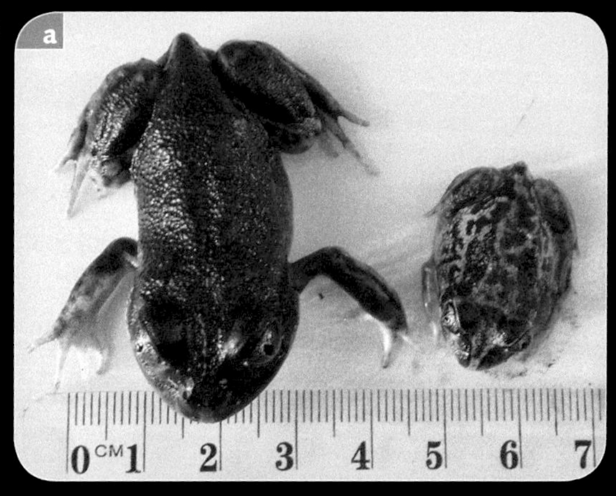

Plate 15.4 Most amphibians have aquatic, fish-like larval stages metamorphosing into terrestrial adults with moist skins. (a) The moaning frog takes its name from its distinctive call. Adult and subadult are shown. (b) The Turtle Frog of south-western Australia lays eggs with tough, water-resistant membranes for development in moist soil. (c) Magnificent tree frog from north-western Australia. (Source: a, b Mike Bamford; c Lochman Transparencies)

Plate 15.5 Reptiles have a waterproof scaly skin, internal reproduction and internal respiratory surfaces to reduce water loss and allow reproduction on land. (a) Red-bellied black snake. (b) Large varanid. (c) Thorny devil. (Source: a, b Lochman Transparencies; c Mike Bamford)

5. frontal section (FS) – while sagittal sections divide animals into left and right halves, frontal sections run crosswise from 'ear to ear', at right angles to sagittal sections. They can rarely divide an animal into equal halves because the anterior and posterior surfaces of animals are nearly always very different. You will note that frontal sections can in many circumstances be equivalent to TSs.
6. oblique sections (OS) – these are sections not parallel with the transverse, sagittal or frontal planes.

Despite their diversity of shapes and sizes, all animals share these features:

- They are all multicellular eukaryotes, lacking the supporting cell walls found in plants and fungi. Instead, they use extra-cellular structural proteins and specialised cell junctions for support and association, supplemented in most species by muscle cells.
- At some stage in their lives they can move from place to place.
- They are heterotrophic, ingesting other organisms in whole or in part and digesting them internally, rather than synthesising their own organic compounds from inorganic matter.
- Diploid adult animals produce haploid sex cells (gametes) from special reproductive organs, the gonads. Testes produce sperm and ovaries produce ova.
- New individuals develop from diploid embryos, which form after the fusion of sperm and ova. A unique family of regulatory genes called *Hox* genes controls the development of animal embryos in all except one group, the sponges.

All animals also face five common problems in surviving and propagating their genes:

1. Being heterotrophic they must obtain food, which involves special structures for locating, ingesting and digesting it.
2. They must access the oxygen needed for cellular respiration and distribute it to their cells.
3. Water balance and temperature must be maintained, so cellular reactions can proceed in solution at an appropriate rate.
4. The nitrogenous waste products of metabolism must be excreted, because the accumulation of even small amounts of these may impede cellular respiration.
5. They must reproduce successfully.

The solutions to these problems used by different animals relate to the environment, lifestyle and size of each species. We will therefore review these general principles before surveying the wide range of animal diversity in Chapters 14 and 15.

Influence of the environment

The physical properties of salt water, fresh water and air have important implications for the anatomy, physiology and behaviour of the animals inhabiting them. The interactions between organisms and their environments are explored in detail in Chapters 18–23.

Here we give a brief introduction to the main features of marine, freshwater and terrestrial environments and their effects on animal life.

The marine environment

Almost half of all animal phyla are restricted to the marine environment. One reason is that it is the most stable of the three main types of environment. Wave action, tides, ocean currents and upwellings ensure that there is little fluctuation in the concentrations of dissolved gases, so although oxygen concentrations are only about 2.5% of those in air they do not fall as low as those in some stagnant freshwater environments. There is no danger of respiratory surfaces drying by evaporation because they are immersed in water, so the external surface of the body or specialised gills exposed in the surrounding water can be used for oxygen uptake. Dissolved salts are also evenly distributed. Sea water has a similar salt concentration to the tissue fluids of many marine animals, so they can maintain water balance easily because there is no tendency for water to enter or leave the body via osmosis. Small organisms and food particles drift suspended in water and many marine animals specialise in collecting such particles, a process known as suspension-feeding. On land, only web-building spiders 'filter' food from the surrounding medium. Sea water also provides a resource for excreting nitrogenous (nitrogen-containing) wastes arising from protein metabolism. Ammonia is the basic nitrogenous waste in all organisms and, with the abundance of water available, marine animals such as bony fish commonly excrete it directly. Although ammonia is highly toxic, there is no shortage of water to dilute it.

Water temperature is much more stable than air temperature, because large amounts of energy are needed to raise water temperature. The narrow temperature ranges experienced by marine animals are less demanding on their physiology than the larger ones experienced on land. Similarly, the pH (acidity) of salt water varies much less than that of fresh water. It is maintained close to 8.1 by the presence of bicarbonate ions. Some physical conditions, however, do vary, such as the decline in light and the rise in water pressure with increasing depth. These changes require specialised adaptations from animals living in the ocean depths.

The buoyancy of sea water supports animal bodies, as well as allowing energy-efficient drifting on ocean currents. This is why the largest animals ever known, both living and extinct, have been marine. However, water is also more viscous than air, meaning that the molecules stick together more readily. This creates resistance to organisms moving through it, so marine animals, especially small ones, may need special physiological or behavioural adaptations for movement or for suspension-feeding.

Lastly, the marine environment is ideal for reproduction involving external fertilisation and development, so the great majority of marine animals use this form of reproduction. Sperm and ova can be shed into the water where they can mix for fertilisation, with no threat of desiccation or salt imbalance. The embryos of marine animals often develop into larvae, which may look quite unlike the adult. Larvae are specialised for feeding, growth and dispersal, rarely requiring large egg yolks to nourish their development. As they grow, they increasingly resemble adults, a process known as metamorphosis.

The freshwater environment

Freshwater environments provide good buoyancy for the animals that live in them, an excess of water so nitrogenous wastes can be excreted as ammonia and no danger of water loss by evaporation over respiratory surfaces. Apart from these points, however, they are much more variable than marine environments. Freshwater bodies may vary greatly in turbidity, velocity and volume, and these factors may change rapidly in response to heavy rain. Drought may also be a factor, sometimes leading to the complete drying of a water body.

Salt concentrations may also be very variable in fresh water. In general, freshwater environments have lower salt concentrations than those found in animal tissues, so animals tend to gain water by osmosis. The excess water must be removed while retaining important salts in the tissues, using a process called osmoregulation. However, in times of drought, evaporation may concentrate salts so that an inland water body may become saline, not fresh. In such cases, freshwater animals must prevent excess water loss by osmosis into the surrounding saline water (Box 13.2). In addition to this natural variability of the freshwater environment, freshwater animals may face other, externally imposed problems. Rivers and streams are often repositories for a range of human wastes and this pollution may have significant effects on freshwater life (Box 13.3).

Box 13.2 Osmotic regulation in brine shrimp

Brine shrimp are small crustaceans of the family Artemiidae, which live in salty lakes and other inland saline water bodies. The most widespread species is *Artemia salina*, originally described from salt works in Lymington, England, but which is now found (either naturally or as an introduced exotic) throughout the world. *Artemia* are a relatively recent introduction to Australia, which has its own, endemic genus of brine shrimp, *Parartemia*. Brine shrimp are harvested in large numbers from the wild because they are a commercially important source of food for fish larvae, for both industrial aquaculture and aquarium hobbyists.

Brine shrimp have an extremely wide salinity tolerance. They can live in water with a salt concentration as low as 4 g/L (about one-tenth the concentration of sea water) or as high as 200 g/L (about five times the concentration of sea water). Brine shrimp are able to cope with such large changes in salinity by maintaining the osmotic pressure of their haemolymph (blood) relatively independent of the external environment. They do this by active transport of ions and water from the environment, through the gut and into the haemolymph, and by active excretion of ions across the branchiae (gills) into the environment. This continuous, active transport of ions requires energy, and the more extreme the salinity the greater the energy expenditure. Survival, growth and reproduction all decline at salinities above 70 g/L, as more and more energy is required for ion transport and less is available for other bodily functions.

Box 13.3 The effect of pollution on aquatic environments

Water's properties as a solvent have made it a favourite medium for disposing of human wastes for a long time. In modern societies, domestic, industrial and agricultural wastes are discharged in vast quantities into waterways and coastal regions. The main pollutants and the problems they cause are:

- *inorganic nutrients.* Nitrogen and phosphorus pollutants from sewage and fertiliser run-off stimulate abundant growth of plants and algae. When these organisms die, bacteria decompose their bodies and in the process create an oxygen-depleted or hypoxic dead zone from the ocean bed upward to within a few metres of the surface. Free-swimming marine animals avoid these areas, but bottom-dwelling animals are killed. About 60 large coastal areas around the world are recognised as dead zones, including an area of about 17 000 km^2 in the Gulf of Mexico, into which pollutants are drained from the Mississippi River.
- *organic compounds.* These are chemicals that contain carbon atoms. Some, such as pesticides or herbicides, are toxic in their own right, whereas others may mimic the actions of natural hormones. The free-living embryonic and larval stages of many marine animals may be especially vulnerable because their small surface area to volume ratios facilitate uptake of pollutants and they are undergoing critical developmental processes that may be disrupted by organic pollutants.
- *inorganic chemicals.* These chemicals do not contain carbon. They do not degrade easily, so they may persist in the environment for long periods or accumulate in organisms. Many, such as lead and mercury, are toxic.
- *radioactive materials.* When unstable isotopes of some elements decay, they spontaneously emit energy either as rays or as subatomic particles. While all living things are exposed to background radiation normally, significant increases in this from exposure to radioactive materials may cause genetic damage.
- *sediments.* Suspended soil sediments from stormwater drains or agricultural run-off may enter waterways and be carried some distance before finally settling. In large quantities they may smother bottom-dwelling aquatic life, while even smaller quantities may clog gills or interfere with suspension-feeding.
- *heated water.* Many industries use water as a coolant and when this has been heated it is often discharged into waterways, raising local water temperatures. The increased temperature may interfere with reproductive cycles and oxygen availability (less oxygen dissolves in warm water than cool water) and alter metabolic processes.

Freshwater animals are much less likely than marine animals to produce free-floating eggs, because they can be swept away by rapid currents. Instead, eggs are usually retained by the parent until they hatch, or attached firmly to the substrate. Larvae are rarer than in the marine environment and the eggs are often provided with a large yolk for nutrition.

The terrestrial environment

The terrestrial environment is home to humans, so it can be hard to accept that it is the harshest of the three main environments. To begin with, compared to water there is little buoyancy, so all terrestrial animals require some form of skeleton strong enough to enable support and movement on land. Extremes of temperature are much more common, requiring biochemical, physiological or behavioural adaptations to cope with them.

All land animals also face the problem of significant water loss by evaporation. To control this, their body coverings must offer a more permeable barrier between internal and external environments than is common in marine and freshwater animals. It is also critical for terrestrial animals to locate their respiratory surfaces inside the body, or to protect them from desiccation if they remain external. Surfaces permeable to oxygen and carbon dioxide are also permeable to water, so evaporative water loss would be great if respiration occurred on the outside of the body. If respiratory surfaces are internal, with only a small external opening, evaporative water loss can be reduced. The few terrestrial animals, such as earthworms, that use the body surface for respiration protect it with mucous secretions to reduce water loss. Behavioural adaptations such as burrowing may also be important in reducing evaporation across the external body surface (Box 13.4).

Box 13.4 Life in a burrow

(Contribution from Mark Newton)

The arid-zone scorpion, *Urodacus yaschenkoi*, is a large scorpion (reaching a length of 7 cm) found throughout the interior of Australia. In this harsh environment, with extremes of temperature and low, unpredictable rainfall, the scorpion is one of the commonest of animal inhabitants. The arid-zone scorpion is remarkably fixed in its habitat, being found almost exclusively on sandy soil. This is related to the way that it has adapted to life in the desert – by inhabiting a deep and complex burrow.

The burrow of the adult scorpion consists of an entrance, with a width of about 2–3 cm, leading into a straight tunnel, constructed at an angle of around 30° to the ground. This entrance tunnel varies in length, being about 1 cm long in adult burrows, but longer in the burrows of younger animals. The entrance tunnel is followed by a series of spirals that vary in number and direction of rotation, with up to nine spiral turns being recorded. The burrow ends in a terminal living chamber, which is large enough for the animal to move around in and sometimes has an additional side chamber. The terminal chamber is the deepest point of the burrow and may be anywhere from 20 cm to 1 m below the surface of the ground. During construction of its burrow, a scorpion may excavate up to 400 times its own weight of sand.

Scorpions spend most of the day in the terminal living chambers of their burrows, usually emerging in the evening to catch their prey. By measuring temperature, humidity and moisture content at different soil depths, it has been found that below 40 cm in depth, the soil microenvironment is relatively constant, with a temperature of 20–25°C and a relative humidity above 80%. At the surface of the ground, temperatures may reach

continued ›

Box 13.4 continued ›

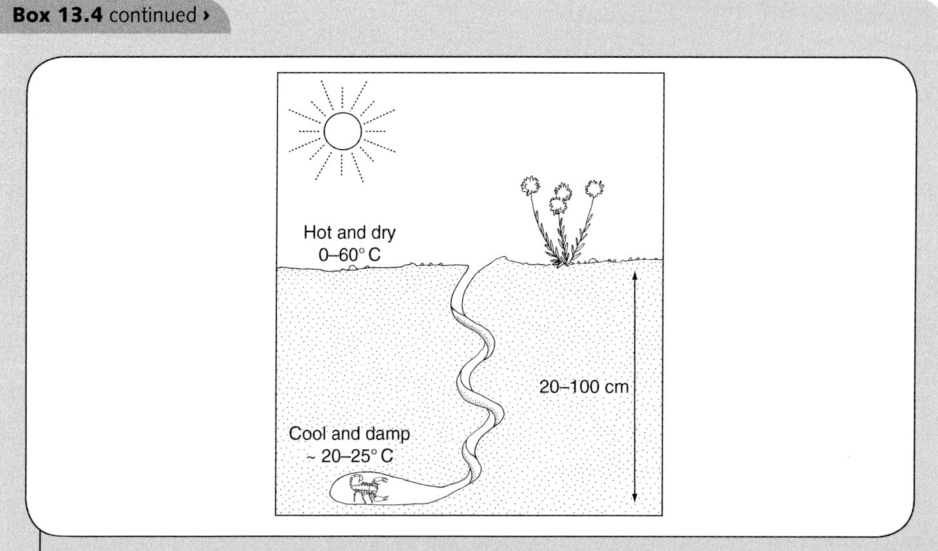

Figure 13.4.1 The arid-zone scorpion, *Urodacus yaschenkoi*, escapes the extremes of the arid-zone climate by constructing a deep, spiral burrow. (Source: From drawing by Mark Newton)

50–60°C during the heat of the day, with a relative humidity of less than 10%. In the burrow the scorpion escapes the severest extremes of the arid zone climate and burrowing is a key behavioural adaptation for desert dwelling in this species (Figure B13.4.1).

Land animals can also conserve water when excreting nitrogenous wastes. It would be extremely wasteful to excrete ammonia because of the large volume of water required to keep it dilute and non-toxic, so energy is expended to convert ammonia to urea (in mammals) or uric acid (reptiles and birds). Despite the energy cost, these compounds are less toxic so less water is needed to remove them. Uric acid demands the highest energy expenditure, but its toxicity is very low and little water is lost in its excretion.

Lastly, fertilisation must be internal in terrestrial animals, because the gametes could not survive and fuse without water. This in turn may require elaborate courtship behaviour. Young may be born alive, or eggs laid enclosed in a protective coat or located in a moist spot. The eggs cannot obtain water-soluble nutrients from their surrounding environment, so all the nutrients necessary for development must be available within the egg. Larvae are rare in terrestrial animals, but insects are a significant exception.

Influence of animal lifestyle

Broadly, animals can be grouped into those with a sedentary or passive, drifting lifestyle and those that are actively moving. Simple though it is, this distinction matches two very different animal body plans (Figure 13.2).

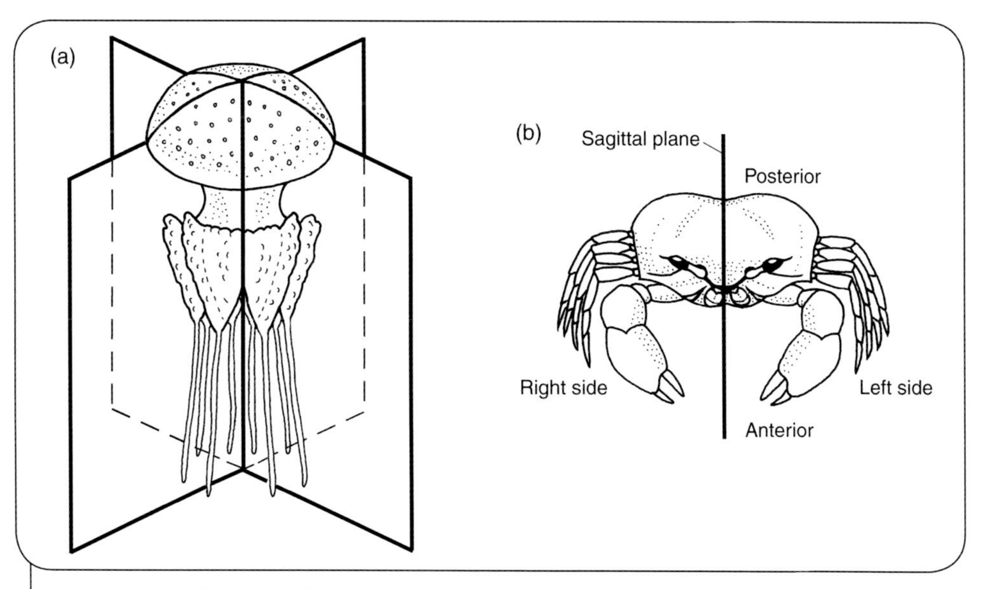

Figure 13.2 Two basic animal body plans. (a) Sedentary or drifting animals, such as this jellyfish, are radially symmetrical – any longitudinal plane will divide the animal into two equal halves. (b) Actively moving animals, such as this crab, are bilaterally symmetrical – only one plane (the sagittal plane) will divide the animal into two equal halves. (Source: B. Cale)

Sedentary or passively drifting animals show radial symmetry, in which the body parts radiate out from the centre. Any imaginary line dividing the animal vertically through the centre will produce two halves that are mirror images of each other. Such an animal has a top and a bottom, but it does not have right or left sides. This is an effective body plan for an animal that spends much of its life in one place. Radially symmetrical animals may feed by trapping passing prey, eating the substrate they rest on, or suspension-feeding. Skeletons or spines may be present for support or to discourage predators. A special problem they may face is obtaining oxygen in very still waters, because unless water is moving constantly past them they may exhaust the oxygen in the water close by. Regularly beating appendages may create water currents to overcome this problem.

In contrast, actively moving animals are bilaterally symmetrical. There is only one imaginary line that can divide the animal along its long axis and produce two halves that are mirror images. Therefore such animals have right and left sides, a distinct head end (anterior), a distinct tail end (posterior), an upper surface (dorsal) and a lower surface (ventral). Bilateral symmetry, often accompanied by a concentration of sense organs into a head at the anterior end, is well suited to active animals needing to process information from the direction in which they are moving. They may be fast-moving predators, or move more slowly and feed on plants or dead and decaying matter.

Whether an animal shows radial or bilateral symmetry also reflects a basic pattern of its embryonic development (Figure 13.3). When a sperm and an ovum unite, they form a diploid zygote. This divides by mitosis, forming a hollow ball of cells called a blastula. In most animals, this is followed by the gastrula stage in which one side of the blastula

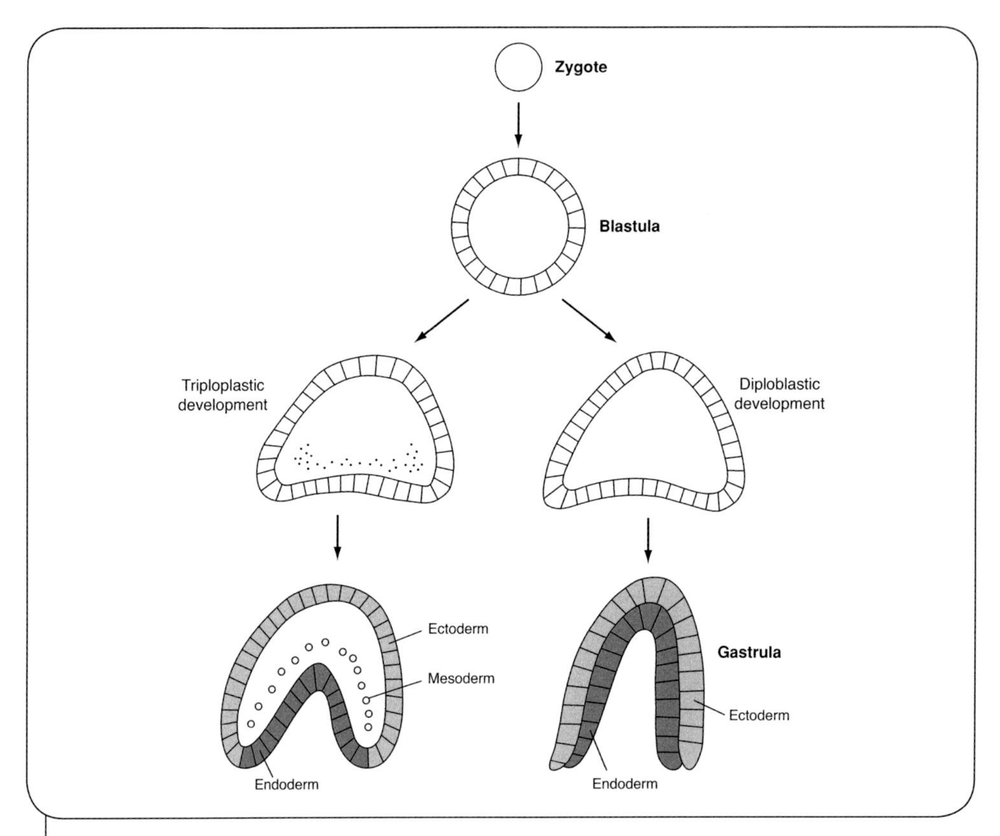

Figure 13.3 Gastrulation and tissue formation. The blastula, a hollow ball of cells arising from the zygote, folds inwards (invaginates) to form a gastrula. In (a) diploblastic animals, two primary germ layers (endoderm and ectoderm) arise during this process, whereas in (b) triploblastic animals, mesoderm develops between endoderm and ectoderm. Each of these cell layers may give rise to many tissues performing specific tasks within the animal body. (Source: B. Cale)

folds inwards, producing an internal sac that ultimately becomes the digestive tract. The cell layer lining the internal sac is known as the endoderm, and the second cell layer that lines the outer surface is called the ectoderm. In many animals a third cell layer, called the mesoderm, develops between the ectoderm and the endoderm. Each of these cell layers may give rise to tissues, which are groups of specialised cells, often isolated from other tissues by membrane layers, which perform specific tasks within the animal. Different tissues organised as a group to perform specific functions are called organs, and organs with related functions form systems. Animals having only two cell layers, the endoderm and the ectoderm, in the early embryo are called diploblastic and are all radially symmetrical. The great majority of animals have a mesoderm as well and are called triploblastic. Triploblastic animals all show bilateral symmetry. There is one exception to this radially symmetrical diploblastic/bilaterally symmetrical triploblastic dichotomy. A primitive group of animals, the sponges, live a sedentary adult lifestyle, but their bodies show no symmetry at all (they are asymmetric). Their embryonic development is also unique. After fertilisation, the blastula does not develop into a gastrula and no specialised tissues form.

Influence of animal size

A crude approximation to the volume of an animal is the cube of its length (length × length × length), and the surface area is about the square of its length (length × length). Therefore the ratio of surface area to volume (SA:V) falls as length increases (Figure 13.4). This simple mathematical relationship has profound consequences for the body plans of animals. Small animals may use diffusion across the body surface as a simple, effective means of gas exchange, excretion and transportation of materials within the body. However, as size increases there is insufficient surface area to support the larger volume of cells. The solution adopted by larger triploblastic animals is to develop circulatory systems to move dissolved substances around the body, and to fold and coil organs into fluid-filled internal body cavities, thereby packing the large surface areas of these organs into a compact space. For example, if the human small intestine was unfolded, it would extend for about 7 metres.

The fluid-filled body cavity is known as a coelom if it is completely lined by tissue of mesodermal origin (the peritoneum), or a pseudocoelom if not (Figure 13.5). Animals with

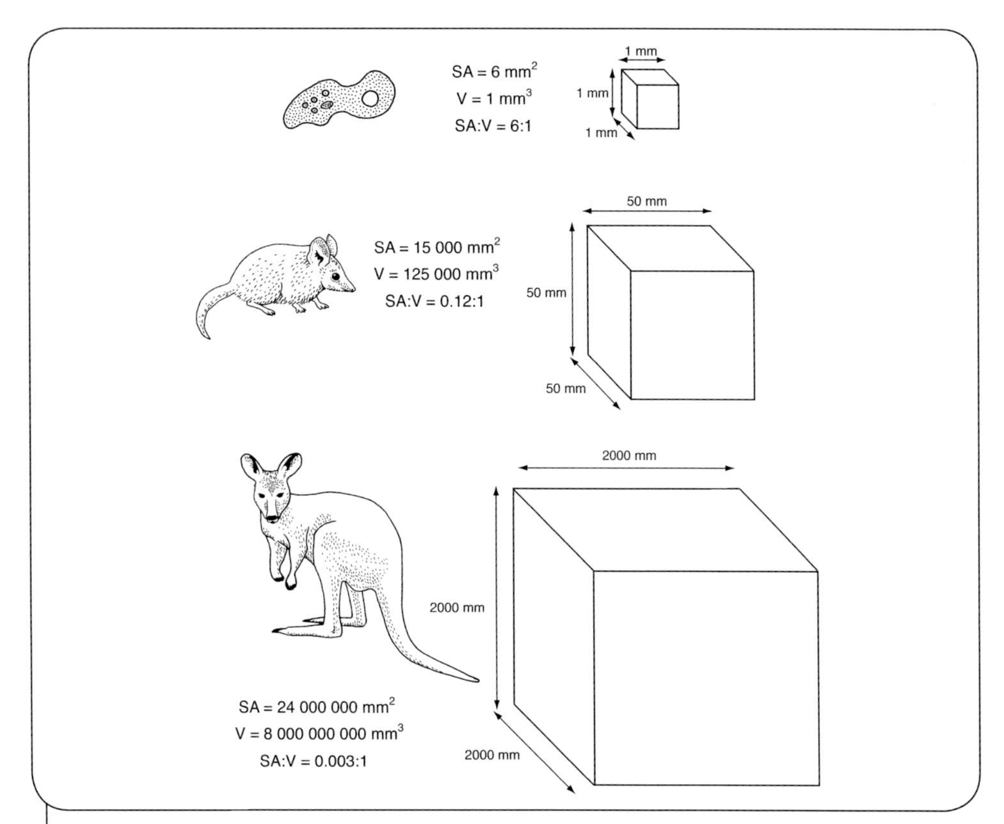

Figure 13.4 As the size of animals increases, their ratio of surface area to volume decreases, with important consequences for the efficiency of gas exchange, excretion and the internal transport of materials. (Source: B. Cale)

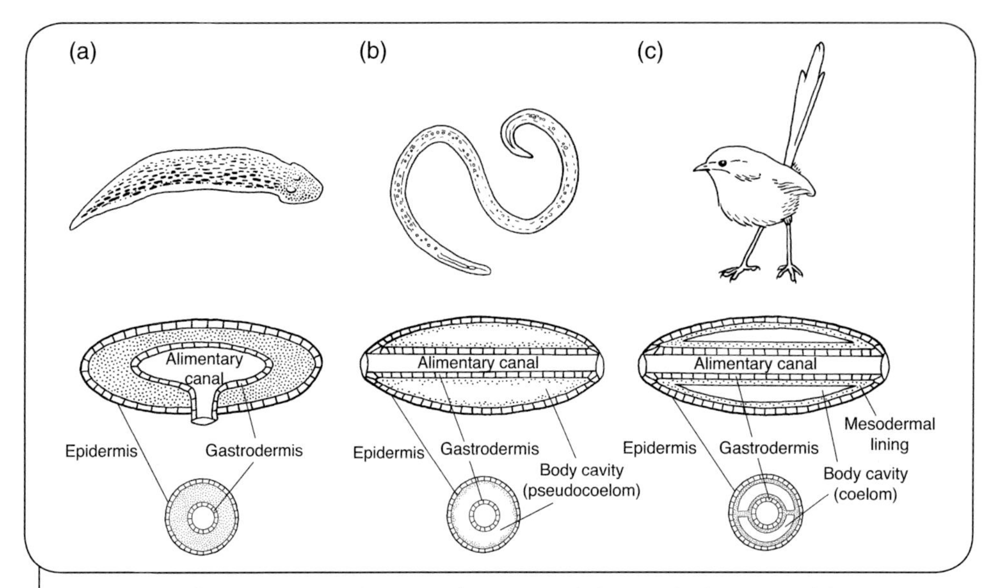

Figure 13.5 Body cavities in animals. (a) Acoelomate animals have no body cavity and the space between the digestive tract and the epidermis is filled with unspecialised cells. (b) Pseudocoelomate animals have a body cavity, the pseudocoelom, between the digestive tract and the epidermis. (c) Coelomate animals also have a body cavity between the digestive tract and the epidermis, and this is lined completely with cells of mesodermal origin. (Source: Adapted from an original drawing by Olivia Ingersoll published in Boolootian, R.A. and Stiles, K.A. (1981). *College zoology.* Macmillan Publishing Company and Collier Macmillan Publishers. 10th edn)

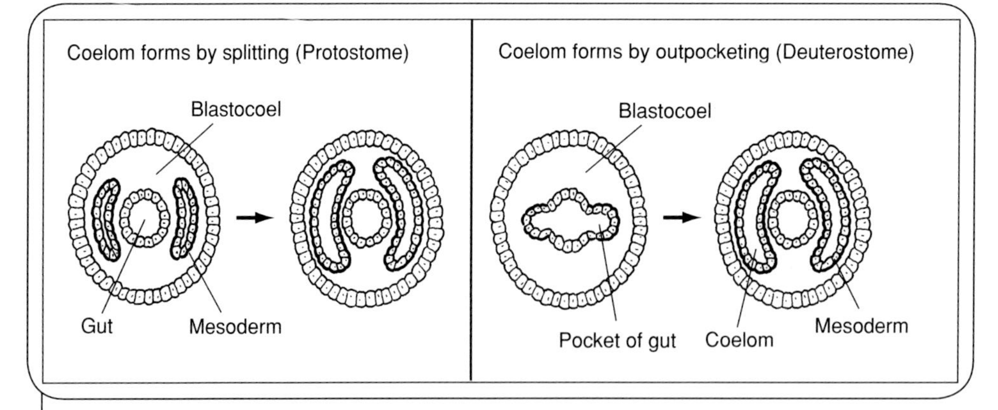

Figure 13.6 Embryonic development in protostomes and deuterostomes. In protostomes the blastopore, the opening formed when the blastula folds inwards during development, becomes the animal's mouth; the coelom forms from a mesodermal split. In deuterostomes, the blastopore becomes the animal's anus and the coelom arises from outpocketing of the embryonic gut. (Source: Adapted from Hickman, C.P. Jr., Roberts, L.S. and Larson, A. (2003). *Animal diversity*. McGraw-Hill, Boston. 3rd edn)

a coelom are further divided based on the development of the embryonic digestive tract. In protostomes, the opening formed when the blastula folds inward ultimately becomes the animal's mouth and the anus forms secondarily, whereas in the deuterostomes the original opening becomes the anus and the mouth forms secondarily. The coelom of protostomes forms by the splitting of bands of mesodermal cells surrounding the developing gut. In deuterostomes, the coelom arises from outpocketing of the embryonic gut (Figure 13.6). Regardless of the means of origin, the coelom provides space for folded and coiled organs of large surface area and enables increases in overall body size.

Animal body plans and classification

Adaptations of animals to different environments all show variations on a limited number of body plans. The most important distinctions are those between:

- animals with true tissues and those without them
- radially symmetrical and bilaterally symmetrical animals
- animals with or without a body cavity
- animals with a pseudocoelom and those with a coelom
- protostomes and deuterostomes.

These distinctions can be used to produce a phylogeny of the major phyla of the animal kingdom (Figure 13.7). The first branch splits the sponges, which lack true tissues, from the rest of the animals. Those with true tissues are then divided on the basis of radial or bilateral symmetry. The bilaterally symmetrical animals split into three groups: the flatworms, which lack a coelom (acoelomate); the nematodes, which have a pseudocoelom; and the remaining phyla, which have a true coelom. Those with a coelom then divide into those with deuterostome development and those with protostome development. Such a phylogeny is a subjective judgment based on careful studies of body plans.

As with protists and plants, recent developments in molecular techniques for studying the nucleotide sequences of genes coding for ribosomal RNA (rRNA) supported some features of the traditional phylogeny in Figure 13.7 and challenged others. The new results are controversial, but one widespread view is shown in Figure 13.8. The main similarities between the molecular and the traditional phylogeny lie in the unchanged positioning of the sponges, the cnidarians and the deuterostome phyla (the echinoderms and the chordates). The classification of the other groups is very different. The molecular phylogeny groups the acoelomate, pseudocoelomate and protostome phyla as protostomes. Within this group, the arthropods and the nematodes are placed together as Ecdysozoa, while the annelids, flatworms and molluscs form the Lophotrochozoa. While this classification no longer emphasises distinctions based on body cavities, it still has a morphological basis. The animals grouped in the Ecdysozoa all have a rigid body covering, which is moulted periodically as they grow. Many of the Lophotrochozoa have a specialised feeding structure called a lophophore at some stage of their lives, while others in the group have a particular type of larva called a trochophore.

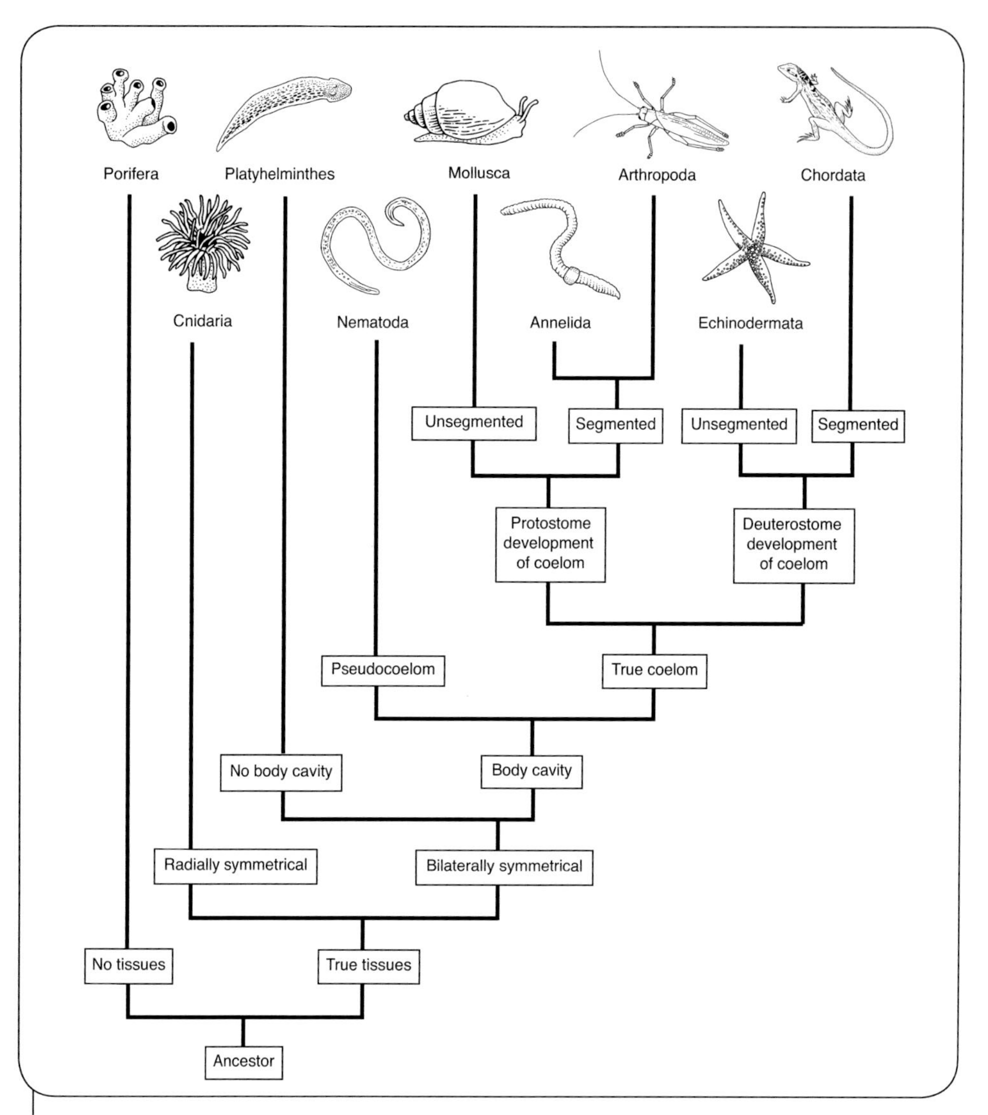

Figure 13.7 Traditional phylogeny of animal phyla based on the key transitions in animal body plans: the evolution of tissues, bilateral symmetry, a body cavity, deuterostome development and segmentation. (Source: B. Cale)

The differences between the two phylogenies illustrate the ongoing research and debate concerning the true evolutionary relationships underlying animal diversity. However, proponents of both trees share the common view that evolution is responsible for the great diversity of animal life and that all living animal groups show features reflecting their evolutionary history. In subsequent chapters we use the traditional phylogeny (Figure 13.7) when discussing the major animal phyla, although parts of it are likely to be overturned when final consensus on interpreting the molecular data is reached.

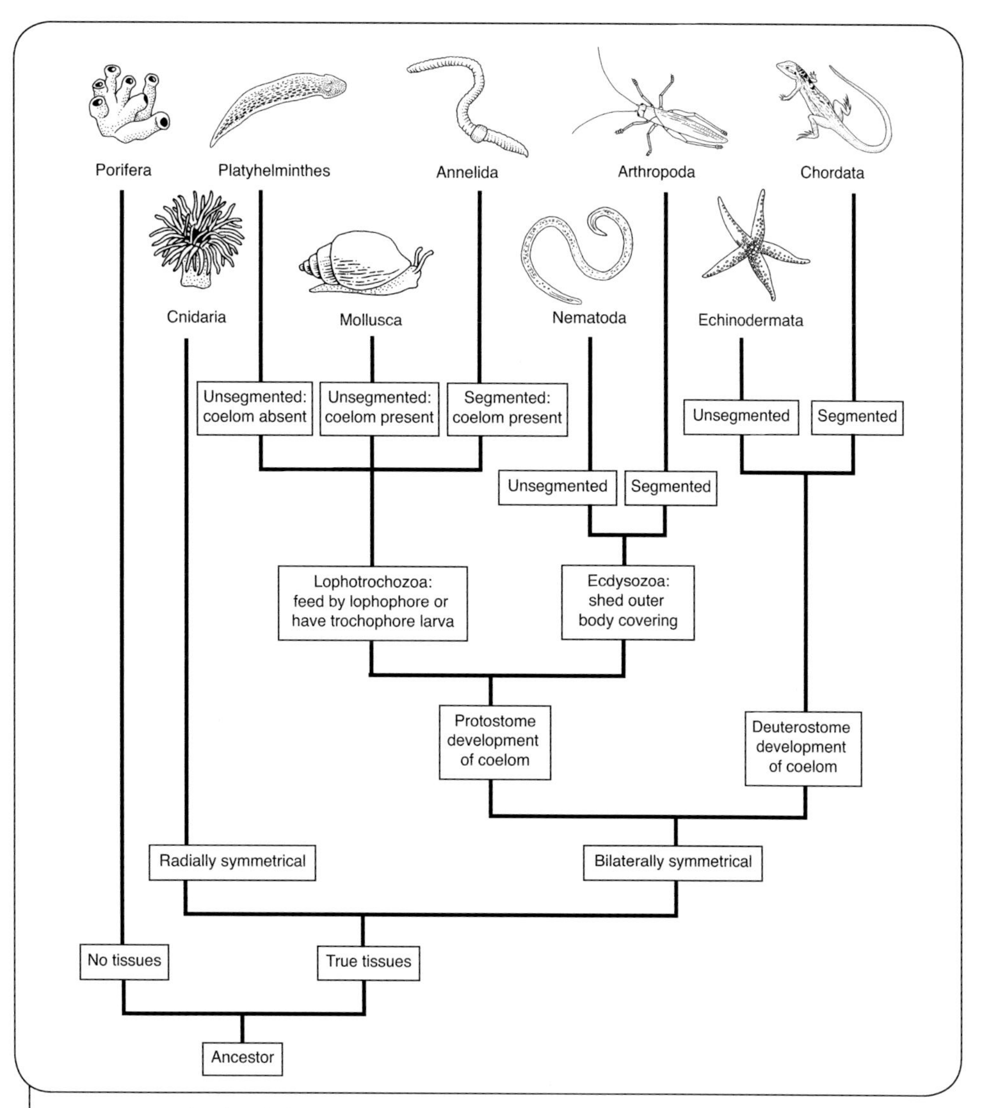

Figure 13.8 Molecular phylogeny of animal phyla. This phylogenetic tree, based on ribosomal RNA gene sequences, has some similarities to the traditional phylogeny, but also important differences, particularly in grouping animals with different body cavities. (Source: B. Cale)

Chapter summary

1. Animals are multicellular, heterotrophic and mobile organisms that develop from embryos arising from the fusion of gametes.
2. All animals must overcome five main problems: finding food, finding oxygen, maintaining water balance, removing nitrogenous wastes and reproducing. Animals' solutions to the problems of life depend on the environment, lifestyle and size of each species.
3. Marine animals benefit from a buoyant, stable and isotonic environment that provides an ideal medium for external fertilisation and development, often involving larvae. They excrete their nitrogenous wastes as ammonia.
4. Freshwater animals also benefit from buoyancy and the opportunity to excrete nitrogenous wastes as ammonia. However, they must cope with changes in the extent, turbidity and velocity of the water body and sometimes marked fluctuation in dissolved oxygen concentrations and salt levels. Osmoregulation is needed because salt levels are usually higher within the animal than outside it. Larvae are less common than in the marine environment.
5. Terrestrial animals must support their bodies in the absence of significant buoyancy. Water loss is a significant problem that is overcome by a resistant body covering, placing respiratory surfaces inside the body and excreting nitrogenous wastes as urea or uric acid. Fertilisation is internal.
6. Radially symmetrical animals have a top and a bottom, but no right or left sides. They are often sedentary or drifters. They are diploblastic, arising from two cell layers in the embryo. Free-moving animals often have a distinct head, where sense organs are concentrated, and a bilaterally symmetrical body. They are triploblastic, arising from three cell layers in the embryo.
7. As animals grow longer, their volumes increase more rapidly than their surface areas, so they can no longer rely on diffusion across the body surface for gas exchange and removal of metabolic wastes. Instead, they develop fluid-filled body cavities in which large organs can be folded or coiled to provide the necessary surface area in a compact package.
8. Phylogenies, whether derived from studies of animal body plans or from data on the molecular similarities of different animal phyla, highlight features believed to reflect the evolutionary history of animal groups.

Key terms

asymmetric
bilateral symmetry
blastula
coelom
deuterostomes
diploblastic
dorsal
ectoderm
endoderm
gastrula
protostomes
pseudocoelom
radial symmetry
triploblastic
ventral
zygote

Test your knowledge

1. What are the five problems of life faced by all animal species?
2. Name the principal nitrogenous waste product of each of the following animals:
   - freshwater fish
   - seed-eating bird
   - kangaroo.

   Relate the properties of the nitrogenous waste product to the environment of each animal.
3. Explain, with diagrams, the difference between radial and bilateral symmetry.
4. Explain, with diagrams, the differences between protostome development and deuterostome development.
5. Why is the terrestrial environment described as the harshest of all environments for life?
6. Outline the main differences between a traditional phylogeny of the animal kingdom and one based on similarities in ribosomal RNA.

Thinking scientifically

A friend announces excitedly that a new, large species of predatory marine animal has just been discovered. Before she can say anything more, you amaze her by listing several features which you say are likely to be true of the biology of this new animal. What features could be on this list?

Debate with friends

Consider the traditional and molecular phylogenies of the animal kingdom. Debate the topic: 'There is no practical reason to study molecular phylogenies'.

Further reading

Gordon, M.S. and Olson, E.C. (1995). *Invasions of the land: the transitions of organisms from aquatic to terrestrial life*. Columbia University Press, New York.

Gross, M. (1998). *Life on the edge: amazing creatures thriving in extreme environments*. Perseus Books, Cambridge MA.

Pechenik, J.A. (2005). *Biology of the invertebrates*. McGraw-Hill, Boston. 5th edn, Chapter 1.

14

Life on the move II – the spineless majority

Alan Lymbery and Mike Calver

Invertebrates – out of sight, out of mind

Land and Water Australia's 2002 terrestrial biodiversity assessment painted a grim picture for Australia's terrestrial fauna. It was estimated that 27 species of mammals, 27 species or subspecies of birds, one reptile species and four frog species have become extinct in Australia since European settlement, with a further 253 species or subspecies currently threatened with extinction. Yet these figures, sobering as they are, are a massive underestimate of the extinction crisis facing Australia's wildlife. The terrestrial biodiversity assessment considered only vertebrates (animals with backbones). However, the 6000 described species of vertebrates make up only about 6% of described animal species in Australia compared to about 100 000 described species of invertebrates (and there are at least 200 000 invertebrates still to be described).

The lack of attention paid to the conservation of invertebrates is not because they are not endangered. The problem is lack of knowledge. In a scientific sense, we have much to learn about the taxonomy, biology, ecology and conservation status of invertebrates. Invertebrates are very often ignored or actively discouraged by the general public. However, they are critical components of ecosystems, providing integral links in the food chain as well as essential ecosystem services such as plant pollination, soil aeration, organic decomposition and pest control.

Chapter aims

In this chapter we introduce you to the main phyla of invertebrate animals. Each phylum has a common evolutionary history and common features that have evolved to cope with environmental challenges (Chapter 13). This is the body plan and our coverage of each phylum begins by describing it. Next, we describe the organs and organ systems responsible for movement (skeletal and muscular systems), feeding (digestive system), transporting food and oxygen internally (respiratory and circulatory systems), excreting waste products (excretory system), coordinating bodily activities (nervous and endocrine systems) and reproducing (reproductive system).

Phylum Porifera – the sponges

General description

Sponges (Plate 14.1) are mainly marine (about 5000 species) and sometimes freshwater (about 150 species) invertebrates found throughout the world. There are about 1500 Australian species. They are usually classified into three classes:

1. the Calcarea, with calcareous skeletons
2. the Hexactinellida or glass sponges, a deep-sea group with siliceous skeletons
3. the Demospongiae, with skeletons made of protein or silica, but in a different form to that of glass sponges.

Sponges have a simple body plan not organised into tissues or organs. They are tube-like, with a body wall in three functional layers (Figure 14.1). The inner layer consists of flagellated choanocytes (collar cells) lining an internal cavity or atrium. In simple sponges, the atrium is a single tube, but in more complex forms infoldings of the atrium wall produce many flagellated chambers for feeding (Figure 14.1). The atrium opens by one or more oscula (singular: osculum). The outer surface consists of a layer of flattened cells called pinacocytes, similar to the epithelial cells forming the outer layer of other multicellular animals. Between these two layers is a gelatinous, protein-rich matrix called the mesohyl, containing amoeboid cells (amoebocytes) and skeletal material.

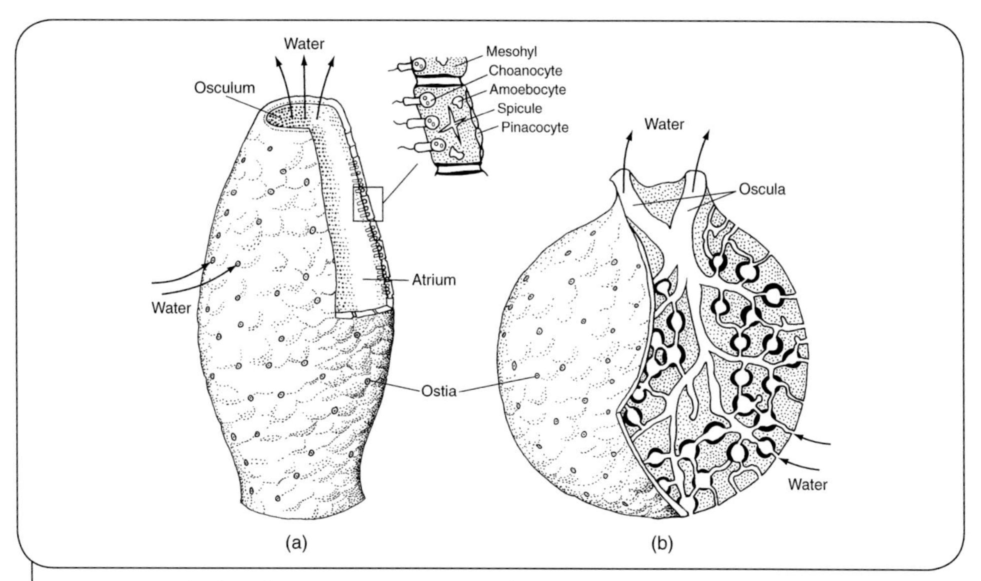

Figure 14.1 Body plan of a sponge. (a) Simple tube form. (b) More complex form with feeding chambers. The internal cavity or atrium is connected to the outside by an osculum. Surrounding the atrium are three body layers: choanocytes, mesohyl and pinacocytes. (Source: B. Cale)

Skeleton and locomotion

Although larval sponges are free swimming, adults attach to rocks, often living colonially on the ocean floor. Shape is maintained by a skeleton composed of minute needles (spicules) of calcium carbonate or silica, fibres of a coarse protein called spongin, or both.

Feeding and digestion

Sponges filter feed. The choanocytes lining the inside of the sponge beat their flagellae, creating a current drawing water in through numerous small pores (ostia) in the external wall. Plankton and other small organisms are filtered by pinacocytes or choanocytes as water flows into the atrium before being expelled through the oscula. A complex body form, with many feeding chambers, increases the efficiency of filter feeding, allowing a larger body size.

Reproduction

New sponges may form asexually from buds. Sponges have a great capacity for regeneration, and any fragment containing both amoebocytes and choanocytes can reconstitute a new individual. Sponges also reproduce sexually, with differentiated amoebocytes producing eggs and sperm. Most sponges are hermaphroditic (the same individual produces both sperm and ova), but some have separate sexes (dioecious). The fertilised eggs are usually retained within the parent until flagellated larvae develop and swim out through the osculum. After a short planktonic stage they settle and develop into adults.

Phylum Cnidaria – the jellyfish and their relatives

General description

There are about 9000 cnidarian species, nearly all marine, with a few found in fresh water. Australia has about 1700 described species. We recognise five major classes:

1. the Hydrozoa or hydroids, mostly marine and colonial species, although some, such as *Hydra*, are solitary and live in fresh water
2. the Scyphozoa or jellyfish, which are exclusively marine, spending most of their lives floating near the surface of coastal waters
3. the Cubozoa or box jellyfish, cuboidal jellyfish with similar life histories to scyphozoans
4. the Anthozoa, a large marine group, including sea anemones, corals, sea fans and sea pens, and
5. the Ctenophora or comb jellies, a marine group quite different to other cnidarians and often classified in a separate phylum.

Cnidarians represent a major evolutionary transition from sponges. They are radially symmetrical and their bodies comprise distinct tissues, but not true organs. They are

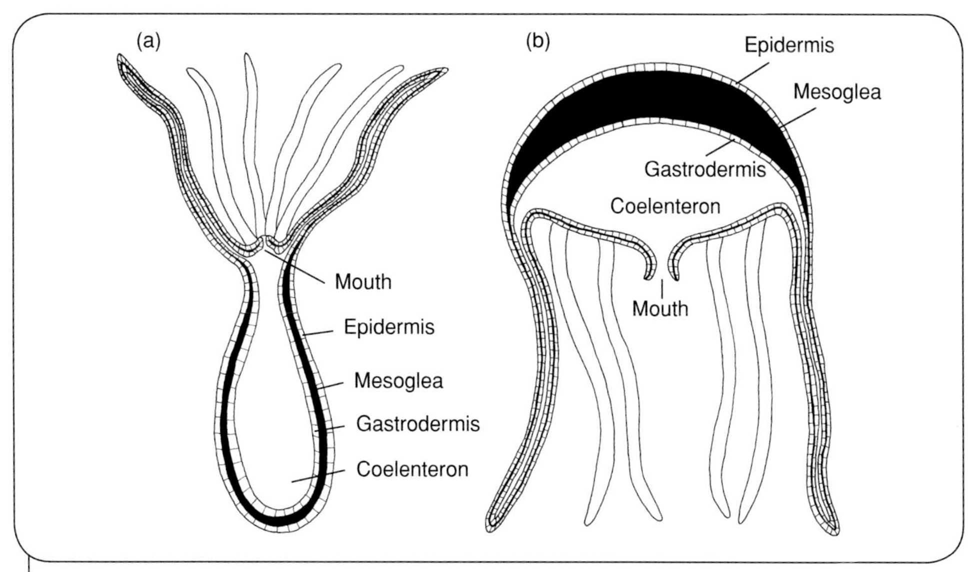

Figure 14.2 Body plan of a cnidarian. (a) Polyp form. (b) Medusa form. The central cavity or coelenteron is surrounded by three tissue layers: gastrodermis, mesoglea and epidermis. (Source: B. Cale)

diploblastic (see Chapter 13), and the inner endoderm and outer ectoderm differentiate into the adult tissues.

The basic body plan is a sac with a central gastrovascular cavity or coelenteron (Figure 14.2). The opening, fringed with tentacles, serves as both mouth and anus in all cnidarians except comb jellies, which have additional anal pores. The coelenteron is lined by a gastrodermis, formed from endodermal cells, whereas the body is covered with an epidermis, formed from ectodermal cells. Between the gastrodermis and epidermis is a layer of gelatinous mesoglea. There are two variations on this body plan: polyp and medusa (Figure 14.2; Plate 14.2). Polyps are tubular and usually attached to a substrate, with the mouth upwards. Medusae are bell-shaped and free floating, with the mouth underneath. In many cnidarian species, particularly in the class Hydrozoa, polyp and medusa stages alternate in the life cycle, whereas in other species one stage is reduced or absent (this is an alternation between two diploid forms of the same organism and thus different from haploid and diploid generations in plants). In scyphozoans, cubozoans and ctenophores, the medusa predominates, whereas anthozoans have only a polyp phase.

Skeleton and locomotion

Corals and colonial hydroids are the only cnidarians with solid skeletons (Plates 14.3 and 14.4). In soft corals this is an endoskeleton of calcareous spicules embedded in the mesoglea, whereas hard corals and some colonial hydroids secrete an exoskeleton. The massive calcium carbonate exoskeletons secreted by hard corals form coral reefs (Box 14.1).

Box 14.1 Corals as ecosystem engineers

Ecosystem engineers are organisms that alter their physical environment and create living spaces or niches for other organisms. In many ways, therefore, ecosystem engineers are important determinants of local biodiversity. Corals are ecosystem engineers *par excellence*.

Coral reefs are composed of the calcium carbonate secretions of coral polyps, built up over many years as each generation of coral polyps enlarges the reef by building on the skeletons of its ancestors. Only the surface layer of a coral reef is composed of living polyps. Coral reefs are centres of marine biodiversity; individual reefs may harbour up to 400 species of coral, 400 species of sponges, 4000 species of molluscs and 1500 species of fish.

Coral reefs contain such a rich diversity of life because they provide the essential ingredients of space, light and nutrients. Physically, reefs provide substrate and attachment points for benthic organisms, such as sponges, molluscs and crustaceans, and shelter and breeding sites for active organisms such as fish. Corals live only in shallow, well-lit seas, so abundant light is present for photosynthesis. The coral polyps constantly secrete calcium, which is essential for photosynthesising organisms such as phytoplankton. In addition, abundant populations of filter-feeding organisms such as sponges continually clean the water, enhancing water quality for all organisms.

Whereas the polyps of corals, sea anemones and colonial hydrozoans are mostly sessile, those of some solitary hydrozoan species, such as the common freshwater genus *Hydra*, glide on a basal disk or move by looping – bending over to attach to the substrate by their tentacles and then somersaulting to a new location. Some species of anemones swim by rowing with their tentacles or by strong bending movements of the body. Free-swimming medusoid stages have various means of locomotion. Specialised muscular cells ring the margin of the bell of jellyfish; rhythmic contractions against the mesoglea or fluid in the gastrovascular cavity expel water from beneath the bell, propelling the animal upward or forward. Ctenophores or comb jellies contain eight comb-like plates of fused cilia, which beat in synchrony to propel the animal.

Feeding and digestion

Cnidarians are predators, using stinging nematocysts to capture prey. Nematocysts are characteristic of all cnidarian classes except Ctenophora, in which only one species (*Euchlora rubra*) produces its own nematocysts.

Nematocysts are produced in specialised cells called cnidocytes. Each is a coiled, thread-like tube, often barbed and inverted within a capsule (Figure 14.3). When the nematocyst is triggered it discharges explosively, propelled by osmotic pressure into the prey (or into a predator or competitor, for nematocysts are used defensively too). Some nematocysts contain toxin that immobilises the prey and produces a stinging sensation in people. Some stings, like those of the box jellyfish *Chironex fleckeri*, can be fatal to people.

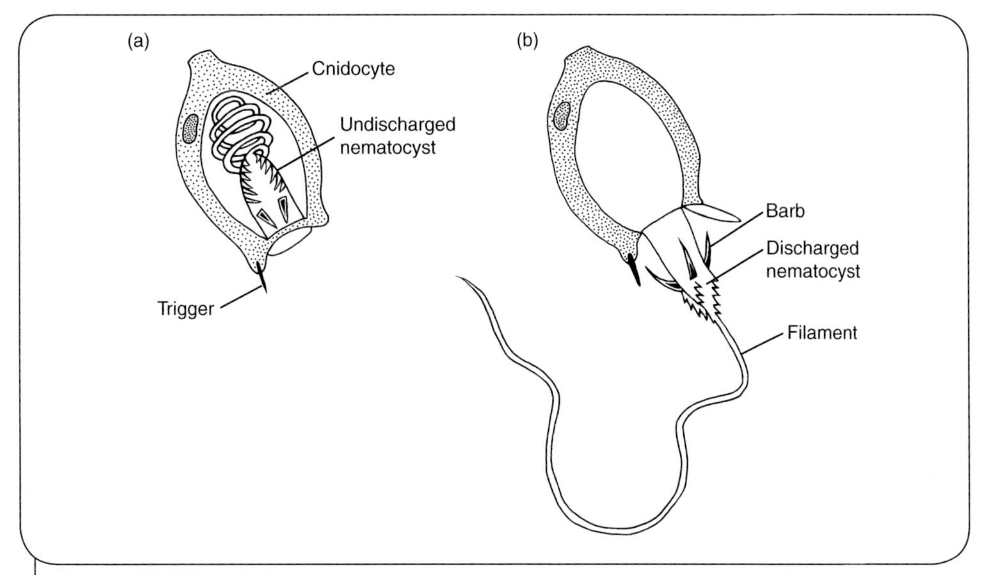

Figure 14.3 Structure of a nematocyst. (a) Undischarged. (b) Discharged. Nematocysts are produced in specialised cells called cnidocytes. (Source: Adapted from an original drawing by Olivia Ingersoll published in Boolootian, R.A. and Stiles, K.A. (1981). *College zoology*. Macmillan Publishing Company and Collier Macmillan Publishers. 10th edn)

Although all cnidarians are carnivorous, some species, particularly corals, have intracellular algae (zooxanthellae), which provide additional nutrition. Almost all reef-building (hard) corals possess zooxanthellae. The coral use the oxygen and food produced by photosynthesising algae to grow, reproduce and form their exoskeleton. The algal pigments also colour the coral. Stressed corals may expel their zooxanthellae, whitening and eventually starving the corals (coral bleaching; see Chapter 18, Box 18.1).

Cnidarians have extracellular digestion. Digestive enzymes secreted into the coelenteron break ingested prey into smaller fragments engulfed by cells lining the cavity. Cnidarians can therefore eat animals much larger than individual cells.

Respiration, circulation and excretion

Because most of the living cells of cnidarians lie very close to the animal's surface, complex organs of respiration, circulation and excretion are unnecessary. Oxygen is absorbed from the surrounding water through the epidermis or gastrodermis, with direct diffusion to underlying tissues. Metabolic wastes, largely carbon dioxide and ammonia, diffuse outwards. Larger waste particles are carried by amoeboid cells into the coelenteron and expelled along with undigested food through the mouth in hydrozoans, jellyfish and anthozoans, or through two anal pores in comb jellies.

Nervous and endocrine systems

The contractions of cnidarian muscles are controlled by nets of nerve cells over the surface of the mesoglea, but there is no central control of nerve function. All cnidarians possess epidermal sensory cells, perceiving light, chemical stimuli or mechanical stimuli. There are no endocrine glands, but some peptides and other unidentified compounds appear to be important in regulating growth.

Reproduction

Polyps often reproduce asexually by budding off new polyps or medusae. Colonial hydrozoans and corals are long chains of asexually produced polyps. All cnidarians can regenerate body parts. Sexual reproduction usually occurs in the medusoid form, except in anthozoans, where medusae are absent and polyps reproduce sexually. Sexes are usually separate (dioecious), although sea jellies are hermaphroditic (male and female gametes produced in the same individual). Eggs and sperm are produced in simple gonads. Typically, gametes are released into the coelenteron and then expelled for external fertilisation. Fertilised eggs give rise to ciliated, free-swimming planulae larvae, which are common in plankton and may be dispersed widely in oceanic currents. In some anthozoans, fertilisation is internal and larvae are retained to the polyp stage.

Phylum Platyhelminthes – the flatworms

General description

Platyhelminths or flatworms live in marine, freshwater or terrestrial habitats. There are about 20 000 free-living and parasitic species worldwide (about 1600 Australian), from four classes:

1. the free-living Turbellaria, also called planarians
2. the Monogenea, which are nearly all ectoparasitic (living on the outside of their host) on aquatic vertebrates
3. the Trematoda, or flukes, which are all endoparasitic (living inside their host), often with complex life cycles involving both invertebrate and vertebrate hosts
4. the Cestoda, or tapeworms, which are also endoparasitic with two or more hosts in the life cycle, at least one of which is usually a vertebrate.

Platyhelminths are bilaterally symmetrical, having distinct anterior and posterior ends, and dorsal and ventral surfaces. They are also triploblastic, with a third germ layer (mesoderm) forming between ectoderm and endoderm during embryonic development (see Chapter 13). This allows the development of specialised organs for digestion, excretion or reproduction.

Platyhelminths are ribbon-shaped and dorsoventrally flattened. Their bodies are solid, with no internal cavity except for an endoderm-lined digestive tract, with one opening (Figure 14.4a). Tapeworms are the most specialised, with many adaptations to parasitism. They usually consist of an attachment organ called a scolex, behind which is a growth zone or neck, budding off segment-like proglottids. Each proglottid has a complete set of reproductive organs and more reproductively mature proglottids are pushed further back as new ones form in front of them (Figure 14.4b).

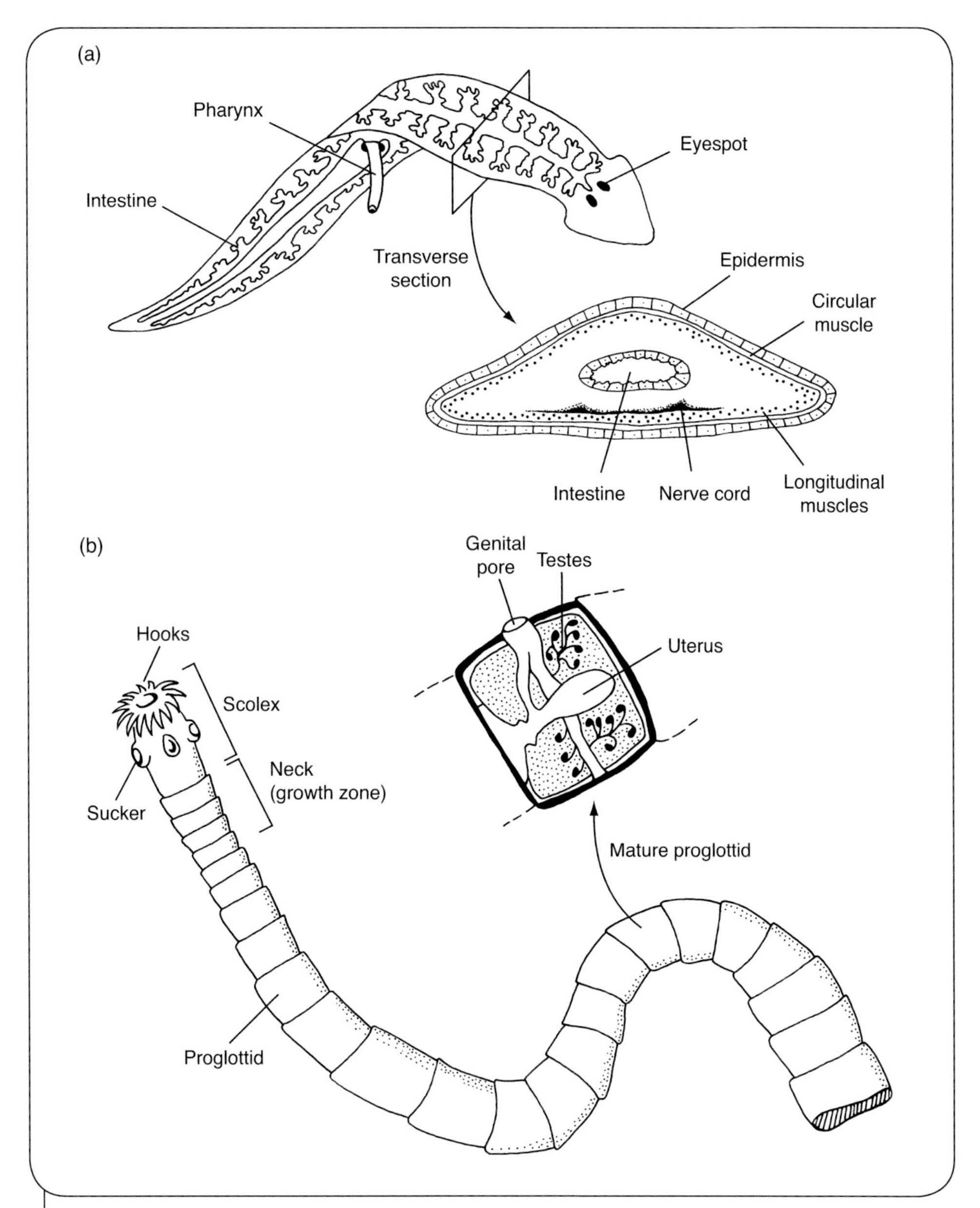

Figure 14.4 Body plan of a flatworm: (a) typical planarian structure; (b) typical tapeworm structure. The planarian body plan is similar to that of monogeneans and flukes, whereas the tapeworm body plan is highly specialised. (Source: Adapted from an original drawing by Olivia Ingersoll published in Boolootian, R.A. and Stiles, K.A. (1981). *College zoology*. Macmillan Publishing Company and Collier Macmillan Publishers. 10th edn)

Skeleton and locomotion

Platyhelminths are soft-bodied, with no internal or external skeleton. Parasitic platyhelminths (monogeneans, trematodes and cestodes) have a resistant outer body covering or tegument. Cestodes also secrete a protective cuticle against the host's digestive enzymes. In turbellarians the epidermis, particularly on the ventral surface, is ciliated and these cilia are used for creeping. In terrestrial species, ventral mucus-secreting glands provide lubrication for ciliated locomotion. Muscular contractions are used for more rapid movement in all four classes. All flatworms have a layer of circular muscles (operating in a radial axis) and a layer of longitudinal muscles (operating in the longitudinal axis). Alternate contractions of these muscle layers in different sections of the body create undulations, passing backwards along the body and pushing the animal forwards.

Feeding and digestion

Most free-living turbellarians are predators or scavengers. Their simple gut, which may be branched to increase the surface area for digestion, has a single ventral opening. A muscular, ectoderm-lined pharynx extends back from the mouth and may be everted from the mouth for feeding. The pharynx leads to an endoderm-lined intestine. Prey or organic debris is swallowed whole and digested by intestinal enzymes, or partially digested externally. Intestinal cells then engulf the food particles.

Monogeneans live and feed on the outside of their host, usually on the skin and gills of fish or other aquatic vertebrates. Their posterior adhesive organ bears suckers or hooks for attachment. Trematodes and cestodes are endoparasites, living in and feeding on the internal tissues of their host. Trematodes attach with suckers, usually one oral and one ventral. The scolex of cestodes also bears suckers and often hooks. Monogeneans and trematodes have a mouth, pharynx and intestine for feeding and digestion, but cestodes have no gut and absorb food directly through their tegument.

Respiration and circulation

Platyhelminths lack respiratory and circulatory systems. Their thin bodies and branched digestive tracts enhance diffusion to underlying tissues.

Excretion

Most metabolic wastes diffuse directly through the epidermis or into the gut for expulsion through the mouth, but there is an excretory system consisting of a network of fine tubules running through the body. They open to the outside through excretory pores and within the body end in organs known as flame cells. Water and waste products enter the flame cells and beating cilia move them through the tubules to the excretory pores.

Nervous and endocrine system

The simplest turbellarians have a nerve net similar to cnidarians. More advanced turbellarians and the parasitic groups have a central nervous system, with a concentration of nerve cells at the anterior end (cerebral ganglion) and two or more long nerve cords, often joined by transverse connections, running the length of the body. Sensory organs such as chemoreceptors and tactile receptors are spread over the body. Turbellarians and the free-living stages of parasitic groups may have eyespots containing concentrations of photoreceptors. Neurosecretory cells in the cerebral ganglion or main nerve cords produce neuropeptides acting at sites remote from the point of secretion. This gives a simple endocrine system (where ductless glands secrete specialist chemicals, hormones, for transfer to another tissue where they will have a specific effect).

Reproduction

Adult turbellarians reproduce asexually by budding or by transverse fission, where the animal divides into roughly equal halves and each half regenerates to form a complete animal. Trematodes and cestodes may also reproduce asexually, usually by specialised budding in the larval stages.

Sexual reproduction also occurs in platyhelminths. Most are hermaphroditic, although some are dioecious. Fertilisation occurs internally, sometimes by hypodermic impregnation of sperm directly through the body wall, but more often using copulatory organs to transfer sperm from male to female ducts.

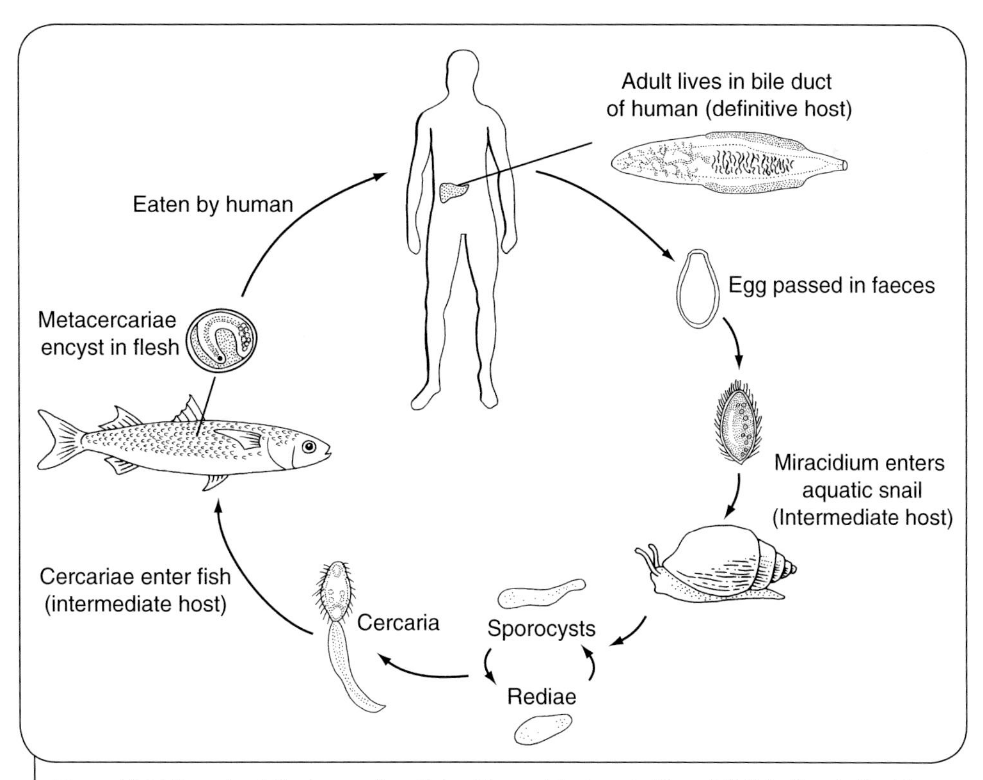

Figure 14.5 Life cycle of the human liver fluke, *Clonorchis sinensis*. The adult fluke lives in the bile duct of people, or sometimes other mammals. Eggs are passed in the faeces and hatch in water to a larval form called a miracidium. The miracidium penetrates an aquatic snail, where it transforms into a sporocyst, which in turn gives rise to many rediae through a process of asexual reproduction. Rediae produce cercariae, which leave the snail and bore into the body of a fish. Within the muscles of the fish they encyst as metacercariae. If the fish is eaten by a person, the metacercariae become young flukes which burrow out of the intestine and find the bile duct. (Source: B. Cale)

In turbellarians, eggs are usually laid in cocoons and hatch into miniature adults. The eggs of ectoparasitic monogeneans hatch into a ciliated larva (oncomiracidium), which then attaches to the same host or goes through a free-living stage before attaching to a new host of the same species. After attachment, the larva undergoes marked changes in structure (metamorphosis) to the adult form. Endoparasitic trematodes and cestodes have complex life cycles involving two or more host species. The adult parasite lives and reproduces sexually in the definitive host, whereas larval stages occur in one or more intermediate hosts.

Trematodes typically have a vertebrate definitive host, with adult flukes living in the digestive tract, blood or other tissues. Eggs are usually shed into water with the host's faeces and develop into a ciliated larva that infects an aquatic mollusc. The sequence of developmental stages in the mollusc intermediate host is variable, but usually involves asexual multiplication through a series of larval stages, ending with a tailed, free-swimming stage that leaves the mollusc and infects either another intermediate host or a definitive host. Figure 14.5 shows the life cycle of the human liver fluke, *Clonorchis sinensis*, a serious parasite of people throughout South-East Asia.

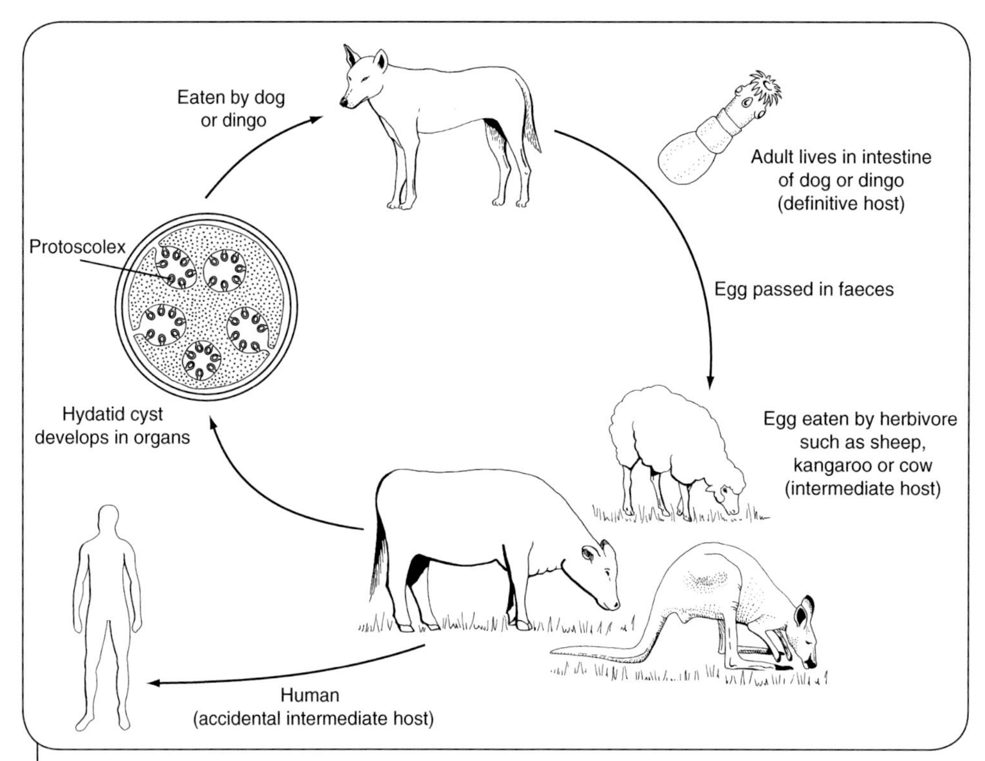

Figure 14.6 Life cycle of the hydatid tapeworm, *Echinococcus granulosus* in Australia. The adult tapeworm is very small, and lives in the small intestine of dogs or wild canids, such as wolves. Eggs are released in the faeces and are then eaten by grazing herbivores, such as sheep, cattle, deer or kangaroos. Humans may also swallow the eggs. The egg hatches in the intestine and the larva burrows through the intestinal wall and finds its way to an internal organ, such as liver or lung. There it becomes a fluid-filled hydatid cyst, which expands in size and asexually produces within itself many larval forms called protoscoleces. If the cyst is then eaten by a definitive host, each protoscolex develops into an adult tapeworm in the intestine. (Source: B. Cale)

Cestodes also have vertebrate definitive hosts, where the adult tapeworm is usually found in the digestive tract. Typically, eggs are expelled with the faeces and eaten by an intermediate host, which may be vertebrate or invertebrate. Asexual reproduction may occur in various body organs in the intermediate host. Larval stages pass through one, two or three intermediate hosts before returning to the definitive host, often when the intermediate host is eaten. Figure 14.6 shows the life cycle of the hydatid tapeworm, *Echinococcus granulosus,* which is found throughout the world.

Phylum Nematoda – the roundworms

General description

This is a large group of about 20000 described species worldwide (about 2000 Australian), with probably 10 to 100 times as many species still undescribed. Nematodes are found in terrestrial, marine and freshwater habitats; many are free living, especially in soil and sediments, whereas others are parasites of plants or animals. Their classification is in a state of change, but molecular studies suggest there are three major classes, all containing free-living and parasitic species:

1. the Chromadoria
2. the Enoplia
3. the Dorylaimia.

Nematodes are triploblastic, bilaterally symmetrical worms with complex organ systems, a gut with both anterior (mouth) and posterior (anus) openings and a body cavity lined on the outer side by mesoderm and on the inner side by endoderm (and therefore referred to as a pseudocoelom, to distinguish it from the coelom of other animal groups, which is lined entirely by mesoderm). Most free-living nematodes are microscopic, but parasitic roundworms can be large. The largest, a parasite of whales, reaches 9 m. In general body plan, all nematodes have elongated, unsegmented, cylindrical bodies tapering to a point at both ends. They are covered by a thick, flexible cuticle secreted by the epidermis. Beneath the epidermis a layer of longitudinal muscles extends the length of the worm (Figure 14.7).

Skeleton and locomotion

The fluid-filled pseudocoelom, bounded by the elastic but deformable cuticle, creates a hydrostatic skeleton. This is essential for effective movement in nematodes, which have longitudinal muscles but no circular muscles to oppose them. The longitudinal muscles on each side of the body contract alternately against the hydrostatic skeleton, shortening first one side, then the other side, of the body and flexing it from side to side. Swimming in open water is possible, but laborious and energetically expensive. More effective movement occurs against a substrate or even the surface tension created by thin water films.

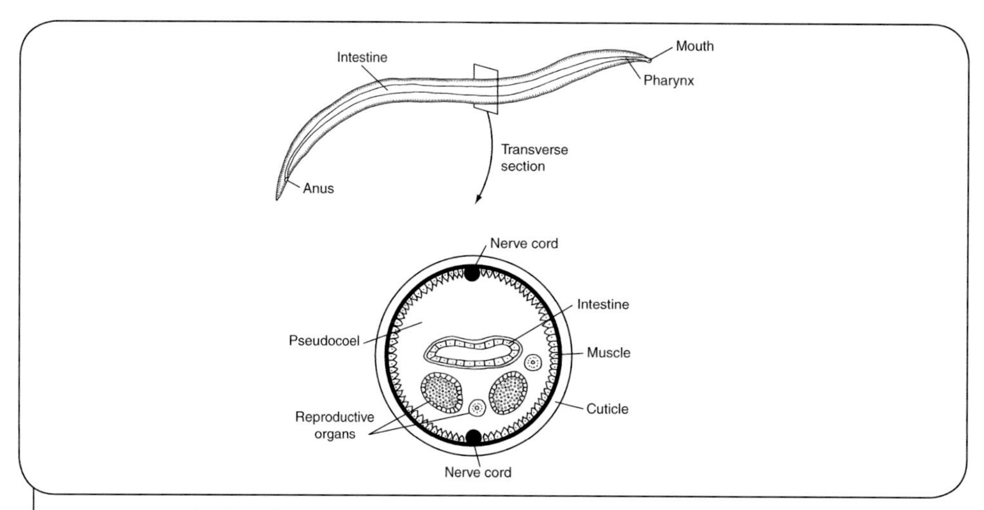

Figure 14.7 Body plan of a roundworm. The cylindrical body shape, longitudinal muscle layer and pseudocoelom are characteristic of all nematodes. (Source: Adapted from an original drawing by Olivia Ingersoll published in Boolootian, R.A. and Stiles, K.A. (1981). *College zoology*. Macmillan Publishing Company and Collier Macmillan Publishers. 10th edn)

Feeding and digestion

Free-living nematodes may be herbivores, carnivores or scavengers. Nematodes are also important endoparasites of plants, invertebrates and vertebrates. All species of higher plants and animals probably have at least one species of parasitic nematode. The digestive system is well developed. The anterior, cuticle-lined mouth is often surrounded by sensitive lips and sometimes equipped with piercing stylets, used by blood-sucking parasitic species to pierce the host's tissues. The mouth leads to a muscular pharynx, also lined by cuticle, which pumps food through the digestive system. Undigested material passes to the cuticle-lined rectum, opening at the anus in females or, together with the gonoduct, at the cloaca in males.

Respiration and circulation

Nematodes lack respiratory and circulatory organs, exchanging oxygen directly through the cuticle.

Excretion

Nematodes usually possess either a large glandular cell (cervical gland), located ventral to the gut, with a duct opening by a pore near the lips; or a complex, H-shaped canal system, with a short ventral canal linking two long lateral canals and opening via a duct at a pore near the nerve ring. These organs are usually referred to as the excretory/secretory system. In some nematodes the system secretes enzymes, whereas in others it is osmoregulatory. The extent to which it has an active excretory function is unknown.

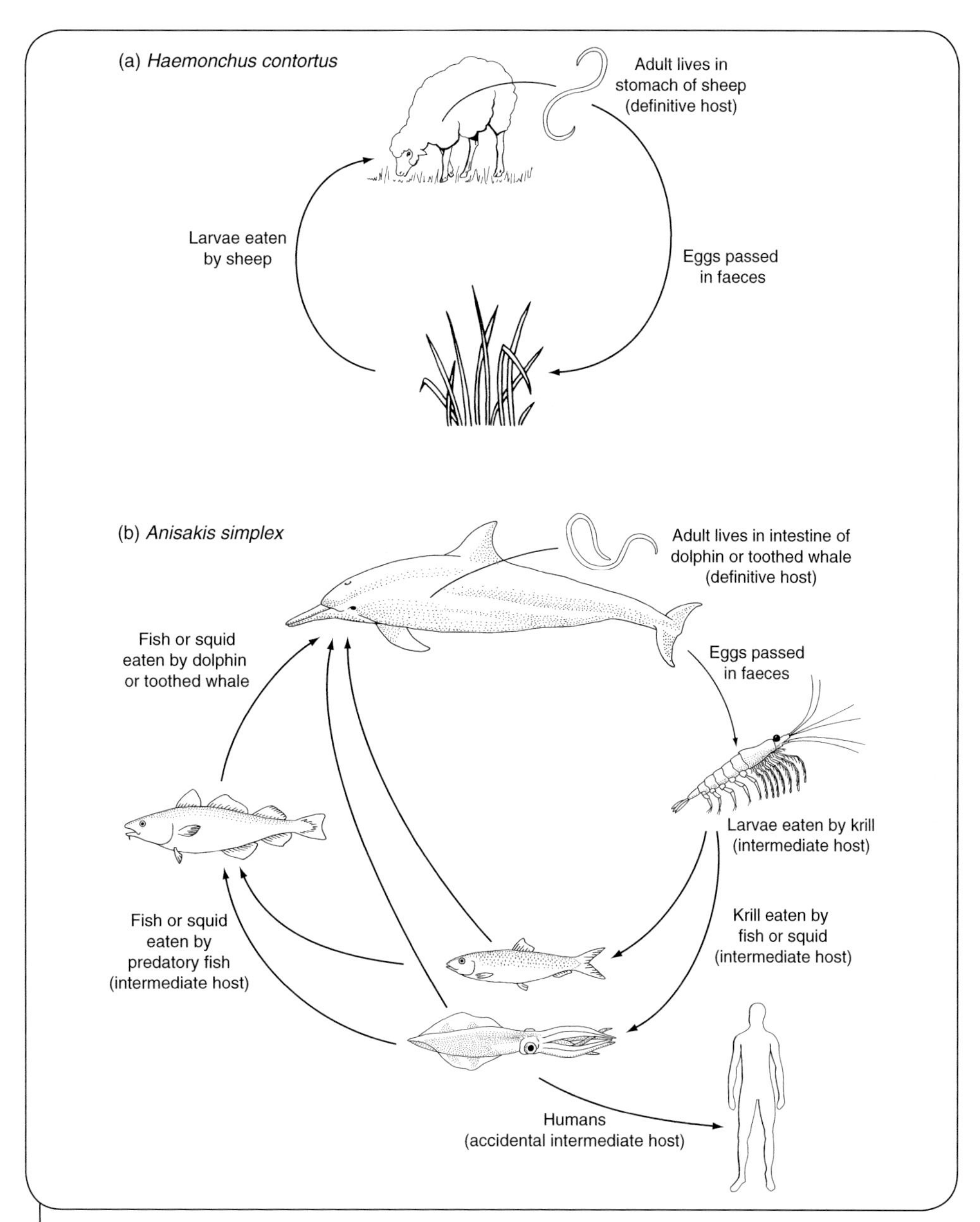

Figure 14.8 Life cycles of nematode parasites. (a) *Haemonchus contortus*, the barber's pole worm, has a simple direct life cycle. Adult worms live in the abomasum (a stomach compartment) of sheep, goats and related animals. Eggs are passed out in the faeces onto pasture, where they hatch into free-living larval stages. The larvae are eaten by grazing animals and pass directly to their preferred site in the abomasum. (b) *Anisakis simplex* has an indirect life cycle. Adult worms live in the intestine of dolphins or toothed whales. Eggs are passed out in the faeces and hatch into larvae in the water. The larvae are eaten by minute crustaceans (krill), which are in turn eaten by squid or fish. The larvae usually live in the body cavity of these intermediate hosts and may be passed up the food chain as larger, predatory fish prey on smaller fish or squid. Eventually fish containing the larvae are eaten by dolphins or toothed whales to complete the cycle. (Source: B. Cale)

Nervous and endocrine systems

The central nervous system consists of a nerve ring around the pharynx, with ventral, dorsal and lateral nerves extending posteriorly. Papillae, sensitive to touch, and chemoreceptors are often found on the body surface. Some aquatic nematodes have paired photoreceptors near the pharynx. All nematodes shed their cuticle periodically as they grow, and neurosecretory hormones control this moulting process.

Reproduction

Reproduction is exclusively sexual. The sexes are usually separate, although hermaphroditism occurs in some soil-dwelling species and parthenogenesis (development of unfertilised eggs) is known. There are complex reproductive organs in both sexes and fertilisation is internal. Either eggs or live young are produced and there are generally four larval stages, all similar to the adult. Parasitic life cycles may be direct or indirect (Figure 14.8). In direct life cycles there is only one host, within which the adult worm resides. Eggs or first stage larvae are released to the environment and the first three larval stages are free living, with the third stage infecting a new host. Nematode parasites with indirect life cycles have a definitive host, in which the adult worm lives, and one or more intermediate hosts harbouring larval stages, with the third larval stage usually infective to another definitive host.

Phylum Annelida – the segmented worms

General description

The annelids comprise about 15 000 known species (about 2300 Australian) from marine, freshwater and terrestrial environments. There are three main classes (Plate 14.5):

1. the Polychaeta, which are mainly marine burrowing and free-living worms
2. the Oligochaeta, which are mainly freshwater and terrestrial, and
3. the Hirudinea or leeches, which are mainly freshwater and terrestrial ectoparasitic species.

Annelids have a true coelom. It is completely bounded by mesoderm, but during embryonic development the mesoderm and endoderm combine in places to form specialised tissues and organs. The coelom facilitates increases in body size. Annelids are one of only three animal phyla to show segmentation (metameric). The body is a series of repeating segments, each with its own muscular, nervous, circulatory, reproductive and excretory systems (Figure 14.9). Segments are usually separated by layers of tissue called septa, isolating the coelomic fluid in adjacent segments.

Skeleton and locomotion

The fluid-filled coelom forms a hydrostatic skeleton surrounded by an inner layer of longitudinal muscles within an outer layer of circular muscles. The basic movement

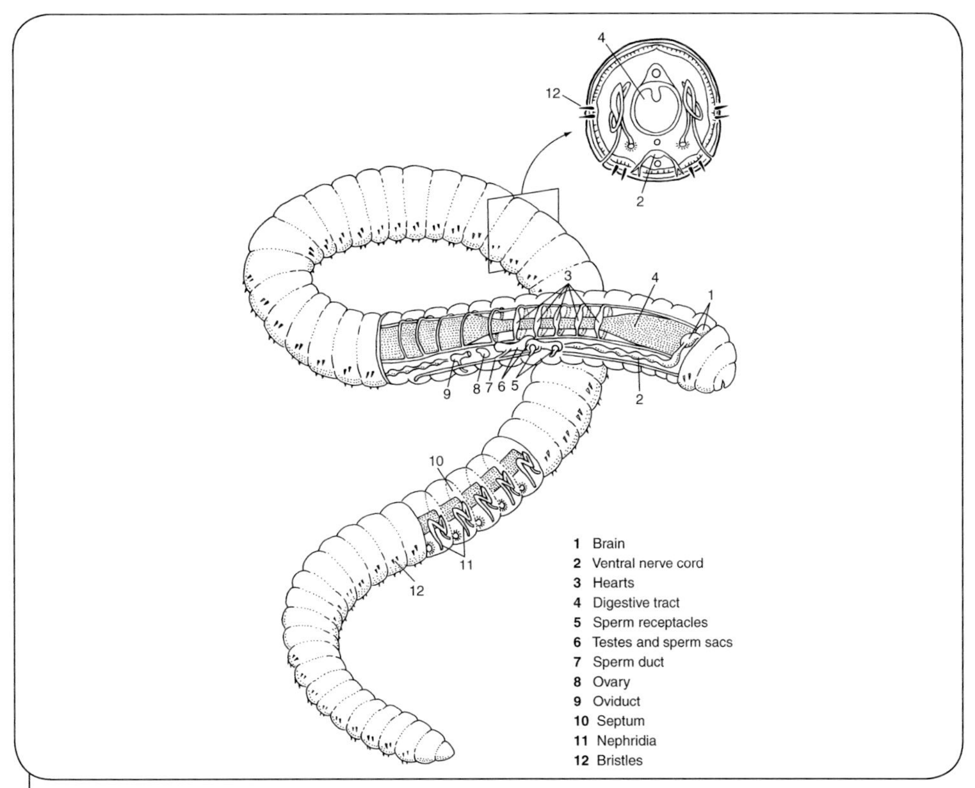

Figure 14.9 Body plan of an annelid. Each segment has its own muscular, nervous, circulatory, reproductive and excretory systems. (Source: Adapted from Wilson, E.O., Eisner, T. Briggs, W.R., Dickerson, R.E., Metzenberg, R.L., O'Brien, R.O., Susman, M. and Boggs, W.E. (1973). *Life on Earth*. Sinauer Assoc., Sunderland, Massachusetts)

pattern in polychaetes and oligochaetes involves localised contractions and relaxations of the circular and longitudinal muscles. These alternately anchor some segments of the body and extend others, enabling forward movement in burrowing and soil-dwelling forms. Active, mobile polychaetes ('errant' forms) lever themselves forward using parapodia, which are thin, flattened outgrowths. Leeches move in a looping motion, using anterior and posterior suckers. While the posterior sucker is attached, the leech extends forward by contracting its circular muscles. The anterior sucker then attaches, the posterior sucker releases and contractions of the longitudinal muscles pull the leech forward.

Feeding and digestion

Annelids may be suspension feeders, deposit feeders collecting organic materials from their substrate, active predators or, in the case of most leeches, ectoparasites. The basic gut structure is linear and unsegmented, linking an anterior mouth with a posterior anus. Modifications include an eversible pharynx in polychaetes and a muscular gizzard for grinding food in oligochaetes. Most leeches wound the host with teeth. A muscular,

Box 14.2 Medicinal uses of leeches

There is a long history of using leeches medicinally. Ancient Greeks applied leeches to poisonous bites to draw out venom, and in the early 19th century drawing blood from patients with medical leeches (*Hirudo medicinalis*) was popular in Europe. The Emperor Napoleon I of France treated his prolapsed piles with leeches, and physician Francois Broussais popularised their medical use to such an extent that dresses decorated with embroidered leeches were fashionable. Modern medical research has found important new applications.

To facilitate bloodsucking, leeches secrete chemicals to dilate veins and prevent clotting. These include hirundin, which neutralises compounds critical in clotting, and hementin, which dissolves existing clots. These substances have potential in cardiovascular medicine. Furthermore, leeches' gentle removal of blood is applied in microsurgery, reducing congestion in tissue grafts.

Modern medical leeches are specially bred, but the fate of *Hirudo medicinalis* in Europe warns of the importance of conserving invertebrates. It was nearly collected to extinction during the 19th century leech mania. Although international trade in wild-collected medicinal leeches is now banned, swamp draining poses a severe threat.

pumping pharynx ingests blood, while associated salivary glands secrete anti-coagulants to prevent clotting (Box 14.2).

Respiration and circulation

The body covering is permeable to gases and water and is used for respiration. This restricts annelids to aquatic or moist terrestrial environments.

The circulatory system is closed, confining blood within vessels and not mixing it directly with fluids surrounding the cells. Blood is carried anteriorly by a dorsal vessel and posteriorly by a ventral vessel. Lateral vessels link them. Five enlarged, anterior lateral vessels serve as pumps or hearts, and the blood often contains specialised oxygen-transporting pigments. In leeches the circulatory system is reduced or absent, replaced by the coelomic fluid.

Excretion

Annelids excrete some metabolic wastes through the body surface, but most are removed by paired nephridia located in the segments. Coelomic fluid is drawn into one end, required substances are reabsorbed and wastes are voided through the nephridipore.

Nervous and endocrine systems

Aggregations of nerve tissue at the anterior end form a brain, from which a solid ventral nerve cord extends the length of the body with swellings, or ganglia, in each segment. Receptors for light, touch, vibration and chemicals occur along the body. Neurosecretory cells stimulate maturation of the gonads, attainment of sexual maturity and regeneration.

Reproduction

Polychaetes reproduce sexually and are dioecious. Unusually, there are no gonads and gametes are produced from peritoneal tissue in adjacent segments. Fertilisation occurs externally and the free-living embryo develops into a trochophore larva, moving by rings of cilia. In contrast, oligochaetes are hermaphroditic with distinct gonads. Sperm are exchanged simultaneously during mating. Subsequently, ova and sperm pass into a protective external cocoon secreted by the clitellum, a specialised region of the body surface. Young emerge fully developed. Leeches are also hermaphroditic, but fertilisation occurs internally. Fertilised eggs develop within nutritive cocoons and there is no larval stage. Asexual reproduction, in which the body divides and each section produces a new individual, occurs in polychaetes and oligochaetes, but not in leeches.

Phylum Arthropoda – the joint-legged animals

General description

Arthropods occur over all major habitats from the oceans to the polar regions (Figure 14.10). Worldwide there are approximately 1 236 000 species (about 93 246 Australian). Their diverse ecological functions include being the major herbivores in most terrestrial ecosystems, pollinating many plant species, providing important human

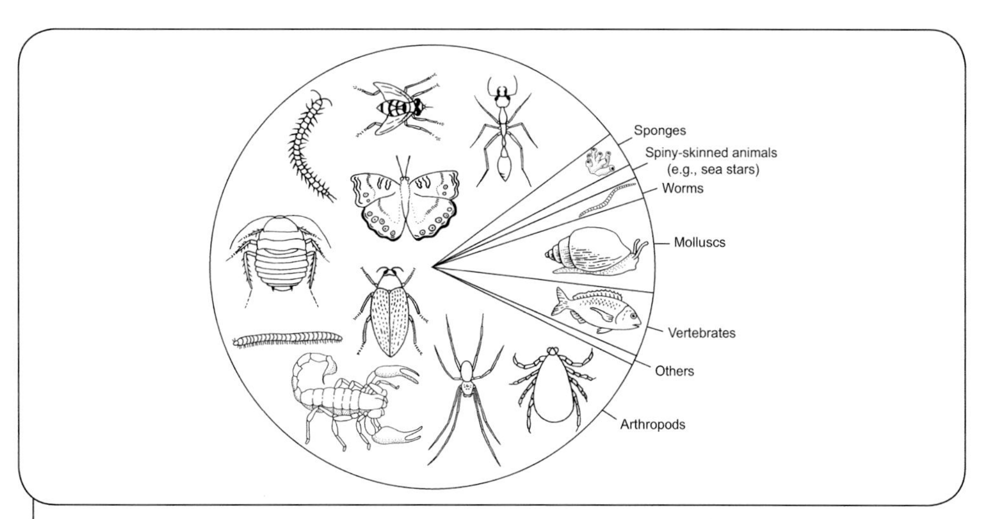

Figure 14.10 The numerical superiority of arthropods. The pie diagram indicates the proportion of all animal species that each group makes up. (Source: B. Cale)

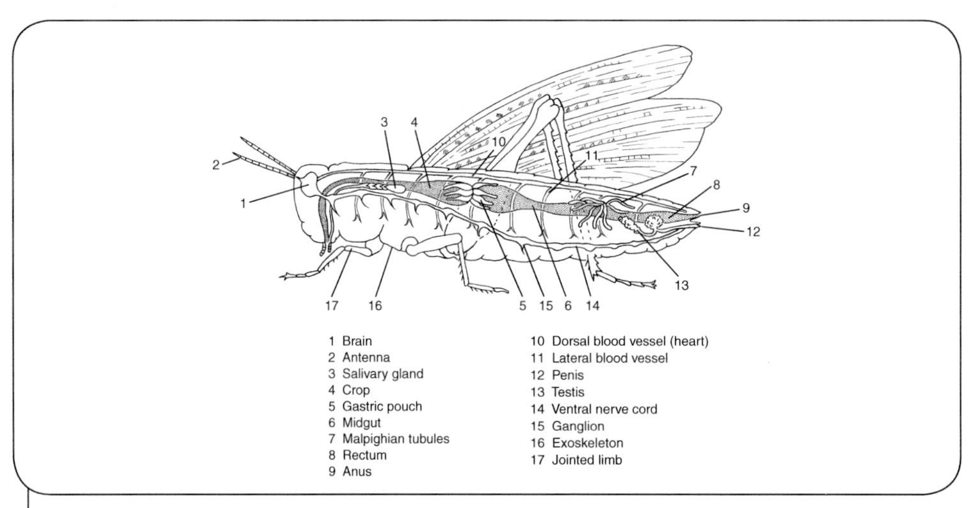

Figure 14.11 Body plan of an arthropod, showing specialised organ systems within a rigid exoskeleton. (Source: Adapted from Wilson, E.O., Eisner, T. Briggs, W.R., Dickerson, R.E., Metzenberg, R.L., O'Brien, R.O., Susman, M. and Boggs, W.E. (1973). *Life on Earth*. Sinauer Assoc., Sunderland, Massachusetts)

foods, competing with humans for food, transmitting diseases and providing critical links in marine food chains. We recognise five main classes (Plate 14.6):

1. the extinct Trilobita
2. the mainly marine Crustacea, including some freshwater and terrestrial species (crabs and woodlice are familiar examples)
3. the Arachnida, a mainly terrestrial group including spiders, scorpions and ticks
4. the Myriapoda, a terrestrial group including centipedes and millipedes
5. the Insecta, which are mainly terrestrial but include many freshwater and a small number of marine species.

The basic body plan is that of a segmented coelomate with an external skeleton and jointed appendages (Figure 14.11). The segments are grouped into specialised regions of the body (tagmatisation). The coelom is reduced and most of the body cavity consists of blood-filled spaces, collectively called a haemocoel.

Skeleton and locomotion

The rigid external exoskeleton consists of a series of plates connected with flexible articular membranes to permit movement. It is primarily composed of chitin, a polysaccharide combining strength, flexibility and light weight. In most terrestrial forms a waxy epicuticle on the outer surface reduces water loss. Many crustaceans impregnate the exoskeleton with calcium salts for greater strength. The exoskeleton cannot expand, so for growth it is shed periodically in a process called moulting or ecdysis.

Jointed appendages may be modified for movement, feeding, reproduction or sensing. One striking example is the wings of insects, formed from thin, lightweight outfoldings of the body wall supported by networks of veins. Flight helps insects to find food or new habitats, exploit temporary resources, escape predators, disperse and exchange genetic material between populations. It is one of the main reasons for their extraordinary species diversity.

Feeding and digestion

Arthropods may be suspension feeders, deposit feeders, herbivores, active predators or parasites. They often have specialised feeding appendages. The gut comprises a foregut for ingestion, mechanical digestion and storage (divided into a cardiac stomach and a pyloric stomach in crustaceans), a midgut with outpockets to increase surface area for chemical digestion and absorption, and a hindgut or intestine for absorbing water and minerals.

Respiration and circulation

Many terrestrial arthropods use pores or spiracles on the external surface connecting to branching tracheae, conveying oxygen directly to the cells. The efficiency of oxygen transport declines rapidly with increasing size, limiting the maximum size of arthropods. Some spiders supplement or replace tracheae with a book lung of parallel air pockets within a blood-filled compartment. Crustaceans respire with gills, usually attached to appendages and ventilated by movement through the surrounding water. All arthropods have open circulatory systems with a dorsal heart and vessels carrying blood to different parts of the haemocoel, where it escapes from vessels and bathes the tissues. The blood may contain respiratory pigments to aid oxygen transfer.

Excretion

Aquatic crustaceans excrete nitrogenous wastes across the gills, often as ammonia. Insects and many arachnids use Malpighian tubules for excretion and water conservation on land. These slender tubes project from the junction between the midgut and the hindgut, exchange fluid with the surrounding blood and concentrate the nitrogenous wastes as uric acid or, in arachnids, guanine, for elimination via the hindgut. Some arachnids supplement or replace Malpighian tubules with coxal glands, concentrating wastes from the blood and excreting them via pores.

Nervous and endocrine systems

Aggregations of nerves at the anterior end form a brain, from which a double ventral nerve cord extends along the body. Sensory organs are highly developed, including compound eyes with a great ability to detect movement, as well as organs for chemical reception, smelling, touching, hearing and balancing. Complex behaviour patterns are common, often associated with feeding or reproduction. Although many of these are inherited, some species learn. True endocrine glands (ductless glands secreting directly into the circulatory system) contribute to many functions, including moulting, growth, and reproduction (Box 14.3).

Box 14.3 Methoprene, an insect growth regulator

In holometabolous insects, development from larva to pupa to adult is regulated by a moulting hormone (ecdysone) and juvenile hormone (JH). Ecdysone stimulates shedding the exoskeleton, and the concentration of JH prior to moulting determines the features of the newly moulted insect. JH concentration decreases as the insect ages, allowing the development of adult features. If JH concentration remains high, the insect does not mature.

The chemical methoprene mimics the action of JH and is sometimes used as an insecticide. It does not kill insects, but prevents them maturing and reproducing. Methoprene is not toxic to humans if ingested or inhaled, and only slightly toxic if absorbed across the skin. This makes it an effective insecticide in situations such as controlling mosquitoes in rainwater tanks, but it can also harm beneficial insects if used in broad-scale crop protection.

Reproduction

Arthropods reproduce sexually and most are dioecious. External fertilisation is common in marine forms, often leading to free-living larvae that undergo metamorphosis. Terrestrial species protect gametes from desiccation using internal fertilisation. In arachnids, development is direct and the young hatching from the eggs resemble miniature adults. The small number of wingless insects also have direct development, but winged insects undergo metamorphosis. In incomplete or hemimetabolous development, eggs hatch into nymphs resembling adults but lacking wings. As they moult and grow, external wing buds develop until, at the final moult, a winged adult emerges. In complete or holometabolous development, eggs hatch into larvae totally unlike the adult. After several moults they enter a non-feeding pupal stage and are extensively reorganised into the winged adult body. This provides the advantages of a larval stage specialised for feeding and growth and an adult specialised for reproduction and dispersal, as well as the possibility of two stages in the life history (egg and pupa) resistant to environmental pressures such as cold or drought.

Phylum Mollusca – the snails and their relatives

General description

With about 70000 living species (about 8700 Australian), the molluscs are one of the largest phyla after the arthropods (Plate 14.7). They are mainly marine, although some gastropods and bivalves occur in fresh water and some gastropods live on land. The main classes are:

1. the Gastropoda, including the familiar snails. Many species within this class undergo torsion as larvae, in which the right and left muscles attaching the shell to the body grow unequally, twisting the viscera through a 180° rotation and bringing

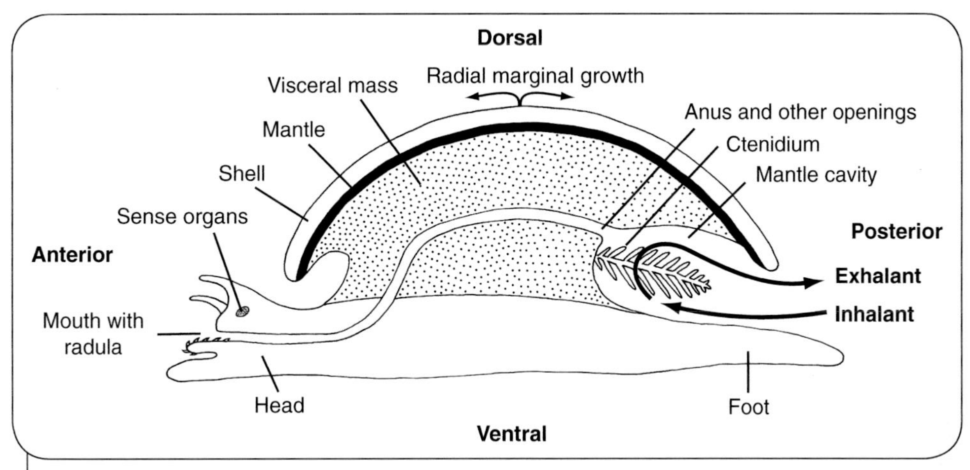

Figure 14.12 Body plan of a mollusc. Although there are many variations on this generalised body plan, all molluscs have a head, a flattened foot and a visceral mass containing the body organs and covered by the mantle. (Source: Adapted from Hickman, C.P. Jr., Roberts, L.S. and Larson, A. (2003). *Animal diversity*. McGraw-Hill, Boston. 3rd edn)

the mantle cavity, gills and openings of the excretory, digestive and reproductive systems to the anterior

2. the Bivalvia, in which the soft body, nourished by suspension feeding, is enclosed between two hinged shells
3. the Cephalopoda, including squids and octopus. The foot is modified into a ring of tentacles and a funnel for expelling water from the mantle cavity.

The basic body plan is a muscular foot beneath a soft visceral mass containing the internal organs (Figure 14.12). The coelom is reduced and the main body cavity is a haemocoel. Two outgrowths of skin from the dorsal surface form a protective mantle and also enclose a mantle cavity between the mantle and the body wall. In many species the mantle also secretes a hard, protective shell. This is internal or absent in most cephalopods. Many molluscs (the bivalves are a significant exception) have a unique, specialised feeding structure called a radula (see below).

Skeleton and locomotion

Muscular attachment to the shell and hydrostatic pressure based on the haemocoel provide skeletal support. In gastropods, muscles attached to the shell and the dorsal mantle move the foot through waves of muscular contraction. Bivalves extend the muscular foot through the gap between the shells using blood pressure. The foot then anchors to a surface and muscular contractions pull the body forward. Cephalopods are agile, propelling themselves by forcefully expelling water from a funnel in the mantle cavity.

Feeding and digestion

The digestive system begins with a mouth, supplemented in most species except the bivalves by a protrusible, tongue-like radula. This bears up to 250 000 teeth drawn across

Box 14.4 Molluscs as bioindicators

The town of Rouen straddles the Seine, a river in France. When assessing water pollution in the Seine estuary downstream from Rouen, biologists found an unlikely ally in the zebra mussel, *Dreissena polymorpha*. This highly invasive freshwater bivalve, native to lakes in south-eastern Russia, has spread through waterways in Europe and North America, outcompeting native invertebrates. As a very efficient filter feeder it also ingests and accumulates environmental pollutants, including polychlorinated biphenyls (PCBs) and polycyclic aromatic hydrocarbons (PAHs). Caged zebra mussels suspended in the Seine estuary rapidly accumulated high levels of these compounds, indicating a significant pollution problem.

Zebra mussels can also filter and concentrate bacteria such as *Escherichia coli*. *E. coli* occurs in the human intestine and its presence in the environment indicates faecal contamination. Caged zebra mussels filter, ingest and concentrate even brief increases in *E. coli* concentrations, so levels of contamination not normally detectable in natural waterways can be identified.

food to scrape, cut or tear particles for ingestion. The predatory cephalopods seize prey with tentacles, often armed with suckers or hooks. They supplement the radula with paired jaws like a parrot beak. Some species also subdue prey with venom, with the bite of the blue-ringed octopus (*Hapalochlaena maculosa* and *H. lunulata*) potentially fatal to humans. In contrast, bivalves are headless suspension feeders, collecting small food particles using their gills. Ingested food passes via a short oesophagus to the stomach, where digestion occurs in associated digestive glands. Undigested wastes pass through the intestine to the anus. Humans exploit the filter-feeding behaviour of bivalves to monitor pollutants and bacterial contamination of waterways (Box 14.4).

Respiration and circulation

Gas exchange is primarily through gills, which may be external or entirely within the mantle cavity and ventilated via water movement through inhalant and exhalant siphons. Terrestrial gastropods breathe air using the highly vascularised lining of the mantle cavity as a 'lung', and the cavity as a whole is sealed to the back of the animal, leaving only one opening, a small pulmonary pore.

Most molluscs have open circulatory systems. However, cephalopods meet their high demand for oxygen for their active, predatory lifestyles with a closed circulatory system. Accessory hearts increase blood pressure at the gills for efficient gas exchange. The oxygen-transporting pigment haemocyanin, which is functionally similar to haemoglobin in vertebrates but contains copper rather than iron, enhances the oxygen-carrying capacity of the blood. The efficient respiratory and circulatory adaptations of cephalopods, combined with the buoyancy of sea water, enable the giant squid (*Architeuthis* spp.) and the colossal squid (*Mesonychoteuthis* spp.) to grow to several metres in length.

Excretion

Molluscs excrete using large, paired, folded metanephridia where there is selective resorption of important water and solutes before the final waste is passed through a pore in the mantle cavity. Aquatic species excrete primarily ammonia or ammonium compounds, but, to conserve water, land snails expend energy to convert these wastes to the drier and less-toxic uric acid.

Nervous and endocrine systems

The basic nervous system comprises an anterior collection of ganglia into a brain with paired ventral nerve cords. Information is provided by photoreceptors of varying complexity, statocysts for maintaining balance and chemical receptors. Cephalopods have highly developed brains, perhaps an adaptation to their predatory lifestyle. They have excellent sight, touch, shape recognition and both long-term and short-term memory. Some cephalopods also learn by copying the behaviour of others. Neurosecretory cells occur in the brains of many molluscs and true endocrine glands occur in some species.

Reproduction

Most molluscs are dioecious, although many gastropods are hermaphrodites. Fertilisation is usually external in bivalves, but some freshwater species brood fertilised eggs. Fertilisation is internal in cephalopods and many gastropods. Many molluscs show metamorphosis in which the egg hatches into a trochophore larva similar to that of annelids. This develops into a veliger larva (which in the gastropods is the stage that undergoes torsion) and finally into the adult. In some species, the trochophore larval stage is passed in the egg and a veliger larva hatches, whereas others have direct development in which both larval stages are either suppressed or passed in the egg, a miniature adult emerging. Internal fertilisation and direct development are important adaptations to terrestrial life in gastropods.

Phylum Echinodermata – the sea stars and their relatives

General description

The roughly 7000 species (about 1200 Australian) of echinoderms are all marine and are remarkable for their extraordinary features (Plate 14.8). To begin with, the radially symmetrical adults arise by metamorphosis from free-swimming bilaterally symmetrical larvae. Also, part of the coelom is modified into a hydraulic water vascular system powering many tiny tube feet (podia) used in movement, gas exchange, excretion and collecting food. Many species also have pincer-like structures, called pedicellariae, on the body surface to remove settling organisms. Echinoderms also have deuterostome development, in which the coelom arises from outpocketing of the embryonic gut

(Chapter 13, Figure 13.6). The only other major phylum with deuterostome development is the Chordata. The major classes of echinoderms are:

1. the Crinoidea, including mobile feather stars and the stalked and sedentary sea lilies
2. the Asteroidea, the well-known sea stars with arms radiating out from a central disk that is not sharply differentiated
3. the Ophiuroidea, or brittle stars, which are also star-shaped, but the arms are narrow and sharply demarcated from the central disk
4. the Echinoidea, or sea urchins, which are commonly globular or disk-shaped and armless. They are protected by strong skeletal plates and movable spines
5. the Holothuroidea, which take the common name 'sea cucumber' from their cylindrical shape with a crown of tentacles surrounding the mouth.

The basic body plan of adult echinoderms is a central disk surrounded by arms with tube feet protruding through an ambulacral groove on the underside. The central mouth faces downwards, as do the tube feet. The surface to which the mouth points is called the oral surface and the surface opposite to this the aboral surface. This is the basic shape of sea stars and brittle stars. If you imagine the arms curling back upwards to meet over the top of the animal to form a ball, the basic shape transforms into the globular sea urchin. If it elongates and rests on its side the sea cucumber shape is achieved, while if it rests head down with mouth and tube feet upwards it resembles a crinoid (Figure 14.13).

Skeleton and locomotion

Echinoderms have endoskeletons of small calcareous plates (ossicles). In sea urchins, these ossicles form a protective test encasing the internal organs and the spines are pronounced and movable, for defence and locomotion. Some sea cucumbers also have large protective ossicles, but in most they are reduced and embedded in a leathery skin.

Echinoderms move using tube feet, flexible arms, movable spines or sometimes a combination of these. Tube feet are the external component of an extensive coelomic compartment forming the water vascular system, connecting to the outside via a small pore called the madreporite. From this a tubular stone canal connects to the ring canal, encircling the mouth. Radial canals then branch into each arm and in turn into lateral canals, connecting to the tube feet. Each tube foot has a muscular internal sac (ampulla) at one end and often a sucker at the other. Muscular action forces fluid from the ampulla into the tube foot to extend it, or back into the ampulla to retract it.

Feeding and digestion

Echinoderms have diverse feeding mechanisms. Many sea stars are predatory, attacking other invertebrates. Suspension feeding is common in brittle stars and crinoids, whereas echinoids use a complex, toothed chewing structure called an Aristotle's lantern to scrape algae. Sea cucumbers either suspension feed using their tentacles, or collect food particles from the substrate.

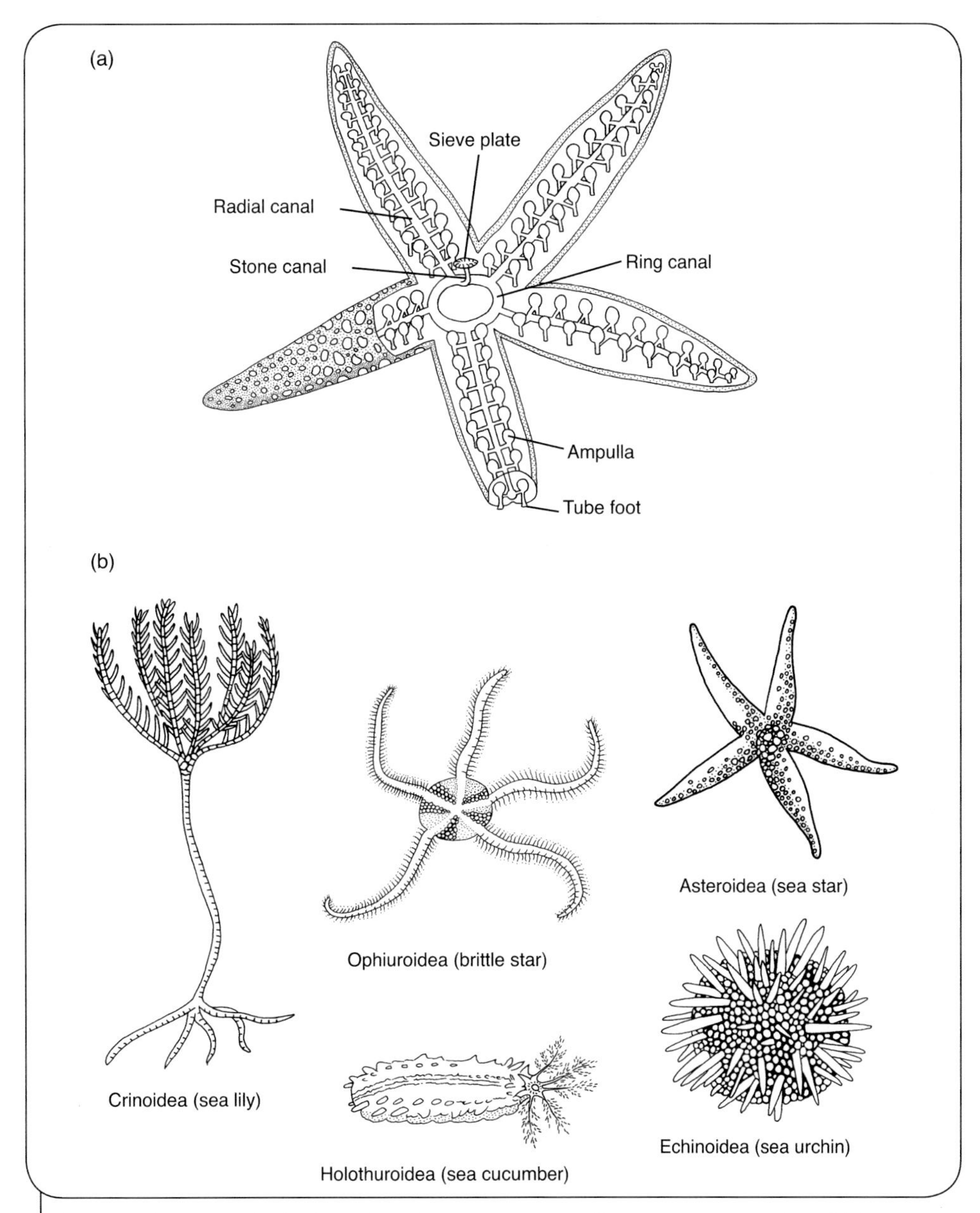

Figure 14.13 (a) Basic body plan of an echinoderm, showing the water vascular system. (b) Comparison of body forms of the five major classes of echinoderms. (Source: Adapted from Boolootian, R.A. and Stiles, K.A. (1981). *College zoology*. Macmillan Publishing Company and Collier Macmillan Publishers. 10th edn. Copyright © Pearson Education Inc. Reproduced by permission.)

Respiration and circulation

Sea cucumbers are the only echinoderms with specialised internal respiratory structures, extending their paired, highly branched, muscular respiratory trees from the hindgut into the coelom. The hindgut pumps water into them, whereas contractions of the respiratory trees themselves circulate and ultimately expel the water. Gas exchange also occurs by diffusion over the surface of the podia – this occurs in crinoids and sea stars as well. Sea stars further supplement this with diffusion across papulae, which are protrusions of the epidermis. Brittle stars respire using ciliary action or muscular pumping to circulate water through clefts in the body called bursae.

Excretion

Larval echinoderms have a simple nephridial system. In adults, ammonia diffuses across the body surface, including the tube feet and possibly the respiratory trees of sea cucumbers.

Nervous and endocrine systems

There is a circular or star-shaped ring of nerves below the mouth, with radial nerves branching into each arm. Simple sense organs detect light and chemicals, orient the body in water currents and right the animal when they have overturned. Endocrine mechanisms occur in sea stars and sea urchins, where the radial nerves produce a gonad-stimulating hormone that induces spawning.

Reproduction

Most echinoderms are dioecious. Spawning is usually followed by external fertilisation and a free-swimming larval stage, although brooding occurs in some polar and deep-sea species where conditions are unsuited to larval development. Echinoderms also regenerate lost body parts.

Conservation of invertebrates

The immense diversity of invertebrates is a substantial conservation challenge. Specialists are often required to identify species, the details of the life history (reproduction, growth and development) of many species are poorly known and advances in molecular taxonomy are still revealing their full biodiversity. Although research on the abundance, distribution and basic biology of many species will help to identify appropriate measures for conservation, immediate actions can ameliorate pollution and the physical destruction of habitats. Marine reserves, especially those that include intertidal habitats, may be especially important for conserving marine invertebrates, and terrestrial reserves are also critical.

Perhaps just as importantly, we need to alert people to the importance of invertebrate biodiversity. Many people do not empathise with invertebrates as much as they do with vertebrates, especially mammals and birds. The problem is one of scale. Invertebrates are small and their complex beauty can often be fully appreciated only through a magnifying glass or microscope. Despite this, it is vital to communicate the essential roles that

Box 14.5 Conserving parasites in captive-bred animals

Captive-breeding programs aim to establish a breeding population of an endangered species in captivity, before returning the offspring to the wild to establish new populations or enhance the size of existing populations. Worldwide, many zoos now have active captive-breeding programs, often exchanging animals to maintain genetic diversity in their captive-bred collections.

When animals are brought into captivity to establish a breeding program, their parasites provide a dilemma. On the one hand, parasitic diseases can be a real threat to the viability of an endangered population, especially one artificially housed in confined spaces. Many zoos therefore treat new arrivals for parasites and maintain their collection in, as much as possible, a parasite-free environment.

There is a danger with this practice, however. Young animals not exposed to natural parasites as they mature do not develop immunity, and on release may succumb to parasitic diseases normally present, but relatively innocuous, in wild populations. In addition, the natural parasite fauna of a species may protect it against invasion by more virulent exotic parasites. Parasites and their hosts often have a long history of coevolution, leading to many complex and subtle interactions between them. If we wish to conserve the hosts, sometimes it may also be necessary to conserve their parasites.

invertebrates play in maintaining healthy and functioning ecosystems through pollination, nutrient cycling and pest control. Even those invertebrates that are usually considered detrimental, such as parasites, may in fact be vital components of balanced ecosystems (Box 14.5).

Chapter summary

1. Each animal phylum has a characteristic body plan, reflecting a common evolutionary history.
2. Sponges are simple aquatic animals, with three layers of cells, but no tissues or organs.
3. Cnidarians have distinct tissues derived from two cell layers, an inner endoderm and an outer ectoderm, that form during embryonic development. They are radially symmetrical animals with two body forms during their life cycle (polyp and medusae).
4. Platyhelminths or flatworms exhibit an organ level of development. They are bilaterally symmetrical and dorsoventrally flattened, with no body cavity. Some platyhelminths are free living, but many are very important parasites.
5. Nematodes or roundworms also have three cell layers, complex organ systems and bilateral symmetry. They have a gut with both mouth and anus and a body cavity, called a pseudocoelom. Nematodes have a thick cuticle, are round in cross-section and unsegmented. There are a very large number of both free-living and parasitic species.
6. The annelid (segmented worm) body has a succession of segments supported by a hydrostatic skeleton based on the coelom. Respiration is often via the body surface, restricting their abundance and distribution on land, although there are many aquatic forms.
7. Arthropods are abundant in marine, freshwater and terrestrial environments. Body segments are combined in larger, specialised groups (tagmatisation) enclosed in an exoskeleton and supported by jointed appendages.
8. The basic molluscan body plan is that of a soft but muscular foot beneath a soft-bodied visceral mass containing the internal organs. Modifications to the visceral mass and the foot give rise to the major structural adaptations of the different classes.
9. Echinoderms are the earliest deuterostomes, in which the coelom arises from outpocketings of the digestive tract. The radially symmetrical adults develop from bilaterally symmetrical larvae and possess a unique water vascular system.
10. Invertebrates provide a wide range of ecosystem services and human foods, justifying their conservation.

Key terms

aboral surface
chitin
choanocytes
circular muscles
closed circulatory system
coelom
corals
coxal gland
definitive host
dioecious
ectoparasites
endocrine gland
endoparasites
exoskeleton
flame cells
ganglion
haemocoel
hemimetabolous development
hermaphrodite
holometabolous development
hormones
hydrostatic skeleton
intermediate host
longitudinal muscle
Malpighian tubules
mantle
mantle cavity
medusa
mesoderm
mesoglea
mesohyl
metameric
metamorphosis

metanephridia
nematocyst
nephridia
oral surface
parapodia
parthenogenesis
pedicellariae
pinacocytes
polyp
proglottid
pseudocoelom
radula
respiratory trees
scolex
segmentation
septa
sessile
spiracles
tegument
torsion
tracheae
trochophore larva
tube foot
veliger larva
water vascular system
zooxanthellae

Test your knowledge

1. In what ways is the cnidarian body plan more complex than that of the poriferans?
2. What special problems are presented by a parasitic lifestyle and how do parasitic flatworms and nematodes overcome them?
3. A cephalopod and a snail look very different, but they are both classified in the phylum Mollusca. What features are shared by cephalopods and other molluscs? What features are unique to cephalopods among the molluscs?
4. The arthropods make up the great majority of animal species and in turn most of the arthropod species are insects. From your knowledge of the biology of insects and the other invertebrates, suggest why there are so many insect species.
5. Early biologists classified cnidarians and echinoderms together in one phylum, the Radiata, because they all had radial symmetry. On what basis are they now placed in two distinct phyla?

Thinking scientifically

Midges are holometabolous insects spending their egg, larval and pupal stages in freshwater ponds and lakes. Winged adults emerge from the pupae, mate and lay eggs in the water. They are often attracted to light and can annoy local residents.

Residents living alongside five different wetlands in a large city are urging their local councils to spray insecticide on the wetlands to control midges. Some environmental officers are concerned that the spraying will kill not only the midges but also aquatic dragonfly larvae that prey on midges. Others believe that the correct dose of insecticide will kill only midges. Using the detailed examples given in Chapter 2, design experiments to determine whether spraying can kill the midges but not the dragonfly larvae.

Debate with friends

Debate the topic: 'It is not necessary to be concerned about the conservation of parasitic animals'.

Further reading

Anderson, D.T. (1996). *Atlas of invertebrate anatomy.* UNSW Press, Sydney.

Barnes, R.S.K., Calow, P., Olive, P.J.W., Golding, D.W. and Spicer, J.I. (2001). *The invertebrates: a synthesis*. Blackwell Science, Oxford.

Pechenik, J.A. (2005). *Biology of the invertebrates.* McGraw-Hill, New York.

Tudge, C. (2000). *The variety of life: a survey and a celebration of all the creatures that have ever lived.* Oxford University Press, Oxford.

15

Life on the move III – vertebrates and other chordates

Mike Bamford

The hoax that wasn't

In 1799, British naturalist George Shaw excitedly studied an extraordinary dried specimen from the newly established colony of New South Wales. The small, otter-like animal, measuring about 35 cm, had a duck-like bill on the furred body of a four-legged animal (Plate 15.9b). He named it *Platypus anatinus*. Until further specimens arrived the following year, however, he worried that the remarkable animal was a surgically prepared hoax. Renowned German anatomist Johann Blumenbach had no such reservations when he received a specimen in 1800. Unaware of Shaw's work in the slow communications of those pre-internet days, he also described the new animal, naming it *Ornithorhynchus paradoxus*. The scientific name was ultimately resolved as *Ornithorhynchus anatinus* and 'platypus' was kept as the common name. The surprises, though, were far from over. Excavations of platypus burrows on river and creek beds revealed that, although furred, it laid eggs and nourished the newly hatched young with milk. The truth of platypus eggs was not confirmed until 1884 when William Caldwell, fresh from excavating a burrow and dissecting an egg, telegrammed from outback Queensland to a meeting of the British Association for the Advancement of Science. His message read 'monotremes oviparous, ovum meroblastic'. The platypus thus blended characteristics of reptiles, birds and mammals – most of the groups of animals with a backbone – in one species.

The platypus is but one of many remarkable Australian backboned animals. Charles Darwin summarised aptly the wonder of European settlers in his observation that one would be excused for thinking that these animals came from a separate creation.

Chapter aims

In this chapter we examine the evolution of the distinguishing features of the phylum Chordata, concentrating on the infraphylum, the Vertebrata (the backboned animals). We describe in detail the main classes of the vertebrates and their adaptations to their environments, because of the ecological significance of the vertebrates.

A small but significant group

The presence or absence of a backbone is a great distinction among animals. Yet animals with backbones are only in one infraphylum, a level of classification between a subphylum and a class; they are the Vertebrata. This infraphylum lies within the subphylum Craniata, one of four subphyla of the phylum Chordata, just one of nearly 30 animal phyla on planet Earth. The vertebrate group is by far the largest within the Chordata with about 50 000 species, but by comparison there are nearly as many species in some families of insects. In terms of species richness and the taxonomic hierarchy, vertebrates are a minor branch of animal evolution.

What vertebrates lack in numbers and taxonomic standing, however, they make up for in size and diversity of form. They include the largest animals and occur from ocean depths to mountain tops, from pole to pole. Some live entirely underground; some spend most of their lives on the wing. In short, the vertebrate body plan challenges the physical limitations restricting the size of invertebrates and allows a dizzying adaptive radiation. The biomass of vertebrates may exceed that of invertebrates, and the individual size of vertebrates makes them a force to be reckoned with in any ecosystem. The vertebrates and their fellow chordates represent a very special clade of animals with some remarkable characteristics. They also include humans.

What makes a chordate?

Although most chordates are vertebrates, there are four living subphyla of non-vertebrate chordates (see Table 15.1 and Figure 15.1). Three of these subphyla, the hemichordates, urochordates and cephalochordates, are referred to as protochordates, and the DNA of the urochordates identifies them as the closest living vertebrate relatives. The fourth non-vertebrate subphylum consists of the enigmatic hagfish, placed with the vertebrates in the superphylum Craniata. Key chordate features shared by protochordates and craniates are:

- a stiff, flexible internal rod called a notochord. This is the precursor to the vertebrate backbone.
- a single, hollow, dorsal nerve chord
- a pharynx pouch at some stage of development
- a mucus-secreting organ, the endostyle (precursor to the thyroid gland in most vertebrates)
- metameric organisation of muscles in segments (metameres). This metameric structure can be seen in the way the meat of a well-cooked fish falls apart into segments.
- a post-anal tail of skeleton and muscle at some stage.

What makes a vertebrate?

The subphylum Vertebrata includes 12 living and seven extinct classes (see Table 15.1). Australia has abundant fossils of the extinct classes, with sedimentary rocks in central Australia and the Kimberley containing some of the best fossils of ancient fish in the world. In addition to the main chordate features, vertebrates have:

Table 15.1 A who's who of living chordates. There are also seven extinct classes of vertebrates.

Phylum	Subphylum	Infraphylum	Class	Number of species
Chordata	Cephalochordata (lancelets)			13
	Hemichordata			85
			Enteropneusta (acorn worm)	
			Pterobranchia (pterobranchs)	
	Urochordata (tunicates or sea squirts)			1300
			Ascidiacea (sea squirts)	
			Thaliacea (salps)	
			Larvacea	
	Craniata	Myxinoida (hagfish)		64
		Vertebrata (vertebrates)		
			Hyperoartia (lampreys)	41
			Chondrichthyes (cartilaginous fish)	> 900
			Osteichthyes (bony fish)	25 000
			Amphibia (amphibians)	5000
			Reptilia (reptiles)	6000
			Aves (birds)	8600
			Mammalia (mammals)	4300

- a basic body plan consisting of a head, a trunk and a post-anal tail
- organ systems including (1) a complete digestive system with mouth and anus, supported by a liver and a pancreas, (2) a closed circulatory system with a ventral heart of two to four chambers, a network of blood vessels and red blood cells with haemoglobin to transport oxygen, (3) an excretory system of paired kidneys, (4) an endocrine system of ductless glands producing hormones to control a range of functions, and (5) a well-developed brain and paired eyes and ears
- an endoskeleton. In its most ancestral form it includes a cranium and an internal, dorsal vertebral column. This is usually bony and incompressible, but flexible. In

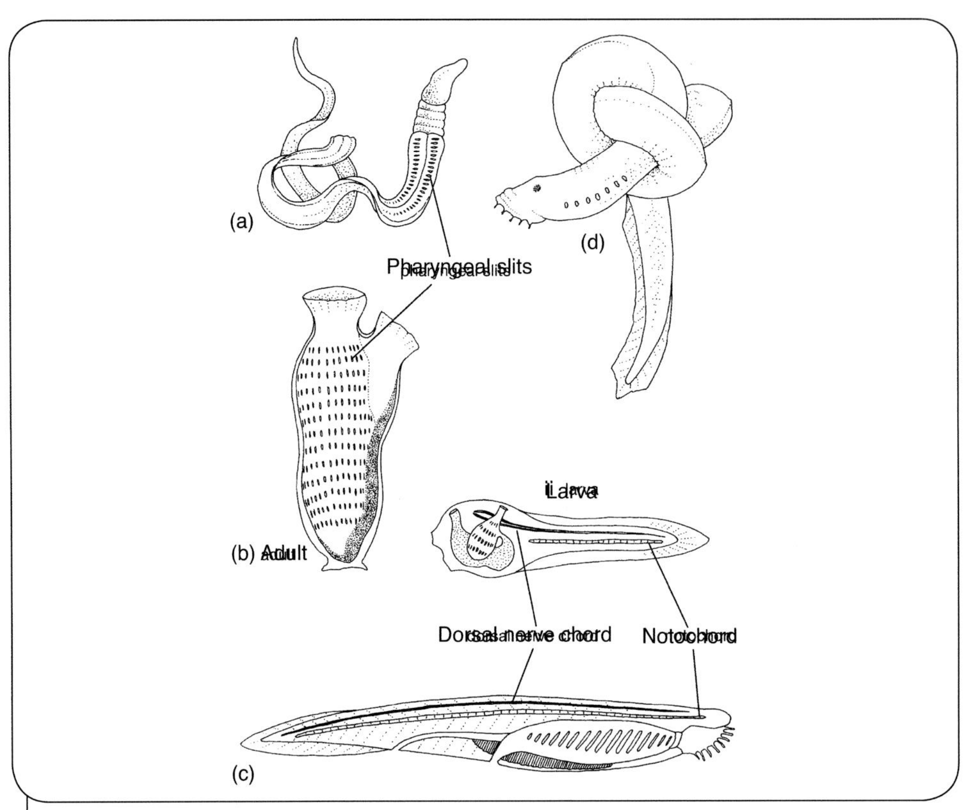

Figure 15.1 Protochordates and non-vertebrate craniates. (a) Hemichordata (acorn worms). These are sometimes considered not to be chordates but are placed in their own phylum with affinities to the chordates. (b) Urochordates (sea squirts and allies). Although the adults look little like other chordates, the larvae possess the characteristic chordate features. (c) Cephalochordata (lancelets). *Branchiostoma* (formerly *Amphioxus*). It is like a tiny fish but has no head, eyes or jaws, and no paired fins. (d) Hagfish. Formerly classed as vertebrates, hagfish are considered to represent a unique chordate lineage. They have a skull but no vertebral column. (Source: M. Bamford)

more evolved forms, the endoskeleton includes a column of bony vertebrae and bones supporting the viscera and limbs and anchoring the muscles.

- an outer integument divided into an outer epidermis and an inner dermis. Modifications of the integument such as hair, scales and feathers characterise different vertebrate groups.
- a neural crest, a line of cells formed on the dorsal surface of the early embryo. Neural crest cells form parts of the nervous system, skeleton and some organs.

Milestones in vertebrate evolution

The features of the vertebrate classes, both extinct and extant, represent milestones in vertebrate evolution reflecting fundamental environmental pressures such as feeding and adaptation to marine, freshwater and terrestrial environments. They are described below.

Development of hinged jaws

Agnathans, fish without true jaws, include all extinct classes of fish and the living lampreys. All other vertebrates are gnathostomes, with true jaws. The ancestral chordates were filter feeders and the mouth was merely an opening to the pharynx. The first vertebrates were also jawless and some may have continued filter feeding. Others fed actively, using teeth embedded into the edge of the mouth and possibly also the tongue, as do modern lampreys. Modifying the gill arches supporting the pharynx into jaws opened up many approaches to capturing and processing food for these animals. The ancestral gill arches provided structural material not only for jaws, but for jaw articulation and ear bones. The evolutionary fate of the anterior gill arches of agnathans is illustrated in Figure 15.2.

Development of legs

Amphibians, reptiles, birds and mammals are tetrapods (four-legged). Even species that have secondarily lost their legs, such as snakes and whales, are tetrapods. All other vertebrates are fish.

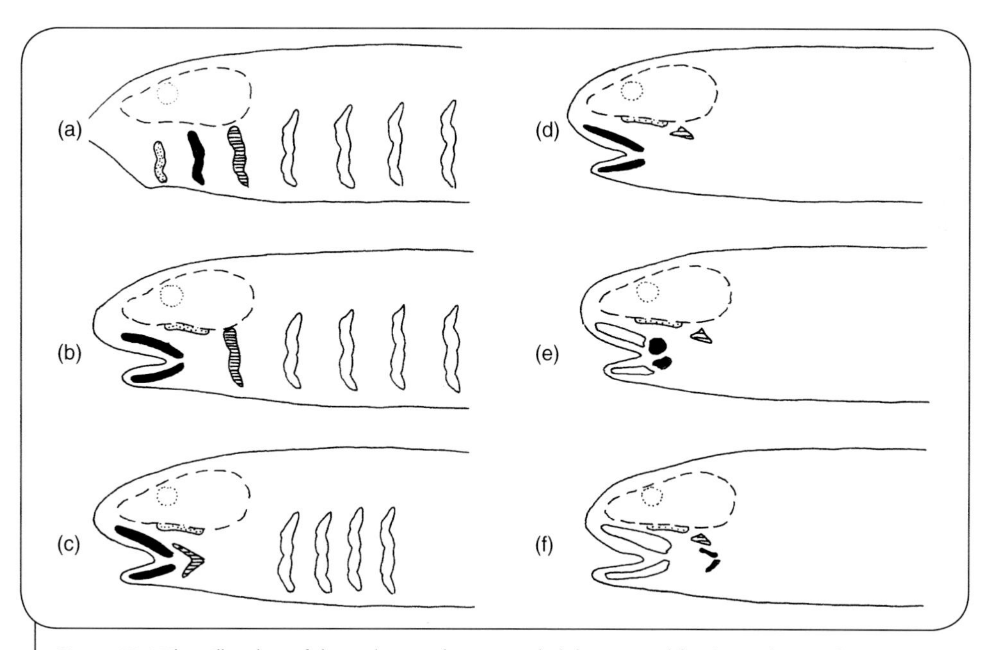

Figure 15.2 The gill arches of the early agnathans provided the material for the evolution of jaws and ear ossicles in later vertebrates. (a) Agnathans. Series of cartilaginous gill arches with associated gill slits. (b) Early gnathostomes. First (pre-mandibular) arch forms rods (trebeculae) on the base of the brain case. Present in all subsequent gnathostomes. Second (mandibular) arch forms the upper jaw (palatoquadrate) and lower jaw (Meckel's cartilage). (c) Modern fish. Jaws derived from second gill arch with third (hyoid) gill arch incorporated into hyomandibular jaw structure. (d) Amphibians. Jaws derived from second gill arch with third (hyoid) gill arch forming an ear ossicle (stapes). (e) Reptiles and birds. Palatoquadrate and Meckel's cartilage, originally derived from the second gill arch, modified to form the quadrate and articular bones of the jaw hinge, with the jaw function replaced by the maxilla (upper jaw) and dentary (lower jaw). Hyoid gill arch forms ear ossicle (stapes). (f) Mammals. Jaw articulation simple, with the lower jaw hinged directly against the base of the skull, and the quadrate and articular forming ear ossicles (maleus and incus respectively). Third ear ossicle (stapes) still derived from hyoid gill arch. The jaw is a very rigid structure, allowing the complex heterodont dentition to function effectively. (Source: M. Bamford)

The tetrapod ancestors were fish adapted to shallow wetland margins. This environment provided shelter from predators that lived in more open waters and was probably warm and rich in food. The wetland margins were low in oxygen, however, so many fish had primitive lungs. Their descendants survive today and modern fish in similar environments often have auxiliary breathing systems as well as gills. The shallow water and dense vegetation favoured fish with low body profiles and sturdy limbs with fingers, rather than fins, to pull themselves along underwater. Colonisation of the land thus occurred *after* the evolution of legs. The earliest movements across land probably occurred when fish trapped in shallow pools dragged themselves to other pools.

Development of the amniotic egg

The amniotic or shelled egg does not dry out on land and was critical in freeing terrestrial vertebrates from the need for water to reproduce. As with the development of legs, the evolution of the amniotic egg may have been driven by conditions in aquatic environments. Eggs and young in aquatic environments are prone to predation. Eggs laid on the margins of wetlands or even on land may have had better chances of survival. The amniotic egg may therefore have been the ultimate adaptation by a largely aquatic animal to protect its young.

The most obvious feature of the amniotic egg is the shell providing support and protection. It may be either rigid, as in birds and some reptiles, or flexible and leathery, as in most reptiles and the egg-laying mammals. There is also a yolk, a massive food supply encased in a membrane continuous with the embryo's gut. Less obvious are the membranes enveloping the yolk and embryo, but they are the strength of the amniotic egg (see Figure 15.3). The semipermeable chorion lies immediately beneath the shell, allowing gaseous exchange. The allantoic membrane encloses the allantoic sac storing waste products, and also forms the respiratory surface with the chorion. The embryo itself

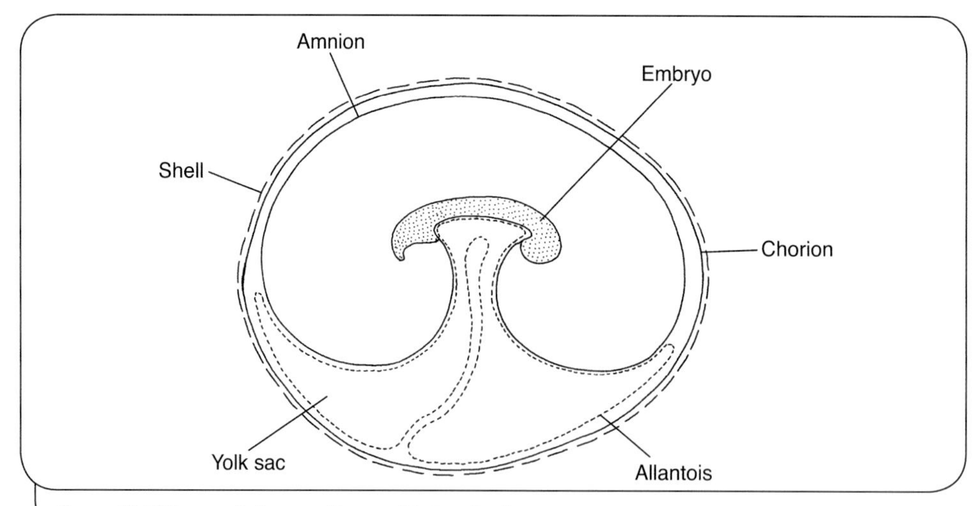

Figure 15.3 The amniotic egg. (Source: M. Bamford)

is encased within an inner membrane, the amnion, enclosing a pool of fluid, a 'personal pond' for the developing embryo. The amnion also presses the allantois flat against the chorion and the inside of the shell for gaseous exchange. Thus the amniotic egg is a self-contained survival capsule that allows embryonic development in an environment hostile to the effectively aquatic embryo, with just enough contact with the outside world for the embryo to breathe. Reptiles, birds and mammals are *amniotes* while fish and amphibians are *anamniotes*.

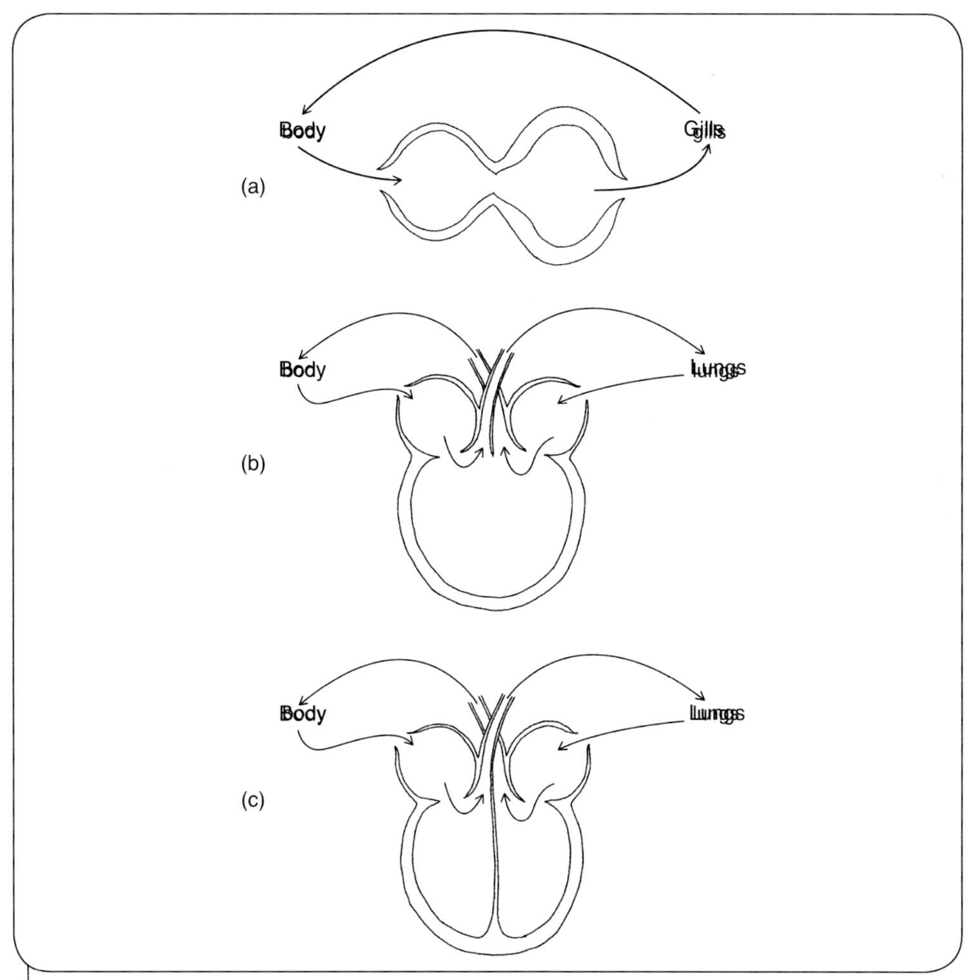

Figure 15.4 Diagrammatic two-chambered (a), three-chambered (b) and four-chambered (c) hearts. In the two-chambered heart, deoxygenated blood from the body passes into the atrium and from there to the ventricle. It is pumped to the gills for gas exchange and then on to the body. In the three-chambered heart, the right atrium receives deoxygenated blood from the body and the single ventricle pumps it to the lungs for gas exchange. Oxygenated blood from the lungs returns to the left atrium and thence to the ventricle for pumping to the body. Some mixing of oxygenated and deoxygenated blood occurs in the ventricle. In the four-chambered heart, deoxygenated blood enters the right atrium and is pumped to the lungs by the right ventricle. Oxygenated blood returns to the left atrium and is pumped to the body by the left ventricle. Oxygenated and deoxygenated blood are kept separate in the heart. (Source: M. Bamford)

Development of three- and four-chambered hearts

A vertebrate has a closed system of blood vessels with a pumping heart. The hearts of most fish have two primary chambers: an atrium for receiving blood and a ventricle for pumping it around the body (lung-fish have a partly divided atrium). Some fish, frogs and most reptiles have three-chambered hearts, whereas crocodiles, birds and mammals have four-chambered hearts (Figure 15.4). In most fish, blood enters the atrium of the heart and is pumped into the larger ventricle. The ventricle pumps blood to the gills, where it is oxygenated, and through to the rest of the body. Pressure is lost, forcing blood through the gills, but the loss is much greater when forcing blood through lungs, in which the blood is constrained in capillaries. As a result, air-breathing vertebrates have a double circulatory system, with deoxygenated blood from the body being received in one atrium, pumped into the ventricle and then to the lungs. It is then received into a second atrium before being pumped around the body by the ventricle. The significance of the number of chambers lies with the efficiency of separating oxygenated blood from the lungs with deoxygenated blood that needs to be pumped to the lungs. In groups with two atria and one ventricle, blood flow is such that little mixing of deoxygenated and oxygenated blood occurs, whereas groups with a four-chambered heart have two ventricles so no mixing occurs.

Development of lungs

Gills are blood-filled vesicles with very large surface areas. Gaseous exchange occurs through diffusion as blood flows through the gills and water flows over them. The evolution of lungs allowed vertebrates to breathe atmospheric oxygen. They were seen initially among fish living in water with low levels of dissolved oxygen. Lungs are effectively modified air bladders (used to achieve neutral buoyancy in most fish). Lungs also diffuse gases through thin membranes, but whereas gills are supported by the water around them, lungs require support to ensure the membranes don't collapse.

An overview of vertebrates: living on water, on land and in the air

The lamprey and other agnathans

The first vertebrates were agnathans (Plate 15.1). Today there are about 40 species, directly descended from the ancestors of all other vertebrates. Importantly, most vertebrate organ systems were established in agnathans. The body is covered with mucus and locomotion is by unpaired fins. The skeleton is made of cartilage and reproduction is external (Box 15.1). Modern lampreys metamorphose from a filter-feeding larval stage living in sediment to a free-swimming adult with a toothed, muscular tongue. Agnathans are ectothermic (Box 15.2) and excrete nitrogenous wastes primarily as ammonia, although they produce some urea.

Box 15.1 Reproduction – more than one way to ...

Chordates are almost all dioeceous, although parthenogenesis (development of unfertilised eggs) occurs in some fish and reptiles so that the entire species is female. Most bony fish and amphibians practise external fertilisation, with eggs and sperm shed into the water. However, this is not random, and elaborate courtship rituals have often developed to determine mate selection. External fertilisation usually requires large numbers of eggs and little parental care, although some fish and amphibians protect their young with elaborate systems such as mouth brooding in many fish, and constructing nests on the water's edge by some amphibians. The recently extinct gastric-brooding frog (*Rheobatrachus silus*) of Queensland swallowed its eggs and the young developed in the stomach, which acted as uterus and placenta.

Internal fertilisation is essential among amniotes as the fully developed amniotic egg cannot be penetrated by sperm. Despite their aquatic environment, cartilaginous fish also practise it. In most amniotes, embryonic development occurs within the egg outside the female's body, but internal fertilisation makes internal development possible. There are two basic forms of internal development: ovoviviparity, when development takes place in the egg within the female, and viviparity, when embryonic development is nurtured by the female. In some viviparous sharks, the developing young feed on unfertilised eggs or even on the other developing young. Internal fertilisation requires the male to get his sperm into the female, but in the tuatara and most birds, there is no penis. Instead, the cloacas (combined genital, excretory and digestive tract openings) of male and female are brought together to allow sperm transfer.

Parental care is almost universal among birds, probably because birds are incapable of flight until adulthood. In contrast, parental care in reptiles is very rare; young reptiles are miniature adults and can function effectively almost immediately. The young of the viviparous bobtail lizard (*Tiliqua rugosa*) eat their own placentas. Parental care is highly developed in mammals because a great deal of physical development is required for the young to be able to fend for themselves.

Box 15.2 Temperature regulation in vertebrates

Body temperature affects the rate at which chemical reactions occur, or whether they occur at all. There are three broad areas involved: body temperature regulation, body temperature variation and resting metabolism.

Body temperature regulation

Animals may regulate their body temperature using either internal or external sources of heat. Ectotherms regulate body temperature behaviourally, moving between warm and cool environments and manipulating their blood flow to fine-tune their temperature

continued ›

Box 15.2 continued ›

regulation. They are typically solar-powered because that is the ultimate source of warmth, although some animals may gain warmth from objects warmed by the sun. Ectothermy has the great strength that an animal's energy demands from its food are low; ectotherms typically eat infrequently and survive long periods without food. This makes them very successful in harsh environments. Ectothemic animals survive in cool environments and can be active at cool times of the day. Geckoes are active on cool nights by suspending metabolic activities requiring high body temperatures. They bask unseen during the day, such as under bark on the sunny side of a tree, raising their body temperatures for digestion. However, without external warmth, ectotherms cannot feed, search for mates and generally get on with living, so they are confined to environments where they can get some warmth.

Endotherms regulate their body temperature internally by generating metabolic heat. Endothermy liberates animals from external warmth, but energy demands are high. Temperature regulation in an endotherm is like keeping a building at an even temperature by running the heaters constantly and turning the cooling system up when it gets too hot. Endothermy allows high levels of prolonged activity, but when food sources are seasonal or unreliable, endotherms must find novel solutions. Some migrate. Small animals may deliberately lower their metabolic rate (torpor) so that energy stored in the body isn't wasted keeping warm overnight. Even then a small bird may lose 10% of its body weight between sunset and sunrise. Hibernation, in which an animal becomes inactive for many months and allows its body temperature to fall to levels that would normally be lethal, occurs in species living in strongly seasonal environments such as the high latitudes of the Northern Hemisphere.

Body temperature variation

Animals maintaining a constant body temperature, often well above the ambient temperature, are homeothermic. Those with widely varying body temperatures, often close to ambient temperatures, are poikilothermic. Fish may be poikilothermic, but some may bask in shallow water (ectothermy) or generate warmth internally through muscular activity (endothermy). Reptiles are rarely poikilothermic, and large reptiles may be inertial homeotherms because their great size ensures low rates of heat loss. The largest of dinosaurs are sometimes referred to as gigantotherms.

Resting metabolism

Animals maintaining high levels of metabolism at rest are tachymetabolic, whereas those that slow their metabolism markedly at rest are bradymetabolic.

Vertebrates and other animals show differing combinations of these three features in their temperature regulation. As a generalisation, mammals and birds are endothermic, homeothermic and tachymetabolic. However, many bats and small birds are effectively poikilothermic and bradymetabolic when they sleep overnight and 'heterothermy' is sometimes used to describe their temperature regulation. Reptiles, amphibians and fish may show variations of ectothermy, poikilothermy and bradymetabolism.

The first gnathostomes – fish with jaws

Jawed fish appear in the fossil record about 430 million years ago. They have paired fins and a lateral line system along their flanks to detect movement and vibration, and are ectothermic. There are two extant classes and one, the bony fish, contains over half of living vertebrate species.

Chondrichthyes (sharks and allies)

The Chondrichthyes have skeletons of cartilage and a tough, leathery skin incorporating small, tooth-like scales (Plate 15.2). Fertilisation is internal and they may lay eggs (oviparous), have the eggs hatch internally and the young emerge (ovoviviparous) or bear live young nourished internally before birth (viviparous). They are an ancient group, with fossils dated to 420 million years ago. They underwent considerable adaptive radiation in the late Palaeozoic, when they were more diverse than modern sharks. Much of this diversity was lost in the Permian extinction event (Chapter 7). There are two very distinct modern subclasses.

Elasmobranchii includes typical sharks, but also skates and rays specialised for bottom feeding (Figure 15.5). True rays swim through undulations of modified pectoral fins; sharks swim with their tails. There are about 900 living elasmobranch species, including the largest living fish, the 15 m whale shark.

Holocephali are the chimaera and rabbitfish (Figure 15.5). They are a small group of only 34 species, found mainly in deep water. Chimaera have a single gill opening, unlike sharks and rays which have up to seven gill openings.

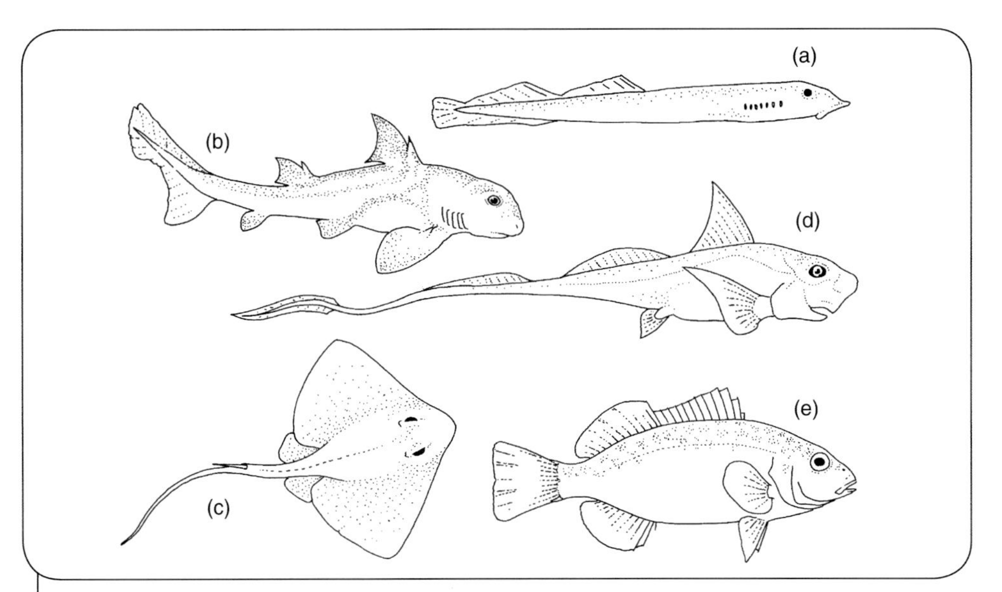

Figure 15.5 The main groups of living fish. (a) Lamprey. Note the lack of lateral fins. (b) Shark. Note the heterocercal tail. The Port Jackson shark (*Heterodontus portusjacksoni*) illustrated is a bottom-feeding species with grinding teeth. (c) Ray. Note the enlarged pectoral fins. The gills are on the underside of the body, with water taken in through the gills and expelled through spiracles located behind the eyes. (d) Chimaera. (e) Bony fish. Note the homocercal tail. (Source: M. Bamford)

The chondrichthyes have no swim bladder for buoyancy, and they reduce weight to compensate with a cartilaginous skeleton and a liver filled with oil to improve buoyancy. The heterocercal (two unequal lobes) tail, with the larger upper lobe containing the continuation of the vertebral column, combined with broad, rigid pectoral fins, imparts lift to counteract the tendency of a swimming shark to sink (see Figure 15.5). Although they are aquatic they excrete their nitrogenous wastes as urea.

All chondrichthyes are carnivorous, but some are specialised for filter feeding (whale shark, manta ray). The cookie-cutter shark bites chunks out of larger fish by latching onto them and then twirling around so that the teeth slice out a circular plug of flesh. The teeth are similar in structure to the scales, both having a bony base and a dentine cap.

Osteichthyes (bony fish)

The Osteichthyes have bony skeletons and a body covering of scales (Plate 15.3). Fertilisation is normally external and larval forms hatching from the eggs may undergo marked metamorphosis during development. They are ectothermic and excrete their nitrogenous wastes as ammonia. Bony fish are slightly younger than the chondrichthyes, appearing in the fossil record about 410 million years ago. Initially they were a minor component of vertebrate biodiversity, but they flourished and are now the most species-rich vertebrate class, with about 25 000 living species. They range in size from less than 10 mm to 12 m, and include such oddities as the seahorses and pipefish, mudskippers and lungfish. There is also a huge variation in life styles and reproductive strategies. Their success comes from:

- a lightweight and complex bony skeleton. Spines on vertebrae support the body and the rib cage supports the viscera.
- a gas-filled swim bladder, which gives adjustable neutral buoyancy, so the fish does not need to expend energy to hold its depth
- neutral buoyancy, which allowed the development of the powerful homocercal (two equal lobes) tail and lightweight, flexible fins based on a membrane supported by fin-rays. Flexible paired fins allow great manoeuvrability, but are impossible in fish lacking neutral buoyancy, because rigid paired fins are required for lift.
- varied tooth structure arising from the tissue types of bones and teeth
- thin but strong and greatly overlapping scales, combining external rigidity and flexibility
- a bony operculum, which protects the gills and creates a distinct gill chamber
- the jaw structure, which is very complex in most modern forms, including a shortening of gape from a simple mouth to a complex jointed structure. Many species have extrusible jaws (see Figure 15.6).

Tetrapods – gnathostomes with legs

The four tetrapod classes (Amphibia – Figure 15.7, Reptilia – Figure 15.8, Aves – Figure 15.9 and Mammalia – Figure 15.10) are distinctive but unified by the demands of living

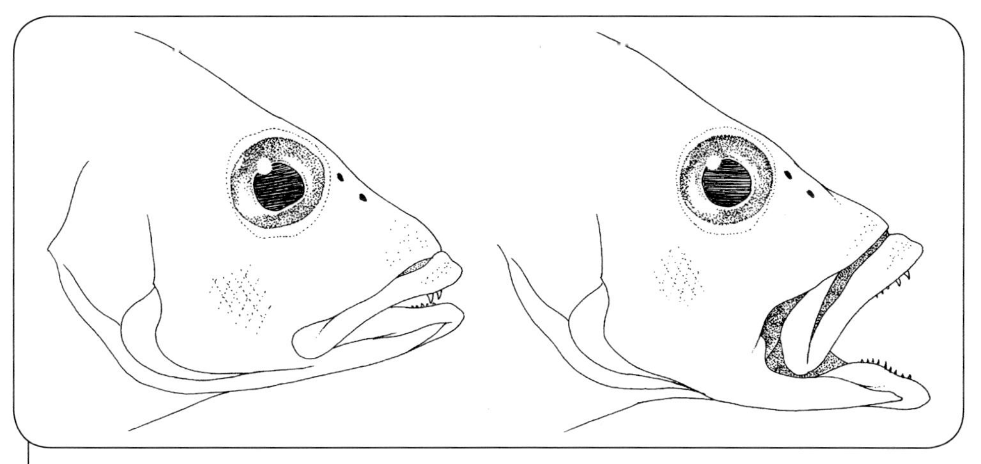

Figure 15.6 Jaw articulation in a bony fish, showing the double articulation in the upper jaw as the mouth is opened. (Source: M. Bamford)

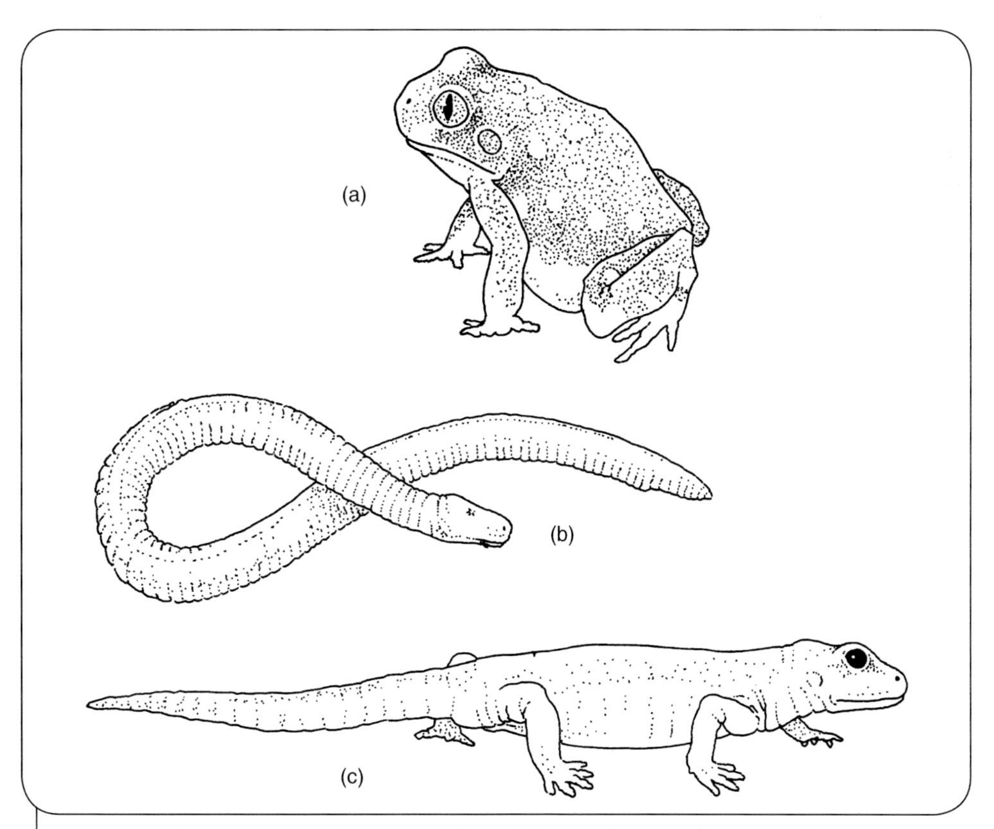

Figure 15.7 The major groups of living amphibians. (a) Frog. The spotted burrowing frog (*Heleioporus albopunctatus*) illustrated is a terrestrial species that goes near wetlands only to breed. (b) Apodan. An entirely limbless, largely subterranean amphibian. (c) Salamander. Similar in appearance to the first terrestrial vertebrates. (Source: M. Bamford)

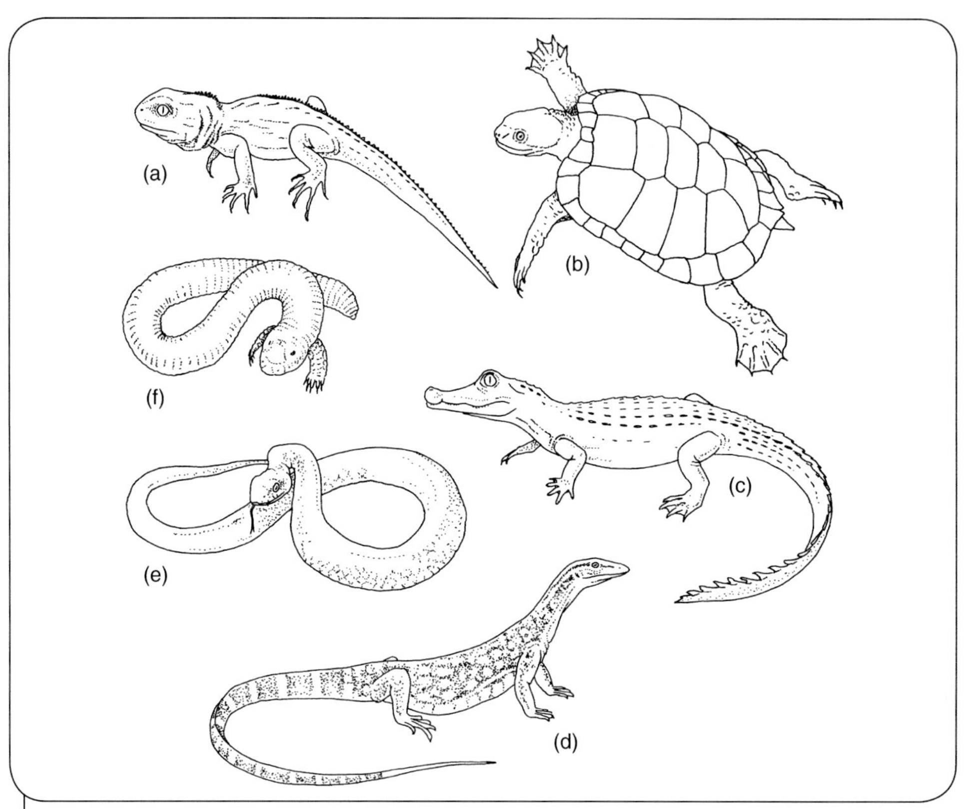

Figure 15.8 The major groups of living reptiles. (a) Tuatara (*Sphenodon punctatus*). Confined to a few island locations in New Zealand and represents an ancient reptile lineage. (b) Tortoise (turtle). The western swamp tortoise (*Pseudemydura umbrina*) illustrated is one of the world's rarest reptiles, with a population that at one stage numbered less than 50 individuals. (c) Crocodile. Illustrated is a juvenile saltwater crocodile (*Crocodylus porosus*), one of the largest living reptiles. (d) Lizard. Illustrated is Gould's goanna (*Varanus gouldii*), a fast-moving predator found across much of Australia. The skull is illustrated in Figure 15.11. (e) Snake. (f) Amphisbaenian. Illustrated is the Mexican worm lizard (*Bipes biporus*), one of the two-limbed amphisbaenians. The short but powerful forelegs assist with burrowing. Many other amphisbaenians are limbless and burrow with the head. (Source: M. Bamford)

in terrestrial environments. The following sections discuss the key characteristics and origins of each class.

Amphibia (frogs, toads, newts, salamanders and caecilians)

'Amphibian' means 'double life' and accurately describes these vertebrates (Plate 15.4). Almost all species have an aquatic, fish-like larval stage that metamorphoses into a usually terrestrial adult that is not at all fish-like. Other key features are a moist, glandular skin lacking scales, a mainly bony skeleton and a skull that articulates with the vertebral column via two rounded condyles (the ball part of a ball-and-socket joint). In fish, the skull fuses with the vertebral column, reptiles and birds have only one condyle, whereas mammals, like the amphibia, have two. Amphibians are ectothermic. Unlike other tetrapods, they are anamniotic. In many species the aquatic larvae excrete their nitrogenous wastes mainly as ammonia, but shift to excreting urea as adults.

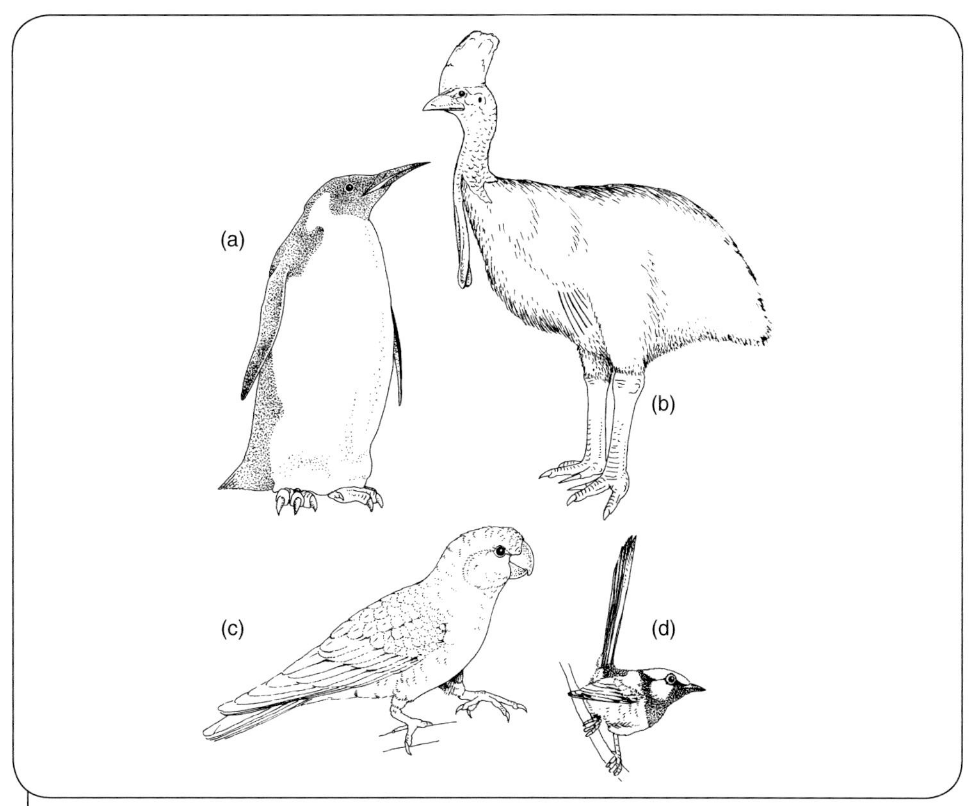

Figure 15.9 Examples of birds. (a) Emperor penguin (*Aptenodytes forsteri*). A flightless bird with the wings modified as flippers to allow the bird to fly through the water. (b) Southern cassowary (*Casuarius casuarius*). A flightless terrestrial bird with the wings reduced to stumps that support spines. (c) Glossy black cockatoo (*Calyptorhynchus lathami*). (d) Red-winged fairy-wren (*Malurus elegans*), one of the order Passeriformes, which includes over half of living bird species. (Source: M. Bamford)

The moist skin of amphibians is an important respiratory surface, accounting for about 30% of gaseous exchange, and it works under water and on land. This is a great advantage for an amphibious tetrapod. The disadvantage is that it is also an avenue for water loss. The gland-rich, oily skins of amphibia are designed to reduce water loss, but are not entirely successful.

Amphibia are almost exclusively predators as adults, but the larvae of some groups may be omnivorous or herbivorous, grazing on algal and bacterial films. Metamorphosis can therefore involve a considerable change in the digestive tract as well as the growing of legs, the development of lungs and the absorption of gills and tail.

The class Amphibia is usually divided into three subclasses: the extinct Labyrinthodontia and Lepospondyli, and the living Lissamphibia.

The modern Amphibia (the subclass Lissamphibia) contains three orders: Anura (frogs, toads and allies), Caudata (or Urodela, the newts and salamanders) and Apoda (or Caecilians, pan-tropical, generally limbless amphibians that are aquatic or burrow in damp soil).

Figure 15.10 The major groups of living mammals. (a) Monotremes, such as the platypus (*Ornithorhynchus anatinus*). (b) Marsupials, such as the woylie or brush-tailed bettong (*Bettongia penicillata*) carrying nesting material in its prehensile tail. (c) Eutherian mammals, such as Mitchell's hopping mouse (*Notomys mitchelli*), a rodent native to southern, semi-arid Australia. The order Rodentia is the most species-rich order of the Eutheria. (Source: M. Bamford)

Reptilia (tortoises and turtles, lizards and snakes, crocodiles, tuatara and numerous extinct groups)

In many ways, reptiles are the success story of the amniotes. From their appearance about 340 million years ago until only 65 million years ago, they were the most abundant and diverse vertebrates on Earth. Even among living terrestrial vertebrates they are outnumbered in species only by the birds, a group recently evolved from the reptiles. Modern reptiles range in size from lizards less than 50 mm in total length to crocodiles of over 6 m and snakes (the anaconda of South America and the reticulated python of South-East Asia) of over 10 m. Australia has a particularly diverse lizard fauna (Box 15.3).

Reptiles are characterised by a unique combination of features, many of which are adaptations to provide support, reduce water loss and allow reproduction on land (Plate 15.5), although some have returned to the water (Plate 15.6). The bony skeleton provides support and the skull has a single condyle. The skull is also kinetic, meaning that the bones are not fused, allowing for great flexibility so that very large prey can

Box 15.3 The lizards of Oz – destined for diversity

Australia is rich in lizard species. Eric Pianka found that while a single site in arid Australia may support 45 lizard species, sites in similar climatic zones in southern Africa or North America support only a dozen or less. The underlying cause seems to be geological and climatic history. Australian ecosystems are 'ancient' by world standards because vulcanism, tectonic activity (collisions between crustal plates) and glaciation have not occurred for very long periods. There has been climatic change, however, causing changes in the type and distribution of plant communities and therefore of animals too.

Lizard assemblages in Australia have therefore been enriched by periodic colonisation, isolation and speciation. In any one location, there are lizard species with southern affinities and species with northern affinities. For example, of 48 lizard species recorded in one area on the Cape Range of Western Australia, 10 species had a southern distribution and were at the extreme north of their range, six species were from the north, three species were from inland areas and seven species were endemic to the region (occurring only there). Twenty-two species were widespread.

be swallowed (see Figure 15.11). The teeth are all similar (homodont) but occasionally pseudoheterodont, meaning that there is some specialisation of the teeth within the mouth. Such specialisation is more highly developed among the heterodont mammals. Teeth in reptiles are used to capture, but rarely to chew or process food. In some species, teeth are modified to form venom-injecting fangs. Limb reduction and absence are common. In snakes, the body is greatly elongated and organ systems adjusted accordingly, with paired organs (lungs, kidneys and ovaries/testes) modified so that only one of each pair is developed. The waterproof skin is dry, scaly and generally lacks glands. Its outer horny (like fingernails) layer is periodically shed. Respiration is internal using lungs, which also reduces water loss. Fertilisation is internal so water is not necessary for the sperm to swim to the egg, and all reptiles except the tuatara have a penis for transferring sperm to the female reproductive tract. Most species lay amniotic eggs, but live-bearing has evolved independently several times. There is no metamorphosis. Reptiles excrete nitrogenous waste as uric acid. Although this is energetically expensive to produce, it emerges as a white paste so excretion involves little loss of water. Reptiles have only a single bone in the ear and in all except the Crocodilia the heart is three-chambered.

Reptiles are also ectothermic (Box 15.2) and most modern reptiles are predators. Leafy plant material is low in nutrients and difficult to digest for an animal without a high, constant body temperature. Many species regulate their body temperatures behaviourally, basking to absorb heat when necessary or sheltering from the sun to avoid overheating.

The main orders of modern reptiles are the Chelonia (tortoises and turtles), the Crocodilia (crocodiles, alligators, caimans and gharials), the Squamata (snakes, lizards and the Amphisbaenia, a pan-tropical group of mostly limbless, burrowing reptiles

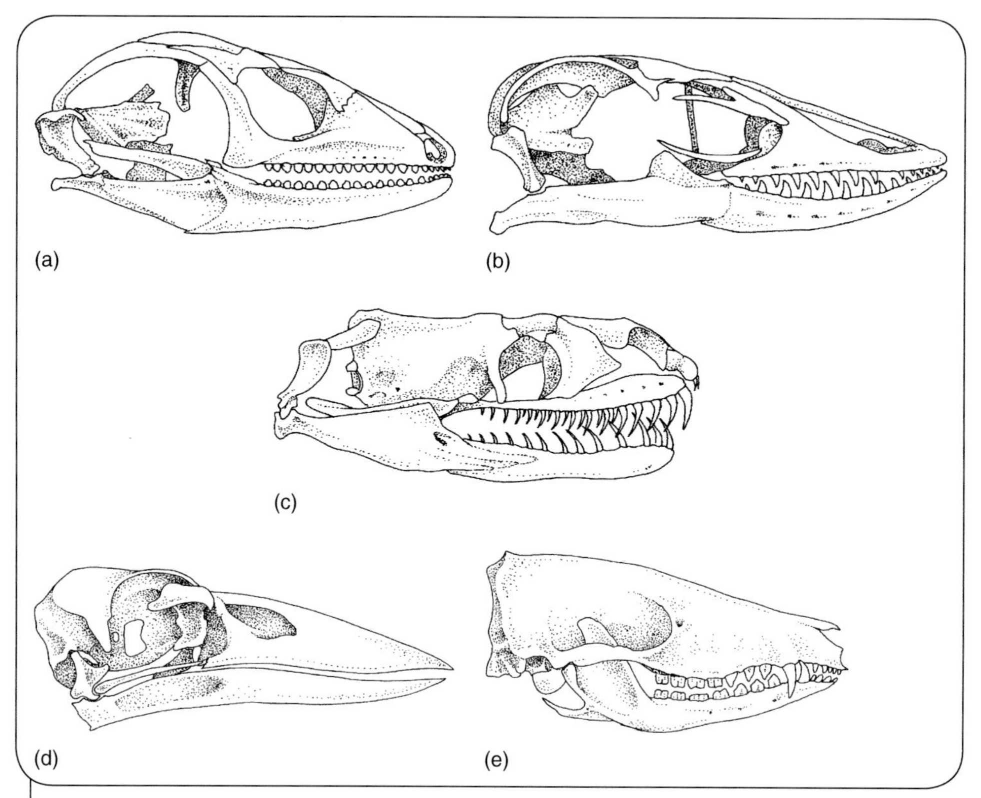

Figure 15.11 The skull of reptiles, birds and mammals. (a) Bobtail skink (*Tiliqua rugosa*), a partly herbivorous lizard with a rigid skull, rounded teeth and a powerful, crushing bite. (b) Gould's goanna (*Varanus gouldii*), a carnivore with stabbing teeth and a kinetic skull to allow large chunks of food to be swallowed. (c) Carpet python (*Morelia spilotes*), a large snake with a highly kinetic skull and very slender, gripping teeth. The skull can be distorted to enable food items much larger than the head of the snake to be swallowed. Note the complex jaw articulation. (d) Laughing kookaburra (*Dacelo gigas*). Jaw articulation is like that of a reptile, but the skull bones are extensively fused and the jaws form a toothless beak. (e) Quenda or brown bandicoot (*Isoodon obesulus*). A typical mammal with a rigid skull, lower jaw hinged directly against the skull and heterodont teeth. (Source: M. Bamford)

reminiscent of the apodan amphibians) and the Rhyncocephalia, an important group during the first period of reptile radiation. The inclusion of the Rhyncocephalia among this modern line-up is being questioned, as the single living species, the tuatara of New Zealand (Plate 15.7), is thought by some authorities to be an unusual squamate. Australia is especially rich in lizards and snakes, with 500 and 170 species respectively. Representation at the family level is poor, however, with only five lizard and six snake families present, compared with 17 and 12 worldwide. This lack of richness at the family level is because Australia's long period of geographic isolation has meant that many families did not colonise the continent.

Aves (birds)

The birds are the most distinctive and successful of modern tetrapods, with as many living bird species as all other tetrapod groups combined (see Table 15.1). Many of their features relate to the demands of flight (Plate 15.8). The most obvious is the modification of the forelimbs into wings (occasionally vestigial or secondarily modified as flippers). The body is covered with feathers (modified scales) that aided thermoregulation in the first birds, and subsequently aerodynamics. There is remarkable consistency in the arrangement of feathers on the body into discrete tracts.

Weight reduction is also important. Instead of heavy teeth, birds have a horny beak of keratin, which is both light and strong. Their bones are often hollow with supporting struts to give strength while reducing weight, and the tail is reduced to a short, muscular stump supporting tail feathers. The clavicle is greatly enlarged as an anchor point for flight muscles and the pelvis greatly elongated and fused with vertebrae to further reinforce the anchorage of flight muscles and support the walking limbs. Birds have more neck vertebrae and more flexible necks than other tetrapods, possibly as compensation for restrictions on body shape imposed by the needs of flight. Weight reduction also extends to the excretion of nitrogenous waste as uric acid. Although this is energetically expensive to produce, it emerges as a white paste so there is no heavy, fluid-filled bladder and there is also little loss of water.

Flying demands considerable energy and birds are endothermic so that their muscles are constantly at the right temperature to power flight. Their feathers are important in temperature control by providing insulation and waterproofing. To meet the high oxygen demands of the flight muscles, the lungs are augmented with air sacs providing additional respiratory surface and a highly efficient, unidirectional flow of air. There is also an efficient, four-chambered heart to deliver oxygenated blood to the flight muscles.

The demands of flight also greatly influence reproduction, parental care and growth. Although most birds lack a penis, fertilisation is internal. All birds are oviparous and this allows for large clutches without the female having to carry the weight of developing young. There is no metamorphosis. Parental care is greatly developed in almost all birds, probably because of the vulnerability of the young before they can fly.

Birds arose from small, bipedal dinosaurs, and feathers probably evolved as a means of regulating body temperature. The origin of flight is uncertain, but elongated feathers on the front legs of a bipedal dinosaur may have improved speed and manoeuvrability, leading to a form of wing that was adaptive but not designed for flight. Birds diversified during the Cretaceous (135 million to 65 million years ago), when most modern orders evolved, and again following the Cretaceous extinction event.

Flight allows birds to exploit habitats in a way not available to other animals. The ability to migrate hundreds and even thousands of kilometres allows birds to use seasonal resources at locations on opposite sides of the planet. About 10% of Australia's birds breed in Asia, many taking advantage of the briefly productive Arctic summer to breed before migrating south to forage during the Southern Hemisphere spring, summer and autumn. Migration requires the ability to sense direction, not necessarily true navigation, and birds are sensitive to the Earth's magnetic field and innately recognise star patterns. In contrast, some species are flightless. The four orders of ratites are flightless, although their wings retain features suggesting they evolved from flying ancestors. Penguins are also flightless. Flightlessness has evolved in representatives of a number of other groups in response to local conditions, particularly on islands.

Birds are mostly predators or eat the most nutritious parts of plants, such as fruits, grain and nectar. Few species are grazers, browsers or folivores because such a bulky diet is restrictive for animals constrained by the weight control needed for flight. Extreme specialisation of the beak and often feet to a bird's diet is common.

There are about 26 living orders of birds (the number varies with taxonomic approach), with over half of living species in one order: the Passeriformes. These are the perching or songbirds, with a rear toe that opposes three forward-directed toes, facilitating perching.

Mammalia (mammals)

Mammals include the largest vertebrates to have ever lived, while modern mammals display tremendous variation in their body form, from entirely marine species to species capable of true flight. Mammals are also distinctive among living vertebrates as the only group with a body covering of hair, to nurture their young on a fluid (milk) exuded from enlarged skin glands, with three bones in the middle ear, with a variety of tooth types (heterodont teeth) replaced as the animal grows (dyphyodont), with seven neck vertebrae, with a muscular diaphragm separating the chest cavity from the abdominal cavity, and with a lower jaw articulating directly against the skull. They excrete nitrogenous wastes as urea dissolved in water, which requires less energy than producing uric acid but is remarkably wasteful for terrestrial animals in which water conservation is often important.

The features of modern mammals reflect their evolutionary origins. Mammal-like reptiles (theropods) appear in the fossil record at about the same time as the earliest reptiles. Following the Permian extinction event (Chapter 7), reptiles diversified spectacularly while mammals persisted only as small forms, although these underwent adaptive radiation through the 'Age of Reptiles', occupying nocturnal niches poorly exploited by reptiles. Endothermy, sensitive hearing, sense of touch often aided by whiskers and sense of smell are consequences of evolving in the shadow of the dinosaurs.

Many mammalian features relate to their endothermy (see Box 15.2). For example, hair insulates and the diaphragm facilitates breathing for sustained activity. Teeth specialised for stabbing, snipping, slicing, grinding and so on can process large quantities of food efficiently to support the high energy demands of endothermy. Mammalian teeth often act against each other and are replaced once as the animal grows to maintain efficiency as the jaws enlarge, but are not replaced continuously, as seen in reptiles, as that would lead to inefficiency from gaps and teeth of uneven size. The direct jaw articulation and solid skull improve the efficiency of mammalian teeth in processing food.

Mammals range in size from shrews and bats of about 2.5 g to the blue whale of up to 136000 kg. The dentition, large size of many species and endothermy allow mammals to exploit plant material largely uneaten by other vertebrates. Specialist herbivores have modified digestive tracts to create chambers where symbiotic bacteria break down otherwise indigestible cellulose. To further the mechanical breakdown of their food, herbivores may regurgitate their stomach contents (chewing the cud) or may eat their own faeces (coprophagy).

Modern mammals can be divided into three subclasses: the Prototheria (monotremes), Metatheria (marsupials) and Eutheria (often erroneously referred to as placental mammals) (Plates 15.9–15.11). Metatheria and Eutheria are sometimes placed together, as infraclasses, in the subclass Theria. The relative competitive advantages of metatherians and eutherians are often debated (Box 15.4).

The Prototheria includes only three living species (platypus and two species of echidna, Plate 15.9), all restricted to Australasia. Their fossils are more widespread and include many early mammals. Prototherians mix characters that are distinctly reptilian and therefore presumably ancestral, in combination with more obviously mammalian

Box 15.4 Metatherians versus eutherians

Some 80–90 million years ago, metatherians or marsupials were present on all continents. They were most diverse in Gondwana and parts of North America, less diverse in Europe and appear to have been a minor component of the mammalian faunas of Africa and Asia. At the time, Gondwana seems to have had the most diverse mammalian fauna, with representatives of the monotremes, a prototherian group called multituberculates, metatherians and eutherians. The oldest definite metatherians come from North America, but this is probably due to lack of rocks of the right age in Gondwana, where the later diversity of mammals suggests most of the major mammal groups evolved. Adaptive radiation of metatherians was greatest in South America and Australia, although a modest radiation also occurred in North America.

Metatherians in South America were a large component of the mammalian fauna from the beginning of the Tertiary (65 million years ago) until less than 5 million years ago, with 40% of living species still present in South America. They persisted in North America until 10 million years ago. The richness of marsupials in North and South America declined when eutherians invaded from Asia via the Bering land bridge, and this has often been taken as an example of eutherian 'superiority'. In fact, all the autochthonous mammal fauna of the Americas suffered from this invasion, including the eutherians. It was a case of the mammal fauna from a large land mass (Asia) displacing the mammal fauna upon invading a smaller land mass. In this case, however, the invading mammals were all eutherians because metatherians were poorly represented in Asia and were not part of the invasion. The impact was particularly severe because the invasion occurred at a time of huge environmental change following the evolution of grasses. Grasslands formed first in Asia about 15 million years ago and rapidly spread into similar climatic zones around the world. It was a eutherian mammal fauna that evolved to exploit this new environment first and consequently had an advantage during the invasion phase.

features. They have the skull and neck of a mammal, the back vertebrae and pectoral girdle of a reptile, features of the brain, kidneys, circulatory and respiratory systems that are mammalian, they lay eggs, and while they nurse their young on milk, the milk oozes from a patch of glands and is lapped up. The young therefore do not strictly suckle. They are hairy, but their body temperature varies more than is usual for a mammal, although this may be a specialisation to cope with adverse conditions. Other specialisations include poisonous spurs on the hind feet, and electrosensitivity, which allows a platypus to catch a freshwater crayfish with its eyes closed by detecting faint electrical discharges in the crustacean's muscles.

Metatherians and eutherians are similar in most respects, with subtle differences between some bones in the skull, the shape of the lower jaw and the location of the tear duct, but a profound difference in their mode of reproduction. Both are viviparous (giving birth to live young), but eutherians rely on a specialised structure called the placenta to nourish the developing embryo. This is similar to placentas seen in some fish, amphibians and reptiles and is not a unique characteristic of mammals. By contrast, metatherians have a very short gestation and a tiny, poorly developed embryo that completes its development externally, usually in a pouch. It has traditionally been assumed that this is an incomplete form of viviparity resulting because marsupials lack a placenta and a trophoblast (tissue isolating the embryo from the maternal tissue and preventing rejection of the embryo as foreign material). In fact, metatherians do have a trophoblast and a placenta does develop in some species. Metatherians therefore could retain the embryo *in utero*, suggesting that external development is an effective alternative.

Conservation of Australia's vertebrates

The uniqueness of Australia's biota, including its vertebrate fauna, is a result of the continent's long isolation. Impacts from human settlement have resulted in massive rates of extinction, especially among the mammals. Threatened vertebrates are listed in the Commonwealth's *Environment Protection and Biodiversity Conservation Act 1999* (Table 15.2).

Habitat degradation and loss

Agricultural clearing has destroyed habitat on a massive scale. In agricultural regions of southern Australia, native vegetation has been reduced in area by more than 95% and remnants are too small to support the original animals. In the pastoral areas that occupy about a third of Australia, grazing by introduced livestock degrades native vegetation, contributing to the decline and extinction of some vertebrates. Wetlands have been degraded or lost through exploitation of water resources and increases in salinity associated with clearing of native vegetation. This affects freshwater fish, frogs and waterbirds in particular.

Table 15.2 Numbers of species and subspecies within each of the vertebrate classes listed as extinct or threatened under the *Environment Protection and Biodiversity Conservation Act 1999* (Cth). See also Chapter 24.

Vertebrate group	Extinct	Critically endangered	Endangered	Vulnerable	Conservation dependent
Fish	1 (in wild)	2 species, 1 subspecies	15 species, 1 subspecies	22 species, 2 subspecies	1 species
Frogs	4 species	2 species	13 species	11 species, 1 subspecies	
Reptiles	0	1 species	10 species, 3 subspecies	34 species, 4 subspecies	
Birds	8 species, 14 subspecies	3 species, 3 subspecies	17 species, 23 subspecies	31 species, 31 subspecies	
Mammals	22 species, 6 subspecies	2 species, 1 subspecies	25 species, 8 subspecies	30 species, 24 subspecies	1 subspecies

Introduced species

Introduced species have caused many declines and extinctions in Australia's native fauna. While problems caused by introduced foxes and rabbits are well known (see Chapters 2 and 16), there are many other examples. Freshwater fish and frogs are threatened by introduced species of fish. The introduced cane toad (*Bufo marinus*) causes local declines of both aquatic and terrestrial fauna, mostly because adults and larvae are toxic to eat (Box 6.4, Box 16.4). Abrupt extinctions of frog species have occurred since the 1980s following the introduction of a pathogenic chytrid fungus (Chapter 10). On the Australian Indian Ocean Territory of Christmas Island, endemic reptiles have declined and some may have become extinct since the introduction of the small, predatory Asian wolf snake (*Lycodon capucinus*) in the early 1990s.

Hunting

Hunting should be the most manageable of impacts, but a policy of persecution contributed to the extinction of the thylacine in Tasmania and hunting severely depleted populations of several whale and seal species in Australian waters. A fishery based on Australia's largest freshwater fish, the Murray cod (*Maccullochella peelii*), collapsed following over-hunting, as have several modern fisheries.

Fire

Fire is a natural factor in most Australian ecosystems. Many vertebrate species cope with fire and are even adapted to the fire-induced changes that occur in the vegetation. As a result, changes to fire regimes following the disruption of traditional Aboriginal societies and the imposition of European attitudes towards fire led to massive impacts on a wide range of species. Using fire in conservation management remains a great challenge in Australia.

Climate change

Climate change can mean that the climatic region that will support a species may shift, expand or contract. This is a concern for species with small natural distributions, and for species surviving in fragmented landscapes where surrounding agricultural land presents an inhospitable barrier.

Chapter summary

1. Taxonomically, vertebrates are restricted to an infraphylum within the phylum Chordata, but in terms of absolute body size, diversity of form and habitats occupied they are an ecologically significant group.
2. Chordate characteristics include a notochord, a ventral nerve cord, pharynx pouches at some stage of development, an endostyle, metameric organisation and a post-anal tail.
3. Vertebrate characteristics include a body plan with head, trunk and post-anal tail all supported by an endoskeleton; organ systems including a well-developed brain and paired sense organs, a complete digestive system, an endocrine system, a closed circulatory system and excretion by paired kidneys; a body covering with many different adaptations; and a neural crest in the embryo.
4. Milestones in vertebrate evolution such as development of jaws, growth of legs, the amniotic egg, respiration using gills or lungs and increases in the number of heart chambers are associated with food processing and adaptations to aquatic, terrestrial and flying lifestyles.
5. Major living vertebrate classes include:
   - classes with aquatic lifestyles
     - jawless fish (Agnatha)
     - cartilaginous fish (Chondrichthyes)
     - bony fish (Osteichthyes)
   - classes with amphibious lifestyles:
     - amphibians (Amphibia)
   - classes with predominantly terrestrial lifestyles:
     - reptiles (Reptilia)
     - birds (Aves)
     - mammals (Mammalia).
6. Significant conservation issues for the Australian vertebrate fauna include competition from and predation by introduced species, wetland degradation, clearing of native vegetation, changes in fire regimes, and climate change.

Key terms

amniotic egg
atrium
bradymetabolic
cartilage
closed circulatory system
ectotherm
endocrine system
endoskeleton
endostyle
endotherm
gnathostome
heterodont
homeothermic
homodont
integument
neural crest
notochord
operculum
oviparous
ovoviviparous
pharynx
placenta
poikilothermic
swim bladder
tachymetabolic
tetrapod
torpor
ventricle
viviparous

Test your knowledge

1. List the major distinguishing characteristics of (a) chordates and (b) vertebrates.
2. List the milestones in vertebrate evolution. How were these associated with the movement of vertebrates from aquatic to terrestrial environments?
3. Compare and contrast aquatic and terrestrial vertebrate classes in terms of (a) their mode of reproduction, (b) their mode of respiration, (c) their body coverings and (d) their means of excreting nitrogenous wastes.
4. Outline the advantages and disadvantages of (a) endothermy and (b) ectothermy for terrestrial vertebrates.
5. What are the main conservation pressures facing Australian vertebrates?

Thinking scientifically

Temperature regulation is vital for the functioning of vertebrates and is achieved by a variety of means. Using terms such as endotherm, ectotherm, poikilotherm, homeotherm, bradymetabolic and tachymetabolic, describe the thermal relations of the following:

- a hibernating echidna
- a crocodile basking on the bank of a river
- a gecko (small lizard) foraging at night
- a brush wallaby
- an inactive lizard
- a Murray cod (large freshwater fish)
- a perentie, Australia's largest lizard, out foraging in daylight
- a marine turtle
- a boobook owl, when active at night and when inactive by day
- a southern bluefin tuna (marine fish).

Debate with friends

Debate the topic: 'Eutherians are a superior line of evolution to metatherians'.

Further reading

Cogger, H. (1992). *Reptiles and amphibians of Australia*. Reed New Holland, Frenchs Forest, NSW. 5th edn.

Dickman, C. and Woodford G. R. (2007). *A fragile balance: the extraordinary story of Australian marsupials*. Craftsman House, Sydney.

Last, P. (1994). *Sharks and rays of Australia*. CSIRO Publishing, Collingwood, Victoria.

Schodde, R. and Mason, I.J. (1999). *The directory of Australian birds – passerines*. CSIRO Publishing, Collingwood, Victoria.

Van Dyke, S. and Strahan, R. (eds) (2008). *The mammals of Australia*. Reed New Holland, Sydney, NSW. 3rd edn.

Theme 4

Applying scientific method – biodiversity and the environment

The main characteristics of organisms from the various kingdoms of life were selected through evolution because they adapt the organisms to their physical and biotic environments. As biologists Marc Johnson and John Stinchcombe point out, we now realise that genetic variation and evolution within a species influence not only that species but also others with which it interacts and the surrounding environment. Thus the study of the interactions within and between species and their environment, collectively called ecology, is also the study of evolution in action. This is the topic of this theme.

We consider interactions between organisms of the same species as they build up a population and then examine the interactions between different species living together and forming a community. Finally, we cover the main characteristics of marine, freshwater and terrestrial environments and how the interactions between the communities within them and the physical environment form characteristic ecosystems. We need different techniques to study the microbes, plants and animals in the ecosystems and we describe some of them, as well as some of the instruments available and the ways of analysing and presenting data.

16

Boom and bust – population ecology

Mike Calver and Stuart Bradley

The bettong, the fox and the rabbit

When Europeans settled Australia, the burrowing bettong (*Bettongia lesueur*) was widespread and abundant. It occurred in western New South Wales and Victoria, all of South Australia, the southern half of the Northern Territory and much of the eastern and north-western reaches of Western Australia, with isolated populations in the woodlands of the south-west. However, it declined rapidly and was probably extinct in New South Wales and Victoria by the end of the 19th century, in South Australia by the 1950s, in the Northern Territory by the 1960s and in mainland Western Australia by the 1940s, persisting only on islands off the Western Australian coast.

While the burrowing bettong declined, the introduced European rabbit (*Oryctolagus cuniculus*) and red fox (*Vulpes vulpes*) expanded rapidly in range and numbers. The first big release of rabbits was in Victorian in the late 1850s to establish populations for hunting. Despite belated attempts at control, they had spread across the southern two-thirds of Australia by the early 20th century, damaging native vegetation and crops and accelerating soil erosion. Foxes were also introduced for hunting near Melbourne, Victoria, in 1845 and 1860. They bred and dispersed rapidly, encouraged by the proliferation of rabbits as food. Foxes now range across much of Australia except the tropical north and many offshore islands, including recently Tasmania. Foxes are a significant cause of decline in native mammals (Chapter 2), an agricultural pest and a potentially important reservoir for rabies should the disease reach Australia.

Population ecology, the subject of this chapter, seeks reasons for phenomena such as the calamitous decline of the burrowing bettong and the rapid increase in the rabbit and fox populations.

Chapter aims

In this chapter we describe the properties of populations, the methods for measuring their key features (density, size, recruitment, deaths and migrations), and how techniques differ for plants and animals. We discuss the factors influencing population growth, and illustrate practical applications of population ecology for fields such as pest control, conserving endangered species and harvesting natural resources.

What is population ecology?

Populations are groups of individuals of the same species living in the same place at the same time. 'Individual' is theoretically a genetically distinct organism arising from a single zygote following sexual reproduction (see Chapter 5). However, in many plants only molecular techniques (Chapter 5) can distinguish between genets (different individuals each arising from a single zygote) and ramets (seemingly distinct individuals arising by asexual reproduction and therefore genetically identical). Therefore studies of plant populations regard both genets and ramets as individuals. The spatial boundaries of populations can be large, such as the human population of the Earth, but populations are studied more easily when restricted to limited regions such as an island, lake or forest.

Population ecologists study the interrelationships between individuals in a population, between the individuals of different populations, and between populations and their environment.

Properties of populations and how to study them

If the boundary of a population is restricted such as for barnacles in a tide pool on a rocky shore, it may be possible to count all individuals and monitor their fates. Mostly this is impractical, because the boundaries of populations are not defined so clearly. In this chapter we describe how to estimate the density, dispersion pattern and size of populations and how to track the survivorship and mortality of individuals within them.

Density

Density is the number of individuals per unit area (e.g. the number of shrubs of a particular species per square metre of a heathland) or volume (e.g. the number of a small crustacean per cubic metre of water in a lake). It indicates local abundance in relation to the distribution of important resources such as water, nutrients or shelter.

Plants

The density of plant populations is estimated commonly by delimiting an area of ground and counting the number of individuals in that area. The delimited area is called a quadrat and is usually square or rectangular. A mean density per unit area (plants m^{-2} ± error) is calculated by establishing multiple quadrats, counting the individuals in each and averaging the counts (Plate 16.1).

One way to decide whether sufficient quadrats have been taken is to calculate a running mean. This is graphed as the mean number of individuals against the number of quadrats

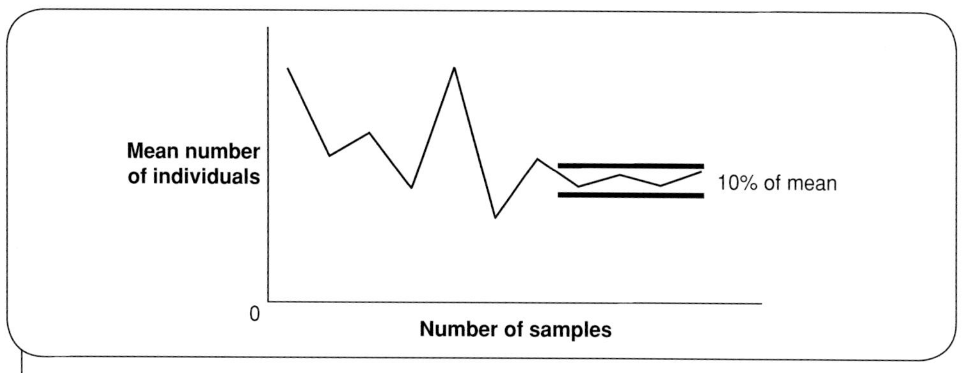

Figure 16.1 Running mean of the number of individuals recorded in increasing number of samples. Sufficient samples are determined when the running mean stabilises at some predetermined range about the mean (perhaps 10% of the mean). (Source: B. Cale)

(see Figure 16.1). The first point is the number of individuals in quadrat 1. The second point is the mean of the individuals in quadrats 1 and 2, the third is the mean of the individuals in quadrats 1, 2 and 3, and so on. Sufficient samples are determined when the running mean stabilises at some predetermined range about the mean (perhaps 10% of the mean).

If the species sampled is a tree, then quadrats may need to be very large to accommodate even one individual. An alternative method is plotless sampling. This may be based on selecting random plants and measuring the distance between them and their nearest neighbours, or selecting random points and determining the distance between them and the nearest trees. A simple protocol is the point-quarter method (Box 16.1).

Box 16.1 The point-quarter method for estimating tree density

The point-quarter method is widely used to estimate the density of trees and other plants too large for assessing with conventional quadrats. It assumes that the trees are distributed randomly. Begin by constructing a notional grid over the study area, similar to the coordinates seen on a map. There is no strict rule about the distance between gridlines, except that they should be far enough apart that the same tree is not measured from more than one point. Number all the intercepts on the grid and use random numbers to select the grid points to sample. For example, if the area being studied measures 1 km × 1 km, gridlines could be set 100 m apart, giving a total of 100 intercepts for the area. Each of these is assigned a number and a sample of them (e.g. 10) is selected by reference to random number tables. At each of the sampling points chosen, the distance to the nearest tree in each of four quadrants is measured (see Figure B16.1.1). The formula for density is:

$$\hat{N}_p = \frac{4(4n-1)}{\pi\sum\left(r_{ij}^2\right)}$$

continued ›

Box 16.1 continued ›

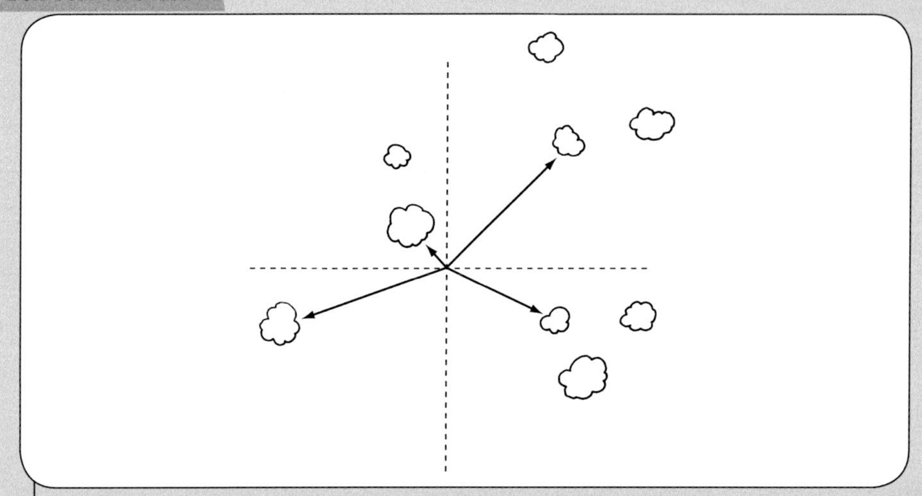

Figure B16.1.1 Measuring the distance to each tree in each of the four quadrants. (Source: B. Cale)

where $\hat{N}_p$ = the density estimate

r_{ij} = distance from random point i to the nearest organism in quadrant j (j = 1, 2, 3, 4; i = 1, . . . , n)

$\pi = 3.14$

n = number of random points

Σ is a symbol meaning 'the sum of'.

Consider an example where density of a tree species is estimated from 10 random points (Table B16.1.1).

Table B16.1.1 Sampling data for estimating the density of a tree species based on the point-quarter method.

	Distance from point to nearest tree (m) in each of four quadrants			
Sampling point	**Quadrant 1**	**Quadrant 2**	**Quadrant 3**	**Quadrant 4**
1	7	11	9	5
2	5	6	7	8
3	14	9	7	12
4	10	10	7	7
5	8	8	9	7
6	11	6	3	7
7	10	11	7	11
8	5	3	8	6
9	7	9	5	12
10	8	6	2	16

Applying the formula above, the density for this tree species is 0.01772 m^2. Converting this to a density per hectare (10000 m^2), gives a density of 172.72 trees per hectare.

Animals

Quadrat sampling is also appropriate for estimating the density of sessile (immobile) animals. However, most animals move so sampling programs attract them to traps or intercept them (Plate 16.2). Attraction traps use food, smells or lights. Interception traps vary from nets to tubes or buckets in the ground (pitfall traps). The main problem with all trapping is that different species and even different sexes and age classes within a species vary in catchability. We therefore cannot justify concluding that different species have different densities based solely on the relative proportions captured. The solution is to present the results as an index (e.g. number of animals caught per traps set) rather than as a density. Indices for the same species may be compared between different times or between different places as indications of relative changes in density (e.g. more animals of this species occur in site A than in site B), but it is usually wrong to regard them as estimates of absolute density.

Alternatively, large or conspicuous animals may be observed rather than trapped (e.g. by bird watching or aerial surveys of large grazing animals) or detected via tracks, diggings or droppings. One generic approach is the line transect in which a survey line is traversed through a habitat. The number of animals seen to a given distance on each side of the line is recorded, together with the estimated distance from the observer to the animal and the angle of the observation to the line (so that we can later calculate the perpendicular distance of the animal from the line and convert the observations to density estimates). If a fixed width is surveyed, if all the animals within the strip are seen, if the presence of the observer does not alter the chance of an animal being present, if no animal is counted twice, and if distances and angles are measured accurately, density estimates can be calculated. These are daunting assumptions, so some population ecologists prefer to present the results as an index (e.g. number of animals seen per transect distance travelled). These indices are usually comparable between sites and times for the same species, enabling estimates of relative densities over space or time.

Dispersion pattern

The dispersion pattern of a population describes how individuals are spread across an area. It provides insights into interactions between individuals, or indicates how they respond to differing environmental characteristics.

There are three main dispersion patterns in space:

- *clumped* – individuals aggregate in patches. This may be because resources are limited or patchy, as would be the case for fungi living on fallen trees. Alternatively, animals may aggregate to breed, for defence against predators or to hunt collaboratively.
- *uniform* – uniform patterns occur when individuals are distributed evenly across a landscape. They often arise because of interactions between individuals such as territorial behaviour in animals or shading by large plants.

- *random* – the location of any individual is uninfluenced by the location of other individuals and all locations are equally likely to be occupied. Random dispersion is rare, with most individuals responding to local environmental conditions or to each other.

Quadrat-based studies enable a simple test to determine if distribution is clumped, uniform or random. The mean number of individuals per quadrat ($\overline{x}$) is divided by the variance (s^2) to give the coefficient of dispersion, CD:

$$CD = \frac{s^2}{\overline{x}}$$

If CD is about 1, the population is distributed randomly. If CD is greater than 1 the population is clumped, and if it is less than 1 the distribution is uniform.

Populations also disperse in time. These changes may occur daily, as when an owl leaves its roost to hunt at dusk and returns before dawn, or at specific points in the life history, such as when young leave a nest or a natal territory. One spectacular example is the annual migration of the Christmas Island red crab (*Gecarcoidea natalis*) from its forest habitat to the sea to breed. Migrations can be tracked most effectively by following tagged individuals (Box 16.2).

Box 16.2 Follow that shark!

The great white shark (*Carcharodon carcharias*) occurs off South Africa, Australia, New Zealand, California and some Pacific islands such as Hawaii. It is one of the most formidable marine predators. Adults grow to over 6 metres in length and weigh over 2 tonnes. The liver of a large specimen may weigh more than an average human adult. Size comes at the cost of a low reproductive rate and it can take up to 9 years for a great white to reach sexual maturity, so populations are slow to recover from substantial losses. Such losses did occur in the 1970s and the great white is endangered. The causes of the decline are uncertain, but may include overfishing both for meat and for sale of the jaws and teeth as souvenirs.

Management of the great white shark was hampered by very limited knowledge of dispersal patterns, but recent satellite-based telemetry techniques enabled researchers to tag sharks and trace their movements over long periods and potentially long distances. The results were startling.

A detailed satellite-tagging study of great white sharks off South Africa in 2002 and 2003 found that some sharks made long-distance coastal return migrations of over 2000 km, returning to their original sites 4–6 months later. Others patrolled over shorter distances or stayed close to one site. Remarkably, one female headed north for about 750 km and then turned directly across the Indian Ocean to Western Australia, arriving near Exmouth Gulf 99 days later. She moved at 4.7 km/h, the fastest sustained speed known in sharks, and did not follow island chains. Nine months after she set out she was back off South Africa, having completed an amazing transoceanic migration.

These results are important in conserving great white sharks, because they indicate potential interbreeding between widely separated populations. Furthermore, the evidence of a female shark making such a long migration overturned previous beliefs that only the males made long-distance migrations. The results also pose a challenge. If great white sharks migrate over such long distances they pass through international waters and come under the jurisdictions of various governments in widely separated coastal waters. If the different authorities do not collaborate in management, efforts to conserve sharks in one locality may be offset by mortality elsewhere.

Dispersal is also important in sustaining metapopulations. These are a series of subpopulations the distributions of which do not overlap, but which exchange some individuals through migration. Each subpopulation responds to its local environment independently of the others. Even if one subpopulation becomes extinct, it may be re-established by migrants from another. Land clearing and urbanisation fragment once-continuous distributions for many species, converting them into metapopulations.

Population size

Estimates of population size can be calculated from density estimates by multiplying the density estimate (e.g. numbers per hectare) by the total area occupied. This is effective for plant populations, but less so for animals because many indices of abundance are problematic. Instead, animal ecologists use removal methods, catch per unit effort methods and mark-release-recapture techniques.

Removal methods

There are several methods based on systematic removal of animals from a population and 'Eberhardt's method' or the 'index-removal' method is the simplest to calculate. For example, suppose you wished to estimate the population of introduced red fox in a nature reserve. First, derive an index of population size, perhaps by driving along tracks through the reserve at night and recording the number of foxes seen with spotlights per kilometre of track travelled. Then set traps throughout the reserve and remove and kill all trapped foxes. Finally, record the population index again. The original population size of the foxes can be estimated by:

$$\hat{N} = \frac{x_1 R}{x_1 - x_2}$$

where $\hat{N}$ = the estimated population size at the start of the study
x_1 = the first index

x_2 = the second index

R = the number of foxes removed.

Statistically, the method works best when most of the animals are seen and a high proportion of them is removed. This is not a problem during commercial harvesting or pest control, but hardly applicable to endangered species!

Catch per unit effort

These methods exploit falling capture rates over a series of trapping events when animals are removed after capture. They only work if no migration occurs, all individuals in the population have an equal likelihood of being caught on a given sampling occasion, and large enough samples are removed to cause decline. If these assumptions hold, a graph of catch per unit effort (e.g. number of animals caught per traps placed) on the vertical axis against accumulated animals caught on the horizontal axis yields a straight line. The *x*-intercept (the point where the catch per unit effort equals zero, at the arrow in Figure 16.2) estimates the original population size. The method is not widely applied because of the restrictive assumptions and the need to remove the captured animals.

Mark-release-recapture techniques

Mark-release-recapture (MRR) techniques are used widely to estimate animal populations and also rates of birth, death and migration. The principle is simple: catch a sample of animals, mark them, release them back into the population and then sample again (Plate 16.3). Assuming that no animals enter or leave the population, all animals are equally catchable, marks are neither gained nor lost and marking does not change

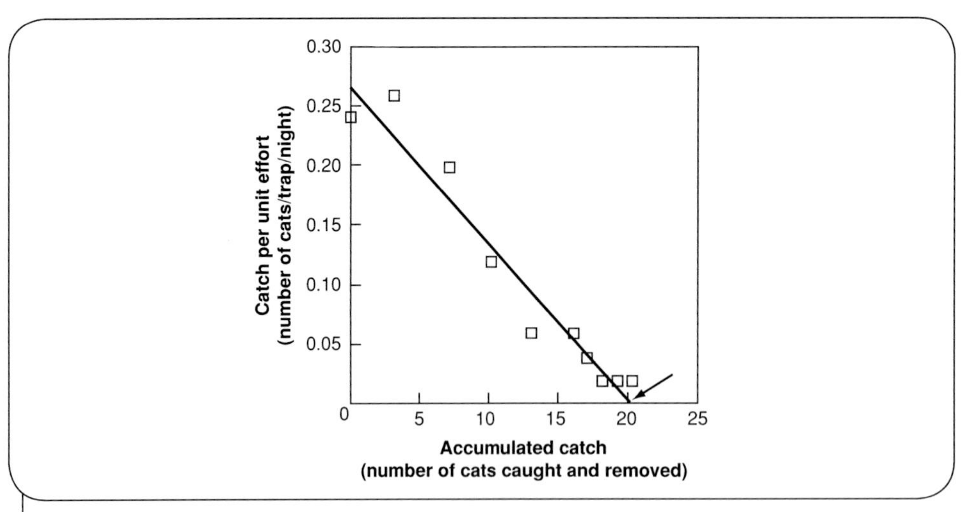

Figure 16.2 A plot of catch-effort data for removing feral cats by trapping on an island (the data are hypothetical). The estimated original population size, indicated by the arrow, is 20 cats. (Source: B. Cale)

catchability, the ratio of marked and unmarked animals in the second sample should be the same as the ratio of the total population to the number of animals captured in the first sample. This is expressed mathematically in the Lincoln index, also called the Petersen method:

$$\hat{N} = \frac{n_1 \times n_2}{m}$$

where $\hat{N}$ = estimated population size
n_1 = the total caught on day 1
n_2 = the total caught on day 2
m = the total caught on both day 1 and day 2 (the marked recaptures).

If animals carry unique marks and sampling continues for a third sampling occasion or more, a detailed capture history against sampling occasion can be compiled for each animal. The Jolly-Seber MRR models use such data to calculate population size estimates over time and to estimate the probability of survivorship of animals between sampling occasions, the number of new animals joining the population and the rate of population change.

MRR analyses are attractive because they quickly turn field data into estimates of population size or other useful statistics. However, the models also make specific assumptions about how data were collected and how animals behaved. If those assumptions are untrue, the estimates may be unreliable. This can be appreciated by re-examining the Lincoln index formula above. Any feature of the sampling design, the marks or the behaviour of the animals that makes a marked animal more or less likely to be caught on the second day will bias the population estimate. If a marked animal is more likely to be caught, m will be higher and the population will be underestimated. If a marked animal is less likely to be caught, then m will be lower and the population will be overestimated.

Some possible problems and their solutions are:

- *Marks harm an animal or change its chance of survival.* If the stress of capture and marking kills some animals after release or if marked animals are more vulnerable to predators, then fewer marked animals will be recaptured and the population will be overestimated. If sampling extends over several occasions, the proportion of animals marked for the first time and later recaptured can be compared with the proportion of animals recaptured on more than one occasion. If initial capture and marking is detrimental, then a higher proportion of animals caught twice will be recaptured at a later time compared with the proportion of animals caught once that are later recaptured.
- *Marked and unmarked animals are not equally likely to be caught again.* A marked animal might be more easily recaptured if capture is by hand and the marked animal is easier to locate. When traps are used, marked animals may be more likely to return to traps to eat the bait provided ('trap-happy' animals). This inflates the recaptures

and underestimates the population. Alternatively, animals may become 'trap-shy' and avoid traps after being caught. Recaptures will fall and the population will be overestimated. One solution is to use different methods of capture on different occasions, or to relocate traps between samples.

- *Marks are either gained or lost*. Sometimes animals are marked by clipping a toe or notching an ear. Some animals may be 'marked' by natural injury, inflating the number of marked animals in a population. Perhaps more common is losing a mark such as a tag or band, so a recaptured animal is not identified. Careful choice of marking method can minimise the chance of a tag appearing 'naturally'. If tags are suspected to be at risk of loss, a 'double tagging' system might be used. Thus if a small spot of paint on a wing is used to mark an insect, two small dots might be used. The number of animals recaptured with only one paint dot gives an idea of the rate of loss of marks.
- *Different subgroups of the population are not equally catchable, regardless of whether or not they are marked.* This could happen, for example, if males and females have different movement patterns. In this case, it is best to estimate the size of each subgroup separately.
- *Sampling is not at regular times, or very little time is left between sampling occasions.* If sampling occasions are close together, there may not be enough time for marked animals to mix randomly in the population. The solution is to adjust the sampling schedule.

Overall, the moral in MRR, as in all population studies, is to think carefully about the assumptions and limitations of techniques, and strive to account for them during data collection and analysis. Comprehensive computer software is available for MRR under various conditions, including cases where calculations are complex by hand.

Quantifying birth, death and survivorship with life tables

Birth, death and age structure

The number of individuals of different ages in a population is its age distribution and is determined by a balance of births and deaths. All continuously breeding populations tend towards a stable age distribution, where the relative proportions of the population in each age group are unchanged and each age group has its own specific birth rate (the number of young produced per number of individuals in the age group) and death rate (the number dying per number of individuals in the age group). Sometimes biologists can age organisms accurately, and age in days, weeks, months or years is used to determine the age distribution. Where this is impossible, life-history stages such as egg, juvenile, subadult and adult may be used. Plants are particularly difficult to age and present special problems when some individuals are stunted by shading by faster-growing competitors or regrow from underground parts after a fire. In these cases, age may not be a useful variable and discussion of reproductive rate and death rate by size-classes of plants is more meaningful.

Figure 16.3 Examples of age structures for a hypothetical animal population with a life span of up to 6 years. The reproductive ages are shaded. (Source: B. Cale)

Graphical representations of the age distribution of a population are called age pyramids (Figure 16.3). Populations with a high proportion of young individuals (leading to a broad base in a pyramid) have the greatest capacity for high population growth.

Life tables

Age pyramids give a quick visual impression of age structure, but life tables provide the most detailed information about age composition, birth rate and death rate. The insurance industry developed them to forecast the probability of death of people at different ages and hence the appropriate premiums to charge. Biologists adapted them to very different purposes, including the projection of population sizes and age structures into the future (Box 16.3). Here we consider the main types of life table and their features.

Box 16.3 How quickly did the moa disappear?

When Polynesians reached New Zealand in the late 13th century, birds dominated the terrestrial vertebrate fauna. Most spectacular were the 11 species of flightless moa, ranging from the turkey-sized *Euryapteryx curtus* to *Dinornis giganteus*, which was 2 m high at the shoulder. The new arrivals quickly exploited this abundant food. Numerous archaeological sites show evidence of extensive butchering and cooking of moa. People made clothing from the skins and feathers, and fish hooks, harpoon heads, ornaments and other tools from the bones. They probably ate eggs too and used the empty shells for water containers. Faced with this intense predatory pressure, the moa populations collapsed and all species were extinct before the European discovery of New Zealand. Their demise also meant the extinction of their principal predator, the giant eagle, *Harpagornis*. Maoris were forced to change their diets to shellfish, fish and vegetables. How quickly did this happen?

To answer this question, New Zealand palaeoecologists (ecologists who study ancient ecosystems) Richard Holdaway and Chris Jacomb applied population ecology techniques. If the age profile of a population and the survival and reproductive rates for different ages are known, the size and age composition of the population in the future can

continued ›

Box 16.3 continued ›

be predicted by calculating the number of survivors of each class and their reproductive success. The calculations are commonly applied to females only for simplicity, and the total population projected based on an expected sex ratio. The process is expressed clearly using the notation and methods of matrix algebra and the biological applications are called Leslie matrices.

When studying the moa extinctions, Holdaway and Jacomb wanted to establish the impacts of a small initial human population using low levels of exploitation and also altering some moa habitat. They assumed an initial total moa population of 158 000 (nearly twice as large as other estimates) and an initial human population of 100. They also assumed that humans hunted only adult moa at low levels and ignored predation on moa eggs. Even with these cautious assumptions, the Leslie matrices showed that the moa population was susceptible to even small increases in adult mortality. A range of models exploring different rates of human population growth and habitat loss all predicted a time to extinction of less than 160 years. Allowing for errors in radiocarbon dating of archaeological deposits, the suggested brief period to moa extinction is supported by archaeological data. It may be one of the swiftest extinctions of large animals known.

The cohort life table follows a group of individuals all born within a narrow time interval throughout the life span. An alternative is the static life table, which is a cross-section of a population at a single time constructed from animal sampling. Often only females are followed in life tables because they are the reproductive individuals and only their history needs to be known to project future populations. An example of a static life table is given below.

Data are recorded in a series of columns (Table 16.1). The first column indicates the age class interval (x), the second the number of individuals entering each successive interval (N_x) and the third the proportion of the original cohort entering each successive interval (l_x). The next column records d_x (the number of individuals dying in a particular age class), followed by the proportion of individuals dying in each age class (the age-specific death rate). To complete the life table there is a column for m_x (maternity, or the number of female young produced per female). To adjust for mortality in each age class, m_x is multiplied by l_x to give $l_x m_x$, the mean number of females born in each age group, adjusted for survivorship.

Summing the $l_x m_x$ values gives R_0, the net reproductive rate for the population. This is the number of female progeny produced by each female that survives to maturity. The females in a population will replace themselves if $R_0 \geq 1$.

Another important population property calculated from a life table is survivorship, often represented in a survivorship curve plotting the logarithm of the number of survivors of each age interval against age. If we generalise these by plotting the logarithm of the percentage of survivors against the percentage of life span, we can compare survivorship curves for very different types of organisms (Figure 16.4).

Table 16.1 Life table for vixens (females) from a fox population in rangeland in Western Australia. (Source: Modified from Marlow, N.J., Thomson, P.C., Algar, D., Rose, K., Kok, N.E. and Sinagra, J.A. (2000). 'Demographic characteristics and social organisation of a population of red foxes in a rangeland area in Western Australia.' *Wildlife Research*: 457–64; CSIRO Publishing <http://www.publish.csiro.au/nid/145/issue328.htm>).

Age interval in years (x)	N_x (number living at start of age interval x)	l_x (proportion entering age interval x)	d_x (number dying in age interval x)	q_x (probability of dying in age interval)	m_x (number of female young produced)	$l_x m_x$ (number of female young produced, corrected for survivorship)
0	94	1.000	43	0.457	0	0
0–0.3	51	0.543	31	0.608	0	0
0.3–1.3	20	0.213	7	0.352	2	0.426
1.3–2.3	13	0.138	1	0.072	4	0.552
2.3–3.3	12	0.128	6	0.500	4	0.512
> 3.3	6	0.064	6	1.000		
						$R_0 = \sum l_x m_x = 1.49$

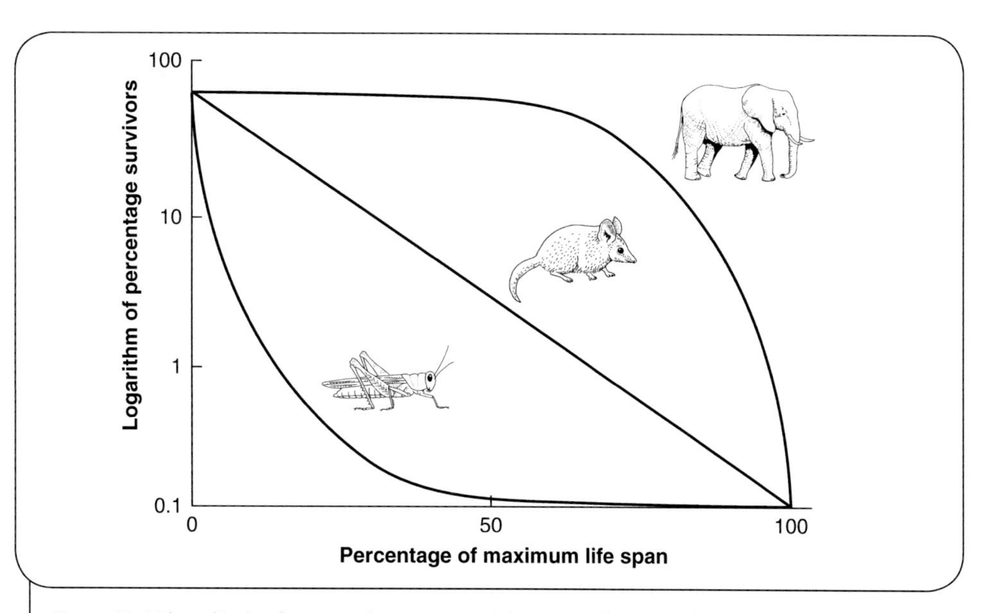

Figure 16.4 Three kinds of survivorship curves and the types of animals that follow them. (Source: B. Cale)

Three broad types of survivorship curve can be recognised:

1. *type I* – infant survival is high and most mortality occurs later in life. These species usually produce few offspring and show high levels of parental care. Many large mammals, including humans, are examples.
2. *type II* – mortality is about constant over the life span. It is characteristic of some small mammals, reptiles and some invertebrates.
3. *type III* – juvenile mortality is high, followed by a low death rate for individuals surviving to maturity. Many species producing huge numbers of eggs, most of which do not survive, fit this curve.

What influences population growth?

Describing the properties of populations is only part of the picture. For practical purposes such as harvesting natural resources, controlling pests and recovering endangered species, we also need to know what influences population changes. One of the best ways to approach this is through the application of mathematical models.

Exponential growth models

An old legend claims that an Indian king's games master invented chess. The delighted king asked him to name his reward. He asked for some wheat: one grain for the first square on the board, two for the second, four for the third and so on for all 64 squares on the chess board. When the royal mathematician calculated the final amount the king was horrified by the answer: about 1.84×10^{19} grains, probably more wheat than exists in the world.

The increase in the number of wheat grains from square to square on the chessboard is an example of exponential growth. This can also occur in populations increasing at their maximum rate under ideal conditions, where the whole population multiplies by a constant factor in each generation. It is represented graphically as a J-shaped curve (Figure 16.5). Mathematically, differential calculus is used commonly to express population growth as a rate at an instant in time, the equation being:

$$\frac{dN}{dt} = rN$$

where dN/dt = the change in population size over a very short time interval
N = the initial population size
r = the intrinsic rate of increase, the average contribution of each individual to population growth.

Those without calculus may find the equation easier to understand as the approximation $G = rN$, where G is the number of new individuals added at each time interval (growth rate of the population), N is the initial population size and r is the intrinsic rate of increase, the average contribution of each individual to population growth.

Given that $dN/dt = rN$, if r remains constant the number of new individuals added per unit time depends on the population size, N. A large population will add more individuals over a specific time period than a small population, which is why the exponential growth curve becomes steeper with time.

Many rapidly reproducing species such as bacteria, many invertebrates and small mammals have much higher r values than larger, slower-breeding animals such as the rhinoceros. Human r values are low, but because the global population is large increases in absolute numbers over a given time are substantial. Increasing human populations place growing demands on many natural resources and these will restrict the final population size. Thus the exponential growth model has limited application because it ignores these restrictions.

The logistic growth model

In the late 19th century, the Englishman Thomas Malthus argued that population increase outruns food production and that surplus population is checked by starvation, disease and war. Modern biologists incorporate these effects in the logistic growth model, which predicts that exponentially growing populations eventually encounter limiting factors constraining further growth. This results in an S-shaped curve for population growth (Figure 16.5).

The equation for the logistic curve is:

$$\frac{dN}{dt} = rN\frac{(K-N)}{K}$$

(those without calculus may prefer to approximate by substituting G for dN/dt). The new term in this equation is K, the carrying capacity or the maximum population size the

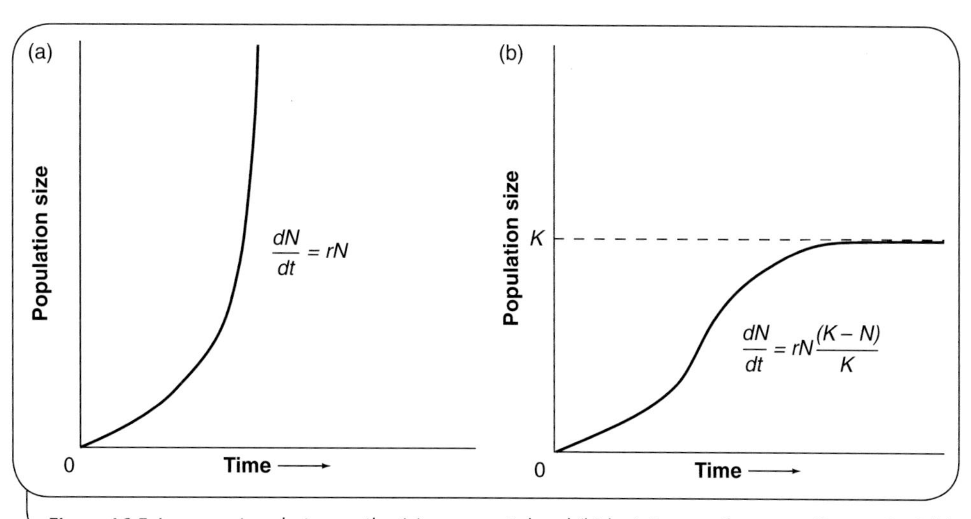

Figure 16.5 A comparison between the (a) exponential and (b) logistic growth curves. (Source: B. Cale)

environment can sustain. To see how carrying capacity produces the characteristic S-shaped logistic growth curve, look closely at the equation above. The left-hand component $dN/dt = rN$ is the exponential growth equation. The right-hand component corrects for carrying capacity. When the population is small $K - N$ is small, so the correction is close to K/K or 1, and the population increases at close to an exponential rate. However, as the population increases $K - N$ is smaller and $(K - N)/K$ is considerably less than 1, slowing growth.

If limiting factors decrease birth rates, increase death rates, or both as population densities rise, they are density-dependent and regulate population density. This most often occurs through competition for limited resources. Gardeners apply this principle when thinning seedlings, ensuring that remaining plants are unlikely to be restricted by limited water, nutrients or light. Gardeners growing many plants of the same species may encounter two other density-dependent problems. Disease spreads more easily in large populations and aggregations are often more attractive to predators.

When limiting factors are removed from a population, as may happen when a plant or animal is introduced to a new country, its numbers may increase until it becomes a pest. For example, veldt grass (*Ehrharta calycina*) is a minor component of the vegetation of South Africa, but introduced populations are a major weed across southern Australia. One unfortunate consequence of the rapid expansion of introduced species can be displacement of native species by competition or predation. This is a particular problem in isolated places such as the Galapagos Islands, about 1000 km west of Ecuador in South America, where eradication of introduced species is a high priority for management. This is believed to be a significant factor in these islands retaining about 95% of their pre-human biological diversity.

Environmental factors such as fires, floods and droughts may also determine population numbers, but their impacts are often independent of population size. Organisms respond on an evolutionary scale to the balance of density-dependent and environmental factors through their life histories.

Population growth relates to life history

Life history refers to the important events in an organism's life from birth (or germination in the case of a plant), through dispersal (if it occurs), feeding habits, reproduction and finally death. Reproductive success is important in evolution, so traits such as when an organism breeds, how many young it produces and the extent of care it gives them are influenced strongly by natural selection (Chapter 6). This can be seen in the population biology of different species.

Species living in unpredictable and sometimes crowded environments may maximise reproductive success by maturing rapidly and producing many offspring. They are often small, short-lived, poor competitors but excellent dispersers, mainly living beneath carrying capacity. This strategy is called *r*-selection, after the term *r* in the logistic growth curve, because there is strong selection to maximise the intrinsic rate of increase. The contrasting strategy is called *K*-selection, after *K* or carrying capacity in the logistic growth curve, because there is strong selection to stabilise at carrying capacity. These organisms mature later in life, producing fewer offspring often receiving high levels of parental care. They tend to be larger, long-lived and good competitors, and maintain populations near carrying capacity.

Applying population ecology

There are at least three main reasons for managing populations of wild plants and animals: pest control, sustainable harvest of natural resources and increasing population sizes of threatened species. Principles of population biology are important in all of these.

Biological control

Humans deliberately introduced many organisms to environments far from where they evolved. Some were domestic animals, whereas others, such as the European rabbit released in Australia, were for sport. Many other organisms were introduced accidentally with the baggage of travellers and settlers. For example, Polynesian rats (*Rattus exulans*) stowed away in the canoes of Polynesian seafarers and were dispersed throughout the Pacific. Some of these new arrivals, freed from limiting factors in their native environments, multiplied rapidly, displacing indigenous species or damaging crops. One solution is biological control, where a predator or pathogen of the pest in its original environment is introduced to limit the pest population. Prospective biological controls are tested thoroughly before release to ensure that they are specific for the pest species, will not change to a native prey when the target species is depleted, or become a problem in other ways (Box 16.4). Although most attempts do not succeed, biological control remains attractive because it is self-maintaining and valuable for the control of environmental weeds in inaccessible areas or where annual spraying is too expensive.

One current biological control study in Australia targets the introduced slender thistles (*Carduus pycnocephalus* and *C. tenuiflorus*). These short-season winter annuals from southern Europe and North Africa were introduced to southern Australia over 100 years ago and are invasive pasture weeds, displacing more-palatable livestock feed and contaminating fleeces. They have been declared noxious weeds in most Australian states and are also significant environmental weeds in native vegetation.

Biological control is attractive because of low ongoing maintenance after establishment, lack of side effects, and its applicability in remnant native vegetation as well as pasture. The plant-feeding crown weevil (*Trichosirocalus mortadelo*) is a promising candidate. It is specific to thistles and in preliminary studies halved plant size and reduced seed production by 70%.

Box 16.4 Now we've got those beetles by the . . .

One of Australia's most unfortunate experiences in biological control began in 1935, when about 100 cane toads (*Bufo marinus*) were imported into Queensland to control the native greyback cane beetle (*Dermolepida albohirtum*) in sugarcane plantations. Adult beetles eat sugarcane leaves while larvae attack the roots, stunting or even killing plants. The cane toad, a native of Central and South America, was known for its voracious appetite and wide range of prey. It had already been introduced to several Caribbean and Pacific islands to control insect pests. In anticipation of its success against the greyback cane beetle, one grower chortled: 'Now we've got those beetles by the . . .'

The optimism was misplaced. The cane toad failed as a biological control and has become a serious pest. It out-competes native amphibians and preys on a wide range of native organisms, while adults and tadpoles are toxic and kill many native predators. This exerts a strong selective pressure on native fauna (Box 6.4). The toads spread rapidly both south and north-west at rates of over 25 km/year. By 2001 their north-west expansion reached Kakadu National Park in the Northern Territory and by 2003 their southern expansion reached New South Wales.

The cane toad introduction illustrates how an introduced species spreads rapidly when freed from limiting factors applying within its natural distribution. Today, potential biological controls are scrutinised thoroughly before release to ensure that they can control the pest as intended and are most unlikely to become pests themselves.

Harvesting

Many human activities such as fisheries, forestry and game management harvest natural resources. The challenge is to maximise yield while not driving the population to extinction. Although this is a complex area, some generalisations can be made based on an understanding of population ecology:

- *r- and k-selected species* – *r*-selected species such as prawns are adapted to rapid population growth and, if conditions are favourable, replace harvesting losses rapidly. By contrast, recovery of *k*-selected species from even modest harvesting losses is protracted. Many species harvested by humans are long-lived and therefore threatened by substantial harvesting.
- *the logistic growth curve* – under the logistic growth curve, species grow fastest at intermediate densities because these are closest to the exponentially growing part of the curve. The maximum sustainable yield for a population is the largest harvest that can be taken while maintaining a harvested species at an intermediate population size. In practice, harvest levels are set below the maximum sustainable yield to allow for unexpected natural catastrophes that may alter population numbers.

However, maximum profit may occur by driving one resource to extinction and then turning to another. The social issues in balancing these pressures are discussed in Chapter 25.

Recovering endangered species

Species decline when mortality outweighs recruitment. A critical task in recovery of an endangered species is to identify and reverse the causes of the increased mortality before reintroducing the species into parts of its former range where it has become extinct. One example is the management of endangered Australian small macropods (wallabies and tree kangaroos). Increasing evidence implicates introduced predators, especially the red fox and to a lesser extent the feral cat, in the declines (Chapter 2) (Plate 16.4). The challenge for wildlife managers seeking to reintroduce them to their previous range is to either control the foxes before reintroduction or to teach the threatened species to avoid the foxes.

Small macropods usually shelter by day and feed at night. Their response when threatened by a predator is often to dash for shelter and then freeze until the danger passes. This behaviour is effective against visually hunting predators swooping from above, such as birds of prey. However, the small macropods may not recognise novel predators such as the red fox as dangerous and hence become easy prey.

Studies carried out mainly by New Zealand animal behaviourist Ian McLean and his colleagues showed that small macropods can learn that foxes are dangerous and remember the lessons for at least several weeks. In one study, captive rufous bettongs (*Aepyprymnus rufescens*) were exposed to a stuffed fox in their cage (the pre-test). The fox suddenly started jumping (it was controlled as a puppet) and a dog was introduced simultaneously to the cage to chase the bettongs (the training). The dog and fox were then removed from the pen. The next day the stuffed fox was introduced alone to the cage as a post-test. Following training, bettongs in the presence of the stuffed fox ate less, were more watchful and kept as far way from it as possible. Such training may help reintroduced animals avoid predators.

Chapter summary

1. Populations are groups of individuals of the same species living in the same place at the same time. Population ecologists study the interrelationships between individuals in a population, between the individuals of different populations, and between populations and their environment.
2. Density is the number of individuals per unit area or volume. In plants it is commonly studied using quadrats or plotless sampling methods. Quadrats may be used for sessile animals, but indices of abundance are usually calculated from trapping or observational data.
3. The dispersion pattern of a population describes the way individuals are spread across an area. Dispersion patterns may be clumped, uniform or random.
4. A metapopulation is a series of subpopulations the distributions of which do not overlap, but which exchange some individuals through migration.
5. Estimates of population size can be calculated from density estimates by multiplying the density estimate (e.g. numbers per hectare) by the total area occupied. Alternatively, removal methods, catch per unit effort and mark-release-recapture techniques can be used.
6. The number of individuals of different ages in a population is called its age distribution. It is determined by a balance of births and deaths. All continuously breeding populations tend towards a stable age distribution, in which the relative proportions of the population in each age group are unchanged and each age group has its own specific birth rate and death rate.
7. The cohort life table follows a group of individuals all born within a narrow time interval throughout the life span. An alternative to the cohort life table is the static life table, which is a cross-section of a population at a single time.
8. Exponential growth occurs in populations increasing at their maximum rate under ideal conditions. The whole population multiplies by a constant factor in each generation.
9. The logistic growth model predicts that exponentially growing populations eventually encounter one or more limiting factors that constrain further growth.
10. *r*-selected species are often small, short-lived, poor competitors but excellent dispersers, with population densities mainly below carrying capacity. *k*-selected species mature later in life, producing fewer offspring that often receive high levels of parental care.
11. Population ecology is applied in biological control, harvesting of populations and recovery of endangered species.

Key terms

biological control
carrying capacity
catch per unit effort
clumped distribution
coefficient of dispersion
cohort life table
density
dispersion pattern
exponential growth
feral
genet
K-selection
life history
limiting factors
logistic growth
mark-release-recapture
maximum sustainable yield
metapopulation

plotless sampling
population
population ecology
quadrat
r-selection
ramet
random distribution
static life table

Test your knowledge

1. Define the terms 'density' and 'dispersion' and explain why they are important in population ecology.
2. Describe the main techniques used to study the densities of plant and animal populations.
3. What important assumptions are made when using the different methods for estimating the sizes of animal populations?
4. Distinguish between a cohort life table and a static life table. What kinds of information are recorded in a life table? What statistics can be calculated from a life table?
5. Explain the differences between the exponential and logistic models of population growth.
6. Describe briefly some of the main applications of population ecology.

Thinking scientifically

Apply your understanding of scientific method (Chapter 2) and population biology (this chapter) to design a field trial to test the effectiveness of the crown weevil in controlling slender thistles.

Debate with friends

Debate the topic: 'Anyone who is not a vegetarian has no grounds to object to the resumption of commercial whaling once it can be shown that numbers are sufficient for a sustainable harvest'.

Further reading

Attiwill, P. and Wilson, B. (eds) (2006). *Ecology: an Australian perspective.* Oxford University Press, Oxford. 2nd edn.

Krebs, C.J. (1998). *Ecological methodology.* Benjamin Cummings, San Francisco. 2nd edn.

Smith, R.L. and Smith, T.M. (2001). *Ecology and field biology.* Benjamin Cummings, San Francisco. 6th edn.

17

Living together – communities and ecosystems

Barbara Bowen and Philip Ladd

Along the dingo fence

During the late 19th and early 20th centuries, frustrated pastoralists in South Australia, New South Wales and Queensland fenced their properties to stop the Australian dingo (*Canis lupus*) killing their livestock. Eventually links formed between individual fences, creating a continuous barrier now covering over 5000 km. Known variously as the 'Dingo Fence', 'Dingo Barrier Fence' or 'Wild Dog Fence' it is of wire mesh standing 1.8 m high, with a further 30 cm buried. There is a 5 m wide cleared buffer on each side and the entire structure is well maintained. On the New South Wales side of the fence where sheep are the main livestock, dingoes are controlled and numbers are low, whereas on the South Australian side dingoes are tolerated alongside cattle husbandry. Unintentionally, the fence initiated a large-scale experiment, allowing biologists to assess the biological consequences of removing a large predator.

Alan Newsome and his colleagues from the Commonwealth Scientific and Industrial Research Organisation (CSIRO) studied vertebrate abundances on either side of the fence by counting animal tracks at stock watering points. One striking finding was that the introduced predator the red fox (*Vulpes vulpes*) was present on both sides, but in higher numbers in the absence of dingoes. Dingoes eat foxes and drive them away, explaining the differences in fox numbers across the fence. Foxes threaten several native Australian mammals (see Chapters 2 and 16), including many burrowers and diggers important in soil turnover, nutrient cycling and dispersing plants. It may be that, by regulating fox numbers, dingoes are protecting native fauna that in turn modify the environment to the benefit of many soil organisms.

The case of the Dingo Barrier Fence shows how relationships between organisms determine the range and relative abundance of species.

Chapter aims

In this chapter we describe the range of relationships occurring between populations of organisms coexisting in communities. Communities, together with the physical environment, form ecosystems within which there are important flows of energy and nutrients. As the biota itself modifies the environment, a changing pattern of communities may occur over historical or geological time. Natural forces or humans can cause major damage to ecosystems, and we describe the situations under which they recover, or decline.

Characteristics of biological communities

In Chapter 16 we saw how interacting individuals of a particular species form a population. In turn, assemblages of populations of different species interacting together form communities. Within these communities each species has a specific role or niche, which is the aggregate of its use of the living and non-living resources of its habitat. The fundamental niche of a species includes the full range of environmental conditions in which it can grow and reproduce. However, species rarely occur without other species and the realised niche is the segment of the fundamental niche to which a species is restricted by interactions with others.

All biological communities ranging from coral reefs to deserts and tropical rainforests share common characteristics, including species richness and diversity, dominance/ keystone species, and trophic relationships.

Species richness and diversity

Different communities are formed by a different composition of species (Box 17.1). This composition is often summarised by the species richness, which is the number of different species in a community, and by the species diversity, which combines both the number of different species and their relative abundance (Box 17.2).

Box 17.1 Quantifying the species composition of a community

Plants

You are already familiar with the use of quadrat sampling to estimate population densities (Chapter 16). It is also appropriate to determine the number of species in a community. In this case all the species in the quadrat are recorded by counting them or, more commonly, by estimating their area of coverage within the quadrat. Accurately assessing cover is very difficult, especially for inexperienced operators. Therefore standardised scales such as the Domin scale (Table B17.1) are used.

continued ›

Box 17.1 continued ›

Table 17B1.1 Domin scale categories used to measure relative abundance of species in a quadrat.

Cover (%)	Density	Domin symbol	Transformed Domin value
< 5	1–2 plants	+	**0.04**
< 5	3–5 plants	1	0.2
< 5	6–10 plants	2	0.4
< 5	11–30 plants	3	0.9
< 5	31–100 plants	4	2.6
5–10		5	3.0
11–25		6	3.9
26–33		7	4.6
34–50		8	5.9
51–75		9	7.4
76–100		10	8.4

The optimum quadrat size can be determined by constructing a species area curve. A successive series of quadrats of increasing size is marked out and a cumulative species list recorded. The cumulative species curve levels as the area sampled (quadrat size) increases, indicating the optimum quadrat size (Figure B17.1.1). This is often related to the structure of the vegetation. In a forest a much larger quadrat is needed than in a low herb field.

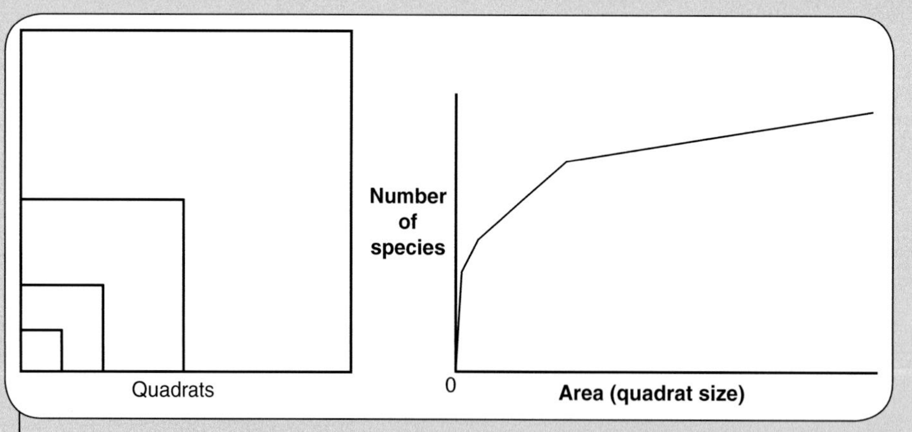

Figure B17.1.1 Graph showing the increasing cumulative number of species recorded as the size of the quadrat increases. Note that the number of new species recorded starts to level off at larger quadrat sizes. (Source: B. Cale)

Animals

As explained in Chapter 16, quadrat sampling is also appropriate for sessile animals, especially those on rocky seashores. Animal sampling programs usually involve trapping. If the aim is to determine the number of species occurring in a particular area, species accumulation curves are plotted to determine if trapping intensity is sufficient to detect most species. These plot the cumulative number of species detected (vertical axis) against the cumulative number of individuals caught (horizontal axis) (Figure B17.1.2). When the curve levels, it is likely that sufficient animals have been caught to represent the number of species in the area.

The ease of catching different species and seasonal factors may complicate the assessment. For example, several species of Australian mammals are reluctant to enter enclosed metal Elliott traps, but are caught more readily with pitfall traps (holes sunk in the ground). Therefore, a mix of traps is recommended. Sampling restricted to only one season may miss animals inactive at that time. If a complete species list is important, sampling should occur across the seasons.

Assessing the similarity of two communities

The similarity of two communities is often compared using a coefficient of community similarity, *C*. This ranges from 0 (no species in common between the two communities) to 100 (the two communities are identical in species composition). The relevant equation is:

$$C=\frac{2W}{a+b}(100)$$

where a = the number of species in community A
b = the number of species in community B
W = the number of species present in both communities.

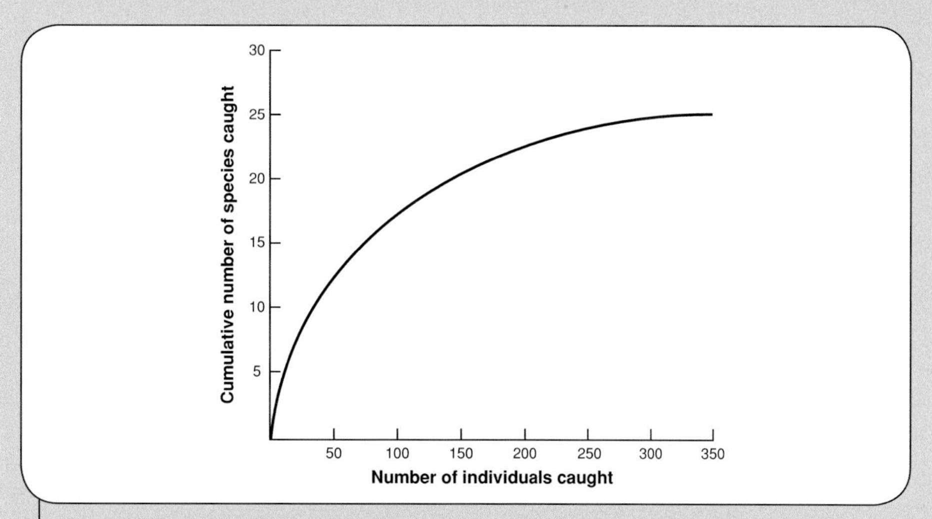

Figure B.17.1.2 A species accumulation curve for animal-trapping data. (Source: B. Cale)

Box 17.2 Calculating species richness and species diversity

Two of the basic attributes of a community are the number of species present (species richness) and a combination of the number of species present and their relative abundances (species diversity). Although these concepts are simple, there are three important issues. First, species richness and species diversity are not the same as biodiversity, which explicitly includes the species present and also genetic diversity and the diversity of ecosystems (Chapters 1 and 24). Species richness and diversity are easier to measure than biodiversity as a whole, but we need to be aware of the simplification. Secondly, measuring species richness and diversity assumes good taxonomy, so we recognise different species when we see them (Chapters 6 and 8). This is generally true for vertebrates or vascular plants, but is much less true for invertebrates or non-vascular plants. Thirdly, traditional measures of species richness and species diversity treat all species as equivalent, but this is not always appropriate. For example, should a species of kangaroo (one of the 14 species in the genus *Macropus*) contribute the same amount to a diversity measure as a species that is unique in its genus, such as the platypus (*Ornithorhynchus anatinus*)? To overcome this problem a number of new 'taxonomic diversity' measures, which attempt to measure diversity as the amount of taxonomic distinctness in a community, have been proposed. Lastly, it is important to specify the boundaries of the community being studied.

Species richness

If it is possible to count all the species in a community, then species richness is clear and unambiguous. However, most figures for species richness are based not on a complete count of species present, but on estimates of species richness based on samples. Species richness rises with the number of samples taken and, although this levels off with higher sampling effort, there can be problems when the species richness of two communities is compared based on unequal sampling effort. For example, if 400 individual insects are collected from community A and 2000 from community B and the number of species in each sample identified, it is probably invalid to compare the species richness of the two communities based on these samples. Instead, a statistical technique called rarefaction is used to standardise the data from the different communities to a common sample size so that species richness can be compared validly. However, this will be valid only if the sampling methods used in each community were the same, because different sampling techniques may sample very different species.

Dominance/keystone species

The activities of only a few of the many species in a community regulate the presence and relative abundance of the others. Sometimes these species are obvious, such as trees in a forest, and they are called dominant. For example, in a eucalypt forest the tree canopies rarely overlap and comprise the top third of the forest. Below these are smaller trees, shrubs, herbs and ferns forming layers down to the ground. The density of tree canopies influences the lower layers, because shade restricts access to sufficient sunlight for photosynthesis. Keystone species are usually present in lower numbers than dominant species, but their actions allow many other species to thrive. For example, the cassowary, a large flightless bird, disperses the seeds of many north-east Queensland rainforest plants. Marine examples include the tiger sharks at Shark Bay in Western Australia that protect seagrass habitats by eating dugongs and green sea turtles grazing the seagrass.

Trophic relationships

Autotrophs use solar energy to produce their own organic compounds from inorganic compounds. In turn, they are eaten by heterotrophs that are food for other heterotrophs. These feeding relationships determine the flow of energy and the cycling of nutrients within communities.

All these community characteristics emphasise one important point: communities are defined by interactions between different species of organisms.

Interactions in communities

Communities are characterised by six main types of interactions between species (Table 17.1). In combination, they determine the range and relative abundances of species in a community.

Table 17.1 Types of interactions occurring between two species.

Term for the interaction	Description of interaction
Neutralism	Neither species is affected by the other species.
Commensalism	One species benefits from the interaction and there is no effect on the other species.
Amensalism	One species is disadvantaged whereas the other is unaffected.
Mutualism	Both species benefit from the interaction.
Predation	One species benefits whereas the other species is disadvantaged.
Competition	Both species are disadvantaged by the interaction.

Neutralism, commensalism, amensalism and mutualism

Superficially, neutralism (no interaction) seems widespread. For example, how could a eucalypt tree and a worm in the soil interact? Possibly, leachate from fallen eucalypt leaves may influence soil pH, which in turn affects the worm. Subtle interactions such as this often appear as commensalism or amensalism.

Shading by trees is a good example. It could improve the microenvironment for understorey plants by preventing overheating in summer. If the understorey plants benefit and the trees are unaffected, the relationship is commensal. However, if the understorey is disadvantaged by reduced light for photosynthesis and the trees are unaffected, the relationship is amensal.

Lichens illustrate mutualism, where close, mutually beneficial associations of a fungus and an alga allow them to be classified as one species (Plate 17.1a–c). The fungus shelters the alga and provides mineral nutrition, and the alga photosynthesises. Other examples are associations between cyanobacteria and cycad roots, and the formation of nitrogen-fixing nodules by *Rhizobium* bacteria on the roots of pea plants (legumes, Plate 17.1d–e).

Predation

In predation, one species benefits at the expense of another. There are three main types:

1. *carnivory* – animals (predators) eating animals (prey) (or, more rarely, plants eating animals) where the prey is killed
2. *herbivory* – animals eating plants, but not necessarily killing them
3. *parasitism* – small animals (parasites) feeding on larger animals or plants (hosts), or plants (parasites) using other plants (hosts), where the host is harmed, but not necessarily killed.

Predators have physical and behavioural adaptations to locate and capture prey (Plate 17.2). Many, such as the death adder and python, are camouflaged. Others set traps. Some trapdoor spiders spin an array of fine threads from their retreat to detect approaching prey, and oysters pump water through their mantle cavity and sieve out zooplankton prey. Carnivorous plants offer glistening fluid droplets that mimic nectar, but are sticky and capture invertebrates. They usually grow in nutrient-poor soils and obtain nitrogen and phosphorus from digested prey. Sun dews (*Drosera* spp.) are carnivorous and widespread in southern Australia's nutrient-poor heathlands (Plate 17.2a–b).

In turn, prey species are adapted to avoid capture (Plate 17.2). Primary defences operate whether or not a predator is nearby. These include camouflage (Plate 17.2f) and chemical defences making the animal poisonous or distasteful. Sometimes these species are brightly coloured so they are not mistaken for palatable species. If different poisonous species have similar colours then they all reduce predation because predators learn that the colours mean 'danger'. This is Mullerian mimicry. For example, the unpalatable monarch butterfly (*Danaus plexippus*) and the unpalatable viceroy butterfly (*Limenitis archippus*) resemble each other. Other species capitalise on the unpalatability of other animals by closely mimicking the form or colour of the distasteful species, although they are palatable themselves. This is Batesian mimicry. One example is the harmless juvenile of the African lizard (*Heliobolus lugubris*), which mimics the noxious 'oogpister' or 'eye squirter' beetles that emit pungent chemicals when disturbed.

Secondary defences operate when a predator is nearby. Solitary animals may run to escape, enter a prepared retreat such as a rabbit into a warren, or even fight. Animals in herds may flee, aggregate for defence or mob a predator.

Herbivory seems simpler than carnivory because plants cannot run away, but plants vary in palatability and have defences to discourage herbivory. Different parts of plants have different food quality. Leaves are low in energy and nutrients, but seeds are much more rewarding (Plate 17.2d). The most nutritious parts are cell contents high in nitrogen and phosphorus, but largely indigestible cellulose makes up the bulk. Many herbivores have mutualisms with bacteria in their guts that decompose the cellulose so their digestive systems can absorb the nutrients. This is particularly true for termites, an important part of Southern Hemisphere arid communities, where they decompose plant debris.

Many plants deter grazers with thorns or toxic chemicals. The spines on many acacia species are obvious deterrents. Eucalypts have distasteful oils and tannins, preventing some animals from eating their leaves. However, koalas subsist almost entirely on eucalypt leaves and many Australian invertebrates consume them with impunity, so resistance can evolve. An example of a toxin is sodium fluoroacetate, produced by species in the Western Australian pea plant genus *Gastrolobium* (the toxic plants referred to at the beginning of Chapter 3).

Pollen and nectar feeding are forms of herbivory approaching mutualism. Plants encourage animals to visit their flowers with nectar or sometimes pollen (Plate 17.2c). In return, the animal unwittingly carries pollen to another flower where pollination and seed set occur. If the flower visitor transfers pollen inefficiently, the animal obtains food at the plant's expense and the interaction is not mutualistic. Some plants may have as few as one preferred pollinator, so in the case of rare plants such as the Corrigin grevillea (*Grevillea scapigera*) it is important to understand pollination biology.

Parasitism is the most common lifestyle on earth – all free-living species have at least one parasitic species which attacks them. Parasites are generally smaller than their hosts and rarely kill them, although they may reduce their fitness (see Chapter 14). Viruses and pathogenic bacteria (Chapter 9), which infect both plants and animals, may be regarded as parasites. In general, animals parasitise plants or other animals and plants parasitise plants (Plate 17.3), but plants are sometimes parasitic on animals. Orchid pollination by sexual deception is one case where plants might be considered to parasitise animals. Some orchids emit a chemical similar to the pheromone (sex hormone) of a female thynnid wasp. Males are attracted to the orchid and part of the orchid flower (the labellum) resembles the abdomen of a female wasp. The labellum is hinged and when the male attempts to clasp it he is flipped into the anthers of the flower. Pollen sticks to the wasp and is carried to the next flower the wasp is deceived into visiting, where it is delivered to the female part of the flower. There is no benefit to the wasp and the extra exertions no doubt take a toll on his fitness for actual matings.

All types of predatory interactions (carnivory, hebivory and parasitism) may cause an 'arms race' between exploiter and exploited species, where prey or hosts try to protect themselves, but predators or parasites adapt to circumvent these innovations. A term for this is the 'red queen' hypothesis of evolution. This term is derived from the section in Lewis Carroll's book *Alice in Wonderland* where the Queen of Hearts tells Alice to run as fast as she can, but Alice finds she has to run as fast as she can just to stay in the same place. It seems in the animal and plant kingdoms that arms races never end with a winner – perhaps there is a lesson in this for human affairs!

Although being eaten or parasitised is disadvantageous for individual organisms, predation in some communities enables the coexistence of more species. This occurs because some dominant species may suppress the growth or survival of other species in the same habitat. If predators reduce the numbers of the dominant species, less robust species may prosper, increasing species richness. In contrast, herbivores may sometimes consume all the palatable plant species in a community so that unpalatable species become dominant – this is one cause of rangeland degradation from overgrazing.

Competition

Interspecific competition occurs because of a shortage of resources needed by two species and is negative for both. The competition may be by direct contact such as fighting for territory or over a particular type of prey – interference competition. Exploitation competition is when two species access a resource but do not directly come into contact. The more efficient species benefits most from exploiting the resource to the detriment of the other species, but is still adversely affected.

No two coexisting species can be completely equal competitors for a single, limited resource, because one would have to completely exclude the other. This gives rise to the competitive exclusion principle – no two species can be complete competitors for a limiting resource, so coexistence means that each species uses resources slightly differently or that some other factor such as predation keeps populations below the level where the resource becomes limiting.

Consequences of competitive interactions for species

Five thousand years ago the predatory thylacine (marsupial wolf or Tasmanian tiger) occupied all Australia, including Tasmania, before disappearing from the mainland about 3000 years ago. One hypothesis for its extinction is competition following the introduction of the eutherian dog, the dingo, about 3500–4000 years ago. According to this hypothesis, the dingo out-competed the thylacine on the mainland but could not reach Tasmania, where thylacines survived until exterminated by Europeans in the early 20th century. A second hypothesis suggests competition as well as direct hunting by humans were responsible. Aboriginal populations on the mainland increased and expanded in range about 4000 years ago, supported by innovations in hunting technology. Both Aboriginals and thylacines hunted similar prey, and this increased competition may have driven the thylacine extinct.

From community to ecosystem

An ecosystem includes the organisms of a community and the non-living environment they interact with. Ecosystems vary from large estuaries to ephemeral pools on granite outcrops after heavy rain. With very few exceptions (see Box 4.1), they are driven by the solar energy that autotrophs fix into the energy of chemical bonds during photosynthesis. Energy and matter then flow or cycle through the ecosystem as organisms interact with each other and with their abiotic environment. The pattern of flow of energy and matter through the ecosystem is known as ecosystem function.

Flows of energy through food chains and food webs

Energy flows through all organisms in the ecosystem through their feeding relationships. Feeding relationships within an ecosystem form food chains that aggregate into food webs (Figure 17.1, Box 17.3). Each step in a food chain is a trophic level. At the base are the primary producers, the plants. Animals eating plants are primary consumers, or

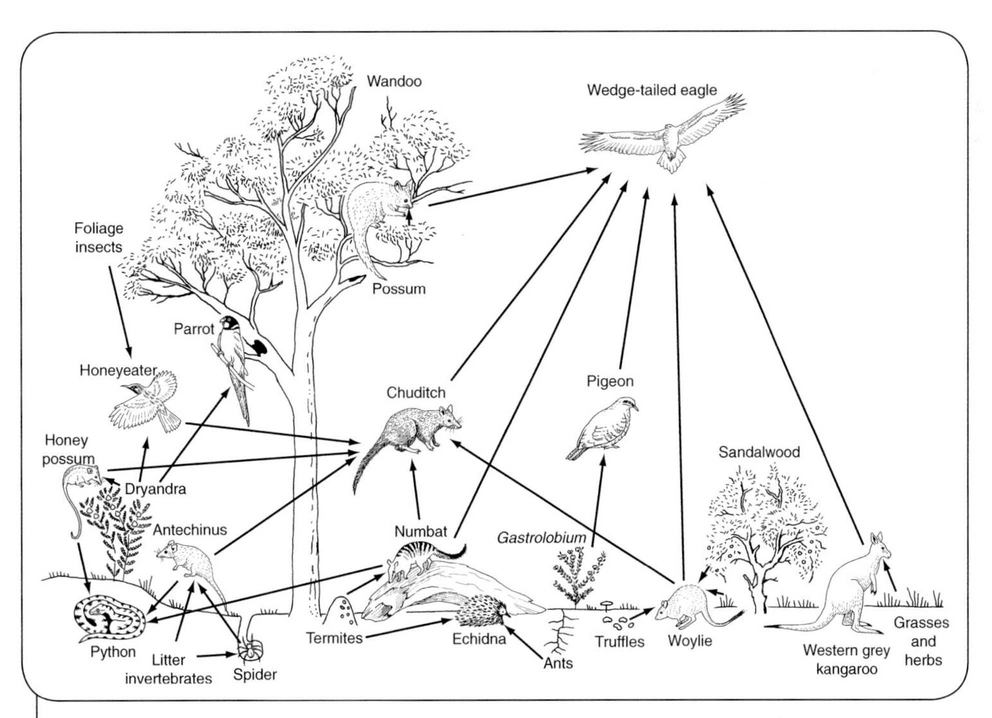

Figure 17.1 Simplified food web in the wandoo (*Eucalyptus wandoo*) woodland of Western Australia. (Source: B. Cale)

Box 17.3 Unravelling the food web

Understanding feeding relationships is critical to describing interrelationships within communities and also in tracing nutrient cycles. Some of the most important techniques to study these include:

- *exclusion experiments* – if predators can be excluded from some prey populations (experimental treatments) but given ready access to others (controls) then changes in the prey population in the presence or absence of the predator can be monitored.
- *stomach contents* – colloquially known as 'last supper biology', this involves trapping, killing and gutting animals and identifying food remains in their digestive tracts. Killing animals may raise ethical objections (Box 2.1), so non-destructive methods such as stomach pumping or using emetics (drugs inducing vomiting) may be valid alternatives.
- *faecal analysis* – droppings can be collected either by searching in the field, or by trapping animals and holding them until they produce a sample. One problem with this method is that only prey with hard body parts or tough plant cells that resist digestion are likely to be detected.
- *immunological methods* – proteins from putative prey are injected into laboratory animals, which then produce specific antibodies against them. Stomach samples from predators are then screened with antibodies for the presence of prey proteins. The method is effective for animals such as blood feeders, or for invertebrates that ingest prey fluids.
- *tracers* – dyes, particles or radioactive tracers may be used to label putative prey animals and predators later screened to determine if they have accumulated any tracer.
- *stable isotopes* – these are isotopes (forms) of an element that are not radioactive and behave chemically in the same way as the most common isotope for that element. The ratios of two common isotopes of carbon, ^{13}C and ^{12}C, vary in different groups of plants. The tissues of grazing animals reflect the types of plants they eat, so examination of the isotope ratios gives information on their diets.

more specifically grazers or herbivores. Animals eating primary consumers are secondary consumers or predators. There can also be tertiary consumers and quaternary consumers (Figure 17.1). Irrespective of the length of the food chain, the final link is the detritivores or decomposers, which feed on the dead bodies or wastes of other organisms, returning their nutrients to the soil for recycling. This is a simplified scheme. For example, omnivorous animals such as the honeyeaters in Figure 17.1 eat both plants and animals and therefore are both primary and secondary consumers.

In general, the number of links in a food chain does not exceed four because the transformation of energy from level to level is inefficient. A herbivore, for example, converts energy in the organic compounds of plant food to ATP for its own use, losing

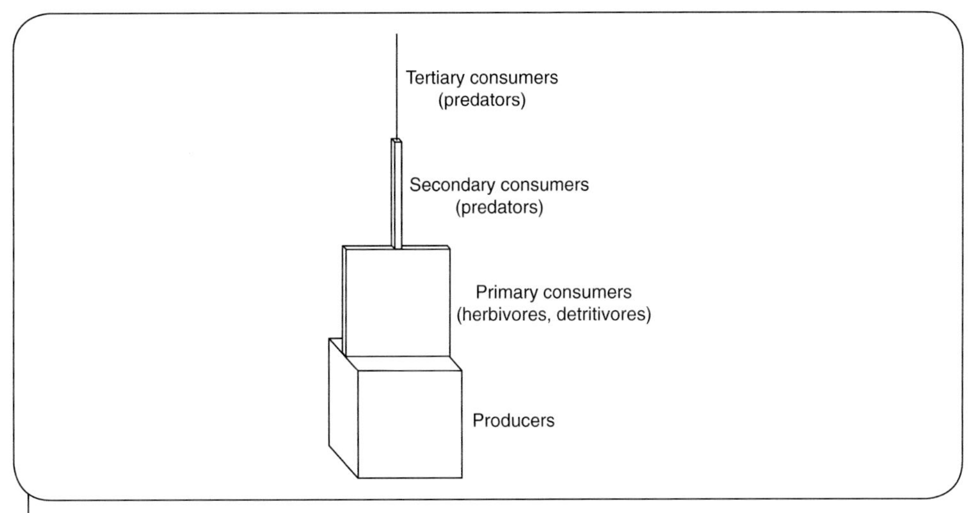

Figure 17.2 Diagrammatic pyramid of net production or energy. Tertiary consumers are only a very small proportion in comparison to the producers and contain only a very small proportion of the energy fixed by the producers. (Source: B. Cale)

energy as heat in the process. As a rule of thumb, only 10% of energy in the bodies of producers is transferred into the bodies of primary consumers, and only 10% of that to the secondary consumers. Very quickly the system runs out of energy. This gives rise to the idea of a pyramid of energy, numbers or biomass (Figure 17.2). The broad base of the pyramid consists of producers, but the apex of top-order carnivores is small. They are relatively rare because they need a large base of productivity to support them. Overall, energy transfer in food chains and webs is open ended. It enters from the sun and is gradually dissipated in chemical bond energy and heat. From the Earth's point of view this does not matter, as the sun is effectively an unceasing energy source. The situation with matter is different because limited supply makes recycling important. Even communities apparently isolated in underground caves or aquifers illustrate these key principles of energy flow and nutrient cycling (Box 17.4).

Box 17.4 Out of sight, out of mind – stygofauna communities

Recently an Australian mining company had to revise its environmental review and management plan to take account of the need to conserve some organisms referred to in the press as 'cave spiders'. This term does not adequately describe the diversity of aquatic invertebrate communities that can be found in groundwater and caves: the stygofauna. The Cape Range National Park near Exmouth has stygofauna of special value.

continued ›

Box 17.4 continued ›

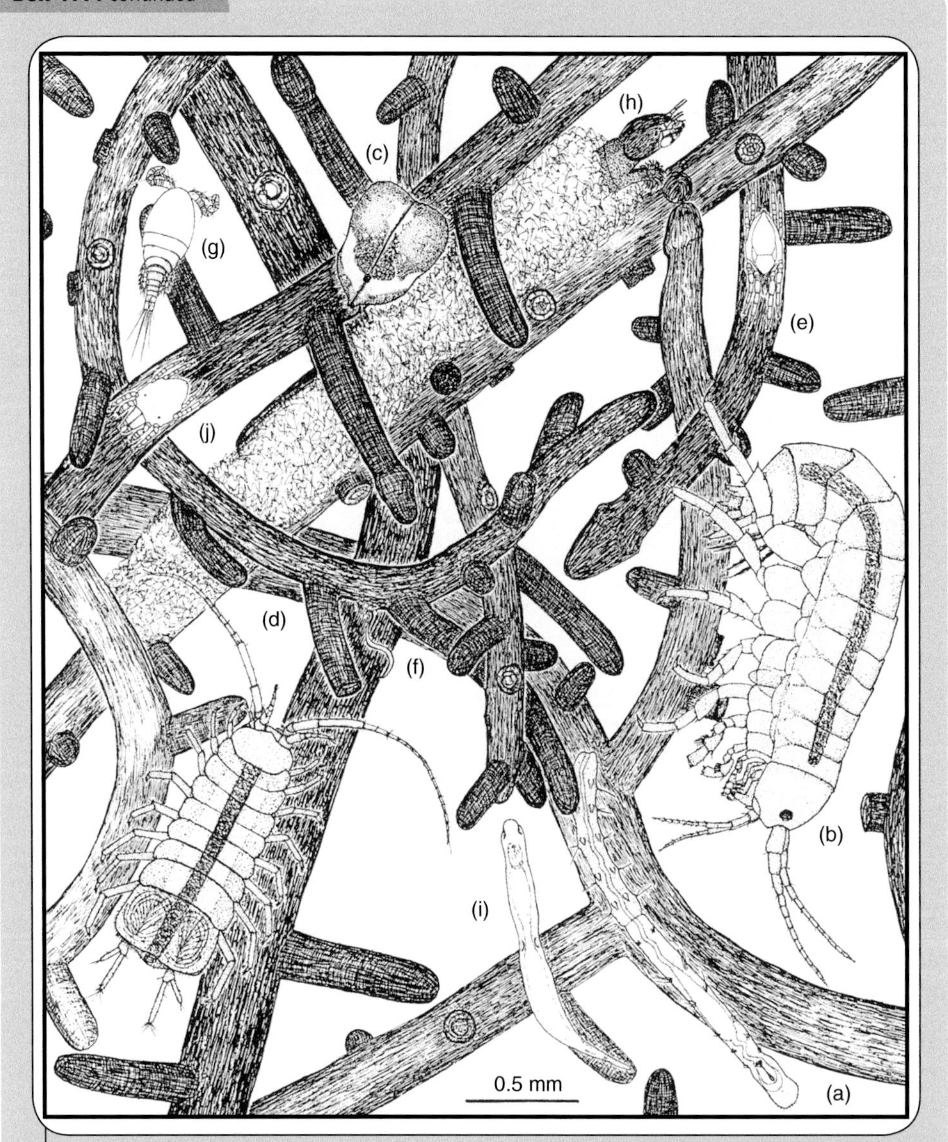

Figure B.17.4.1 Stygofauna within root mats of tuart (*Eucalyptus gomphocephala*) in the stream of Cabaret Cave, Yanchep National Park, Western Australia. (a) Aeolosomatid worm (annelid). (b) *Austrochiltonia subtenuis* (amphipod). (c) *Gomphodella* (ostracod). (d) Janirid isopod. (e) *Lobohalacarus weberi* (of the order Acarina). (f) Nematode. (g) *Paracyclops* sp. (h) *Polypedilum* sp. (Diptera) (i) *Stenostomum* (platyhelminth) (j) *Soldamellomx monardi* (Acarina). (Source: E. Jasinka)

Despite its arid surface, the Australian continent is surprisingly rich in groundwater. The Great Artesian Basin, for example, extends for 1.7 million km^2 under north-eastern Australia. Many Australian river systems rely to some extent on groundwater recharge, and river water and groundwater may be exchanged at many points along a river channel. Human water extraction or disruption of hydrological cycles through damming and irrigation is threatening many of these systems, and in some cave systems water has to be pumped in to prevent the extinction of these ecosystems due to low water tables.

In dark, underground aquifers there is no opportunity for photosynthesis, and energy enters stygofaunal communities in the bodies and waste of organisms coming from the land surface. In caves where root mats penetrate cave lakes, there may be particularly high numbers of taxonomically diverse species (Figure B17.4.1). In Europe, where they are studied extensively, 40% of the crustaceans are estimated to be stygobitic (groundwater dwelling). The Australian stygofauna is poorly known, but limited surveys suggest that it is at least as rich as that in other areas of the world.

Nutrient cycling

Any matter that is required for the nourishment of an organism is called a nutrient. Nutrients are obtained from other organisms or from the environment. Furthermore, the Earth is finite. There is some small gain of material from the cosmos in the form of meteorites and possibly comets that might hit the Earth, but it is so small that nutrients must be recycled.

Nutrients in terrestrial ecosystems are derived from the soil and the atmosphere, and are accumulated selectively by organisms. Living things need ample water, carbon, nitrogen and phosphorus, somewhat less calcium, sodium and potassium, and very small amounts of micronutrients such as iron, magnesium, molybdenum, zinc, boron, selenium and copper. All these nutrients are part of biogeochemical cycles. We now consider three examples of nutrient cycles: the nitrogen cycle, the water cycle and the carbon cycle.

The nitrogen cycle

Nitrogen is stored mostly in the atmosphere (which is 70% nitrogen gas) rather than the soil (Figure 17.3). Nitrogen gas is stable because strong electrical bonds join its two nitrogen atoms. However, most organisms need single nitrogen atoms to make proteins, not the two bonded in nitrogen gas. Lightning in electrical storms can separate nitrogen atoms, but it is minor compared with an important biological route. Microbes converting atmospheric nitrogen to ammonia (NH_3) are widespread in ecosystems, and the ammonia in turn is changed into forms such as nitrates that plants absorb. Higher plants often harness nitrogen-fixing microbes in mutualisms where the microbe fixes nitrogen while the plant fixes carbon. Cycads (Chapter 11) and the water fern *Azolla*

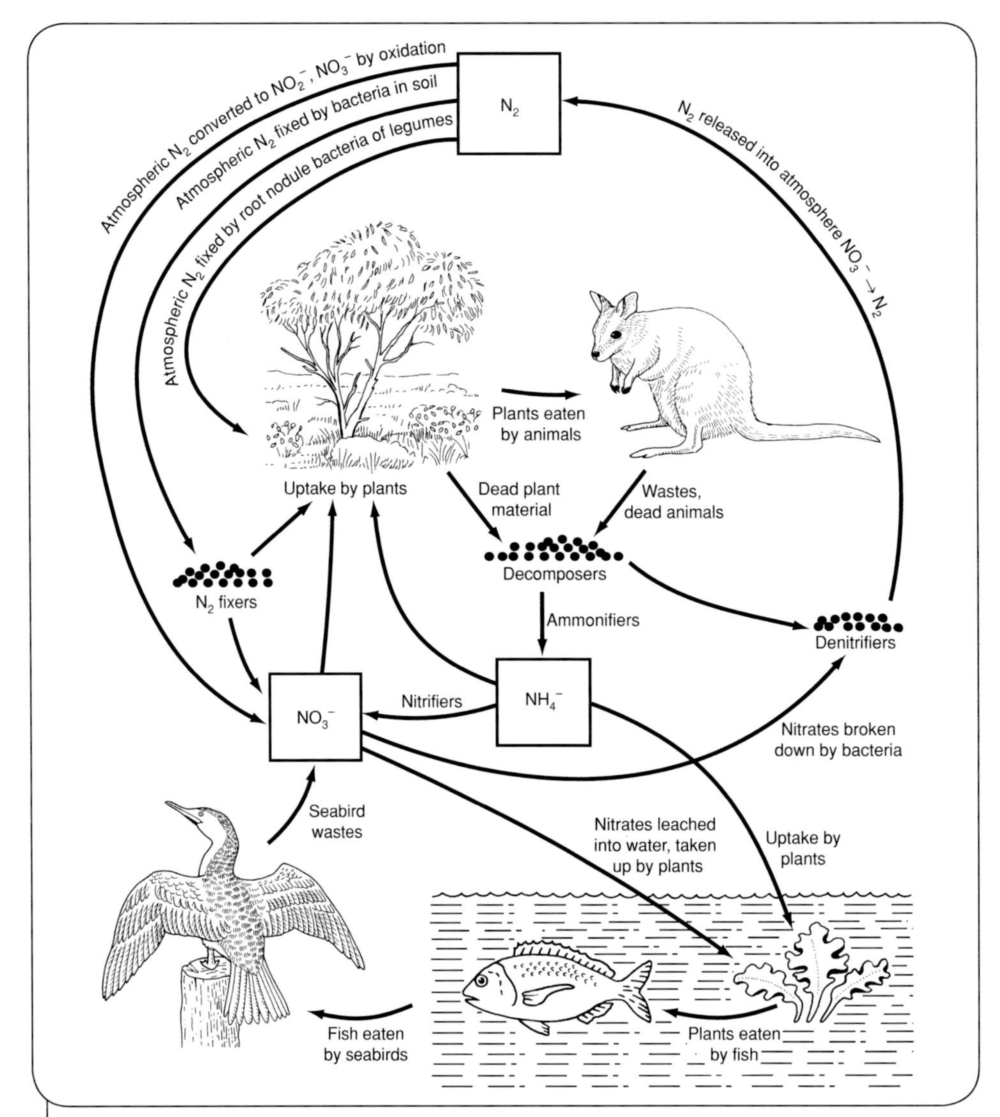

Figure 17.3 Diagrammatic representation of the nitrogen cycle. (Source: Adapted from D. Milledge, in Recher, H.F., Lunney, D. and Dunn, I. (1986). *A natural legacy: ecology in Australia*. A.S. Wilson Inc.)

provide microhabitats for nitrogen-fixing cyanobacteria and benefit from access to the fixed nitrogen. A more widespread cooperation is between the nitrogen-fixing bacteria *Rhizobium* spp. and legumes (Plate 17.1e). The *Rhizobium* forms an infection (nodule) on the legume roots where it fixes nitrogen, while the legume provides sugars.

Nitrogen fixation is vital in agriculture and natural ecosystems. In Australia, fire volatises much of the nitrogen in the vegetation, removing it from the system. However, legumes (peas and acacias) are common in most Australian fire-prone systems and through their mutualism with *Rhizobium* the nitrogen lost in fires is soon restored.

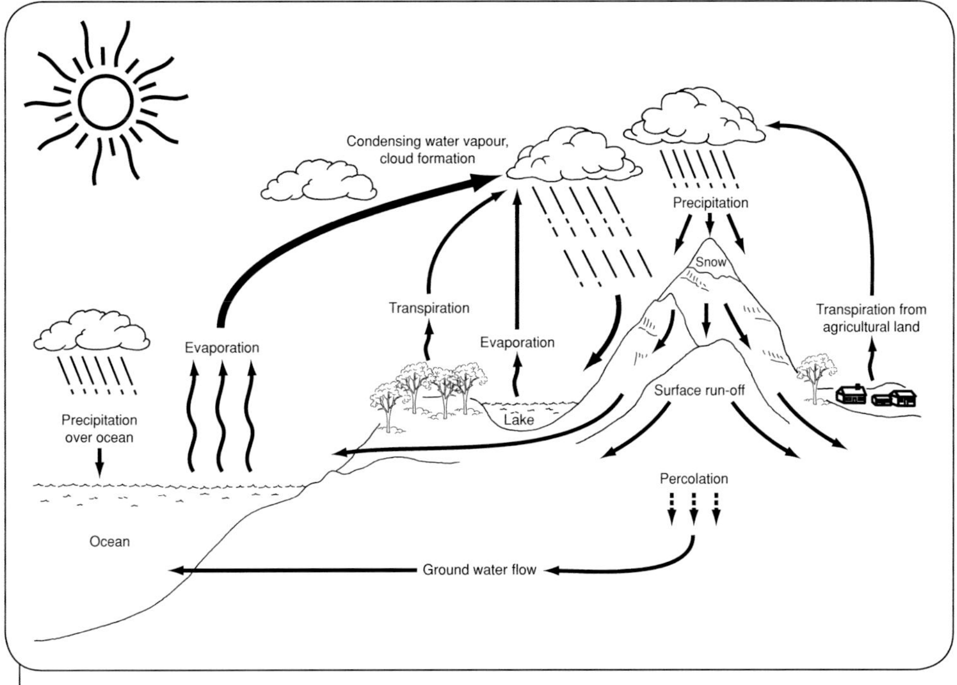

Figure 17.4 Diagrammatic representation of the water cycle. (Source: B. Cale)

The water cycle

Precipitation (rainfall), evaporation and transpiration, all driven by solar energy, determine the availability of water to living things (Figure 17.4). Over the oceans, water transfer to the atmosphere by evaporation exceeds precipitation. Clouds transfer the excess water over the land, where it falls as rain. Where rainfall over land exceeds water loss to the atmosphere by transpiration or evaporation, water collects in lakes, rivers or groundwater and ultimately flows back to the sea to complete the cycle. Humans affect the water cycle through land clearing that disrupts transpiration and by pumping groundwater supplies to the surface for irrigation.

The carbon cycle

Carbon is the key component of all organic molecules. Its availability is regulated in a cycle with biotic reservoirs in organisms and abiotic reservoirs in the atmosphere, the oceans and fossil fuel deposits (coal and petroleum) (Figure 17.5). Respiration and photosynthesis are mainly responsible for cycling carbon between the biotic and abiotic reservoirs. Photosynthesis fixes atmospheric CO_2 in organic compounds, which are broken down in respiration by both autotrophic and heterotrophic organisms to return CO_2 to the atmosphere. In recent times, humans burning fossil fuels greatly increased CO_2 release to the atmosphere, with possible serious consequences (Box 4.5).

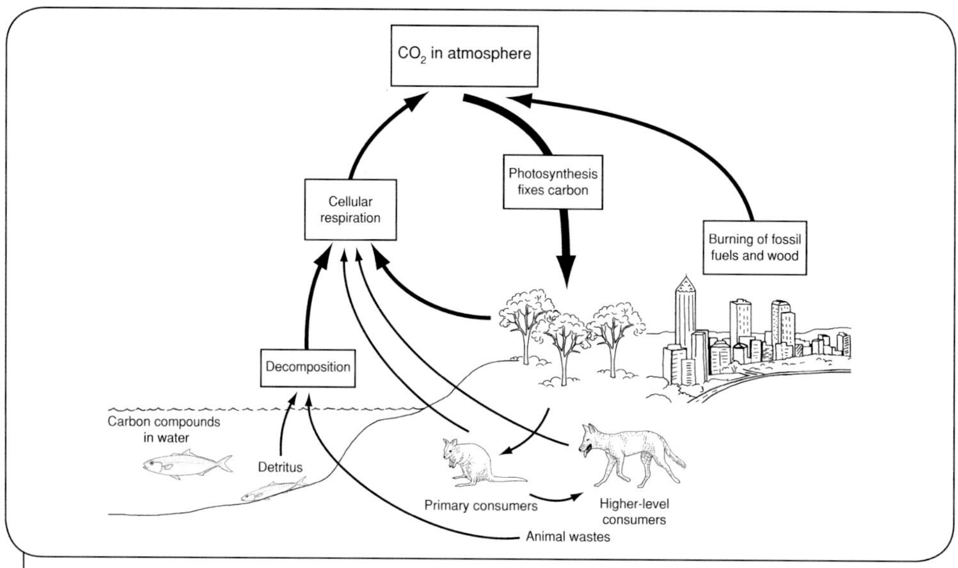

Figure 17.5 Diagrammatic representation of the carbon cycle. (Source: B. Cale)

Nutrient capital in ecosystems

The nutrient capital of an ecosystem occurs in compartments: the living biological component, the non-living biological compartment, the atmosphere, the soil and the underlying rock. Nutrients move between these compartments. Nutrients in plants are lost in fallen leaves, dead branches, bark, unfertilised flowers and fallen fruits and seeds. These combine with animal wastes and corpses to form litter gradually incorporated into the soil or the bottom of a lake in aquatic systems by detritivores. There are two main steps:

1. *comminution (reducing the size of particles)* – mostly, small invertebrates fragment the litter. In wetter forest in the eastern highlands of Australia lyrebirds are integral too. They scratch through the litter looking for invertebrates (their staple diet) and act as composting machines, fragmenting, turning and aerating the litter.
2. *mineralisation (hydrolysis and oxidation)* – the increased surface area after comminution accelerates hydrolysis and oxidation by microbes and soil fungi. Much of the litter resists decomposition, especially in many nutrient-poor Australian systems, where plants withdraw many useful nutrients before shedding parts. In addition, plant secondary compounds, initially intended to protect the living plant, also inhibit microbial breakdown. Fortunately several fungal groups are able to decompose resistant plant chemicals such as lignin and waxes. Litter decomposes only during cool, moist times providing suitable conditions for invertebrates, soil microbes and fungi to grow.

Bioaccumulation and biomagnification

If an organism absorbs a substance at a greater rate than it loses it through excretion or degradation, then that substance bioaccumulates in its tissues (Box 17.5). Predators

Box 17.5 Bioaccumulation in crops

For wheat cropping to be successful in many areas of Australia, nutrients (fertilisers) are added when the crop is sown each year because the soils are poor in nutrients. Phosphorus is particularly important and much of the Australian phosphate fertiliser comes from Christmas Island.

Christmas Island phosphate rock arises from a combination of seabird guano (faeces) and coral rock formed over hundreds of thousands of years. Seabirds have been using Christmas Island as a roosting place since it appeared as a coral atoll above the ocean on a subsea volcano. They eat fish and their guano has a high level of phosphate and some cadmium, a toxic heavy metal. Although individual birds contribute only a very small amount of cadmium, it accumulated over millennia together with the phosphate. Following decades of adding this phosphate to Australian cropland, the cadmium in the soil accumulated to a point where plants absorb it and it reaches unacceptably high levels in some of the grain. The low rainfall of most of the cropping areas means that the cadmium will be leached very slowly from the soil, so the problem is persistent.

ingesting these organisms also ingest the concentrated substance. If they cannot degrade or excrete it, the substance remains in their tissues. To meet their energy needs the predators must ingest many prey, so the substance reaches higher concentrations in their tissues. This process is repeated at higher trophic levels. Such concentrations of substances through a food chain are called biomagnification. Some organic molecules such as pesticides that do not biodegrade may be below toxic levels in primary consumers but accumulate to toxic levels in a secondary predator. This is the case with the insecticide DDT, which accumulated to high levels in top predators such as the bald eagle in America. It caused shell thinning in eagle eggs, which broke easily, causing breeding failure and bringing the species to the edge of extinction. Biomagnification can also occur with inorganic substances such as heavy metals (Box 17.5).

Response to disturbance – succession and resilience in communities

Communities constantly suffer external disturbances such as drought, fire and flood that may change their composition in a process called succession. This may return a community to its predisturbance state, or to a different composition. A community's capacity to recover from disturbance to its original state is called its resilience. These important properties can be seen most clearly in plant communities.

Few parts of the natural environment lack plants. Take the example of a sandy beach, an unstable, harsh environment. It is salty from the sea water, dry from exposure to air and with little water-holding ability because of its open, porous nature. However, beyond the wave zone are small sand dunes where plants grow, accumulating more beach sand blowing inland. This is a pioneer community. On a coastline accumulating sand the beach gradually moves seaward and the small dune behind it builds in height and in distance from the waves. The propagules of more plant species establish on the developing dune and vegetation thickens, forming a second community. Soon the coloniser plants of the pioneer community are shaded by taller growing invaders that eventually exclude the less competitive pioneers. Gradually, as the beach gets further away and the site becomes more protected from salty winds, other plants enter the second community and form a third type of community. This progressive formation of different communities is termed succession and is the way bare ground is colonised. The earlier species prepare the environment for later species. If you stood in one place for many hundreds of years, different communities would pass by as one matured and was replaced by another. In this system (also called the primary succession model) an association of plants eventually develops where species regenerate within their own community, becoming a climax community as long as there is no disturbance.

Disturbance is sudden marked change, but not complete destruction, of a community. It includes natural fire, floods, cyclones, volcanic eruptions, tsunamis and human-mediated activities such as overgrazing. Typically, the number of species in an area falls following disturbance. If the disturbance is not sustained, species begin to return. In the early stages of recovery there are often many species in an area, but some will eventually be eliminated by competition. Thus the number of species is lowest when there is a high frequency of disturbance and also when there are very long intervals between disturbances. Intermediate levels of disturbance may lead to the greatest range of species present. This is called the intermediate disturbance hypothesis. Resilient communities recover from the disturbance, but in other cases disturbance leads to secondary succession differing from the primary succession pathway. In Chapter 26, we explore the recovery and restoration of communities following human-induced disturbances.

The importance of fire as a disturbance in Australian communities

In Australia fire has long been an important disturbance, so most Australian plant and animal communities are fire adapted (Plate 17.4). Hurricanes and floods rearrange the organic components of a community, but fire converts organic molecules to inorganic and other organic molecules and energy.

Fire influences community structure in many ways. On the east coast of Australia and in parts of the Northern Territory closed forests with dense canopies occur in moist environments where fires are infrequent. These forests contain many co-dominant tree species as climax vegetation – the species regenerate within the community. This type of vegetation is uncommon in Australia, where communities are commonly event-dependent systems, depending on disturbance for regeneration.

Tall eucalypt forests with open canopies are icons for the Australian community, as seen by strong debates about conserving old-growth forest. Open forests contrast with closed forests by having few dominant canopy species (usually eucalypts), but species-rich understoreys. Most of these forests occur in high-rainfall areas of the east and south-west and depend on catastrophic disturbance (wildfire) for establishment. Without fire closed forest species invade under the eucalypts and eventually take over, while the eucalypts senesce and die. In general, eucalypts do not produce seedlings in mature undisturbed forests while closed forest species can do so. Thus, without disturbance, closed forest is the climax type of community in areas of very moist climate.

In general, eucalypt forests depend on fire for regeneration and maintenance (Plate 17.4). However, intervals between fires vary with climate. In the moist mountains of the eastern highlands and Tasmania fire intervals are very long, but they are shorter in drier forests. Open forests contrast with closed forests by having few dominant canopy species (usually eucalypts) but species-rich understoreys. After a fire, plants may resprout from resistant structures or leave seed that germinates in the burned environment. These two strategies, called 'resprouter' or 'seeder', occur in varying proportions depending on fire frequency (Plate 17.4c, d). Frequent fires favour resprouters because seeder species might not have long enough to mature and produce sufficient seeds before the next fire. In intermediate fire intervals both seeder and resprouter species occur in similar proportions. Where there are long fire-free periods seeder species outnumber resprouters (e.g. the tall open forests).

Human-modified ecosystems

Although ecosystems developed by humans may seem very different from natural systems, similar processes occur within them.

Agriculture

Most crops derive from pioneer plants that colonised disturbed habitats where nutrients are initially abundant. They capitalise on these by rapid growth, producing propagules dispersed to the next disturbed area before nutrients decline. The next plants to colonise live longer than the pioneer species, accumulating nutrients in their plant biomass. In our agricultural systems we expend energy maintaining succession at an early stage. We remove the nutrient capital as the crop (e.g. wheat or rice) and add more nutrients as fertiliser for the next crop. In doing this repeatedly we may add toxins (Box 17.5). We may also change soil structure. In natural systems the soil stores water, oxygen, carbon dioxide and nutrients. Litter increases water-holding capacity, releasing nutrients to plants and clumping soil in small aggregates, increasing spaces for holding water and oxygen. In agricultural systems the litter component is usually low.

Eutrophication

Eutrophication refers to nutrient enrichment of ecosystems. In wetlands and rivers it causes algal blooms such as in the Swan River in Western Australia and the Darling River in New South Wales. Warm, salty and nutrient-rich waters encourage cyanobacteria which, when they die and decay, decrease the oxygen content of the water. If the cyanobacteria produce toxic chemicals, fish and aquatic invertebrates die. Eutrophication is usually controlled by limiting nutrient input or flushing the nutrients to dilute their concentration (Chapter 20).

Urbanisation

Pioneer species, including many weeds, are favoured in urban systems. Weeds colonise initially but there are no secondary successional plants to follow on unmanaged land, so a depauperate weed community dominated by only a few species remains. Few animals persist in such communities. For example, in weed-dominated communities there are few species available for flower visitors. Honeyeater birds, so prominent in Australian bushland, lack resources in urban weedy environments or even gardens of introduced exotic plants. Domestic pets, especially dogs and cats, are maintained at high densities and may predate or harass native fauna.

Some urban areas retain remnants of the original vegetation as public open space. Managing these areas is challenging. Ecologically, there is a relationship between the number of species and the area available, and smaller areas support fewer species. Certain animals such as those at the top of food chains or with large home ranges cannot persist in small areas. Small remnants also have larger perimeters for their areas than similarly shaped larger ones, causing problems when weeds or exotic animals invade across the long boundaries (Chapter 24).

Trophic cascades

Humans often remove large predators from natural ecosystems through sport hunting or to protect livestock or people. This may cause significant impacts. For example, following removal of large fish from some North American lakes, their prey increased in numbers and fed heavily on zooplankton. Zooplankton numbers fell, in turn easing grazing pressure on phytoplankton that increased rapidly, reducing water quality.

On land, the principle was seen in reverse when wolves were reintroduced into Yellowstone National Park in North America in 1995–1996, after a 70-year absence. Browsing animals soon avoided areas of the park where they could be ambushed easily by wolves. As a result, in these areas cottonwood and willow trees that had been heavily browsed by herbivores in the past began to recover. These changes help in managing regeneration of vegetation in the park.

Such cases where top predators suppress the abundance of their prey, thereby leading to changes in the population of a species at a lower trophic level, are called trophic cascades. They are important in maintaining species diversity in a range of ecosystems.

From ecosystem to biosphere

The biosphere includes all parts of the Earth inhabited by living organisms. It is formed by the interlinked ecosystems of the oceans, land and freshwater environments throughout the world. These ecosystems are explored in more detail in Chapters 18–23.

The biosphere supports human existence through many 'ecosystem services', valued in a 1997 study at a staggering US$16–54 trillion ($10^{12}$) per year. These ecosystem services include such things as climate regulation, water supply, soil formation, waste treatment and food production. For example, manipulating the forests of a catchment influences the amount of water that can be used for domestic water supply. Production of many crops depends on pollination of flowers, accomplished mostly by free-range insects. Healthy aquatic systems prevent outbreaks of mosquitoes and midges, restrict the transmission of insect-borne diseases and maintain good human health.

The biosphere is increasingly influenced by human activities and this is expressed in such problems as ocean pollution, the possibility of an enhanced greenhouse effect (Box 4.5), air pollution and overexploitation of resources in general.

Declining Australian ecosystems

The ethos of early European Australians was to clear the bush for prosperity and to create a familiar 'European' landscape. As one correspondent to a Brisbane newspaper put it: '... the settler's first instinct was to shoot every strange animal and sink his axe into every unfamiliar tree ...'. The scale of the clearing was staggering. For example, about 25.7 million hectares of land in New South Wales were ringbarked or partially cleared between 1893 and 1921, which, as Stephen Jackson pointed out, exceeded the total land area of all of England, Scotland and Wales. This attitude is not behind us, because at the turn of the century land clearing in Australia was estimated at over 550 000 hectares per year, the fifth-highest of all nations in the world. This rampant clearing causes problems such as erosion, the decline of tree health in paddocks, dryland salinity, nutrient discharge into waterways, loss of soil structure, acid sulfate soils, loss of biodiversity in remnant bushland and threats to the survival of some native species, and facilitates the depredations of weeds and feral animals.

As an example, declines in tree health may be caused by hydrological changes, inappropriate burning, insect attack or disease. In forest areas of south-western Australia, dieback – death of woody plants – is caused by the soil pathogen *Phytophthora cinnamomi* (Plate 10.1). Hygiene programs now stop the pathogen spreading on earth-moving equipment or landfill from infected sites. Rural tree dieback in many areas of Australia (Plate 17.5b) can be caused by increased nutrient levels under isolated paddock trees where stock shelter. The nutrient content of the foliage rises, attracting insect herbivores. The birds that might control them are lacking because there are few understorey plants to provide a diversified food source. The increased insect grazing defoliates the tree. The tree resprouts but further insect damage leads to poor condition and eventually death. In many paddocks, trees kept for shade are now ageing. Solutions to the problem can be to plant more trees, retain remnant vegetation on the farm to support native bird populations and fence stock away from trees.

Clearing and farming also have impacts on aquatic and marine habitats, especially when sediments or nutrients from agricultural lands are flushed into the sea. The impacts on the Great Barrier Reef off eastern Australia are discussed in Chapter 27. Maintenance of our vast marine resources is crucial in particular to coral reefs, considered by the International Union for Conservation of Nature (IUCN) to be one of the life-support systems essential for human survival. These are just a few examples of why communities need to function at optimal levels.

Chapter summary

1. Biological communities are assemblages of populations of plant and animal species living together.
2. Each species occupies its own niche in a community, although this may be restricted by competition with other species.
3. Species richness refers to the number of species in a community, whereas species diversity combines both the number of different species and also their relative abundance.
4. Communities are characterised by interactions between species, which may be positive, negative or neutral. Types of relationships include neutralism, commensalism, mutualism, predation, competition and amensalism.
5. An ecosystem is made up of the living organisms within a community and their non-living environment.
6. Food chains transfer energy initially derived from the sun and fixed by producers. On average only 10% of the energy is transferred through each level of a food chain. Unlike energy, nutrients derived from the soil and the atmosphere are recycled through food chains. Interconnected food chains form food webs.
7. A pioneer community is the initial community of colonising species. They make the habitat suitable for the progressive formation (succession) of other communities until a stable climax community is established. If a climax community is disturbed, there is a redevelopment called secondary succession.
8. Species must reproduce for the community to continue. Event-dependent systems depend on disturbance events for regeneration.
9. The total of all the interconnected ecosystems of the Earth is called the biosphere. The biosphere provides us with ecosystem services that are essential to human survival and prosperity. Human alteration of natural ecosystems causes huge environmental problems including erosion, salinisation, nutrient discharge into waterways, loss of soil structure and loss of biodiversity. Solutions lie in scientific approaches and public awareness.

Key terms

amensalism
Batesian mimicry
bioaccumulation
biomagnification
biosphere
carnivory
climax community
commensalism
community
competition
competitive exclusion principle
consumer
detritivores
ecosystem
ecosystem services
eutrophication
event-dependent systems
exploitation competition
food chain
food web
herbivory
interference competition
intermediate disturbance hypothesis
interspecific competition
intraspecific competition
keystone species
Mullerian mimicry
mutualism
neutralism
niche (fundamental and realised)
nitrogen fixation
pioneer community

predation
primary succession
resilience
secondary succession
stygofauna
trophic level

Test your knowledge

1. What are the common characteristics of all communities?
2. Distinguish between species richness and species diversity.
3. List the six main types of interactions between organisms in a community and indicate which of the participants benefit, are harmed or are unaffected.
4. Explain the difference between primary and secondary succession.
5. How do feeding relationships regulate energy flow and nutrient cycling within communities?
6. Give examples of how human disturbances alter natural communities.

Thinking scientifically

Pine plantations have a rather gloomy, sterile appearance, but may support more animal diversity than nearby farmland. How would you go about testing this idea?

Debate with friends

Debate the topic: 'Ecotourism is loving our natural habitats to death'.

Further reading

Attiwill, P. and Wilson, B. (eds) (2006). *Ecology: an Australian perspective.* Oxford University Press, Oxford. 2nd edn.

Bond, W. J. and van Wilgen, B.W. (1996). *Fire and plants.* Chapman & Hall, London.

Costanza, R., d'Arge, R., deGroot, R., Farber, S., Grasso, M., Hannon, B., Limburg, K., Naeem, S., O'Neill, R.V., Paruelo, J., Raskin, R.G., Sutton, P. and van den Belt, M. (1997). 'The value of the world's ecosystem services and natural capital' *Nature* 387: 253–60.

Krebs, C.J. (1998). *Ecological methodology.* Benjamin Cummings, San Francisco. 2nd edn.

Pimm, S. (2001). *The world according to Pimm: a scientist audits the earth.* McGraw-Hill, Boston.

Smith, R.L. and Smith, T.M. (2001). *Ecology and field biology.* Benjamin Cummings, San Francisco. 6th edn.

18

Marine habitats

Eric Paling

Journey to the bottom of the world

The average depth of the world's oceans is 4 km, and about 75% of their water is at a depth of 1 km or more. These depths include the two most extensive marine habitats on Earth: the deep ocean floor and the water column above it. They are dark, cold and at great pressure, with low concentrations of nutrients and dissolved oxygen. There is no photosynthesis in the absence of light, so oxygen and organic material must descend from above. Despite this, just over 10% of known fish species feed and breed in these habitats.

The fish in the water column at these depths are characterised by sparse populations and little movement. They seize any prey passing by or lured to them by photophores (light-producing organs). They are large-mouthed and sharp-toothed to maximise chances of success, but to conserve energy their bodies have weak muscles, no scales and substantial reductions in many internal organs. Population densities are so low that finding mates is a significant problem. Female anglerfish release pheromones (chemical attractants) to lure males, which then attach to her as ectoparasites. In some species the sex organs of males and females do not mature until an ectoparasitic male is attached. Females then produce millions of eggs.

The fish found on the deep sea floor are more active and often more abundant than those above, feeding on each other, organic material descending from above and invertebrates on the ocean floor. Many have elongated shapes, possibly to increase the length of the lateral line system and improve sensory perception in the dark. They attract mates acoustically, by pheromones or by producing light.

Deep-ocean fish illustrate clearly how characteristics of the physical environment such as light and substrate determine the abundance and distribution of marine organisms. In turn, organisms show specific adaptations to find food and mates within the physical constraints of their immediate environment.

Chapter aims

This chapter introduces the main physical characteristics of the marine environment and how they interact with the biota to produce specific marine habitats. Chapter 19 returns to adaptations and lifestyles suiting organisms to specific habitats.

Life began in the ocean and about 70% of the Earth's surface is covered with sea water. Over 90% of the world's living biomass is contained in the oceans and humans harvest a mere 0.2% of its total production. Marine sources provide around 16–20% of the animal protein eaten by the world's population overall, but in Asia it is estimated that about 1 billion people rely on fish as their primary source of protein.

In 1992, a special United Nations Conference on Environment and Development recognised the importance of coastal waters. It adopted an 'Oceans Declaration' to control degradation of the marine environment, develop its potential to meet human nutritional needs and promote its integrated management and sustainable development. Following this, several large marine areas (LMAs) were defined to assist management. These encompassed coastal areas from river basins and estuaries to the seaward boundaries of continental shelves, enclosed and semi-enclosed seas, and the outer boundaries of major currents. Presently 64 LMAs are defined. They are large (greater than or equal to 200 000 km^2) and characterised by distinct bathymetry (depth), hydrography, productivity and trophic interactions. Annually, LMAs produce 95% of the world's fish catch and contain most of the ocean's degraded habitats and coastal pollution. They are the focal point of global efforts to reduce degradation of marine resources and coastal environments from pollution, habitat loss and overfishing.

To learn about the major marine habitats, it is first necessary to consider the physical factors shaping them. We will therefore examine important concepts in physical oceanography before describing briefly the main marine habitats that determine the distribution of organisms. Finally, we consider the kinds of substrates supporting marine producers and the communities they sustain.

Physical features of the oceans

Two of the most important characteristics of sea water are its salinity (the concentration of dissolved salts) and temperature. These work together to control its density and variations in density help drive oceanographic processes and mixing.

Salinity and circulation

Despite being salty, the ocean actually contains 96.5% fresh water by weight. The rest comprises dissolved salts and gases, organic materials, undissolved particles and pollutants. Sea water has a concentration of dissolved salts of 35 gL^{-1} (or parts per thousand, ppt). About 80 elements occur in sea water, but only 11 ions are abundant. These are (in grams per kilogram of water): chloride (19.135), sodium (10.760), sulfate

(2.712), magnesium (1.294), calcium (0.413), potassium (0.387), bicarbonate (0.142), bromide (0.067), strontium (0.008), boron (0.004) and fluoride (0.001). The rest are in trace quantities (less than 1 part per million).

Salinity influences water movements through its effect on the physiology of organisms and water density. All marine organisms are influenced by salinity, primarily by osmosis (discussed in Chapters 3 and 19). Density increases with salinity. Denser water is heavier and 'sinks', whereas fresher water 'floats'. Surprisingly small differences in seawater density have profound effects on ocean mixing and currents. We can see the effects of density on mixing by examining cases where fresher water meets sea water.

The first case is where rivers meet the sea. This is often in estuaries, although sometimes river water flows some distance into the sea, as with the Amazon River where outflow is so intense that fresh water can still be measured a few kilometres into the Atlantic Ocean. In an estuary, there is normally a gradient between the up-river, or fresher end, and the saltier seaward end. In a well-mixed estuary, this gradient is smooth and organisms are distributed according to their salinity tolerances. In less-mixed conditions the estuary may consist of saltier water sitting at the bottom and fresher river water above, the interface being termed the halocline. This prevents the bottom water from mixing with the surface water, reducing oxygen in the bottom layer. In winter, fresher river water may form a surface layer running from the estuary into the sea, carrying nutrients from farmlands.

Another example of the effects of density upon mixing is in ocean wastewater outfalls. Outfalls use a pipeline on the ocean floor at about 20 m depth with a series of diffusers at the end to release relatively fresh water from treatment plants. As the water emerges, it is so buoyant in relation to sea water that it rapidly rises to the surface, 'dragging' in the surrounding sea water to dilute it.

Finally, density effects can cause environmental 'catastrophes'. Corals spawn at very specific times of the year (see Box 19.1). In Western Australia, corals spawn on an outgoing tide, eight or nine nights after the March full moon. The buoyant, fertilised coral eggs float on the surface. If it rained on one of these nights, the fresher water would sit above the denser sea water and it is likely that the eggs, adapted to 35 ppt sea water, would be killed by the osmotic shock of coping with fresher water (0 to 5 ppt).

Temperature

There are three ways in which temperature influences the physical and biological properties of marine habitats: its broader influence on the distribution of organisms, its effect on physiology (Chapter 19) and its influence on seawater density and salinity. Certain groups of organisms are restricted to particular temperature regimes. Mangroves, for example, usually occur where air temperatures are above the 24°C isotherm. Corals and coral reefs grow optimally between seawater temperatures of 23 and 29°C, although many non-reef-building corals grow in colder temperatures. Seagrasses are divided into tropical and temperate suites. The most obvious macroalgae, giant kelps, best develop in cold climates.

Temperature influences seawater density directly by heating or cooling and indirectly by evaporation. Warmer water is less dense, whereas colder water is denser and sinks. The interface between the layers may be sharp and is termed the thermocline. In enclosed water such as estuaries or coastal bays, the sun heats the surface layer. That layer floats on the denser underlying water, promoting water-column stability. At night the surface water cools. Cooled sufficiently, it becomes denser and sinks, promoting mixing. In coastal areas with high evaporation and low rainfall, such as Australia or South Africa, sea water in narrow inlets or estuaries evaporates, increasing salinity and producing dense water that encourages mixing.

Light

Sunlight heats water and affects its density, and also drives primary production. The penetration and availability of light, coupled with water clarity, determines where plants and coral can grow. The depth to which sufficient light penetrates to allow biota to thrive is the euphotic zone, often abbreviated to photic zone. This is usually 10 to 20 m, but may range from a few metres to 50 m depending on turbidity (caused by suspended particles).

Not only does the intensity of light decrease with water depth, but different wavelengths of light are absorbed at different rates. The different photosynthetic pigments present in green, brown and red algae (Chapter 10) result in maximum photosynthesis in different parts of the spectrum, leading to a general correlation with the depths to which each can grow. Although green, brown and red algae may all be found in shallow waters, green algae have highest absorption and maximum photosynthesis in the red part of the spectrum. This is the first to be absorbed by water, so green algae are found no deeper than 30 m. In contrast, brown algae, which have maximum absorption from the green sector, can grow at greater depths, whereas red algae, which have highest absorption from the blue sector, are the only ones that occur in the pale blue light of the deepest parts of the ocean (2000 m or more below sea level).

Tides

Many habitats, including mangroves and rocky reefs, are profoundly influenced by tides, regular rises and falls in the depth of the water column. To comprehend tides, we have to consider two forces, centrifugal motion and gravity, and the balance between them. At the equator, our planet spins at 1670 kmh^{-1} (a staggering 464 ms^{-1}). So why do people not fly off into space? The answer, of course, is gravity. It is these two forces (centrifugal and gravitational) that combine to produce the Earth's tides.

Consider the Earth/moon system and pretend (for simplicity) that the Earth is entirely covered by water. The Earth and the moon both rotate as a combined system. The centre of gravity of the spinning Earth is at its centre, as is that of the moon, but the combined centre of gravity for the Earth/moon system is about 2100 km below the Earth's surface (Figure 18.1a). Gravitational forces depend upon the distance between the Earth and the

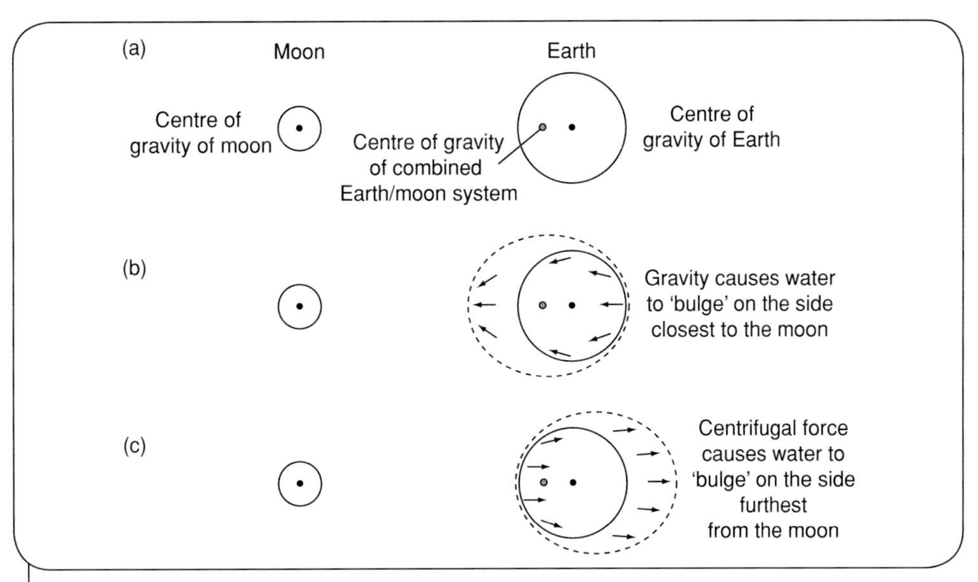

Figure 18.1 How tides are caused. (a) The centres of gravity of the Earth, moon and combined Earth/moon system. (b) The bulge caused by the moon's gravity and (c) the bulge caused by the centrifugal force of the Earth/moon system. (Source: B. Cale)

moon. The gravity on the side of the Earth closest to the moon is greater than that on the other side. This causes water to 'bulge' on the side of the Earth nearest the moon (Figure 18.1b).

In terms of centrifugal motion, the forces produced by the spinning Earth/moon system are weakest on the side of the Earth facing the moon and strongest on the opposite side. This causes water to 'bulge' on this side (Figure 18.1c). The combined gravitational and centrifugal forces produce two bulges in the water on the Earth on each side (Figure 18.2). This is called the 'tidal bulge'. On our theoretical Earth, we can now see why, in any particular location, you would expect two high tides and two low tides each day (or 'semidiurnal' tides). People at point A (Figure 18.2) would experience a low tide at time zero. Because the Earth rotates, 6 hours later they would have travelled one-quarter of one revolution, and would experience a high tide. In another 6 hours they would have travelled half a revolution and would experience another low tide (and so on until they return to their starting point).

Another influence on tides is the interaction between the Earth and the sun. Although it is much further away than the moon (150 000 000 km compared to the moon's 383 000 km), the sun is more massive (27 million times). The combined gravitational and centrifugal forces between the Earth and the sun also cause a tidal bulge. The Earth/moon and Earth/sun systems both combine to affect tides. This is most easily observed when we have spring and neap tides. Spring (or high) tides occur when the Earth, moon and sun are in alignment. The tidal forces are additive and they combine to produce high tides (Figure 18.3). When the moon is perpendicular to the axis of the sun, the two forces are working against each other and we experience lower (or neap)

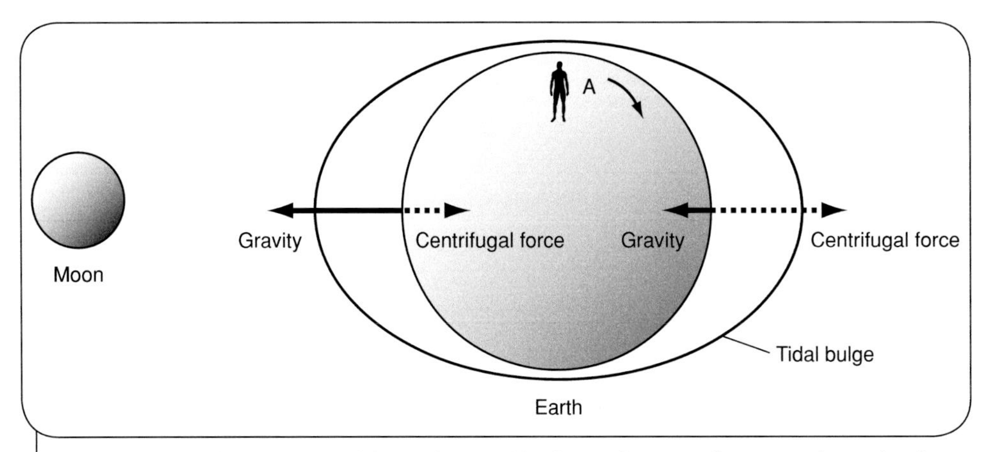

Figure 18.2 The tidal bulge around the Earth caused by the combination of gravity and centrifugal force. (Source: B. Cale)

tides (Figure 18.3). As the moon rotates about the Earth this causes rises and falls in tidal height over 14 days (Figure 18.3).

Topography, bathymetry and barometric forces also affect the tides at any particular location. These effects mean that not all high tides occur twice a day as in our theoretical example (Figure 18.2). High tides can occur once per day or they can display a mixture of timing behaviours. The effects of bathymetry are most obvious on wide continental shelves. A tide acts like a wave and shoals up to produce a very high tide. In northern Western Australia, for example, the shelf reaches Indonesia, and this area experiences very high tides (up to 10 m).

'Barometric' tides are not really tides but are the effects on water level caused by high and low pressure systems that influence the weather. A high pressure system sitting over

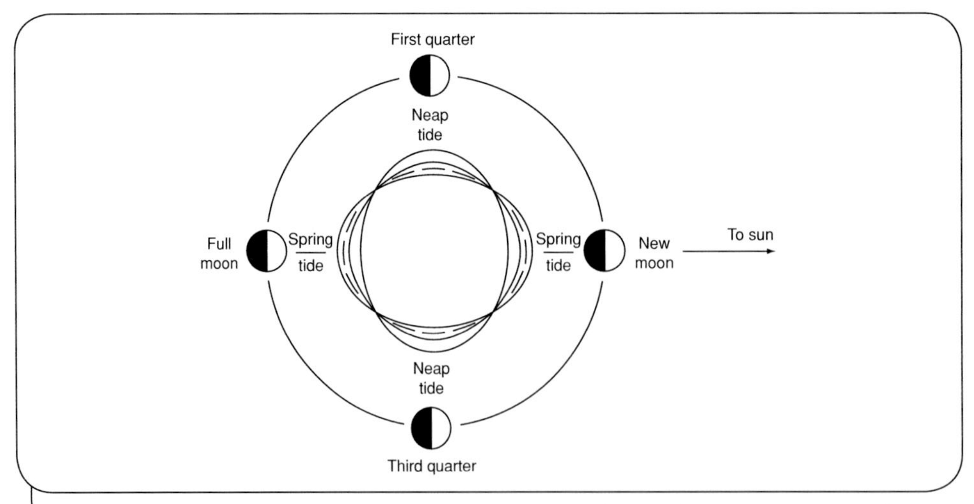

Figure 18.3 The interaction between the sun and the moon influences the tides. (Source: B. Cale)

the ocean pushes the water down. Low pressure systems 'suck' the water up. For every 1 hPa of pressure, seawater level varies by 1 cm. In a region that has a pressure range from 980 hPa to 1020 hPa, water will move by up to 40 cm. Some areas such as Fremantle in Western Australia, with low astronomic tides (about 0.5 m), may have barometric pressure water-level variations of equal magnitude.

Other factors affecting seawater level and motion

Wind setup

Wind also affects ocean water levels. When wind blows, it moves water in the same direction. This is called 'wind setup' and is most obvious during storms. Cyclonic winds (up to 270 kmh^{-1}) can cause storm surges up to 20 m. This, combined with the usual barometric rise caused by the low pressure system (up to 1 m in severe cases), can raise the water level even further.

Waves

Waves are temporary fluctuations in sea level usually caused by wind. 'Swell' is generated by storms at sea distant from the land. Water-level variations radiate away from the influence of the winds at the storm centre and form into long-period waves. The period of a wave is the distance between its crest and that of the next wave. In swell, the period is usually greater than 10 seconds. When swell waves reach the coast they rise to form waves that can be surfed. 'Sea' waves are generated within the influence of the wind. They are typically chaotic, with short periods (1–2 seconds) and usually of low amplitude. In Australia we also call sea waves 'chop'. These two waves are recognised in boating forecasts, which report a swell and a sea height. If the two are high and they combine, they form quite spectacular coastal waves.

Tsunamis are waves developing after underwater earthquakes, volcanoes or landslides. The ocean floor shifts, displacing water, which rapidly moves away from the epicentre. In the open ocean, tsunamis might hardly be noticed because their heights are only half a metre but they travel at tremendous speeds (up to 700 kmh^{-1}). When they reach the coast, they rise into massive waves that surge up to 30 m and cause colossal destruction, as most recently seen in the Boxing Day tsunami of 2004 off Indonesia's Aceh coast. This tsunami killed at least 275 000 people and was caused by one of the largest earthquakes recorded in human history. Tsunamis most commonly occur in the Pacific Ocean.

Coriolis effect and upwelling

The Coriolis effect is caused by the spinning Earth deflecting water that is in motion. It causes moving water to veer to the right in the Northern Hemisphere and to the left in the Southern Hemisphere. The Coriolis effect near the coast causes water to rise from the deep ocean. 'Upwellings', as they are called, bring cold, nutrient-rich water to

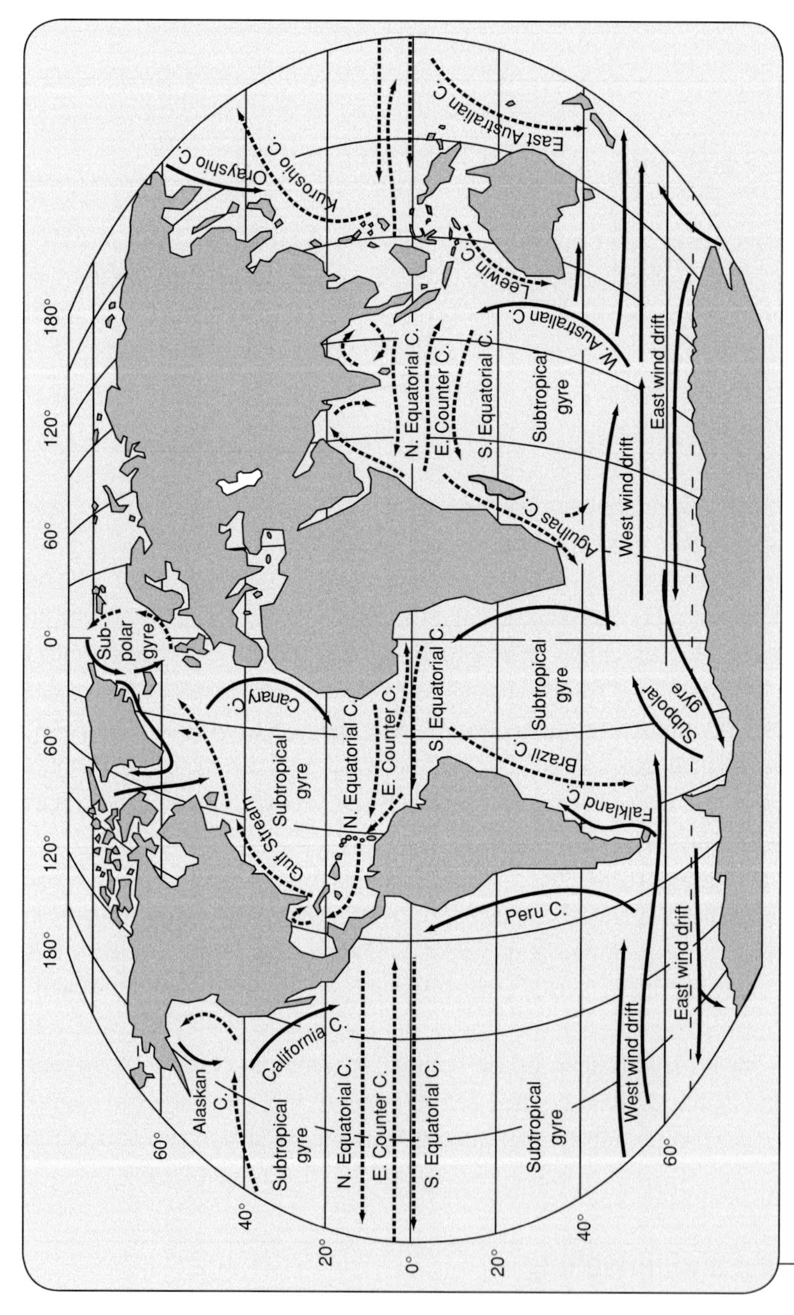

Figure 18.4 Oceanic wind-driven surface currents during the Southern Hemisphere summer (February and March). Dotted lines are warm currents and solid lines are cold currents. (Source: Adapted from Thurman, H. V. (1996). *Essentials of oceanography* Prentice-Hall Copyright © Pearson Education Inc. Reprinted by permission.)

the surface on coasts. This stimulates phytoplankton growth, benefiting other creatures higher up in the food chain, and may support substantial fisheries (Plate 18.1). Pelagic (open ocean) seabirds such as the shearwaters of New Zealand and southern Australia congregate around upwellings, feeding on pilchards and krill. Humans harvest fish such as sardines and anchovies. Uniquely, significant upwelling does not occur on Western Australia's coast because it is suppressed by the Leeuwin current (Figure 18.5). This is a warm water current, anomalous compared to those along other western coasts, that in most years flows southward parallel to the coast, bringing warm, nutrient-poor waters from the equatorial region. It is driven by relatively high water levels in the Indonesian archipelago, and raises water levels in the south by as much as 20 cm. This warm current may decrease marine productivity off the west coast, but it allows coral reefs to grow as far south as Perth in Western Australia.

Major marine habitats

Open-water and coastal habitats

The open ocean forms a habitat for a huge range of organisms because it comprises most of the sea. The pelagic zone, strictly speaking, is the water column of the open sea, although the term is now often used to describe the water column of the sea in general and thus this habitat can be either close to the coast or in the open ocean.

The open ocean receives few external nutrient inputs because it is far from the land from which nutrients are derived. It is usually deep, and waste material and particulate nutrients from the land sink before recycling. It is often termed oligotrophic (poor in nutrients). In contrast, coastal waters may receive high nutrient input from land run-off, erosion, or sediment resuspension in energetic environments. These areas are called eutrophic (high in nutrients) and they are highly productive.

Most human impacts, as well as our fish resources, are associated with the high biodiversity of coastal regions. Thus much of our discussion on marine habitats will focus upon coastal regions (Plate 18.2). Once the constraints of light, temperature, salinity and water motion have been considered, the primary characteristic of marine habitats is the kind of substrate present. Substrate availability is so important that we should focus upon the different substrates available before we discuss the organisms associated with them. This provides a convenient platform to consider marine primary producers in detail. Simplistically, substrates are divided into hard and soft. Hard substrates include rock, coral and artificial surfaces such as jetty pylons, rigs, boats and even plants and animals. Soft substrates are grouped into sand and mud.

Rocky substrates

Rocky substrates vary from easily weathered limestone and volcanic material to lithified dunes, often in tropical areas where sand is rapidly cemented, to more resistant material such as (commonly) granite and (occasionally) basalt. We can characterise reef substrate by depth or tidal influence. Rock forms high (or intertidal) reef, low reef and pavement. Low reef is permanently underwater and pavement forms a veneer comprising a shifting mosaic of rock and sand. Rocky reefs often include networks of caves and crevices. Insufficient light reaches these areas to support plant life, but they are often rich in sponges, ascidians and other animals.

Coral reefs

Many animals, such as oysters and serpulid tubeworms, form hard surfaces, but the best examples are those that form coral reefs. Although corals grow from the poles to the tropics (and usually require some rock to first colonise), most reef-building corals are restricted between latitudes 30° south and north of the equator. Coral reefs are common in all tropical waters with water clean enough for them to grow in. They are usually restricted to depths less than 4 m because they have mutualistic algae (known as zooxanthellae) living within them and providing sustenance (Box 18.1). Fringing reefs grow on coasts in regions where there is restricted fresh water or sediment input, and barrier reefs grow parallel to the coast 10 to 160 km offshore.

Rubble

Rubble consists of broken or water-worn material. Although not an extensive substrate (except perhaps in coral reef zones), it is important in certain areas. For example, the

Box 18.1 Coral partners – the zooxanthellae

Zooxanthellae, which are algae, live mutualistically within coral polyps, and assist them by producing nutrients through photosynthetic activities that provide useful carbon and enhance calcification. The host polyp provides a protected environment and carbon dioxide for the algae to grow. This combination allows corals to compete with other organisms on the reef, such as macroalgae, because the corals can effectively photosynthesise in the day and feed at night by predation. The colour of coral reefs is derived from the zooxanthellae living within them, not from the clear tissue of the polyps.

Many corals require warm sea water in which to grow. This sometimes causes problems when temperatures increase, because zooxanthellae are adversely affected by high temperatures or high light intensity. Although the actual mechanisms are still hypothetical, it is believed that at high temperatures the enzyme systems of the zooxanthellae become disrupted, particularly those protecting against oxygen toxicity (a by-product of their photosynthesis). When this happens, the corals expel them to avoid tissue damage and subsequently they lose their colour. Hence, this phenomenon is known as 'coral bleaching'.

islands of the Houtman Abrolhos in Western Australia consist largely of coral rubble supporting mangroves.

Artificial substrates

Artificial substrates such as the hulls of ships, oil rigs and jetty pylons are common, and can form important habitats. In the Gulf of Mexico, for example, up to 80% of the hard substrate available is provided by oil rigs. In the same area there are at least 50 boats sunk by submarines in World War II. Many aquaculture operations, such as those growing algae, mussels and oysters, provide substrates such as ropes or nets for organisms to grow upon until they are harvested.

Soft substrates

Soft substrates occur on sandy beaches, estuarine embayments and deeper areas. Depending upon wave energy and depth, they are colonised by several organisms. The size of the sediment particles can be as important for biota as its origin. Marine sediments are classified as calcareous or terrigenous. The first is usually biogenic, derived from hard parts of living creatures such as shells, and consists of calcium carbonate. Terrigenous sediments are often finer (they can be sands, silts or clays) and are derived from the land via river drainage, erosion or wind. These sediments are derived from the breakdown of rocks such as quartz or feldspar.

Sediment particle size is significant with regard to both water motion and habitat. The standard definition of sand is particles ranging in size from very coarse (2 mm in diameter) to very fine (0.06 mm). Silt is next, with a grain size of 0.06 to 0.004 mm, followed by clay (smaller than 0.004 mm). The difference in settling velocity of these particles is dramatic. Sand takes around 30 minutes to settle to the bottom in 50 m of water. Silt takes two and a half days and clay takes around 8 months. Sediment affects water-column clarity and its movement affects the growth of coral, seagrass and mangroves.

Sand

Sand is the most common substrate on beaches and in coastal areas with energetic hydrodynamics. The grain size is determined by the intensity of water motion. Sheltered embayments have finer sands and rougher areas more coarse sands, although both can occur on the same beach in different seasons. Sand particle size becomes finer the deeper the ocean, because of reduced wave energies. Although sand may initially appear lifeless, microscopic organisms including bacteria, invertebrates and algae live on and between sand particles. Deeper and more stable sandy areas are colonised by seagrasses and various soft-bottom communities.

Mud

Silt and clay are at the lower end of the sediment grain-size spectrum. Mud is the sediment component combining these two materials, although it may also contain a small amount of sand. Because it is so fine and easily suspended in shallows, mud is associated with deeper areas. Below 30 m, marine sediments usually consist of very fine sand or mud.

Estuaries can be much calmer than oceanic areas and often their substrates consist of fine, terrigenous sediments at shallow depths. Some tropical regions, particularly those with high rainfall, have muddy coastal marine sediments and turbid coastal waters. The tropics are usually associated with high rainfall that flushes terrigenous sediment, along with fresh water, into marine environments. This is often exacerbated by agriculture, land clearing and associated erosion. One of the reasons that fringing coral reefs do not occur in such areas is the combination of fresh water and mud, which impede coral growth.

Lack of light, often associated with muddy sediments in deeper ocean areas, means that the only colonising organisms are animals, often termed 'soft bottom macrozoobenthos'. Another sediment component in these areas is organic detritus, such as that derived from macroalgae and animal waste, especially from more productive coastal habitats. Two key differences between hard and soft substrate environments are the lack of vegetation in soft substrate environments and the fact that organisms usually live in soft sediment, not on it. Grain size therefore has a large effect on which organisms actually colonise it.

Marine producers and marine communities

Marine organisms are often described according to their location within the marine environment. Organisms that are found on the sea floor are called benthic, whereas those in open water are called pelagic. Benthic animals may use the sea floor quite differently to pelagic animals. Infaunal animals live completely in the sediment, for example, and epifaunal creatures live on other animals. In the water column, different environments are recognised. 'Neuston' refers to creatures at the water's surface, and plankton and nekton to those living within the water column (Figure 18.6). Planktonic organisms are unable to maintain their position independently of water movements, whereas nektonic organisms are able to move independently of water currents through active swimming.

Primary production is the process whereby sunlight and nutrients are converted into biomass, which forms the basis of food chains. Plants and algae containing chlorophyll are important oceanic primary producers, although some bacteria also use pigments and sunlight to generate biomass.

Biomass and productivity

It is sometimes thought that larger organisms must be more important than smaller ones in an ecosystem. The biomass of an organism, whether plant or animal, is the amount of

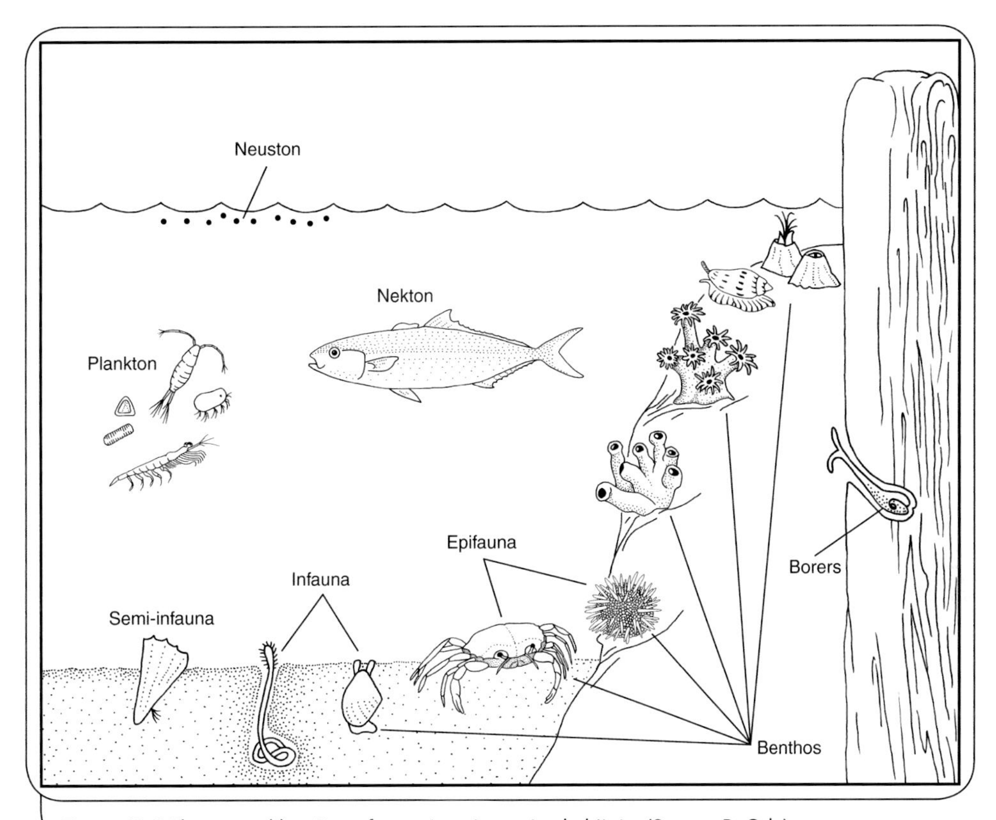

Figure 18.5 The general location of organisms in marine habitats. (Source: B. Cale)

living matter it contains, and is often represented as the amount of material per unit area or volume, for example the number of whales in a hectare of ocean surface, the quantity of seagrass per square metre of ocean floor or the amount of phytoplankton per litre of sea water.

Productivity, on the other hand, is the amount of living material produced by an organism as it grows. When you mow your lawn in summer, you remove the lawn's production. Because you mow the lawn, its biomass might stay very much the same throughout the year, but it is more productive during summer. Primary productivity is the amount of material produced by autotrophs per unit area per unit time. This is a rate, for example grams produced per square metre in a year or growth in a day. Productivity is also referred to as 'turnover'. Secondary productivity refers to animal productivity.

Knowing and appreciating the difference between biomass and productivity, and the fact that they can vary dramatically between organisms, is paramount when considering ecological function in the ocean. As a general rule, large organisms may be slow in their turnover (or production). Smaller organisms (microscopic algae or bacteria) often turn over very rapidly indeed and may even double their biomass in a few hours.

Algae

Marine algae can be functionally divided into those consisting of single cells or filaments (the phytoplankton or microphytobenthos) and larger species known as macroalgae. Phytoplankton and microphytobenthos are found in the water column or on soft sediments, whereas macroalgae almost always occur on hard substrates (rocky reefs, artificial substrates, seagrass leaves or other algae).

Phytoplankton

Phytoplankton occur in the water column and usually consist of single-celled algae or sometimes filaments, both moved mainly by tides and water currents. Although individual organisms may not have high biomasses, they do have rapid turnovers and they may also cover vast expanses. Microscopic algae are at the base of the food chain and are consumed by animal plankton (zooplankton) before they too are consumed by larger predators. In coastal regions influenced by upwellings, the phytoplankton respond to the nutrients and their growth begins the food chains supporting major fisheries. There are many different groups of phytoplankton and they are probably the biggest consumers of carbon dioxide on Earth (see Box 10.2).

Microphytobenthos

Microphytobenthos are a group of algae, often diatoms, growing on soft substrates such as sand or mud (hence the term 'benthos'). Sometimes the coloured layer seen on top of the sand in areas with relatively stable sediment is caused by these organisms. They are food for creatures living in and around the sand. They are more often found in shallow warm areas with high light and respond rapidly to favourable conditions. Sediments that are stable because of a seasonal reduction in water motion may be rapidly colonised by these organisms.

Macroalgae

The macroalgae are a diverse group including representatives from the red (Rhodophyta), brown (Phaeophyta) and green (Chlorophyta) algae (Chapter 10). Whereas a few macroalgal species are free floating in the water column (e.g. *Sargassum* in the Sargasso Sea), most attach to a hard substrate (Figure 18.6). This is usually rocky reefs or pavement, but any hard or stable substrate will do. Jetty pylons or seagrass leaves are often covered in algae.

In general, macroalgae can be loosely classed into functional groups based upon the robustness of their thalli. The green algae tend to be less robust species that respond rapidly to light, temperature and nutrients. They often grow in shallow, intertidal areas and have structures based upon filaments or simple sheets. The brown algae, which include kelps, are more robust. Kelps grow in extremely rough areas with cooler water temperatures. The giant kelps off the Californian coast grow up to 6 m long in water 4 m deep, and form an extensive habitat occupied by over 1000 species of creatures including sea otters. Red algae consist of a number of different forms and many groups grow at greater depths than green or brown algae.

Macroalgal beds are very important in the world's oceans. They provide a valuable habitat, are directly grazed by many organisms and contribute significant organic material to detrital pathways. Large kelps are often three to five times more productive than seagrasses. Off the coast near Perth (Western Australia), two species of brown algae (*Ecklonia radiata* and *Sargassum*) produce twice as much organic matter as the extensive seagrass meadows in the region.

Figure 18.6 Reef macroalgae. (Source: Photograph by E. Paling)

Seagrasses

Seagrasses are true aquatic angiosperms, most spending the greater part of their lives completely submerged (Figure 18.7, Plates 18.3 and 18.4), although some may be exposed at low tide. Seagrasses grow on soft substrates (sand or mud) in the ocean or in areas influenced by salt water such as estuaries or rivers. Because they require a stable substrate, water movement is one of the main physical factors limiting their distribution. Most seagrasses are restricted to calm areas, or depths of water in which water motion does not disturb the sediment. Thus sand movement, wave energy and desiccation limit the upper range of seagrass distribution. The lower range (the deepest depth to which seagrass can grow) is limited by light. This very much depends upon the clarity of water, but generally seagrasses will grow to depths that receive 10% of the land irradiance. This can vary from just a few metres in very turbid areas to up to 50 m in clear water.

Figure 18.7 Seagrass. (Source: Photograph by E. Paling)

Seagrasses evolved around the mid-Cretaceous (65 million years ago) and probably arose from some form of salt marsh plant. There are 12 genera of seagrass and around 35 species in the world's oceans and estuaries. Australia has a good representation of seagrass with 11 genera and over 25 species. Seagrasses have a worldwide distribution and it is useful to divide them into those of tropical and temperate zones.

Like terrestrial grasses, seagrasses have a rhizome and root system that grows underground and from this arise either leaves or stems. They are adapted to an aquatic environment. Their leaves have much thinner cuticles than land plants do, and they have no stomates (the pores that allow gas exchange in land plants), and so can transfer dissolved gases directly across the leaf epidermis. The ocean is always hydrodynamically active, and the flexible leaves or stems of seagrasses can withstand the constant motion of being waved back and forth without snapping. Their pollen survives underwater, and because sand on the ocean floor is not well oxygenated, they usually have roots and rhizomes filled with air-containing tissue, and oxygen diffuses from the roots into the sediments.

Seagrass meadows can be very extensive and highly productive. The largest seagrass meadow in the world, the Wooramel Bank, is located in Shark Bay, Western Australia, and has an area of 1030 km^2. Seagrasses can produce up to 2 $kgm^{-2}y^{-1}$ of organic matter (compare this with 3 $kgm^{-2}y^{-1}$ for tropical rainforests). The leaves contain tannins, ensuring that all but the smallest (and softest) species are unpalatable to large grazers. Very few animals eat seagrass in temperate areas, but in the tropics the softer, smaller species are eaten by dugongs, manatees and green turtles.

As well as providing a rich source of organic material to grazers and detritivores, seagrasses perform many other important functions. Seagrass leaves reduce water motion, preventing erosion and trapping and binding sediments and organic detritus. They also provide stable surfaces for colonisation by epiphytes (non-parasitic algae and animals that grow on the leaves or stems of other plants). These epiphytes provide a rich food source for a wide variety of fish and invertebrates. As they trap and bind organic material, seagrasses also help fix nutrients. Finally, the three-dimensional structure of seagrass meadows helps provide habitat and shelter (e.g. nursery environments) for many fish and other organisms.

Seagrass ecosystems were believed to be fragile because of their decline in areas of the world where high nutrient inputs occurred. However, research in the last decade showed that seagrasses are quite dynamic systems able to either naturally expand into suitable areas or be rehabilitated (Box 18.2).

Mangroves

Mangroves are angiosperms that, like seagrasses, grow on soft substrates (sand or mud) in coastal areas influenced by salt water (Figure 18.8, Plate 18.5). Unlike seagrasses, they are emergent trees or shrubs. There are about 70 species in 19 genera from 16 to 20 different families. There is good evidence that mangroves arose in the mid-Cretaceous when the angiosperms were evolving, and that they were primarily derived from an

Box 18.2 Rehabilitation of seagrass meadows

Most international efforts to rehabilitate seagrass meadows have been carried out at small scales using manual methods. In the Northern Hemisphere, the success of these methods was largely because of the properties of the seagrass species used and the calm areas planted. In Australia, seagrass transplantation has largely failed, mainly because of the need to establish plants in high wave-energy areas. Western Australia is one of the few places in the world where seagrasses grow in the open ocean. The ECOSUB program was initiated to allow transplant of a unit of seagrass that was large enough to withstand considerable wave energy and restore large areas. This required a mechanical device to extract and plant seagrass (Figure B18.2.1).

Figure B18.2.1 Transplanting seagrass. (Source: Photograph by E. Paling)

The ECOSUB system was the first mechanical seagrass transplantation device developed, it moves the biggest transplanted unit (or sod) in the world (0.5 m^2 in area), and it was involved in the world's largest area of seagrass rehabilitation ever considered. Over 5000 sods were transplanted by ECOSUB modules. Functioning, transplanted seagrass meadows have now been on the ocean floor off Perth for more than 5 years and further work is under way to refine mechanical planting.

area just north of Australia. They display what is known as 'convergent' evolution, meaning that they arose from several different ancestors, but share a range of attributes because of adaptation to living in salt water on soft substrates.

Mangroves form one of the most productive ecosystems in the world. They are able to produce some 3 kg of organic matter per square metre per year (equivalent to a tropical rainforest). High production must be sustained by a large supply of nutrients, which are incorporated into mangrove biomass and organic products, and there is evidence that mangroves help to support adjacent offshore ecosystems, including fisheries.

There are various constraints to mangrove distribution along the coasts. The first is temperature. Most mangroves occur within the 24°C isotherm, although there are some notable exceptions in Western Australia, South Australia and New Zealand. All mangroves require a substrate of mud or fine sand in areas with tides usually exceeding a daily range of 1 m. These environments usually possess gently sloping shorelines. Most

Figure 18.8 Mangroves. (Source: Photograph by E. Paling)

mangroves cannot withstand the full force of the ocean and often occur in protected embayments or tidal creeks. One species (*Avicennia marina*) is most often found at the seaward edge of mangroves because of its ability to cope with water movement. It is also temperature tolerant and grows in colder areas.

Mangroves often grow better in salinities less than sea water. This can be attributed to their ancestry; all probably originated from estuarine trees that moved towards the sea. However, few mangroves can survive pure fresh water, requiring some salt in the water to grow effectively. The final constraint to mangrove distribution is propagule dispersal. Amenable ocean currents move propagules to areas in which they can grow. This is made clear by the observation that, although introduced mangroves grow well in Hawaii, they do not occur there naturally because of the lack of favourable ocean currents.

Mangroves perform many useful functions in their environments. As they reduce water motion they allow fine particles to settle within their root systems and build new land. They form a refuge or nursery area for many fish species and provide valuable habitat for birds, bats and insects. Finally, the great South-East Asian tsunami of 2004 showed that mangroves protect coastlines from storm and wave damage. Unfortunately, this was noted because many Asian countries have removed coastal mangrove forests to use the areas for aquaculture.

Salt marsh

A salt-marsh plant could be likened, in a way, to a 'temperate mangrove'. These plants usually occur in greater numbers outside the 24°C isotherm, but require a tidal range

Figure 18.9 Salt marsh. (Source: Photograph by E. Paling)

of around 0.5 m or more, a soft substrate such as sand or mud and protected shores (Figure 18.9). Salt marsh plants can be either shrubs (chenopods) or emergent monocots such as reeds and rushes. Shrubs tend to occur in the tropics and in temperate regions. In the tropics, these shrubs occur behind mangroves in areas that are sufficiently high in elevation to be not too salty. In temperate areas they occur on the banks of rivers and embayments up to the water's edge.

Salt-marsh reeds or rushes occur predominantly in the Northern Hemisphere. Spartina marshes (*Spartina alterniflora*) form extensive habitats, particularly on the eastern coast of North America and in Europe. In Australia other suites of marsh-grass species occur, but not to the same extent as in the Northern Hemisphere. Salt marshes are very important providers of primary productivity to near-shore ecosystems where their detritus forms the base of the food web.

Bacteria

Bacteria can take up almost any form of material, use it for growth and, in the process, make nutrients available for other organisms. They were the first organisms to develop in the oceans, have occupied at least half of the Earth's biological history, and are probably responsible for around half of the ocean's primary production. For example, blooms of the cyanobacterium *Trichodesmium* can cover 40 000 km^2.

Bacteria drive the nitrogen, phosphorus, carbon and sulfur cycles in the water column and sediments of the ocean. For example, bacteria transform nitrogen, which cannot be processed by other organisms, into accessible forms. A normal marine community could consist of bacteria converting ammonium (and other forms of nitrogen) to nitrite, bacteria converting nitrite to nitrate, plants taking up dissolved inorganic nitrogen, plants and animals releasing ammonium and organic nitrogen when they decay, and animals excreting ammonium and other nitrogenous wastes.

Bacteria also drive the detrital cycle. They decompose organic carbon (both particulate and dissolved) that cannot be ingested by larger organisms, including liquid wastes or material from animals and plant cells. Microscopic flagellates and ciliates consume these bacteria, helping to recycle organic matter back into the marine food web.

Chapter summary

1. The salinity and temperature of sea water determine its density, and this has an important role in determining the circulation and mixing of water in the open ocean. Salinity and temperature are also very important determinants of the distribution of marine organisms.
2. Light intensity is reduced at depth in water, and the photic zone (where there is sufficient light for algae, seagrasses or corals to grow) depends on water clarity.
3. Tidal oscillations of water level are caused by the gravitational pull of sun, moon and Earth, interacting with centrifugal forces generated by the spinning of the Earth. The Coriolis force, also caused by the Earth's rotation, leads to upwellings of cold, nutrient-rich water to the surface of the ocean.
4. Marine habitats may be divided into open water or coastal. The primary characteristic of coastal habitats is the kind of substrate present. Hard substrates include rocky reefs, coral reefs, rubble and man-made objects. Soft substrates can be divided into sand or mud.
5. Marine organisms may live on the sea floor (benthic) or in the water (pelagic). Benthic animals are often divided into those living completely in the sediment (infaunal) and those living on other animals (epifaunal). Pelagic organisms may be divided into those living at the surface of the water (neuston) and those living in the water column; water, column organisms may drift with water movements (plankton) or actively maintain their position by swimming (nekton).
6. Productivity is the amount of living material produced by an organism as it grows, and biomass is the amount of living matter present. Plants containing chlorophyll are the major marine primary producers.
7. Phytoplankton and microphytobenthos are microscopic algae which occur in the water column or on top of soft sediments, and are at the base of the marine food chain. Macroalgae usually grow attached to hard substrates, such as submerged rocks, including reefs, and man-made objects such as jetty piles.
8. Seagrasses are aquatic flowering plants which grow rooted in sediment, when sufficient light is available for their survival. Seagrass meadows can be very extensive and form highly productive ecosystems.
9. Mangroves are woody higher plants with leaf canopies projecting above the water surface, usually rooted in mud in warm areas of the world subjected to regular tidal inundation. They build new land, form a refuge for fish and provide valuable habitat for terrestrial animals. Salt marshes, made up of salt-tolerant shrubs and rhizomatous plants (rushes or reeds) rooted in sand or mud and projecting into the air, usually occur in cooler climates than mangroves do.
10. Bacteria occur widely in marine habitats, playing key roles in the decomposition of organic matter. Bacteria drive the nitrogen, phosphorus, carbon and sulfur cycles in the ocean's water column and sediments.

Key terms

artificial substrates
barometric tides
benthic
chenopods
Coriolis effect
epifauna
halocline
infauna
mangroves
nekton
neuston
osmosis
pelagic
photic zone
plankton
salt marsh
seagrass
terrigenous
thermocline
tide

Test your knowledge

1. List those major physical features of marine environments that are important in controlling the abundance and distribution of organisms.
2. How do salinity and temperature influence ocean circulation?
3. Explain how tides are created by a combination of the centrifugal force arising from the spinning of the Earth, and the gravitational pull of the Earth, the moon and the sun.
4. What is the Coriolis effect, and what influence does it have on marine life?
5. What is primary production, and what types of organism are the primary producers in marine habitats?
6. What are the main physical and biotic features of (a) seagrass habitats, (b) mangrove habitats and (c) salt marsh habitats?

Thinking scientifically

If humans harvest only 0.2% of the ocean's annual productivity, why are so many fisheries depleted or on the point of collapse?

Debate with friends

Debate the topic: 'Conservation of seagrass meadows is no longer important because damage can be rehabilitated'.

Further reading

Ballestra, L. and Descamp, P. (2007). *Planet ocean: voyage to the heart of the marine realm.* National Geographic.

Brearley, A. (2005). *Ernest Hodgkin's Swanland: estuaries and coastal lagoons of south-western Australia.* University of Western Australia Press, Perth.

Fox, A. and Parish, S. (2007). *Wild habitats: a natural history of Australian ecosystems.* ABC Books/Steve Parish Publishing, Sydney/Archerfield, Queensland.

19

Marine lifestyles

Eric Paling and David Ayre

The hitchhiker's guide to the oceans

Shipping practices rewrote the travel guides for many marine organisms because many of them are well suited to producing stowaways, forming exotic populations far beyond their normal ranges. With increased shipping to Australia, marine creatures that have never lived here before are arriving as dispersive spores or larvae in ships' ballast water or directly as adults attached to their hulls.

Luckily, not all introduced marine species are harmful and only half a dozen or so are true pests. They range from several toxic dinoflagellates (planktonic creatures secreting poisons) to tube worms, macroalgae and sea stars. Many pests successfully compete for substrate or food used by native creatures. For example, the European fan worm (*Sabella spallanzanii*) fouls hard surfaces and commercial oyster beds, competing with native animals for food in the water column. The northern Pacific sea star (*Asterias amurensis*) is a voracious predator, eating many native animals including commercial and non-commercial shellfish.

How did these creatures become pests? The answer is quite simple. The Earth's oceans have two temperate regions, in the Northern and Southern Hemispheres, divided by warm equatorial waters. The warm water around the equator is a natural barrier to migration. First, it is warm and may exceed the temperature tolerances of the cool-temperate organisms and secondly the surface ocean currents are restricted to each hemisphere with limited exchange between them. Organisms transported to a new hemisphere often find themselves in a utopian environment lacking predators and parasites, and they out-compete other organisms. Once established, they are often difficult or impossible to remove.

Chapter aims

In this chapter we explore the lifestyles of marine organisms. Salinity, temperature, light (or depth) and water motion affect the distribution of marine habitats and the organisms inhabiting them (Chapter 18). Given these physical conditions, marine organisms must maintain temperature regulation and water balance, dispose of wastes, obtain food and oxygen, and reproduce. We discuss solutions adopted by marine organisms to these life problems. On a cellular level, most plants share the same adaptive responses as animals, so animals are the focus of this chapter.

Temperature regulation

Apart from the overarching effects of temperature mentioned in Chapter 18, most marine animals must cope with changes in seawater temperatures. Animals are usually divided into two extremes: poikilotherms and homeotherms (see Chapter 15, Box 15.2). Subtidal invertebrates and most fish are poikilotherms. As the external temperature increases, so too does their metabolic activity (Figure 19.1). Every process or reaction in an organism has a rate and it will vary with temperature. The change in rate is defined in terms of a 10°C change in temperature and this value is known as Q_{10}. This value is calculated as the respiration rate at the higher temperature divided by the respiration rate at the lower temperature, raised to the power 10 divided by (the higher temperature – the lower temperature). For example, if a coral respires 20 molecules of CO_2 per minute at 25°C, and 40 molecules CO_2 per minute at 35°C, then its $Q_{10} = (40 \div 20)^{10/(35-25)} = 2$. So for every 10°C increase in temperature, the rate doubles. Metabolic activity increases up to a point where cellular processes and tissues cannot survive. This does not necessarily have to be a high temperature: the upper lethal temperature of the Antarctic ice fish (*Trematomus* sp.) is 6°C. One of the advantages of poikilothermy is that an animal needs less food and can survive long migrations and prey shortages that

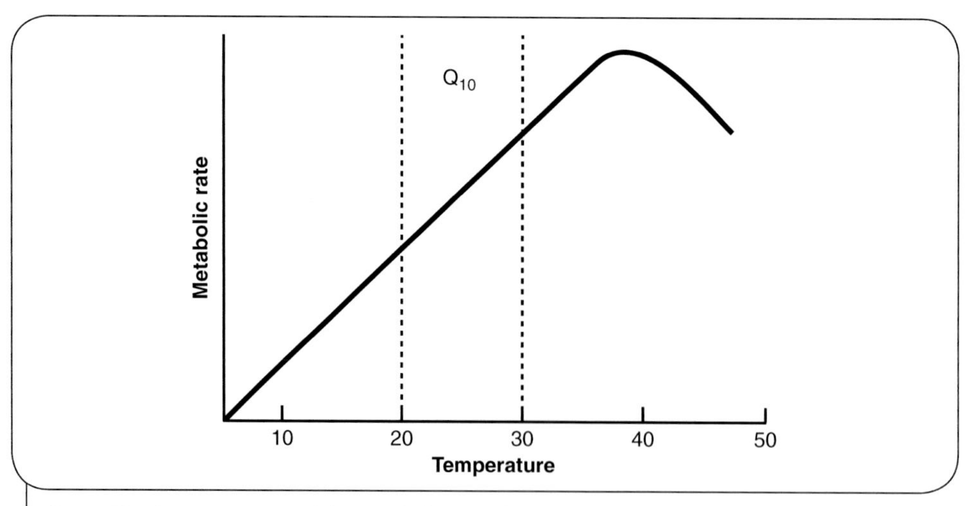

Figure 19.1 A representation of the metabolic rate of poikilotherms. (Source: B. Cale)

Plate 15.6 Although adapted to life on land, some reptiles have returned to an aquatic or semiaquatic lifestyle. (a) Mating sea snakes. (b) Turtle. (c) Freshwater crocodile. (Source: a, b John Huisman; c Lochman Transparencies)

Plate 15.7 The tuatara from New Zealand is of great interest because it shares many features with the first, ancestral reptiles. (Source: Lochman Transparencies)

Plate 15.8 Birds show many adaptations to flight, including their wings, feathers (for insulation and aerodynamics), weight reduction in bone structure, and beaks of light, strong keratin. (a) Osprey in flight. (b) Black swan. (c) Eclectus parrots nesting in a tree hollow. (Source: a–c Lochman Transparencies)

Plate 15.9 The prototherians or monotremes have the back vertebrae and pectoral girdle of a reptile and the fur of a mammal. They lay eggs and nurse their young on milk. The milk oozes from a patch of glands and is lapped up. (a) Echidna. (b) Platypus. (Source: a Mike Bamford; b Lochman Transparencies)

Plate 15.10 The metatherians or marsupials have a very short gestation and a tiny, poorly developed embryo that completes its development externally, usually in a pouch. (a) Koala. (b) Leadbeater's possum. (c) Squirrel glider. (d) Numbat. (e) Tasmanian devil. (f) Common brushtail possum. (g) Pouch young of western grey kangaroo. (Source: a–g Lochman Transparencies)

Plate 15.11 The young of eutherian mammals are usually born more highly developed after a longer gestation. (a) Domestic cat. (b) Elephants and humans. (c) Bottlenose dolphin. (d) Giraffes. (e) Zebra. (f) Rhinoceros. (Source: a M. Waters; b, d–f B. Dell; c H. Finn)

Plate 16.1 The frame in the foreground is a 1 m × 1 m quadrat for sampling vegetation density. In this example, multiple readings would be taken at intervals along the taped transect line. (Source: P. Ladd)

Plate 16.2 Many different kinds of traps are available for animal sampling. (a) A walk-in trap. Small animals encounter the low fence, move along it and are caught in the cage. (b) A pitfall trap (left) and an Elliott trap (right). Small animals encounter the fence, move along and either fall into a lined hole (pitfall trap) or enter a metal box (Elliott trap) where the door closes behind them. (c) Cage traps are larger than Elliott traps, but work on the same principle. (d) Trapped animals are often transferred to calico or canvas bags where they are held until measured and tagged. This chuditch is being released. (e) Fine mist nets suspended between poles are used to trap birds. (f) Flying invertebrates may also be caught using suspended interception traps. (g) Cameras, operated by a motion sensor, can provide a record of animals nearby. (Source: a–c, f–g M. Bamford; d R. Armistead; e (C) Lochman Transparencies)

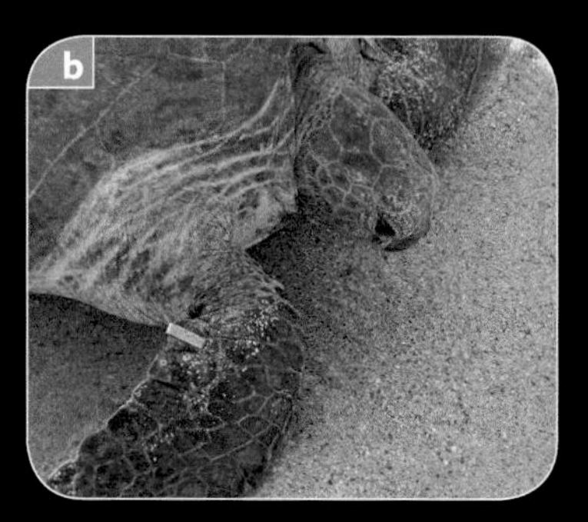

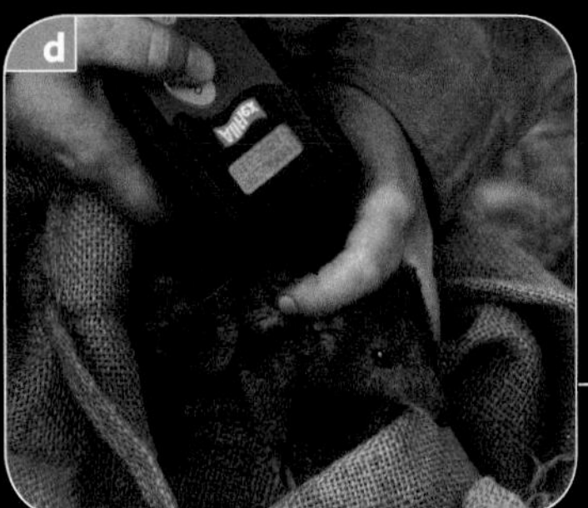

Plate 16.3 Trapped animals may be tagged so they can be identified if caught again. (a) Leg bands are used to tag birds. (b) This green turtle is identified by a metal tag on its flipper. (c) Animals may be powdered with a fluorescent dye so their movements can be traced. (d) Microchip implants are effective, long-lasting tags. (Source: a, b, d M. Bamford; c R. Armistead)

Plate 16.4 The introduced red fox and the feral cat are major predators of native Australian fauna. (a) Red fox. (b) The stomach contents of a feral cat include many small reptiles. (c) Extensive baiting is used to control foxes. Here baits are being prepared. (d) Shooting may also be used for control. These foxes were killed in a shooting campaign. (Source: a Lochman Transparencies; b M. Bamford; c, d L. Twigg)

Plate 17.1 Examples of mutualistic associations. (a) Foliose lichen growing over crustose lichen on granite rock. (b) Foliose lichen growing on a dead twig in a rainforest. (c) Light microscope section of lichen thallus showing the algal cells (stained red), loose fungal hyphae in the middle layer and a more dense layer of fungal hyphae at the base. (d) Small *Macrozamia communis* seedling with coralloid roots (arrow) that contain cyanobacterial cells. (e) Nodules containing *Rhizobium* bacteria on the roots of the clover *Trifolium*. (Source: a–e P. Ladd)

Plate 17.2 Example of organism interactions – herbivory and predation. (a) Upright *Drosera* with glistening glandular hairs that trap insects. (b) The rosette *Drosera erythrorhiza* also has dense glandular hairs on the broad leaves. (c) A *Nomia* bee species visiting a flower of *Senna* sp. to obtain pollen for her larvae. (d) Sandy inland mouse (*Pseudomys hermannsburgensis*) consuming legume seeds – a good source of energy. (e) Feral cats (*Felis catus*) are very destructive of native Australian wildlife, in this case a galah (*Cacatua roseicapilla*). (f) Cryptic grasshopper on red sand (Source: a R. Froend; b B. Dell; c T. Houston; d–f Lochman Transparencies)

Plate 17.3 Plant parasites. (a) The large Western Australian Christmas tree (*Nuytsia floribunda*) is parasitic on the roots of other plants. (b) The annular root attachment (haustoria) of *Nuytsia* attaching to the fine root of a *Banksia attenuata*. (c) Flowers of the parasite *Balanophora fungosa* in rainforest in north Queensland. (d) Flowers of the parasite *Pilostyles hamiltonii* appearing on the stem of its host, a species of *Daviesia*. (e) The aerial stem parasite *Amyema gibberulum* (a mistletoe) with red flowers on a species of *Melaleuca*. (Source: a–e P. Ladd)

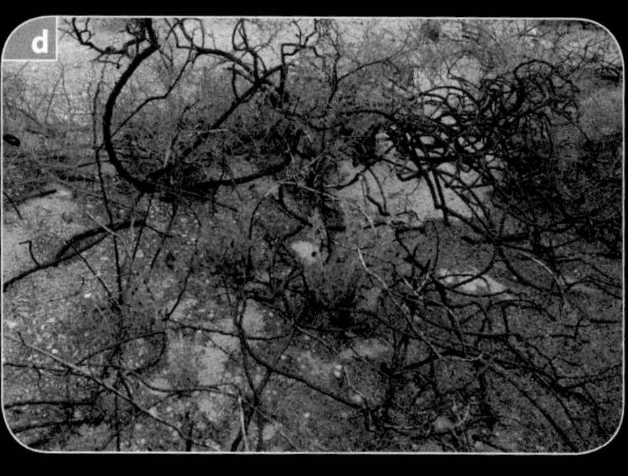

Plate 17.4 Fire in the environment. (a) Fire converts the chemical energy stored in plant material into heat and produces ash rich in inorganic minerals from the plants, (b) Rural die-back on the western slopes of the Great Dividing Range. This eucalypt displays death of the ends of branches and may ultimately die. The causes are complex and one may be a change in fire regimes. (c) Many plants can survive fire by resprouting (e.g. *Nitraria shoberi* after a fire on the Nullarbor Plain). (d) Many species that cannot survive fire produce abundant seed that is stimulated to germinate after a fire (eg. *Banksia lanata*). Cones on the dead stems were the source of the seeds that have germinated under the skeleton of the banksia. (Source: a–d P. Ladd)

Plate 17.5 (a) Areas of the Western Australian wheat belt that once supported freshwater lakes have been severely degraded by increases in salinity killing many species that used to inhabit the wetland areas. (b) Rural tree decline may be due to a number of ecological factors associated with increases in salinity, insect attack and even changes in soil nutrients. Here, the swamp gum (*Eucalyptus rudis*) shows evidence of canopy decline, although canopy regeneration by epicormic regrowth is also evident. The causes of this are still not well understood. (Source: a R. Hobbs; b P. Ladd)

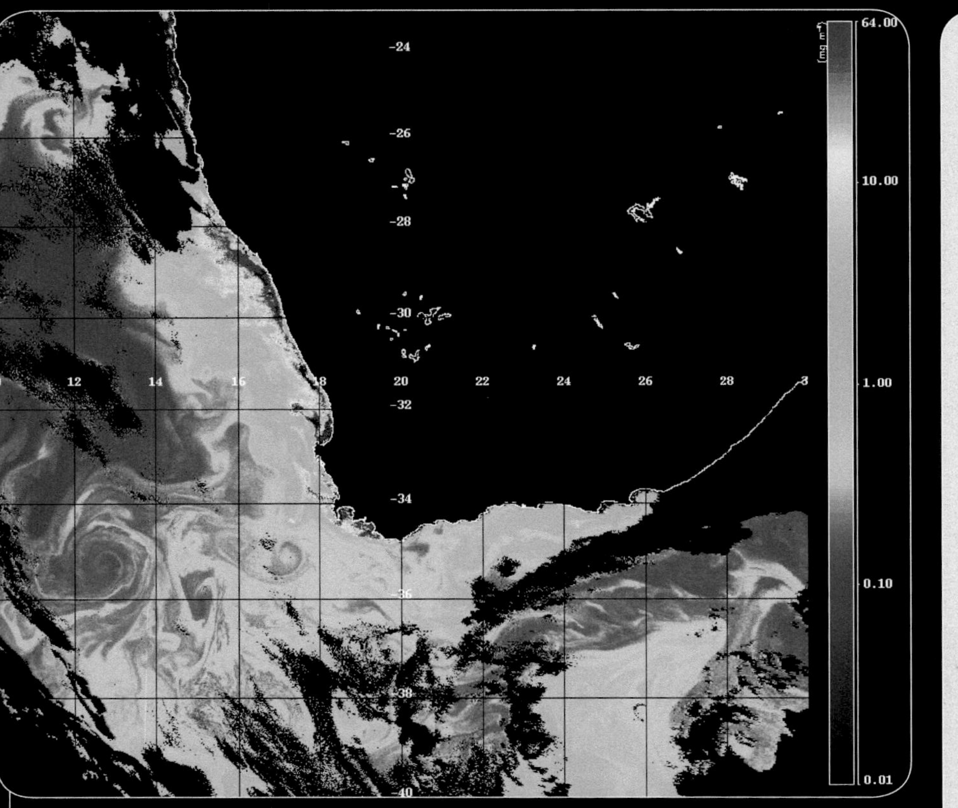

Plate 18.1 Chlorophyll concentrations measured by a satellite using the Sea-viewing Wide Field-of-View Sensor (SeaWiFS). The high concentrations are associated with upwellings on the coast caused by the Coriolis effect. (Source: NASA)

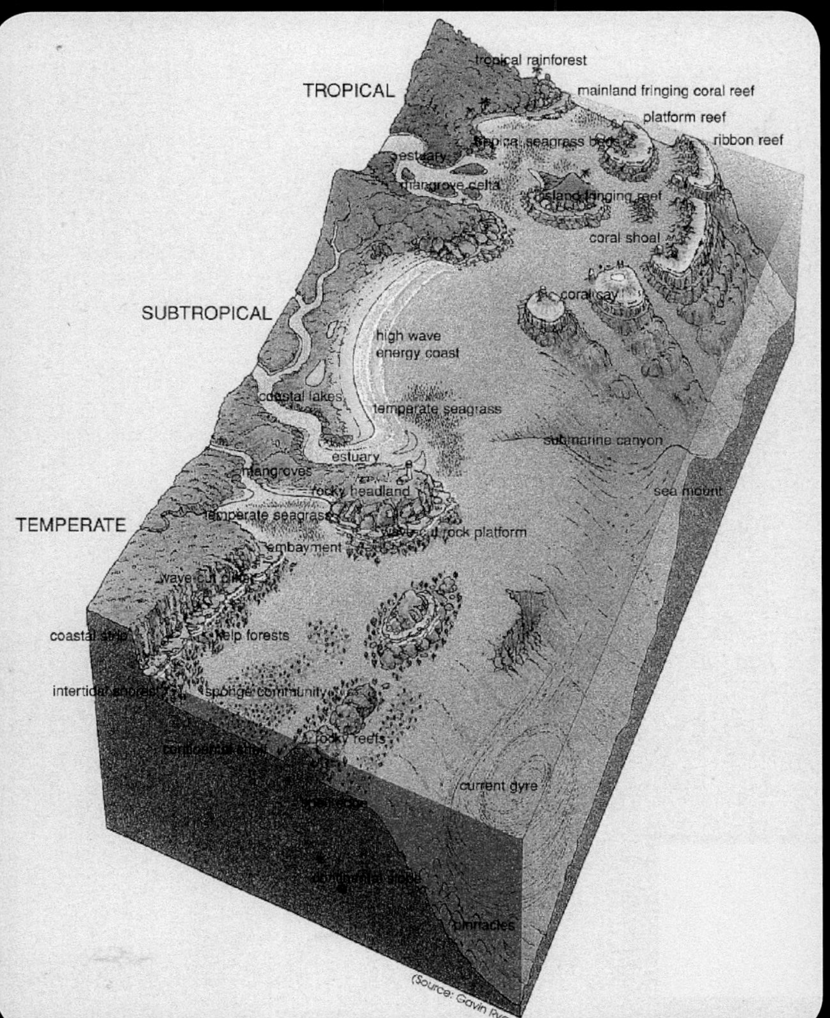

Plate 18.2 The range of marine habitats from temperate to tropical regions. (Source: Department of the Environment and Heritage, Canberra)

Plate 18.3 Seagrass meadows are important marine habitats. (Source: E. Paling)

Plate 18.4 Dr Lotte Horn measuring the photosynthetic capacity of seagrasses in the environment. (Source: E. Paling)

Plate 18.5 Mangroves are angiosperms that, like seagrasses, grow on soft substrates (sand or mud) in coastal areas influenced by salt water. They are among the world's most productive ecosystems. (Source: E. Paling)

Plate 19.1 The nasal salt gland in the southern giant petrel discharges from the nostrils. For an explanation of its function, see Figure 19.3. (Source: © Peter Scoones Naturepl.com)

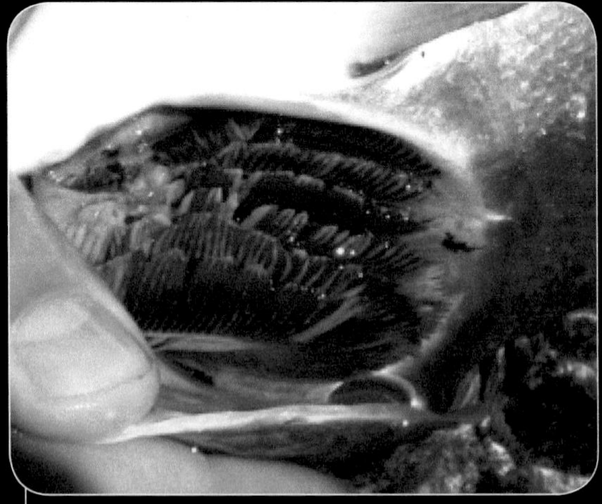

Plate 19.2 Fish gills showing the bright red appearance caused by the blood pigment haemoglobin. (Source: J. Negus, Tennessee Wildlife Resources Agency)

Australia has a huge diversity of aquatic ecosystems because it covers a number of climate zones and has variable topography. Being the driest inhabited continent, water occurs often only temporarily on the landscape, leading to a variety of adaptations by aquatic organisms to survive the dry spell. The type of aquatic ecosystem that will form in any area can be predicted through knowledge of the interaction between the geomorphology (geology, soils, relief) and climate (rainfall, temperature). This is true of wetlands throughout the world, but the diversity of types in Australia makes a valuable illustration.

In cold climates such as in Tasmania, where rainfall exceeds evaporation, permanent deep lakes (7: Dove Lake at Cradle Mountain, Tas.) and permanent streams (8: Strahan, Tas., here stained by tannins from the surrounding rainforest) occur on the mountainous tablelands. In flat alpine areas, evaporation is so low that bogs (see Figure 21.2) form (6: Mt Wellington, Tas.). In more temperate climates, high evaporation during summer results in water levels fluctuating seasonally, creating quite different environments at low and high water, for example 12: a seasonally flooded swamp near Harvey, WA and 5: the Murray River, NSW/Vic. The seasonally and episodically flooding waters of inland rivers are vitally important to river function as they spill across the floodplain, rejuvenating billabongs (cut-off lakes) and bringing nutrients to the river from the floodplain.

In the mostly flat, arid centre of Australia, rainfall is episodic, with sufficient rain to run off into rivers and lakes only every few years or even tens of years. Evaporation far exceeds rainfall, so wetlands flood then dry until the next rainfall event. A visit to the interior might reveal the Todd River at Alice Springs, NT (9) to be a small waterhole or a raging torrent or you may see a huge lake (spreading over several square kilometres) or a pan of salt (e.g. 10: Lake Baandee, WA). Even estuaries in this area are dry for much of the time, as a bar forms across the mouth and the water slowly dries out (11: Stokes Inlet, WA). Salt lakes can occur in temperate climates. Lake Pearse (13) is a hypersaline, meromictic lake on Rottnest Island, WA, formed through the concentration of salts of marine origin. The wonderful pink colour of these lakes is due to an alga *Dunaliella*, which contains very high concentrations of beta-carotene.

In tropical regions, rainfall is higher than in temperate climates, but very seasonal due to the monsoons, resulting in a distinct wet and dry season. The rivers flood across the floodplain after huge rainfall events, carrying large quantities of silt. Even when they return to their sinuous banks (1: South Creek, Port Hedland, WA) they are usually very turbid (2: South Alligator River, NT). In these flat, northern regions the whole landscape becomes a wetland for several months of the year before drying back to a low, dry woodland (3: Kakadu, NT). In the tropical rainforest of the Daintree, Qld, streams flow down the hills onto the flatter coastal plain where they form mangrove forests (4).

Plate 20.1 Inland aquatic environments in different parts of Australia. (Source: Relief map of Australia from <http://www.fotosearch.com/DGV077/200199991-001/>; temperature map and rainfall map adapted from the Australian Bureau of Meteorology; other photographs J. Davis and J. Chambers)

Limnologists who stayed too long …

Bog
Bogs form in cold, wet climates where rainfall exceeds evaporation. Water stays for long periods on the ground, encouraging the growth of aquatic plants such as sphagnum moss. As the plants die they form a layer of peat beneath the live plants; this layer maintains a high water level through capillary action. Humic acids that leach from the dead moss cause the water to become acidic and tea coloured. Humic acids are antiseptic in nature and inhibit the growth of bacteria. As water in the peat is stagnant, there is little input of oxygen for the respiration of microbes. In the anaerobic environment, decomposition is very slow, and the moss and anything else that falls into the bog tends to be preserved. Tollund Man (above) was killed and thrown into the Bjaeldskovdal bog in Denmark in the Iron Age, 300–400 years BC, and is remarkably well preserved over 2000 years later. Tollund Man is held in the Silkeborg Museum in Denmark.

Salt lake
Salt lakes are a common feature of arid lands where evaporation far exceeds rainfall. For the majority of the time they are often dry. For example, Lake Eyre in South Australia is 9300 km^2 in area and is fed by a huge (over 1 million km^2) inland catchment that feeds into the centre of the continent rather than out to sea (arheic drainage). In inland Australia rainfall is highly episodic, resulting in massive flooding every few years or even decades. Then the water will stay around for up to years while the hot desert sun evaporates the water, leaving salts washed from the land to desiccate into a salt crust up to 290 mm thick. Lake Eyre filled only twice in the last century, in 1950 and 1974. Aquatic plants and animals in salt lakes have to be able to cope with increasing salinity as the lake dries and tolerate the long periods when there is no water in resting phases.

The importance of climate and hydrology to the formation and functioning of aquatic ecosystems can be clearly illustrated by comparing a bog and a salt lake. The geomorphology of these two types of wetlands is similar: both occur on relatively flat land. However, bogs form in cold, wet climates where rainfall exceeds evaporation, resulting in water staying on the landscape for long periods of time. In contrast, salt lakes usually occur in hot, dry climates where evaporation exceeds rainfall, resulting in the rapid loss of water to the atmosphere, concentrating any salts that might be present in lake water.

Plate 20.2 Limnologists who stayed too long (Source: Silkeborg Museum, Denmark (Tollund Man) and Axel Kayser (skeleton)).

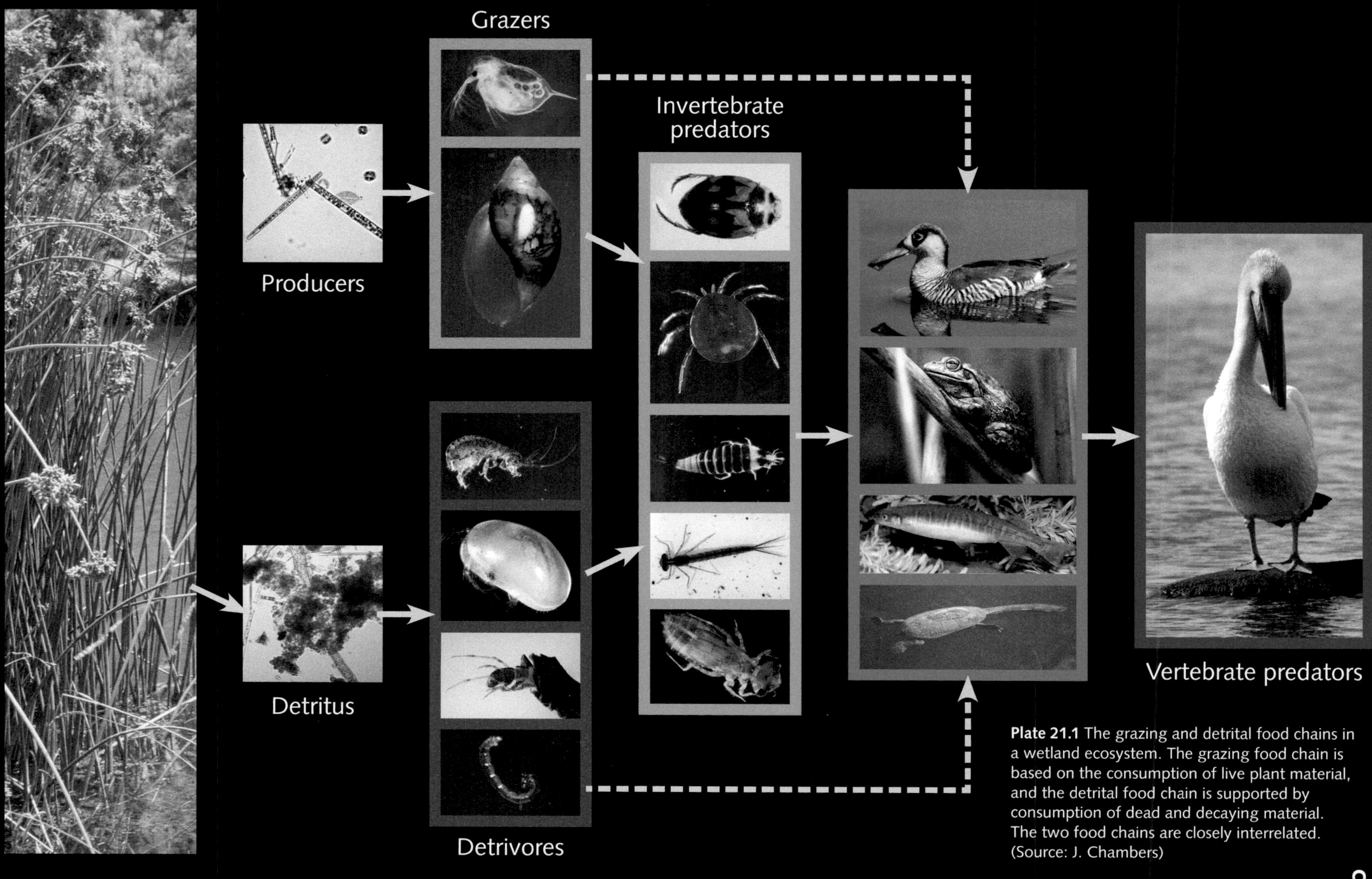

Plate 21.1 The grazing and detrital food chains in a wetland ecosystem. The grazing food chain is based on the consumption of live plant material, and the detrital food chain is supported by consumption of dead and decaying material. The two food chains are closely interrelated. (Source: J. Chambers)

Plate 21.2 (a) Low-order streams (1–3) are shallow and narrow, with steep slopes and high velocities, and substrates ranging in size from large boulders to cobbles to pebbles. This example is from Australia. (b) Mid-order streams (4–6) are wider and deeper, substrates range from cobbles to pebbles to sand, and the stream gradient decreases. This example is from New Zealand. (c) High-order streams (6–12) are much wider and deeper, substrates are finer (often sands, silts or muds) and, although the gradient may be low, velocities are often fast and flows high. This example is from Sarawak. (Source: a–c J. Chambers and J. Davis)

would hinder homeothermic animals. However, a disadvantage of poikilothermy is that an organism needs to have multiple chemical pathways, some for cooler temperatures and others for warm.

At the other end of the temperature regulation scale are the homeothermic marine birds and mammals, maintaining a constant temperature using internal processes. Ocean water temperatures are usually less than 27°C, so these animals lose heat continuously, for they have body temperatures at, or greater than, 38°C. Apart from blubber, which rarely occurs in birds, many animals rely on countercurrents (Figure 19.2). As blood flows in arteries through a limb, or towards the outside of the body, it loses heat to surrounding tissues and, near the extremities, the external environment (Figure 19.2a). This would chill blood returning through the veins and dramatically reduce the core body temperature. If, however, arteries and veins occur close to each other (i.e. in a countercurrent flow), heat transfer occurs directly between them (Figure 19.2b). Blood flowing outward through an artery loses heat to surrounding tissues and the returning blood in the veins. Thus heat loss is minimised. An excellent example of this principle is a dolphin flipper (Figure 19.2c). The main artery occurs deep within the limb and the surrounding veins gain heat from it when returning to the body. Despite these mechanisms, homeothermic animals must eat frequently to maintain their metabolism.

We have examined the extremes of temperature regulation but this is a rather simplistic view, because some animals have an intermediate ability to adjust their body temperatures independently of the environment. Although they are effectively poikilotherms, many intertidal invertebrates move to optimise their temperatures to become warmer or cooler than the surrounding environment. Fast-swimming fish such as tuna and swordfish generate heat via muscle activity or fat burning (thermogenesis), retaining heat deep in their bodies and preventing heat loss through their gills. Recently, a salmon shark from Alaskan waters was found to have an internal temperature 20°C higher than the surrounding water. Its internal swimming muscles were the cause.

Maintaining water balance and removing liquid waste

Animals in terrestrial and aquatic habitats face the constant challenge of maintaining their salt and water balances. Terrestrial and marine animals lose water to their environments and must conserve and replace it. Additionally, marine organisms must, like their terrestrial relatives, remove metabolic wastes by elimination (the production of faeces) and/or excretion of salts and toxins (usually through urine). Some solutions are uniquely aquatic, whereas others foreshadow terrestrial mechanisms. For many terrestrial organisms, maintaining water balance (osmoregulation) and waste removal are connected. This is because the number of charged ions in solution within the cells determines a key component of the composition of tissue fluids, their osmolality (a measure of the number of moles of osmotically active particles in solution). These ions can be either inorganic (predominantly Na^+ and Cl^-) or organic wastes from cellular metabolism (ammonia, urea and uric acid).

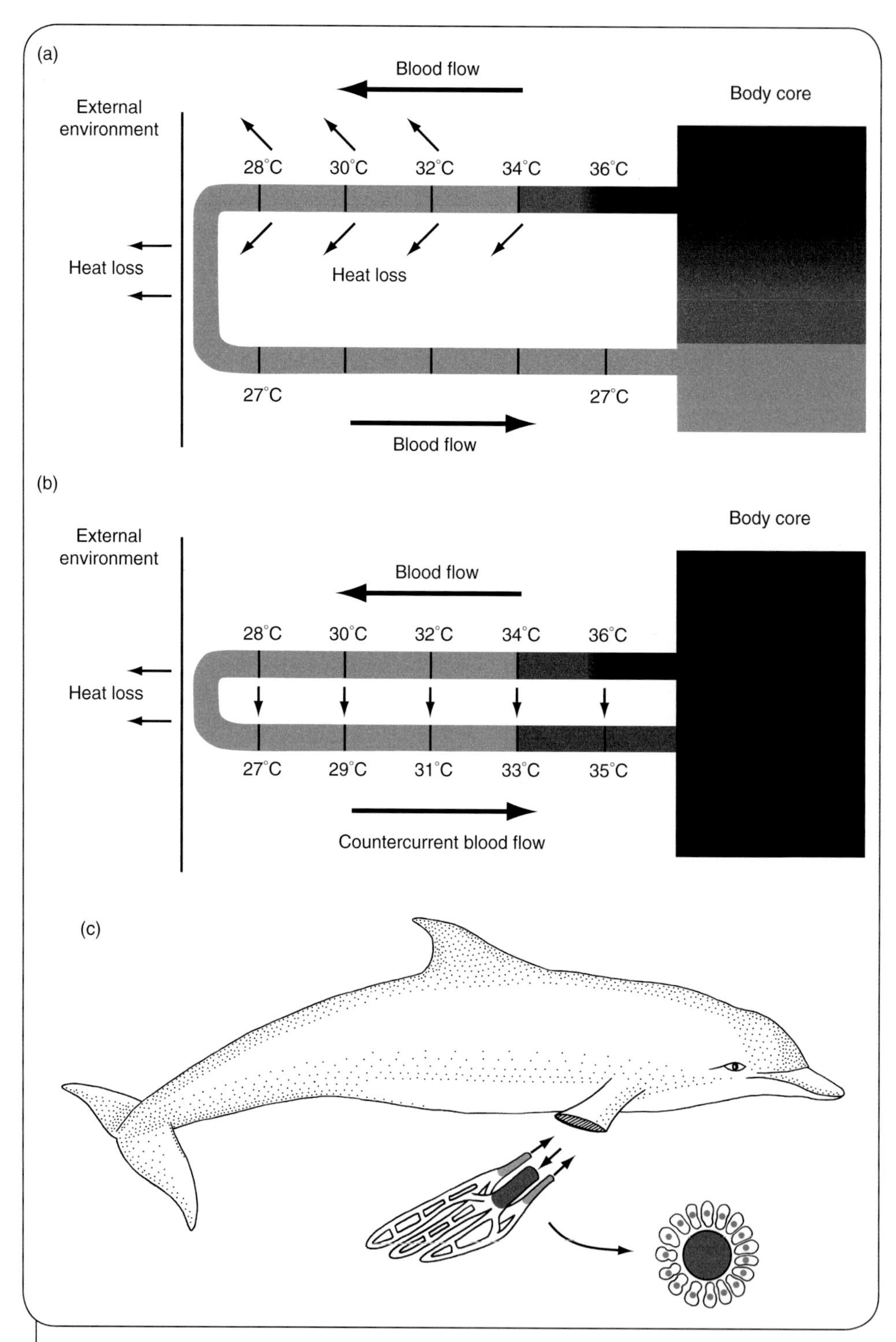

Figure 19.2 (a), (b) Countercurrents in temperature regulation (see the text for further explanation). (c) Countercurrents in a dolphin's flipper. (Source: a, b B. Cale; c Adapted from Schmidt-Nielsen, K. (1970). *Animal physiology*. Prentice-Hall Inc. Reprinted by permission of Pearson Education.)

Semipermeable membranes surround animal cells, allowing some small molecules and water to pass through. Water moves in response to the osmotic gradient and enters cells with high osmolality or leaves cells with low osmolality. Maintaining either the osmotic pressure or the rate at which water enters and leaves cells is crucial; cells may shrink if too much water leaves, or expand and burst if excess water enters (see Chapter 3). The most energetically efficient solution is to be an osmotic conformer, with cell contents at the same osmolality as the surroundings. However, life is not quite that simple because not all cell contents are simply interchangeable. Some, such as nitrogenous wastes, are highly toxic and must be removed, whereas the relative concentrations of others are crucial to cell function. In particular, cell functions such as muscle contraction and conduction of nerve impulses depend upon maintaining a charge difference across the cell membrane by regulating the concentration of particular ions inside the cell. Cells achieve these concentrations by actively pumping specific ions in or out (see Chapter 3).

Marine organisms are bathed in sea water and face unique challenges and opportunities for osmoregulation and waste removal. Marine vertebrates (birds, reptiles, mammals and fish) are osmoregulators, maintaining concentrations of dissolved substances inside their bodies different from those of the environment. The birds, reptiles and mammals evolved on land and returned secondarily to the sea, and their solutions to osmoregulation and waste excretion are based on that evolutionary history.

Birds and reptiles have highly developed salt glands, using a countercurrent flow system, far more efficient than their kidneys in removing excess salts (Figure 19.3, Plate 19.1). Their kidneys are reptilian in nature (even the bird ones) and, although they can remove nitrogenous wastes, they are poor at concentrating salts. With salt glands, sea birds, turtles and marine iguanas have little or no need for fresh water and the marine environment really poses no special problems for waste removal. A similar dependence on nasal glands is seen in arid-land birds such as the ostrich and North American roadrunner.

Marine mammals such as whales, dolphins and seals face a severe shortage of fresh water, but they solve the problem with efficient kidneys producing concentrated urine. These kidneys are large for their body size in relation to land animals and anatomically complex. Kidneys use three mechanisms to filter blood of its wastes: filtration, reabsorption and secretion. All occur in specialised nephrons. Ion pumps (proteins that assist in pumping salts against concentration gradients) increase salts in urine by countercurrent concentration. As there is rarely fresh water to drink, marine mammals rely on metabolic water produced by their cellular metabolism and water obtained from food. They also lose much less water in respiration than terrestrial mammals do.

The truly marine vertebrates, elasmobranches (cartilaginous fish such as the sharks and rays) and teleosts (bony fish) have different solutions to the problems of osmotic balance and excretion. Sharks and rays tolerate high levels of nitrogenous wastes as urea and have high circulating salt concentrations. Their blood is slightly hyperosmotic (more concentrated) than surrounding sea water. They protect their tissues against the toxic effects of urea using trimethylamine oxide, which helps stabilise proteins. Because of the concentration gradient they passively absorb fresh water across their gills (Figure 19.4a).

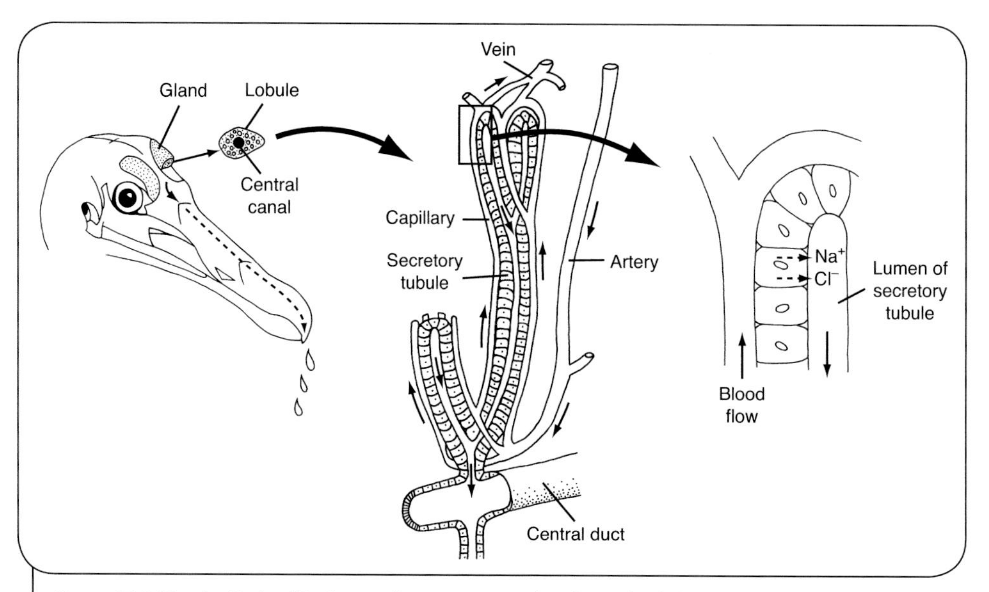

Figure 19.3 Nasal salt gland in the southern giant petrel. When a bird drinks sea water, sodium enters the blood plasma from the intestine and the solute concentration of the blood plasma increases. This causes water to move out of cells (osmosis), increasing the extracellular fluid volume. The increases in blood plasma solute concentration and extracellular fluid volume stimulate salt gland secretion. (Source: Left: Schmidt-Nielsen, K. (1960). 'The salt-secreting gland of marine birds.' *Circulation* 21: 955–67. Middle: The late E. Mose, by permission of his son. Right: Adapted from Randall, D. et al. (2002). *Eckert Animal Physiology*. WH Freemean & Co. 5th edn.)

Thus they have a ready supply of water for excretion. They also use their gills to excrete salts and other wastes via diffusion. Their kidneys and rectal gland, an additional kidney-like organ located at the end of their alimentary canal, remove other waste products.

In contrast, bony fish have internal salt concentrations about one-third of the surrounding sea water and tend to lose water via osmosis and gain salts by diffusion (Figure 19.4b). They drink sea water constantly, excrete salts across their gills using active transport and use their kidneys to pass small amounts of dilute urine.

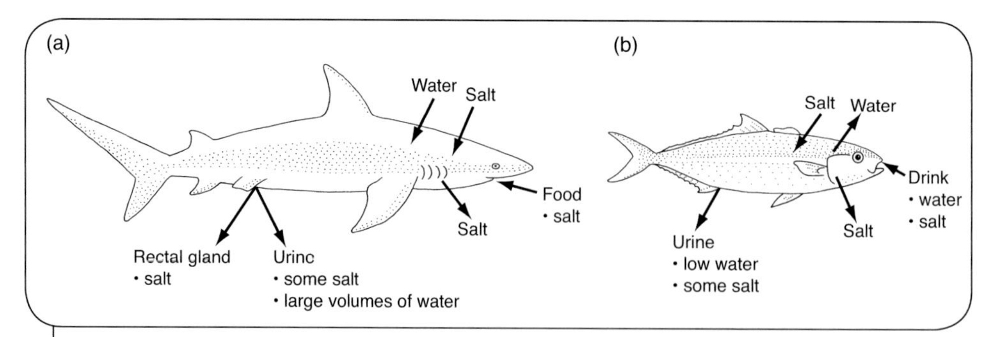

Figure 19.4 Different solutions to osmotic regulation in the ocean. (a) Elasmobranchs such as the sharks and rays have slightly hyperosmotic blood, whereas (b) teleosts, or bony fish, have hypoosmotic blood. (Source: Evans, D. H. (2002). *Osmoregulation by vertebrates in aquatic environments*. Macmillan Publishers Ltd, Nature Publishing Group)

Most marine invertebrates are simple osmoconformers, maintaining an internal ion concentration close to that of the environment. Because the body and extracellular fluids are isotonic to sea water there is rarely a net movement of water. Marine osmoconformers include echinoderms, cephalopods, sipunculids, ascidians, coelenterates, polychaetes, and most marine molluscs and crustaceans. All regulate internal solutes and may adjust solutes using proteins (known as osmotic buffering). However, even when tissues are isosmotic, ion concentrations vary between tissues and water. For example, the concentrations of sulfate and magnesium are usually reduced, whereas potassium and calcium concentrations are usually higher in the tissues than in sea water. Ions are regulated across cell membranes in the gills, digestive tract and excretory organs.

Although sea water is mainly of constant salinity, in environments such as estuaries, animals meet salinities different to their own internal concentrations. If an animal moves into a lowered salinity it becomes hyperosmotic (in relation to the surrounding sea water) and gains water, losing solutes and increasing cell volume. In these circumstances, animals reduce membrane permeability and produce dilute urine copiously. In increased salinities, the animal tends to lose water, gain salts and decrease cell volume. It copes by reduced membrane permeability and active excretion of salts.

Food intake and removing solid wastes

For most Cnidaria (corals and sea anemones), digestion occurs within an all-purpose chamber (the coelenteron), which also houses their gonads and often serves as a brood chamber. They have no digestive tract, but simply draw food into this chamber through an oral sphincter (see Chapter 14). Digestive enzymes are released and surrounding cells absorb nutrients. Indigestible material is ejected through the oral sphincter. Some sea-anemone-like creatures, such as the jewel anemone *Corynactis australis*, evert mesenterial filaments (normally used for digestion within the coelenteron) to cover and digest prey too large to swallow.

Many marine animals, including fish, many molluscs and crustaceans, that either graze on algae or ingest large prey have a complete digestive tract. Food is ingested through the mouth, digested within a gastrointestinal tract and eliminated through the anus as faeces. Some organisms with such tracts (e.g. sea stars) digest material externally by temporarily forming a sealed chamber outside their bodies, so that inedible material is not internalised. Japanese pygmy squid (*Idiosepius paradoxus*) use the extreme example of this technique when, like a terrestrial spider, they inject digestive enzymes into the body of the prey and then suck up the digested flesh.

Marine animals such as ascidians, polychaete worms, crustaceans, basking sharks and whales strain food from massive volumes of water. They either take this material into their gastrointestinal tract for digestion and elimination (in the case of whales and basking sharks), or filter appropriately sized particles and reject other solids. Basking sharks may pass more than 1000 tonnes of water per hour through their mouths, filter it through gill rakers and pass it out through the gill slits. The filtered material that is swallowed can weigh as little as 1 kg of crustaceans per hour. Some of these filtering mechanisms mean that much material is rejected before being taken into the gut. For example, tube-building polychaete worms capture small prey on crowns of tentacles covered in cilia and mucous. Particles are sorted on the basis of size with small particles being taken to the mouth, medium particles stored for tube construction and large particles completely rejected. The structure and developmental origin of filtering mechanisms vary widely from a modified pharynx (ascidians), sticky tentacles (fan worms), specialised gills (bivalves) and bristly legs (barnacles).

Reproduction and dispersal

Marine organisms are potentially a population geneticist's dream. Here it really does seem reasonable that synchronised spawning and external fertilisation create the random mating required by theoretical models such as the Hardy-Weinberg equilibrium (Chapter 6). The mass-spawning corals have molecular mechanisms deterring self-fertilisation (even though they are hermaphroditic) and potentially the eggs of any given colony could be fertilised by a vast number of genetically different fathers. Coral larvae (planulae) typically develop in the plankton for several days before they are competent to settle and metamorphose into miniature coral colonies. This form of marine reproduction ensures genotypically diverse larvae and disperses them widely, thus increasing the chances that at least some are well suited to the conditions of the reefs where they settle. The big questions in marine population biology, however, are: (1) is this normal for the majority of marine organisms? (2) how do marine species form breeding aggregations or synchronise their spawning? (3) does the structure of marine populations match our expectations based on the apparent tactics of reproduction and dispersal?

The diversity of reproduction

So, are synchronised reproduction and widely dispersed, genotypically diverse larvae the normal tactics for marine organisms? It's tempting to imagine that this is the case, because living in a watery medium provides the potential for long-distance dispersal and many marine species have larvae that survive for long periods without settlement. The presence of many species on isolated islands provides good evidence that long-distance dispersal occurs.

In fact, however, marine organisms have a huge range of reproductive modes and far more species have adaptations promoting highly localised settlement of genotypes than was once expected by marine biologists. The most likely explanations are that copies of local genotypes that have been successful within a particular habitat out-compete genotypes that come from outside; and, whereas the ocean provides a wonderful medium for dispersal, long-distance travel as a tiny larva is dangerous and usually unsuccessful.

Consequently, many marine organisms are modular. Corals, bryozoans, ascidians, sponges and sea stars are made up of many genetically identical daughter polyps or zooids produced asexually. At the extreme, these modules are physiologically interconnected and function as a single individual. This may mean that a colony expands to occupy a large habitat patch as a single organism made up of thousands of genetically identical units. This strategy involves no risk of dispersal, but it could be argued that this is really growth rather than reproduction – the one organism has just become larger. However, modular marine organisms also have strategies to make genetically identical copies of themselves that can detach, which clearly is asexual reproduction. The importance of this process seems emphasised by the fact that it has arisen many times within the same phyla, particularly within the Cnidaria. In this group, asexual reproduction occurs through essentially passive means such as colony fission or fragmentation, where parts of a large colony become separate – perhaps because a central section is killed by sediment – and surviving sections form daughter colonies. Fragmentation also occurs though the breakage of staghorn corals by storms or cyclones.

Active asexual reproduction can occur through formation of polyp balls, miniature colonies that develop on a soft stalk of tissue until large enough to survive by themselves when they drop off and roll away. Sea anemones reproduce asexually by dividing in half (fission) to form two similar-sized polyps or many smaller polyps through pedal laceration (where a small portion of a polyp's foot detaches and forms a new individual). In all of these cases, the new individual is physiologically distinct and in the case of the anemones can walk to another site. Perhaps most surprising, however, is that both corals and sea anemones use asexual reproduction to produce brooded babies released as free-swimming planula larvae (corals) or miniature anemones. Generally, we don't know how this is achieved but, in the case of anemones, both males and females produce broods and it seems unlikely that sex cells are involved. One anemone (*Bunodeopsis medusoides*) is known to swallow its own tentacles. These become pinched off (autonomised) and develop into miniature anemones within the adult's coelenteron or gastric cavity (Figure 19.5). Typically, however, even those species using asexual reproduction to maintain local populations also use sexual reproduction at least occasionally, and virtually all marine organisms must therefore solve the problem of finding a mate.

Finding a mate

Some hermaphroditic marine animals (e.g. barnacles), like their terrestrial counterparts (e.g. tapeworms) may hedge their bets. They reproduce with others when mates are available or by themselves (self-fertilisation) when they are not. This is potentially a huge advantage to colonists of unoccupied reefs or islands. However, one of the clear disadvantages of living in the vast expanses of the ocean is that, in order to mix your genes (reproduce sexually), it may be necessary to travel long distances to find a mate. Even for marine organisms living in close association, separation by water poses a big problem, especially if gametes can be swept away by waves or currents. Animals with mobile adults, including many fish, solve this by forming annual spawning aggregations,

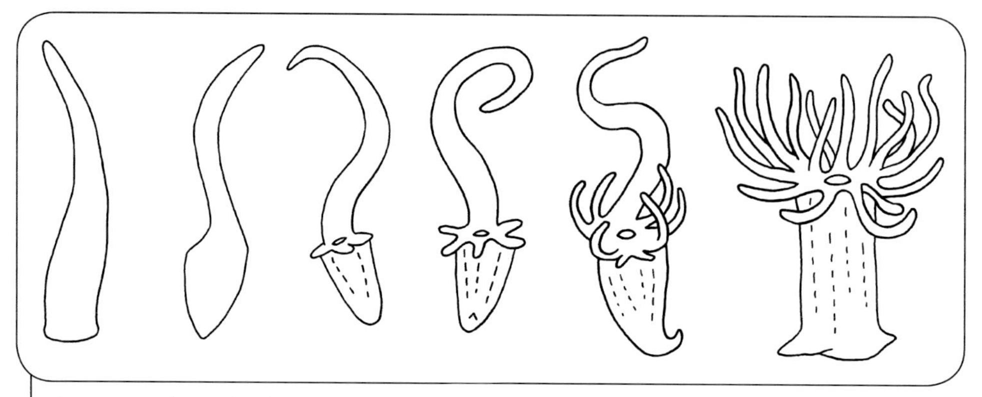

Figure 19.5 The method by which the anemone *Bunodeopsis medusoides* reproduces asexually. It swallows its own tentacles, which become pinched off (autonomised) and these develop into miniature anemones within the adult's coelenteron or gastric cavity. This figure (from left to right) shows the development of a tentacle into a small anemone. (Source: Modified from Cuttress, R. E. (1979). 'Bunodeopsis medusoides Fowler and Actinodisous Fowler, two Tahitian sea anemones: redescription and biological notes.' *Bulletin of Marine Science* 29: 96–109)

determined by factors such as water temperature and the lunar cycle. For example, eastern Australian long-finned eels (*Anguilla reinhardtii*) move from lakes and streams to the ocean, and migrate thousands of kilometres to aggregate in their spawning grounds off the coast of New Caledonia. Their young (elvers) then spend almost a year in the ocean before returning to the streams.

Marine organisms with sedentary adults and no opportunity for direct contact must solve the problem of watery isolation in other ways. Broadcast (or mass) spawning, guided by lunar synchronisation, can be successful if potential mates occur in sufficient densities. Again, the broadcast spawning of corals on the Great Barrier Reef provides a spectacular example of the success of this approach (Box 19.1), but it also highlights one of the potential problems for broadcast spawners. On any given reef, thousands of colonies may shed their egg and sperm bundles within a few minutes to hours. To prevent wasting gametes, colonies must spawn at different times to other closely related species. This problem has been addressed in part by related species evolving a strict timetable, so that spawning periods are separated within the several days that make up each mass spawning event.

Box 19.1 Synchronised spawning in corals

Many marine animals have external fertilisation, with sperm and eggs shed into the water where they subsequently mix. One of the most spectacular examples of external fertilisation is the synchronised spawning of corals. In many corals this occurs annually in response to seasonal and lunar cues. On the Great Barrier Reef, spawning begins during spring and early summer, just after a full moon, and continues for about a week.

Millions and millions of gametes are shed into the water from a number of different coral species. Fertilisation between sperm and ova from the same species occurs and, within a day, fertilised eggs develop into swimming larvae called planulae. Predators eat most of the planulae, but some descend after a few days at the surface to find a suitable site to form a new colony. The precise timing of spawning, just after a full moon, during periods of low tides and rising water temperatures, increases the chances that settling larvae will remain within the reef and not be carried away to unsuitable habitat.

After settling, the coral changes form from the pelagic larval stage to the benthic juvenile stage. The juvenile coral lays down a basal plate to attach itself to the substrate and secretes a protective skeleton. The tiny coral polyp is now ready to bud new polyps and extend the coral reef (Figure B19.1.1).

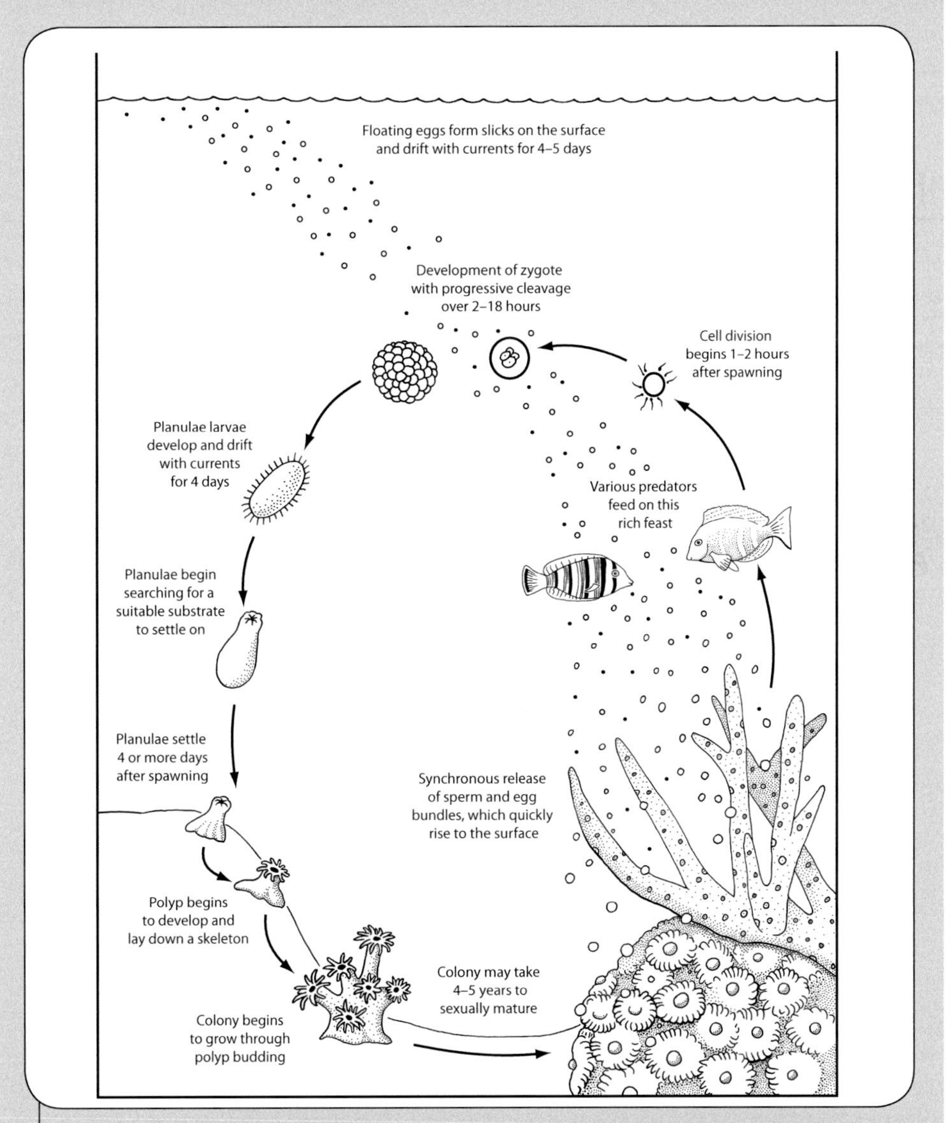

Figure B19.1.1 Generalised life cycle of a broadcast-spawning hard coral. (Source: B. Cale)

On rocky intertidal reefs, where opportunities for spawning may be limited by wave action and desiccation, many species reduce the risks associated with reproduction by internal fertilisation. For example, many marine gastropods are mobile, form mating aggregations and use internal fertilisation essentially as if they were land snails. However, in contrast to their terrestrial counterparts (which simply develop *in situ* within a yolky egg), their offspring may be brooded and develop wholly or partially within egg capsules (often cannibalising their siblings or nurse eggs) before either emerging to crawl away or completing development in the plankton. Others have completely planktonic development with either feeding or non-feeding larvae. Other less mobile taxa rely on physical adaptations in their search for a mate. The secret of success for barnacles is to have the greatest penis length relative to body size of any animal. In order to mate, a barnacle can uncoil its penis and search its immediate neighbourhood for a suitable mate without moving (failing that, barnacles are hermaphroditic and can fertilise themselves). It is possible that other taxa have more flexibility in their reproductive strategies than we have yet discovered (Box 19.2).

Does the structure of marine populations match our predictions?

Molecular tools allow us to examine the genetic structure of marine populations (Chapter 5). One of the most surprising outcomes is the discovery that many marine populations are highly genetically subdivided. That is, different alleles occur with very different frequencies in separate subpopulations. This result clashes with the expectation

Box 19.2 Flexible reproduction in a sea anemone

Biologists from the University of Melbourne and the University of New South Wales have recently reported that, in Victorian rock pools, the intertidal sea anemone (*Oulactis muscosa*) is seemingly able to reproduce asexually. It also appears able to be, like corals, a synchronised broadcast spawner, releasing eggs that remain within a viscous matrix until fertilisation. In this case, however, synchronisation occurs within pools through the release of pheromones (chemical triggers) and females in pools with few or no males are at a severe disadvantage, with markedly lower fertilisation success. It seems that *O. muscosa* populations may be maintained in part by clonal replication, and success in broadcast spawning is achieved in all-or-nothing, big-bang spawning events that are highly localised.

However, this strategy contrasts markedly with the same species of anemone occurring in the northern part of its range. In this case *O. muscosa* is largely found on flat vertical and horizontal wave-swept surfaces, where retention of gametes and chemical signals must be far more problematic. We still do not know how reproduction is achieved in this case, but genetic studies by biologists at the University of Wollongong revealed no evidence of clonality, and anemones under these circumstances show no distinct spawning period.

that marine species could form large homogeneous populations that are strongly connected by larval dispersal. It is true that many species of subtidal and intertidal organisms, with larvae that are thought to survive in the plankton for long periods (e.g. several limpets, ascidians and echinoderms), show little or no genetic differentiation over hundreds and sometimes thousands of kilometres of Australian shores. There are perhaps as many that do not.

A part of the explanation lies in the great variation in dispersal potential. Many species with no known means of dispersal rely on the direct development of 'crawl away' larvae. For example, populations of the sea star *Patiriella exigua* show massive levels of genetic differentiation even among neighbouring rocky headlands. Other species have planktonic larvae and, although they show little or no genetic differentiation among extensive sets of subpopulations, clearly display the effects of barriers in the sea. For example, the area around the New South Wales and Victorian border sees the convergence of the southward-flowing East Australian Current and the northward-flowing Bass Strait Cascade. Several marine species are known to show genetic discontinuities to the north and south of this region, presumably because their larvae are caught in long-lived eddies and cannot progress north or south.

The genetic structure of coral populations is a paradox. Broadcast-spawning species have life histories adapted to ensure widespread dispersal as their larvae are only competent to settle after several days. This contrasts with other, brooding coral species that have internal fertilisation and produce larvae that are competent to settle immediately on emergence. Typically, the brooders show striking evidence of genetic differentiation within reefs (even over hundreds of metres), although genetic surveys have detected little additional variation among populations separated by hundreds or thousands of kilometres. This suggests that brooded larvae can and often do settle close to their parents, but those larvae that are washed off their reef may survive for long periods and travel large distances before settlement. Broadcast spawners also show striking levels of genetic subdivision within reefs (which begs more detailed study), but seem less well able to disperse among widely separated reefs. It is a major research challenge to determine which life-history characters really equip marine species for widespread dispersal and to try to use this information to plan the size and structure of marine reserves (Box 19.3).

Although fish are probably seen as the organisms with the greatest capacity for dispersal, realised dispersal patterns are often a very poor fit to predictions. In bony fishes there is passive dispersal of larvae, and larval fish may remain in the plankton for periods ranging from minutes to months. Ocean currents also influence the dispersal of adult fishes, but they are typically powerful swimmers that may migrate over vast distances and often in surprising directions.

Conventional tagging studies are inefficient because they rely upon recapturing fish, and the only information obtained is the distance between points of tagging and recapture. Archival tagging, satellite tagging and molecular genetic analyses offer greater power to track dispersal of both adults and larvae. Archival tags can

Box 19.3 Development of Marine Protected Areas (MPAs) in Victoria

Whereas national parks and nature reserves have a long history on land, the idea of reserving sections of ocean for conservation is new. The task is daunting because of incomplete knowledge of the distribution of many marine animals and the diversity of lifestyles and dispersal strategies used by marine animals.

Nevertheless, since 2002 just over 5% of Victorian coastal waters have been declared high-level protection Marine Protected Areas (MPAs), where exploiting living natural resources and extracting most minerals are forbidden. The MPAs recognise the biological importance of Victoria's 2000 km of coastline and aim to protect important marine habitats and species, significant natural features important to biota, and areas of cultural heritage and aesthetic value. Protected areas were chosen to be comprehensive, adequate and representative (often known as the CAR system). The protected areas must protect the full range of habitats and communities in the region (comprehensive), be sufficiently large to allow natural ecosystem processes to occur and to minimise external disturbance (adequate) and include the full range of biodiversity within their boundaries (representative). This achievement, significant by international standards, was the culmination of over 30 years of lobbying, scientific research and political debate.

The core of the debate centred on the relative values of marine conservation and protection of Victoria's two key fisheries, abalone and rock lobster. These species are highly localised, so MPAs had the potential to remove important areas from the fisheries altogether. The main stakeholders supporting the MPAs included conservation groups such as the Victorian National Parks Association, the Australian Conservation Foundation, the Australian Marine Conservation Society and many local groups spread along the Victorian coastline. Many senior government bureaucrats were also supportive and ultimately the two major political parties, the Australian Labor Party and the Liberals, also decided that the MPAs were valuable. However, there was opposition from bodies representing commercial and recreational fishers and also from the National Party, which has traditionally supported agriculture and natural-resource-utilisation industries.

Key factors in the final outcome were the persistence of the conservation lobbyists, the completion of the necessary scientific surveys to identify key habitats, active involvement of 'champion' scientists to argue the merits of MPAs, the dissemination of sound information to the public and active public debate. These ultimately convinced the politicians of the feasibility and acceptability of action.

record months of information for variables such as temperature and salinity that is retrieved when the fish is recaptured. Satellite tags are an expensive option, but track large fish by receivers on orbiting satellites. This approach has recently tracked the extraordinary dispersal of individual white sharks around thousands of kilometres of the Australian coastline and even the travels of one shark back and forth between Australia and South Africa (see Box 16.2). Equally surprising was the finding that North American tuna, expected to travel south to breed in the warm waters of the Gulf of Mexico, instead migrated north to breed in the cold waters off Maine. Molecular genetics and otolith (inner ear structures) microchemistry allow the tracking of larval movements (by comparing either the genotypes or chemical signatures of captured fish with those of fish resident in known areas) and have revealed both surprisingly long- and short-distance dispersal. Paternity analysis has also been used to demonstrate that, on Australia's Great Barrier Reef, a third of larval clownfish settle within a few hundred metres of their parents.

Respiration

All marine animals respire and must procure oxygen to survive. However, oxygen is poorly soluble in sea water and its solubility decreases as the water warms (Figure 19.6; see also Chapter 20 for a discussion of oxygen in the aquatic environment). In fact, it is so poorly soluble that its concentration in the water is measured in parts per thousand, and its usual concentration is about 8 mg L^{-1}. Added to this difficulty is the fact that oxygen concentrations are not uniform across marine habitats. Sediments, particularly fine ones, often lack oxygen and the only organisms capable of surviving are specialised bacteria. Most organisms therefore require an effective means of getting oxygen into their bodies. How they do so depends very much upon their size.

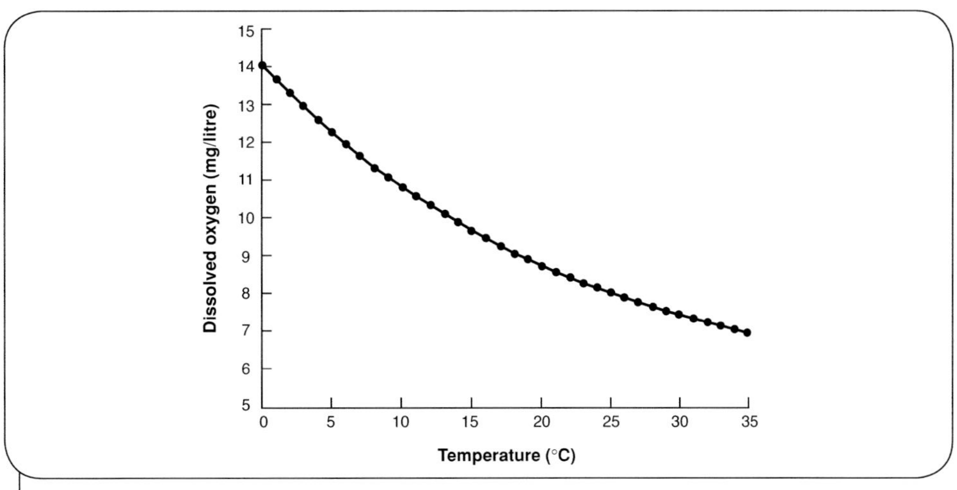

Figure 19.6 Oxygen solubility in water at different temperatures. (Source: B. Cale)

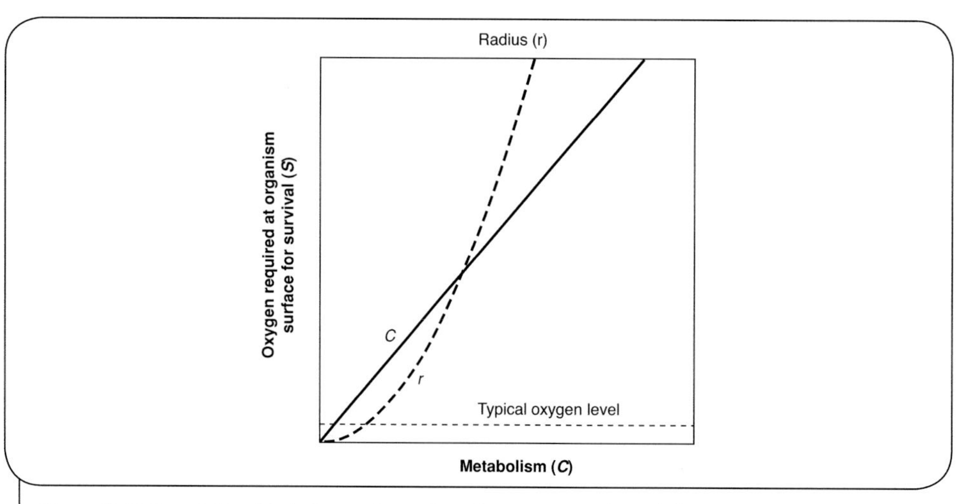

Figure 19.7 The relationship between the radius of an organism (*r*) and the oxygen required at its surface (*S*) to support its metabolism via diffusion (*C*). As can be observed from the graph, normal oxygen levels in sea water (the dashed horizontal line) are sufficient only to support very small organisms. (Source: B. Cale)

Oxygen diffuses from surrounding sea water across membranes and into the cells. Smaller organisms use diffusion alone but, because of the laws of physics, they are unable to grow much beyond about 3 mm in thickness or diameter (Chapter 3). This is illustrated by comparing the amount of oxygen an organism needs to survive with its radius (Figure 19.7). In the graph we can see that, as either the oxygen-consumption rate (or metabolism) increases or an organism's size (radius) gets bigger, a larger amount of oxygen is needed at its surface for diffusion to be able to provide enough oxygen to survive. As can be seen from the graph, the typical oxygen level in the water is simply too low for any creature that is either large or has a high metabolic rate. So for an organism to survive by using diffusion alone, it must either be very small or have an extremely low metabolism.

Larger animals, or those with high metabolisms, develop respiratory structures to meet oxygen requirements. Most animals achieve this with gills that pass water over blood-rich capillary beds in specially designed anatomic structures (Figure 19.8, Plate 19.2). Water passes over the gills in a direction opposite to the flow of blood, to achieve countercurrent flow and maximise oxygen transfer via diffusion (Figure 19.9). Gills fail in air because they cannot support themselves, and collapse, reducing surface area and causing suffocation.

Having gills, however, does not solve problems of the low oxygen concentration in the water and oxygen saturation in the blood. Imagine the water outside a blood vessel in a gill and the blood vessel itself. Diffusion is driven by the difference between the concentrations of oxygen. In any given situation, we would expect the concentrations to be much the same. In other words, the blood will be saturated for oxygen and no more will transfer across. This is where proteins can help. Blood pigments are very effective

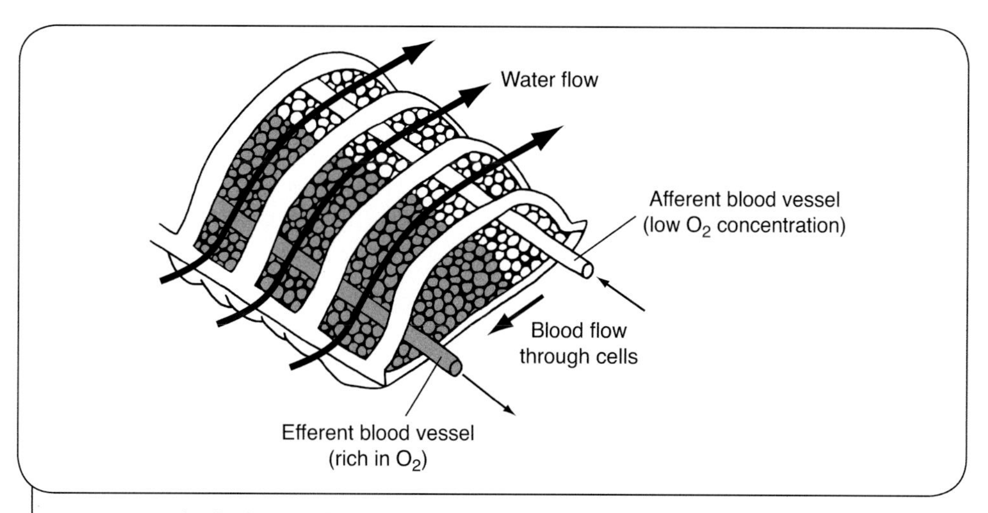

Figure 19.8 Fish gills showing the countercurrent flow of blood against water across the gills. (Source: Adapted from Schmidt-Nielsen, K. (1970). *Animal physiology*. Prentice Hall Inc. Reprinted by permission of Pearson Education)

at binding oxygen to themselves. As they do so, they lower the relative concentration of oxygen in the blood, keep the concentration low and maximise the transfer of oxygen from the water into the blood. By having proteins that bind oxygen, an animal can increase the capacity of its solutes for oxygen by up to 100 times. Several oxygen-binding proteins are used in the marine world. Some molluscs and arthropods use copper-based ones such as haemocyanin. Iron-based proteins including haemoerythrin, chlorocruorin

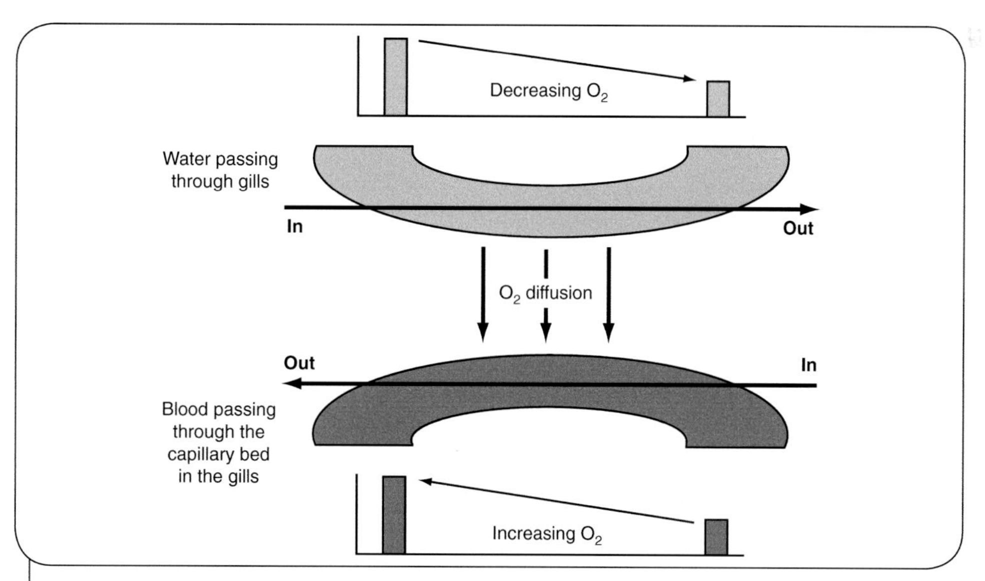

Figure 19.9 Countercurrent flow of blood and water passing across fish gills. The oxygen content of the water decreases as it flows across the gills (top), caused by diffusion into the blood flowing in the opposite direction (bottom). (Source: B. Cale)

(used by polychaetes) and haemoglobin (used by echinoderms, flatworms, annelids and some arthropods and molluscs) are more common.

Marine mammals, birds and reptiles use lungs, not gills, because they have returned to marine lives after a long evolutionary history on land. Their specialist adaptations include lowering their metabolic rates when diving to reduce the need for oxygen, having large blood volumes with high oxygen-carrying capacity, and high concentrations of the protein myoglobin in their muscles to store oxygen. Sperm whales hold the record for the deepest-diving air breathers, descending to depths of up to 2200 m and remaining underwater for up to 2 hours.

Chapter summary

1. Marine reserves are important for protecting biodiversity in the ocean. These reserves, if they are to function efficiently, must be developed with a sound understanding of the lifestyles and dispersal patterns of the marine organisms they contain.
2. Marine organisms must maintain temperature regulation and water balance, dispose of wastes, obtain food and oxygen, and reproduce.
3. In terms of temperature regulation, animals are divided into poikilotherms and homeotherms. Poikilotherms cannot regulate their body temperature and their internal temperature conforms to that of the external environment. This group includes all subtidal invertebrates and most fish. Homeotherms, including marine birds and mammals, regulate their body temperature. They rely on the principle of countercurrents to avoid losing heat continuously to the external environment.
4. Maintaining the correct water balance (osmotic pressure) is important for animals as their cells cannot change volume without suffering damage. Marine vertebrates (birds, reptiles, mammals and fish) are osmoregulators, maintaining a concentration of dissolved substances inside their bodies different from that of the surrounding environment. Birds, reptiles and mammals use organs and tissues that have undergone long periods of evolutionary development on the land. Birds and reptiles have highly developed nasal glands, using a countercurrent flow system, which are more efficient than their kidneys in removing excess salts. Marine mammals have efficient kidneys producing concentrated urine. Sharks and rays tolerate high levels of nitrogenous wastes in the form of urea and have high blood-salt concentrations. In contrast, bony fish have internal salt concentrations about one-third of that of the surrounding sea water.
5. Most marine invertebrates are simple osmoconformers, maintaining internal ion concentrations close to those of the environment.
6. Although most cnidarians capture and digest their food in an all-purpose coelenteron, most marine animals have a complete digestive tract, with food taken in through the mouth and solid wastes excreted through the anus. Many animals, including ascidians, crustaceans, worms, basking sharks and whales, have filtering mechanisms that allow them to strain food from massive volumes of water.
7. There is a wide range of reproductive strategies in the ocean, and currents assist widespread dispersal. There are highly mobile adults (fish, dolphins) or, in sedentary organisms, vast numbers of tiny planktonically dispersed larvae that may remain in the water column and travel great distances for months before settling. Alternatively, other organisms such as some corals and sea squirts produce swimming larvae that typically settle within metres of their parents.
8. Smaller organisms obtain oxygen by diffusion, but are limited in size to 3 mm in diameter. Larger animals have developed respiratory structures (gills) to meet their oxygen requirements. All gills involve the passing of water over blood-rich capillary beds in specially designed anatomic structures, using the principle of countercurrents to enrich the blood with oxygen. Countercurrents are very useful, but they cannot overcome the problem

of oxygen saturation. To solve this, many animals use blood pigments to bind oxygen and increase the oxygen-carrying capacity of their blood by up to 100 times.

Marine birds, reptiles and mammals use lungs, not gills, because they returned to the sea after a long evolutionary history on land.

Key terms

archival tag
asexual reproduction
blood pigments
broadcast spawning
brood
countercurrent
diffusion
dispersal
epifauna
fission
gills
hermaphroditic
homeotherm
larva
metabolic rate
metabolic wastes
nasal glands
nitrogenous wastes
osmoconformers
osmoregulation
osmoregulators
planula
poikilotherm
polyp
Q_{10}
synchronised spawning
thermogenesis
water balance

Test your knowledge

1. Compare and contrast the temperature regulation mechanisms of a poikilotherm such as a marine jellyfish with those of a homeotherm such as a dolphin, a marine mammal. What are the advantages and disadvantages of each method?
2. Explain the principle of countercurrent exchange and give examples of its use by marine animals in temperature regulation and respiration.
3. Explain, using marine examples, the differences between osmoregulators and osmoconformers.
4. What dispersal strategies are used by marine animals and how are these accommodated when designing marine reserves?
5. Marine mammals, birds and reptiles all returned to the sea after long periods of evolution on land. In what ways does this make their osmoregulation and respiration different to those of fish, which are strictly marine vertebrates?

Thinking scientifically

One suggested explanation for outbreaks of the crown-of-thorns starfish on the Great Barrier Reef was that starfish were being introduced from elsewhere in ships' ballast water. What data could you collect to determine whether the populations on the Great Barrier Reef were likely to be introduced?

Debate with friends

Debate the topic: 'Commercial and recreational fishing interests should be the primary consideration when designing marine reserves'.

Further reading

Edgar, G.J. (2001). *Australian marine habitats in temperate waters.* Reed New Holland, Sydney.

Fox, A. and Parish, S. (2007). *Wild habitats: a natural history of Australian ecosystems.* ABC Books/Steve Parish Publishing, Sydney/Archerfield, Queensland.

Nybakken, J.W. and Bertness, M.D. (2004). *Marine biology: an ecological approach.* Benjamin Cummings, York. 6th edn.

20

Inland aquatic environments I – wetland diversity and physical and chemical processes

Jane Chambers, Jenny Davis and Arthur McComb

Trouble in Kakadu

Water is a primary requisite for life. As human populations expand, demand for fresh water increases exponentially. Only about 2.5% of the world's water is fresh and of this 2.4% (or 99.6% of the total fresh water) is frozen in glaciers, in permanent snow cover or hidden in ground water. Only 0.1% of the world's fresh water flows freely in aquatic systems. Not surprisingly, there is most conflict between people and the environment where water is most readily available to humans: in inland aquatic ecosystems.

Human use of water and, at times, ignorance of how aquatic systems function have led to wetlands disappearing at alarming rates globally. Inland aquatic ecosystems are among the most productive ecosystems on Earth, supporting an enormous biodiversity as well as providing water for much terrestrial life. Aquatic ecosystems maintain water quality, support biodiversity and underpin the Earth's ecology. We harm them at our own expense.

In the Northern Territory of Australia, Kakadu's World Heritage-listed freshwater wetlands cover 195000 hectares. Three million waterbirds feed and breed there and they are also home for turtles, frogs and fish. The wetlands are valued for ecological importance on a global scale, because they support many migratory waders. However, within the original boundaries of the National Park is a uranium mine with potential to pollute downstream wetlands. To ensure that the wetlands are not adversely affected requires a very good understanding of how they function and how pollutants might be transported and affect wildlife. This is just one example, but it typifies the importance of understanding and conserving valuable wetland ecosystems.

Chapter aims

In this chapter we describe the main environmental factors determining the biota that inhabit different types of inland water bodies: light, temperature, dissolved oxygen, pH, salinity and nutrients. We explain how to measure these parameters, and how changes caused through human impacts are detrimental to water quality and consequently to the biota and the health of the ecosystem.

The diversity of inland aquatic systems

Inland aquatic environments are highly diverse. They are sometimes called 'freshwater' systems to differentiate them from marine environments, but in inland Australia many lakes are more saline than the ocean! The International Union for the Conservation of Nature defines wetlands as: 'Areas of marsh, fen, peatland or water, whether natural or artificial, permanent or temporary, with water that is static or flowing, fresh, brackish or salt including areas of marine water, the depth of which at low tide does not exceed six metres'. This may seem a long definition, but it was refined at numerous international meetings to encompass aquatic ecosystem diversity. We will use it in this chapter, although we will not extend beyond the coastal margin, because this area is covered in Chapter 18. With such diversity, how is it possible to understand and manage these systems?

It is possible because the fundamental component of every aquatic system is water. The crucial nature of water is sometimes overlooked in light of environmental concerns such as waterbird habitat loss or fish kills resulting from algal blooms. Yet every physical, chemical and biological process in aquatic ecosystems is affected by hydrology, and any modification to hydrology may have catastrophic results. The integral nature of hydrology means that if we understand the inflows, outflows and dynamics of water within an aquatic system we are well on the way to understanding how it works.

Hydrology and wetland diversity

The formation and attributes of inland waters (e.g. lakes, swamps, rivers and ephemeral water bodies) depend on climate (rainfall and temperature) and geomorphology (geology, soils and relief). A brief overview of the diversity of inland aquatic environments in different parts of Australia highlights interactions between climate, elevation and geomorphology (Plate 20.1).

The direct impact of climate and geomorphology is expressed in a wetland's hydrology. The hydrology of a water body is summarised by a water budget (Figure 20.1) describing the different ways in which water enters or leaves aquatic systems. The components of a water budget include rainfall (or precipitation), evapotranspiration (the loss of water to the atmosphere via evaporation and transpiration by plants), surface flow (through channels and by overland flow), groundwater inflow and outflow (including subsurface movement and deeper ground water), tides and storage (or volume of water held within the water body). The relative importance of each of these parameters promotes different types of wetlands.

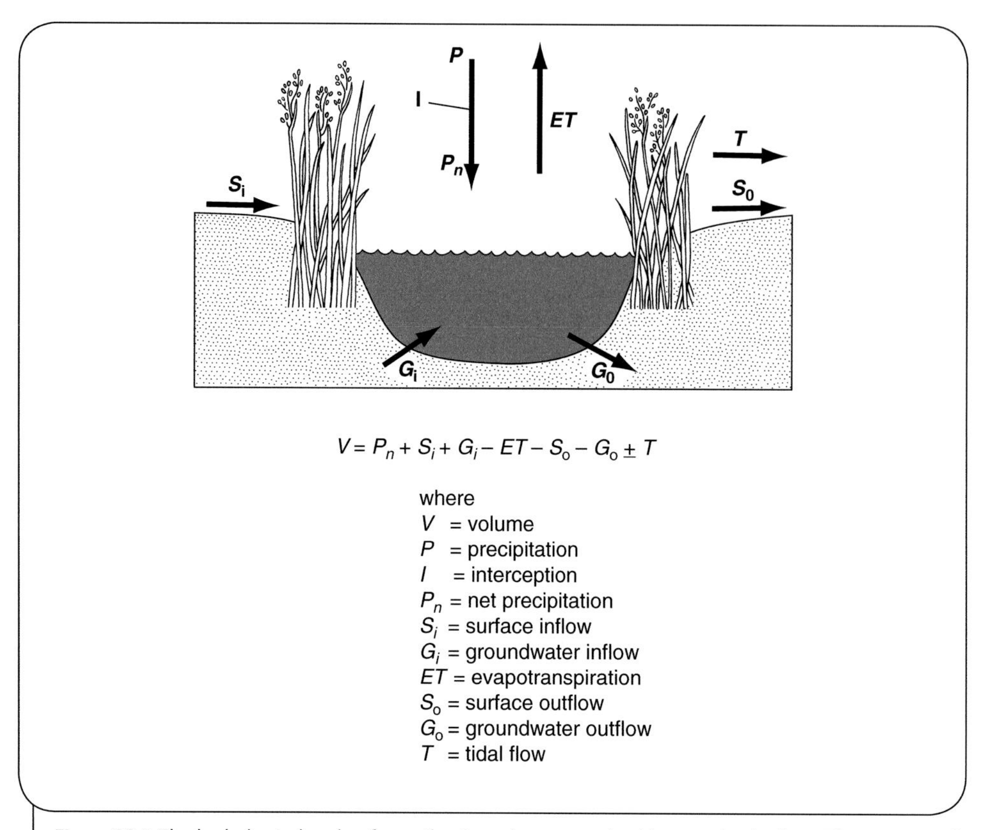

Figure 20.1 The hydrological cycle of a wetland can be summarised by a water budget. The amount of water in a wetland is equal to the water flowing in by net precipitation (= precipitation – interception, by plants for example), surface, tidal and groundwater inflows minus water being lost to the wetland by evapotranspiration (= evaporation + transpiration by plants), surface, tidal and groundwater outflows. (Source: B. Cale)

In Figure 20.2 we see the different interactions of hydrology with the landscape, creating different wetland types. For example, in tidal reaches adjacent to rivers and estuaries, tides dominate the formation of marshes. A semidiurnal pulse of water flowing across flat to gently sloping land sets up a zonation of inundation and possibly salt concentration, which is reflected in the plants and animals living there. In depression wetlands (swamps or lakes), several hydrologic parameters can dominate: rainfall in perched systems (for example, on Fraser Island in Queensland); groundwater flow through the landscape, such as occurs on the coastal plains of south-western Australia; surface water inflows via creeks and rivers (such as the salt lakes of the Australian interior); or all factors combined. The driving force is that the wetland occurs in a depression and forms the end point of flow from the surrounding catchment. In flat landscapes, where rainfall markedly exceeds evapotranspiration, bogs form. Water remains on the soil for long periods, promoting the hydrophytic plants such as mosses (e.g. *Sphagnum* sp.). They maintain the water table through capillary action in the plant mass and

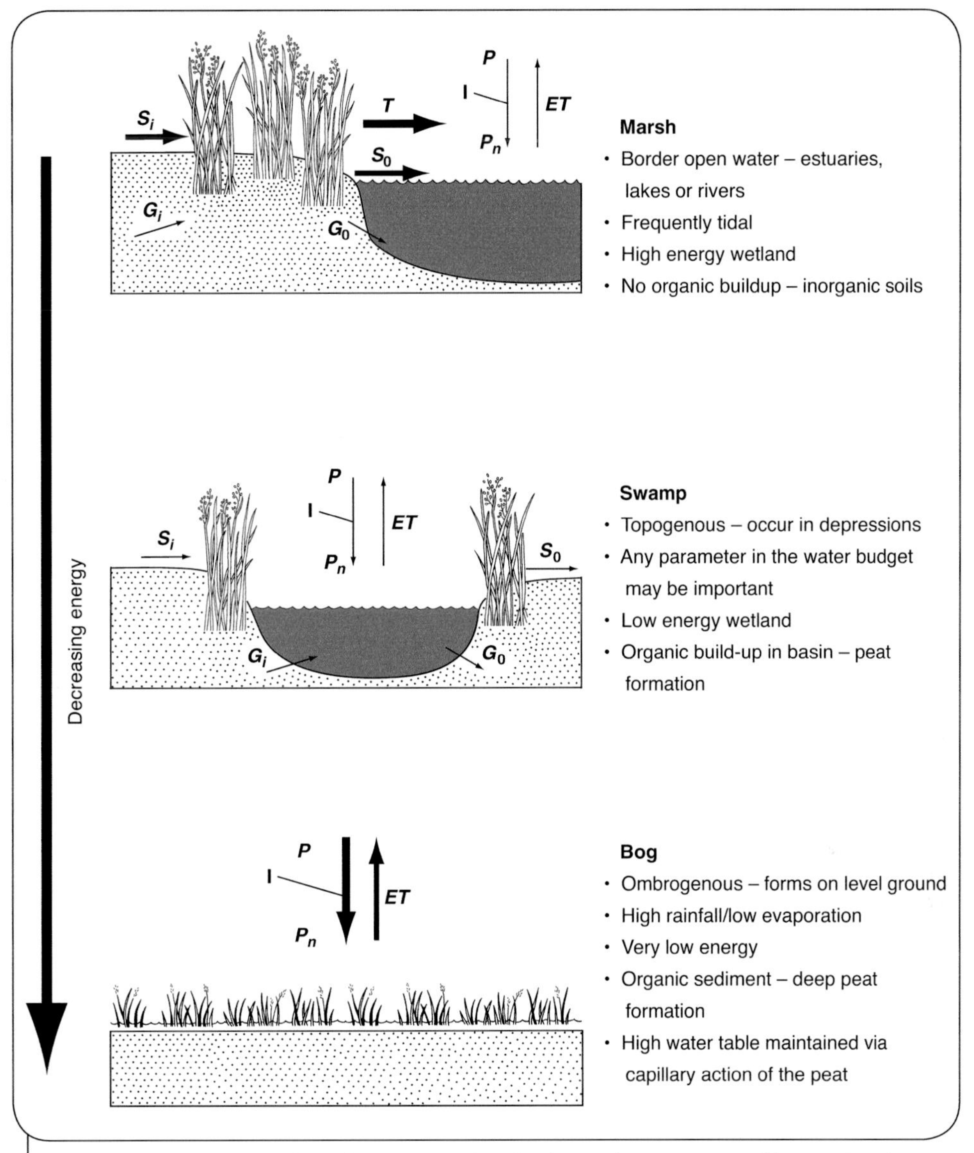

Figure 20.2 Water budgets for different types of wetlands (marshes, swamps and bogs). The relative importance of the hydrological parameter in the formation of the wetland is indicated by the thickness of the arrow. For example, the dominant hydrological parameter in marshes is the tide. See Figure 20.1 for legend. (Source: B. Cale)

the detritus accumulating below the plants. Managing any of these systems requires understanding the proportional contribution of the different hydrologic parameters to the wetland.

If we look at the hydrologic budget for a swamp (Figure 20.2) and imagine it on sloping terrain, the water will flow downhill as a stream or even a river. This is a primary distinction between aquatic systems: the difference between standing water bodies (lakes, wetlands and marshes) on relatively flat land or in a depression in the landscape, and running waters (rivers and streams) flowing down slopes. This simple distinction determines how these systems develop and function and divides them conceptually. Standing waters are called lentic systems, running waters are called lotic systems (think lo-tic: flowing). For this reason, the ecology of these two types of aquatic systems is discussed separately in Chapter 21.

The rate or periodicity of flow in an aquatic system further drives wetland formation. The hydroperiod describes the frequency of inundation in a wetland (Figure 20.3). Some wetlands have permanent waters varying little in depth (e.g. groundwater-maintained lakes). The water remains still and possibly stagnant (still and unmixed for long periods). These wetlands have a long hydroperiod. Others, such as tidal marshes, have frequent changes in flow rate, direction and water depth and a rapid hydroperiod. The level of

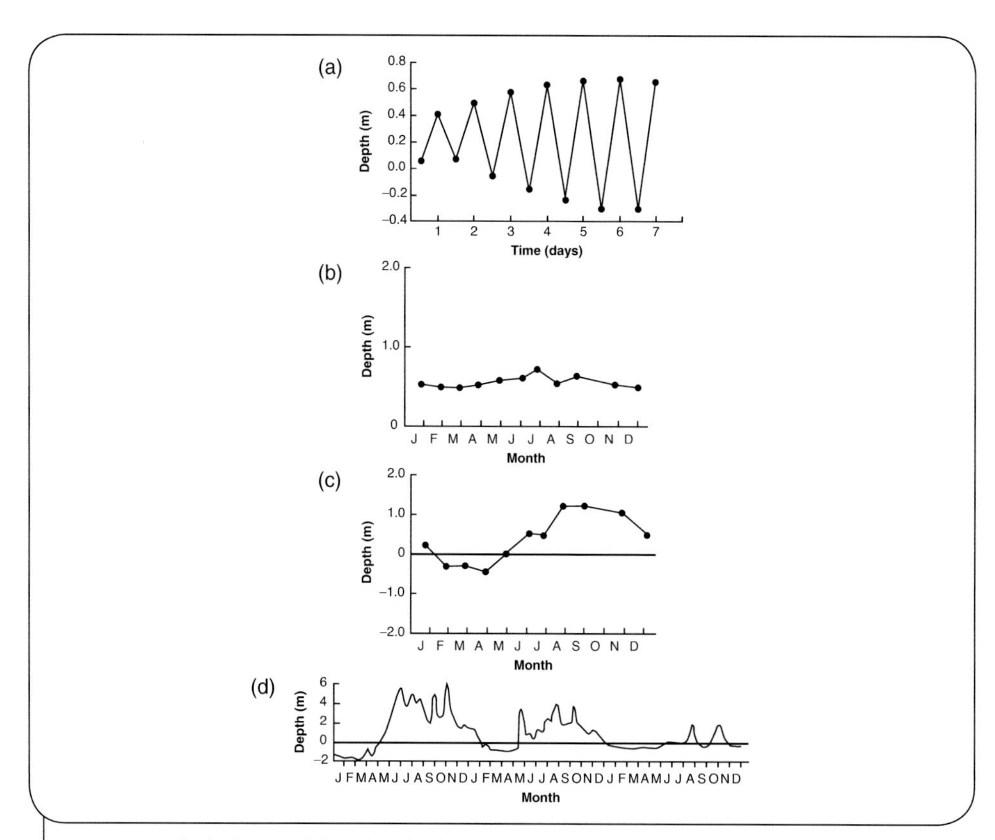

Figure 20.3 The hydroperiod describes the frequency of inundation in a wetland. (a) The rapid hydroperiod of tidal salt marsh receiving daily inflows and outflows. (b) The almost constant water level of a groundwater-fed wetland. (c) A wetland in a Mediterranean climate showing flooding with winter rainfall and seasonal drying in summer–autumn. (d) A billabong (a cut-off oxbow lake), which receives inflow only when the river floods across its flood plain. Note the differences in water level as well as frequency. The depth of water affects the type of plants and animals that can live there. The hydroperiod affects physical, chemical and biological parameters in a wetland (see Figure 20.4). (Source: B. Cale)

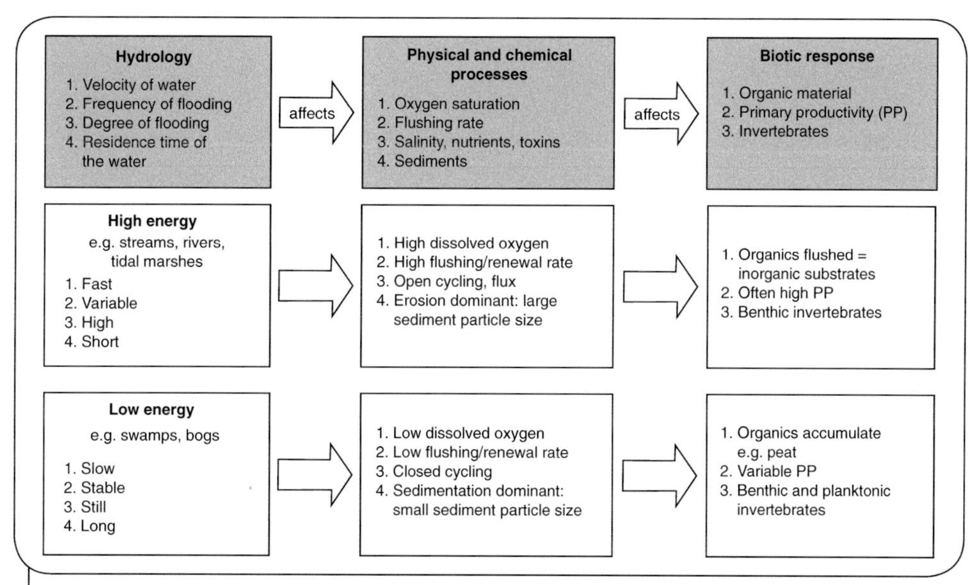

Figure 20.4 The hydrology of a wetland affects physical, chemical and biological parameters. In high-energy systems, fast-flowing water, rapid hydroperiods and short residence times result in mixing of air and water (increasing oxygen saturation), rapid movement and exchange of solutes in and out of a wetland and scouring of river beds, whereas low-energy, still water keeps all material within the confines of the wetland. (Source: B. Cale)

energy in the system is defined by the time water remains in the wetland (residence time), the rate of flow, and the change in water depth and volume in the wetland. The level of energy directly affects the physical, chemical and biological parameters of the wetland (Figure 20.4).

Human impacts

Removing water for human use can put severe stress upon lentic and lotic waters. In the past, water was seen purely as a resource for human needs. Wetlands were drained, rivers were dammed or water diverted to provide consistent supplies for humans. When the human population is low or the water quantity is high, impacts are small, but as competition for water increases, so too does the impact on the timing, duration and quantity of water available to natural systems. A relatively new concept is that there must be an allocation of water to the environment, enabling aquatic ecosystems to function and support the ecosystems and resources on which humans rely. This is the concept of 'environmental flows'. Increasingly, environmental flows are allocated in water-resource policies as awareness of interactions between hydrology and ecosystem health permeate governments, managers and the community. Sustainable water usage requires balancing the needs of farmers, industry and communities on the one hand and, on the other, the need to ensure an adequate and timely flow of good-quality water to maintain healthy aquatic ecosystems.

Environmental factors

Water quality is a prime determinant of what can live in an aquatic ecosystem. Several environmental parameters, including light, temperature, dissolved oxygen, pH, salinity and nutrients, determine water quality. They vary with hydrology, geomorphology and biotic inputs at different scales: between different regions, between different wetlands or even between different areas within a wetland. This creates complex mosaics of environmental and community types. Understanding how a wetland works requires awareness of the variability and interactions between physical, chemical and biological parameters.

Light

The sun is the primary energy source in aquatic ecosystems. The amount of light available controls photosynthesis and thereby the rate of plant growth. Hence the amount of light directly affects the productivity of the plants and (indirectly) the animals within aquatic systems, as plants form the basis of the food chain.

The amount of light within the aquatic environment is much less than in the air. As light passes through water, it is absorbed by the water itself, by dissolved substances in the water which cause colour (e.g. humic acids), and by particles in the water which cause turbidity. Particles can be either biotic (e.g. phytoplankton) or abiotic (e.g. sediment, clay and silt). These factors attenuate (or reduce) the amount of light with increasing water depth. Attenuation is described by the vertical attenuation coefficient (K_d) of the water, calculated by the equation:

$$K_d = \frac{\ln(I_0) - \ln(I_z)}{z}$$

where I_0 = irradiance at depth 0

I_z = irradiance at depth z

z = depth

K_d has the units m^{-1}, so it is describing the rate at which light decreases with depth, and is used to compare light attenuation in different water bodies. Irradiance is measured using a light or quantum meter, which measures the energy of photons in $\mu Es^{-1}m^{-2}$. The unit μE stands for microeinstein. A whole einstein is defined as 1 mole of photons (a mole being 6.02×10^{23} atoms), so it is actually measuring the number of photons or light energy.

The primary reason for measuring underwater light is to determine the light available for photosynthesis, and it is important to remember that water does not absorb light equally across all wavelengths. The important wavelengths of light for photosynthesis (photosynthetically active radiation, PAR) are between 400 and 700 nm, corresponding to light's visible spectrum. The pattern of light absorption by chlorophyll a (the dominant photosynthetic pigment in plants) is shown in Figure 20.5.

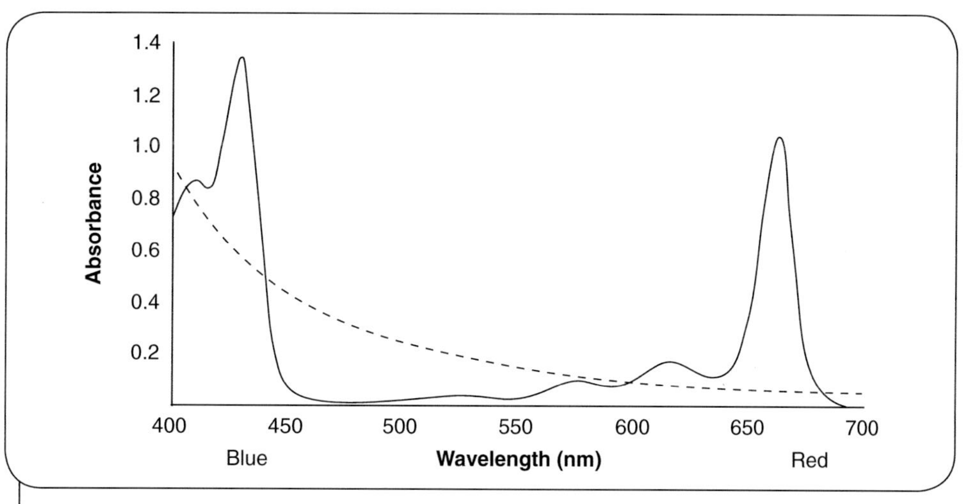

Figure 20.5 The solid line on this graph shows the relative absorption of different wavelengths of visible light by chlorophyll a. The dashed line shows the absorption by dissolved organic colour (humic substances). A comparison of the two shows that in tannin-stained water the dissolved organic carbon absorbs light in the important blue wavelengths, making it unavailable to chlorophyll and photosynthesis, thus reducing plant growth. (Source: B. Cale)

The importance of the attenuation coefficient (K_d) is that the less light transmitted through the water, the shallower the depth of the water body that supports plant growth. The depth to which plants can grow is the euphotic zone (Z_{eu}). It extends to about 1% incident light. The relationship between the attenuation coefficient and the euphotic zone is described by the equation:

$$Z_{eu} = \frac{4.6}{K_d}$$

The depth of the euphotic zone (often abbreviated to 'photic zone') can be estimated using a Secchi disc. Originated by Fr Pietro Angelo Secchi in 1865, this is a 0.2 m diameter disc either white or, more commonly, shaded in alternate black and white quadrants. It is lowered by an attached string until it becomes just invisible. The measurement at that depth is affected by the observer's eyesight, the time of day, and shadow and water characteristics, but it approximates about 5% incident light. The lower limit of the euphotic zone is approximately three times the Secchi depth.

The two main parameters reducing light transmission through the water column in inland aquatic systems are colour and turbidity.

Colour

In its broadest sense 'colour' describes factors imparting colour to the water, such as the yellow/green or red/brown of phytoplankton blooms or the muddy brown resulting from soil particles. However, in limnology (the study of inland waters), colour in natural water bodies is attributed to dissolved substances, the most common of which is dissolved organic material. This imparts a yellow or tea-coloured staining (depending on concentration) and is caused by tannins and humic acids derived from decomposed vegetation. In fact, the colour of the tea you drink is caused by the same tannins that occur in wetlands and streams, but is derived from the leaves of a species of camellia.

Colour reduces light penetration, particularly in the blue wavelengths (400–475 nm, Figure 20.5), resulting in the red colour being reflected and giving that characteristic tea colour. This skews the wavelengths available for photosynthesis and so, apart from reduction in overall light, the imbalance of wavelengths also reduces photosynthetic capability.

Colour is measured using a spectrophotometer at a certain wavelength. Various units, including hazen units or gilvin, are used. Gilvin is measured at 440 nm (the wavelength of maximum absorbance by the humic substances – see Figure 20.5) and has the units m^{-1}.

Turbidity

Turbidity is a measure of suspended solids in the water column, for example, clay, silt and algal cells. Suspended material reduces light penetration as light is absorbed, reflected or deflected by the particles (Figure 20.6). The particles may be biotic (e.g. phytoplankton) or abiotic (e.g. clay). The higher the turbidity the less light is available for photosynthesis, and hence the productivity of the aquatic ecosystem decreases (Figure 20.6). High turbidity is particularly a concern for rooted aquatic plants, which require light to reach the bottom in order to survive. The level of natural turbidity is a function of soil type, slope in the surrounding catchment and flow rate. Steep slopes, clay soils and high rainfall may result in high turbidity, especially in tropical rivers.

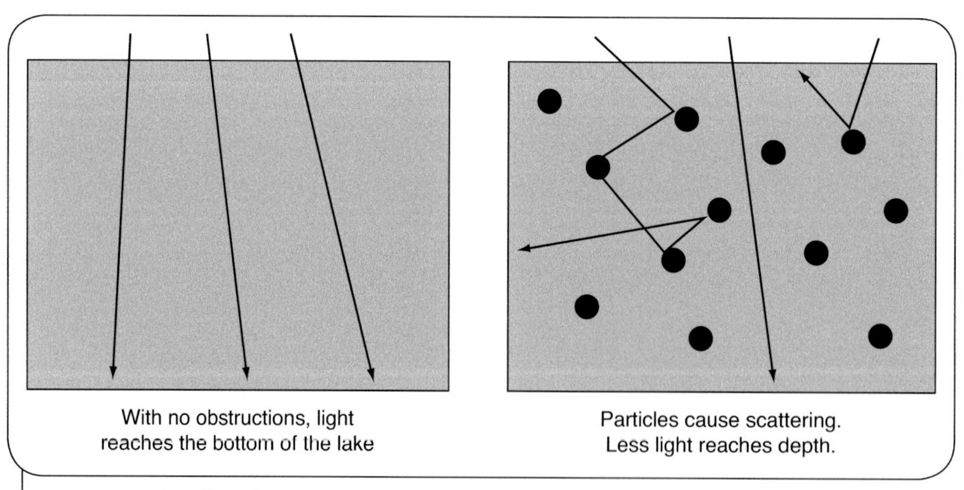

Figure 20.6 The effect of turbidity on light transmission through water. In clear water, where there are no obstructions, light passes to the bottom of the wetland. In turbid water, each time a light beam comes in contact with a particle it may be absorbed, reflected or refracted, scattering the light in different directions. The more particles in the water, the less light continues downwards through the water column. (Source: B. Cale)

Turbidity is measured by passing light through water and measuring the proportional reduction using a turbidity meter. There are various units, the most common being NTU or nephelometric turbidity units, an arbitrary unit related to standardised suspensions of the polymer formazin. Total suspended solids (TSS), the component that causes turbidity, are measured by filtering a known volume of water through a pre-weighed filter paper and determining the weight of sediment per unit volume of water, for example gL^{-1}. There is no direct relationship or conversion possible between TSS units and NTU.

There is often some confusion between observing turbidity and colour. People sometimes think if water is darker it must have higher turbidity. The difference is that water may be clear but have dissolved colour, whereas turbidity clouds the water. Put water in a glass beaker and place your finger behind the beaker: if the water is coloured, the outline of your finger will be sharp but darker, and if it is turbid your finger outline will be blurred.

Human impacts

Increases in turbidity result from increased sediment loads caused primarily by land-clearing that results in erosion (Figure 20.7). Apart from reducing light levels,

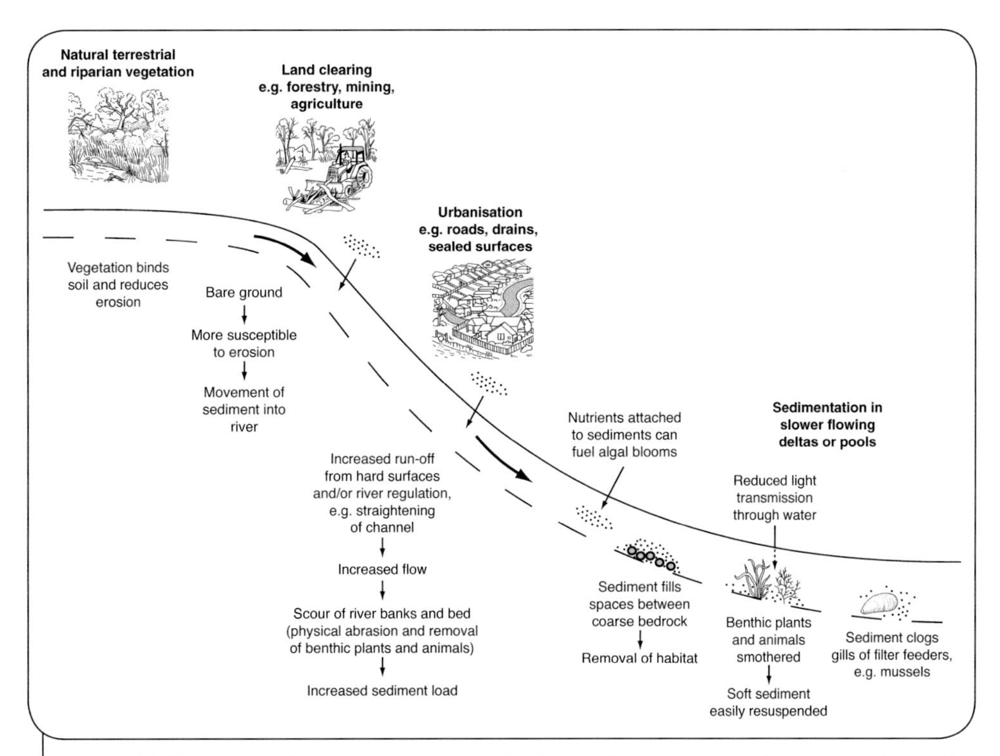

Figure 20.7 The impact of land use on erosion and sedimentation in a river. In the upper reaches of the catchment, land clearing increases flow and erodes the river bed and banks. The increased sediment load resulting from this erosion, often with attached nutrients, is deposited when the river slows in river pools or in flatter deltas before it reaches the sea. (Source: B. Cale)

abiotic sediment can severely affect invertebrates that filter water for food or oxygen. The silt provides no sustenance and clogs filtering mechanisms and gills, potentially causing death. Sedimentation also alters habitats and can smother benthic vegetation (Figure 20.7). Careful land management together with maintaining riparian buffers (vegetation along the margins of rivers) are the best means of reducing erosion and sedimentation.

Increases in turbidity may also be caused by biotic particles such as algal cells (phytoplankton). This reduces light in the water column and can kill benthic vegetation. This in turn can increase abiotic turbidity because the sediment is no longer bound by a root mass (see Chapter 21 on plants and wetland function) and is readily suspended by wind and currents. An increase in phytoplankton is caused by input of nutrients to the aquatic system (see 'Human impacts' under 'Nutrients' below).

Removal of native vegetation, particularly trees and sedges fringing wetlands, can also reduce the colour of a wetland. Removing vegetation reduces sources of humic acids (decomposing plant material) and also removes shading of the water column. Humic acids are then broken down by photo-oxidation. The resultant clear water is often subject to algal blooms, as the light limitation is removed and humic wetlands may be nutrient rich.

Temperature

The temperature of inland aquatic water is not buffered by volume as is water in the ocean and therefore it varies more. Temperature may change greatly daily or seasonally, and does not necessarily decrease uniformly from the top to the bottom of the water column. This may cause stratification (described below) with serious effects on wetland physical, chemical and biological processes. Temperature affects the rate of metabolic processes, which in turn affect productivity. Growth and blooms of algae, for example, occur in the warm, high-light conditions of spring and summer, whereas during cool winters plants may be dormant or die back to resting states. Most organisms have an optimum range of temperature that determines their geographic ranges. Temperature also affects growth and respiration by influencing oxygen availability. Hot water holds less oxygen than cold water (see below).

Temperature is measured using a thermometer in the SI units of degrees Celsius (°C). It is important to consider the time of day when measuring water temperature as temperature fluctuates over a diurnal cycle.

Human impacts

Temperature becomes important when conditions change. For example, clearing trees from a wetland decreases shading, increasing water temperature and algal growth. Animals often have specific temperature cues for spawning or reproduction and if water temperature is changed through thermal (hot or cold) water pollution (e.g. the release of cold water from the bottom of dams, or heated water from industrial processing plants) it can have serious implications for their life cycle.

Dissolved oxygen

Oxygen comprises 21% air (by volume) but in water the amount is 35 times lower, at only 0.6%. The availability of oxygen, required by all aerobic organisms for respiration, is therefore critical. The amount of oxygen in the water depends on the atmospheric pressure at the air–water interface, the temperature, and the concentration of other dissolved substances in the water. This is because the amount of oxygen held in water depends on the atmospheric pressure exerted. If this increases, more oxygen dissolves. However, the total pressure of gases in the water is the sum of all partial pressures exerted by each gas, so if there are many gases or solutes in the water, proportionally less oxygen dissolves. The partial pressure of oxygen is directly proportional to the number of molecules, or the concentration, of the gas.

Increasing temperature decreases the pressure within the water and so, as temperature increases, dissolved oxygen concentration decreases. The concentration of dissolved substances is generally measured as the salinity of the water. As salinity increases, the dissolved oxygen concentration decreases. The effects of temperature and salinity on dissolved oxygen are shown in Table 20.1. Oxygen is measured as a concentration usually as mgL^{-1} or ppm, using an oxygen probe. The meter must correct for salinity and temperature. Some oxygen meters are designed only for fresh water and are inaccurate in salt water.

As dissolved oxygen concentrations are so variable (depending on temperature and salinity), the amount of oxygen in water is often expressed as % saturation. Percentage saturation is the ratio between the concentration of oxygen measured in a water body at any one time divided by the maximum possible concentration of oxygen at the given salinity and temperature. For example, the maximum dissolved oxygen concentration at 20°C and 0 ppt salinity is 9.1 ppm (Table 20.1). This is 100% saturation of oxygen at this temperature and salinity.

Various factors increase or decrease dissolved oxygen concentration. Oxygen enters the water by the physical mixing of air and water resulting from turbulence, currents, wind mixing or waterfalls. You can see entrainment of oxygen occurring when you see bubbles of air caught under water such as at the bottom of a waterfall. Under these conditions oxygen saturation is 100%. Oxygen also enters the water via photosynthesis. If plants and algae are photosynthesising in water that is already well mixed, greater than

Table 20.1 Dissolved oxygen saturation based on temperature and salinity calculated at sea level.

Temperature (°C)	Salinity				
	0 ppt	10 ppt	20 ppt	35 ppt	60 ppt
0	14.6	13.7	12.8	11.5	9.7
10	11.3	10.7	10.0	9.1	7.8
20	9.1	8.6	8.1	7.4	6.4
30	7.6	7.1	6.7	6.2	5.4
40	6.4	6.0	5.7	5.2	4.5

100% saturation (supersaturation) can occur. When there is an algal mat on the bottom of a wetland you can often see bubbles of oxygen beading the mat in the middle of the day, when photosynthesis is highest.

Oxygen is lost from the water during mixing with the air, especially at times of supersaturation, but, more importantly, through the respiration of aerobic organisms including invertebrates, fish or aerobic microbes. 'Aerobic' simply means organisms are reliant on oxygen for respiration; in contrast, anaerobic organisms can respire in the absence of oxygen and use nitrate or other molecules. Lentic systems, rather than streams, are susceptible to deoxygenation because they are still. If protected from wind by trees, which reduce wind mixing, they are particularly susceptible. Oxygen can be rapidly stripped from still water if there is a large amount of organic material present, as microbial decomposition of the organic matter uses oxygen. This is particularly so at night when photosynthesis stops and there is no oxygen source to replenish supply. In these cases anoxic conditions can occur (anoxic: absence of oxygen), particularly in deeper water, because of the distance from the replenishing atmosphere. Fish and other aerobic organisms are killed. Different organisms require different amounts of oxygen. Large animals such as fish are most susceptible to low-oxygen conditions. As a general rule, fish cannot survive in water with less than 30% oxygen saturation.

pH

The pH of water is the potential of hydrogen, a measure of its acidity or alkalinity on a scale from 0 to 14. Specifically, it describes the logarithm of the reciprocal of hydrogen ion concentration in gram atoms per litre. High concentrations of hydrogen ions (H^+) cause the water to be acidic and have pH values less than 7. A low concentration of hydrogen ions (or a high concentration of hydroxyl ions, OH^-) causes the water to be alkaline, or basic, and have pH greater than 7. Water is neutral at pH 7 (Figure 20.8).

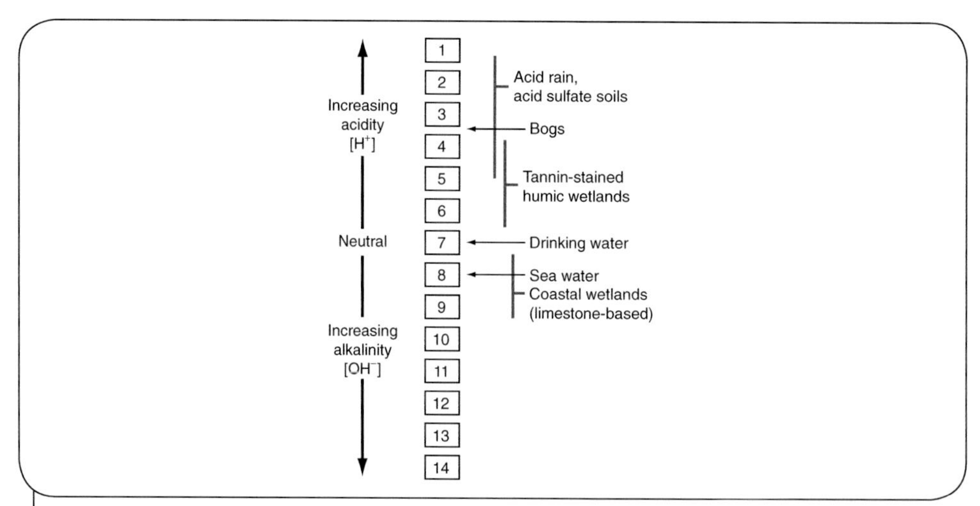

Figure 20.8 The pH scale from 1 to 14, showing the range of pH in natural and human-affected wetlands. (Source: B. Cale)

The pH of water is affected by solutes from the surrounding environment. For example, humic acids from the decomposition of vegetation (see 'Colour' above) cause acidity (pH 4–6), but only when they are not bound to wetland sediments. In granitic sandy soils or where peat formation is deep, humic acids are influential. Very low pH results in low wetland metabolism. Decomposition processes are slow, preserving organic material in the peat. Human bodies over 1000 years old have been found perfectly preserved in peat bogs in Europe (e.g. Tollund Man – Plate 20.2). Conversely, limestone results in waters being alkaline. Most coastal wetlands and estuaries underlain by limestone have a pH between 8 and 9. Increases in pH also occur during high photosynthetic activity because the reduction in carbon dioxide in the water column affects the carbonate equilibrium.

pH is routinely monitored in limnological studies, but small changes in pH rarely have large effects on biological systems. Organisms are usually adapted to the natural pH range within their environment. Extremes of pH (whether acid or basic) limit the productivity of most plants and animals. Chemical transformations on which organisms depend (e.g. carbon availability – see below) may also be affected by pH.

pH is usually measured via an electrode, the voltage of which changes as the hydrogen ion concentration changes, but a rough approximation can be made using colour reagents or papers (e.g. litmus paper).

Human impacts

Changes in pH because of anthropogenic factors can be catastrophic. The two main concerns are acid rain and acid sulfate soils. Acid rain occurs when gases, particularly sulfur dioxide (SO_2) and nitrogen oxides (NO_x^-), react in the atmosphere with water, oxygen and other chemicals to form acids. This is common in industrialised areas of Europe and North America, but less so in the Southern Hemisphere. In the USA, about two-thirds of all sulfur dioxide and a quarter of nitrogen oxides come from electric power generation burning fossil fuels such as coal.

Acid sulfate soils are common in coastal wetlands (e.g. salt marshes) where there are high concentrations of sulfur, iron and organic material forming pyrite (FeS_2) under anoxic conditions. Under natural conditions they are not a problem, but when human activity drains or modifies the hydrology of the wetland exposing the sulfides to air, oxidation occurs, producing sulfuric acid.

Acid rain and acid sulfate soils have similar effects. The pH of the wetland becomes highly acidic (pH of 1–2 has been measured) and arsenic and metal (e.g. aluminium, iron and lead) concentrations become toxic because metal solubility increases at low pH. This kills fish and plants (especially from aluminium and manganese toxicity), reduces aquatic biodiversity and creates predator-free habitats where pests such as mosquitoes (which tolerate low pH) breed. This increases the risk of mosquito-borne diseases. Not only the immediate wetland may be affected, but water flow may contaminate ground water.

Salinity

Salinity describes the salt concentration of the water. Salts are comprised of cations (positively charged ions) and anions (negatively charged ions). Inland waters vary in ionic composition. Four cations (sodium, potassium, magnesium and calcium) and three anions (chloride, sulfate and bicarbonate) account for almost all of the total ionic concentration or salinity.

In many areas, in particular Europe and North America, fresh waters are usually dominated by calcium and bicarbonate. In Australia, sodium and chloride dominate in all inland waters (both fresh and saline), although bicarbonate can be dominant or co-dominant with chloride. Sodium chloride dominance is largely caused by the atmospheric supply of oceanic salts to inland waters (much of the salt comes from sea spray transported on rain and dust). In some areas of inland Australia salt lakes may result from former marine sediments or saline ground water.

Salinity is measured as salt concentration in mgL^{-1} (ppm) or more typically in gL^{-1} (ppt). However, a salinity meter generally uses the capacity of a salt solution to conduct electrons. Ions in the water conduct an electrical charge. This is known as conductivity. Particularly in 'fresh' waters where salinities are low, salinities may be expressed as conductivity in μScm^{-1} or $mScm^{-1}$. Conductivity can be converted to salinity. The conversion factor depends on pressure, temperature and the type of salts present. In Australia, where sodium chloride is the dominant ion, a close approximation of salinity can be obtained using the equations:

$$\mu S/cm \times 0.64 = mgL^{-1} \text{ (ppm) or } mScm^{-1} \times 640 = mgL^{-1} \text{ (ppm)}$$

Salinities vary in inland waters from fresh (0 ppt), to various concentrations resulting from the dilution of sea water (35 ppt), to salt lakes, which may have salinities up to saturation point (360 ppt for sodium chloride at 20°C) (Table 20.2). Fresh water is defined as having a salinity of less than 0.5 ppt (about 800 μScm^{-1}); the upper limit for drinking water set by the World Health Organization. The ecological boundary between fresh and saline water is often debated and depends on the part of the world in which you live. Organisms living in permanent fresh water are much more susceptible to increases in salinity than those in parts of Australia where seasonal variation in water depth results in evapoconcentration of salts annually. Most freshwater organisms show stress above

Table 20.2 Typical salinities of different aquatic ecosystems: units and concentrations. Conductivities are not provided for sea water, salt lakes and marshes, as conductivity is generally only used to measure salinity of less than 5000 mS/cm.

Aquatic system	Salinity (ppt)	Conductivity (mS/cm)
Fresh	0	0
Brackish	2	3125
Sea water	35	NA
Estuarine salt marsh	20–60	NA
Salt lake	20–360	NA

1 ppt (about 1500 μS/cm), and 3 ppt (about 5000 μS/cm) is considered the upper limit of freshwater ecosystems.

Changes in salinity affect osmoregulation. Organisms adapted to fresh water have a higher osmotic potential (salt concentration of their body fluids) than the surrounding water, although most have a lower concentration than comparable marine organisms. As explained earlier (Chapter 13), this means there is a tendency for water to be gained by osmosis and for ions to be lost by diffusion. Freshwater organisms are hyperosmotic regulators, maintaining body salts by excreting hypo-osmotic urine with respect to the body fluids and uptaking ions actively.

Most freshwater animals cannot hypo-osmoregulate (they cannot maintain their osmotic pressure below that of the surrounding water) and are usually restricted to salinities below 1 ppt. Animals inhabiting highly saline waters are usually hypo-osmoregulators (for example, the Australian brine shrimp *Parartemia*) or osmoconformers where conformity between body fluids and the external medium is maintained over a wide range of salinities. Aquatic animals tolerating wide variations in salinity are called euryhaline and those with a limited tolerance are called stenohaline.

Human impacts

Increased salinity following anthropogenic landscape modifications is often called secondary salinisation. There are two forms. Dryland salinity is caused by the removal of native, perennial, deep-rooted vegetation and its replacement by shallow-rooted annual crop species (Figure 20.9a). The widespread loss of deep-rooted vegetation reduces the amount of evapotranspiration and the surface capture of rainwater, causing increased run-off and recharge. These hydrologic changes cause groundwater tables to rise, bringing to the surface salt stored in the soil through millennia of input from salt-laden winds and rain. The rising watertable also waterlogs surface soils. Secondary salinisation is widespread in Australia, mostly (70%) occurring in south-western Australia. The effect on wetlands is twofold: an increase in salinity and deeper, more permanent water.

Irrigation salinity is caused by the build-up of salt near the soil surface where drainage is inadequate. Irrigation water recharges the ground water, bringing stored salt to the surface. This can be exacerbated by fertiliser addition and evapotranspiration of surface water from the irrigated area concentrate salts over time (Figure 20.9b).

Aquatic ecosystems are vulnerable to secondary salinisation because their low-lying position makes them susceptible to both groundwater intrusion and floods of saline surface water. This kills freshwater organisms and reduces biodiversity. Only organisms with a high salinity tolerance survive. Within the water column there is a shift from mainly species of freshwater macrophytes to fewer species of salt-tolerant macrophytes (both angiosperms and algal charophytes) up to about 45 ppt and then a dramatic shift to algal mats, which can survive up to 300 ppt. Invertebrate communities shift from insect-dominated populations in fresh water to more salt-tolerant crustacea, but always with a decrease in diversity with increasing salinity. Few organisms survive the stress of high salt content.

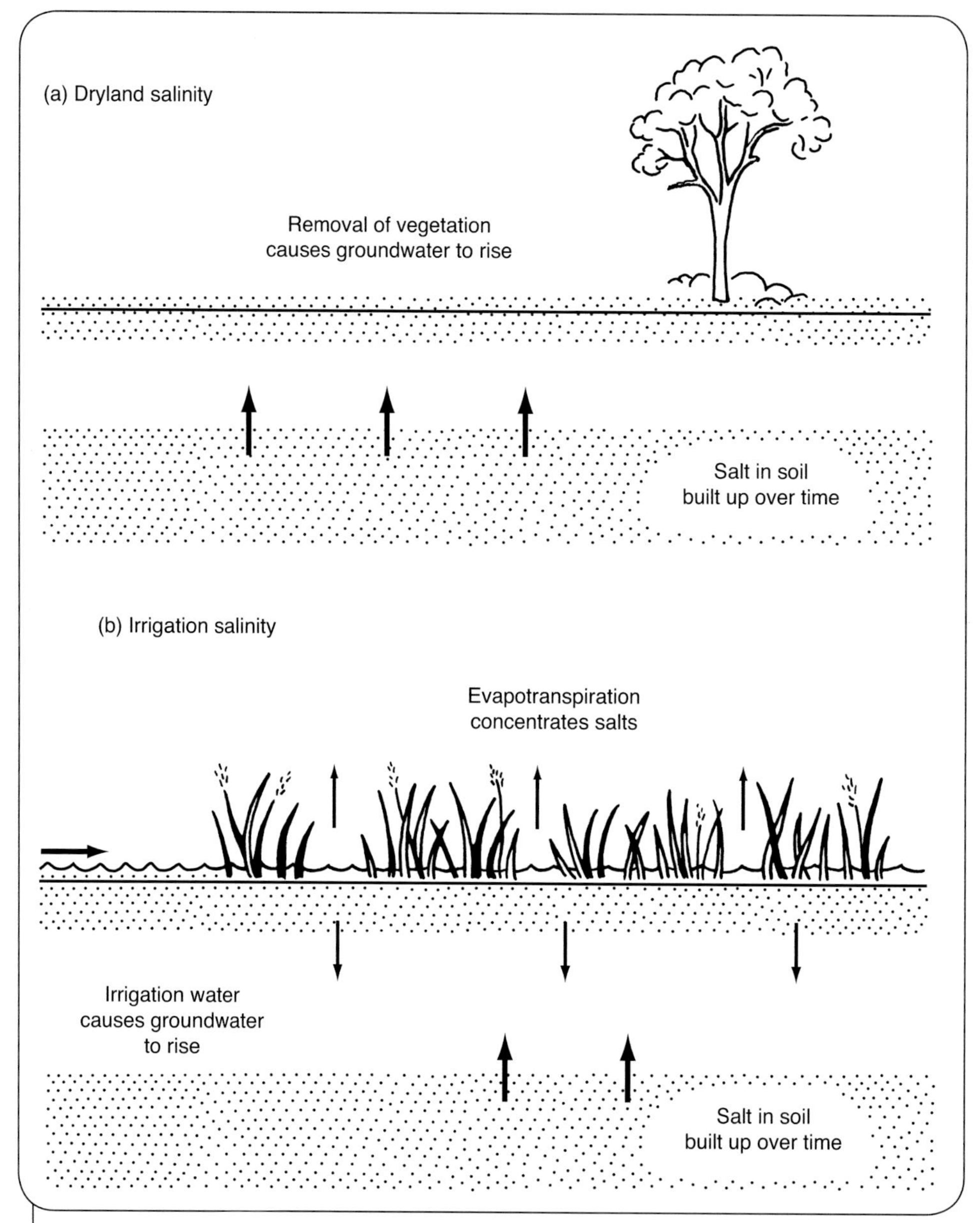

Figure 20.9 Salinisation caused by (a) dryland and (b) irrigation salinity. Dryland salinity is caused by the replacement of deep-rooted native vegetation with shallow-rooted annual cropping species which use less water, resulting in a rise of the groundwater table bringing salt stored in the soil profile to the surface. Irrigation salinity is caused when increased inputs of surface water result in a similar rise in the groundwater table. Evaporation of slightly saline irrigation water at the soil surface can exacerbate surface salinities. (Source: B. Cale)

Nutrients

Nutrients are essential for plant growth (Table 12.3), yet few make up more than 1% of the dry weight of a plant. Their availability controls plant productivity and thereby the productivity of the whole food chain. If we look at the composition of plant tissue we can see which nutrients are required in the greatest amounts for growth. Many aquatic plants are composed of up to 99% water (hydrogen and oxygen) that is plentiful in wetlands. Next is carbon – carbon dioxide (CO_2) from the atmosphere dissolves into water to form carbonic acid (H_2CO_3). This rapidly dissociates into carbon dioxide in solution, bicarbonate (HCO_3^-), carbonate (CO_3^{2-}) and traces of carbonic acid, the proportion depending on the pH of the water (Figure 20.10). The different forms of carbon are important, because all plants can use free CO_2 but only larger plants (macroalgae and flowering aquatic plants) can use bicarbonate. Carbonate is not available for plant growth. The availability of carbon then is dependent on the equilibrium between these different forms of carbon, controlled in turn by pH. Carbon dioxide is freely available from the atmosphere, so carbon availability rarely limits plant growth.

The remaining nutrients that may contribute more than 1% of the dry weight of a plant are nitrogen, phosphorus, sulfur, magnesium, calcium and potassium. The concept of 'limiting factors' suggests that if one of these nutrients is in a lower supply than the others, it will limit the rate of plant growth. In aquatic ecosystems, the nutrients in low supply are usually nitrogen or phosphorus. There are several forms of nitrogen and phosphorus (Table 20.3), but the inorganic nutrients are used for plant growth. The inorganic forms of nitrogen are oxidised nitrogen (NO_x^- – comprising nitrate (NO_3^-) and nitrite (NO_2^-)) and ammonium (NH_4^+). Nitrate is generally found under aerobic conditions (Figure 20.11) and frequently enters wetlands in run-off from urban and agricultural land. It derives from fertilisers and nitrogen-fixing *Rhizobium*/legume associations. Ammonium forms under anaerobic conditions (Figure 20.11), primarily from the breakdown of organic matter or nitrogenous wastes (urea). The inorganic form of phosphorus is orthophosphate (PO_4^{3-}), which also enters aquatic systems as phosphatic fertiliser in run-off.

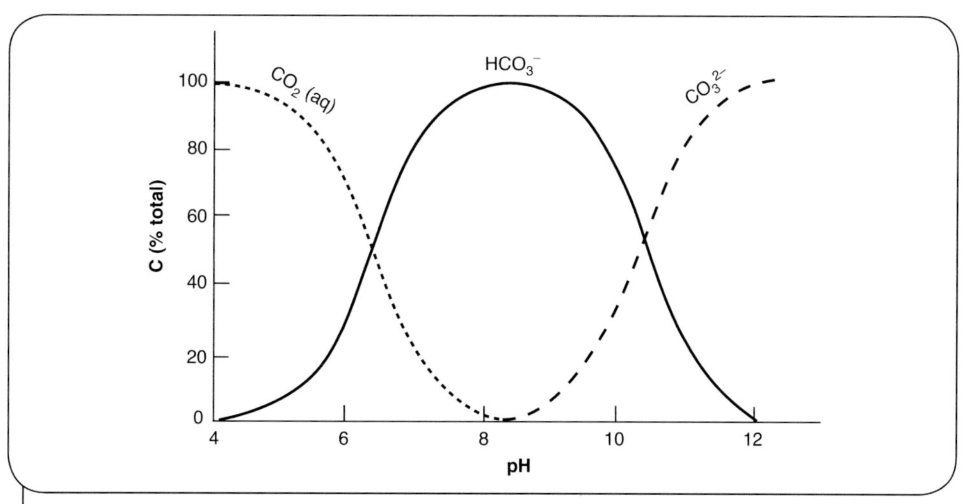

Figure 20.10 The proportion of dissolved carbon as carbon dioxide (CO_2), bicarbonate (HCO_3^-) or carbonate (CO_3^{2-}) as a function of pH of the water. (Source: B. Cale)

Table 20.3 Forms of nitrogen and phosphorus.

Form	Nitrogen	Phosphorus
Gas	Nitrogen (N_2) Ammonia gas (NH_3)	None
Dissolved inorganic	Oxidised nitrogen (NO_x^-), comprising nitrate (NO_3^-) and nitrite (NO_2^-) Ammonium (NH_4^+)	Orthophosphate (PO_4^{3-}) (the actual molecule) Soluble reactive phosphorus (SRP) (the term includes all dissolved phosphorus smaller than 0.45 μm)
Dissolved organic	N bonded to carbon molecules (e.g. urea, $CO(NH_2)_2$)	P bonded to carbon molecules (C–O–P, C–P bonds)
Particulate or sediment bound	Organic (live or dead biotic material) Inorganic (e.g. clay-bound)	Organic (live or dead biotic material) Inorganic (e.g. clay-bound, apatite or non-apatite) Apatite P – bound in crystal lattices of apatite grains (e.g. $Ca_5(PO_4)_3OH$). Not available for algal growth (acid soluble). Non-apatite P – adsorbed, complexed with iron and other compounds (e.g. ferric hydroxy phosphates such as $Fe_3(PO_4)_2.8H_2O$). Available for algal growth (alkali soluble)

Both nitrogen and phosphorus undergo several transformations in the aquatic environment as described by their biogeochemical cycles (Figures 20.11 and 20.12). In lentic wetlands, this cycling is primarily maintained within the water body but in rivers water is moving downhill as the nutrients cycle, resulting in nutrients spiralling between water and sediment as they travel from the headwaters to the river outflow.

Nitrogen has a more complex cycle than phosphorus, as some organisms (particularly the cyanobacteria) are able to 'fix' nitrogen gas from air. The atmosphere is 78% nitrogen gas (N_2) and provides plentiful supplies of this potentially limiting nutrient for organisms able to use it. This external supply also makes it difficult to control plant growth through limiting nitrogen, as a reduction will provide a competitive advantage only to nitrogen-fixing organisms. There is no gaseous supply of phosphorus, so this nutrient is often the focus of management programs to reduce algal growth. Many transformations of nitrogen and phosphorus are microbially mediated and occur only under either aerobic or anaerobic conditions (Figures 20.11 and 20.12). Nitrogen fixation is actually an anaerobic reaction, but cyanobacteria in oxygenated water fix nitrogen by carrying out the transformation in sealed anaerobic cells known as heterocysts.

Some of these transformations are redox (reduction/oxidation) reactions in which electrons are lost from one substance (oxidation) and accepted by another (reduction). The tendency of a substance to gain electrons and be reduced is its redox potential. Substances with higher redox potentials have a greater tendency to gain electrons. Redox

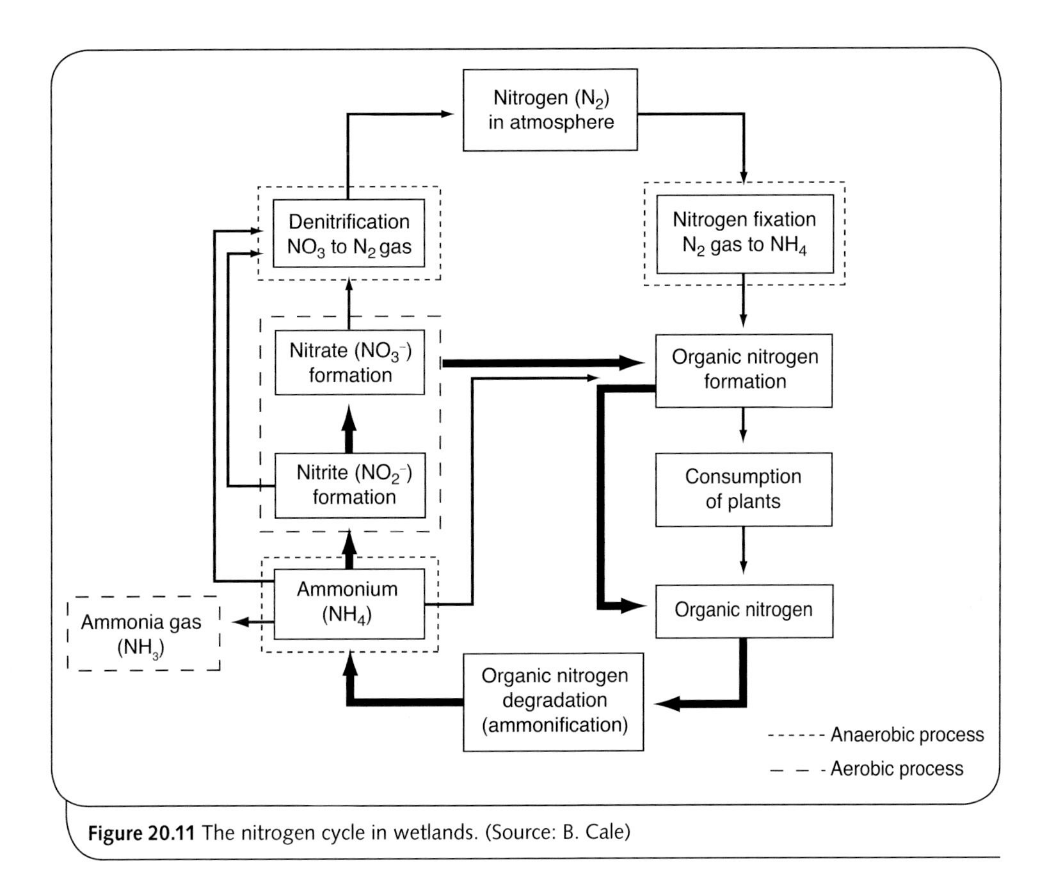

Figure 20.11 The nitrogen cycle in wetlands. (Source: B. Cale)

reactions always occur together and they release energy used by microbes for respiration. Aerobic respiration is a redox reaction releasing the greatest energy per molecule of oxygen, but anaerobic organisms can use nitrate when oxygen is not present and its consequent reduction (called denitrification) results in a loss of nitrogen from the aquatic

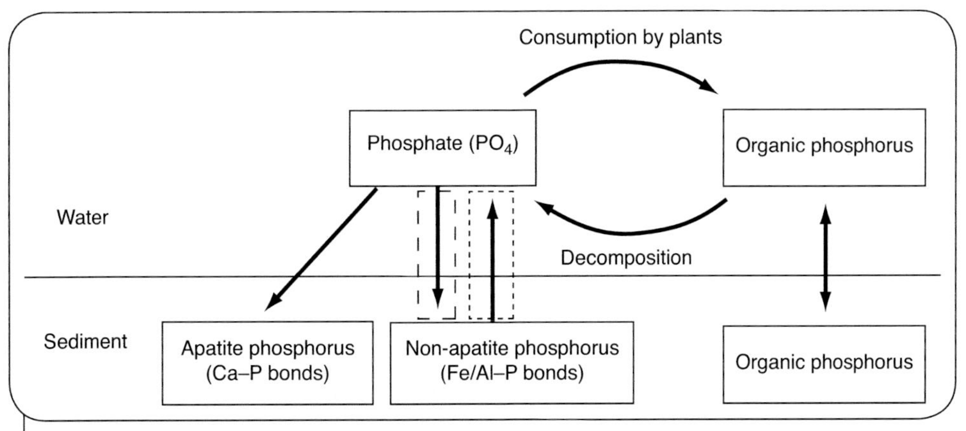

Figure 20.12 The phosphorus cycle in wetlands. Organic phosphorus may be in the form of live (e.g. phytoplankton) or dead (e.g. detritus) material. Aerobic processes are outlined in a large dashed line, anaerobic processes in a dotted line. (Source: B. Cale)

system in the form of nitrogen gas. Other bacteria (*Nitrosomonas* and *Nitrobacter*) oxidise ammonium to nitrite and nitrate. This process is called nitrification. Under very low redox potentials (100 to –100 mV) bacteria use iron (Fe) as an electron acceptor. If the iron is bonded to phosphorus, this transformation renders it soluble, releasing orthophosphate into the water column. As an aside, at even lower redox potentials microbes use sulfate or carbon dioxide, which is why you often smell 'rotten egg gas' (hydrogen sulfide) or methane in wetlands.

Analysis of nitrogen and phosphorus is usually carried out colorimetrically using a spectrophotometer. Chemicals that react to produce a colour, the intensity of which is directly proportional to the concentration of the element measured, are added. 'Total' nitrogen or phosphorus, which includes all forms of the nutrient (inorganic, organic, dissolved or particulate) requires that the bonded forms in organic material or on particulates are first broken down to a soluble form, which can be measured colorimetrically. This is done by digesting with a strong acid after filtering to remove larger particles or molecules. For example, water to be analysed for orthophosphate must first be filtered using at least a 0.45 μm filter paper. In reality, most of this phosphorus is usually available for plant growth. The fractionation of nitrogen and phosphorus in its different forms can be achieved by subtracting inorganic measurements from the 'total' analysis to determine the organic or particulate component. Digestion for total nitrogen results in ammonium as its breakdown product; this is known as total Kjeldahl nitrogen (TKN). To calculate total nitrogen, TKN must be added to the concentration of nitrate in the water.

Eutrophication

Lentic water bodies can be divided on the basis of their organic-matter production. Oligotrophic lakes are usually characterised by greater depth, a poorly developed littoral zone and low concentrations of nutrients such as nitrogen and phosphorus. There are few organisms present and very little organic matter. The water is usually clear and well oxygenated. Eutrophic lakes are shallow and rich in organic matter and nutrients. Oxygen depletion may occur during summer because of the accumulation of dead organic matter and rapid oxygen use by aerobic decomposers.

As a lake ages, it steadily fills with sediment and organic matter from catchment erosion and biological activity. It becomes more eutrophic with age and eventually progresses from an open water system, to marshlands, to dry land. For a deep oligotrophic lake such a process may take millions of years, whereas the life span of shallow wetlands may be much shorter.

Human impacts

One of the primary human impacts on aquatic ecosystems is the acceleration of eutrophication by adding nutrients such as fertilisers from urban or rural land uses, and waste water from sewage or industry. Eutrophication that would naturally take thousands of years compacts into years or decades. This is often termed cultural eutrophication. The main symptom of nutrient enrichment is excessive plant growth. In the early stages this may mean a rapid growth of flowering plants, but these will be overtaken by epiphytic plants (see Chapter 21) and macroalgae (which are better adapted to absorb nutrients directly from the water column). They in turn will be out-competed by phytoplankton and cyanobacteria (especially when nitrogen is limiting). Through this succession, the biodiversity and functionality of the ecosystem will decrease. Many cyanobacteria are toxic. Excess plant material decomposes, stripping oxygen from the water, killing fish and other aerobic organisms. Only hardy species survive. Deoxygenation promotes ammonium- and phosphorus-release from the sediment, further fuelling algal growth.

The best solution for cultural eutrophication is reducing nutrient run-off from the catchment. In some cases the sediments of the water body become so enriched that, even if no further nutrients were added, internal nutrient cycling would maintain eutrophic conditions for many years. Common management strategies include oxygenation (to prevent anaerobic phosphorus release and support aerobic organisms), dredging and removal of enriched sediment, treating the water with chemicals to bind phosphorus (e.g. alum or modified clays such as Phoslock™) or improving flushing of the water body (such as building a channel out to sea as was done in the Peel-Harvey estuary in Western Australia).

Stratification

Stratification of layers of water of different densities can occur in aquatic systems as in marine systems (Chapter 18). Differences in density can be caused by temperature or water chemistry, primarily salinity.

Water is most dense at 4°C (Figure 20.13). Increases in temperature above 4°C decrease water density. Density also decreases as water freezes, as we know from the way

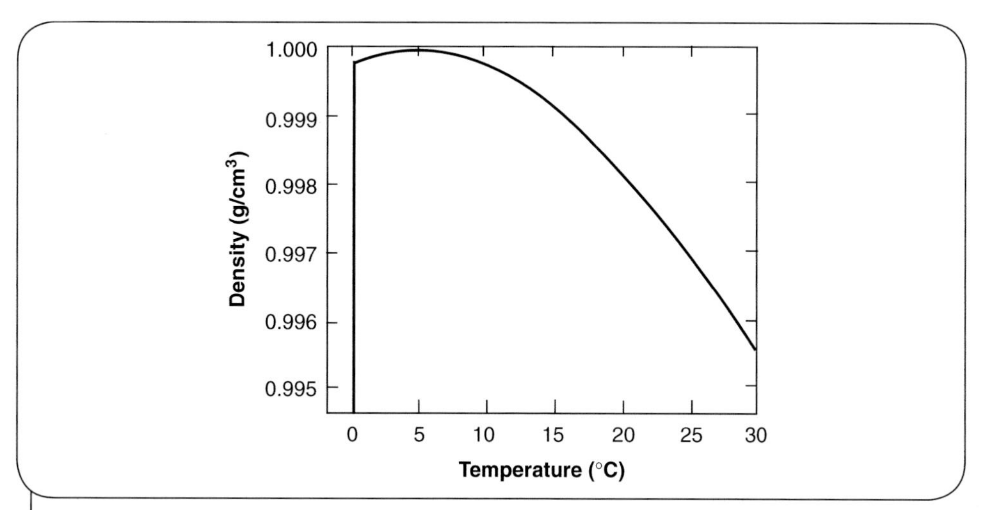

Figure 20.13 The density of water is related to temperature. Water is at its most dense at 4°C. (Source: B. Cale)

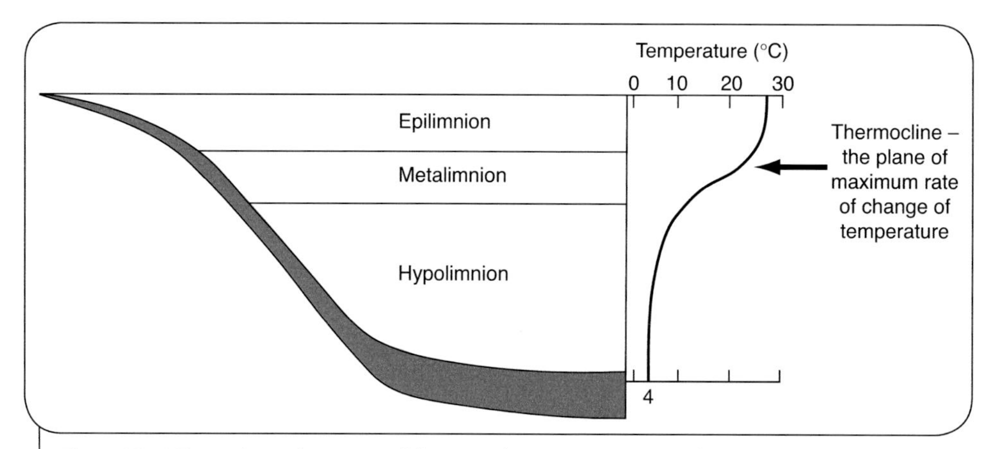

Figure 20.14 Thermal stratification in a lake. Less-dense warm water (epilimnion) overlies denser, colder water (hypolimnion). The boundary between the two is called the metalimnion (the region in which the thermocline occurs). (Source: B. Cale)

ice blocks float in summer drinks. Thermal stratification in temperate regions occurs where sun-warmed surface waters stay at the top of a lake and do not mix with the denser, colder water underneath. It is generally a feature of deeper lakes (or reservoirs). The upper warmer layer is the epilimnion and the deeper colder layer is the hypolimnion. Between them is a narrow region of rapid temperature change, known as the thermocline (Figure 20.14). This type of stratification occurs once a year over summer and is known as monomictic stratification (Figure 20.15a).

The nature and extent of thermal stratification is determined by depth, climate and degree of exposure to prevailing winds. The action of wind across a lake surface generates horizontal and vertical currents, which mix the water; this may either break down thermal stratification or prevent its establishment. Where this occurs, a water body is considered to be fully mixed. This is generally the case in shallow standing waters such as wetlands.

In cold climates, stratification occurs twice: once during summer, as described above, and then in winter, when the lake freezes. Complete circulation (mixing) occurs in spring and autumn. This is a dimictic pattern (Figure 20.15b). In winter, water at 4°C sinks beneath cooler water, which freezes at the surface at 0°C. Plants and animals survive below the ice in water between 0°C and 4°C. In spring, as the ice melts, the surface water warms to 4°C and again sinks below the cooler water, forcing the bottom water to the surface. This is known as the spring overturn and it brings nutrients from the bottom of the lake to the surface and oxygen to the bottom. In summer, warmer surface waters stay on top and do not mix with the cooler waters underneath. During autumn, the air and surface-water temperatures fall, eventually leading to an autumn overturn. Lakes in the temperate regions of the Northern Hemisphere generally undergo two overturns each year. In the Arctic and parts of the tropics, there is usually only one overturn. In Australia, dimictic lakes are rare and restricted to alpine regions. The two commonest patterns are polymictic and warm monomictic. Polymictic lakes are generally well mixed and stratification usually lasts for

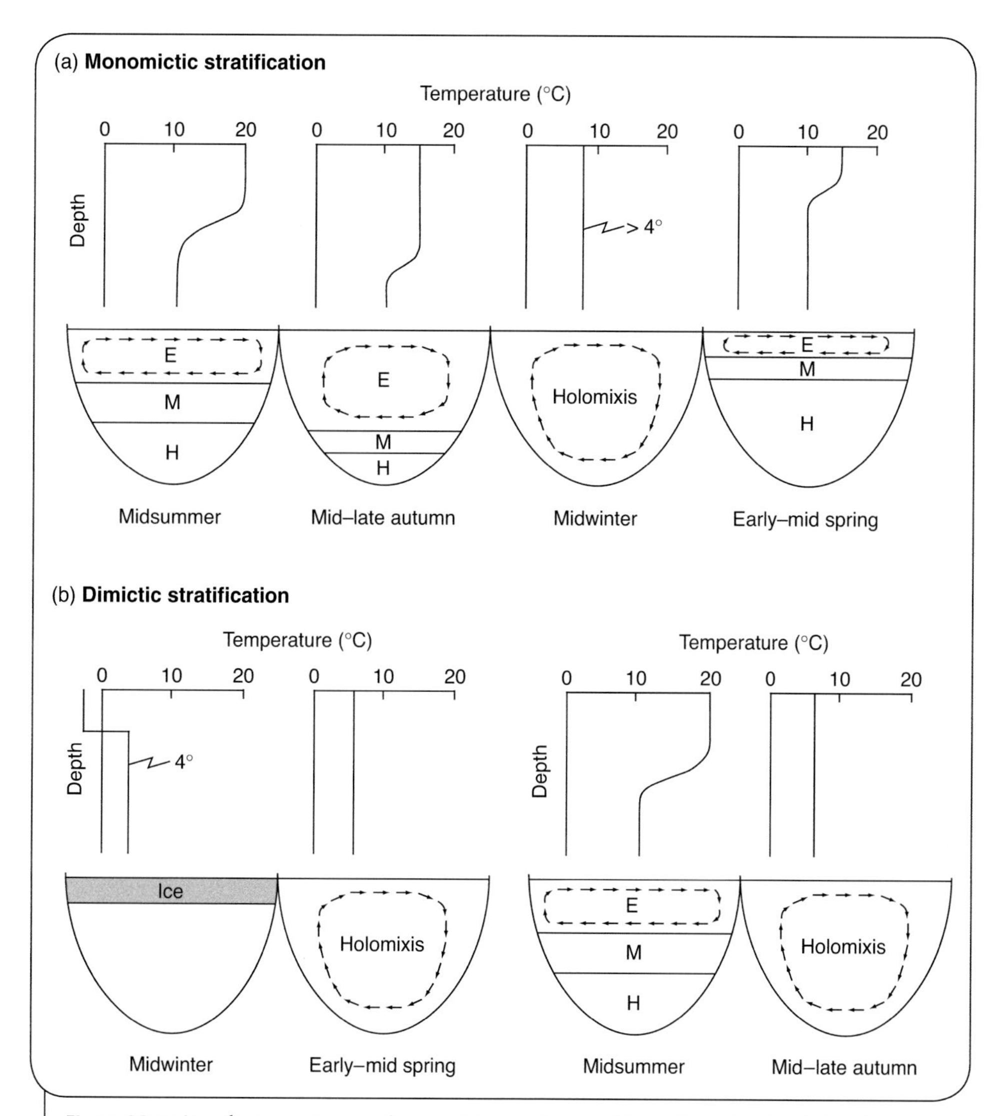

Figure 20.15 Stratification into an epilimnion (E), metalimnion (M) and hypolimnion (H) and mixing (holomixis) in regions where there are seasonal changes in temperature. Stratification and mixing in regions (a) where water does not freeze in winter (monomictic – stratification occurs once per year) and (b) where ice does occur in winter (dimictic – stratification occurs twice per year). (Source: B. Cale)

days or less in any one season, if it occurs at all. Tannin-stained systems, for example, can undergo rapid heating at the surface because of the dark colour of the water. They stratify during the morning but then, in these generally shallow systems, the radiation of heat overnight causes complete mixing. In warm monomictic lakes (Figure 20.15) stratification develops in summer, but in other seasons water mixes completely.

Consequences of stratification

Once thermal stratification is established, secondary chemical stratification usually occurs. The amount of oxygen in the hypolimnion falls and at the height of summer it may become anaerobic. The amount of CO_2 in the hypolimnion rises and the pH falls below that of the epilimnion. If the temperature gradient is steep and the hypolimnion becomes completely deoxygenated (through respiratory use and non-replenishment of oxygen), the vertical distribution of many organisms will be sharply cut off at the

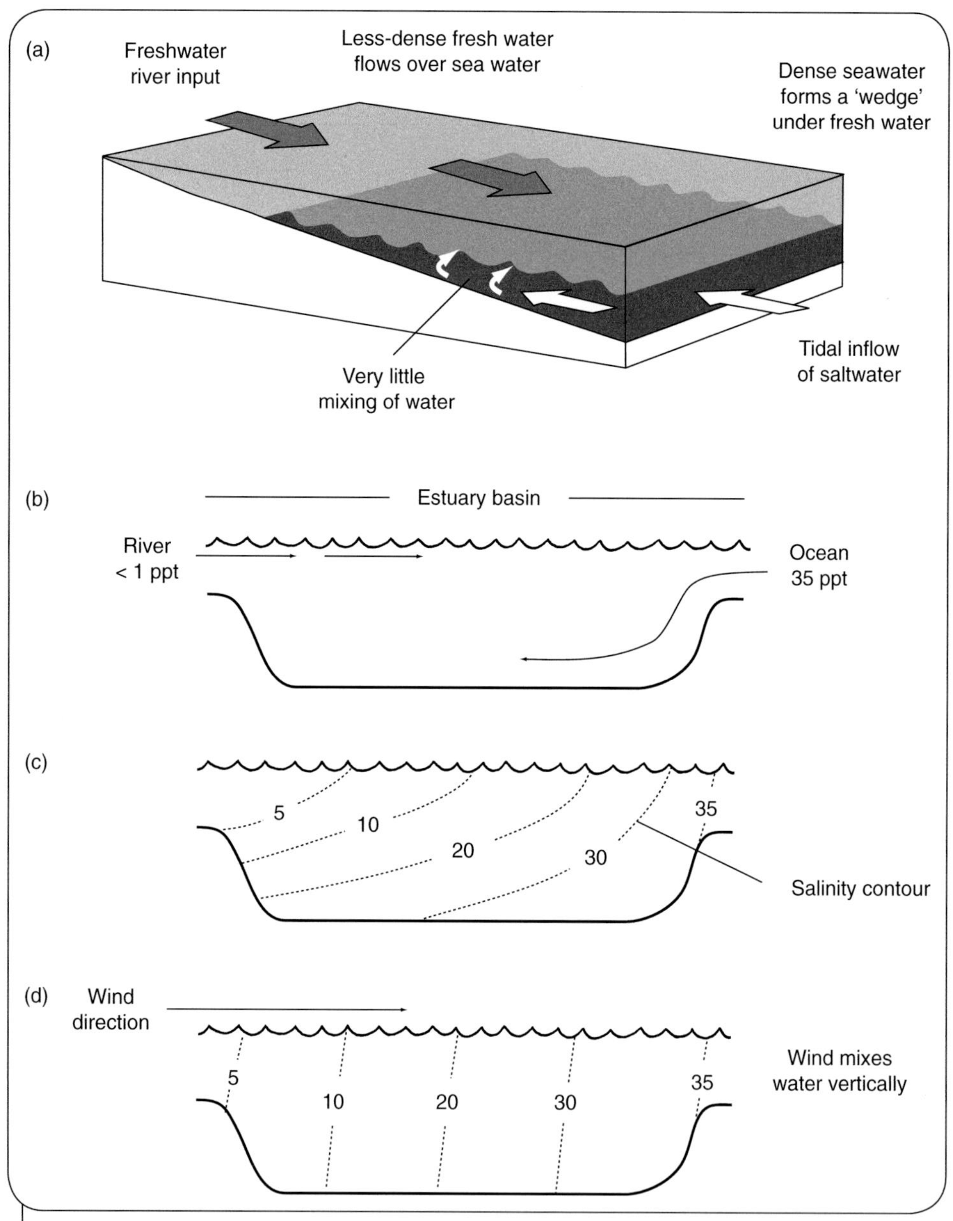

Figure 20.16 Stratification in estuaries. (a) When less-dense fresh water from the land meets the dense salty ocean water, stratification occurs. (b) The fresh water flows over the ocean water. (c) Tides can push salt water kilometres upstream under freshwater flow, resulting in wedge-shaped salinity contours from ocean salinities (35 ppt) to fresh water (0 ppt). (d) Strong winds can result in these wedges forming vertical blocks of different salinities. (Source: B. Cale)

thermocline. If oxygen is not severely depleted, then some organisms may choose to remain in the cooler waters of the hypolimnion. In thermally stratified lakes, nitrogen and phosphorus may be more abundant in the hypolimnion than in the epilimnion. This results from accumulating dead organisms falling into the hypolimnion, decomposing and releasing nutrients from the sediments under anoxic conditions. The overturn is important in redistributing these elements to the upper waters.

Chemical stratification

Density differences caused by salinity are much greater than those caused by temperature. As a rule of thumb, a change of 4 ppt salinity is equivalent to a 10°C temperature change. The most common salinity stratification occurs in estuaries where fresh water from the river meets the tidal inflow from the sea (Figure 20.16).

Meromictic lakes

In meromictic lakes some water remains partially or wholly separate from the main mass of water; that is, there is a permanently stagnant layer of water on the bottom. The upper region of such a lake which mixes completely is termed the mixolimnion. The lower stagnant region is the monimolimnion. The boundary between them is the chemocline. Usually there is a marked difference in salinity between the two regions and the chemocline (comparable to a thermocline) is the place of maximum rate of change of salinity. Meromictic water bodies provide the theoretical basis of 'solar ponds' where heat is trapped in the denser saline water of the monimolimnion.

Chapter summary

1. Inland water bodies can be described as lentic (standing water: lakes and wetlands) or lotic (flowing water: rivers and streams). The type of wetland results from the interaction of climate (rainfall and temperature) and geomorphology affecting hydrological inputs and outputs into the aquatic ecosystem.
2. Understanding wetlands requires awareness of the variability and interactions between physical, chemical and biological parameters. Several environmental parameters vary between water bodies including light, temperature, dissolved oxygen, pH, salinity and nutrients. These modify water quality and thereby its suitability for particular organisms.
3. *Light*: The light available in the water column controls photosynthesis and thereby the productivity of the plants and animals within aquatic systems. Light transmission is reduced by dissolved substances (such as tannins). Particles suspended in the water column, whether inorganic (silt, clay) or organic (phytoplankton) in origin, also intercept light. This is turbidity.
4. *Temperature*: Temperature may change greatly daily or seasonally, and does not necessarily decrease uniformly from the top to the bottom of a lake or wetland. This may cause stratification (the separation of a lake into discrete layers of different density) with significant effects on physical, chemical and biological processes.
5. *Dissolved oxygen*: Air has 21% oxygen but in water oxygen comprises only 0.6%. The amount of dissolved oxygen depends on the atmospheric pressure at the air–water interface, the temperature, and the concentration of other dissolved substances. The amount of dissolved oxygen (DO) is reduced as temperature or salinity increases.
6. *pH*: The pH of a wetland is affected by solutes entering from the surrounding environment. Extremes of pH limit plant and animal productivity, so anthropogenic causes of pH change, such as acid rain and acid sulfate soils, are catastrophic. Highly acidic conditions increase solubility of arsenic and metals (aluminium, iron, lead and so on), killing fish and plants.
7. *Salinity*: Salt concentration of inland water ranges from fresh up to the saturation point (360 ppt for sodium chloride at 20°C) in some salt lakes. Most freshwater organisms show stress above 1 ppt (about 1500 μS/cm), and 3 ppt (about 5000 μS/cm) is the upper limit of freshwater ecosystems.
8. *Nutrients* required for plant growth are important for controlling productivity in aquatic ecosystems. The nutrients required in the greatest amounts are carbon, nitrogen, phosphorus, sulfur, magnesium, calcium and potassium. In aquatic ecosystems, the nutrients in low supply are usually nitrogen or phosphorus. Wetlands naturally increase in nutrient content over time. This is eutrophication and naturally takes thousands of years. Human inputs compact the time scale of eutrophication into years or decades (cultural eutrophication). The main symptom is excess plant growth. This decomposes, stripping oxygen from the water and killing fish and other aerobic organisms.

Key terms

acid rain
acid sulfate soils
aerobic
ammonium
anaerobic
anoxic
attenuation coefficient
bicarbonate
biodiversity
carbon dioxide
carbonate
colour
cultural eutrophication
decomposition
dimictic stratification
dissolved oxygen
epilimnion
euphotic zone
euryhaline
evapotranspiration
fen
hydrologic budget
hydroperiod
hypolimnion
Kjeldahl nitrogen
lentic
light
limnology
lotic
meromictic stratification
monomictic stratification
nitrogen fixation
nutrient cycling
nutrient spiralling
nutrients
oligotrophic
orthophosphate
particulates
pH
phosphorus
salinisation (primary and secondary)
salinity
saturation
stenohaline
stratification
total nitrogen
total phosphorus
turbidity

Test your knowledge

1. List the main physical and chemical parameters that determine the water quality in a wetland.
2. What are the effects of acid sulfate soils on wetland ecology?
3. What factors of the aquatic environment reduce the depth to which light penetrates into the water?
4. What human activities are likely to cause an increase in the salinity of a water body in an area with saline ground water at depth?
5. Why is the availability of phosphorus often of more interest than the availability of nitrogen in fresh waters subject to eutrophication?
6. What is meant by the thermal stratification of a water body and what factors might disrupt such stratification?
7. Explain the difference between lotic and lentic inland waters.

Thinking scientifically

In 1994, a channel was cut between the eutrophic Peel-Harvey Estuary in Western Australia and the Indian Ocean to improve flushing and reduce nitrogen and phosphorus concentrations. This reduced algal blooms. Unfortunately, in the 2000s there are now algal blooms in the lower sections of the rivers feeding the estuary. Which of the following statements represents a valuable scientific explanation of this observation? (You can choose more than one.) Use the explanations you choose to suggest a management plan for controlling eutrophication of the estuary.

- The change in hydrology may result in the tides pushing water back up the rivers and preventing the nitrogen and phosphorus in the rivers entering the estuary basin.
- People in the catchment must be adding more nutrients to the rivers upstream in the catchment.
- Cutting a channel did not address the source of the problem: nitrogen and phosphorus run-off from the catchment – long-term control of eutrophication requires the source be removed.
- Cutting a channel to the sea would increase the salinity in the estuary and make it unsuitable for the fresher water species that originally lived there. They are now restricted to the rivers.

Debate with friends

Consider the diversity of wetlands shown in Plate 20.1 and their relationship to rainfall, temperature and geomorphology. Debate the topic: 'Successful creation of new wetlands by humans will occur only if the climate and geomorphology would naturally support that type of system in that location'.

Further reading

Boulton, A.J. and Brock, M.A. (1999). *Australian freshwater ecology: processes and management.* Gleneagles Publishing, Glen Osmond, South Australia.

21

Inland aquatic environments II – the ecology of lentic and lotic waters

Jane Chambers, Jenny Davis and Arthur McComb

Created wetlands

Imagine being asked to advise a mining company about implementing their vision to convert an area devastated by mining into a visually attractive waterbird habitat that would become an essentially self-sustaining ecosystem in perpetuity, and a showcase for wetland management, rehabilitation after mining and public education. Two of the authors of this chapter were involved in just such an advisory committee for a company that is mining mineral sand rich in rutile and ilmenite near the town of Capel in Western Australia. Mining had created an undulating landscape in the sandy soil, and where this dipped below the water table temporary or permanent lakes were created, giving rise to a 'created wetland'.

The task bristled with problems. Was the water table constant or seasonal? Were there sufficient nutrients to maintain biological production in the wetlands, or was there so much that undesirable algal blooms would form? What food chains and food webs could be established, and would these meet the objective of sustaining water bird populations?

This chapter provides information needed to address such questions, which are relevant not just to local problems such as managing the Capel Wetlands Centre, but to much broader questions of water quality, and management of rivers, lakes and scarce water resources.

Chapter aims

In this chapter we explain how the physical parameters of aquatic systems, described in the previous chapter, result in different types of habitats in still or flowing waters. We describe plant and animal adaptations to these aquatic habitats, and the interlinked grazing and detrital food chains. Understanding these physical and biological processes is crucial for the development of management techniques appropriate for aquatic ecosystems.

The importance of aquatic ecosystems

Aquatic ecosystems are either fresh or saline water bodies in a terrestrial setting, and are complex environmental systems supporting considerable biodiversity. The species present differ in tropical, temperate and arid zones. Animal species range from monotremes (the platypus) to marsupial (swamp wallaby) or placental mammals (water rats), reptiles (crocodiles, turtles), amphibians (frogs), fishes and invertebrates. The invertebrates alone comprise many different groups – worms, snails, mussels, shrimps, crayfish, spiders, water mites, beetles, bugs, dragonflies, damsel flies and many others. Aquatic vegetation includes both single-celled organisms (phytoplankton) and more complex, multicellular algae, ferns and angiosperms (macrophytes). The presence of a large number of species of plants and animals depends on water quantity, permanence and depth, water quality and the presence of a range of habitats. Different habitats provide different food resources and refuge areas for many aquatic organisms.

Unfortunately there are many threats facing aquatic systems. These include altered flow regimes (by damming and irrigation), loss of habitats, nutrient enrichment, salinisation, acidification, other forms of pollution, invasion by exotic species, and potential impacts arising from climate change.

With such a myriad of impacts, maintenance of biodiversity in aquatic ecosystems might well seem an unattainable goal – how can you ensure so many species with different needs are provided for? The key is understanding how organisms and their environment interact. Both lentic systems (lakes and wetlands) and lotic systems (rivers and streams) can be divided into separate zones, habitats or microhabitats characterised by certain environmental parameters. Within these zones and habitats, organisms will have a smaller range of interactions than those within a whole ecosystem, and thus the requirements of those organisms can be better predicted and managed.

One of the main divisions in aquatic ecology is between the standing or comparatively still water of lakes and wetlands and the flowing waters of rivers and streams. The influence this has on environmental parameters has been discussed in the previous chapter. In this chapter we consider how aquatic organisms interact with their environment and discuss the ecology of lentic and lotic ecosystems separately.

The ecology of lentic environments (lakes and wetlands)

Habitats in wetlands and lakes

The sun is the main energy source for standing waters. As light passes through a body of water it is gradually absorbed and scattered. Light is used by the plankton near the surface. Down to a certain depth, which depends on the clarity of the water, there is enough light to enable oxygen production by photosynthesis to equal or exceed the amount of oxygen consumed by organisms for respiration. This is known as the compensation depth.

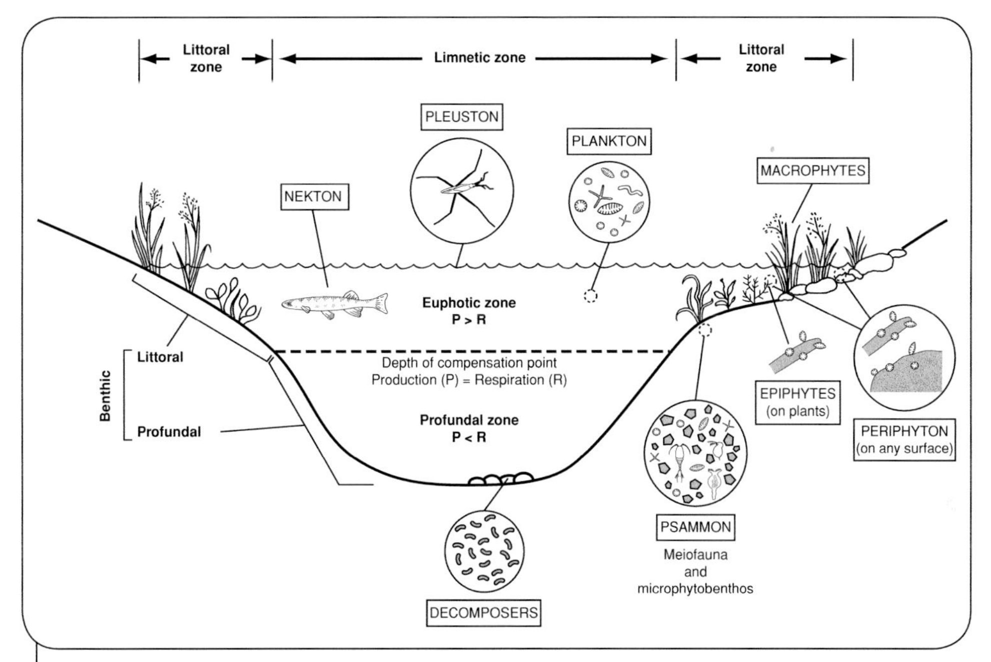

Figure 21.1 Ecological zones and habitats in a lentic wetland. A primary distinction between these zones is whether they are in the euphotic zone (where production is greater than respiration and plant growth dominates) or the profundal zone (where production is less than respiration and decomposition dominates). (Source: B. Cale)

The region above this depth, in which more organic matter is formed than consumed (production exceeds respiration), is known as the euphotic zone. Below this depth, decomposition is the dominant process (production is less than respiration) and this region is known as the profundal zone (Figure 21.1).

This zonation has a profound impact on the distribution of aquatic organisms. Since plants can grow only where there is light, plant growth in lakes is generally confined to surface waters. The shallow margins of a lake, where light penetrates all the way to the bed, is known as the littoral zone. This region of the lake often supports rooted aquatic plants and a diverse fauna of invertebrates, amphibians and waterbirds.

As in the marine environment (Chapter 18), aquatic habitats can be considered as a number of zones, each of which supports a different array of organisms. The open surface waters of the lake are called the limnetic zone. Although all of the littoral zone is, by definition, in the euphotic zone (see Figure 21.1), the limnetic zone is deeper and is divided into euphotic (surface) and profundal (bottom) zones. The dominant organisms in this region are floating algae (phytoplankton), zooplankton, macroinvertebrates and fish. The organisms present in the limnetic zone are generally considered to belong to two main groups: the plankton and the nekton. Plankton consists of microscopic phytoplankton and zooplankton, which drift passively in the water column and are generally dependent on water currents for transport. The term 'nekton' encompasses swimming organisms that are independent of water currents and are able to move in any direction. The most obvious group of nektonic organisms is the fish, but it also includes actively swimming insects such as notonectids (backswimmers) and crustaceans (shrimps).

In the limnetic zone, and below the compensation point, is the profundal zone. The bed of the lake, whether it occurs in the littoral or profundal zone, is known as the benthic zone, and organisms living in or on the bed are, like organisms living in similar habitats in the ocean, collectively termed the benthos. The most important source of energy for the benthic region of the profundal zone is dead organic matter falling from above. Microbial breakdown of this organic material provides a rich food source for detritivores (organisms that feed on detritus: see 'Aquatic food webs' below). The decomposing microbes (or decomposers: bacteria and fungi) use up oxygen in the process of converting dead organic matter into inorganic nutrients. As a consequence, in lakes with highly productive surface waters, decomposition processes may be so great as to result in little oxygen being left in the profundal zone. Therefore many organisms that live at depth are adapted to tolerate low oxygen conditions. Two of the most abundant macroinvertebrates in this region are the oligochaetes (aquatic worms) and larval chironomids (larval midges of the class Insecta: family Diptera – the flies). In this region various detritivores, both fish and invertebrates, consume dead organic matter.

Organisms may be adapted to one of several microhabitats within lakes. These include those attached to substrates in the lake, collectively known as periphyton; inhabiting the air–water interface, known as the pleuston; and colonising the spaces between sediment particles, called the psammon. Within these microhabitats there are specialisations. Periphyton, sometimes also called biofilm, can be divided into the types of substrate to which the organism attaches; for example, epiphytes are organisms (usually algae in aquatic habitats) which attach to larger algae or plants. Organisms using other substrates including rock, sediment or the shells of animals are known as epilithic, epipelic or episammic, and epizooic organisms respectively. Within the psammon are the microscopic plants (sometimes known as microphytobenthos) and the microscopic animals known as meiofauna. Adaptations to such a wide variety of habitats promotes the huge biodiversity of organisms in aquatic environments.

Aquatic food webs

Two processes important to wetland food webs are productivity and decomposition (Chapter 17). Productivity can be divided into primary and secondary production. Primary production is represented in a wetland by the growth of algae in the water column, aquatic plants in the littoral zone, and sedges, shrubs and trees fringing the wetland. Secondary production refers to animal productivity, and decomposition refers to the breakdown of dead organisms. Decomposition is usually microbially mediated (e.g. by bacteria, fungi or protozoans) and is an essential process that recycles nutrients from dead plants and animals back into the ecosystem.

Wetland food webs comprise two interconnected food chains: a grazing food chain and a detrital food chain (Figure 21.2, Plate 21.1). The grazing chain comprises those organisms that feed on algal cells or aquatic plants, and these in turn are eaten by other animals, the second-order consumers or predators. Grazing organisms include

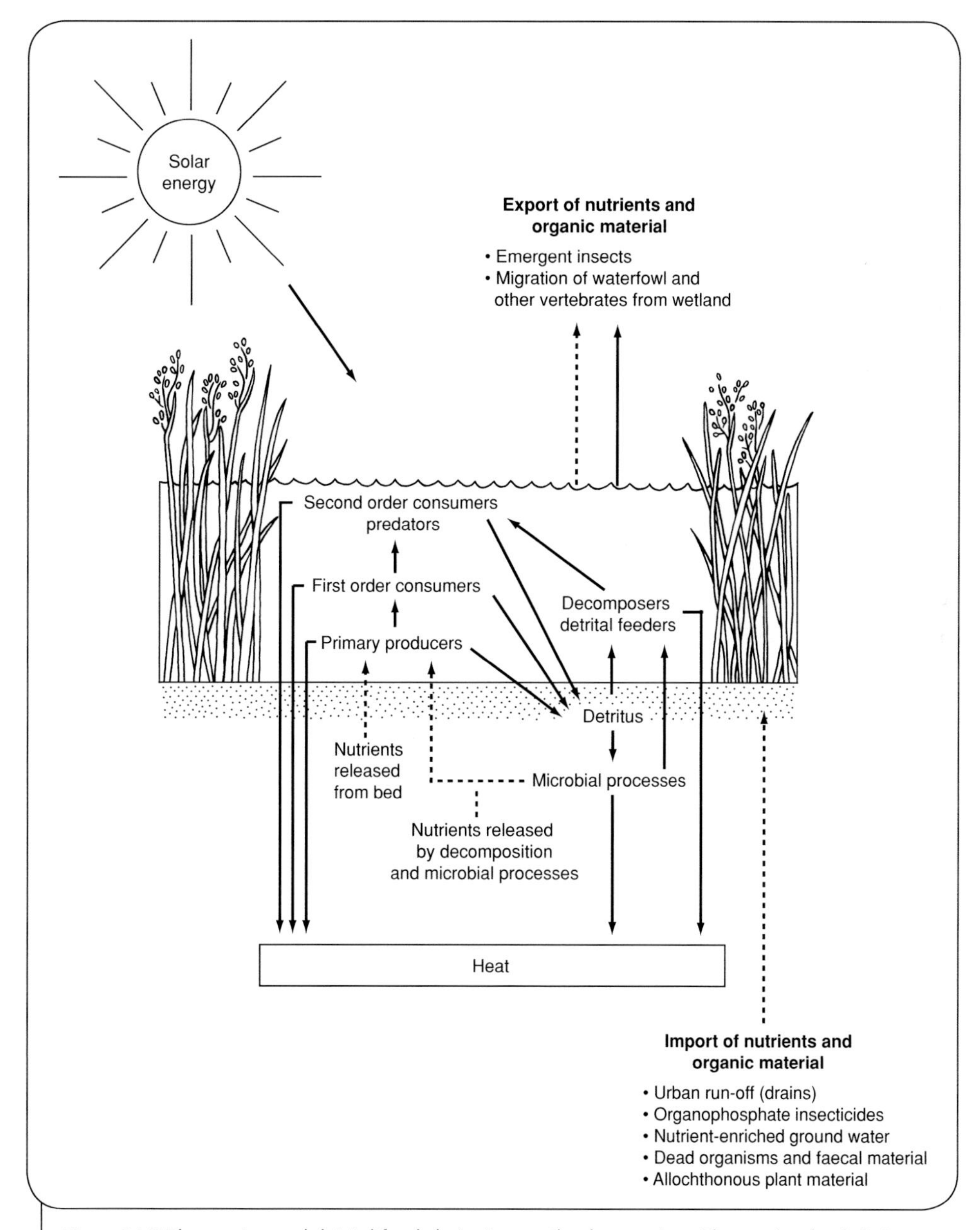

Figure 21.2 The grazing and detrital food chains in a wetland ecosystem. The grazing food chain is based on the consumption of live plant material, and the detrital food chain is supported by consumption of dead and decaying material. The two food chains are closely interrelated. (Source: B. Cale)

many of the microcrustaceans that make up the zooplankton, such as *Daphnia* (the water flea) and aquatic snails. The detrital food chain comprises those organisms that feed on dead and decaying plant and animal material, and so this chain is responsible for the processes of decomposition that take place within a wetland. The detrital food chain is often the dominant food chain in wetlands, especially in shallow aquatic systems where macrophytes (see below) dominate. Macrophytes are often too large, tough (containing lignin) or unpalatable (containing antiseptic tannins) for faunal consumption, and their productivity cannot be incorporated into the food chain until it has been broken down by microbes. The detrital food chain can also dominate when nutrient enrichment unbalances the natural grazing pathways maintained by algal growth (see below). Common detrital feeders include the larval chironomid or midge, several crustacean groups (e.g. ostracods, amphipods and isopods), the aquatic oligochaetes (worms) and insects such as mayflies and caddis flies. Predatory invertebrates include damsel flies and dragonflies, aquatic beetles and bugs, water mites and flatworms. Vertebrates at the top of wetland food chains include waterbirds, fishes, turtles and frogs.

Both food chains play important roles in the functioning of wetland ecosystems. The grazers maintain good water quality by feeding upon algae, and the detritivores process the decaying organic matter of the lake bed, performing a vital recycling function.

Consequences of altering wetland processes and food webs

In stable ecosystems (in this context 'stable' is used to indicate systems that do not fluctuate unduly in terms of species composition and abundance from established seasonal patterns), the processes of primary and secondary production proceed such that the input and output of materials and energy are generally equivalent. When processes become unbalanced, the effects are often considered to be of considerable nuisance or harm to humans. However, it must be noted that it is usually human impacts that have caused this disturbance in the first place. Impacts known to have deleterious effects on wetland function include interference with water regimes, the removal of fringing vegetation, the input of excessive nutrients and pesticides, and the introduction of exotic biota.

As a general rule, disturbed aquatic ecosystems have a lower biodiversity (number of species) and often an increased abundance (number of individuals) of one or a few species. In eutrophic or polluted lakes, detrital food chains become increasingly dominant and many invertebrate and even top predators (e.g. fish) may be absent.

As noted above, grazing organisms such as *Daphnia* play an important role in maintaining good water quality within a wetland by feeding on suspended algal cells. However, when algal blooms become very large, as a result of the presence of excessive nutrients or because they are comprised of toxic species such as cyanobacteria (blue-green algae e.g. *Microcystic* or *Anabaena*), the numbers of grazing organisms will be unable to process all the algal material and reduce the blooms to tolerable levels.

As the blooms die off and decompose, oxygen concentrations within a lake may become seriously depleted as a result of intensive microbial action. Decomposing algal blooms provide a large food source for detritivores, and the growth of larval midges is particularly favoured. As a consequence, the management of midge problems at a wetland must involve some treatment of the cause of the problem: the entry of excessive nutrients into the lake and ensuing algal blooms. Decomposing blooms may also result in offensive odours and contribute to the incidence of waterbird deaths from botulism.

The presence of pollutants, such as pesticides, and introduced species (e.g. the mosquito fish, *Amusia affine)* can also result in poor water quality through their effects on wetland food chains. Many grazing organisms, such as *Daphnia,* are particularly susceptible to pesticides. These larger members of the zooplankton may also be selectively grazed by mosquito fish. Both impacts will result in a decrease in the numbers of grazers and, as a consequence, the capacity of the wetland to process algal blooms will be diminished. Pesticides, toxic algae and low dissolved oxygen may also result in the death of larger predatory invertebrates such as larval dragonflies and beetles. Invertebrate predators are often an effective means of control of nuisance organisms such as larval midges, and so their absence from a wetland may result in greater problems with nuisance midge swarms than would otherwise occur.

Adaptations to aquatic life: aquatic plants and wetland function

Plant communities are the basis of all life in a wetland, as they provide a food source and a range of habitats for aquatic and terrestrial fauna. The vegetation also forms an important structural component, stabilising and aerating the sediment and providing a filtering mechanism for material passing into the wetland.

In shallow aquatic systems, large aquatic plants or macrophytes (flowering plants, ferns and structurally complex algae such as charophytes) have the greatest biomass (amount of plant material per unit area) and in consequence have a great impact on wetland structure and function. Even in rivers or deep lake systems, the fringing vegetation or riparian zone has a disproportionate effect on ecological function relative to its size. This is because the fringing vegetation grows at the overlap (ecotone) between the terrestrial and aquatic environments and thereby buffers the aquatic system from terrestrial inputs. Sometimes the fringing vegetation is described as a buffer zone around a wetland, but in fact the fringing vegetation is an integral part of the wetland and it cannot function correctly without it. The boundary of any aquatic ecosystem is always placed at the landward edge of the plants exhibiting adaptations to waterlogging.

So what adaptations do plants have for an aquatic environment? The aquatic environment is very different from the terrestrial one. In many ways the aquatic environment is benign – water is usually in plentiful supply and provides a support mechanism that is not available to species living on land. There is a huge diversity of plants from nearly every taxonomic group, from simple unicellular to the most complex algae (the charophytes), bryophytes, ferns and fern allies, and angiosperms. The only group with no truly aquatic representative is the gymnosperms. Despite this diversity, most of the plant biomass in wetlands is comprised of angiosperms. Land plants evolved from algal ancestors and thus aquatic plants represent a reinvasion of the aquatic environment.

The so called 'lower plants' of the evolutionary tree, such as the algae, require water to survive and are perfectly adapted to the aquatic environment, having thin cell walls through which dissolved nutrients and gases may readily pass. Some merely float in the water currents, whereas others have active mechanisms to ensure their position in the photic zone, such as adjusting buoyancy using gas vesicles.

To develop larger size requires greater adaptation, as exemplified by the flowering aquatic plants (Figure 21.3) The main limitations of the aquatic environment are reduced light passing through the water column limiting photosynthesis to the euphotic zone, reduced gas availability (oxygen and carbon dioxide) in the water, and reduced nutrient availability in the water column relative to that of the sediment. This latter limitation has prompted the success of the aquatic macrophytes – most have roots in the sediment, providing access to a greater nutrient supply than do their waterborne counterparts, and they attain a larger size and play an important role in the structure and function of aquatic systems.

The ability of macrophytes to buffer against changing nutrient concentrations is partly a result of the way they absorb nutrients. Macrophytes primarily absorb their nutrients from the sediment, and by doing so effect a number of physical and chemical processes which limit the availability of nutrients in the water column, consequently limiting the occurrence of phytoplankton.

The process of sedimentation is greatly increased in macrophyte beds, as their structure reduces current flows by as much as 90%. Nutrients attached to particulates in the water column are returned to the sediment, increasing the nutrient status of the sediment immediately adjacent to the plant roots. Macrophytes are able to take up more nutrients than are required for growth. This ability is termed 'luxury uptake', and allows macrophytes to become very effective nutrient sinks during growth periods (usually the warm spring/summer months), reducing nutrient concentrations in the water column, and thus nutrient availability for algal growth. As a result of these processes, increasing sedimentation will in most cases result in a clearer water body, encouraging the extension of macrophytes into deeper water. Sedimentation also results in any pollutants or other materials attached to particulates being precipitated to the sediment, removing impurities from the water and improving water quality. Macrophytes further maintain clear water by reducing erosion. The dense mat of roots and rhizomes bind sediments, impeding scouring by water currents or wave impact, and reducing sediment suspension into the water column. Fringing plants also intercept rainfall, reducing erosion from raindrop impact (which can be significant in areas that dry seasonally), and improve percolation of water into soil by channelling flow down their stems and along the outer surface of their network of roots.

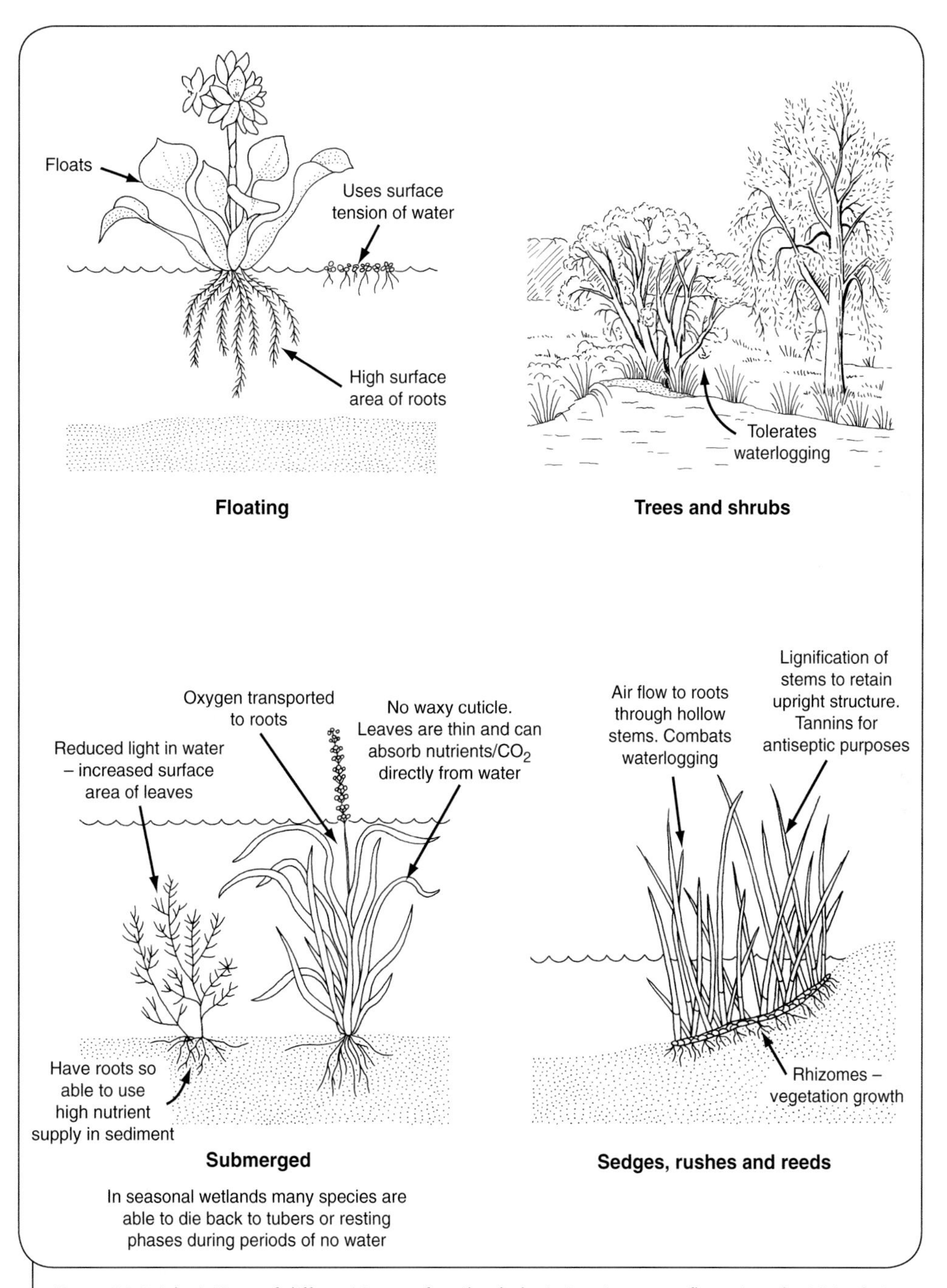

Figure 21.3 Adaptations of different types of wetland plants (angiosperms: flowering plants) to their habitat. Plants that incur only occasional waterlogging have a comparable morphology to terrestrial plants, but significant adaptations are required for those species that live entirely underwater. (Source: B. Cale)

In waterlogged sediments, oxygen is often at very low concentration or absent altogether (anoxic), and aquatic plants transfer oxygen from their leaves to the roots to allow them to respire and grow. Some of this oxygen leaks from the roots to produce a thin film of oxygenated or aerobic soil around each root. This film is called a rhizosphere. The rhizosphere provides a more benign environment for the plant roots, allowing plants to colonise where otherwise metals and other elements might be in toxic concentration. Importantly, it provides an environment for the proliferation microbes, which are vital to nutrient and other cycles in a wetland.

Macrophytes also provide a safe haven for macroinvertebrates to escape from predators, facilitating their survival. Interestingly, they provide increased habitat and protection for phytoplanktonic grazers such as *Daphnia*. By doing so the population of these grazers is maintained, placing pressure on phytoplankton in macrophyte beds. This has been found to keep macrophyte beds free from phytoplankton blooms even at high nutrient concentrations.

Adaptations by animals to the aquatic environment

The primary adaptation to living in an aquatic environment is driven by the need for oxygen. As in marine animals, the most common adaptation is the development of gills that filter oxygen from the water. Gills are found in all fish, larval insects (e.g. dragonflies), crustaceans (e.g. freshwater crayfish) and amphibians (the tadpole stage of frogs, the adults of which have lungs). However, not all aquatic animals have gills. Many insect larvae have novel ways of obtaining oxygen. Diving beetles capture bubbles of air from the surface, which they carry underwater to use as their supply of oxygen. Mosquito larvae hang by their tails from the surface film of the water, accessing air through discs in their tails; water scorpions similarly live at the air–water interface and have snorkel-like tubes extending from their tails.

As noted above, an important feature of fringing vegetation is that it occurs in the ecotone between aquatic and terrestrial environments. As such, it shelters species from both environments, as well as some that exist only in the ecotone. Many animals use both environments to complete their life cycle. This has the benefit of using different environments for habitat and food, so that juveniles and adults are not in competition for resources. Numerous aerial insects (e.g. dragonflies, mosquitoes, midges and black flies) have aquatic larval stages, and emerge into the air as adults. Most frogs have the first stage of their life as tadpoles in the aquatic environment and then emerge into the terrestrial environment as adults.

In arid areas of Australia, temporary and episodic wetlands cover a greater area than do permanent systems. Many organisms are adapted to exploit these temporary environments. Early colonisation strategies enable species to avoid predators. For example, larval mosquitoes rapidly colonise shallow pools in salt marshes, avoiding predators (fish) in deeper water. Tadpoles are most abundant in shallow pools for similar reasons. Many animals have adaptations to allow survival during dry periods. Some species of frogs

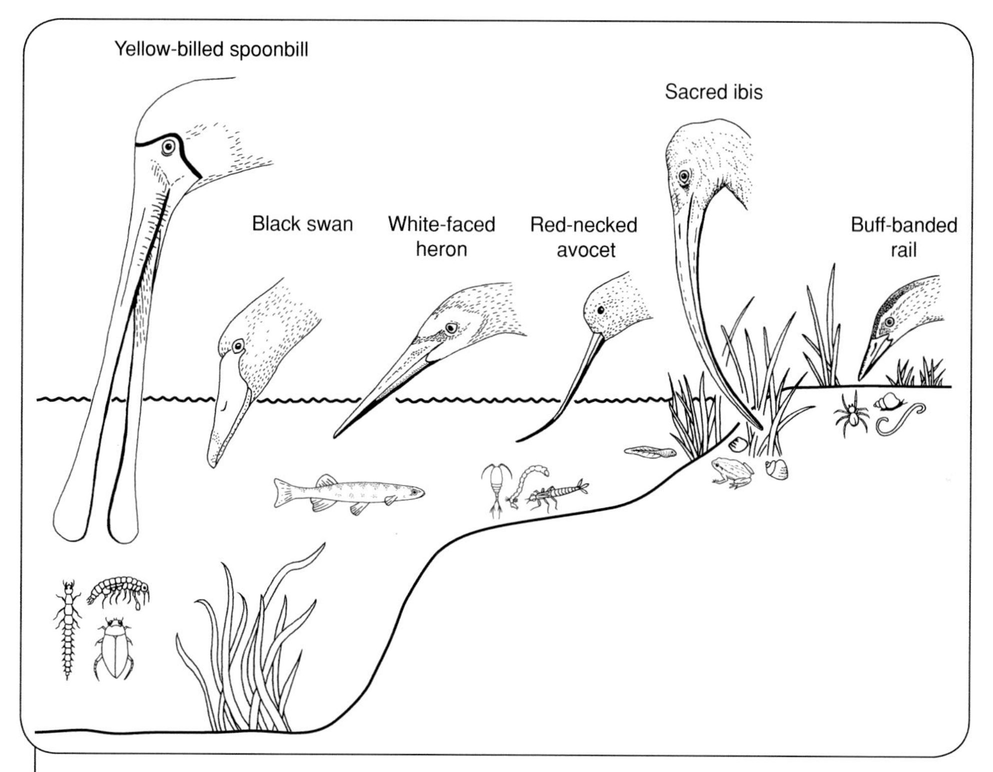

Figure 21.4 One wetland will provide a multiplicity of habitat types. Numerous species of waterbirds may be foraging in a single area, utilising different food resources. A close look at their beak types shows they are carefully adapted to obtain different food resources that occur in different habitats. (Source: B. Cale)

(e.g. desert burrowing frog, *Limnodynastes spenceri*) will burrow in the ground and await wet conditions to return, when they rapidly mate and breed. The Australian lungfish (*Neoceratodus forsteri*) has both gills and lungs. and when ponds dry up it can gulp air and survive a few days with very little water. Its African counterpart, *Protopterus annectens*, burrows in the mud and secrets a mucous cocoon in which it survives until reflooding reinstates its aquatic habitat.

Apart from aquatic organisms, lakes and wetlands provide a vital source of water for terrestrial animals. Waterholes provide a common source of water and food. Waterbirds in particular use a multitude of habitat and food types in wetlands. Waterbirds are highly specialised to optimise their success at obtaining nourishment. For example the size, shape and strength of a waterbird bill is suited to the type of food it eats. Bills variously tear meat, spear fish, crack hard seeds, catch insects, gather aquatic plants, probe into mud for tiny shellfish, or filter tiny creatures from mud. Even the length of the bill can be important to the type of food it takes (Figure 21.4).

The ecology of lotic environments (rivers and streams)

The main features which distinguish lotic (flowing water) systems from lentic (standing water) systems are a linear morphology and unidirectional flow. Water flows downhill and, ultimately, to the sea (with some notable exceptions in inland Australia where rivers flow inland to flood Lake Eyre and other episodically filled lakes). Flowing waters in Australia can be broadly classified into three groups: permanent rivers and streams which always contain water, temporary rivers and streams which contain water only seasonally (depending on the pattern of rainfall), and episodic or ephemeral rivers and streams (these occur in semiarid to arid areas, where rainfall is unpredictable and contain water only after major rainfall events such as cyclones).

Unlike lakes, where ecological zones are based on water depth and the extent of light penetration, lotic systems change longitudinally as the rivers become larger, deeper and faster flowing as you move downstream. As a consequence, a useful approach to hydrological classification is that of stream order, often called the Strahler method after the first limnologist to propose it. First-order streams have no tributaries, second-order streams are formed by the confluence of two first-order streams, third-order streams are formed from the confluence of two second-order streams, and so on (Figure 21.5). This type of classification has limitations, but is useful because it effectively summarises a number of physical features of streams and rivers that are directly related to stream ecology (see the later section on the river continuum concept).

Generally low-order streams (1–3) are shallow and narrow (you can jump across them), with steep slopes and associated high velocities, and substrates ranging in size from large

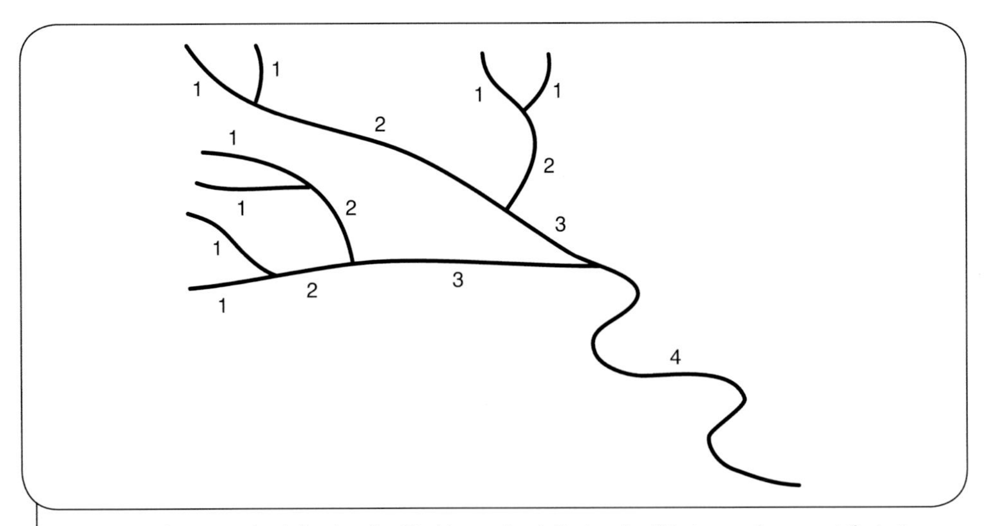

Figure 21.5 Stream order following the Strahler method. First-order (1) streams have no tributaries, but are found in the headwaters of a stream. Second-order (2) streams are formed at the confluence of two first-order streams, third-order (3) streams are formed at the confluence of two second-order streams, and so on. Knowing the order of a stream allows you to picture its size. The outlet of the Amazon River is stream order 12. (Source: B. Cale)

boulders to cobbles to pebbles, depending on the geology of the area (Plate 21.2a). Mid-order streams (4–6) are wider and deeper (you can usually wade across them), substrates range from cobbles to pebbles to sand, and the stream gradient decreases (Plate 21.2b). High-order streams (6–12) are much wider and deeper (you would need a boat to cross them), substrates are finer (often sands, silts or muds) and, although the gradient may be low, velocities are often fast and flows high because of the amount of water being carried (Plate 21.2c).

All the tributaries that make up one river system collectively drain the catchment (also known as the river basin or watershed). Recognition of the integral relationship between a river system and its terrestrial catchment underpins much of the current catchment-based approach to land management in Australia.

The unidirectional flow, which characterises lotic systems, often results in biological communities that are quite different to those of lentic systems. Rivers and streams are often physically harsh environments characterised by high water velocities. Floods may scour the substrate and reshape stream channels. In contrast, droughts may result in the river having little or no flow, high temperatures and low oxygen concentrations that may be equally stressful for organisms living there.

Adaptations for life in flowing waters

Fast-flowing waters tend to be high in dissolved oxygen, but the drag on organisms inhabiting these high-energy environments means that much energy may have to be expended to avoid being swept downstream. As a consequence, many organisms exhibit morphological, behavioural and physiological adaptations to reduce drag; for example, many fish are highly streamlined whereas others make use of patches of slower flow created by woody debris (snags) or indentations in stream banks to rest or move upstream. Many aquatic invertebrates avoid direct exposure to flow by burrowing into the riverbed or living within the crevices and dead spaces created between boulders and cobbles. Small size and dorso-ventral flattening enable organisms such as mayflies (Insecta: Ephemeroptera), water pennies (Insecta: Coleoptera) and flatworms (Turbellaria: Platyhelminthes) to inhabit the region of reduced flow associated with the stream bed. Some stream-dwelling caddis flies (Insecta: Trichoptera) construct cases of small pebbles that act as ballast. Suckers, hooks and other anchoring or friction-reducing devices may also be present in lotic invertebrates.

Phytoplankton and zooplankton communities are virtually absent except in large or high-order rivers where extensive reaches of relatively still water may be present. Most of the algae of flowing waters are attached to rocky substrates or large woody debris (periphyton). To maintain themselves in the current, aquatic macrophytes are anchored by strong root systems, but they may be scoured from the riverbed under flood conditions, and are rarely found under persistent, high-flow conditions.

Many organisms exhibit life-history strategies that enable them to cope with the floods and droughts characteristic of Australian rivers and streams. Some may have desiccation-resistant eggs, or survive droughts by burrowing into the damp riverbed. Flood conditions may act as a physiological trigger to spawn in some fishes. Quick-hatching eggs and widespread dispersal may also help maximum use of resources when water is present.

Ecological models of lotic systems

Why do we need models?

Unlike lentic systems, where discrete zones based on depth and light penetration can be clearly defined, the changes in a flowing stream are often more variable in terms of the predictability of the suite of environmental parameters found at any one location. To cope with this variability and classify the river into manageable habitats or subunits, a number of models have been developed.

The well-defined longitudinal changes in river systems have provided a starting point for ecological classification. An early European approach recognised zones based on temperature (ranging from cool, spring-fed headwaters to warmer lowland zones), whereas more-recent models recognise rivers and streams as integral parts of catchments (drainage basins) with well-defined pathways and mechanisms for the processing and transport of organic material. The importance of terrestrial inputs of plant material (in the form of leaves and wood) to the energy base of streams also forms an important part of models such as the river continuum concept.

The river continuum concept

The river continuum concept is a useful conceptual model that describes the longitudinal changes in organic matter inputs and energy pathways from headwaters (low-order) through to lowland (high-order) sections of a river system. An important underlying premise of this model is that the invertebrate biota reflects the nature of organic matter inputs. Functional feeding groups are defined by the food resources used by stream macroinvertebrates:

- *Shredders* feed upon large pieces of organic matter (coarse particulate organic matter, greater than 1 mm in diameter) such as leaves, wood and other plant material that has fallen into the stream. They include some mayfly nymphs (Insecta: Ephemeroptera) and caddis larvae (Insecta: Trichoptera).
- *Collectors* utilise small particles of organic matter (fine particulate organic matter, less than 1 mm in diameter) either by filtering from the passing water or gathering from deposits in stream sediments. Collectors include freshwater mussels (Mollusca: Bivalvia), larval black flies (Insecta: Simuliidae) and larval midges (Insecta: Chironomidae).
- *Scrapers* are adapted to removing attached algae, especially where it grows on rock or log surfaces in the current. Examples of this group include snails (Mollusca: Gastropoda), some caddis-fly larvae (Insecta: Trichoptera) and water pennies (Insecta: Psephenidae).
- *Predators* feed upon other invertebrates and possess specialised structures and/or behaviour for the capture of prey. Predatory aquatic invertebrates include the dragonflies, damsel flies, aquatic beetles, bugs and water mites.

According to the river continuum concept (Figure 21.6), headwaters are characterised by heterotrophy (due to shading by riparian vegetation), a significant input of coarse particular organic matter from terrestrial sources, and a dominance of shredders and collectors. Mid-order streams are characterised by autotrophy (more light falls on the streambed in wider reaches where the influence of riparian shading is less), increasing amounts of fine particulate organic matter from upstream, decreasing amounts of coarse particulate organic matter from terrestrial sources, and collectors and grazers are dominant. High-order systems are characterised by heterotrophy (greater depth and turbidity reduce the light available for photosynthesis), large inputs of fine particulate organic matter from upstream, and a dominance of collectors.

Associated with the river continuum concept is the concept of nutrient spiralling, which is similar to the concept of nutrient cycling developed for lakes, in that nutrients move from the biotic to the abiotic environment and back. The concept of 'spiralling' adds a longitudinal component to cycling, where nutrients move downstream as a result of leakage from 'inefficient' upstream processes.

Other models

Considerable debate still exists as to how generally applicable the river continuum concept is, especially to Australian systems where eucalypt leaves provide a more refractory (difficult to break down) source of organic material than do deciduous plants, and shredders are often rare. Riparian shading, which promotes heterotrophy, does not occur in streams in alpine regions above the tree line, nor in arid areas where vegetation is sparse.

The highly variable and unpredictable flow regimes of many Australian lotic systems may negate many of the predictions of the river continuum concept. Disturbance, particularly in the form of floods and spates, can create a mosaic of habitats (both spatially and temporally) that may have a greater effect in organising stream communities than longitudinally related factors. The importance of disturbance in structuring stream communities is described in the 'patch dynamics concept', which recognises that rivers are a series of discrete patches rather than a single longitudinal system. Patches can be caused by a change in substrate, a change in nutrient availability, a change in riparian vegetation, floods or droughts.

The flood pulse concept extends the river continuum concept by considering the lower reaches of large undisturbed rivers with extensive flood plains (as typified by the flood plains of the Amazon and the rivers of the Kakadu region in northern Australia). The flood-pulse concept suggested that the lateral exchange of organic matter between flood plain and the channel, brought about by large overbank floods, is more important to ecosystem processes than the longitudinal delivery of fine particulate organic matter from upstream, as proposed by the river continuum concept. For some large rivers with well-defined channels and stable substrates, it is the input of material from the immediate edges and in-stream primary production that accounts for most of the organic material present. This model, the riverine productivity model, probably also

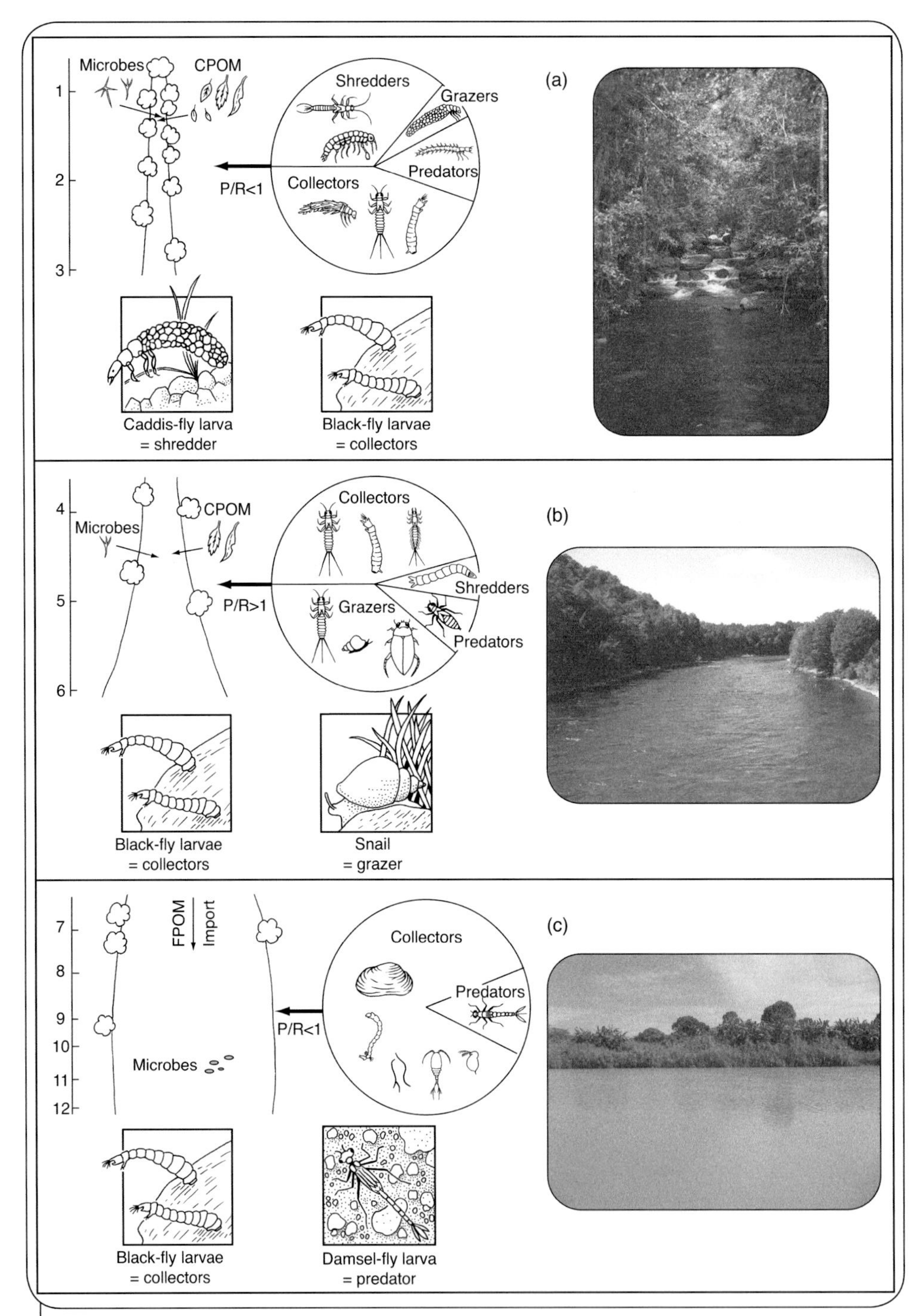

Figure 21.6 The river continuum concept. (a) The top third (stream order 1–3) is the headwaters characterised by heterotrophy, (b) the middle third (stream order 4–6) is the mid-order streams characterised by plant production and grazing, whereas (c) the bottom third (high-order streams 7–12) is once again characterised by heterotrophy. CPOM and FPOM are coarse and fine particulate matter respectively. P is photosynthesis and R is respiration. (Source: B. Cale)

best describes the sources of organic material to lentic systems (wetlands). For rivers that flow through catchments which have been heavily modified by human activities, the ecosystem can be linked to the type of land use. Lotic systems in forested regions are characterised by steep slopes, fast flows and low levels of organic detritus due to high transport rates. In contrast the streams and rivers characteristic of agricultural areas have lower gradients, slower flows and higher levels of organic detritus.

Chapter summary

1. Lentic environments can be divided into ecological zones based on the depth of water and the penetration of light. In the euphotic zone, there is adequate light to enable the production of oxygen by photosynthesis to exceed that consumed by respiration. Production is dominant in this zone. Below the compensation depth, in the profundal zone, respiratory use of oxygen exceeds that produced by photosynthesis. Decomposition is dominant in this zone.
2. Within either of these zones are a number of microhabitats described by the location of organisms in the zone. These include periphyton (attached to substrates) in the lake, pleuston (at the air–water interface) and psammon (in spaces between sediment particles).
3. Awareness of the zone, microhabitat and environmental parameters in which an organism lives enables a better understanding of the adaptations and requirements of each species. The greater the variety of microhabitats, the greater the biodiversity.
4. Wetland food webs comprise at least two interconnected food chains: a grazing food chain and a detrital food chain. The grazing chain comprises those organisms that feed on algal cells or aquatic plants, which in turn are fed upon by other animals. The detrital food chain comprises those organisms that feed on dead and decaying plant and animal material. Both food chains play important roles in the functioning of wetland ecosystems. The grazers maintain good water quality by feeding upon algae, whereas the detritivores process the decaying organic matter of the lake bed, performing a vital recycling function.
5. Plants, particularly macrophytes, have a suite of adaptations that allow them to colonise all parts of a wetland both within the euphotic zone and in the ecotone between the aquatic and terrestrial environments. Aquatic plants play a vital role in wetland function, improving water quality, recycling nutrients, stabilising sediments and providing a variety of habitats for aquatic organisms.
6. Most adaptations displayed by aquatic animals focus on obtaining sufficient oxygen for respiration. Structures to reduce drag and maintain position within a current are also important. Innovative ways to tolerate drying in temporary wetlands also fuel the wide biodiversity of aquatic life.
7. The main features that distinguish lotic (flowing water) systems from lentic (standing water) systems are a linear morphology and unidirectional flow. Fast-flowing waters tend to be high in dissolved oxygen, but the drag on organisms inhabiting these high-energy environments means that energy and adaptations are needed to avoid being swept downstream.
8. Unlike lentic systems, changes in a flowing stream are often more variable in terms of the predictability of the suite of environmental parameters found at any one location. To cope with this variability and classify the river into manageable habitats or ecological zones, a number of models have been developed. These are the river continuum concept, the patch dynamics concept, the flood pulse concept and the riverine productivity model.

9. The river continuum concept uses stream order classification, which describes the size and location of river sections from the headwaters to the mouth, and the concept of functional feeding groups, which describes the way in which invertebrates use food resources.

Key terms

adaptation
benthos
biodiversity
biofilm
biomass
compensation depth
decomposers
decomposition
detrital food chain
detritivores
ecological models
ecological zones
ecotone
epilithic
epipelic
epiphytes
episammic
episodic wetland
epizooic
euphotic zone
food web
functional feeding groups
grazing food chain
habitat
lentic
limnetic zone
littoral zone
lotic
macrophytes
meiofauna
nekton
periphyton
plankton
pleuston
productivity
profundal zone
psammon
riparian
river continuum concept
sedge
stream order
temporary wetland

Test your knowledge

1. Why are macrophytes important to wetland function? What functions do they perform?
2. What aspects of aquatic environments promote biodiversity?
3. List the components of a detritus-based food chain.
4. How might a detritus-based food chain interact with another food chain to form a food web?
5. As you move from the surface waters to the bottom of a deep lake, what environmental factors will change, and what kinds of organisms might be present?

Thinking scientifically

One of the ways in which river flooding is controlled is to remove snags (dead tree trunks and branches that have fallen under the water) that impede flow. After a period of time following desnagging, fishermen on the Murray River in Australia noticed a reduction in the number of native fish in the river. Some think the snags may be an important nursery habitat for juvenile fish, whereas other believe snags might provide a habitat for organisms that the fish eat. Describe how you would design an experiment to determine whether there is a link between fish populations and snags.

Debate with friends

Fresh water is an essential and diminishing resource for both humans and aquatic ecosystems. Removal of water for human use can reduce flow and flooding in rivers or reduce the water depth of wetlands. In more arid areas, seasonal wetlands might dry earlier than otherwise. All of this reduces and modifies the habitats available for aquatic organisms and thereby affects biodiversity. Debate the topic: 'It is essential to maintain environmental water allocations even when water shortages affect humans'. Consider the implications for humans of reducing aquatic biodiversity.

Further reading

Boulton, A.J. and Brock, M.A. (1999). *Australian freshwater ecology: processes and management*. Gleneagles Publishing, Glen Osmond, South Australia.

Brearley, A. (2005). *Ernest Hodgkin's Swanland: estuaries and coastal lagoons of south-western Australia*. University of Western Australia Press, Perth.

Fox, A. and Parish, S. (2007). *Wild habitats: a natural history of Australian ecosystems.* ABC Books/Steve Parish Publishing, Sydney/Archerfield, Queensland.

McComb, A.J. and Lake, P.S. (1990). *Australian wetlands*. Angus & Robertson, Sydney.

22

Terrestrial habitats

Philip Ladd and Mark Garkaklis

Earth, rain and fire – a trinity of the Australian landscape

Aboriginal Australians used fire to improve productivity of the landscape and to clear land for easier travel or hunting. Different parts of the landscape were burned at different frequencies and at different times, depending on soil characteristics and climate.

In grasslands on the fertile basalt plains in Victoria the starchy root of the daisy *Microseris scapigera* was an important food for the Koori people. The daisy was encouraged with frequent burning to lower competition from the dominant kangaroo grass (*Themeda australis*). In contrast, nutrient-poor heathlands were burned less frequently to stimulate ephemeral grasses and herbs that attracted grazing animals.

Plant reproductive traits in fire-prone environments are attuned to the season when fire occurs. In the southern heathlands fires occur mainly in summer or autumn at intervals of 10–30 years. Many species have seed banks either in persistent fruits or 'cones' (serotinous seed banks) or in the soil. Serotinous cones release seeds when the plant is burned and killed. Smoke chemicals frequently break seed dormancy so soil-stored seeds germinate when moisture is available (Chapter 12). This happens soon after summer or autumn fires. However, if fires occur in spring, seeds released are exposed to possible predation and desiccation for over 6 months before rainfall. Spring fires favour 'resprouter' species.

In the monsoonal north, winter is the dry, fire season. Accumulation of coarse annual grasses supports frequent fires and the reproductive traits of the savanna species differ from southern species. Canopy-stored seed is uncommon and many species are annuals or resprouters after fire.

Interactions involving soil fertility, biomass production, and the timing of the dry season and of fires have shaped vegetation types in Australia and elsewhere in the world, particularly in the Southern Hemisphere.

Chapter aims

In this chapter we describe the major factors influencing the Australian climate and how soil is formed, because these factors determine the occurrence of plant and, in turn, animal species. We introduce the concept of biomes (climatically determined groups of plants and animals spread over wide areas) and describe the typical biomes of Australia developed from the biota provided by evolutionary history.

Patterns of life on land

The terrestrial ecosystems of the world are distributed in patterns, both within and between continents, which are products of climate, soils and evolutionary history. It is important to understand the roles of these factors in explaining distribution patterns, and to appreciate the similarities and differences between Australian terrestrial ecosystems and those of other regions.

The climate

Influences of isolation and seasons

The Southern Hemisphere is characterised by oceans, and the Northern Hemisphere by continents. Southern hemispheric climates are influenced by oceanic circulation that, in turn, is regulated by the large ice mass on Antarctica.

The isolation of the southern continents began many millions of years ago in the Cretaceous epoch, when southern Africa, America and Antarctica separated from Australia (Chapter 7) and the oceans surrounding Australia grew with time. As Antarctica moved over the South Pole, ice accumulated and gradually the cold Antarctic circumpolar current developed, together with the 'roaring forties' wind system controlling weather movements from Antarctica towards Australia. Whether you regard Australia as the world's largest island or smallest continent, it is uniquely a large land mass isolated by sea in a moderate climatic zone (Figure 18.4).

The differential heating of the Earth throughout the year pushes the atmospheric circulation systems in the Southern Hemisphere south in the summer and north in the winter. Thus monsoonal systems cross the equator in summer, bringing rain to northern Australia, while high pressure systems bring dry weather to southern Australia. Conversely, in winter the pressure systems move north and the westerly system moves up across the southern continent, bringing rain to the south-west and south-east while the north is dry. Central Australia is generally dry because the high pressure system moves back and forth over the continent, blocking any rain-bearing systems. Most rain in central Australia comes from strong low-pressure outbreaks (cyclones) moving south in summer and also from patchy thunderstorms. In contrast, the western mountains of Tasmania, the south island of New Zealand and the mountains of Chile are always in the path of the wet westerly wind systems and have high rainfall almost all year.

Influence of oceanic circulation on terrestrial climate

While the cold waters of the Southern Ocean circulate from west to east they also influence areas further to the north. As described in Chapter 18, dense cold water sinks below warmer surface water and forms a subocean current. Such currents flow north-west, rising toward the surface as they reach the continental shelves on the western coastlines of Africa and South America. The nutrient-rich water increases marine productivity, but reduces productivity on land. The water off shore is very cold so there is low evaporation and hence little production of rain-bearing clouds. These parts of the continents are some of the most arid environments on Earth, with very low terrestrial productivity because of lack of rain.

A cold current flowing along the western edge of continents is the norm for all except Western Australia, where the warm Leeuwin current, coming from the north between Australia and the Indonesian archipelago flows down the west coast, forcing the cold current off shore (Figure 18.4). This benefits the terrestrial environment. As the water off Western Australia is warm, evaporation from the sea surface creates clouds, bringing rain further north than would be the case off the coast of South Africa or South America. Mean annual rainfall in the mid-west of Australia is more than twice as high as rainfall at an equivalent latitude on the west coast of Africa. Agriculturally, Western Australia benefits from the warm Leeuwin current. Although there are no high mountains to produce orographic rainfall (rainfall related to altitude) on the west coast, the warmer seas compensate to some extent. However, there can be no escaping the large influence that mountains have in determining rainfall patterns around the world.

Landscape and climate

Orographic relief influences the Southern Hemisphere climatic macropattern. The east coast of Australia has higher rainfall on average than the west coast because the Great Dividing Range causes air travelling towards eastern Australia to rise, cool and condense water vapour into rain-bearing clouds. Similarly, New Zealand and South American mountains interact with westerly winds, producing very moist environments on the western side. Two glaciers on the South Island of New Zealand (the Fox and Tasman glaciers) are the glaciers closest to the equator that almost reach sea level, maintained by heavy snows in the South Island mountains. The high mountains forming the spine of New Guinea cause moist air to rise, bringing high rainfall. Rain falls throughout the year and some peaks are high enough to have miniglaciers on their summits. Even the small increase in altitude of the Darling Range in south-west Western Australia is reflected in increased rainfall.

ENSO – El Niño southern oscillation

The influence that the oceans hold over the world's weather patterns is emphasised by the El Niño-Southern Oscillation (ENSO) climatic pattern, producing droughts in some parts of the Southern Hemisphere and devastating floods in others. The driving force is the waters of the central Pacific Ocean called the Pacific warm pool. In some years the Pacific warm pool expands and blocks the cold upwellings on the west coast of South America. This dramatically decreases the productivity of the ocean in this area and hence there is low food supply for breeding albatrosses and other seabirds on Pacific islands, so populations of these birds crash along with their food sources. The warmer seas on the eastern side of the Pacific Ocean increase evaporation feeding into the atmospheric circulation pattern at the equator (Walker circulation). This promotes very heavy storms on the South American coast, causing flooding and landslides (the high rainfall is called an El Niño event because it comes at Christmas time – *niño* is 'boy child' in Spanish). In contrast, on the Australian side of the Pacific the south-east trade winds supplying summer rainfall to north-eastern Australia weaken, causing severe drought in eastern Australia. The alternative (La Niña) years are times of a smaller Pacific warm pool so the upwellings strengthen, increasing marine productivity on the west coast of South America. The Walker circulation is moved to the western Pacific Ocean, producing good rainfall years on the east coast of Australia. In Western Australia there is no direct link with the ENSO system, but some evidence suggests links with the southward flowing Leeuwin current of the following year.

El Niño years may be more frequent because of global warming. The mean annual temperature of the Earth has risen by 0.6 degrees over the last 100 years, causing sea level rises, melting of the polar icecaps and mountain glaciers, higher frequency of severe storms such as hurricanes and more extreme drought.

Soils

In terrestrial systems the soil, an amalgam of living and non-living components, is a primary determinant of plant communities and the animals they support because it is the nutrient source for most primary producers.

Soil formation and structure

There are five interlinked soil-forming factors:

1. *substrate* – in most cases, this is igneous or sedimentary rock. Igneous rocks form from silicate minerals (a combination of various elements and silicate, SiO_4^-) at very high temperatures in the Earth's interior. Silicate minerals such as feldspars can contain magnesium, calcium and sodium whereas mica has iron, magnesium and potassium as well as silicon. These minerals weather to form softer clay minerals. Sedimentary rocks are formed from eroded sediments of other rocks accumulated at the base of water bodies. In time, the weight and pressure of accumulated sediment

compresses the lower layers into rock. The structure and mineral composition of rock influences how quickly it will disintegrate under weathering, because some minerals degrade more easily than others.

2. *climate* – temperature and humidity are the most important factors. Bare rock is broken down over long periods by physical processes such as expansion and contraction of the rock crystals during temperature changes, and by the action of weak carbonic acid formed from water and carbon dioxide. Where temperatures fall below freezing, water between rock crystals will form ice that expands and gradually can break the boundaries between different rock crystals. Cumulatively, these processes are called weathering.
3. *biota* – living things continue the process of breaking down rocks. For example, lichens (Chapter 10) grow on rocks, probing with their anchoring hyphae and etching the rock with organic acids. Furthermore, the wastes of organisms and the decay of their bodies after death add organic compounds and inorganic nutrients to the developing soil.
4. *slope* – soil cannot accumulate on steep slopes.
5. *time* – weathering and colonisation by organisms take long periods of time.

We can classify the soil into two horizons: the A horizon (top soil) and the B horizon (subsoil). The A horizon is where most nutrient recycling occurs and it is the key to productivity. The B horizon grades into the weathered rock and accumulates some materials washed down from the A horizon. The B horizon may also add minerals to the A horizon nutrient store as deep roots gather nutrients from the weathered substrate.

Soils and plant communities

Within a uniform climatic area, soils across a landscape can change markedly, supporting different plant communities. As an example we follow a 100 km transect (a line across country) along the Victorian coast from Bacchus Marsh due south to Anglesea (Plate 22.1). The striking changes in soil type and vegetation are related to the diversity of rock types across the landscape.

Near Bacchus Marsh the soils are formed from Permian sediments. These are low in nutrients and prone to erosion in cleared areas. Going south from Bacchus Marsh we cross basalt plains derived from volcanic eruptions millions of years ago, interspersed with Tertiary sedimentary deposits. Weathering of the basalt produces a clayey, high-nutrient soil that cracks severely when dry, limiting support for tree roots. Parts of the basalt plains supported only grasslands of kangaroo grass (*Themeda australis*) and herbs before they were cleared for grazing. Swampy areas supported thickets of lignum (*Muehlenbeckia cunninghamii*) and there were sparse woodlands of red gum (*Eucalyptus camaldulensis*), particularly near streams. Soils on the Tertiary sediments on the other hand are sandy and poor in nutrients, but they support low woodland to low open forest with a grassy understorey dominated by manna gum (*Eucalyptus viminalis*) and casuarinas (*Casuarina stricta, C. littoralis*).

Further towards Geelong, in the Barrabool Hills, the rocks are Cretaceous sediments similar to those underlying the Otway Ranges to the west. These rocks, despite being sedimentary, form a rich, loamy soil. Almost no native vegetation remains in this part of the transect and the only hint of what used to occur are isolated *E. viminalis*, *Acacia melanoxylon* and blackthorn or South Australian Christmas bush (*Bursaria spinosa*), often beside roads or on cliff sides.

Beyond the Barrabool Hills the transect again crosses Tertiary sediments, with podsolic soils – a sandy A horizon and clayey B horizon. On this soil the manna gum–casuarina woodland with grassy understorey occurs again, as on the Tertiary sediments south of Bacchus Marsh. Closer to the coast the Tertiary sediments are heavily weathered, losing nutrients and developing red colouration. Vegetation on these soils is low woodland of stringy-barks such as *Eucalyptus obliqua* and *E. baxteri* with a heath understorey or treeless heathland dominated by the heather family (Ericaceae, including *Epacris* and *Leucopogon*), teatrees (*Leptospermum*), banksias (*Banksia*), low-growing casuarinas and several sedges (Cyperaceae) and orchids (Orchidiaceae).

On the recent sand dunes at the sea at Anglesea coastal teatree (*Leptospermum laevigatum*) introduced from further east and moonah (*Melaleuca lanceolata*) are the dominants, as well as grasses including marram grass (*Ammophila* spp.) introduced for sand stabilisation, and a native grass, coastal poa (*Poa poiformis*).

This transect shows a great range in vegetation and changes in species dominance over an area with almost uniform rainfall, but marked changes in soil types. Such changes occur globally in all climatic zones where soil types differ.

Evolutionary history

Whereas climate and soils determine the distribution and abundance of species, the taxonomic mix of species occurring on different land masses and their specific adaptations are in turn products of evolutionary history. Recall that about 250 million years ago the continents were united in the super landmass Pangaea. This ultimately separated into the supercontinents Laurasia in the north and Gondwana in the south. The separation of Gondwana in turn gave rise to several land masses including Australia, Antarctica, Africa, India, Madagascar, South America and New Zealand (see Chapter 7). Thus for most of the last 50 million years Australia has been isolated from the other land masses of the world. During the ice ages, lower sea levels created land bridges between Australia and Papua New Guinea, across which some organisms may have dispersed. However, for much of the period isolation was complete and, with the exception of a small number of long-distance migrants, Australia's biota evolved alone under conditions of low soil fertility and increasing aridity.

As it drifted northwards, the Australian continent shared no land boundaries with other tectonic plates to raise mountains or create substantial seismic or volcanic activity.

Thus the land became aged and eroded. Parts of the Pilbara in Western Australia are the oldest exposed land surfaces in the world, and the continent as a whole is extremely flat. The average elevation above sea level is 440 m, which is the lowest average elevation of all continents. Indeed, the highest Australian peak (Mt Kosciuszko) is equal in elevation to the average elevation above sea level of the Antarctic continent. Steady erosion removed nutrients from the soil, and these were not replaced by volcanic eruptions or glaciers grinding down existing rocks.

Australia's northward drift gradually changed the climate. The flat landscape, with no high mountain ranges to intercept clouds and encourage rainfall, gradually became drier with most rainfall restricted to coastal belts. Northern regions receive rainfall mainly in summer, whereas southern regions have mainly winter rainfall. The centre of the continent became a large arid zone receiving little, unpredictable rain. More than half of Australia has a median rainfall of less than 30 cm a year. This situation is exacerbated by Australia's susceptibility to ENSO, which contributes to oscillating cycles of drought and heavy rain.

Australian flora

The evolution of the plants was driven by low soil fertility and aridity. Plant adaptations to cope with poor soils also confer some advantages to aid surviving in dry conditions, preadapting these plants to the arid conditions ultimately prevailing over much of the continent.

The Australian flora includes some very ancient remnants from the time when all the world's land was united in the supercontinent Pangaea (for example, the Wollemi pine, Chapter 11). More recent, but still very old, species trace their lineage to Gondwanan founders. Nevertheless, the long period of evolution in isolation gave rise to many unique plant species, so that about 85% of Australian plants are endemics, meaning that they occur in Australia and nowhere else in the world.

The endemic species are not spread evenly over the continent. Now that Australia's northward drift has carried it close to South-East Asia, the north of the continent is within reach of long-distance dispersal of plant propagules by wind, animal vectors or ocean currents. Some tropical plant groups colonised northern Australia, so that the proportion of Australian endemic species is lower in the north than it is in the south. For example, in Western Australia about 14% of species in the tropical Kimberley are endemic whereas in the south-west the figure is closer to 80%.

Australian fauna

Animals also showed extraordinary adaptations to cope with extreme conditions. For example, the hopping gait of kangaroos is extremely energy efficient at mid to high speeds, and species of water-holding frogs can survive for many years below ground before emerging to breed following a brief period of rain.

Australia's animals derive from two broad evolutionary sources: the descendants of those species that were on the continent when it separated from Gondwana and more recent colonisers that migrated to Australia when its northward drift brought it close to South-East Asia. The case of the mammals illustrates the significance of this history.

About 90% of Australia's mammal species are endemic, yet much of that endemism is restricted to the marsupials. Fossils indicate that marsupials evolved in North America and spread into Europe, South America and Australia. Following the separation of the continents, marsupials became extinct in Europe and North America, but they diversified in South America and Australia. When Australia separated from Gondwana, marsupials predominated in the mammal fauna and the few early eutherian mammals present became extinct. Other elements of the mammal fauna were introduced more recently as Australia drifted northwards towards South-East Asia. Insect-eating bats are believed to have entered Australia about 50 million years ago, and fruit bats probably arrived only within the last 6000 years. Australia's recent proximity to South-East Asia also led to entry by rodents about 15 million years ago, while the dingo (*Canis lupus*) was introduced by seafarers to northern Australia about 3500–4000 years ago. The dates of introduction of fauna are of more than academic interest, because predation, competition and habitat change caused by introduced species may influence the abundance and distribution of other fauna. For example, the wolf-like carnivorous marsupial the thylacine declined to extinction in mainland Australia, following competition from the dingo and competition and predation from Aboriginals (Chapter 17).

We will now see how climate, soil and biota unite to form characteristic distributions of life across the continent.

Biomes

Biomes are ecological communities of wide extent, with a particular structure and characterised by climate and soil. The same biome (grasslands, for example) may be found in various parts of the world, but contain plants and animals from different taxonomic groups. The life form of the plants (trees, shrubs, climbers and so on) (Table 22.1) is determined largely by the climate, which poses physical stresses of temperature and water availability which determine how long plants can photosynthesise each day. Animals are integral to biomes as well. Clearly, there are no koalas in a grassland or buffalo in a dense rainforest, so the forms of the plants determine the types of animals to some extent.

There are several ways to describe and compare biomes. One is to compare the main plant species occurring in the biomes (vegetation floristics). However, these species vary globally, especially between the Northern and Southern hemispheres, so comparisons are difficult unless one is familiar with all the species. A second system uses the structure of the vegetation (forest, grassland, heathland and so on). This is more universal, because similar-looking vegetation occurs in similar climates around the world.

Table 22.1 Simplified structural classification of vegetation.

Growth form	Height range	Foliage cover classes (%)		
		100–70	69–30	< 30
Trees	> 10 m	Closed forest	Open forest	Woodland
Shrubs	< 10 m	Closed scrub, shrub or heathland	Scrub, shrub or heathland	Open scrub, shrub, heathland
Grasses/sedges	< 2 m	Closed grassland	Grassland	Open grassland
Herbs	< 1 m	Closed herbfield	Herbfield	Open herbfield
mosses	< 0.5 m	Closed mossland	Mossland	Sparse mossland

In Australia, biomes are classified by the height of the vegetation and the projective foliage cover (the area of ground overshadowed by the foliage) of the tallest layer (Table 22.1). Plant height and cover depend mainly on the efficiency of photosynthesis, which in turn relates to climate. In productive sites there is a high likelihood of a positive carbon balance, but in stressful environments carbon gain may be small and plants remain low. This approach could be extended beyond Australia, but Northern Hemisphere terminology is influenced by where some vegetation types occur (e.g. taiga in Russia), and by the importance of deciduous forests (where trees lose all their leaves at one time) that are largely absent in Australia (Table 22.2). We now consider the important Australian biomes.

Table 22.2 Comparison of biome concepts for northern and southern viewpoints.

Northern Hemisphere	Southern Hemisphere	Critical climatic variables
Tropical rainforest	Closed forest	High moisture and even temperature
Temperate deciduous forest	No extensive representation in the south, small areas in South America, temperate closed forest	Cold, high-nutrient soil
Temperate evergreen forest	Open forest	Moderate rainfall and temperature, fire
Savanna	Woodland (temperate) Savanna (tropical)	Pronounced dry season, fire important in the tropics
Taiga	Not represented	Severe cold, but with brief warm summer
Shrubland	Heathland, scrub, mallee	Fire frequent, low-nutrient soil
Grassland	Hummock or clump grassland	Environmental extremes of drought, cold, waterlogging, fire
Tundra	Herbfield, low closed heathland	Cold season
Desert	Grassland, succulent heathland	Drought

Closed forests

In the humid tropics, conditions for plant growth are ideal and trees grow densely with a closed canopy, forming closed forest with an understorey of shade-tolerant herbs and ferns (but see Box 22.1). Palms may be prominent and in north-eastern Australia conifers often emerge above the canopy of angiosperm trees. Conifers are more prominent towards the drier end of the climatic gradient occupied by closed forest (Plate 22.2b) and angiosperms dominate towards the wetter end (Plate 22.2a). In Australia, closed forest species are often killed by fire, favouring fire-tolerant eucalypts and the creation of open forest.

Box 22.1 Closed forests of the temperate Southern Hemisphere

Closed forests are not restricted to the tropics. They extend to any area where soils are well aerated, temperatures are equable and moisture is not limiting. The Gondwanan biogeographical connections are very strong in closed forests of the Southern Hemisphere. *Nothofagus* (southern beech) is a key element in temperate closed forests in the Southern Hemisphere in montane eastern Australia, New Guinea, Tasmania, New Zealand, South America and, in a minor way, in New Caledonia. Conifers are frequently associated with closed forests, often as emergents, but also as co-dominants. *Agathis australis* (kauri) dominates forests in the north of the North Island of New Zealand while *Podocarpus* and *Dacrydium* species occur further south (Plate 22.2d). In New Caledonia there is a high diversity of *Araucaria* species and these grow as emergents above closed forest angiosperm species (Plate 22.2c). In Chile *Nothofagus* is associated with the tall conifer alerce (*Fitzroya cupressoides*, Plate 22.2f). *Nothofagus* does not occur in Africa, but the conifer *Podocarpus* does, and is associated with lowland and highland closed forest. Although there are taxonomic similarities between the southern continent forest dominants, the factors controlling the forest form may not be the same in the different areas. Some appear to be event-dependent systems, whereas others are the climax type of vegetation.

In the Chilean closed forests, regeneration of alerce and some *Nothofagus* species requires a landscape-wide disturbance such as fire or landslides associated with earthquakes, so it is an event-dependent system (see Chapter 17) just like some Australian open forests (see below). In the wet climate parts of Chile long disturbance-free intervals lead to dominance by other closed-forest species, often in the Myrtaceae and laurel families with close relatives in the forests of north-eastern Australia.

In south-eastern Australia, including Tasmania, the highland closed forests are dominated by *Nothofagus* (Plate 22.2e). These forests tend to be the climax type of vegetation in the absence of disturbance. However, as in the tropics, fire has a large controlling influence on where closed forest can survive, as the species are not very fire-tolerant. Lowland closed forest has taxa more similar to those in the tropical rainforests. One of the iconic species that used to be common in central New South Wales, but is now rare due to logging exploitation, is red cedar (*Toona australis*), which can be deciduous – an uncommon trait in such forests.

In some parts of the eastern Australian coastal mountains, such as the Atherton Tableland in north-eastern Queensland, rainforest clearing allowed productive farming on fertile soils derived from nutrient-rich basalt rocks. However, in other places closed forest grows on nutrient-poor soils. The nutrient capital is mostly contained in the forest and its topsoil, and soil nutrients are quickly exhausted by agricultural production after clearing.

Many animals in closed forests are tree-dwelling (arboreal), such as tree kangaroos, cuscus and possums. Bird groups include pigeons, bowerbirds, riflebirds and the beautiful birds of paradise. Important flightless birds such as the cassowary disperse plant seeds in their droppings. Invertebrates are extremely diverse. In South-East Asia, Africa and South America, primate groups (monkeys and apes) are an important closed-forest element.

In the Northern Hemisphere, deciduous forests are major vegetation formations (Plate 22.3a) and our scheme of height and canopy cover classifies them as closed forests. However, the dominant trees lose their leaves in winter to cope with lack of water. They occur on nutrient-rich soils, so plants can afford to lose leaf nutrients every year. Australia has only one winter deciduous forest tree – *Nothofagus gunnii* from wet environments in Tasmania. However, in southern South America there are winter deciduous trees, again in the genus *Nothofagus*. Large areas of the cold temperate Northern Hemisphere replace deciduous forests with a taiga dominated by evergreen conifers. The closest Australia gets to taiga is in wet areas of Tasmania where sparse conifers are emergent over heathland (Plate 22.3d).

The next step towards drier sites from closed forest is open forest. The influence of water in driving the formation of forest versus other biomes is summarised in Plate 22.4.

Open forests

Australia's open forests are dominated by one tree genus – the fire-adapted *Eucalyptus*. In the cool, temperate regions of southern Australia where soil water does not generally limit growth, some of the tallest species that dominate tall open forest are event dependent (Plate 22.3f). They have poor fire resistance, but establish well from seed after severe fires with a return time of about 300–500 years (mountain ash *E. regnans*, Sydney blue gum *E. salign* and karri *E. diversicolor*). The understorey is composed of smaller trees from various families, tree ferns (in eastern Australia), ground ferns, liverworts and mosses.

In drier areas eucalypts are not as tall and are mostly resistant to fire, resprouting from epicormic buds in the stem or, after very severe fires, from the base (e.g. messmate *E. obliqua,* jarrah *E. marginata* and snow gum *E. pauciflora*). The understorey comprises other fire-resistant species including wattles (*Acacia* spp.), the protea family (Proteaceae, *Banksia*, *Hakea*, *Grevillea*, *Persoonia*) and the pea family (Fabaceae). The wattles and peas replenish soil nitrogen after fires through their association with the nitrogen-fixing bacterium *Rhizobium* (Chapter 9).

Fauna in these forests comprises arboreal possums (gliders, Leadbeater's possum) and the koala (Plate 22.3e), as well as ground-dwelling wallabies, wombats, carnivorous dasyurids (chuditch, quoll) and smaller marsupials such as *Antechinus* and native rodents. Important birds include cockatoos, lyrebirds, parrots, honeyeaters and tree creepers. Many of the bird species feed on the diverse invertebrates.

Woodlands

As the water availability gradient becomes drier, the trees become sparser. The tree form changes from forest form (tall, long-boled with narrow crown) to open-crowned trees with shorter boles and spreading canopies. This is the woodland formation or savanna. As the canopy is less dense the understorey is diverse with shrubs, herbs and grasses. Eucalypts are prominent – in the east, poplar box (*E. populneus*), yellow box (*E. melliodora*) and ironbark (*E. crebra*) are common. In the west, salmon gum (*E. salmonophloia*, Plate 22.3b) and wandoo (*E. wandoo*) are common. However, in drier conditions other groups such as *Acacia, Casuarina* and conifers (*Callitris*) become important. In the north, the savanna occurs in a monsoonal climate extending into Papua and Timor. Here eucalypts are prominent (*E. miniata, E. tetradonta*) as well as the boab (*Adansonia gregorii*, especially in Western Australia, Plate 22.3c), *Terminalia* spp. and ironwood (*Erythrophleum chlorostachys*). Tall grasses such as *Sorghum plumosum* and many others dominate the understorey.

Some woodland tree species are deciduous in areas of winter drought, although some species retain the leaves if the dry season is not severe. Examples of deciduous trees are the baobabs in north-western Western Australia, boabs of Africa (*Adansonia*), and many acacias in Africa (e.g. *Acacia nilotica*). Few eucalypts (about 20–25 species) in northern Australia are deciduous (e.g. *E. tectifera* and *E. grandifolia*).

Fire is frequent. Summer rainfall promotes luxuriant grass growth that dries in winter to a high fuel load ignited by lightning. Humans, once they reached Australia, learned to burn the understorey to encourage regrowth palatable to grazing animals. This interaction of fire and savannas is worldwide.

African savannas have a very diverse grazing fauna and an equally diverse carnivorous fauna to eat them, but Australia's larger animals (megafauna) disappeared in the last 100 000 years (see Box 22.2). At the time of European settlement the only large herbivores of the savannas were kangaroos (principally the red kangaroo, *Macropus rufus*). The only large carnivore on the mainland was the dingo (*Canis lupus*), which reached Australia about 3500–4000 years ago. There was a greater diversity of smaller marsupials such as

Box 22.2 Why are there few large animals in Australian savanna?

Australian monsoonal savanna has far fewer large animals than the African savanna although the vegetation is structurally similar. The answer may lie in soils and climate. The African savanna soils are nutrient rich whereas the Australian soils are relatively nutrient poor. The rainfall in African savanna is more reliable because the savanna systems extend into the equatorial region, whereas in Australia the extension of the savannas is south into the desert rather than north into more humid environments. In Africa there are extensive areas available for migration so animals can follow the seasonal productivity, but in Australia the sea prevents migration to the north, and to the south the land is very arid. The contrasting latitudinal positions of the continents and the differences in soil nutrients may be the reasons for the faunal disparity.

the bilby (*Macrotis lagotis*), mala (*Lagorchestes hirsutus*) and several wallaby species. The insect fauna in Australian woodlands is also abundant. Termites and ants are crucial to recycling nutrients and are important foods for specialised native fauna such as numbats and echidnas.

Non-tree-dominated formations

Regions with extreme conditions of temperature, moisture, soil nutrient availability or exposure to toxins such as salt or other chemicals in the soil support few, if any, trees. In these regions plants of lower stature become dominant (Plate 22.4).

Grasslands

Grasses have radiated to occupy many inhospitable environments as well as some of the most productive (Plate 22.5). In many cases they are subordinate to large plants, forming the understorey stratum. However, when given the chance by less than ideal (from a tree's point of view) environmental factors such as fire, drought, soil toxicity or low temperature, they dominate the vegetation. Selected species also provide the staple food source for the most humans and grazing animals.

Grasslands of several types occur in the arid centre of Australia. On sandy soils hummock grasslands are characteristic. Sparse clumps of spinifex (*Triodia* spp., colloquially known as 'porcupine grass') initially form a very prickly hummock, but grow to a doughnut form with either bare ground or other plant species in the centre (Plate 22.5d). These grasses are one reason why Australian deserts are resilient to damage from overgrazing by feral animals. They are so unpalatable that they retain at least a skeleton of the vegetation even when other species are eaten out. However, the hummocks are very flammable so fire is an integral part of even the most unproductive environments in central Australia.

On more clayey soils tussock-type grasses occur (including *Astrebla* spp.). In dry times there may be few other plants obvious between the tussocks, but after rain there are many annuals, particularly daisies.

While these formations dominate vast areas of inland Australia, extending south to the Great Australian Bight, special formations occur in limited areas such as beside ephemeral water courses, in depressions or on rocky hillsides. Trees may be present. River red gum (*E. camaldulensis*) is the most widespread eucalypt in Australia, occurring along dry rivers throughout the arid zone (Plate 22.5c). Water plants such as the aquatic fern, nardoo (*Marsilea hirsuta*), occur in wetland depressions forming after storms. Eucalypts and acacias may grow in inhospitable landscapes if their roots reach a soil water source. Acacias are more drought resistant than the eucalypts, and many species survive with access to very little soil water for long periods.

In the eastern Australian alps and Tasmania, grasslands occur above the timber line and in frost hollows. The timber line in Australia is not very well defined, but is correlated with the 10°C summer isohyet. Above the timber line the growing season is not long enough for plants to develop into trees. Thus grasses, herbs and low-growing shrubs dominate. Grasses (mostly *Poa* spp.) form tussocks interspersed with perennial herbs and occasional low shrubs.

Grasslands dominated by *Poa* tussocks or the related genus *Austrostipa* occur along the coast where salt-laden winds and sandy soils present a difficult combination of environmental factors for most plants (Plate 22.5g). In more saline areas saltgrasses (*Distichlis distichophylla*) or dropseed grasses (e.g. *Sporobolus virginicus*) may form dense mats. The true *Spinifex* (not to be confused with the desert plants from the genus *Triodia* commonly called spinifex) occurs as a pioneer plant on coastal dunes, sending out runners that stabilise the loose sand as the start of coastal primary succession (see Chapter 17). Finally, in waterlogged areas such as on the edge of coastal freshwater lakes the common reed (*Phragmites communis*) forms dense swards of culms up to 2 m tall, often with sedges and bulrush (*Typha* spp.).

The close relationship between savanna and grassland means that similar animal species occur in both biomes. Kangaroos are predominant in Australia, but elsewhere ungulates and ruminants are the most important groups – in North America the central plains were the range for bison (buffalo), the European steppes had horses and camels, and Africa has both groups.

Scrub and heathlands

Chenopod shrublands occur on clayey soils and in salt marshes in near-coastal sites. They comprise sparse shrubs in the Chenopodiaceae, the main genera being *Atriplex*, *Maireana* and *Sclerolaena*. These communities abound in annuals, including grasses and herbs, after rain, but are at other times rather bare in between shrubs (Plate 22.5h). The Sturt's desert pea (*Swainsona formosa*) may form a spectacular focus of red and black flowers against silvery foliage in a wet spring.

Low shrubs dominate the vegetation where soils are nutrient poor (Plate 22.5a, b). These can be very rich in species and are termed kwongan (Western Australia) and wallum (Queensland). The main families characteristic of these heathlands are Proteaceae, Restionaceae, Ericaceae (Epacridaceae) and Myrtaceae. The species may look similar, most having very small leaves, and the high diversity is revealed only at flowering. Similar heathlands occur in the Cape of South Africa – called fynbos, with very similar plant forms and plant families being dominant.

Heathland and scrub occur from sea level to the mountains on nutrient-poor soils. Where water availability is not a limiting environmental problem, the vegetation may be very dense. Temperature may be a limiting factor at high altitudes. In Tasmania there are areas of bolster moor (Plate 22.5f). The plants form dense, compact, low, rounded canopies (cushion plants) that resist wind resistance and warm up more quickly in the morning than do more open structures. The same plant structures occur in New Zealand and in Patagonian heathlands under subantarctic climatic conditions.

Herbfields

Herbfields are uncommon in Australia, confined mainly to alpine areas or granite rocks. These two areas have very different environmental conditions. On granite rocks the herbfields comprise mosses and annuals that are only functional during the moist winter and spring. During the dry season they exist as seeds or spores or become dormant. In alpine areas temperature, not water, is the limiting factor. The plants only function during the growing season of late spring, summer and early autumn – exactly the opposite of the case on the granite outcrop.

A more extreme example is the moss bogs of alpine areas, where *Sphagnum* moss is the controlling life form. *Sphagnum* contains many times its own weight in water and thus controls the hydrological conditions of the vegetation. Plants growing in the moss beds cope with severe waterlogging, and are often of the same form as plants in heathlands. This may in part be related to the poor nutrient conditions of the moss beds where carbon-to-nitrogen ratios are very high and not conducive to productive growth. Bogs and moss mat communities are very sensitive to trampling, which causes erosion and decreases the water-holding capacity of catchments.

In very exposed conditions in the alps a combination of herbfield and very diminutive heath is termed 'fjeldmark' or 'fellfield' (Plate 22.5e). This is very low vegetation in areas where there is poor snow cover in winter, when plants are exposed to high winds and very low temperatures. The harshness of this sort of environment leads to very poor productivity and low, ground-hugging vegetation.

Changing biomes

Biomes/vegetation types are presented here as discrete entities, but this is an oversimplification. There are rarely sharp boundaries in nature and one biome merges into another over 10–100 km, depending on the characteristics of the landscape and the soils that support the vegetation. The understorey of a savanna is, in many cases, composed of the components of a grassland – remove the trees and the area would be a grassland. Similarly, the understorey composition of a eucalypt forest is similar to some heathlands. Despite the usefulness of recognising biomes as a way of summarising the patterns of vegetation and faunal distribution, plants and animals still have their own range of distribution mediated by environmental tolerance and interactions with other species (their niche – see Chapter 17).

Although the environmental factors shown in Plate 22.4 can be considered the primary factors influencing the change from tree-dominated to non-tree-dominated systems, fire can cause similar changes. A change in fire regime can cause a shift to another vegetation type. As was emphasised for savannas, frequent fires can suppress tree proliferation. Similarly, an increase in fire frequency may force a change from forest to heathland. In the unusual situation of a *Sphagnum* peat bog drying during a drought that is followed by a fire, the peat may burn for a long time, changing the hydrology of the area and allowing invasion by heathland species, which may replace the bog species. Clearly, the interaction between vegetation and environmental factors may be complex.

Currently the natural biomes of the Earth are increasingly disrupted and some have largely been converted to farmland, cities or rangeland. These human-modified systems are often monocultures or weed dominated. They still have processes similar to natural systems, but in most cases are much less complex and very much less diverse than the natural formations.

Chapter summary

1. Climatic patterns result from atmospheric circulation systems caused by differential heating of the Earth by the sun. Oceanic currents also influence the atmosphere. Where these currents originate determines their temperature, which in turn affects the amount of water that evaporates to form clouds and hence provide rainfall.
2. The biological systems on different continents are the product of the environmental factors that influence them now and have done so over the millions of years of evolutionary development of biota.
3. Plants develop specific life forms in response to climate, so the structure of vegetation may be similar in similar climatic zones even though composed of unrelated species.
4. Animals, too, in similar climates may be structurally similar even though the species on different continents are not from the same taxa.
5. In the past, migration between areas was dependent on the isolation or connectedness of land masses, which ensured slow interchange between areas separated by oceans. Human transport increased the interchange of flora and fauna substantially, bringing benefits and problems.
6. Biomes are climatically determined groups of plants and animals spread over wide areas. In Australia biomes are classified by the height of the vegetation and the projective foliage cover (the area of ground overshadowed by the foliage of the tallest layer). In the Northern Hemisphere terminology is influenced by where some vegetation types occur (e.g. taiga in Russia) and by the importance of deciduous forests largely absent in Australia.
7. Major biomes in Australia are the closed forest, open forest, woodland, grassland, scrub and heathland, and herbfields.

Key terms

A horizon
B horizon
biome
closed forest
deciduous
El Niño
endemic
ENSO
fjeldmark
fynbos
heathland
herbfield
hummock grass
kwongan
La Niña
Leeuwin current
mallee
moss bog
open forest
orographic relief
resprouter
savanna
serotinous seed banks
silicate minerals
taiga
timber line
upwelling
vegetation floristics
vegetation structure
Walker circulation
wallum
woodland

Test your knowledge

1. How do isolation and landscape contribute to the Australian climate?
2. How are soils formed?
3. Using examples, explain how soil characteristics determine vegetation structure.
4. How has isolation shaped the evolutionary history of the Australian biota?
5. What are biomes? How does their classification in Australia differ from that in the Northern Hemisphere?
6. Biomes are described as discrete entities in the world. Is this justified?

Thinking scientifically

The Hamelin Bay mallee (*Eucalyptus calcicola*) grows naturally on sandy soils, whereas the Cowcowing mallee (*E. brachycorys*) grows naturally on loamy clay. Design an experiment to determine whether or not soil type determines the distribution of these species, or if another environmental factor is responsible.

Debate with friends

Debate the topic: 'Introduced species that have been in Australia for less than 10000 years should be eradicated'.

Further reading

Augee, M. and Fox, M. (2000). *Biology of Australia and New Zealand.* Pearson Education Australia, Sydney.

Fox, A. and Parish, S. (2007). *Wild habitats: a natural history of Australian ecosystems.* ABC Books/Steve Parish Publishing, Sydney/Archerfield, Queensland.

Fox, M.D. (1999). 'Present environmental influences on the Australian flora' *Flora of Australia.* 2nd edn. Vol. 1: 205–49.

Groves, R.H. (1999). 'Present vegetation types' *Flora of Australia.* 2nd edn. Vol. 1: 369–402.

Vandenbeld, J. (1988). *Nature of Australia.* Collins/Australian Broadcasting Corporation, Sydney.

23

Terrestrial lifestyles

Philip Ladd and Mark Garkaklis

Life on the rocks

Granite rock formations are often prominent in Australian landscapes, for example in Girraween National Park in Queensland, Wilsons Promontory in Victoria and Mt Franklin in Western Australia. In arid areas waterholes in granite outcrops were significant in summer for Aboriginals (Plate 23.1).

There are considerable similarities between granite outcrops across a wide range of climates. All have poor soils of restricted volume, severe fluctuations between wet and dry conditions, low levels of soil nutrients, high insolation (light intensity) and exposure to strong winds. However, as in any harsh environment, species with specialist adaptations survive and even thrive.

Algae and lichens colonise the rock surface and may be overgrown by mosses. These plant forms and some ferns and flowering plants (such as the pincushion (*Borya* spp.) and the feather flower (*Verticordia staminosa*)) can desiccate completely in the dry season, but revitalise and function again hours after rewetting. More commonly, plants die down to seed or rootstock in the dry period. The insectivorous sundew (*Drosera* spp.), which catches and digests insects for extra nutrients, particularly nitrogen, survives summer as an underground tuber. Other plants such as the elbow orchid (*Spiculea ciliata*) are succulents, storing water to persist for some time into the summer. Shrubs and trees on granite outcrops use water conservatively most of the year and need access to rock cracks for water over the dry season.

Many invertebrates flourish in temporary pools on rocks and survive the dry season as eggs in the sediment. Terrestrial invertebrates are often flattened to find protection under rock plates. Few vertebrates live in these dry conditions. One, the rock dragon (*Ctenophorus ornatus*), does not need to drink for 80 days.

All these traits for survival under dry conditions are found in plants and animals occurring in the wider terrestrial landscape, but they are more accentuated in the harsh conditions on the rocks.

Chapter aims

Australia has large areas where conditions are hostile to life: temperatures are excessively high or low, water availability is low and markedly seasonal, and soils are low in nutrients. In this chapter we examine the adaptations evolved by plants and animals that allow survival under these environmental conditions.

Living on land

Animal and plant distributions are controlled strongly by their environmental tolerances. Animals can move to suitable environments, sometimes travelling long distances, particularly if they fly. Many bird species migrate thousands of miles between Siberia and southern Australia to remain in continual summer. Whales migrate from Antarctica to the northern waters of the east and west coast of Australia, and every winter bogong moths (*Agrotis infusa*) migrate from the black soil plains in northern New South Wales and southern Queensland to the Australian Alps to find a cooler summer climate. However, many animals remain in harsh environments without migrating: natural selection has adapted them to the adverse conditions. The keys to survival lie in strategies adopted for temperature control (thermoregulation), maintaining water balance and obtaining vital nutrients. Plants must stay where they germinate and the plant body must tolerate extremes of weather to survive and reproduce. However, if environmental conditions change, plant seeds may disperse to newly favourable areas outside their previous range. Over many generations this may cause a shift or 'migration' in the range.

Thermoregulation

Organisms exchange heat with their environment by:

- *conduction* – the transfer of heat between bodies that are touching, such as an animal and a sun-warmed rock. The heat flow is from the hot body to the cold one.
- *convection* – the transfer of heat by a flow of air or water, such as a breeze or a stream current
- *evaporation* – energy is required to transfer a liquid to a gas. When water evaporates from an organism the energy comes from the organism's heat, so the organism is cooled.
- *radiation* – radiation in the form of electromagnetic waves transfers heat between bodies that are not touching. On a hot summer's day a plant will gain heat from solar radiation and will also radiate some of its own heat into the environment.

Balancing these factors with internal heat production from metabolic processes is essential in maintaining body temperature.

Water balance

Organisms cannot survive an overall gain or loss of water from their cells because biochemical functioning is impaired. Balancing water intake and loss is vital, because water is also important in thermoregulation via evaporative water loss and, in animals, to remove nitrogenous wastes.

The most successful land plants, the gymnosperms and the flowering plants, have vascular tissue for water transport (Chapter 12) and, as we will explore later, specialist adaptations that reduce water loss during photosynthesis. Insects and vertebrates, the most successful terrestrial animals, have external body coverings such as exoskeletons (insects) or fur (mammals), feathers (birds) and scales (reptiles), which reduce evaporative water loss. Internal reproduction and other adaptations such as the amniotic egg protect fertilised eggs and embryos (see Chapters 14 and 15). Insects, birds and reptiles excrete nitrogenous wastes as a uric acid paste, which, while costly in energy, saves water. Mammals excrete nitrogenous wastes as a urea solution (urine) that can be concentrated to high levels without toxicity, so some desert mammals lose little water in their urine. Many terrestrial animals also shelter from extreme temperatures in burrows, which also protect against radiation and reduce evaporative water loss.

Key nutrients

Plant health requires 16 essential inorganic nutrients (Table 12.3), which animals obtain from the plants they eat. Six of these key nutrients are called macronutrients because they are needed in large amounts and are key components in important organic molecules (see Chapter 3). The others are micronutrients and are often components of enzymes. Enzymes are not consumed in the reactions they catalyse, so only small amounts of these nutrients are needed, although their absence can cause nutritional disease. The weathering of the Australian continent stripped nutrients from the soils so that many Australian plants show adaptations to enhance nutrient uptake (Chapter 12). Nutrient levels in soils also influence animals indirectly (Box 23.1).

Box 23.1 Soils and ecosystems

The terrestrial ecosystems we see around us are largely dictated by climate and soils. Evolutionary processes operating over long time frames result in a suite of plant and animal species adapted for the environment they inhabit. Climate and soil type are the first two criteria considered by a biologist wishing to translocate rare plants to a new location. The importance of soils and climate in influencing the evolution of species and communities is emphasised when systems undergo rapid change.

In this region of Australia the south-flowing Leeuwin current brings warmer waters down the west coast, dispersing tropical fish and coral species to lower latitudes and establishing coral reefs at their southernmost limits in the Indian Ocean. Tropical and subtropical seabird species also benefit from the Leeuwin current. In recent decades the range of several seabird species has extended southwards and they have established large colonies on islands where once they were unknown, while in other areas their colonies have undergone rapid expansion. One such species is the wedge-tailed

continued ›

Box 23.1 continued ›

shearwater or mutton bird (*Puffinus pacificus*). It is colonial, nests in burrows and is absent from its colonies for several months. During the breeding season birds forage by day and spend their nights in frenzied burrow building, burrow maintenance and territorial calling. They also bring with them something from the ocean – nutrients in the form of their digestive wastes, called guano.

The soils of most islands off the west coast of Australia are derived from the limestone of which the islands are composed. The soils are extremely nutrient poor, and the vegetation is adapted to an environment where nutrients and rainfall are limiting. However, within seabird colonies the soils are very different because of the guano deposited and the nutrient input increases the productivity of the plant communities.

Animal species too are affected by this increase in nutrients. Two islands in Western Australia contain relict populations of the endangered dibbler (*Parantechinus apicalis*), a small carnivorous marsupial. Dibblers, as with some other genera of small carnivorous marsupials such as *Antechinus*, are renowned for their unusual life history. After a vigorous mating season all the males die, leaving females and their young to grow and develop in an environment with lower competition.

This is certainly the case for dibblers on Boullanger Island where, in most years, all males die after breeding, so males in the population are all no older than 1 year. On Whitlock Island, 500 m to the west, the situation is quite different. Male dibblers on Whitlock Island live up to 4 years. The fundamental difference between the islands is the large number of seabirds breeding on Whitlock that are absent from Boullanger. Soil nutrients are up to 18 times higher, invertebrate productivity is greater, and the dibblers are in better condition. Soils really do have a strong influence on the nature of the world's ecosystems, and ultimately have a great influence on the evolution of the life histories of plants and animals that inhabit them.

Having established these general principles regarding life on land, we move on to consider specific examples.

Adaptations to cold environments

Adaptations to cold in plants

Where key resources for growth are scarce, as in extremely cold conditions, plants are generally small (Plate 23.2), with a low ratio between photosynthesis and respiration. Respiration uses up almost all the energy captured in photosynthesis, leaving little for cell construction.

The juvenile tissue (apical meristems) at the shoot and root tips is the most sensitive and easily damaged part of plants. The above-ground meristems are protected by overlapping young leaves, but species growing in particularly cold (or dry) climates produce special, resistant buds aiding survival during unfavourable seasons. Indeed, in 1934 C. Raunkiaer divided plant species into a range of 'life forms' based on the position of the apical bud in the tissues above or below ground, reflecting the degree of protection afforded these vulnerable meristems (Table 23.1).

Table 23.1 Comparison of plant life form and plant growth form classification

Life form	Description	Growth form	Description
Phanerophyte	Woody plant with perennial shoots higher than 0.5 m	Tree	Usually single stem with elevated canopy
		Palmoid	Woody plant with rosette of leaves at top of usually a single stem, may be clumped stems
		Tall succulents	Green columnar succulent stems > 0.5 m, usually no leaves
		Shrub	Woody plant usually multistemmed over 0.8 m with canopy developed from close to the ground
		Climber, scrambler	Woody plants that use upright plants for support
		Hemi-epiphytes	Woody plants that germinate in trees and eventually establish a root system to the ground, or plants that germinate on the ground, but grow onto plants and disconnect their roots from the ground
		Epiphyte	Woody plant completely established on another plant, but not parasitic
		Woody hemiparasite	Woody plants which attach to branches or roots of other plants by haustoria
Chamaephyte	Woody plant with mature stems to 0.5 m, shoots may seasonally grow higher, but die back to 0.5 m	Dwarf shrub	Woody plant to 0.8m tall with canopy developed from close to the ground
		Short succulents	Globular or prostrate succulent stems < 0.5 m

continued ›

Table 23.1 (Continued)

Life form	Description	Growth form	Description
Hemicryptophyte	Low growing plant with shoots close to the ground or periodically reduced to buds close to ground level	Cushion	Woody or herbaceous plant with tightly packed foliage close to the ground
		Tussock	Herbaceous – many erect leaves produced from a basal clump or multiple meristems
		Herb – various leaf arrangements from basal or semibasal leaves or erect leafy stems	Herbaceous plants, generally low growing some may be climbers
Geophyte	Seasonal reduction of the shoot to a storage organ below ground	Herb – short basal leaves	Low-growing herb with a bulb or similar underground part.
Therophyte	Annuals that survive the unfavourable season as seeds	Herb – various leaf arrangements from cushion, basal or semibasal leaves or erect leafy stems	Herbaceous plants, generally low-growing, some may be climbers
		Herbaceous hemiparasite	Herbaceous plants that attach to roots of other plants by haustoria
Helophyte	Shoot systems produced above a water surface, but seasonally reduced to bud below the water	Herb	Herbaceous forms annual emergent above the water surface
Hydrophyte	Aquatic plants always fully submerged or floating on the water surface	Herb	Aquatic plants always submerged or sometimes leaves float on the surface

Seasonal changes also influence plant life cycles. Most terrestrial biota function best in a narrow range of environmental conditions. Winter in high latitudes is too cold for most species to function well, so many survive the cold season as dormant propagules such as seeds.

In high latitudes, seasonal changes include not only low temperatures, but also changes in day length. Low temperatures limit biochemical reactions in both plants and animals, and short days limit photosynthetic production in plants and hence food production for animals. As pointed out in Chapter 22, some plants lose their leaves with the onset of winter. Those that are not deciduous survive the cold season by making metabolic changes to prevent the formation of ice crystals, which would destroy cell components and kill cells. Snow cover insulates plants and prevents damage from very low temperatures. Eucalypts that survive –15°C in the Australian snowfields do not survive when planted in the Northern Hemisphere where temperatures are similar but there is no snow cover and the ground freezes.

At high altitudes environmental conditions are similar to those at high latitude. At high altitudes in low latitudes, temperature fluctuates greatly between day and night (diurnally), but day length does not change greatly over the year. In such places plants are not deciduous, but tend to have particular growth forms (Table 23.1). Species from several unrelated families developed a cushion form, an example of convergent evolution where natural selection leads to similar adaptations to a common environmental problem in distantly related organisms. The cushion plant form (Plate 23.2b) presents a streamlined shape to strong winds, preventing dehydration and also warming the plant up more quickly than if it was more open growing. Solar heat is absorbed by the leaves and re-radiated to the surroundings. Nearby leaves absorb the radiation and hence the whole plant uses the total energy input efficiently. Hairy leaves also aid heat retention. As the growing season when plants can make positive carbon gains decreases at high altitudes or latitudes, the size reached by mature plants also decreases, creating a timber line (Plate 23.2c) (Chapter 22).

Adaptations to cold in animals

Some animals, especially some bird species, respond to seasonal cold by migrating to warmer areas, and others have adaptations to survive the winter. In many bears and some small mammals and reptiles, fat accumulates during summer and is used over winter during a long-lasting state of inactivity, lowered metabolic activity and lowered body temperature called hibernation. This saves on energy costs for temperature regulation. Some Australian small marsupials (e.g. Leadbeater's possum) display similar behaviour, but over shorter periods of hours or at most days, in which case it is called torpor.

Other animals remain active during winter or combine periods of activity and torpor. For example, the tiny 50 g mountain pygmy possum (*Burramys parvus*) lives at 1340 m altitude in the alpine regions of the Great Dividing Range in New South Wales and Victoria, including the Kosciuszko National Park. In winter the area is covered in deep snow and the possum may be active in tunnels and spaces between the piles of boulders of glacial moraine, where, because of the insulation of the snow, the temperature is above 0°C. It intersperses periods of up to 20 days torpor between bouts of activity, adopting a 'furry ball' position with the tail curled and ears tucked in so there is minimal surface exposed to the air (Plate 23.6). Males and females live in separate habitats for most of the year and development associated with ski fields is threatening some populations (Box 23.2).

Perhaps the most striking adaptations to cold occur in the emperor penguins (*Aptenodytes forsteri*) of the Antarctic. They endure temperatures as low as –30°C in the Antarctic winter

Box 23.2 The Mt Hotham 'tunnel of love'

(Contributed by Rodney Armistead)

The mountain pygmy possum (*Burramys parvus*) (Plate 23.6) was believed extinct before its rediscovery in 1966. It is now known to occur in isolated populations in the Kosciuszko National Park, but the fossil record indicates it was once distributed widely through south-eastern Australia. Its preferred habitat is boulder fields capped by shrubby heathland, where it feeds on insects, fruits and seeds (some of which it stores). In summer migratory bogong moths shelter in crevices between the boulders and are a big part of its diet. Males and females normally live in separate regions, with the females at higher altitudes than the males. After the spring snow melt (around October), males move into the female regions and mate. They then return to lower altitudes and the females raise four young to independence before the onset of the winter.

Unfortunately, the mountain pygmy possum's habitat is popular for ski resorts. At Mt Hotham a major highway, the Alpine Way, had been built between the high-altitude breeding ground and the male habitat at lower altitude. This caused the adult and the juvenile male possums to linger in the 'female' breeding habitat through summer and autumn (and, we presume, winter, though no possum has been trapped in winter). This decreased female survival over winter and reduced the size of litters. In an attempt to restore normal migration, an artificial tunnel of rocks was constructed under the road to link the habitats. It has wire grills to exclude predators and cameras to record use of the tunnel. Possums were moving through the tunnel within an hour of its completion. The sex-age composition of the high-altitude population reverted to that of undisturbed areas and breeding success increased. The tunnel was popularly dubbed the 'tunnel of love'. Similar rock-filled tunnels allow continuity of the possum colony at Mt Blue Cow, where it is crossed by ski runs. Here success has not been so great as at Mt Hotham, perhaps because the noise and vibration of the snow-grading machinery arouses the animals too frequently from their hibernation.

Despite conservation efforts, these animals remain threatened because of climate change effects on their habitat and on their preferred food source, the bogong moth. Furthermore, the bogong moths transport toxic substances such as arsenic from agricultural regions into alpine areas (Chapter 27).

and fierce, cold wind gusts add a substantial chill factor. One defence against the severe conditions is huddling. When many penguins aggregate, the surface area of the group is reduced, lowering heat loss. Feathers provide effective wind and water insulation, and the birds have thick layers of fat, which afford heat insulation. Tucking flippers close against the body reduces surface area, and black feathers on the back absorb solar radiation. Breathing cold air and contact between the feet and the ice are potentially significant causes of heat loss. In both cases, this is reduced through countercurrent blood flow systems similar to those in some marine animals (Chapter 19). Blood flowing to the feet or the nasal passages passes its heat to blood returning from these areas to the body core, so little core heat is lost to the environment.

Adaptations to hot, dry environments

Most rainfall in Australia is restricted to coastal belts, and the centre is a large arid zone receiving little, unpredictable rain. More than half of Australia has a median rainfall of less than 30 cm a year. This situation is exacerbated by Australia's susceptibility to the El Niño-Southern Oscillation (ENSO) (Chapter 22). Nevertheless, plant and animal life thrives in the hot, dry conditions. Periodic water shortage has been an important driving force in plant evolution in much of Australia, but especially in the monsoonal north, the semiarid and arid interior, and the regions of Mediterranean climate in the south. Apart from the drought escapers (short-lived species), all plants have mechanisms to deal with water deficit.

Plant adaptations for reducing heat load and for conserving water

In comparison to most land animals, plants generally have a very high surface-area-to-volume ratio to collect sunlight for photosynthesis. This also means they may have a high heat load. Conduction can dissipate only a small amount of heat because the transfer must occur from the plant to the air, which is not very efficient. Re-radiation may be important and many plants of hot, dry environments have needle-shaped leaves. The tip of the leaf acts as a point source radiator unloading heat, but the rate of heat loss depends on the rate of conduction to the tip, which may be quite slow.

The main method of heat dissipation is evaporative cooling, facilitated by a range of leaf structures (Plate 23.3). Water has a high latent heat of vaporisation (the amount of heat required to convert liquid water into vapour without a change in temperature), so most of the heat absorbed by leaves is dissipated during transpiration. As explained in Chapter 12, plants encounter a large problem because they must maintain a moist internal environment for biochemical reactions in the cells. Most structurally simple plants, such as mosses and liverworts, have poor control of water loss. They are restricted to moist environments, or function only in seasons when conditions are moist. Higher plants from arid environments have adaptations limiting water loss while still allowing evaporative cooling, such as thick cuticles on their epidermal cells, leaves with rolled edges in spinifex (*Triodia*), or a layer of hairs on the leaf surface where the stomates occur (Plate 23.3). All these innovations maintain a slightly moister microenvironment around the stomates.

As a generalisation, there is a decrease in leaf size from humid environments to drier environments. This is best seen on a transect from the north-eastern Queensland rainforests to the inland deserts. In the rainforest the plants have large leaves (macrophylls) whereas in the desert plants have small micro- or nanophylls. The small leaves often have sharp tips and recurved leaf edges, and may have hairs under the leaf – all adaptations to heat transfer and water retention. The root-to-shoot-biomass ratio in dry-country plants is much higher than in wetter climates – plant investment in root production enhances access to soil moisture, while there is a reduction in the proportion of above-ground material that must be supplied with moisture.

Specialised plant forms associated with low water availability

In some arid parts of Australia, a few plants – including the wax vines or porcelain flowers (Asclepiadaceae: *Hoya*) and succulents such as the purslanes or rose mosses (Portulacaceae: *Portulaca*), and members of the genus *Crassula* – evolved an extreme water-saving strategy by opening their stomates at night when the heat load is low and closing them during the day. Carbon for photosynthesis enters the plant at night as CO_2, is converted into malate, and stored in cell vacuoles. During the day, stomates are closed, the CO_2 is released from the malate and conventional photosynthesis using light energy refixes it into sugars (Box 4.4). This is known as crassulacean acid metabolism (CAM). Having the stomates open during the cool of the night means that more CO_2 can be used per unit of water transpired.

As well as having good water-use efficiency, these plants often store water in their tissues, which become succulent. Pineapples, cacti and some orchids are also CAM plants. Succulents (particularly in the family Cactaceae) are common in the North American deserts, and extend down into the arid parts of Chile. Many species have spines, which are very reduced leaves. These re-radiate heat and deter herbivores. An unrelated group, the family Aizoaceae or 'living stone' plants evolved in the Namibian desert in southern Africa. They do not grow as robustly as cacti and lack spines, relying more on camouflage to escape grazing animals in the rocky landscapes. The succulent growth form is rare in Australia where only some species in the family Chenopodiaceae have succulent leaves. Some have lost their leaves, and photosynthesis occurs in the succulent stems (Plate 23.3a). In the Australian deserts the more common growth forms are plants with very small, often spine-like leaves: grasses, annuals and shrubs that are often multistemmed (mallees).

The mallee growth form of eucalypts is important in several Australian communities. Mallees have several thin trunks arising from a lignotuber, a swelling of tissue produced initially in the seedling, just above the cotyledons. It has dormant buds that can produce new shoots if the top of the plant is removed or severely damaged by grazing, drought or fire. Lignotubers are found in many plant species, but the mallee form of eucalypts is more common in inland Australia than in other parts of the world. Another common form seen in grasses in the arid zone is the porcupine or spinifex growth form.

Some trigger plants (*Stylidium* spp.) that grow on exposed sandy soils elevate their shoots a few centimetres above the surface by using stilt roots. This can reduce the temperature of the air around the sensitive shoot meristem by a few degrees below the lethal temperature.

Animal adaptations for reducing heat load and for conserving water

The best place to see animal adaptations for water conservation is the desert. Desert frogs are a striking example. Australian desert frogs (*Cyclorana* spp.) survive the driest times in 'aestivation' – a period of reduced metabolic activity analogous to hibernation – which they undertake in the soil in wet sacks they secrete around their bodies. After rain they emerge and follow a typical frog lifestyle during the brief wet conditions. Other animals have less extreme methods. Many desert animals avoid the heat of the day by sheltering in a burrow (e.g. the bilby – *Macrotis lagotis*, mala – *Lagorchestes hirsutus*, and the scorpion *Urodacus yaschenkoi*, Box 13.4) and feeding in the early morning or late in the evening. This is called crepuscular behaviour and is one reason why kangaroos are so vulnerable on the roads at the end of the day. Other desert animals are active at night (nocturnal). However, in the driest areas behavioural adaptations are not enough, and animals have physiological adaptations for water retention. Australian mammal species such as the quokka (*Setonix brachyurus*) and the woylie or brush-tailed bettong (*Bettongia penicillata*) limit water loss by concentrating their urine. Some species derive most of their water from their food. The woylie's diet is dominated by moist underground fungi, whereas native rodents can obtain sufficient water as a by-product from the metabolism of the seeds they eat.

Adaptations to low nutrients

Inorganic nutrients are also essential to build the framework of plant cells. Within the soil solution nutrients are usually at very low concentration, though this varies with the composition and origin of the soil. Soil from rocks that have a lot of inorganic nutrients have a higher concentration of these nutrients than do soils derived from nutrient-poor substrates such as silica sand (beach sand). Roots take up the nutrients from this dilute solution against a concentration gradient, because there are usually many more ions in the plant cells than in the soil moisture.

The nutrients required by plants have different solubilities in the soil moisture. Potassium, nitrogen, sodium and sulfur are soluble and usually available. Forms of calcium and phosphorus are less soluble. Some soil components have a very high affinity for some of the nutrients required by plants. In particular, phosphorus may be tightly bound to iron and aluminium compounds, and to clay particles. Australian soils are particularly poor in phosphorus and the small amount present is largely unavailable unless the plants have developed some of the adaptations described below.

Specialised root forms

A wide range of specialised roots optimise nutrient uptake in poor soils (Plate 23.4c, d). Roots grow from an apical meristem, and behind the apex epidermal cells elongate quite spectacularly into unicellular root hairs 1–2 mm long. These may be present for several millimetres along the young root (Plate 23.4d), but are shed from the older root, so absorption occurs only near the root tip. Root hairs provide a large surface area for absorbing nutrients, and some families of plants show great proliferation of root hairs, which aid nutrient uptake from nutrient-poor soils. In the sedge family (Cyperaceae) and some southern rushes (Restionaceae), dauciform (parsnip-shaped) roots with dense root hairs are produced on lateral roots (Plate 23.4c). In many species of Proteaceae (the banksia family) globular clusters of lateral roots called 'proteoid roots' or 'cluster roots' form during the growing season, especially on roots in the relatively nutrient-rich litter layer near the soil surface (Plate 23.5). They trap soluble nutrients leaching from the litter horizon and secrete organic acids that release phosphorus from insoluble compounds for uptake by the root.

Low phosphorus concentration in the soil encourages dauciform and proteoid roots. They increase surface area for nutrient uptake, and exude organic molecules, such as citrate, that have a chemical affinity for iron and aluminium complexes. The citrate can exchange for phosphate on soil particles, freeing phosphate for absorption by root hairs. This exchange mechanism and high surface area are found in plants in many parts of Australia and the Cape of South Africa where there are soils of very low nutrient content. These plants are killed by high phosphorus fertilisers.

Mutualisms for nutrient acquisition

Other mechanisms improving nutrient acquisition by land plants involve mutualism. One example is the association between the roots of legume plants and *Rhizobium* spp. bacteria to form root nodules (Chapters 10 and 17). The plant has harnessed the nitrogen-fixing ability of the bacteria to acquire nitrogen that would otherwise be unavailable in the atmosphere. The wattle on the Australian Coat of Arms is an example of a nitrogen-fixing tree. Over 600 species of *Acacia* (wattle) occur in Australia. Up to 75 kg of nitrogen can be fixed per hectare per year by nitrogen-fixing plants, so they are important components of many Australian ecosystems and, particularly after fire, are instrumental in re-establishing the nitrogen economy of the community. Another example is the association between cycads and the cyanobacteria *Nostoc* and *Anabaena*, which provide these plants with nitrogen. In the she-oaks the main nitrogen-fixing organism is *Frankia*.

Instead of a mutualism with bacteria or cyanobacteria, the roots of much of the Australian flora develop in association with beneficial fungi. The fungal component of these fungus roots or mycorrhiza (Box 10.6) scavenges for scarce nutrient reserves and passes nutrients and water to the host plant. For each metre of root length there may be up to 500 m of fungal hyphae in the soil.

Predation – parasitism

A parasite gains all or some of its nutrients and fixed carbon from the host. Hemiparasites are plants with green leaves growing attached by specialised haustoria (singular: haustorium) to the roots or stems of another species, whereas holoparasites lack photosynthetic tissue and rely on other plants for food.

Examples of hemiparasites are the mistletoes and the Western Australian Christmas tree (*Nuytsia floribunda*) (Plate 17.3 a, b). Mistletoes lack roots and parasitise the branches of hosts such as eucalypts, she-oaks and acacias. They may have a leaf morphology similar to that of the host plant.

Angiosperm holoparasites are rare, but Australian examples include the Western Australian *Pilostyles hamiltonii* and species of *Balanophora* that are seen only when their flowering stems emerge from the soil in north Queensland (Plate 17.3c). Further remarkable species are the underground orchids *Rhizanthella gardneri* in Western Australia and *R. slateri* in New South Wales. More is known about the Western Australian species that parasitises roots of broom honey myrtle (*Melaleuca uncinata*) by linking into their mycorrhizal fungi. The orchid has no above-ground parts. Instead, tulip-shaped bracts surround a head of small flowers that cracks open the soil to allow access for pollinating insects.

Predation – plant carnivory

To supplement their nutrition, several families of Australian plants capture insects. Carnivorous plants have leaves (and, in swampy places, roots) modified to capture animals usually less than 5 mm in size. In the sundew group (*Drosera*), common on granite outcrops, the leaves have glandular hairs that attract and ensnare small invertebrates. The glandular hairs gradually move and enfold the victim, secreting digestive enzymes that release nitrogen and phosphorus to be absorbed by the plant. In this case the leaves absorb the nutrients and lack a thick cuticle. Pitcher plants have leaves modified into a jug with a slippery lip, capturing insects that stray near the mouth of the jug and fall in (Plate 23.4a, b). The pitcher has a solution containing digestive enzymes and, as with the sundews, the nitrogenous and phosphorus compounds are taken up through the leaf cells and into the plant. Most carnivorous plants act as passive traps, but the venus fly trap, native to the swamps of Florida, has an active capture mechanism. The modified leaf resembles an open book, with a shiny surface and trichomes (enlarged hairs) in the centre of each side. Invertebrates are attracted to the shiny surface. If they contact the trichomes the 'book' snaps shut, trapping and digesting the victim. Ironically, carnivorous plants rely on insects to pollinate their flowers, yet they also 'eat' many of them.

Sclerophylly and low nutrients

The Australian flora has a high proportion of plants with tough or sclerophyllous (leathery) leaves, and leaves with spiny ends and edges. At one time it was thought these are adaptations to the arid Australian environment, as the tough leaves were considered an adaptation to reduce water loss. However, similarly tough leaves are often found in areas where water is not an important limiting environmental factor (such as the Amazon basin), yet nutrients are often in short supply. One of the early pioneers of Australian physiological plant ecology, Noel Beadle, hypothesised that the harsh form of the leaves of many Australian plants was an evolutionary response to nutrient-poor soils (particularly the low phosphorus concentrations) and only secondarily an adaptation to aridity. This is supported by fossil evidence. Plants with leaves similar to the present-day sclerophyll types were present in Australia during the Tertiary period when the climate was much wetter.

Sclerophylly is important in low-nutrient environments as a deterrent to herbivory. Where nutrients are abundant, growth is rapid (assuming weather conditions are suitable) and plant biomass harvested by animals is replaced rapidly. In contrast, in nutrient-poor environments plant investment in leaves is relatively expensive, plants retain their leaves for longer and they have adaptations reducing herbivory.

Loss to herbivores is reduced in several ways. One is a poisonous or deterrent chemical so the leaf is unpalatable to animals. Poisonous biochemicals such as fluoroacetate (the toxin in *Gastrolobium* species; see Chapter 3), cyanogenic glycosides (found in some *Grevillea* species) or tannins deter herbivores from eating leaves. Alternatively, leaves may deter herbivores by having a high proportion of structural tissue and relatively little photosynthetic tissue. It is the photosynthetic tissue that contains most of the nutrients animals need, not the structural material of cellulose, lignin and tannins. Not only does this make the leaf unpalatable to animals, but it gives it structural rigidity too, reducing damage from wind buffeting.

Animal adaptations to low-nutrient environments

Plants are not uniformly attractive to herbivores and most of them, given the chance, are selective in what they eat. Kangaroos seek out the most palatable species of herb and grass if there is plenty of food to choose from; in rangelands management, these are referred to as 'ice-cream plants'. After fire the new shoots of many species are highly nutritious and kangaroos concentrate their grazing on areas of regeneration. Plants that are unpalatable as adults are often much more acceptable as seedlings. As an example, the regeneration of a rare wattle species of inland Western Australian heathland (*Acacia chapmanii*) depends on fire. As an adult the acacia is spiky and unpalatable, but its seedlings are much softer. If areas regenerating after fire are not fenced to exclude grazers almost no acacia seedlings survive. In heathland unburnt for a long time the few grass clumps are grazed very close to the ground, while adjacent spiky shrubs are untouched and can produce seed.

Diet also depends on the time of year. In northern Australia the wallaroo or euro (*Macropus robustus*) and rock wallaby (*Petrogale brachyotis*) have grass-dominated diets in the wet season when grass forage is most nutritious, but in the dry season eat a higher proportion of herb, shrub and tree leaves.

One way of coping with poor food quality is to be a generalist herbivore. Large species such as kangaroos and emus range over a wide area, selecting the most nutritious forage. The emu is one of the main seed dispersers for plants with fleshy fruits in Australian heathlands. It can consume a variety of propagules from *Macrozamia* seeds poisonous to humans, to fruits of the quandong (*Santalum acuminatum*). It will also consume foliage of some species, and occasionally invertebrates.

Another way to cope with poor-nutrient fodder is to specialise on a nutritious, reliable food. The honey possum (*Tarsipes rostratus*) occurs in south-western Australia in nutrient-poor heathland. Its diet consists entirely of nectar and pollen, and it survives on this diet because there is sequential flowering through the year of many nectar-producing plants such as banksias and dryandras. The koala is also a specialist, feeding exclusively on leaves from some species of eucalypt. This adaptation is aided by certain gut bacteria, which break down the nitrogen-poor leaf fodder produced by the eucalypts.

The ability of larger herbivores to select a mixed diet from a diversity of plants also enables animals to ingest small amounts of toxins with impunity – just as a small amount of chilli in a vegetable curry is very tasty, but a dish consisting entirely of chillies would be inedible. However, many animals, particularly invertebrates, are specialists, living on one or a few species. Over evolutionary time mechanisms to cope with plant toxins may evolve in some animal species. Most of the native south-western Australian mammal fauna is resistant to poisoning by fluoroacetate through a long association with plants in the genus *Gastrolobium*, which contain the poison.

Thus there are many ways of coping with the smorgasbord of foods provided by plants growing in nutrient-poor soils. In some cases specialisation is the key, but the more widespread approach is for the animal to move from one food type to another, and to select the most nutritious option available at the time.

Adaptations to soil toxins

Although land plants in most soils have difficulty in obtaining nutrients, soils in some areas contain more ions of some elements than is necessary or healthy for plants. The most common excess of soil ions comes from solutions of common salt (NaCl). This has a twofold effect on plants. First, saline soils create problems for plants in water uptake. The high osmotic potential (ability to hold onto water) of a salty soil means that it is harder for a plant to extract water than it would be from a non-saline soil. Nevertheless, many plant species are adapted to saline soils, either in coastal areas or in dry inland areas, especially in central Australia.

Salt bush (*Atriplex* – from the drier areas of Australia and other continents) secretes salt. It has on the leaf many swollen 'bladder' cells, which are filled with salty solution and eventually collapse to form a scurfy salt crust over the leaf. This mechanism removes salt from the internal leaf environment. Other species develop succulence. In this system salts are chemically 'pumped' into the cell vacuole and hence do not interfere with the other organelles or the cell cytoplasm. Such mechanisms to cope with excess salt are energy expensive, so plants that possess the means to survive in a saline environment cannot compete well with non-salt-tolerant plants in a non-saline environment.

In certain soils other elements may be in toxic amounts. In serpentine soils derived from basic igneous rocks, there are often excessive amounts of heavy metals such as nickel (Ni), copper (Cu), cobalt (Co) and manganese (Mn). Although these elements may be useful in micro amounts, in large concentrations they become toxic. However, there are plants adapted to survive on such toxic soils. One New Caledonian species (*Sebertia acuminata*) grows on soils with a high nickel content. The plant has blue sap because of the high concentration of nickel compounds it contains. Tolerance involves reducing uptake of heavy metals across the plasma membrane of cells, repairing cell proteins damaged by the heavy metals, or storing the heavy metals inertly in the cell walls, within vacuoles or in chemical compounds in the cytoplasm.

As in the case of saline-tolerant plants, such species are poor competitors with normal species on non-toxic soils. Experience with revegetation of mine wastes showed that races of plants that can survive on toxic soils can be bred from plants that grow on non-toxic soils; but only after generations of selective breeding.

Chapter summary

1. Terrestrial systems are characterised by more extreme environmental fluctuations than most aquatic systems. Highest productivity in the terrestrial environment occurs in the most benign environments, and falls as the environment becomes drier, colder and/or has less available nutrients.
2. As land plants are confined to the place where they germinate, they will either have adaptations that allow them to survive or they will not persist in that environment. Animals may have the opportunity to migrate.
3. Both animals and plants may survive extreme cold as resistant stages (eggs or seeds). Animals may enter torpor or hibernate.
4. In hot environments the direct effect of high temperatures may be lethal. However, plants and animals have a variety of mechanisms including re-radiation, and evaporative cooling to dissipate heat load.
5. High temperatures also affect the availability of water. Plants use modified leaves or life cycles to cope with low water availability. Animals may reduce water loss by sheltering in burrows, aestivation over the dry period, being active at night and/or producing concentrated urine.
6. Most higher plants acquire their inorganic nutrients from the soil, where nutrient concentrations are normally very low. Different root modifications increase the surface area of the roots, increasing their molecular exchange abilities and aiding plants in searching for scarce soil nutrients.
7. Mutualisms between higher plants and nitrogen-fixing microbes (*Rhizobium*) and/or fungal taxa that form mycorrhiza are important ways in which plants are adapted to survive on nutrient-poor soils. Some plants are parasitic on other plants and gain their nutrients at the expense of the host.
8. In some soils, chemical concentrations may reach toxic levels. Plants may cope using glands to secrete the toxins (e.g. salt glands) or metabolic pathways that prevent the excess nutrient load from damaging normal cell biochemical processes.
9. Plants in low-nutrient environments tend to have leaves with longer life spans than those on plants in nutrient-rich environments. Physical features such as spines, and tough cuticles, or secondary metabolites that make leaves unpalatable or poisonous, protect their leaves from grazing.
10. Herbivorous animals select plant foods that maximise the range of nutrients while minimising intake of toxins.
11. The quality of soil influences the productivity of plant communities, and so the structure of animal communities and the life histories of some animal species.

Key terms

aestivation
CAM plants
conduction
convection
crassulacean acid metabolism
dauciform roots
evaporative cooling
haustoria
hemiparasite
hibernation
holoparasites
lignotuber
macronutrient
mallee
micronutrient

migration
osmotic potential
plant growth form
plant life form
proteoid root
radiation
sclerophylly
thermoregulation
timber line
torpor
urea

Test your knowledge

1. Describe, with examples, how (a) animals and (b) plants may be adapted to cold environments.
2. Describe, with examples, how (a) animals and (b) plants may be adapted to hot, dry environments.
3. In an arid area, what are the advantages and disadvantages of a succulent growth form versus a form such as a twiggy shrub?
4. What adaptations of plants maximise nutrient uptake when concentrations of soil nutrients are low?
5. What factors determine diet selection in herbivorous mammals?

Thinking scientifically

Design an experiment to test the hypothesis: 'Carnivorous plants such as *Drosera* can obtain all the nitrogen they need for survival and reproduction through carnivory'.

Debate with friends

Debate the topic: 'The main factor determining the biodiversity of a region is the fertility of its soils'.

Further reading

Cornelissen, J.H.C., Lavorel, S., Garnier, E., Diaz, S., Buchman, N., Gurvich, D.E., Reich, P.B., ter Steege, H., Morgan, H.D., Van Der Heijden, M.G.A., Pausas, J.G. and Poorter, H. (2003). 'A handbook of protocols for standardised and easy measurement of plant functional traits worldwide' *Australian Journal of Botany* 51: 335–80.

Mansergh, I. and Broome, L. (1994). *The mountain pygmy-possum of the Australian Alps.* Australian Natural History Series, UNSW Press, Sydney.

Specht, R. L. and Specht, A. (1999). *Australian plant communities: dynamics of structure, growth and biodiversity.* Oxford University Press, Oxford.

Vandenbeld, J. (1988). *Nature of Australia.* Collins/Australian Broadcasting Corporation, Sydney.

Wolfe, K.M., Mills, H.R., Garkaklis, M.J. and Bencini, R. (2005). 'Post-mating survival in a small marsupial is associated with nutrient inputs from seabirds' *Ecology* 85: 1740–6.

Theme 5

The future – applying scientific method to conserving biodiversity and restoring degraded environments

In introducing the discipline of conservation biology Gary Meffe, C. Ronald Carroll and Martha Groom stated three unifying principles:

1. Evolution unites all biology (the evolutionary play).
2. The ecological world is dynamic (the ecological theatre).
3. Humans must be included in conservation planning (people are part of the play).

Your study so far in *Environmental Biology* should leave you in no doubt about all three maxims, but we have not yet discussed how to integrate the human element with your understanding of evolution and ecology.

In this final theme of *Environmental Biology* we apply your knowledge and understanding from the earlier themes to solving applied conservation problems or to redressing environmental damage. Good scientific method is at the core of all solutions, complemented by a thorough understanding of biodiversity and the interactions of species through the application of evolutionary and ecological principles.

24

The science of conservation biology

Grant Wardell-Johnson and Pierre Horwitz

Rainbows in the swamp

In 1983, two new frog species were discovered in the high rainfall region of south-western Australia. The white-bellied frog (*Geocrinia alba* – Plate 24.1) was found to occur over about 100 square kilometres, much of which had been cleared for agriculture. Fortunately, many sections of creek on private land remain uncleared and some of these sites retain populations of white-bellied frogs (Plate 24.2). The orange-bellied frog (*G. vitellina*) (Plate 24.3) was even more restricted, limited to only 6 square kilometres in just six adjacent creek systems in State Forest to the north of the Blackwood River. Within this range, there are only about 20 ha of suitable habitat. These two frog species are among the most geographically restricted vertebrates on mainland Australia. Their restricted distributions, low population densities and high habitat specificity mean that they are particularly prone to extinction.

Are such rare species unusual? Are the *G. alba* and *G. vitellina* populations stable or in decline? If they are in decline, what is the cause and can management lead to population recovery? How can their habitat be protected, including, for example, the threatened *Reedia spathacea* ecological community, which occurs in the same areas (Plate 24.4)? These types of questions are addressed by conservation biology.

Chapter aims

This chapter covers the science and biology required for conserving species and ecological communities that are rare and threatened, or declining. We first describe the approaches required for managing and conserving threatened species, then examine the ecological community and the wider issues of reserve selection, protected area management, and management outside reserves. This leads into Chapter 25, which addresses the interaction of society and biology and develops a philosophy for conservation biology which encompasses much more than biology.

What is conservation biology?

Conservation biology is a new, integrated science developed in response to the human-induced, worldwide wave of habitat destruction and extinction (see Chapter 7) widely recognised and acknowledged since the 1970s. Although extinction is a natural phenomenon, the concern is that rates of extinction and threats to species have increased markedly with globally significant changes occurring in recent decades. Conservation biology assesses biodiversity, which includes the genetic variation within organisms, the range of species in the environment, and the full spectrum of the world's ecological communities and their associations with the physical environment – the landscape. It investigates human impacts on species and ecological communities and develops practical approaches to prevent extinctions, maintain genetic variation within species, and protect ecological communities and associated ecosystem functions such as nutrient cycling (see Chapter 17). This may involve determining the best strategies for protecting threatened species and ecological communities, maintaining genetic viability in small populations, designing and managing reserves, and reconciling conservation concerns with other needs.

Conservation biology arose because the existing applied disciplines of resource management such as forestry, fisheries biology, wildlife management and agriculture were either not comprehensive enough to address the critical threats to biodiversity, or considered only a narrow range of species or goals. Taxonomy, ecology, population biology and genetics form the core of conservation biology. However, because much of the biodiversity crisis arises from human pressures, conservation biology also incorporates social science disciplines, such as law, sociology, environmental ethics and economics. Thus conservation biology is truly interdisciplinary, with the problems it addresses often needing social change to be effective.

Conserving populations and species

How big is the problem?

About 500 animal extinctions and 600 plant extinctions have been recorded worldwide in the last 400 years, but these figures are underestimates. A species is recorded as extinct only if it is already named and cannot be found in the wild in an intensive search, or after an extended period (normally decades). Thus, there are no extinction data for the great number of unnamed species and even among the named species there is a heavy bias towards some groups, such as vertebrates and flowering plants. Many named species are not monitored at all.

Are these extinction rates high? Studies of the fossil record estimate a 'background extinction rate' of 0.1 to 10 species from all multicellular organisms annually in geological time. Even allowing for difficulties in estimating both past and present extinction rates, data for well-known taxa such as mammals, birds and molluscs are currently well above this background rate. This indicates an abnormally high rate of extinction. The extinctions are rapid in geological timescales of thousands of years, occur worldwide and cover many different groups of plants and animals. Therefore they may be on a par with the five mass extinctions in geological history (Chapter 7), after each of which recovery has taken 10 million to 100 million years. Such a rapid rate of extinction may eliminate the biological resources and ecosystem functions supplied by many species, not to mention the aesthetic pleasure many people enjoy in biodiversity or the ethical enormity of denying future generations access to the range of life forms we enjoy (Chapter 25). The problem is grave, but as you will see in the following sections it is not necessarily a product of the rarity of species.

Rare and threatened species

Rarity

Rare types of microbes, plants and animals occur in all ecosystems; in fact most species are rare (Figure 24.1). Most rare taxa have particular adaptations that allow them to persist at low abundance, or in very few locations. A good assessment of the likelihood of extinction of a species involves assessing the kind of rarity being considered.

The concept of rarity has three 'dimensions':

1. rare in terms of the size of the local population, or the abundance of individuals (alpha rarity)
2. rare in terms of how specialised a species is for its habitat (beta rarity)
3. rare in terms of its geographic range or distribution (gamma rarity).

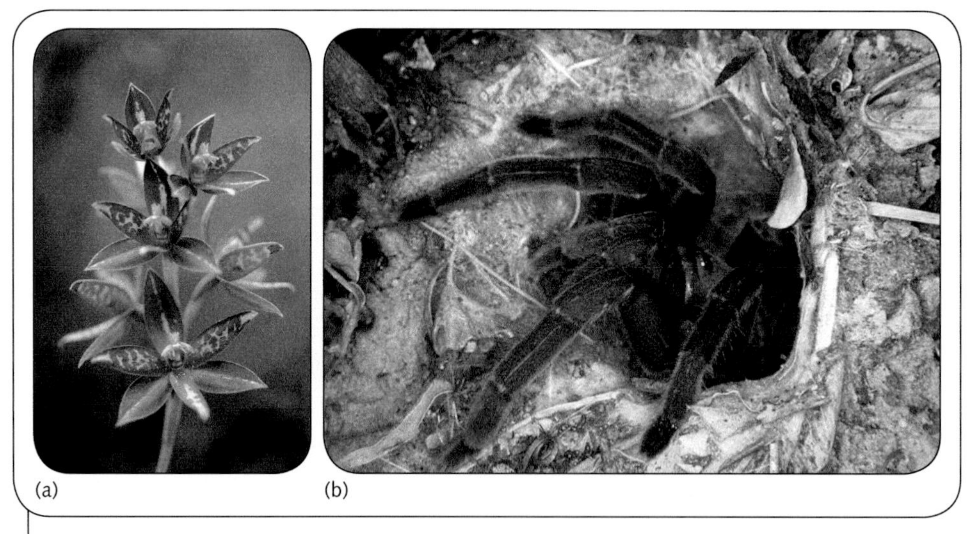

Figure 24.1 Examples of rare species. (a) Jackson's sun orchid, *Thelymitra jacksonii*, is a locally rare species (Walpole, Western Australia) occurring in generalised habitat (sandy-clay soils near winter-wet depressions) restricted to near Walpole in south-western Australia. (b) The bird-eating spider *Selenocosmia* sp., a locally rare species in specialised habitat (the margins of rainforest) that is widespread (west to north-east Kimberley, Western Australia). (Source: G. Wardell-Johnson)

Rarity for any species can be defined by low levels of any one of these dimensions. This leads to different forms of rarity (Plates 24.5, 24.6, 24.7, 24.8) which are (from most to least extreme):

- locally rare in specialised habitats that are geographically restricted (the most extreme form of rarity)
- locally rare in generalised habitat that is widespread
- locally rare in generalised habitat that is restricted geographically
- locally rare in specialised habitat that is widespread
- locally common in specialised habitat that is restricted geographically
- locally common in specialised habitat that is widespread
- locally common in generalised habitat that is restricted geographically.

Common species are locally abundant, widespread and occurring in generalised habitat. Thus the term 'rare' may have different meanings and different management implications.

Species having any of the seven forms of rarity can have persistent stable populations, so rarity does not necessarily mean a species is threatened or lead to conservation concern. Instead, threatened species are those perceived to be at risk of extinction according to specific criteria laid down by the International Union for Conservation of Nature Red List of Threatened Species. These consider factors such as rapid decline, small or declining range, population size and projections of population trends.

Threats

A threat is any biological or environmental process that places a species at risk of extinction. It may be a single devastating event such as a fire, cyclone or flood, or an incremental or ongoing process such as habitat change, salinisation, introduced species, or inappropriate fire regimes (Plates 24.9, 24.10, 24.11, 24.12). Some of the most important of these (such as clearing of native vegetation, and various introduced plants, animals, diseases and other organisms) are listed as threatening ecological processes by state or Commonwealth agencies (Table 24.1). These lists are continually revised. However, virtually any disturbance (or in some cases a lack of disturbance!) or change to a habitat has the potential to threaten. Therefore a catalogue of threats is meaningless without assessing how they can be avoided or alleviated.

Evaluating rarity and threats

The 'small-population paradigm' deals with the risk of extinction that follows from low numbers. It is one side of the conservation biology coin. The other, the 'declining population paradigm', is concerned with the processes by which populations are driven to extinction. These two sides therefore focus on ways of detecting, diagnosing and halting population decline, indicating when a species is threatened with extinction, the exact threats and the actions needed to remove them. In the next section we examine how small populations may put a species at risk, how the technique of population

Table 24.1 Key threatening processes listed currently by the Australian Commonwealth under the *Environment Protection and Biodiversity Conservation Act 1999*.

Category of threat	Threatening process
Introduced species	Competition and land degradation by feral goats (effective 16 July 2000)
	Competition and land degradation by feral rabbits (effective 16 July 2000)
	Loss of biodiversity and ecosystem integrity following invasion by the yellow crazy ant (*Anoplolepis gracilipes*) on Christmas Island, Indian Ocean (effective 12 April 2005)
	Predation by exotic rats on Australian offshore islands of less than 1000 km^2 (100000 ha) (effective 29 Mar 2006)
	Predation by feral cats (effective 16 July 2000)
	Predation by the European red fox (*Vulpes vulpes*) (effective 16 July 2000)
	Predation, habitat degradation, competition and disease transmission by feral pigs (effective 6 August 2001)
Diseases/pathogens	Dieback caused by the root-rot pathogen (*Phytophthora cinnamomi*) (effective 16 July 2000)
	Infection of amphibians with chytrid fungus resulting in chytridiomycosis (effective 23 July 2002)
	Psittacine circoviral (beak and feather) disease affecting endangered psittacine species (effective 4 April 2001)
Harvesting	Incidental catch (bycatch) of sea turtle during coastal otter-trawling operations within Australian waters north of 28 degrees south (effective 4 April 2001)
	Incidental catch (or bycatch) of seabirds during oceanic longline fishing operations (effective 16 July 2000)
Pollution	Injury and fatality to vertebrate marine life caused by ingestion of, or entanglement in, harmful marine debris (effective 13 August 2003)
	Loss of climatic habitat caused by anthropogenic emissions of greenhouse gases (effective 4 April 2001)
Clearing	Land clearance (effective 4 April 2001)

viability analysis can be used to decide on actions in the face of threats and how an understanding of risk and threats can lead to assessments of the conservation status of a species.

The biology of small population sizes

Minimum viable population (MVP) and population viability analysis (PVA)

Information on rarity and threat alone are not enough, because decision makers might ask: 'How low can a population go before it reaches a point beyond which it will definitely become extinct?' The answer to the question lies in probability, because organisms live in an uncertain world. For instance, what are the chances of the remaining females in a population finding a male, or the last 20 individuals of a species surviving a cyclone or a drought? The *minimum viable population* or MVP is the smallest population size that can survive such crises. This requires a quantitative estimate of the size that a population must be to assure survival at a given level of probability for a specific time frame. Conservation biologists also try to predict what will happen to a population faced with real-world uncertainties. Their computer models are called PVA – *population viability analysis.*

PVA deals with the risks that small populations face simply because of their size. These include demographic uncertainties such as the chance of death of all remaining females in a dioecious, sexually reproducing population (or all breeding individuals in an asexually reproducing population), loss of genetic diversity and the effects of chance factors such as environmental uncertainty and catastrophes. All these factors interact, and PVA sets out to establish the relationships between them to predict the probability of extinction under specified conditions.

Demographic uncertainties

The smaller the population, the less likely we are to be able to predict when successful mating will occur. What happens when 10 dioecious flowering plants (female and male flowers on separate plants) are left in a population and five die? If all of the remaining five are female (or male) and do not reproduce vegetatively, then extinction follows. Thus specific demographic characteristics such as age and sex distributions, as well as chance events, become crucial when populations are small.

Genetic uncertainties

What is the minimum population size necessary to avoid significant reductions in the genetic diversity of a population? This is an important question, because genetic diversity provides the raw material necessary for a population to evolve in the face of environmental change. The main factor responsible for the loss of genetic diversity in small populations is genetic drift; random changes in allele frequencies from one generation to the next tend to lead to the loss of rare alleles and the fixation of common alleles in the population. As we saw in Chapter 6, genetic drift is greatest in small populations and is enhanced when there is variation in breeding success because of unequal sex ratios or the reproductive dominance of certain individuals in the population. The fitness of individuals may also be reduced by inbreeding in small populations. Again, as you will remember from Chapter 6, mating between genetically related individuals may lead to inbreeding depression because of the expression of deleterious, recessive alleles in their offspring. Mating between relatives is much more likely in a small population than in a large population, simply because there is less choice of mates.

Environmental uncertainty

Environmental uncertainty refers to variability in population growth rates caused by factors external to the population. Cyclones, floods, droughts, landslides and abnormally high tides may all threaten a population's survival regardless of its abundance in a given location. However, widely distributed populations are at less risk if the threat varies between different locations.

Population viability analysis (PVA) in practice

PVA uses computer modelling to integrate demographic, genetic and environmental data and predict the likely future population size under different management options. The models require detailed background data, so attempting a PVA identifies gaps in information and major threatening processes as well as highlighting trends in population behaviour. PVA also identifies characteristics of a population that allow it to persist over extended periods and quantifies interacting and potentially cumulative factors that influence population dynamics. Managers use PVA to test hypotheses for observed fluctuations in population size, compare options for reserve designs and provide an objective framework for decision making and planning for species recovery (Box 24.1).

Box 24.1 PVA for *Geocrinia* species in south-western Australia

How best should we manage the small populations of the white-bellied and orange-bellied frogs from south-western Australia? PVA can answer this question, given thorough demographic information and the risk of mortality posed by threats such as fire and habitat loss.

Detailed studies of the population ecology of these species found that survival rates to metamorphosis are among the highest observed for all frogs and toads, while adult survival rates are among the lowest. The juvenile life stage is prolonged, with individuals first breeding at a minimum age of 2 years. Recruitment and population size fluctuate markedly between years and populations. Recruitment from the juvenile to the adult life stage is the main driver of population size, with such recruits forming the largest age-class of adults in every year. The frogs also disperse poorly, which makes it difficult for them to found new populations.

Large wildfires may destroy a population, while smaller ones restrict adult body size and condition, and delay breeding and maturation. Land clearing or draining swamps can also destroy populations.

PVA models integrating these data found that the probability of extinction within the next 100 years is low for the white-bellied frog and moderate for the orange-bellied frog. However, both white-bellied and orange-bellied frogs are predicted to have significant range reductions, with only a few large populations persisting far into this century. Given that the species have poor dispersal, improving the overall area of suitable habitat is less effective as a conservation strategy than increasing the protection of core habitat with large populations.

The example of PVA for the white-bellied and orange-bellied frogs demonstrates how PVA can be used in prioritising management actions for conserving species.

Putting it all together – assessing conservation status

After careful assessment of rarity and risk, a species can be given a conservation status, which summarises the likelihood that it may become extinct over a given time. Sometimes these categories are defined by legislation such as the Australian Commonwealth's *Environment Protection and Biodiversity Conservation Act 1999*, or legislation of individual Australian states. However, the most widely recognised international classifications are those of the International Union for the Conservation of Nature, or IUCN (Table 24.2). The position of a species in this classification is determined by very specific quantitative criteria incorporating aspects of its biology and distribution, as well as the results of any detailed PVAs completed (Table 24.3). Of course, classifying conservation status is only the beginning, and management of threatened species must follow.

Managing threatened species

Managing threatened species may be by *ex situ* conservation (where the species is managed away from the location of the threat), *in situ* conservation (where the species is managed in its habitat), or both. *Ex situ* actions are usually highly interventionist. They assume that the long-term survival of the species can be enhanced only by keeping individuals away from where they are threatened. *Ex situ* actions include the establishment of seed banks or gene banks for the long-term storage of genetic material, and captive breeding programs. *Ex situ* actions always involve translocation, the moving of individuals, their genes or their seeds (see Box 24.2).

Table 24.2 The IUCN categories for threat and endangerment of species, in order from the most to the least severe. (Source: Adapted from Rodriguez, A.G.L. et al. (2006). 'The value of the IUCN Red List for conservation.' *Trends in Ecology and Evolution* 21: 71–6)

IUCN Red List categories of conservation status	
Extinct (EX)	A taxon is extinct when there is no reasonable doubt that the last individual has died.
Extinct in the wild (EW)	A taxon is extinct in the wild when it is known only to survive in cultivation, in captivity or as a naturalised population (or populations) well outside the past range.
Critically endangered (CR), endangered (EN) or vulnerable (VU)	A taxon is one of these when the best available evidence is used in an assessment against the criteria and it is thought to be facing a high to extremely high risk of extinction in the wild.
Near threatened (NT)	A taxon is near threatened when it has been evaluated but does not meet the criteria but is close to qualifying or likely to qualify for a threatened category in the near future.
Least concern (LC)	A species does not meet listing under a higher category of threat (for widespread and abundant taxa).
Data deficient (DD)	A taxon is data deficient when there is inadequate information to make a direct, or indirect, assessment of its risk of extinction based on its distribution and/or population status.
Not evaluated (NE)	A taxon is not evaluated when it has not yet been evaluated against the criteria.

Table 24.3 The assessment for Critical, Endangered or Vulnerable is made using the following criteria (simplified overview) (Source: Adapted from Rodriguez, A.G.L. et al. (2006). 'The value of the IUCN Red List for conservation.' *Trends in Ecology and Evolution* 21: 71–6)

Criterion	Critical	Endangered	Vulnerable	Qualifiers and notes
A1: Reduction in population size	≥ 90% (loss of individuals)	≥ 70%	≥ 50%	Over ten years/three generations (whichever is the longer) in the past, where causes of the reduction are clearly reversible AND understood AND ceased
A2: Reduction in population size	≥ 80% (loss of individuals)	≥ 50%	≥ 30%	Over ten years/ three generations (whichever is the longer) in the past, future or combination. Causes of reduction not necessarily understood.
B1: small range (extent of occurrence)	< 100 km^2 (range)	< 5000 km^2	< 20000 km^2	Plus two of: (a) severe fragmentation and/or few locations (1 ≤ 5, ≤ 10); (b) continuing decline; (c) extreme fluctuation
B2: small range (extent of occurrence)	< 10 km^2 (range)	< 500 km^2	< 2000 km^2	Plus two of: (a) severe fragmentation and/or few locations (1 ≤ 5, ≤ 10); (b) continuing decline; (c) extreme fluctuation
C: Small and declining population	< 250 (mature individuals)	< 2500	< 10000	Continuing decline either: (1) over specified rates and time periods; or (2) with (a) specified population structure or (b) extreme fluctuation
D1: Very small population	< 50	< 250	< 1000	Mature individuals
D2: Very restricted population	N/A	N/A	< 20 km^2 area of occupancy or ≤ 5 locations	Capable of becoming Critically Endangered or even extinct within a very short time frame
E: Quantitative analysis	≥ 50% in ten years/three generations (whichever is longer)	≥ 20% in twenty years/five generations (whichever is longer)	≥ 10% in 100 years	Estimated extinction risk using quantitative models (eg. Population viability analysis)

Box 24.2 Translocation of the Monarto mintbush

Translocation is the deliberate transfer of individuals of an endangered species from a nursery or breeding facility into the wild, or from one site where they occur naturally to another. The aim may be to re-establish one or more wild populations following control of the original causes of their decline, or to establish new populations to lessen the chance that the species could be destroyed by a single catastrophe. In many cases, it is appropriate to carry out small-scale experimental translocations first to establish what factors must be controlled to maximise chances of success.

Figure B24.2.1 The Monarto mint bush. (Source: M. Jusaitis)

Manfred Jusaitis of the Department for Environment and Heritage in South Australia applied experimental techniques to the problem of translocating the threatened plant, the Monarto mintbush (*Prostanthera eurybioides*) (Figure B24.2.1). This low, spreading shrub is known only from two small, disjunct populations in the mallee region of South Australia, and translocation would increase the number of individuals in the wild and broaden the genetic diversity of the wild populations. Monarto mintbush grows in shallow, sandy loams associated with granite outcrops and it was unclear how important it was for plants to be near outcrops or to be sheltered from intense sun or prevailing winds. To test the influence of these factors, 10 Monarto mintbush seedlings were planted at each of three sites within the range of one of the field populations at Monarto, South Australia:

- Site 1 was open, exposed and rocky.
- Site 2 was rocky, but partly sheltered by *Melaleuca uncinata* bushes.
- Site 3 had no rocky outcrops, but a thicker cover of *M. uncinata*.

Survival and growth of the seedlings were monitored over 7 years.

The conditions at Site 1 (open, rocky) proved much more suitable for Monarto mintbush than the others. Eight of the 10 seedlings there survived the first year and were still alive at the end of the 7-year trial, growing steadily over the whole time. By contrast, only two of the seedlings at Site 2 (rocky but partly sheltered) survived the first year. They were still alive at the end of the experiment, but their growth rate was slower than those at Site 1. All plants at Site 3 died within 1 year. Therefore rocky, open sites are most suitable for translocation of Monarto mintbush.

This example illustrates the value of using experimental techniques to evaluate different options in translocation. However, it also illustrates how replication and uses of large numbers of test organisms are constrained where the species of interest is rare or endangered.

Where biological material or individuals are moved and released into 'the wild' conservation biologists must pay close attention to the release site. An introduction program involves moving captive-bred or wild-collected animals and plants to areas outside their historical range. This may be appropriate when the environment within the known range of a species has deteriorated to the point where the species can no longer survive there, when return to the original site is impossible because the factor causing the original decline is still present or when the species occurs in a limited area subject to catastrophes such as wildfire.

In situ actions associated with threatened species centre on identifying and controlling threatening processes (Plate 24.13). This may involve protecting or restoring habitat, controlling wildfire, culling introduced predators and competitors or the reintroduction of captive-bred animals or plants (Box 24.2). Reintroductions remain very popular because they involve just about everyone – zoos, wildlife agencies, researchers, local communities and the media. They are the high-profile tip of the threatened species iceberg. Reintroductions are unlikely to be effective unless the factors leading to the decline of the original wild populations are clearly understood and eliminated, or at least controlled.

Limitations to single species conservation approaches

Early work in conservation biology concentrated on species (and other taxa such as subspecies, varieties or particularly threatened disjunct populations). Understanding decline and recovery was conceptually simple given a recognisable entity (such as the white-bellied frog, the Corrigin grevillea or the Monarto mintbush (Box 24.2)) to work with. However, ecosystems and the ecological communities within them provide the environment in which these species occur and the interactions between different organisms and the physical environment are often complex and unpredictable. For example, keystone species are those that, although present in small numbers themselves, are essential for the persistence of many other species in the community. A keystone species could be an animal that disseminates the seeds of many plants (such as the cassowary in North Queensland rainforests), or a predator that prevents prey populations from increasing and competing with each other (such as the sea star *Pisaster* in intertidal environments on the North American coast). Therefore rather than concentrating on species, should we not be considering ecological communities to ensure that we do not miss any critical components or interactions? One alternative to the single species approach is to consider threat from the perspective of the threatened ecological community (Figure 24.2). The rest of the chapter considers these issues of community conservation.

Conserving ecological communities

Recognising a community is easiest where there are marked differences in various features of the environment across the landscape, creating clear boundaries between species

Figure 24.2 Ecological communities vary in size from a few centimetres such as this vernal pool (a) to thousands of hectares. However, even very extensive communities can be threatened. Two examples are the softwood scrub from south-east Queensland (b) and the brigalow (c), both of which occur on agriculturally desirable lands and have largely been cleared. By contrast the extensive chenopod shrublands of the Nullarbor Plain (d) are well catered for in the reserve system. (Source: a–c G. Wardell-Johnson; d © Marie Lochman, Lochman Transparencies)

assemblages. Some assemblages, with typical species reliant on similar environmental conditions, occur predictably within a definable geographic range. The abiotic features of the habitat in association with common, typical, structurally important or key species are used to define or name such communities. They can be almost any size from centimetres (e.g. root mat communities in caves, Box 17.4) to thousands of kilometres in extent (e.g. mulga woodlands in north-western Australia or hummock grasslands in northern Australia).

Establishing protected areas

Protecting habitats that contain intact ecological communities is the most effective way to conserve biodiversity. The first step lies in designating land use for nature conservation and bringing the land under the control of a relevant government department or other nature conservation agency. The momentum to establish protected areas increased during the 20th century, reaching a peak in the 1970s, with another activity peak over the last 10 years (mainly due to private organisations). Steps in establishing new protected areas include identifying those taxa and ecological communities that are the highest priorities for conservation; determining those areas that should be protected to meet conservation priorities; and linking new conservation areas to existing conservation networks.

Choosing areas to protect

A vital task in establishing protected areas is determining which areas to protect – in other words, how to distribute limited resources between regions identified as priorities for biodiversity conservation. The prioritisation of global conservation efforts by international organisations is achieved by various methods including consideration of areas of high endemism, biodiversity hotspots and priority ecoregions.

An endemic is found in a particular area and nowhere else, so levels of endemism are often considered in conservation. Species can be endemic to a very small area such as a single mountain or to a much larger area such as a continent. Many islands and isolated mountain chains are noted for having high levels of endemism and are therefore important sites for conservation. Biodiversity hotspots are areas both rich in endemic species and under severe threat. Examples include the fynbos communities of South Africa, the Mediterranean-climate region of south-western Australia, the island of Madagascar and the Indonesian rainforests.

Unfortunately, the data required for detailed reserve design are often limited, even in areas of high public interest. Where decisions on park boundaries have to be made quickly, biologists may make rapid biodiversity assessments that involve vegetation mapping, species lists and identifying species of special concern. Another approach is to base decisions on general principles of ecology and conservation biology, such as protecting elevation gradients that contain diverse habitats, providing some large parks, protecting representative habitats in different climate zones, and protecting biogeographical areas that have many endemic species.

One way to determine the effectiveness of ecosystem and ecological community conservation programs is to compare biodiversity priorities with existing and proposed protected areas. This identifies gaps requiring new protected areas. This was done informally by establishing national parks in different regions with distinctive ecological communities, but more systematic conservation planning processes such as gap analysis are now used. In gap analysis:

1. Data are compiled on the species, ecosystems and physical features of the region or conservation unit.

2. Conservation goals are identified such as the amount of area to be protected for each ecosystem or the number of endemic species.
3. Existing conservation areas are reviewed to determine what is protected already and what is not (identifying gaps in coverage).
4. Additional areas are identified to help meet the conservation goals (filling the gaps).
5. These additional areas are acquired for conservation and a management plan is developed and implemented.
6. The new conservation areas are monitored to determine if they are meeting their stated goals. If not, the management plan can be changed or additional areas acquired to meet the goals.

Gap analysis is used internationally, nationally and locally to identify representative ecological communities (Box 24.3).

Box 24.3 Where to establish protected areas: an example from the Kimberley region of Western Australia

The tropical rainforests of the Kimberley in Western Australia occur as small isolated patches from a few tree crowns to about 100 hectares in area across the region (Figure B24.3.1). Although these patches are rich in biodiversity, it was difficult to determine whether the areas reserved encompassed the region's biodiversity or whether areas outside the reserve system also contributed to biodiversity conservation.

Figure B24.3.1 The tropical rainforests of the Kimberley in Western Australia usually occur in sheltered, seepage or riverine sites as small isolated patches across the region. (Source: G. Wardell-Johnson)

A regional biogeographic survey was carried out to determine how well the conservation reserves in the region represented the biodiversity of its rainforests. It also identified the

continued ›

Box 24.3 continued ›

optimum positions for any additional reserves necessary and assessed likely future trends in the persistence of rainforest patches under current land management regimes.

Ninety-five rainforest patches were surveyed for perennial plants, birds and land snails to represent the biological scale and environmental range of the rainforests. Twenty-eight species assemblages were distinguished. In 1991, the existing and proposed reserves encompassed the compositional diversity of only 17 of them. The optimal positions for the 12 additional reserves necessary to encompass the diversity were then determined.

These findings led to additional reserve allocation for biodiversity conservation and identified other management needs. For example, fire reduced rainforest area and its impact was exacerbated by cattle. In the drier areas, rainforest patches only persisted in perennially moist refugia that were fire- or cattle-proof.

This survey-based approach in Kimberley rainforest patches provided an exemplar in reserve design, allocation and management in Australia.

Designing networks of protected areas

There are many issues to consider in reserve design. How large must a reserve be to protect species and ecological communities? Is it better to have a single large reserve or many small reserves? How many individuals must be protected to maintain a viable population? What is the best shape for a reserve? When several reserves are created, should they be close together or far apart? There are additional problems related to the lifestyles of marine organisms when designing marine conservation areas (Box 19.3).

How big should a reserve be?

One of the most publicised early debates in conservation biology occurred over whether species richness or biodiversity is maximised in one large reserve or in many smaller ones of an equal total area, known as the SLOSS (single large or several small) debate. The proponents of large reserves argue that only large reserves can support sufficient numbers of large, wide-ranging, low-density species to maintain long-term populations and that they can also support large numbers of species. Large reserves also minimise the ratio of edge to total habitat. For example, a reserve measuring 1 km × 1 km has a perimeter of 4 km and an area of 1 km^2, giving an edge-to-area ratio of 4:1. A reserve of 5 km × 5 km has a perimeter of 20 km and an area of 25 km^2, giving an edge-to-area ratio of 0.8:1. Species averse to conditions at the edge will have more habitat available to them in large reserves. Furthermore, pest species wandering into the reserve from surrounding areas are unlikely to penetrate far from the edge and so will not have an effect over much of a large reserve.

On the other hand, the rate at which the number of new species is added to a reserve tends to decline as total area increases (for the mathematically inclined, the number of species increases with the logarithm of area). Furthermore, habitat varies geographically, so that many well-spread reserves will encompass a greater range of habitats and more species. Creating more reserves, even if they are small ones, decreases the possibility of a single catastrophe destroying a species or ecological community. Thus the optimum size of a reserve will depend very much on the conservation aims it is meant to achieve.

What shape should a reserve be?

The optimum shape is rounded, rather than a long, linear reserve because this maximises the ratio of the area of the reserve to its perimeter and minimises edge effects. However, most reserves have irregular shapes because land acquisition is usually opportunistic rather than by design. Internal fragmentation alters the climate, provides entry points for invasive species, produces additional edge effects, and creates barriers to dispersal.

Habitat corridors

Habitat corridors are narrow strips of habitat used to link isolated protected areas into a single system. Corridors that facilitate natural patterns of migration will probably be the most successful at protecting species. Creating corridors to protect expected migration routes and range shifts under global climate change may also provide some level of insurance. However, there are some potential disadvantages to corridors. For example, large edges are formed leading to attendant edge effects, and diseases, pests and predators may also disperse along the corridors.

The presence and density of many species will be affected by the size of the habitat patches and their degree of linkage. Probably the greatest challenge in designing systems of protected areas is to anticipate how this network will be managed. In many cases, the actual management of the protected areas will be more important than the size and shape of the individual protected areas.

Managing protected areas

Protected areas have different objectives depending on their legal status, establishment history and individual characteristics. Regardless of their objectives, active management is often required in all kinds of protected areas. The most effective parks are usually those where the managers have the benefit of clearly defined goals and information provided by research and monitoring programs, as well as funds to implement management plans.

Small reserves, especially those found in long-settled areas and large cities, will generally require more active management than large reserves because they are often surrounded by an altered environment, have less interior habitat, and are more readily affected by introduced invasive species and human activities. Even in large reserves, active management may be required concerning fire, visitors, introduced species and so on. Simply maintaining the park boundaries may not be sufficient except in the largest and most remote areas.

Monitoring is an important component of reserve management, with the types of information gathered depending on the goals of park management. It is important to consider the monitoring process broadly, so that the process, objectives, outcomes and outputs are all scrutinised for effectiveness.

Adaptive management is a very effective way to achieve this. It treats different management actions in space and time as experimental treatments to increase understanding of the system being managed. In passive adaptive management a panel of experts guides management decisions, drawing on the results of monitoring and research as they come to hand. As an active process, adaptive management uses interventions designed deliberately as research studies with the different management options tested. Thus management and research unite in a feedback loop, with the results of the trials informing management and changes in management policy subject to ongoing testing. Interested community groups should also participate in planning and decision making because values are inevitably involved and the decisions often have widespread social implications (Chapter 25). Adaptive management also applies outside reserves where public lands used for water catchments, timber production, recreation and so on are also managed for biodiversity conservation.

Management outside protected areas

According to even the most optimistic conservation predictions more than 85% of the world's land will remain outside of protected areas so management of the lands adjacent to, or near, protected areas is often essential for the protection of biodiversity. Organisms don't necessarily respect human boundaries and may readily pass into and out of reserves. Thus adaptive management also applies outside reserves where public lands used for water catchments, timber production, recreation and so on are also managed for biodiversity conservation.

Public and private resource managers are moving away from assessing the management of an area as the maximum production of goods (such as volume of timber harvested) or services (such as the number of park visitors), to take a broader perspective including the conservation of biodiversity and the production of ecosystem services such as timber or water catchments. Rather than each government agency, private conservation organisation, business or individual landowner acting in isolation, ecosystem management seeks cooperation to achieve common conservation and resource production objectives. It is associated with several themes including:

- recognising connections between all levels in the ecosystem hierarchy – from the individual organism to the species, community, ecosystem, region and globe
- ensuring viable populations of all species, ecological communities and successional stages and healthy ecosystem functions
- monitoring significant components of the ecosystem, gathering the needed data, and using the results to adjust management

- changing the rigid policies and practices of land-management agencies towards interagency cooperation and integration at the local, regional, national and international levels, and cooperation and communication between public agencies and private organisations
- minimising external threats to ecosystems while maximising sustainable benefits
- recognising that humans are part of ecosystems and that human values influence management goals.

These management guidelines imply greater public involvement and consideration of social components in the conservation-planning exercise than has been considered in the past. This leads into cultural conservation biology – the subject of the next chapter.

Chapter review

1. Conservation biology is a new, integrated science developed in response to the human-induced, worldwide wave of habitat destruction and extinction widely recognised and acknowledged since the 1970s.
2. Conservation biology investigates human impacts on species and ecological communities and develops practical approaches to prevent extinctions, maintain genetic variation within species, and protect ecological communities and associated ecosystem functions.
3. Extinction is a natural phenomenon, but the concern of conservation biology is that contemporary extinction rates are much higher than background extinction rates determined from the fossil record.
4. Rare species occur commonly and species that are considered threatened are only those at risk of extinction according to specific criteria laid down by the International Union for the Conservation of Nature Red List of Threatened Species.
5. The 'small-population paradigm' deals with the risk of extinction that follows from demographic, genetic and environmental consequences of low numbers. The 'declining population paradigm' is concerned with the threatening processes by which populations are driven to extinction. These paradigms focus on ways of detecting, diagnosing and halting population decline.
6. Population viability analysis (PVA) uses computer modelling to integrate demographic, genetic and environmental data and predict the likely future population size under different management options.
7. Species may be conserved outside their natural environments (*ex situ*) by removing plants or animals for breeding in botanic gardens, zoos or wildlife parks, or in their natural environment (*in situ*) after identification and control of threatening processes.
8. Communities are conserved by selecting areas for reservation. Gap analysis, which compares a range of landscape features against those represented in existing reserves, can identify areas needing protection.
9. Adaptive management, which treats different management actions in space and time as experimental treatments to increase understanding of the system being managed, is a very effective way to choose management options for reserves.
10. Effective conservation also requires the management of land outside reserves for conservation purposes as well as other uses, such as production of renewable resources and recreation.

Key terms

adaptive management
biodiversity
biodiversity hotspots
captive breeding
conservation biology
conservation status
corridors
edge effect
endemic
ex situ conservation
feral
hummock grassland
in situ conservation
introduced species
introduction
keystone species
minimum viable population (MVP)

population viability analysis (PVA)
rarity
reintroduction
SLOSS debate
threat
threatened ecological community
threatening process
translocation

Test your knowledge

1. Construct a table to demonstrate the seven forms of rarity. Place the following organisms within this table: the bird-eating spider and Jackson's sun orchid (Figure 24.1), the humpback whale, the orange-bellied frog, the Corrigin grevillea and Leadbeater's possum.
2. Explain the difference between the 'small-population paradigm' and the 'declining population paradigm'.
3. What genetic, demographic and environmental problems may be encountered by small populations?
4. How can population viability analysis be used to assist in the management of a threatened species?
5. Explain the processes involved in deciding the conservation status of a species.
6. What limitations are there to using only a threatened species approach to conservation?
7. What is gap analysis and how can it be used to determine the location and size of conservation reserves?
8. What issues must be considered in setting the size and shape of conservation reserves?
9. What is implied by adaptive management? What monitoring requirements are necessary to achieve effective adaptive management?

Thinking scientifically

Consider the SLOSS debate. What type of rare species or biotic pattern will be best conserved by several small reserves, and what type of rare species or biotic pattern will be best conserved by a single large reserve?

Debate with friends

The accidental introduction of the yellow crazy ant (*Anoplolepis gracilipes*) to Christmas Island in the early 20th century killed up to 20 million of the island's famous land-dwelling red crabs (*Gecarcoidea natalis*) by predation and by competition for burrows. Other ground-dwelling animals such as lizards and a range of invertebrates also declined. However, a reduction in fauna numbers meant that seedlings normally eaten by them had a better chance to survive and grow, changing the character of the island's forests. In the light of this information, debate the topic: 'Management should attempt to eradicate the yellow crazy ant from Christmas Island'.

Further reading

Bradstock, R., Auld, T.D., Keith, D.A., Kingsford, R.T., Lunney, D. and Silversten, D.P. (1995). *Conserving biodiversity: threats and solutions.* Surrey Beatty & Sons, Chipping Norton, NSW.

Caughley, G. (1994). 'Directions in conservation biology' *Journal of Animal Ecology* 63: 215–44.

Forman, R.T.T (1995). *Land mosaics: the ecology of landscapes and regions.* Cambridge University Press, New York.

Huxley, A. (1992). *Green inheritance: the World Wildlife Fund book of plants.* Four Walls Eight Windows, New York.

Konteleon, A., Pascual, U. and Swanson, T. (2007). *Biodiversity economics: principles, methods and applications.* Cambridge University Press, Cambridge.

Lindenmayer, D. and Burgman, M. (2005). *Practical conservation biology.* CSIRO Publishing, Collingwood, Victoria.

Margules, C.R. and Pressey, R.L. (2000). 'Systematic conservation planning' *Nature* 405: 243–53.

Nielsen, J. (2006). *Condor: to the brink and back: the life and times of one giant bird.* HarperCollins, New York.

Roberts, J. (2007). *Marine environment protection and biodiversity conservation: the application and future development of the IMO's particularly sensitive sea area concept.* Springer-Verlag, Berlin-Heidelberg.

25

Cultural conservation biology

Pierre Horwitz and Grant Wardell-Johnson

The case of the Pedder galaxias

Lake Pedder in south-western Tasmania was the largest glacial outwash lake in the Southern Hemisphere (Plate 25.1). In 1972 the Tasmanian state government submerged it under 50 m of water to create a huge dam to generate hydro-electricity. They also sanctioned release into the dam of brown trout (*Salmo trutta*), a non-native predator, for recreational fishing. These controversial actions contributed to the emergence of the environmental movement in Australia and may even have spawned the first green political party in the world.

The Pedder galaxias (*Galaxias pedderensis*) (Plate 25.2) was endemic to the original Lake Pedder. Following impoundment it initially increased in abundance before declining dramatically in the 1980s, probably because of predation and competition from climbing galaxias (*G. brevipinnis*) and trout. In 1992 the Pedder galaxias was formally recognised as 'endangered' and likely to become extinct in 5 years if no action was taken.

The Inland Fisheries Commission of Tasmania studied the life history, diet, conservation status and management of the Pedder galaxias and found the species in only three of 117 creeks in the area and only 68 individuals in the two creeks surveyed intensively.

Three management options for saving the species were (1) protection of existing populations from 'threatening processes' (predation and competition from brown trout and climbing galaxias), (2) captive breeding or (3) moving a portion of the wild population to a place without the threatening processes.

The third option was deemed technically feasible and highly preferred. Lake Oberon (Plate 25.3) in the Western Arthur Range of Tasmania, met stringent selection criteria covering habitat, food availability and accessibility for a translocation site. In late 1991 the species was translocated and the lake now has a healthy population. The recovery actions stopped the threat of extinction.

The 'solution' raises questions, however. Lake Oberon had an ecological community that had evolved, since the last glacial age, without predatory fish. The population of the Pedder galaxias has caused habitat alterations and disturbance to trophic interactions. It is possible that other species including plankton were also accidentally introduced to the lake. What would be worse: not using this option and risking losing a species, or changing the ecology of Lake Oberon forever? Who should make such decisions?

Chapter aims

Conservation biology encompasses much more than biology – it has cultural, political and economic aspects. In this chapter we explore the cultural nature of conservation biology, including a rationale for biological conservation, the question of who should take conservation action and how the cultural and scientific components of conservation biology unite in the important task of planning the recovery of threatened species.

Conservation biology and society

As we saw in Chapter 24, conservation biology involves understanding the biological and ecological characteristics of the endangerment of organisms and their communities and the behaviour of humans in this process. From these characteristics solutions that often have a scientific or technical basis are generated. But the solutions always have a social context. The example of the Pedder galaxias highlights the way biological components that require protection from human activities now must be identified, and how human motivation and behaviour must be taken into account. Thus conservation biology must merge two ways of thinking that are traditionally separate from one another. Ecology is classed as one of the natural sciences, whereas human behaviour, motivations, perceptions and attitudes are dealt with in the social sciences.

Conservation biologists are almost always trained in the natural sciences, but to achieve their goals must work with the social sciences – philosophy, history, business, administration, psychology, law, sociology, anthropology and so on. The problem of endangerment is usually defined and identified in the natural sciences. Its solutions will come from both the natural sciences *and* the social sciences. In fact, part of the solution will require redefining the problem as one that is at least cultural, political and economic in scope.

The cultural nature of conservation biology

Biological features of the landscape that are highly restricted in their distribution are frequently, if not always, regarded as being of special concern from a conservation viewpoint. They are vulnerable to extinction as a result of a local catastrophe, or the genetic and demographic consequences of having a small population size. Management efforts to protect species or communities invariably involve local people or traditional owners of the land. Their commitment to, and involvement in, conservation projects is essential for success. At one extreme, residents may be familiar with species, or characteristic features of species or ecological communities, that are unique to their area or form a critical element in their culture. A place may become recognised as 'the only known location of ...', ecotourism established around the feature, and the feature elevated as an icon or flagship symbol for local identity or action. If local people identify with a particular species or community, then there may be psychological implications if that species is threatened or made extinct, or if habitat features are degraded. At another extreme the residents may be unfamiliar with the unique elements and defining ecological features of the landscape. If they are not convinced of the value, they are unlikely to support conservation efforts. Residents may be hostile to the reintroduction of species they perceive as a threat, such as wolves or bears.

Another example of the social nature of conservation biology relates to an intriguing impasse when, for instance, a species plays a role in the culture of traditional people, but that species is threatened by the activities of other social groups. An example is the dugong (*Dugong dugong*) in northern Australia. Some populations are threatened by the degradation of seagrass habitats and by entanglement in fishing nets, but indigenous people consider hunting dugong to be an important expression of identity. Whose culture is more important: the western notion of protecting all individuals to 'save a species' (which would thereby exclude traditional hunting), or the pursuit of traditional activities by indigenous peoples, which would be seen by others in Australia as cultural favouritism? Such cases are usually resolved through the complex interaction of community involvement, international treaties and government legislation, with conservation biologists potentially involved at all stages of the process.

Why should human cultures conserve biological diversity?

Utilitarian, aesthetic, ecosystem services and ethical arguments

In this book so far we haven't answered at length the critical question: 'Why should we put time and effort into preventing species from going extinct?' How do conservation biologists justify to a broader society giving priority to biodiversity conservation, with its attendant issues of managing rarity, vulnerability, endangerment and threats to biological entities?

The most common response is *instrumental* and *utilitarian*: humans benefit through using organisms for food, clothing, shelter, pharmaceuticals, agriculture, resources of genetic diversity and many other possibilities. These uses are not always obvious. One striking example is the discovery that some animal toxins have powerful antibacterial properties, potentially making them effective drugs. The venom of the wasp *Vespa magnifica* contains peptides (sequences of amino acids) toxic to a wide range of bacteria and fungi but not harmful to human red blood cells, so it could be a rich source of useful drugs. A peptide in the venom of the scorpion *Leiurus quinquestriatus* binds selectively to a range of human cancers, including a form of brain tumour, but not to other body cells. Synthetic versions of the peptide are now linked to radioactive isotopes of iodine to deliver radiation directly to cancer cells. Smoke bushes belonging to the genus *Conospermum* found in Western Australia have been found to produce a chemical ('conocurvone') that has great potential in anti-HIV-AIDS treatment (Figure 25.1). But what if the critical species had been driven to extinction before their medicinal properties were known? Conservation biologists therefore apply the precautionary principle: *Where there are threats of serious or irreversible environmental damage, lack of full scientific certainty should not be used as a reason for postponing measures to prevent environmental degradation*. Extinction is irreversible, so we should take steps to prevent biodiversity loss.

Other forms of instrumental benefits are *non-material* or *aesthetic*. A biological entity may have cultural, spiritual and aesthetic value, giving pleasure or significance simply by existing. This can enrich recreational pursuits such as fishing, hunting and wildlife tourism. Conserving the species preserves all of these tangible benefits to humanity. In short, it is worth saving biological entities from extinction because they are part of a system that enriches our lives, takes away our pain, and provides for the existence of the next generations (intergenerational equity).

An *ecosystem services* argument may also be applied. Thus ecological entities such as individuals and populations of species, as well as the ecological communities in which they occur, provide services to all life on the planet. These include clean oxygenated air, atmospheric moisture and clean water, pollination, dispersal of seeds, decomposition, nutrient cycling, flood control and productive soils, lakes and oceans. This argument is

Figure 25.1 Smoke bushes of the genus *Conospermum* produce a chemical, 'conocurvone', that has potential in anti-HIV-AIDS treatment. (Source: G. Wardell-Johnson)

especially powerful regarding many small and apparently insignificant organisms that are critical in the biogeochemical cycles that sustain ecosystems and ultimately human life. They are, as biologist Edward O. Wilson stated, 'the little things that run the world'. Without them, the complex ecosystems of today could not exist.

A final argument involves moral and ethical reasoning. *Bio- [or eco-] centrism* argues that all *individual* living beings have intrinsic value distinct from their value to humans and therefore humans have an ethical obligation towards them. Part of that moral responsibility includes their right to continue to flourish as a species, in an evolutionary sense. Under this view, the onus is for any person wishing to exploit other organisms to justify the action and show how it will not endanger the exploited species.

Answering the critics

As valid as the arguments in the previous section sound, they will still not convince everyone. Three sophisticated counter-questions, and their possible answers, are presented below.

Extinction is a natural process, so why should we prevent it from occurring?

The answer to this question has two main parts. First, as humans are also part of nature, so one could just as easily argue that our evolved concern for fellow species and acting to prevent them from becoming extinct is as natural as the extinction process itself. Reference to what is and what isn't natural is a false dichotomy if humans see themselves as part of the whole, rather than separate from it.

Secondly, humans with their power of rational reasoning can recognise that contemporary extinction rates are well above the 'background extinction rates' known from the fossil record, that these losses are depriving humanity of a wide range of natural resources and that human activity is responsible. The litany of human actions threatening species and pushing them towards the brink of extinction is as long and varied as human development itself. It includes the over-use of resources, fragmentation of habitat through land clearing, and polluting the environment with toxins, sediments and excess nutrients. Some of these human activities combine to change weather patterns locally and climate regionally and globally. Climate change, including shifting patterns of wind and rain and differing temperature regimes, is now regarded as a significant threat to the future of global biodiversity.

Surely every species isn't crucial to ecosystem functioning? Would it matter if we lost just a few? Are any species redundant?

In May 1968, a gas explosion on the 18th floor of the Ronan Point apartment building in London caused a partial collapse that ultimately killed five people. Investigators blamed a lack of structural redundancy in the modular construction of the building. When one of the modules failed in the explosion, others dependent on it failed as well, causing the collapse. If other 'redundant' measures of support were designed into the building in case the main support failed, it might have survived.

American ecologist Shahid Naeem argued that redundancy is just as important in ecosystems. Species that perform similar functions in nutrient cycling or energy transfer form interdependent functional groups such as autotrophs or decomposers. Failure of one of these groups would lead to collapse of the entire ecosystem. However, because there are usually multiple species in each group, extinction of one species need not lead to collapse, especially given the possibility of colonisation by species with similar functions in neighbouring ecosystems. Unfortunately, human-caused extinctions can lower the species diversity and hence the redundancy within functional groups, while habitat destruction and fragmentation reduce the chances of recolonisation from neighbouring systems. This loss of redundancy increases the risk of ecosystem collapse. The problem is acute in human-created 'modular ecosystems' such as agroecosystems and managed forests where species diversity is reduced. Thus losing any species potentially diminishes a functioning whole.

Restricting or banning use of some natural resources will cause economic hardship in communities depending on that industry. Why should governments care more about the wellbeing of animals and plants than about their own citizens?

Fortunately, restricting or even ending an industry based on using natural resources need not spell disaster for employment and local communities. One strong example is the whaling station at Cheynes Beach near the port of Albany in south-west Western Australia. Its closure in 1978 ended nearly two centuries of whaling in Western Australia. At the time of closure the Cheynes Beach Whaling Company operated four whaling vessels that hunted mainly south of the continental shelf, taking predominantly sperm whales and the occasional blue whale. Humpback whales were taken from 1952 until 1962. The primary product, whale oil, was produced by rendering the carcass. Before the development of petroleum-based alternatives in the 20th century it was highly prized in industry as a lubricant and an additive in cosmetics and pharmaceuticals. Other products included baleen (from humpback whales) for umbrellas and corsets, ivory from sperm whale teeth, ambergris for perfumes (from an intestinal secretion in sperm whales) and protein powder for enriching stock food. By 1978 there were growing protests against whaling, but the primary reason for closure was economic because the price of whale oil was falling, operating costs were rising and three of the four ships needed replacing at an estimated cost of $6 million each. The site today is a popular tourist attraction, with opportunities to tour the site and whaling facilities, an extensive museum, a viewing tower converted from an old whale oil storage tank and theatres and galleries (Figure 25.2). Boat trips to see migrating whales are popular. Far from being the end of Albany, the closure of the whaling station created the opportunity for a vibrant tourism industry.

Considering the environment when planning resource use may also lead to management that more equitably distributes the benefits of resource uses. The idea is not new and is a major tenet underlying long-standing resource use industries. For example, in 1920 Charles Lane Poole, an early Western Australian Conservator of Forests, argued for these steps to protect the state's forest resources:

Figure 25.2 Twenty-five metre skeleton of a pygmy whale on display in the 'Giants of the Sea' exhibit at Whale World, the museum on the site of the old Albany whaling station. (Source: Whale World, Albany Western Australia)

> Some foresters who have visited this State have been so disheartened by the condition of affairs they have found that they have said that there will be no forestry in Western Australia until the last tree has been cut down. I do not hold this pessimistic view, but consider that, by a publicity campaign, the democracy will realise the wealth that the forests represent. It is true that trees to-day have no votes, but when the people develop a forest *conscienceness* [emphasis added] the position will be entirely altered, and they themselves will see to it that the forest policy is maintained and the forests are used for the benefit of the community as a whole for ever, and not for the benefit of the few sawmillers, timber hewers, and timber merchants of to-day. (Source: Lane Poole, C. E. (1920). *Statement prepared for the British Empire Forestry Conference, London*, p. 34. Perth, Western Australia: Western Australian Government Printer.)

Consideration of biodiversity when planning resource use can also contribute to the goal of ensuring that the whole community, not just select interest groups, benefits from resource use (see Plate 25.4).

The 'who' of conservation biology actions

Who should take action on conservation problems: should it be governments, community groups, scientific professionals, industry or some combination of all?

Should conservation actions be left to the government?

This depends on the resources required for conservation action and whether the government feels both responsible for, and able to act for, the species and ecological communities under threat of extinction. Sometimes governments may prefer to deny that a problem exists and in some cases local people willingly take action into their own hands.

On other occasions the scale of the problem may be so massive that government resources are insufficient to resolve it, so private funds are sought from philanthropic sources and the corporate world becomes involved. For instance, Bush Heritage Australia is one of a number of private organisations that have set themselves the task of purchasing land, and managing it exclusively for nature conservation. They receive some support from the government, but most of their funds come from donations from the Australian public.

Another situation arises where an endangered species or ecological community occurs on private land. Here the government can suggest what needs to be done, but needs the landowners' support for that request to be enacted. In general, landholders respond positively to nature conservation efforts as long as they are not perceived as a threat to their livelihood or long-term objectives for their land. Measures that are viewed to be economically unfavourable, alienating the land from its owner, or decreasing its market value, are seldom readily accepted. Let us now consider in more detail the role of non-government and government organisations in conservation.

Community participation in conservation initiatives

In Australia the importance of heartfelt community-based volunteer actions to help endangered species cannot be understated. Landcare activities, along with 'Friends of …' groups (focusing on, for example, a wetland, a species or a landscape feature) are instrumental in raising awareness and conducting actions that serve the environmental interest (Figure 25.3). In agricultural landscapes, areas of attention include clearing of native vegetation and habitats for threatened species, addressing activities leading to degrading of landscape features, and other matters such as soil erosion and water pollution (including salinisation, eutrophication and sedimentation of waterways).

These informal or semi-formal landcare activities are symptomatic of a general rise of environmentalism between 1970 and 1990. They have persisted throughout the 1990s and into the 21st century, and there has been a series of government policy directives that have sought to capture their goodwill and make them more broadly based across the Australian continent. Principal among these was an alliance in the late 1980s that saw the Australian Government, the National Farmers' Federation and the Australian Conservation Foundation declare 'A Decade of Landcare'. This and subsequent governments (through the agency of the sale of government companies such as Telstra and establishment of the Natural Heritage Trust) have sought to fund on-the-ground and community-based activities for environmental objectives, including attention to threatened species.

Figure 25.3 Bush Heritage Australia is an example of a private organisation that manages land for nature conservation. This shows a field day where participants can examine the replanting of former agricultural land in south-east Queensland. The work on this particular site was supported by the Commonwealth's Natural Heritage Trust. (Source: G. Wardell-Johnson)

Local communities participating in actions for the public good, such as conservation or landcare activities, should be seen as assets by professional and/or government agencies rather than as threats. But a key point about participation of communities in conservation action is that it involves not just *what they do*, but *how they are allowed to do it*. Participation means different things to different people and to avoid potential conflict, disappointment or 'burn out' it is important to clearly state, or for community members to agree on, exactly what commitments are being made and what goals are to be achieved (see Box 25.1).

Box 25.1 Project Aware on the Rocks

'Project Aware on the Rocks' is an ambitious, broad-scale project involving a range of stakeholders including local councils, community groups and relevant government and university departments in halting the decline of marine plants and invertebrates in the intertidal zone along the coastline near Sydney, New South Wales. It seeks to raise community awareness, build partnerships among the diverse stakeholders and encourage a cooperative, community-based approach to conservation. The project has produced trained community volunteers to disseminate a conservation message, sustained community and media interest in coastal conservation, and a range of valuable educational outcomes, including tours of rock platforms.

At the scientific level, ecologists from the Institute of Marine Ecology at the University of Sydney helped evaluate the impact of the removal of intertidal invertebrates on intertidal communities. The program set up three Intertidal Protected Areas around Sydney, in which collection of invertebrates as fishing bait was prohibited,

continued ›

Box 25.1 continued ›

and two control sites. Members of the public, with oversight from professional ecologists, helped to collect data on the species abundance and diversity in the protected and unprotected areas. This type of cooperation between ecologists and members of the public is useful for the dissemination of ecological knowledge and expertise. It also ensures that volunteer-based data collection is focused on testing important questions in conservation, conducted with sound scientific designs to maximise value for effort and analysed expertly to draw the best conclusions.

Specific achievements in this case included training volunteers as conservation advocates, heightening and sustaining community interest, involvement of scientists and government authorities and preparation of educational materials. This illustrates how scientists and the general public can work together to solve conservation problems.

Questions that may help conservation ecologists and community members recognise the strengths or weaknesses of government-facilitated community involvement include:

- *access to process*: are community members able to contribute to *how* the action is planned and performed?
- *power to influence process and outcomes*: are community members able to *exert control* over the objectives and goals, and contribute in a meaningful way?
- *structural characteristics to promote constructive interactions*: are dealings with government convoluted, cumbersome, overly bureaucratic?
- *access to information*: is all information readily available?
- *personal behaviours*: is constructive behaviour facilitated in the process?
- *adequate analysis*: is sufficient time and effort given to analyse the information?
- *enabling of social conditions necessary for future processes*: does the process leave behind a willingness of people to contribute in further activities?

Box 25.2 shows how some, but not all, of these questions can be answered using the case of public involvement in the conservation of the white-bellied and orange-bellied frogs discussed in Chapter 24.

Box 25.2 Public involvement in the management of the white-bellied and orange-bellied frogs

A Recovery Team was established consisting of scientists and government officials, but it was clear that public involvement was also required. Action was needed to take care of the stream zone habitat where the frogs occur, and in most cases the stream zone habitat was owned by local farmers. A recovery kit, newsletter and schools kit were produced by the Recovery Team. Frog spotlight nights were held, and were well-attended. The most important step was the protection of stream zones from clearing or stock intrusion, with most farmers implementing a fencing program with financial assistance from government. An appreciation of the broader benefits of the conservation of these species, developed to the extent that some landholders *wanted* a frog swamp on their land. The first 'Land for wildlife' signs in Australia were also produced for landholders through this plan. Signposting the protection of riparian zones on private land was seen as a way of promoting a conservation ethic in the broader social community. There are ongoing negotiations with private property owners for the protection of populations by fencing off the habitat from grazing.

Sometimes individuals or organisations are resistant to the notions that their actions, or inactions, are responsible in some way for the continuing decline of species and/or ecological communities. Under these circumstances a normal response is to challenge and even deny such claims, resulting in conservation debates characterised by public bickering. Conservation biologists must learn how to establish a meaningful dialogue using techniques of listening, sharing ownership of projects, public debate, dispute resolution and diplomacy if they believe that the mindset of a group must change to protect nature, or arrest biodiversity loss (see Box 25.3). So that a dispute does not get out of hand, with accusation and counter-accusation, a willingness to debate openly, and to agree (even about where they disagree!) should become an early goal of conservation action. Disputes can also extend to differing opinions between scientists. In this case, Jamie Kirkpatrick's suggested code of conduct for debate between scientists is a useful guide (see Box 25.4).

Box 25.3 Rules for dialogue

Some basic rules for dialogue (adapted from Gang, P. (2000). *Rules of dialogue*. The Institute for Educational Studies Integrated Studies Program, Endicott College, Massachusetts) are:

- Treat the dialogue as an exploration of ideas in a non-adversarial manner.
- Remember that *everyone* (facilitators, general community, expert advisors, agency representatives) is an equal participant.
- Take time to 'think and reflect' before responding to another's comments.
- Make good use of all the tools for learning: sites for informal discussion, common themes, participant's passionate interests and technical help.
- Be creative: pour in diagrams, pictures, sketches, photos, key points, summaries and fresh ideas, without the need for a finished form. That can come later.

continued ›

Box 25.3 continued ›

Throughout the dialogue:

- Commit yourself to the process.
- Listen and speak without judgment.
- Identify your own and others' belief assumptions.
- Acknowledge the other speakers and value their opinions.
- Balance inquiry and advocacy.
- Relax your need for any particular outcome.
- Listen to yourself and speak when moved.
- Take it easy, enjoy.

Box 25.4 What happens when scientists disagree?

Representatives of government, community groups and corporations are not the only ones to find themselves in dispute over conservation policy. Scientists themselves may disagree over the interpretation of data and the correct procedures to be adopted in a recovery plan. This can be disconcerting for those who believe that objective consideration of data should always lead unbiased scientists to the same conclusion. However, practising scientists know that debate and disagreement fuel good science. It is often by responding to criticism that views are tested rigorously and arguments refined.

How should criticism be given and taken by scientists in the public arena of debate over conservation issues? Jamie Kirkpatrick from the University of Tasmania proposed a four-step code of conduct for scientists entering such public debate:

1. Scientists should hold their opinions honestly (being a 'hired gun' to advocate a particular viewpoint is unethical), be very clear why they hold them (including a declaration of interests such as sources of research funding), and acknowledge their limitations.
2. There is no place in public debate for impugning the character and motives of others. This is the logical fallacy called *ad hominem*, which means attacking opponents personally and not their arguments. It may be true that opponents are drunks, womanisers, in receipt of a grant from industry or members of a political party, but those are not justifications for discounting their reasoned arguments.
3. Scientists should communicate openly as part of the debate. It is very productive to identify areas of agreement (often surprisingly substantial) before considering the disagreements. In particular, proponents arguing scientifically ought to be able to identify what key data or experiment would resolve the issue one way or the other. People not prepared to change their minds on the basis of new data are not arguing scientifically.
4. The debate should be open and public, with the data and arguments accessible to other interested parties. Publication in the scientific literature is one forum, as are media interviews and letters to newspapers.

The role of international efforts, treaties and conventions

There are at least two good reasons for seeing conservation biology as international in scope. The most obvious is that species, ecological communities and ecosystems do not recognise national political boundaries. Migratory species are a classic example where conservation action must occur in different parts of the world. Trade in species also crosses national boundaries, and if the species needs conservation, as invariably it does when trade operates, then action must occur at both the source and the destination of the trade.

The second reason is that some nation-states are reluctant to engage in action to conserve species or ecological communities of global or local importance. International pressure is sometimes the only alternative open to persuade governments that might otherwise resist internal calls for conservation action.

The international non-government community has played a critical and significant role in getting nation-states to respond to growing concerns over the loss of biodiversity. The growing lists of threatened and endangered species produced by national governments and those in the volumes of the IUCN's Red List of Threatened Species (see Chapter 24) are testament to the IUCN's tireless work to produce common criteria for listing species (Chapter 24). Workers in the United Nations (UNESCO) or non-government international bodies such as the IUCN, Conservation International and World Wildlife Fund (WWF), have helped establish conventions and treaties for nation-states to sign (Box 25.5). Conservation conventions and treaties are quasilegal in the sense that once signed they commit the signatories (nation-states only) to negotiate with each other and to resolve issues in a multilateral way. As part of the signing, member states agree to uphold the principles of the treaty or convention. The conventions are often criticised for being

Box 25.5 Some significant conventions and treaties for conservation biology

The Convention on International Trade in Endangered Species (CITES) was adopted in 1972 and came into effect in 1975. The convention contains three lists of species: (1) species that are endangered or vulnerable as a result of existing or potential trade – commercial trade in these species is prohibited, (2) species that might become threatened as a result of high volume trade and (3) species that are protected in the home country and where international cooperation is required to help protect the species from exploitation.

The International Convention for the Regulation of Whaling (1946) established the International Whaling Commission and committed nation-states to supporting its activities. The convention gradually developed conservation of whales as its foremost imperative. It was this convention, and this commission, that adopted a moratorium on whaling (with the much-publicised exception of whaling for scientific purposes) that came into effect in 1985.

continued ›

Box 25.5 continued ›

Japan-Australia Migratory Bird Agreement (JAMBA) and *China Australia Migratory Bird Agreement* (CAMBA) are examples of Australia's commitment to global protection of species that migrate to and from its shores.

The Ramsar Convention on Wetlands was signed in Ramsar, Iran, in 1971, and is a treaty which provides the framework for national action and international cooperation for the conservation and wise use of wetlands and their resources. The four main obligations on contracting parties to that convention can be summarised as:

1. nominating suitable sites as 'wetlands of international importance' ('Ramsar sites') and ensuring they are managed to maintain their ecological character
2. formulating and implementing national land use planning to include wetland conservation, and to promote the wise use of all wetlands in their territory
3. developing national systems of wetland reserves, facilitating the exchange of data and publications, and promoting training in wetlands research and management
4. cooperating with other nations in promoting the wise use of wetlands where wetlands and their resources, such as migratory birds, are shared.

The Convention on Biological Diversity emerged from the *Rio Declaration on Environment and Development* and came into effect in 1993 and is arguably the most important international conservation instrument to date because of its recognition of the breadth of biodiversity (not just iconic species such as birds or whales). It is also critical because it recognises that most of the world's biodiversity reside in the tropical, more populated and poorer nations of the world. Its objectives are the conservation of biological diversity, the sustainable use of its components, and a fair and equitable sharing of the benefits arising from the use of genetic resources. It argues that this is important due to the 'intrinsic ... ecological, genetic, social, economic, scientific, educational, cultural, recreational and aesthetic values' of biodiversity.

'toothless' in the absence of other measures. Sometimes they are strengthened internally by governments when they pass matching legislation supporting their intent. Laws and regulations provide the 'teeth' for conservation action within any democratic nation.

Legislative and regulatory approaches (state and national)

All Australian states have their own legislation that protects threatened flora and fauna, but the processes they use vary from state to state. There is a tendency for states to align themselves either with the categories and criteria of the IUCN (see Table 24.2), or adopt the same listing processes used by the Commonwealth government. This latter process is outlined in more detail here.

In Australia, threatened species and ecological communities may be afforded protection under the Commonwealth's *Environmental Protection and Biodiversity Conservation Act*

1999 (EPBC Act 1999). Threatened taxa and community listings are made after applying criteria that involve combinations of both the extent and nature of the threat. Such criteria often vary between Commonwealth and state jurisdictions. We shall return to discuss recovery plans in detail, for they form a large component of the expenditure on biodiversity conservation in many government agencies.

At the national level, the EPBC Act 1999 is Australia's foremost piece of legislation dealing with conservation of biodiversity. It comes into effect when a project or development occurs on Commonwealth land, is undertaken by a Commonwealth agency, and/or is likely to have an impact on something of national environmental significance. Matters of 'national environmental significance' are:

- World Heritage sites
- nuclear actions (including uranium mining)
- the Commonwealth marine environment
- national heritage sites
- Ramsar wetlands
- listed threatened species and ecological communities
- migratory bird species.

The last three are directly relevant to this chapter. Ramsar wetlands (Box 25.5) are likely to contain special populations of aquatic plants and animals, and the Commonwealth government accepts its obligation to the Ramsar convention to ensure listed wetlands are protected. The same applies for migratory birds since they are the subjects of JAMBA and CAMBA (Box 25.5).

Threatened species and ecological communities are relevant to the *Convention on Biological Diversity*. The EPBC Act 1999 provides for:

- identification and listing of threatened species and threatened ecological communities
- development of recovery plans for listed species and ecological communities
- recognition of key threatening processes, and
- where appropriate reducing these processes through threat-abatement plans.

A key part of the process is to have species listed/gazetted on a schedule of the Act. Public nominations are called for and they are assessed under the EPBC Act 1999 by a panel of experts, the Threatened Species Scientific Committee, which makes a recommendation to the Commonwealth Minister for the Environment, Heritage and the Arts after consultation with the states and other relevant scientists and parties. Species are ranked from extinct to conservation-dependent (see Table 24.2), and ecological communities ranked as critically endangered, endangered or vulnerable (depending on the degree of risk of extinction in the immediate, near future or medium term respectively). Under the Act, the Commonwealth Minister for the Environment, Heritage and the Arts can either make recovery plans for taxa and ecological communities, or adopt plans prepared by others such as state agencies. Such recovery plans usually cover a 5-year period.

Recovery and threat abatement planning for species and communities

Recovery plans are information and management documents that set out actions to assist threatened taxa and ecological communities within a planned and logical framework. A recovery plan might cover a threatened taxon or ecological community, or several taxa or communities across a region. Alternatively, it might target a key threat to many species or ecological communities in a particular region. The most successful plans are usually those supported by a broad range of interested parties, including state, territory and local government, landholders, scientists and researchers, relevant business or industry groups, and the community in general. A recovery team is required and membership usually determined by the agency with jurisdiction. The team's primary focus is on identifying and undertaking actions to protect and restore habitat, reduce threatening processes, and protect important populations. This usually requires some research and planning to understand the needs or composition of the threatened taxa and ecological communities.

According to the EPBC Act 1999 a recovery plan should include:

- a summary of all information used to list the species or ecological community
- the membership of the recovery team, individuals or groups likely to be affected by the plan, relationship to other relevant plans, and benefits to other species or communities
- realistic, measurable and achievable recovery objectives and timelines, with (for reporting purposes) a way of evaluating whether they have been achieved
- the recovery actions themselves, exactly when and how they will reduce threatening processes, protect and restore habitat, protect important populations or occurrences, and improve knowledge
- a guide for decision makers, identifying matters that governments should consider when making decisions on development actions that may have an impact on the species
- tools to assist implementation, including estimates of the duration and costs of the process, proposed incentives and mechanisms to encourage landholder and/or community participation, including, if appropriate, a communication strategy
- the monitoring, reporting and review processes, including time frames and responsibility.

Many of these features are shown in the example in Box 25.6.

Box 25.6 The recovery plan for the white-bellied and orange-bellied frogs and its outcomes

The dependence of the *Geocrinia* species on undisturbed, permanently moist sites requires special conservation efforts directed at their habitats. A recovery plan for these two species received state government approval and was implemented in 1995 by a team that included members from a university, the state and federal agencies responsible for nature conservation, the local shire and the local Land Conservation District Committee.

The overall objective was to *down-list* the frogs from 'vulnerable' to 'conservation dependent' (in the case of the white-bellied frog) and from 'endangered' to 'vulnerable' (in the case of the orange-bellied frog) within 10 years, by protecting existing populations, and if necessary establishing additional populations. Specific criteria to determine success or failure were established. The first criterion was to determine the number of naturally occurring populations. A longer-term criterion was to secure habitat for all orange-bellied frog sites and at least 75% of white-bellied frog sites. Sustainability of all populations was the criterion for success after 10 years.

To address these objectives the following actions were taken:

- *Survey habitat.* A thorough survey of the potential habitat of these species revealed 43 populations of the white-bellied frog and six populations of the orange-bellied frog.
- *Evaluate land tenure and management needs.* Changing land use and development issues in the Margaret River area threaten the survival of some populations. For example, during 2005 viticulture proposals and associated water storage (dams), drainage and chemical use issues threatened three white-bellied frog populations. Management requirements for populations in each case were determined.
- *Conduct fire management and research.* In collaboration with graduate student research projects, the management agency implemented particular actions to prevent a single fire burning the entire habitat of these species at any one time.
- *Protect habitat.* Most habitat was protected through a fully funded fencing scheme for riparian habitat on private property. In 2002, the government purchased land encompassing the most important remnant populations (1570 ha, 30% of the entire population's range) of white-bellied frogs. Other areas occupied by orange-bellied frogs have been upgraded from State Forest to Nature Reserve and Conservation Parks.
- *Ensure community participation.* See Box 25.2.
- *Monitor populations.* An extensive monitoring program involving counts of calling males at multiple sites throughout the breeding season has been in continuous operation for 12 years, with some 72 white-bellied frog and all 12 orange-bellied frog sites (six creeks) monitored. During 2005 a small but previously

continued ›

Box 25.6 continued ›

unknown subpopulation of white-bellied frogs was discovered. However, the trend is for the number of frog populations to decline each year with a total of 22 local extinctions since 1994.

- *Understand the genetics of the populations.* Genetic variation among white-bellied and orange-bellied frogs was measured using molecular techniques (Chapter 5). The genetic variation between populations is very large, despite the maximum distance between populations of these species being only 18 km for the white-bellied frog and 4 km for the orange-bellied frog. This means that current levels of gene flow are almost zero. A mark-recapture study revealed that 95% of adult male white-bellied and orange-bellied frogs moved less than 20 m over 1 year.
- *Carry out translocations and captive breeding programs.* Translocations are often considered for species with very restricted ranges. Because the orange-bellied frog is so restricted, egg clutches were introduced in the year 2000 to two neighbouring swamps (within 2 km) to the east of its distribution as insurance against habitat loss.

By completing these recovery actions, many additional biodiversity benefits were accomplished. The distribution of the white-bellied frog coincides with a corridor of native vegetation in a generally cleared landscape that links two well-vegetated areas – the Blackwood Plateau and the Leeuwin-Naturalist Ridge. These systems harbour many unusual species, including the distinctive locally endemic giant rush *Reedia spathacea* (Plate 24.4) and many previously unknown aquatic invertebrates (which together make up a threatened ecological community in their own right). Several species of declared threatened or priority flora (*Anthodium junctiforma*, *Acacia tayloriana* and a new species of *Chamelaucium*) and two species of threatened mammal (the native mouse *Pseudocheirus occidentalis* and the western quoll *Dasyurus geoffroii*) occur in the area. There are also numerous threatened aquatic invertebrates including several species of highly restricted freshwater crayfish (*Engaewa* spp.) in the immediate area of the distributions of these two frogs.

The total cost of recovery is not cheap. Apart from the costs borne by the management agencies in terms of staff time, it was estimated that about $42 000 (1993 dollars) over 10 years was used to fund activities associated with the conservation of the orange-bellied frog. The funding required for activities associated with the white-bellied frog was much greater – about $301 000 over 10 years (in 1993 dollars). Much of this was used to fund the many kilometres of fencing now protecting the creek lines occupied by the white-bellied frog.

A new recovery plan prepared during 2006 provided an opportunity to evaluate and review the performance of the previous actions and focus attention on actions required for the next 5–10 years. Although all actions have been addressed by a committed and able recovery team, the program is nevertheless a long way from achieving down-listing of the white-bellied frog, and only making slow progress towards this for the orange-bellied frog.

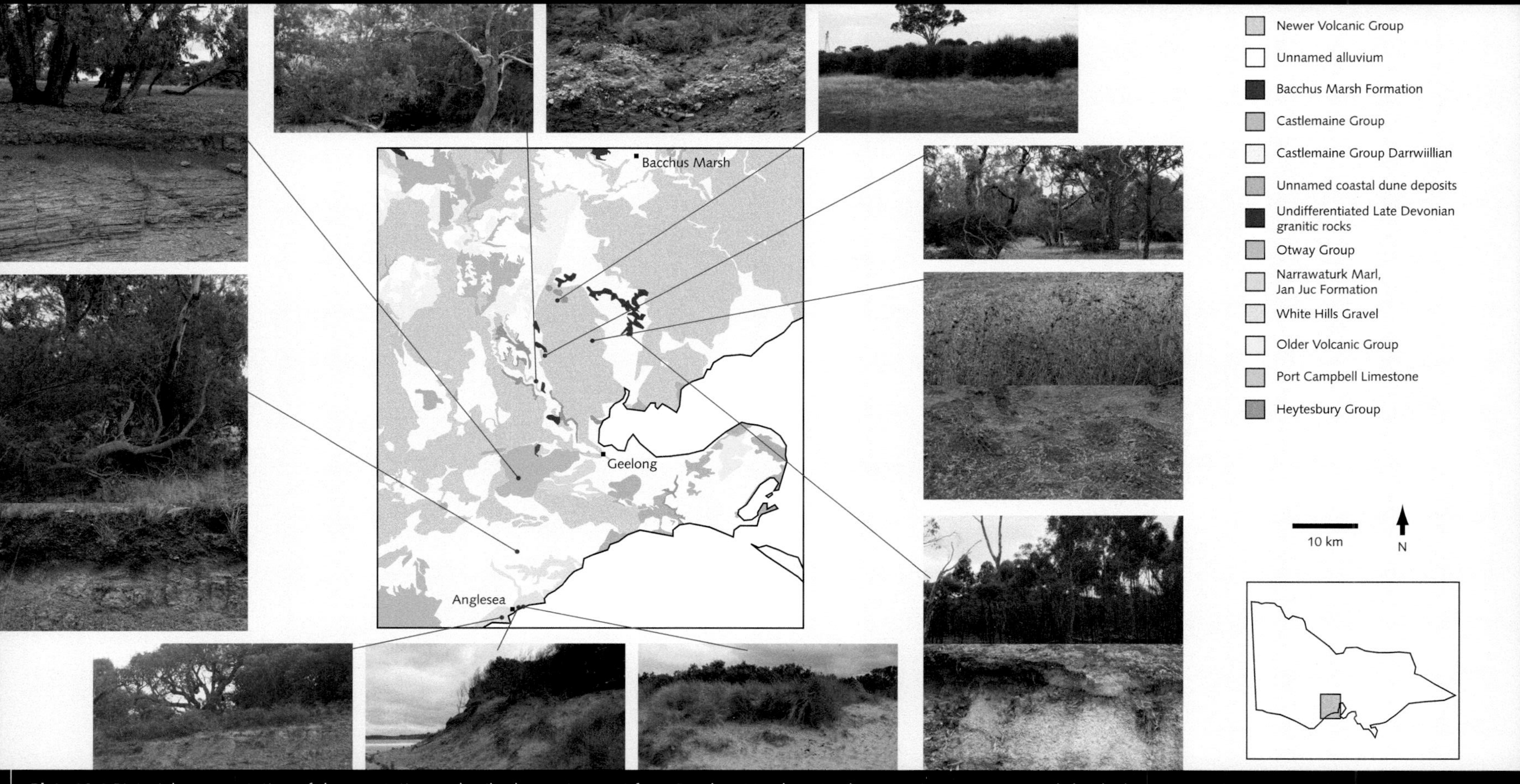

Plate 22.1 Pictorial representation of the vegetation and soils along a transect from Bacchus Marsh to Anglesea in Victoria. From top left (clockwise): *Eucalyptus viminalis* woodland over loamy duplex soil on Cretaceous arkosic sediments; *E. camaldulensis* riparian vegetation on the Moorabol River growing on recent alluvium; podsolic soil on Tertiary sediments over basalt near Bacchus Marsh; lignum (*Muehlenbeckia cunninghamii*) in depressions on basalt-derived soil; *E. viminalis*, *Casuarina littoralis* woodland on alluvial material, probably Tertiary sediments, above the Moorabool River valley; *Themeda australis* grassland on basaltic soil; plantation of *E. cladocalyx* on sandy podsolic soil derived from granite – You Yangs, north of Geelong; marram grass and acacia on sand dunes – Anglesea River mouth; *Leptospermum laevigatum* and marram grass, Anglesea foreshore; *Eucalyptus baxteri* and heathland over podsolic soil developed on lateritised Tertiary sediments – Urquarts Bluff, west of Anglesea; *E. viminalis*, *Banksia marginata* woodland over gravelly podsolic soil developed from Tertiary sediments north of Anglesea. (Source: P. Ladd and M. Garkaklis)

Plate 22.2 Rainforests of the Southern Hemisphere.
(a) North Queensland complex closed forest, Atherton Tableland, Queensland.
(b) Conifer (*Araucaria*) dominated closed forest, Bunya Mountains, central Queensland.
(c) Closed forest on New Caledonia with *Araucaria* emergents.
(d) Podocarp-dominated closed forest, south island, New Zealand.
(e) *Nothofagus cunninghamii* closed forest, Otway Ranges, Victoria.
(f) Alerce (*Fitzroya cupressoides*) closed forest, Largo Sargasso, Chile
(Source: a–f P. Ladd)

Plate 22.3 Woodlands.
(a) Beech (*Fraxinus*) deciduous forest, Apennine Mountains, Italy.
(b) Temperate woodland dominated by salmon gum (*Eucalyptus salmonophloia*), Coolgardie, Western Australia.
(c) Savanna dominated by boabs (*Adansonia gregorii*), north-western Western Australia.
Inset – Tasmanian devil (*Sarcochilus harrisii*) common in Tasmanian open forests.
(d) Taiga-like vegetation in Tasmania, Emergent Athrotaxis over alpine heath, Tarn Shelf, Tasmania.
(e) Koala (*Phascolartos cinereus*), a common tree-dwelling marsupial of open forest in south-eastern Australia.
(f) Tall open forest in south-western Australia dominated by karri (*Eucalyptus diversicolor*) and red tingle (*E. jacksonii*).
(Source: a–f P. Ladd)

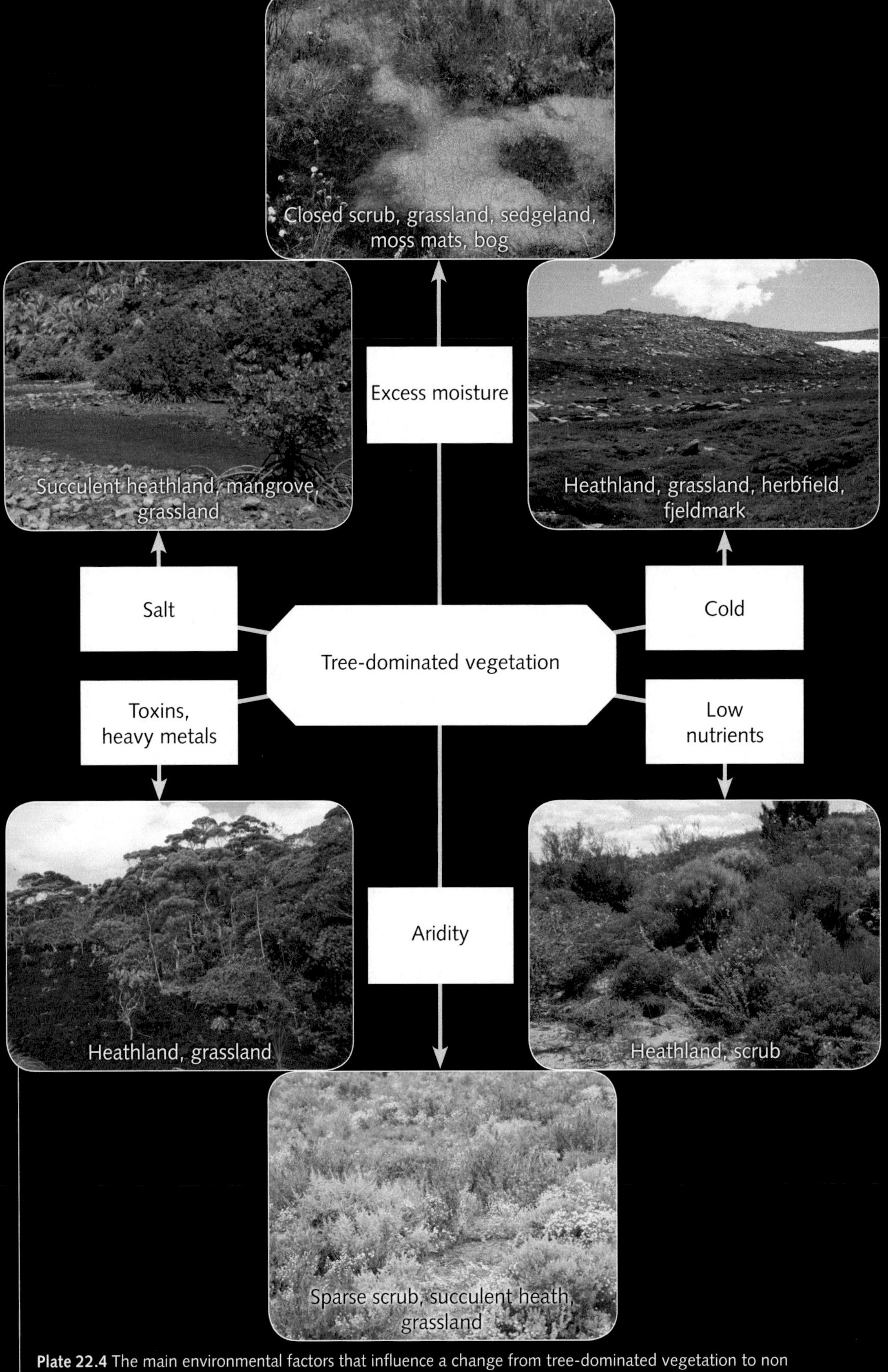

Plate 22.4 The main environmental factors that influence a change from tree-dominated vegetation to non tree-dominated forms. (Source: P. Ladd)

Plate 22.5 Heathlands and grasslands.
(a) Open heath in Fitzgerald National Park, south-western Australia. The large shrub is *Hakea victoriae*.
(b) Open vegetation dominated by mallee eucalypts and porcupine grass (*Triodia*), north-western Victoria.
(c) Arid-zone water course riparian vegetation with *Eucalyptus camaldulensis*, near Flinders Ranges, South Australia.
(d) Spinifex (porcupine) grassland in central Australia, showing the typical hummock form of the grasses.
(e) Fjeldmark (open low heathland) on Mt Kosciuszko, New South Wales.
(f) Open heathland of bolster plants, alpine area of Mt Field, Tasmania.
(g) Tussock grassland of Stipa, coast of Flinders Island, Bass Strait.
(h) Open succulent heathland of *Maireana* and *Sclerolaena*. The red plant is the weed *Rumex vesicaria*, common in inland Australia, New South Wales near Broken Hill.
Inset – Sturt's desert pea (*Swainsona formosa*).
(Source: a–h P. Ladd)

Plate 23.1 Granite outcrop vegetation in the Western Australian wheat belt. Inserts clockwise from right – foliose lichen mosaic on granite, moss and herbfield extending over lichen-covered granite, gnamma hole on granite, the rare shrub *Verticordia staminosa* over a rock pool on granite in spring, *Drosera* herbfield in shallow depression in spring. (Source: P. Ladd)

Plate 23.2 Plant adaptations to cold environments.
(a) Fjeldmark heath, Snowy Mountains, New South Wales.
(b) Cushion plant moor, Tarn Shelf, Tasmania.
(c) Timber line above *Nothofagus* forest, South Island, New Zealand. (Source: a–c P. Ladd)

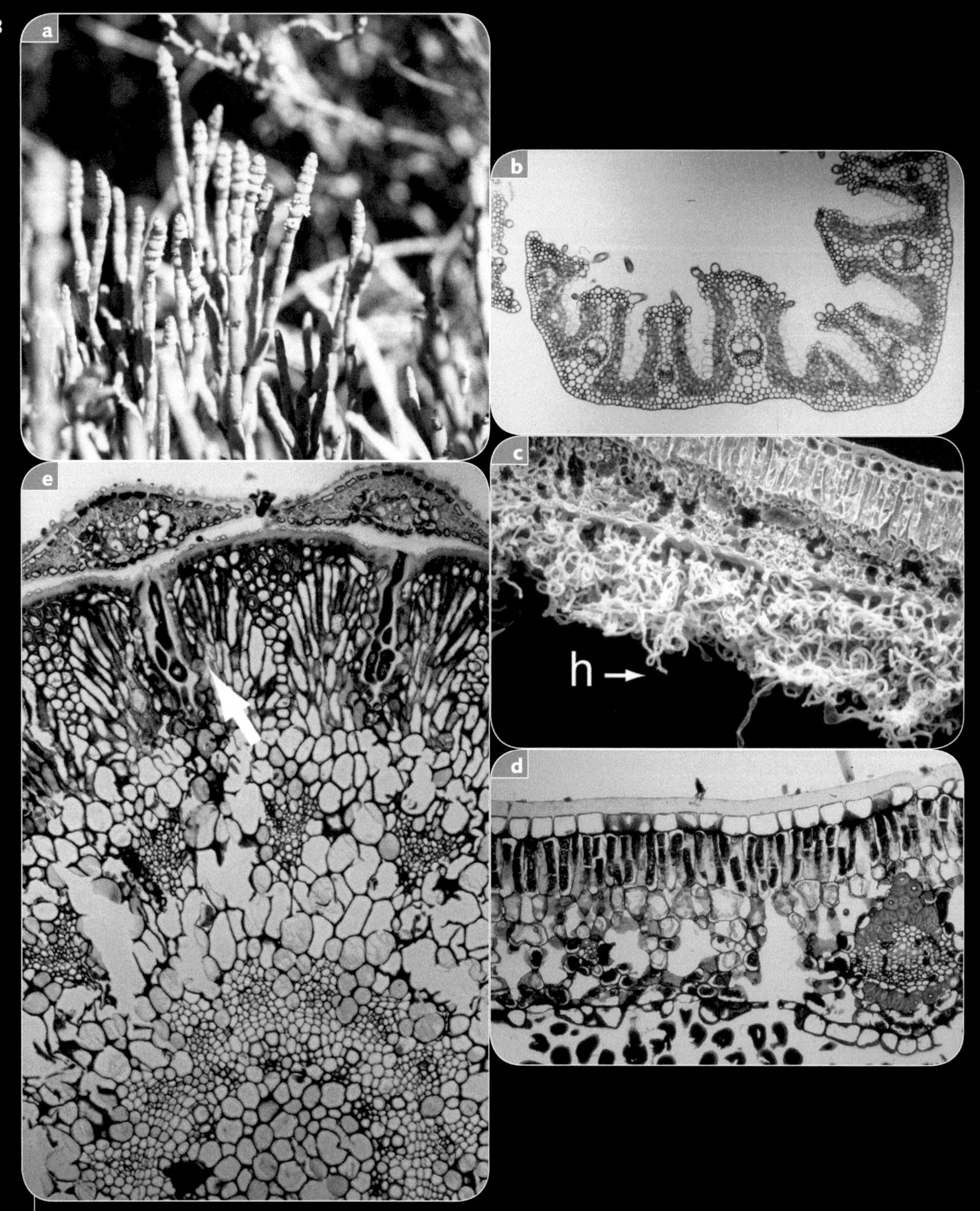

Plate 23.3 Plant adaptations for cooling and for conserving water. (a) *Sarcocornia quinqueflora* stem succulent in coastal saltmarsh. (b) Transverse section of a folded grass leaf from central Australia. (c) Scanning electron microscope image of a transverse section of a *Banksia spinulosa* leaf with dense abaxial hair (h) layer. (d) Transverse section of *Hakea* leaf showing a thick adaxial cuticle at the top of the leaf and sections of abaxial hairs (dark blue) below the leaf. (e) Transverse section of the stem of *Casuarina cunninghamiana* showing outer leaf tips and the photosysnthetic tissue of the outer stem (which is really the leaves fused to the stem). Stomates are located in deep grooves (arrow), which represent the gaps between edges of the leaves fused to the stem. (Source: a–e P. Ladd)

Plate 23.4 Plant adaptations for nutrient intake. (a) Pitcher plant (*Nepenthes* sp.) pitchers. (b) Pitchers of *Cephalotus follicularis* (Albany pitcher plant). (c) Densely hairy dauciform roots of a Restionaceae plant. (d) Root web showing root hairs, from under a stone on alluvial sediments. (Source: a–d P. Ladd)

Plate 23.5 In many species of Proteaceae (banksia family) such as this *Hakea prostrata*, 'proteoid roots' or 'cluster roots' trap soluble nutrients leaching from the litter horizon and secrete organic acids that release phosphorus from insoluble compounds for uptake by the root. White scale bar (bottom right) indicates 25 mm. (Source: M. Shane)

Plate 23.6 The mountain pygmy possum spends up to 20 days in torpor between bouts of activity, adopting a 'furry ball' position with the tail curled and ears tucked in so there is minimal surface exposed to the air. (Source: R. Armistead)

Plate 24.1 The white-bellied frog (*Geocrinia alba*) in its breeding burrow. (Source: G. Wardell-Johnson)

Plate 24.2 Fenced habitat of the white-bellied frog (*Geocrinia alba*) on private land. (Source: G. Wardell-Johnson)

Plate 24.3 The orange-bellied frog (*Geocrinia vitellina*) is one of the most restricted vertebrates known on mainland Australia. (Source: G. Wardell-Johnson)

Plate 24.4 The threatened *Reedia spathacea* ecological community, which occurs in the same areas as the orange- and white-bellied frogs. (Source: G. Wardell-Johnson)

Plate 24.5 Examples of rare Australian plant species.
(a) *Dryandra mimica*.
(b) *Eucalyptus caesia* (flower detail and whole tree).
(c) *Lambertia rariflora*.
(d) *Banksia goodii*.
(Source: a–d G. Wardell-Johnson)

Plate 24.6 Rare, attractive or unusual species are desirable to trade, with conservation implications. This has led to attempts to control both the source and destination of the trade. Two Australian examples are the (a) Albany pitcher plant (*Cephalotus follicularis*) and (b) north Queensland anthouse plant (*Myrmacobius* species). (Source: a, b G. Wardell-Johnson)

Plate 24.7 Red tingle (*Eucalyptus jacksonii*), a locally common species in generalised habitat (granitic terrain in the high and least-seasonal rainfall area of south-western Australia), restricted geographically (to within 10 km of the coast in the southern south-west of Western Australia). (Source: G. Wardell-Johnson)

Plate 24.8 The quokka (*Setonix brachuris*) is a locally common species in specialised habitat (dense riparian vegetation) that is widespread (south-western high rainfall area of Western Australia). (Source: G. Wardell-Johnson)

Plate 24.10 The arum lily is a native of South Africa, but is a serious introduced weed in Australia where it thrives along creek lines, especially in south-western Western Australia. (Source: Lochman Transparencies)

Plate 24.9 Bell miner–associated dieback (BMAD) of trees is associated with increased populations of the native bird, the bell miner (*Manoria melanophrys*). It was once thought to be a localised phenomenon, but it is now causing dramatic change in many thousands of hectares of forest in north-eastern New South Wales. (Source: G. Wardell-Johnson)

Plate 24.11 Fire is a natural and important component of the ecology of many environments. However, where fire is uncommon even a single fire can be a devastating event, as when it occurs in a peat swamp during a dry season. Substantial areas of peat can be destroyed. (Source: G. Wardell-Johnson)

Plate 24.12 Land clearing and drainage of swamps has reduced the habitat of the rare white-bellied frog (*Geocrinia alba*). (Source: G. Wardell-Johnson)

Plate 24.13 Conserving a threatened species may involve intervention to remove a weed or other competitor. This area of national park in northern NSW has become dominated by introduced bitou bush (*Chrysanthemoides monilifera*), so spraying is necessary for control. (Source: G. Wardell-Johnson)

Plate 25.1 The original Lake Pedder in south-western Tasmania. Sometimes old photographs are all that is left of original landscapes. (Source: Tasmanian Inland Fisheries Commission)

Plate 25.3 The highland Lake Oberon. (Source: Brett Mawbey)

Plate 25.2 The Pedder galaxias. (Source: Ron Mawbey)

Plate 25.4 A new use for an old timber mill. This timber mill in Palgarup, Western Australia, is no longer used to cut timber, but has a new use supplying the plants for revegetation schemes. (Source: Grant Wardell-Johnson)

Plate 26.1 (a), (b) River restoration: community groups are transforming highly modified drainage channels into living stream ecosystems. (c) River restoration: humans have generally sought to simplify river systems in the past by channelling and straightening. Restoration efforts are now under way to increase complexity and structure in the river systems, such as by using constructed log dams as illustrated. (Source: a–c Richard Hobbs)

Plate 26.2 Woodland restoration in the Western Australian wheat belt. Intact woodland with a diverse understorey (a) can be degraded by continued grazing so that there are few native species left and (b) bare, compacted soil prevents any regeneration. With effective restoration techniques, tree and shrub species can be re-established (c). (Source: a–c Richard Hobbs)

Chapter summary

1. Conservation biologists are usually trained in the natural sciences, but frequently need to work with people from the social sciences – philosophy, history, business, administration, psychology, law, sociology, anthropology – to achieve their goals.
2. There are strong utilitarian, aesthetic, ecosystem services and ethical arguments for taking action to prevent extinction caused by human activities.
3. Objections to conservation include arguing that extinction is a natural process, that many species may be redundant in providing ecosystem services and that curtailing natural resource use harms the economy. Rejoinders include the concern that contemporary extinction rates greatly exceed background extinction rates, that human engineering gives strong examples of the value of redundancy in preventing system failure, and that changes to natural resource use can stimulate other industries or ensure that wider social benefits flow from using the resource.
4. Local communities value participation in actions for the public good, such as conservation or landcare activities. This should be seen as an asset by professional and/or government agencies.
5. Governments at all levels can participate in biodiversity conservation through enacting legislation, signing and enforcing international agreements, treaties or conventions, and funding scientific research and community conservation initiatives. At the Commonwealth level, Australia's most significant legislation is the *Environmental Protection and Biodiversity Conservation Act 1999*.
6. Recovery plans are information and management documents that set out actions to assist threatened taxa and ecological communities within a planned and logical framework.

Key terms

biodiversity
conservation biology
ecosystem services
instrumental
intergenerational equity
precautionary principle
recovery plan
Red List of Threatened Species
redundant
utilitarian

Test your knowledge

1. A conservation biologist realises that a small (3 mm) roundworm (nematode) requires urgent attention to prevent it from becoming extinct. Which of the following arguments would the conservation biologist most likely *not* use to support the case?
   A It is aesthetically beautiful.
   B It performs an important role in the functioning of the soil ecosystem.
   C It does not occur in the fossil record.
   D Like all other species, it deserves our moral consideration.
   E It is found only on Lord Howe Island.
2. In question 1 (above), with which arguments would the conservation biologist most likely make the strongest case, and why?
3. It is 1991. The Tasmanian government rings you and asks for your evaluation

of the Pedder galaxias recovery plan. What are the five most important points you make in your report? Explain your reasoning for each point.

4. It is 2008. The Tasmanian government rings you and asks for your evaluation of the success of the Pedder galaxias recovery plan. What are the five most important points you make in your report? Explain your reasoning for each point.
5. You are an employee in a government's Nature Conservation Department working on the conservation of a woodland bird community, and your manager asks you what proportion of your work effort should go into working with local landowners. How would you respond?

Debate with friends

Professor Hugh Possingham, a leading conservation biologist, has claimed that species that are nearly extinct are too expensive to save and that time and money is better spent rescuing those with a better chance of survival. He has said: 'The correct metric for evaluating a project is the benefit divided by the cost. Which doesn't always mean the most endangered species are going to be funded because they are too expensive. The Californian condor has been recovered from the brink of extinction but it cost US$10–20 million. That $20 million could have been used to secure large tracts of rainforest to save hundreds of species'. Some animals are more expendable than others, he claimed. Lose another species of beetle or grasshopper and the cost would be low. Do you agree?

Further reading

Bradstock, R., Auld, T.D., Keith, D.A., Kingsford, R.T., Lunney, D. and Silversten, D.P. (1995). *Conserving biodiversity: threats and solutions*. Surrey Beatty and Sons, Chipping Norton, New South Wales.

Chivian, E. and Bernstein, A. (eds) (2008). *Sustaining life: how human health depends on biodiversity*. Oxford University Press, Oxford.

Elliott, L. (1998). *The global politics of the environment*. Macmillan, Hampshire, UK.

Ellis, R. (2004). *No turning back: The life and death of animal species*. HarperCollins, New York.

Hutton, D. and Connors, L. (1999). *A history of the Australian environment movement*. Cambridge University Press, Cambridge.

Jacobson, S.K. (1999). *Communication skills for conservation professionals*. Island Press, Washington DC.

Kirkpatrick, J.B. (1998). 'The politics of the media and ecological ethics' *Ecology for everyone: communicating ecology to scientists, the public and the politicians*. (Wills, R. and Hobbs, R.J. eds) pp. 36–41. Surrey Beatty and Sons, Chipping Norton, New South Wales.

Konteleon, A., Pascual, U. and Swanson, T. (2007). *Biodiversity economics: principles, methods and applications*. Cambridge University Press, Cambridge.

26

Redressing the problem – environmental restoration

Richard Hobbs

Gondwanalink and Atherton to the Alps

In the south of Western Australia, scientists and land managers stand in a paddock pondering a map showing remaining native bushland. Despite some big, globally important reserves such as the Stirling Ranges and the Fitzgerald River National Park, much of the landscape lacks the original native woodland and shrubland. In its place is farmland, often degraded by erosion and salinisation. This is an area in trouble, losing productive agricultural land and conservation value.

The people pondering the map are part of the visionary project Gondwanalink, in which community groups and non-government organisations are developing a grand vision for sustaining agriculture and retaining the conservation heritage of the remaining bushland. Good areas of bushland on private land are bought and put under effective management, areas of degraded farmland are revegetated, and diverse agricultural enterprises are established. Gondwanalink encompasses everything from the tall forests of the south-west to the semiarid woodlands of the goldfields and beyond.

In eastern Australia the similar 'Atherton to the Alps' venture will establish a corridor from the Victorian Alps across 2800 km to the Atherton tablelands in Queensland. It has the support of the New South Wales, Queensland, Victorian and Australian Capital Territory governments, and landholders who wish to conserve parts of their properties can sign conservation agreements. Nobody will be compelled to relinquish land. However, support from landholders is essential because national parks alone will not provide the necessary linkages along the corridor.

A strong motivation for both Gondwanalink and Atherton to the Alps is to provide corridors of connected habitat so that animals and plants faced with climate change can migrate along them. Although there is much uncertainty about how organisms, especially plants, will use the corridor, it is important to provide the opportunity. The projects are examples of the growing emphasis on repair of past damage. Because of ongoing modification of the Earth's ecosystems, it is increasingly important that we learn how to return damaged systems to a more functional or effective state.

Chapter aims

In this chapter we define environmental restoration, describe the different types of ecosystem disturbance that can occur and outline the ecological theories that explain how communities respond to disturbance. The steps involved in restoring disturbed communities are described, with particular attention to ecological processes, vegetation and fauna. We conclude with an assessment of the importance of policy issues in restoration and the relationship between restoration and conservation.

What is environmental restoration and why is it necessary?

Environmental restoration is the repair of damage to the environment. It includes a wide range of activities such as bringing a degraded ecosystem back to an original state, stabilising a landscape and increasing its utility, using plants and microbes to reduce site toxicity and stopping damaging activities to allow an ecosystem to recover without further intervention (Table 26.1). Although in many areas environmental management maintains environmental quality and prevents ecosystem degradation, we face an increasing legacy

Table 26.1 Definitions for concepts used in restoration ecology (Source: After Aronson, J., Floret, C., Le Floc'h, E., Ovalle, C., and Pontanier, R. (1993). 'Restoration and rehabilitation of degraded ecosystems in arid and semiarid lands. I. A View from the South.' *Restoration Ecology* 1: 8–17 and Walker, L.R. and Del Moral, R. (2003) *Primary succession and ecosystem rehabilitation.* Cambridge University Press)

Term	Definition
Reclamation	Actions that stabilise a landscape and increase the utility or economic value of a site. Usually involves amelioration that permits vegetation to establish and become self-sustaining. Rarely uses indigenous ecosystems as a model
Reallocation	Management or development actions that deflect succession of a site from one land use to another, with the goal of increased functionality
Rehabilitation	Actions that seek to repair damaged ecosystem functions quickly (particularly productivity). Indigenous species and ecosystem structure and function are the targets for rehabilitation
Bioremediation	The use of plants and microbes to reduce site toxicity (a special form of rehabilitation)
Restoration (*broad sense*)	Actions that lead to the full recovery of an ecosystem to its pre-disturbance structure and function
Restoration (*strict sense*)	Actions that seek to reverse degradation and to direct the trajectory toward one aspect of an ecosystem that previously existed on the site

of ecosystems damaged by human activities. For example, mining causes intense, but localised ecosystem loss, and agriculture and pastoralism alter ecosystems on massive, even continental, scales. Part of the process of developing sustainable systems is repairing this damage through environmental restoration. Widely practiced by the mining industry for decades, it has developed rapidly in recent years and is applied wherever landscapes and ecosystems are degraded by human activities. More recently, the concept has broadened to include measures taken to reduce degradation in the first place (prevention and conservation).

Environmental restoration applies concepts from many disciplines, particularly ecology, conservation biology and environmental engineering. The key causes of degradation must be identified and clear goals set for implementing and evaluating restoration programs. As with conservation efforts (Chapters 24 and 25), restoration also depends upon adequate financing and incentives stemming from policy decisions at local, state and national levels. Industry, government and community groups must all be involved in setting the goals and implementing the restoration, because projects lacking such wide support will fail. We will begin our study of restoration by considering how ecological communities respond to disturbance, both natural and human-made.

Natural and human-induced disturbance of ecosystems

Disturbances, or episodic destruction or removal of ecosystem components, feature in many ecosystems (Chapter 17). They range from localised events such as animal diggings or individual tree falls to large events such as bushfires, storms and floods. Disturbance often initiates massive ecosystem change before regeneration or recovery. An understanding of the natural roles of disturbance in ecosystems is needed before considering the impacts of disturbance caused by humans and how responses to natural disturbance can be used in restoration.

Inertia and resilience

Ecosystems have a degree of resistance to disturbance, termed inertia. When the disturbance is large enough, this inertia is overcome and the system changes. The degree to which the system recovers from that change is termed resilience. Resilience is made up of a number of characteristics, borrowed from engineering and the physical sciences (Figure 26.1):

- *elasticity*: the rate of recovery following disturbance
- *amplitude*: threshold level of strain beyond which return to the original state does not occur
- *hysteresis*: difference between path in response to disturbance and recovery path
- *malleability*: difference between final recovery level and pre-disturbance level

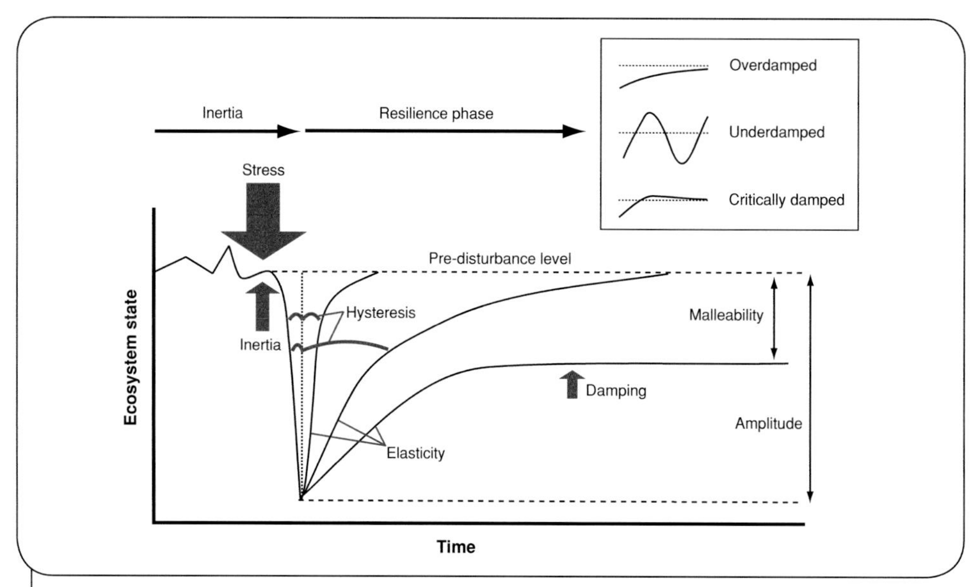

Figure 26.1 The response of an ecosystem to stress and disturbance, illustrating the different elements describing system resilience. (Source: Richard Hobbs)

- *damping*: the degree to which the restoration path is altered by forces that affect the restoration. An overdamped system will take a long time to reach equilibrium, but will not oscillate. An underdamped system will approach equilibrium rapidly, but will overshoot and oscillate. A critically damped system equilibrates more rapidly (it is most resilient).

It is important to note that these concepts relating to resilience are equilibrium concepts, and assume that the ecosystem will return to its original state. However, as we will see below, not all ecosystems can be assumed to be equilibrium systems and a disturbance may cause an ecosystem to redevelop to something different from what was there previously.

Ecological succession

The term 'succession' describes the sequence of species and groups of species present at various times since disturbance (Chapter 17). Pioneer species appear early and tolerate open, often harsh, conditions. Later successional species are either slower growing or appear only after pioneer species modify conditions. Species may appear in a recognisable sequence, or all species may appear shortly after disturbance, but grow and/or assume importance at different rates. Species may facilitate, inhibit or tolerate other species.

Succession theory is based largely on the postglacial landscape of the temperate Northern Hemisphere, where most of the flora (and fauna) are colonising species. These ecosystems experienced catastrophic disturbance from glaciation and volcanism and are therefore composed of species that can cope with disturbance. Australian ecosystems have not experienced glaciation or major volcanic activity in recent geological time, although the changes in fire history 40 000 years ago may have been significant (see Chapter 7). So whereas a predictable post-disturbance recovery period occurs in some Australian ecosystems, in others recovery is unpredictable and results in different ecosystem compositions. This is why restoration of extremely degraded areas such as mine sites in Australia often attempts to seed directly the plants intended to be present in the long term, rather than planning a successional reintroduction.

Thresholds and alternative stable states

Whether a predictable post-disturbance recovery sequence or a change in species composition occurs may depend on the arrival of particular species in the system, the method of management imposed, or climatic and disturbance events during recovery. The ecosystem may reach a state from which little further change occurs. This state is known as an alternative stable state, and indicates that the ecosystem has developed a relatively stable structure or composition that is different from what was present prior to disturbance. An alternative stable state indicates the operation of system thresholds that must be crossed before further system change occurs. The identification of system thresholds is important in assessing appropriate restoration measures.

Impacts of human activities on ecosystems

Human activities affect ecosystems by either modifying the original disturbance regimes (e.g. by changing fire or grazing regimes), or by adding new disturbances such as extensive clearing or introduced species. Sometimes this human disturbance pushes ecosystems beyond the limits of their resilience and active restoration is required.

Ecosystems degrade when human use or alteration changes ecosystem structure and/or function beyond acceptable limits. For instance, vegetation structure may be altered so that it no longer provides adequate habitat for fauna, or the ecosystem may no longer provide clean water, food or fibre. Degradation may change biodiversity or the system's physical or chemical characteristics.

Even with correct assessment and treatment of the problems leading to degradation, it may be impossible to return a system completely to its pre-disturbance state. Natural ecosystems are dynamic and so constantly changing, so it is unlikely that they will return to exactly the same composition and structure as the pre-disturbance system. It is therefore important to set realistic restoration goals that take into account the dynamic nature of ecosystems.

Key steps in environmental restoration

Several steps should be considered in any restoration program to achieve useful outcomes (Figure 26.2). These include setting clear goals with associated success criteria, correctly

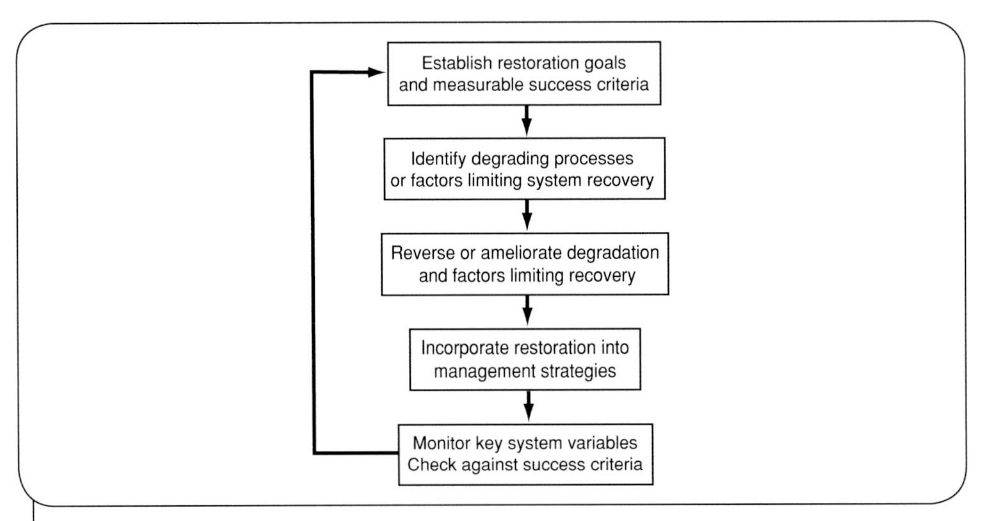

Figure 26.2 Key steps in any restoration program. Each of these steps is essential to achieving successful outcomes, and although they are illustrated as forming a linear sequence here, they in fact interact strongly and the process is iterative. (Source: Modified from Aronson, J., Floret, C., Le Floc'h, E., Ovalle, C., and Pontanier, R. (1993). 'Restoration and rehabilitation of degraded ecosystems in arid and semiarid regions. I. A view from the South.' *Restoration Ecology* 1: 8–17)

identifying the factors limiting system recovery or leading to further degradation, and instigating restoration activities that reverse or ameliorate these factors. These activities should be integrated with broader management objectives and monitored to ensure that progress is made towards the agreed goals.

Setting goals

Many terms are used to describe environmental restoration (Table 26.1 and Figure 26.3). Unfortunately, they are not used consistently and a lot of time can be wasted in discussion about whether a particular activity is restoration or rehabilitation or something else. It is more important to state precisely the actual goals of the activity, as explained below, than to try to decide what best to call the activity.

Clear and achievable goals greatly facilitate choosing restoration options and monitoring progress. Often goals for restoration are set in relation to a particular ecosystem or species composition. Thus a goal for post-mining restoration in the jarrah forest of Western Australia may be the return of the complete forest ecosystem with all the species that were there previously, or it may be simply the stabilisation of mined surfaces and the return of the main tree species. The choice of restoration options needs to be guided by the ecological constraints operating in any given situation, the costs of the different options and the wishes of interested parties such as landholders, government and community groups.

Identifying limiting or degrading factors

The cause of degradation and the factors impeding recovery may be obvious in some cases (e.g. in mine sites), but in others they may be harder to see. They may affect either

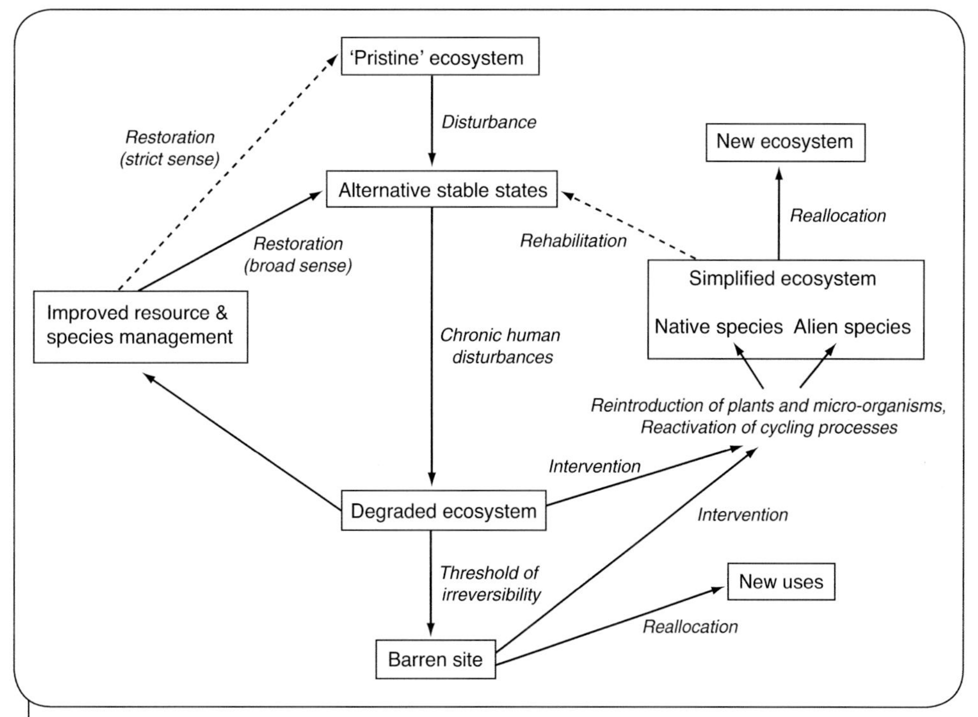

Figure 26.3 Model of ecosystem degradation and possible management responses. (Source: Hobbs, R.J. (1999). 'Restoration of disturbed ecosystems', in *Ecosystems of the world 16. Disturbed ecosystems* (ed. L. Walker), pp. 673–87. Elsevier, Amsterdam)

primary abiotic processes such as nutrient and water retention, or biotic processes such as species recolonisation and survival. Their clear identification is essential to prevent costly mistakes. For example, in a mine-site restoration project, there is no point in replanting the area if there are problems with the stability or chemical composition of the substrate. Similarly, in an area suffering from secondary salinisation through hydrological imbalance caused by vegetation clearance, revegetation of the area may be ineffectual if the broader hydrological condition of the catchment is not also treated.

Incorporating restoration into the management strategy

Once goals are identified and methods determined for overcoming the degrading or limiting factors, the restoration project has to be actually implemented. This involves practical considerations surrounding logistics, budgets and timing of operations, not only early in restoration, but also during ongoing management of the restored site to ensure that the restored system continues to progress toward the restoration goals.

Monitoring progress

Monitoring progress is essential to success, but is often done ineffectively. Restoration projects often have success criteria that meet contractual or legal obligations. Hence tracking progress towards success is important. Monitoring can also identify where the management treatment is unsuccessful and so can be changed. Variables for monitoring must relate to the goals set for the restoration, and be inexpensive and simple to measure. Many of these principles are illustrated in the examples in Boxes 26.1 and 26.2.

Box 26.1 Mine site rehabilitation

The jarrah forest in the south-west of Western Australia is valued for timber production, recreation, protection of water catchments and conservation, especially of flora. Approximately 5% of the forest covers commercial deposits of bauxite, the ore from which aluminium is produced. Open-cut mining for bauxite in the jarrah forest of the Darling Range began over 40 years ago and at present up to 550 ha of forest are mined annually. Rehabilitation of the mine sites has long been a major objective.

Rehabilitation objective

Any restoration project must start with a clear statement of the objective of the program. Most mines in Australia are attempting to return to a native ecosystem. However, many rehabilitation objectives do not incorporate multiple land uses, consider long-term management issues or acknowledge that the rehabilitated ecosystem may not be identical to the original. In the case of the open-cut bauxite mines in the jarrah forest (Figure B26.1.1), the aim is to re-establish a functional forest ecosystem satisfying all the land use requirements normally expected of jarrah forest, including timber production, water catchments, recreation and conservation. Within this broad aim, 5-year improvement milestones are set for achieving better results in difficult rehabilitation areas such as the range of plant species found in rehabilitated sites.

Figure B26.1.1 Mining is essential for the provision of key resources, but also has a dramatic effect on the natural ecosystem. Here bauxite is being surface-mined in the jarrah (*Eucalyptus marginata*) forest of Western Australia. (Source: Richard Hobbs)

Landform design and the soil profile

Mining causes a large-scale disturbance to an ecosystem. Unlike many other types of restoration, mining rehabilitation requires re-establishing the landform and providing a suitable growth medium for plants. This involves the use of heavy machinery and is often the most costly aspect of mining rehabilitation. Open-cut bauxite mining pits are shaped to create a landform blending with the surrounding forest before biotic elements are restored.

Establishing vegetation and fauna habitat

There are many ways to establish vegetation on rehabilitated mine sites including topsoil, seeding, planting, mulching and complete habitat transfer (Figure B26.1.2). Each option has advantages and disadvantages associated with it, and the choice of the best method must be related back to species selection and identification of the rehabilitation objective. Fauna generally return to rehabilitated mine sites as vegetation develops. Vertebrate and invertebrate fauna can therefore be used as bioindicators of the success of mine site rehabilitation.

Stockpiling and returning topsoil is a key feature in rehabilitating bauxite mines. This improves success in establishing native plants from seed or from nursery-grown stock. Low soil-nitrogen levels may limit some plants, but this can be corrected by growing legumes that fix nitrogen in association with soil bacteria (see Chapters 17 and 23). Ongoing research concentrates on ideal ways to establish plants, including 'recalcitrant' species that prove difficult.

Figure B26.1.2 Mine site rehabilitation involves ensuring that the mined surface is stable and that conditions are suitable for the re-establishment of vegetation. Here at Alcoa's Huntly Mine in Western Australia, the restoration aims to return a fully functional jarrah forest. The recently rehabilitated area in the foreground is modelled on the unmined forest in the background. (Source: Richard Hobbs)

continued ›

Box 26.1 continued ›

Table B26.1.1 Spider species richness, species diversity and community similarity at three rehabilitated bauxite mines and two control sites in the jarrah forest. (Source: Adapted from Mawson, P.R. (1986). 'A comparative study of arachnid communities in rehabilitated bauxite mines.' *Western Australian Institute of Technology, School of Biology Bulletin* 14: 1–46)

	Rehabilitated bauxite mine sites			Control jarrah forest sites	
Site	1	2	3	4	5
Age (years)	1	2	8	–	–
Species richness	13	17	36	27	47
Species diversity	0.81	0.60	0.98	1.04	1.18
Similarity to site 5	0.31	0.35	0.48	0.41	–

Fauna are monitored for recovery. Predatory invertebrates such as spiders are useful indicators because a wide range of species in good numbers can re-establish only if there are adequate prey, which in turn require vegetation for food and shelter. Table B26.1.1 shows the results of a study of spider species richness and species diversity at three rehabilitated sites and two control sites. Assessments of the similarity of the spider communities across the sites are also given (see Boxes 17.1 and 17.2 for explanations of species richness, species diversity and community similarity). After 8 years, rehabilitation sites were within the range of the controls for species richness and approaching the controls for species diversity. The older rehabilitated sites were also as similar to the control sites as the control sites were to each other.

Long-term management of rehabilitated mine sites

It is critical that rehabilitated mine sites are managed after the establishment phase. This may include weed control, dieback assessment, fire management and monitoring of nutrient cycling. The mining industry is attempting to define criteria that assess when they have successfully rehabilitated the land and can relinquish leases. These criteria must consider the long-term stability and management of rehabilitated areas.

Box 26.2 River restoration

Human damage to natural waterways is widespread and large sums are spent globally on restoration projects (Plate 26.1). However, few of these are evaluated to determine their success so the best methods to use are still unclear. Here, general issues associated with river restoration are outlined, followed by an assessment of restoration projects in Victorian rivers.

Rivers are open, dynamic systems intimately linked with the immediate riparian area and the broader catchment. Although they are continuous, open systems, natural rivers are also patchy, consisting of many different habitats. River restoration can often be rapid because of the open nature of the river system, with species recolonising quickly once the degrading influences have been dealt with. On the other hand, river restoration also depends on activities in the broader catchment.

The major human disturbances to river systems are:

- *changes in flow regimes*. Water diversion, dams, weirs, culverts, paving, channels or deforestation can change flow regimes. Changes in channel morphology and the destruction of riparian vegetation can also alter flow regimes. These can alter flow velocities, change water levels and affect water quality, sediment deposition, temperature, light and habitat quality. Restoration requires the return of more natural flow rates, removal of impeding structures or the installation of retention ponds or other structures. Reconnection of a river with its flood plain or the reinstatement of riffle/pool sequences may also be necessary.
- *pollution*. Point and non-point pollution can lead to changes in pH, turbidity, nutrients, oxygen levels and concentrations of other chemicals. Appropriate restoration involves reducing chemical, nutrient and sediment inputs – most rivers are fairly resilient and will recover if stresses are removed. Wetlands or buffer strips aid in the uptake of non-point source pollution and in flood control.
- *nutrient enrichment*. Nutrient and organic matter transport from catchments can result in significant nutrient enrichment, which can in turn lead to declining water quality, algal blooms, fish kills and other problems. Restoration involves altering land uses in the catchment and providing buffer strips to filter out nutrients.
- *salinisation*. Stream salinisation is most often caused by altered catchment hydrology brought about by land clearing. This leads to increased salt contents, decline of riparian and in-stream vegetation and loss of diversity. Restoration involves landscape-level treatment of the catchment hydrology and revegetation with salt-tolerant species.
- *erosion*. Erosion can result from the destruction of riparian vegetation, overgrazing, and land clearing or deforestation. Engineered structures can be used to modify stream flows to reduce erosion, and riparian vegetation can be re-established. Improved land-use practices in the riparian and flood-plain areas are required for long-term solutions. For instance, reduced intensity of grazing, fencing of riparian areas and altered logging practices may be required.
- *invasive plants and animals*. Invasive species can be a problem because of weed introductions, livestock grazing, and aquaculture and garden escapes. Restoration usually involves removal or control of the invading species. However, heavy weed infestations may be the result of nutrient enrichment, and without dealing with this problem weed control will be only a temporary success.

Restoration projects in Victorian rivers

Ten regional Catchment Management Authorities (CMAs), established in 1997, manage rivers in Victoria. As part of an extensive international survey of

continued ›

Box 26.2 continued ›

river restoration projects, Shane Brooks and Sam Lake obtained detailed records for the completed restoration projects for four of these areas (Corangamite, Goulburn-Broken, North Central and Port Phillip), covering 2247 projects between 1999 and 2001. It was a difficult task, because many records were lost or incomplete following industry restructuring and poor data archiving. Nearly half the projects were directed at organisms living on the banks of waterways (riparian management), and bank stability, in-stream habitat improvement and channel reconfiguration were also prominent. Although monitoring is critical in determining the success of restoration, only 14% of the total records confirmed that it had occurred (Table B26.2.1).

Table B26.2.1 Types and frequency of restoration projects carried out in Victorian waterways managed by four Catchment Management Authorities (CMAs) between 1999 and 2001 and whether or not they were monitored. (Data from Brooks, S.S. and Lake, P.S. (2007). *Restoration Ecology* 15, 584 91).

Category of restoration project	Percentage of projects[a]	Percentage of projects in each category that were monitored[b]
Riparian management (including livestock exclusion, weed eradication, replanting vegetation)	48.5	17
Bank stability	17	26
In-stream habitat improvement	15	29
Channel reconfiguration	14	11
Stormwater management	3	0
Fish passage	1	5
Water quality management	1	0
Aesthetics/recreation/ education	0.5	0

[a] Based on 2247 records from four CMAs
[b] Based on 1568 records available from two CMAs

Overall, it appeared that despite much effort and good intentions there was little evidence of scientific planning and proper monitoring and evaluation in these restoration projects. Fortunately, the establishment of CMAs in Victoria should improve the situation through mandatory reporting, better archiving of data and increased emphasis on monitoring.

Repairing damaged primary processes

Ecosystem degradation can change an ecosystem's biological component or its abiotic (physical and/or chemical) characteristics. Biotic changes can include loss of species or changes in vegetation structure (spacing of plants, for example), whereas abiotic changes can include changes in hydrology or soil structure or chemistry. Primary ecosystem processes such as the cycling and retention of nutrients, carbon and water may also be damaged, causing more fundamental system changes than simple biotic or abiotic changes. Correct assessment of which ecosystem characteristics have been altered during degradation is essential and reinstating structures and processes to regulate damaged nutrient, water or carbon cycles is often a first step in restoration.

Removal of vegetation cover and degradation of soil structure can influence nutrient cycles through erosion and loss of nutrients, altered hydrology leading to increased run-off, rising water tables, flooding and salinisation. Physical manipulation of the substrate (e.g. modifying soil structure or surface microtopography), introducing physical barriers or revegetation may be necessary. The revegetation may require remediation of the chemical composition of soil or water.

Restoration of primary processes requires attention to the biotic and abiotic components of ecosystems. Attempts to re-establish vegetation on areas where primary processes have not been repaired are likely to fail. On the other hand, attention to the biotic component can also speed up repair of primary processes. For instance, nutrient capture may be enhanced by the re-establishment of mycorrhiza (fungi living in close association with plant roots; see Box 10.6) or the inclusion of plant species with an array of different root architectures.

Directing vegetation change

For natural revegetation to occur following disturbance, the desired plant propagules must arrive by dispersal or develop from an on-site source. The site has to be suitable for the establishment, growth and reproduction of the species. When this happens and a community of plants and animals reassembles, it is referred to as an autogenic recovery. It is the cheapest form of restoration. In some situations, the colonisers may include undesirables: weeds or species that prevent development of the vegetation to a desired state. In others, there may be little or no colonisation because the site is unfavourable or because the native species have low dispersal capabilities. In these cases, assisted recovery is necessary and desired plants are introduced and managed.

Restoration frequently accelerates vegetation development or directs its course to a predetermined goal. In order to do this, site characteristics, plant colonisation and survival can all be manipulated. Where seeds are mobile or dispersed by birds and other animals, they may be effectively dispersed without further intervention. However, this process may be too slow and seed may need to be introduced. In extreme cases, seed germination may be too unreliable and planting of seedlings may be necessary. Once species are established at a site, continued vegetation development depends on their survival and how they interact with other species. Survival can be increased by watering, reducing erosion and reducing damage from herbivory and disease. The development of a functioning biotic community depends on achieving a good mix of species with different life forms such as trees, shrubs and grasses (depending on which sort of community is being restored) and the development of species interactions such as mycorrhizal associations, pollination and seed dispersal.

How biotic communities are built is a central question in restoration ecology, and recently there have been many attempts to consider what factors affect the characteristics of the developing community. Such factors include the timing and extent of arrival of different species on the site, their physiological tolerances, competitive abilities and interactions with other species. These factors may be considered a set of 'assembly rules' governing what species persist on a site. Restoration efforts guided by these assembly rules aim to modify the abiotic and biotic conditions to allow colonisation and persistence of species that will achieve restoration goals.

Fauna and restoration

Restoration projects generally focus on revegetation, assuming that this will provide habitat for fauna to return. However, it is not always clear that this is the case, and closer consideration of the role of fauna in restoration is needed.

Particular animals, known as ecosystem engineers, structure ecosystems and modify ecosystem processes. The removal or introduction of these animals, for example beavers in North America and Europe and large herbivores in Africa, can have dramatic ecosystem effects. In Australia, several medium-sized marsupials alter soil structure by digging for food or shelter, burying fallen leaves and increasing water infiltration rates. This is particularly important for restoration, because these marsupials were eliminated from much of their former range through predation by foxes and cats. So a major conservation program in Western Australia (called Western Shield), which focuses on controlling foxes in order to increase or reintroduce marsupial populations, can also be construed as restoring key ecosystem processes.

Landscape-scale restoration

Landscapes are heterogeneous areas, usually hectares or square kilometres in extent, composed of interacting ecosystems or patches. Landscape ecology studies the patterns and processes operating at this scale, focusing on landscape structure, function and change. The spatial distribution and interrelationships between landscape patches determine functions such as movement of species and water and nutrient cycling and energy flow.

Ecosystem modification and use often lead to dramatic changes in landscape structure and function. The most obvious manifestation of this is landscape fragmentation, which occurs when the original vegetation of an area is cleared and transformed for agriculture or other use. The native vegetation is left as remnants of varying size and degree of isolation, resulting in declining species abundances. In addition, landscape processes such as water flows can be dramatically altered, leading, for instance, to changes in wetland habitats or salinisation.

What landscape-scale interventions are required?

Management interventions need to be based on a sound assessment of the causes of degradation. In the case of landscape fragmentation, the causes of species loss and decline are often the loss of habitat and connectivity. Thus replacing habitat and increasing connectivity are frequently goals for landscape restoration. This is best done by using the existing native vegetation as a skeleton on which to build restoration efforts. Additional habitat or buffer strips can be established next to existing remnants and corridors, or other connecting vegetation can be developed. Restoration actions are best planned in relation to the needs of particular species so that specific recommendations on dimensions of habitat and corridors required can be determined. In addition to habitat re-creation, broader scale revegetation and other activities may be needed to reverse or slow down hydrological changes and influence other landscape-level processes.

Box 26.3 describes an example of landscape-scale restoration. Experimental principles can be used to choose between restoration options (Box 26.4).

Box 26.3 Woodland restoration in south-western Western Australia

Woodlands form part of a vegetation mosaic in south-western Western Australia on the Swan coastal plain and inland from the jarrah forest. These woodlands were once extensive, but have now been heavily cleared for urban development and agriculture. Eucalypt woodlands in the agricultural area now exist mainly as small, isolated patches and many are degraded by prolonged stock grazing, weed invasion and other disturbances. In addition, most woodlands occur in lower parts of the landscape and are under increasing threat from rising saline watertables.

Assessing woodland condition and undertaking restoration

Using a simple assessment protocol, it is possible to go to any patch of woodland and determine what factors may be limiting recovery and regeneration. This assessment focuses on the existing condition of the habitat, the availability of plant propagules and the suitability of the substrate for regeneration. From this assessment, appropriate restoration activities are identified. This may include fencing out livestock, manipulating soil structure, controlling weeds and reintroducing native plant material.

continued ›

Box 26.3 continued ›

Landscape-scale restoration

Patch-based restoration activities, as described above, are often necessary in the heavily modified woodland remnants of the Western Australian wheat belt (Plate 26.2). However, on their own they may be insufficient to ensure the continued persistence of woodlands and their biota. Fragmentation and isolation may also disrupt the movement of biota across the landscape. Disruption of catchment hydrological processes by broad-scale clearing also leads to rising saline watertables that threaten woodland vegetation in low-lying areas. The management response to this may include localised engineering activities such as drainage and pumping, but will also entail landscape-scale revegetation to slow the watertable rise. This will involve large changes in vegetation cover in agricultural landscapes, including a greatly increased use of perennial plant species.

Box 26.4 Using experiments to choose restoration treatments

Sometimes restorationists know what species they want to restore, but are not confident about the ideal method to do so. In such cases, initial trials can be designed as experiments to test the effectiveness of different options.

One informative example of the use of experiments to choose restoration treatments comes from studies of the common blue butterfly (*Polyommatus icarus*) in the United Kingdom. Four restoration schemes were available to restore habitat for the butterfly, involving combinations of soil treatments and sowing of different grasses:

1. control (C): the plot was disturbed periodically by cultivation, but not otherwise sown
2. mound (M): the soil was modified with an elongated mound of fine limestone to make the soil alkaline to encourage specific plants
3. short (S): the plot was sown with a mixture of short-growing grasses and wildflowers
4. tall (T): the plot was sown with a mixture of tall-growing grasses and wildflowers.

The experiment included four blocks (areas of land) with all treatments included in each block. The populations of butterflies were monitored for 3 years (Figure B26.4.1).

When population numbers are summed across the four blocks for each treatment, it is clear that the 'tall' treatment encourages the greatest number of butterflies. However, the results would not be as clear-cut if the experiment was confined to only one block. For example, if the only data considered came from block 3, the conclusion would be that the 'mound' treatment was unsuitable and the other treatments were equivalent. Overall, the example shows the strengths of using experiments to choose restoration procedures and the importance of replicating treatments to ensure confidence in the findings.

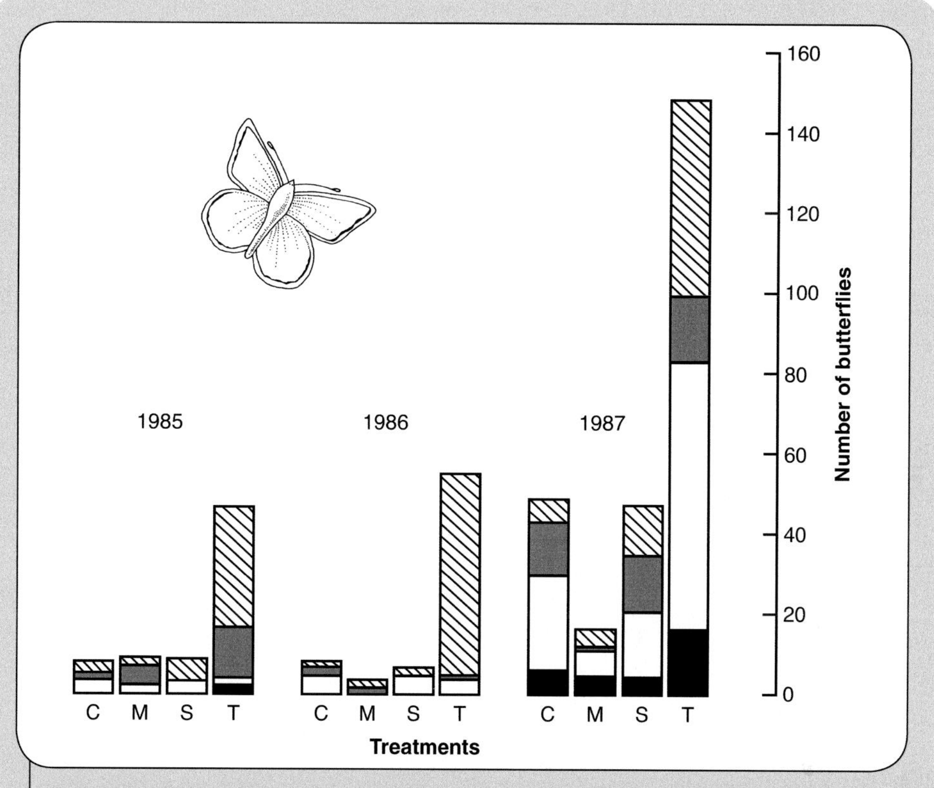

Figure B26.4.1. Populations of the common blue butterfly following four experimental restoration treatments (C, M, S and T, explained in the text) and at four sites shown as different shading in each column. (Source: Adapted from: Davis, B. (1989). 'Habitat creation for butterflies on a landfill site.' *The Entomologist* 108: 109–22)

Working with the community

Involving the community improves the quality of planning and management, reduces conflict, saves time and costs, increases community ownership and trust, and often provides volunteers to achieve on-ground outcomes (see examples in relation to conservation in Chapters 24 and 25).

Stakeholders are individuals, groups and organisations with an interest in a particular management issue or problem. Stakeholders may include local community members, government, industry, researchers and educational institutions, tourists, indigenous people, and special interest groups. Undertaking a stakeholder audit helps identify the various interests and values associated with an area or issue. Such audits involve identifying stakeholders and their values, estimating their relative power, and managing relations with and between stakeholders.

Community involvement is most successful if it is planned. Steps in such planning include identifying the issue/problem, defining the objectives of community involvement, identifying key stakeholders, selecting techniques, helping stakeholders participate, determining the available resources and timetable, making sure the decisions match the original objectives, providing feedback to participants, implementing the techniques, and evaluating the process. A vast array of techniques is available to encourage community involvement. These include citizen advisory groups, workshops, public meetings, commenting on drafts, surveys and information sheets.

Policy instruments

Very often, environmental restoration relies on actions or guidance by government. Policy, as discussed here, is the public policy developed by governments. Policy implementation (via actions called policy instruments) usually requires people to change their behaviour. For example, managing salinity relies on farmers changing irrigation practices and reducing land clearing.

Governments can encourage environmental action with a range of policy instruments including motivational, financial, market-based, self-regulatory and regulatory instruments. Motivational instruments, such as education, aim to change people's behaviour voluntarily. Financial instruments use money to encourage behavioural changes. They are regarded as the most powerful and direct means of encouraging people to participate in biodiversity conservation activities. Examples are grants, and tax and rate relief. Market-based instruments, such as land purchases and tradeable permits, are increasingly popular. Self-regulatory instruments give responsibility to industry groups, companies and individuals. Codes of practice are an example of this approach. The last group, regulatory instruments, centres on legislation and regulation.

The choice of policy instrument is critical to the success of a policy and is based on technical and political considerations including economic efficiency, effectiveness, equitability and workability. Other criteria less frequently considered are the flexibility of the instrument to cope with changing circumstances, its dependability in the face of uncertain information and political acceptability. A complementary mix of instruments is advocated for biodiversity conservation, combining voluntary instruments such as education and grants with regulations (a non-voluntary instrument). The South African 'Working for Water' program (Box 26.5) is an example of implementing government policy to achieve restoration goals.

Box 26.5 Working for water – environment and society

South Africa contains many ecosystem types, including some of the most diverse plant communities in the world. Invasive plants pose a serious threat to plant and animal biodiversity. Invasive plants, particularly woody species such as acacias and hakeas, have become established in over 10 million hectares of land in South Africa. The cost of controlling invasive plants in South Africa has been estimated at more than US$100 million a year over 20 years. If left uncontrolled, the problem will double within 15 years. It is thought that invasive species use 7% of the country's scarce water resources; intensify flooding and fires; cause erosion, destruction of rivers, siltation of dams and estuaries, and poor water quality; and lead to local or complete extinction of indigenous plants and animals.

The Working for Water program

The Working for Water program was initiated in 1995 with a $5 million grant from the South African government. It aims to control invading alien species, and to optimise the potential use of natural resources, through a process of economic empowerment and transformation (Figure B26.5.1). In doing this, the program aims to leave a legacy of social equity and legislative, institutional and technical capacity. It now has 240 projects across South Africa with a budget of US$60 million over three years. Projects involve the removal of invasive species by a number of methods, from mechanical treatment to using biological control.

Linking environmental, social and economic issues

The program, while focusing on the environmental issue of invasive plant species, tackles several important areas simultaneously. By making the link between invasive species and

Figure B26.5.1 A program with the proximate aim of restoring ecosystems by removing invasive plant species also creates employment and alleviates social and economic problems within post-apartheid South Africa. (Source: J. Lawrence)

continued ›

Box 26.5 continued ›

water supply, the immediate relevance of removing invasive species becomes apparent. Working for Water is not just an environmental program, but one that provides jobs, training and opportunities to a wide range of people, especially to women, youth and the disabled. It is also building secondary industries (e.g. crafts) around the removal of invasive species. The program has an extensive education campaign to generate awareness of the issues. The program also incorporates consideration of the provision of childcare for workers, and provides education and guidance in sexual health, in particular HIV/AIDS prevention.

Working for Water has been spectacularly successful and has broad political support for its continuation into the future. If such a program can be successful in South Africa, with all its ongoing social and economic problems, this perhaps provides some hints on how such programs can be implemented elsewhere, including in Australia.

Conservation and restoration

Conservation is the management of human interactions with organisms or ecosystems to ensure that such interactions are sustainable. For biodiversity conservation, the top priority is to retain areas in good condition. The next priority is to repair damaged areas of native vegetation. Finally, and as a last resort, restoration of areas that have been transformed (e.g. by agriculture) may be needed to increase areas of habitat or landscape connectivity. It is much more costly to re-create a natural habitat than it is to protect or repair an existing one. It is also clear that it is impossible to completely restore an area to a former state in terms of species composition and complexity. Restoration therefore forms part of a spectrum of conservation management options, although it is clear that biodiversity conservation is now dependent on some degree of restoration activity in many parts of the world where natural ecosystems have been damaged.

Conservation biology has traditionally focused on preventing the (further) degradation of ecosystems. Conservation biology has therefore often been viewed as reactive and negative, dealing with species and ecosystem declines, whereas restoration ecology has been portrayed as being positive and active because it reverses degradation. This is obviously over-simplistic! Until recently, little interaction between the two fields occurred, but it is increasingly recognised that both fields can benefit greatly from each other. Restoration is not a replacement for good management and prevention of environmental damage and, if these activities were carried out properly in the first place, we would not have the pressing need for restoration activities that we witness today.

Chapter summary

1. Environmental restoration is applied wherever landscapes and ecosystems are degraded by human activities. It draws concepts from ecology, conservation biology and environmental engineering. Important concepts in environmental restoration include ecosystem resilience (the capacity to respond to change), succession (species change over time) and assembly (composition of communities).
2. Disturbance, or episodic destruction or removal of ecosystem components, is natural and integral in ecosystems. It often initiates massive ecosystem change and triggers a period of regeneration or recovery. In some ecosystems, a predictable post-disturbance recovery sequence occurs. In others, especially many Australian ecosystems, recovery is unpredictable and results in different ecosystem compositions.
3. Human activities affect ecosystems by either modifying the original disturbance regimes or by adding new disturbances. If human disturbance pushes ecosystems beyond the limits of their resilience, the ecosystem is said to be degraded and active restoration is required.
4. Restoration should set clear goals with associated success criteria, identify the factors limiting system recovery or leading to further degradation, and instigate restoration activities that reverse or ameliorate these factors. Monitoring progress is essential to success.
5. Restoration involves changes to abiotic processes such as soil structure and biotic processes such as nutrient cycling, revegetation and adding or removing animal species.
6. Replacing habitat and increasing connectivity are often goals for landscape restoration. This is best done by building restoration efforts on the existing native vegetation.
7. Involving the community improves the quality of planning and management, reduces conflict, saves time and costs, increases community ownership and trust, and often provides volunteers to achieve on-ground outcomes.
8. Policy instruments available to environmental restoration include those that are:
   - motivational (e.g. educational)
   - financial (e.g. fines or grants)
   - market-based (e.g. land purchase)
   - self-regulatory (giving responsibility to companies, individuals and so on)
   - regulatory (legislation).
9. In general, it is much more cost-effective to prevent damage than to repair it. Thus restoration forms part of a spectrum of management options for the conservation of biodiversity.

Key terms

alternative stable state
autogenic recovery
bioremediation
disturbance
ecosystem function
ecosystem structure
hysteresis
reallocation
reclamation
rehabilitation
resilience
restoration
stakeholder
succession
threshold

Test your knowledge

1. What is meant by disturbance to an ecosystem? What factors determine how an ecosystem responds to disturbance?
2. Why is it important to set clear goals and success criteria for environmental restoration programs?
3. Using revegetation as an example, explain why successful restoration involves both the biotic and abiotic parts of the environment.
4. Why is landscape-scale restoration necessary?
5. What policy instruments are available to prevent environmental degradation and to restore degraded landscapes?

Thinking scientifically

The woylie or brush-tailed bettong, a native marsupial, often collects and buries the seeds of the sandalwood tree. Caches of buried seeds are used as food stores. In areas where woylies are absent, very few new sandalwood trees are growing. However, where woylies are present many sandalwood trees can be found.

Suggest a hypothesis to explain why new sandalwood trees are more likely to grow if there are woylies in the area. Design an experiment to test your hypothesis.

Debate with friends

Debate the topic: 'Now that degraded ecosystems can be restored, there is no need to prevent further environmental degradation'.

Further reading

Hobbs, R.J. and Harris, J.A. (2001). 'Restoration ecology: repairing the Earth's ecosystems in the new millennium.' *Restoration Ecology* 9: 239–46.

Ludwig, J., Tongway, D., Freudenberger, D., Noble, J. and Hodgkinson, K. (eds) (1997). *Landscape ecology, function and management: principles from Australia's rangelands*. CSIRO Publishing, Melbourne.

McComb, A.J. and Davis, J.A. (eds) (1998). *Wetlands for the future*. Gleneagles Publishing, South Australia.

27

A natural legacy

Mike Calver, Alan Lymbery and Jen McComb
(with contributions by Dan Lunney and Harry Recher)

The case of the toxic moths

In 2001 scientists from the New South Wales National Parks and Wildlife Service (now Department of Environment and Climate Change), La Trobe University and an environmental consulting firm investigated a mysterious death of grass outside a cave in Australia's Snowy Mountains. Heavy rains had washed dead bogong moths (*Agrotis infusa*) from the cave and grass touched by this outwash died. Investigations revealed that arsenic concentrated in the dead bogong moths poisoned the grass. Worryingly, arsenic occurred in the bodies or the droppings of three mammal species eating bogong moths, so arsenic contamination was spreading in the food chain.

The arsenic came from the plains of Queensland and western New South Wales, where bogong moths breed in autumn (Figure 27.1). The grubs, called cutworm caterpillars, eat grasses and crops before pupating in the soil to develop into winged adult moths. Arsenic-based insecticides were used intensively in the early 20th century and some are still available, so cutworm caterpillars probably absorb arsenic when eating crops. The adults migrate to the Snowy Mountains, aestivating during summer in caves until cooler weather when they return to the plains (Figure 27.2).

The bogong migration is amazing and inspirational, part of the natural legacy bequeathed by the Australian environment to its human occupants. By contrast, the moths' residual arsenic toxicity is disturbing, showing how human intervention may squander or spoil rich natural assets. You are now, though, in a position to understand and to solve such environmental problems and to reflect on what the environment and its conservation mean to you.

Chapter aims

In this final chapter we revisit the three case studies introduced in Chapter 1: Leadbeater's possum, the crown-of-thorns starfish and the Corrigin grevillea. We show how theories

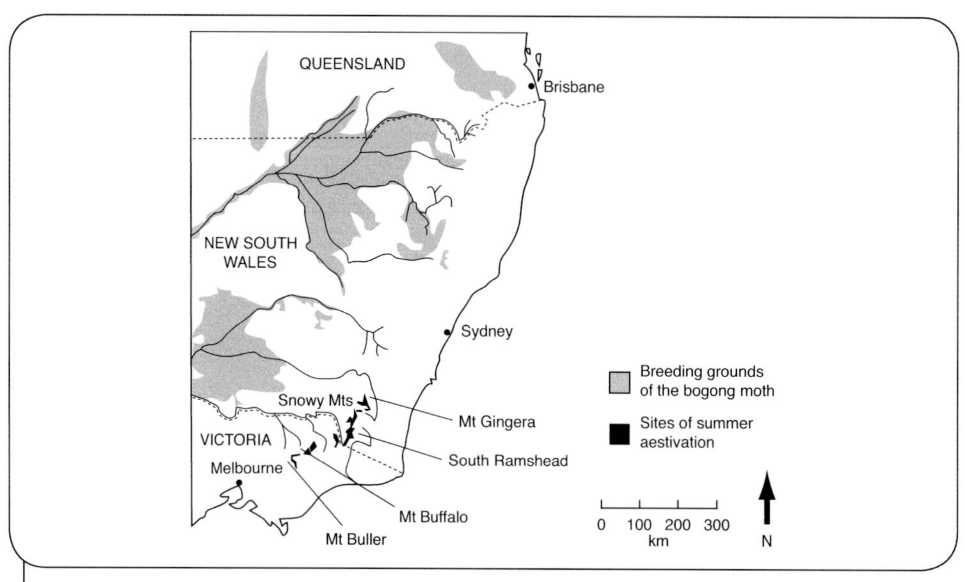

Figure 27.1 Map of south-eastern Australia showing the breeding grounds of the bogong moths and the mountain areas above 1400 m where the adult moths spend the summer. (Source: B. Green)

and techniques described in this book are applied to these cases. To suggest what a career in environmental biology entails, we conclude with views from two eminent Australian environmental biologists. They describe the challenges of their careers and the changes in technology and in society's attitudes they have seen.

Figure 27.2 Adult bogong moths. (Source: Lochman Transparencies)

Revisiting the case studies

In Chapter 1 we introduced three case studies: Leadbeater's possum (*Gymnobelideus leadbeateri*), the crown-of-thorns starfish (*Acanthaster planci*) and the Corrigin grevillea (*Grevillea scapigera*). Leadbeater's possum and the Corrigin grevillea are endangered species under the Australian Commonwealth's *Environment Protection and Biodiversity Conservation Act 1999* and decisions about environmental management are critical to their survival. The crown-of-thorns starfish, the third of the case studies, is not endangered. However, unexplained increases in its population radically change habitat on the Great Barrier Reef, prompting concern as to whether or not human activity might cause outbreaks. Having studied the principles of environmental biology, you can revisit these case studies and apply your knowledge.

Wildlife or wood? Conservation of Leadbeater's possum

Recall that Leadbeater's possum is a small, 120 g marsupial restricted to small areas of mountain ash forests in the Victorian Central Highlands, Australia. It was believed extinct for over 50 years before its rediscovery in 1961, which sparked interest in its conservation. However, the mountain ash forests are valued for timber production, and logging killed possums and eliminated their food and shelter until regeneration occurred. Although food availability returned quickly, some habitat features such as hollows in trees used for shelter did not develop until many years after logging. The challenge was to manage the landscape for both timber production and possum conservation.

The first step in meeting this challenge was improving knowledge of the possum, especially of its population dynamics and role in the forest community (Chapters 16 and 17), before more sophisticated techniques could be applied to predict its response to different management options (Chapters 2 and 24). This work, ongoing for over 40 years, involved numerous biologists and heightens confidence that current management prescriptions give the best chance possible for conserving Leadbeater's possum.

At the time of the rediscovery, ignorance regarding the possum's basic biology caused mistaken ideas about its security, and some incorrect management decisions. For example, while it was recognised that logging for timber and for pulpwood killed possums, it was also believed that habitat regenerated rapidly after logging. Therefore the intimal strategy was to provide reserves excluded from logging, in the belief that they would secure the population and that surplus possums from the reserves would recolonise logged forest when it regenerated. However, there was little information to help choose suitable areas and, because of confidence that the reserves would secure the species, initially there was no attempt to conserve possums on timber production lands as well.

Information on the possum's basic biology, accumulated by the 1980s, shook confidence in these conservation measures. Research established that possums use much energy searching for insects and plant exudates and that they also shelter in hollow trees. Hollows develop in older trees that are removed during logging and are vulnerable to bushfire, so their availability was declining although regrowth after logging or fire provided suitable food. Furthermore, although some areas of the possum's range lie within reserves (including new national parks created to protect water catchments), the overall decline in tree hollows for shelter meant that they were inadequate for long-term conservation unless measures were also taken for conserving possums within timber production areas. Hollow trees, food, some undisturbed habitat and vegetation corridors allowing possums to move between habitat patches were all seen as important in conserving possums, but critical questions such as 'How many hollow trees are needed?', 'How big should reserved patches be?' and 'Will possums use vegetation corridors?' (Chapter 24) remained.

Population viability analyses (Chapter 24) undertaken on Leadbeater's possum in the 1990s answered these questions. They recommended:

- reserving all the remaining old-growth mountain ash forest from logging, because old-growth forest contains the greatest density of tree hollows for shelter
- maintaining corridors of streamside vegetation linking suitable habitat into a network
- reserving substantial areas within timber production blocks to create a patchwork of maturing habitat across the landscape that will eventually provide tree hollows for shelter
- reducing fire risk in all reserved areas and, if fires did occur, prohibiting post-fire salvage logging because fire-scarred surviving trees are likely to develop hollows.

Even with these measures, the population is predicted to decline before increasing as more habitat becomes available.

Legislative and policy changes reflecting changing social attitudes to the environment facilitated the biological studies and the implementation of their findings. Landmark examples include the Victorian State Conservation Strategy (1987), the Victorian *Flora and Fauna Guarantee Act 1988* and its review in 1992, the National Strategy for the Conservation of Australia's Biological Diversity (1996) and the Commonwealth *Environment Protection and Biodiversity Conservation Act 1999*. Their unifying aim is protecting biodiversity and maintaining the ecological systems and processes on which it depends. While they permit using natural resources, they seek to ensure biodiversity is not compromised, in keeping with another Commonwealth policy statement, the National Strategy for Economically Sustainable Development (1992). The legislation and the national strategies reflect social attitudes, and are an example of 'cultural conservation biology' as discussed in Chapter 25.

In summary, the Leadbeater's possum case study shows the critical importance of understanding the biology of an endangered species. Once this groundwork is established, some of the sophisticated techniques from Chapter 24 can be applied to choose management options and plan reserves. All of this is predicated on a social and political will for conservation (Chapter 25).

The starfish, the reef and the scientists

You will remember from Chapter 1 that the crown-of-thorns starfish is a coral predator that was rare on the Great Barrier Reef until first noticed in large numbers in the 1960s. Since then, regular outbreaks have occurred on the Great Barrier Reef and on other coral reefs throughout the Indo-Pacific region. Debate has continued over whether these outbreaks are natural, cyclical events that pose no long-term threat to reefs, or whether they result from new environmental disturbances, and may trigger an irreversible decline in reef size and composition. Although we have no final answer to the ultimate cause of crown-of-thorns outbreaks, there is now a much greater understanding of starfish biology, the nature of outbreaks and the recovery of coral reefs following an outbreak.

Crown-of-thorns starfish are dioecious, with eggs and sperm released into the water over summer and autumn in a very common style of reproduction for marine invertebrates (Chapters 14 and 19). Each female can produce up to 60 million eggs during a spawning season, and starfish can breed for up to 7 years. Males and females gather together to spawn, which leads to high fertilisation rates (the highest measured in the field for any invertebrate). As a result, a small population of starfish can produce large numbers of larvae. Larvae drift in ocean currents for 2–4 weeks, before small juvenile starfish settle onto a reef. Juvenile starfish eat algae until they are about 6 months old, when they start to feed on coral. Although starfish prefer certain species of tabular coral, when they are in large numbers there is intense competition for food and most types of coral will be eaten. During a severe outbreak, the feeding activity of crown-of-thorns starfish can reduce hard coral cover of the reef from 40% to less than 1%.

The rate at which coral reefs recover from outbreaks of crown-of-thorns starfish depends on the extent of reduction of the coral cover and on the rate of recruitment of coral larvae to the reef. Reefs that receive large numbers of coral larvae from ocean currents can have many young corals growing on them a few years after a crown-of-thorns outbreak, whereas reefs that have a limited supply of coral larvae can take 15 years or more to recover from an outbreak. Although coral cover can eventually reach pre-outbreak levels, the composition of corals may be different, because some coral species are better at re-establishing themselves on a damaged reef. If crown-of-thorns outbreaks continue to occur at intervals shorter than the time required for the original coral community to re-establish, then reefs might suffer reduced biodiversity with the consequences that this entails (Chapter 25), even if the size of the coral reef is unchanged.

Although spicules from crown-of-thorn starfish skeletons have been found in old reef sediments, they are difficult to date accurately, so it is still unclear whether starfish outbreaks are a new phenomenon. Some scientists have suggested that outbreaks arise from natural fluctuations in water temperature, salinity or the availability of planktonic food, all of which alter the survival rate of starfish larvae. Other hypotheses point to human influences on starfish numbers. The two most likely human causes of starfish outbreaks are the overfishing of predators, such as giant triton snails (*Charonia tritonis*) and humphead maori wrasse (*Cheilinus undulatus*), and addition of nutrients from the discharge of polluted rivers into coastal reef zones. There is no strong evidence that commercially exploited predators regulate starfish numbers, but it does seem possible that high nutrient loads have contributed to outbreaks. The amount of nutrients reaching the Great Barrier Reef from adjacent rivers has increased substantially since European settlement, as a result of fertiliser use on farmland and urban run-off. Higher nutrient loads increase the growth of algae on reefs, providing more food for developing starfish larvae. Mathematical models have shown that an increased survival of larvae from this increased food supply could lead to regular outbreaks of crown-of-thorns starfish.

In summary, although we do not yet fully understand the causes of crown-of-thorns outbreaks on the Great Barrier Reef, we know a lot more about the basic biology of the starfish, its interaction with coral communities and the response of these communities to starfish predation. This understanding has come from painstaking ecological study that pioneered many new techniques for adapting the basic methods of Chapters 16 and 17 to marine environments, coupled with increased knowledge of coral reefs and associated ecosystems (Chapters 18 and 19) and applications of experimental methods (Chapter 2). Potential human causes of starfish outbreaks have been identified. Although these are not proven, steps have been taken to limit their impact, for example, by protecting starfish predators on the Great Barrier Reef and reducing nutrient outflow into the reef zone. This is an application of the precautionary principle in conservation biology (Chapter 25). There is still much to learn about the crown-of-thorns story, but an encouraging start has been made.

Back from the brink – the Corrigin grevillea

The translocation of populations of the Corrigin grevillea has been an outstanding success to the point that at least one new population appears virtually self-sustaining. The success was achieved through research into the questions raised in Chapter 1, and application of the techniques described throughout this text.

A study of the reproductive biology (Chapter 12) of the Corrigin grevillea showed that the plants have mass flowering and flowers are strongly scented. Pollinators include native insects as well as introduced honeybees. Hand pollination showed that the species was almost completely outcrossing (two individuals interbreeding), producing only a few poor-quality seeds when self-fertilised. Seed set does not seem a problem, but caterpillars and weevils eagerly eat seeds and spraying plants with insecticide at the critical time enhanced seed numbers. Germination of the seed in the laboratory was achieved through scarification (nicking the seed coat in a particular place), and treatment with the plant hormone gibberellic acid or smoke water (water through which smoke has been bubbled). In the field, where there is a seed bank built up, some germination can be achieved through burning or application of smoke to the soil, or cultivation of the soil. A small number of seedlings from these trials and natural recruits have survived to supplement the population. The natural life span of the plants is thought to be no more than 10 years (judged from growth rings on long-dead plants), but many plants live for only 3–4 years. Plants do not resprout after fire and depend on recruitment from seed to replace a population after fire or disturbance.

The species can be grown from soft- or hardwood cuttings using auxin, a plant growth hormone, to induce root growth. In tissue culture, the growth of sterile shoot segments was induced to produce multiple shoots by the application of the plant growth regulators cytokinins and gibberellins, and the microshoots were rooted using the hormone auxin.

As there were so few plants available it was possible to analyse the genetic diversity in the whole species (see the discussion in Chapter 5). A technique called RAPD (randomly amplified polymorphic DNA), which assesses heterozygosity in short lengths of DNA, was used. This showed that 10 plants of the 47 available could be selected to represent 87% of the entire genetic diversity in the species. Clones of these plants were then prepared in tissue culture for the translocation experiment.

As 95% of the bushland in the species range has been eliminated by land clearing, there was not a huge amount of choice of sites for translocation of the species but experimental trials helped determine the best establishment techniques (Chapters 2 and 24). In sites where the species once occurred, soil properties and the range of other species present were recorded and three carefully matched new areas selected. These were freed of weeds, and water tanks and drip irrigation lines set up. It was found that survival, flowering and seed production were higher when plants were irrigated over summer, particularly in dry years. Irrigation was removed or reduced once adult plants were well established.

Although the rarity of the species is largely caused by habitat destruction, it was found that rabbits eat both shoots and roots. Fencing from rabbits was essential for the establishment of the species. Kangaroos, on the other hand, do not appear to damage or eat the plants. With the exclusion of rabbits, the amount of cape weed (*Arctotheca calendula*) increased. This is not a problem for adult plants, but would prevent the establishment of new seedlings. Experiments showed that the herbicide Lontrel® controls weeds without damaging the seed or seedlings of Corrigin grevillea. Applying this herbicide will be important to help populations re-establish after fires.

Some translocated populations have now been in place for almost 10 years and the question arises as to whether they will self-perpetuate and whether succeeding generations will maintain the initial level of genetic diversity. Analysis of the genetic diversity among the seedlings produced by the founder plants revealed some surprises. First, there were only eight clones present instead of 10, with 54% of the founder plants being of one clone – a labelling mistake may have contributed to this as all the clones look similar in tissue culture. Secondly, four clones contributed gametes to 85% of the seedlings. This meant there was a loss of 20% of genetic diversity in a single generation, which if continued could lead to inbreeding (and reduced viability if inbreeding depression results; see Chapter 6). Clearly, manipulation of the frequency of the genotypes in the founder population is needed to ensure that a few genotypes are not over-represented in future generations. This is being addressed by planting under-represented clones.

There are now three translocated populations of Corrigin grevillea in the field comprising a total of some 1800 plants from clonal lines and seedlings. All the original 47 wild plants existing when this project started have died, but some are preserved in cryostorage (frozen). Conservation biologists saved this species from extinction.

Interaction with volunteers such as Kings Park Master Gardeners and local community groups was essential for the success of the project, illustrating some of the principles of community involvement covered in Chapter 25. The Corrigin Landcare Group and Bullaring community provided resources, checked the watering system and monitored growth. Research staff from Kings Park and staff from the Department of Environment and Conservation met with the Landcare group, landholders and schools in the region as the project started and as it progressed. The result of excellent communication was not only a successful translocation project, but a local community with pride in its involvement in the project.

Two encounters with the Australian bush

We conclude our journey through environmental biology with the personal reflections of two biologists with a long connection with the Australian environment. The first, Dan Lunney, is Australian born. The second, Harry Recher, migrated to Australia from the United States of America in 1967 at the age of 29. Despite their different origins, each of these biologists has a deep, personal identification with Australian environments and the Australian biota. Their experiences may help you anticipate the challenges awaiting you as an environmental biologist and the changes you can expect in technology and public attitudes during your career.

Asking new questions

Dan Lunney's career began in 1970 and he has risen to be a Principal Research Scientist in the Department of Environment and Climate Change in New South Wales. He has served as Conservation Officer for the Australian Mammal Society and is one of a very small group of Australians to serve as a member of the International Union for the Conservation of Nature's specialist groups for both bats and marsupials. Dan is an outstanding communicator, writing for both specialist scientific journals and also for the general public. He is a prolific editor of scientific books and journals, playing a major role in peer review (Chapter 2).

(Contributed by Daniel Lunney)

Chapter 1 of this textbook opened with a photograph from another textbook, *Biological science: the web of life*, first published in 1967. It shows a flock of sheep in the shade of a tree in a large paddock. Do take another look at the photo and reread the opening paragraph. You were invited to look at the photograph and compare questions about it selected from the original text with a new, modern set of questions. A primary message of the opening paragraph of this text was contained in two sentences: 'Today, those questions [from 1967] seem less relevant', and 'These new questions reveal a growing concern about the impacts of expanding human populations and the application of new technologies on the natural environment'. Let us return to the original text and consider its opening paragraphs and the question of what has remained relevant.

Figure 27.3 Dan Lunney on the occasion of receiving an Honorary Doctor of Science at Murdoch University, Perth in 2006. (Source: M. Müller)

In *The Web of Life*, the original text reads: 'Look at the photograph [of the sheep in the paddock] carefully. It is a typical view that might be seen by anyone passing in a car along a country road. What you see in it and what you think about depends on your point of view. An artist looking at the scene might note the dappled pattern of light and shade, the texture of the bark, and the haze of the distant horizon. An economist might think of wool and the balance of trade in the economy of the nation.'

The next paragraph opens with: 'What do you think about it?' That question is timeless. The Australian ecologist Charles Birch, one of my lecturers at university and one of the 1967 textbook's authors, often invited students to 'see what everyone else sees, but think what no-one else has before'. Birch was also fond of saying, 'The questions remain the same, and it is the answers that differ'. That seemed to me, as a first-year biology student, to be silly. Yet, I have spent the rest of my working life seeing the sense in it, and answering that question, 'What do you think about it?' It is these metaphysical questions that remain the enduring guide for scientists, particularly those biologists who concentrate on conservation in their working lives.

'Thinking about it' and speaking your mind on environmental matters can arouse strong opposition. Environmental degradation – climate change, overpopulation, habitat loss and the extinction of wildlife, salinity, loss of forests and fisheries, desertification, pest species and pollution – remains the most pressing matter in world survival today and yet we, on the whole, remain more concerned with the balance of trade and the economy of the nation. These are the long-term versus the short-term views of life. Try your hand at expressing your point of view on any of these matters. You will make many firm friends and encounter some bitter opposition. Be ready, read this textbook, and you will have modern science as part of your armoury. However, solving 21st century environmental problems won't involve just science; it will have a human dimension. Let me give a few examples from my working life. Let's go back to the sheep in the paddock as the first example.

My hard-working brother has a sheep property in Yass, just west of Canberra. It is prime merino wool country, but sheep for meat became increasingly important as the droughts bit. The photo in the 1967 textbook could well have been taken on his land. Our father worked as a CSIRO scientist in wool research. He was steeped in the idea that 'Australia has ridden to prosperity on the sheep's back'. You can imagine the shock to my brother and my father when I looked ecologically at the sheep industry in the Western Division of New South Wales in the 19th century. After reading many historical accounts, I concluded: 'Twenty-four mammal species – predominantly the medium-sized, ground-dwelling mammals with a dependence on grass/herbs and seeds – disappeared forever from the landscape of the Western Division of New South Wales in the 60 years from first settlement in 1841. The extinctions can be largely attributed to the impact of sheep, exacerbated in the scarce and fragile refuges... during intense and frequent drought. This conclusion differs from those of many others who pointed to "the impact of feral animals, rather than overgrazing" (particularly the introduced red fox (*Vulpes vulpes*) and the rabbit (*Oryctolagus cuniculus*)) as the cause of mammal extinctions.... The sequence of occupation and land use in the Western Division and the timing of the loss of native mammal species allows the conclusion to be drawn that it was sheep, and the way the land was managed for the export wool industry, that drove so many mammal species extinct. The impact of ever-increasing millions of sheep on all the river frontages, through all the refuges, and across all the landscape by the mid 1880s, is the primary cause of the greatest period of mammal extinction in Australia in modern times'.

My paper was published in an academic journal, but when journalist James Woodford used it in an article on exploding myths in a major Sydney newspaper in 2005, my brother had a hard time explaining things to his fellow wool growers in Yass. What remains is a permanently diminished landscape in western New South Wales, larger in area than Britain. By international standards, it is a shocking outcome environmentally. Yet it stands as a powerful example of the 1967 biology textbook statement: 'What you see in it and what you think about depends on your point of view'. Let me now turn to Australia's largest ecosystem, the arid lands, and how views changed over the last century.

In 1944, the explorer Cecil Madigan captured some of the magic of Australian deserts and the Australian attitude to this land: 'I thought I knew my own country. I spent two years in the Antarctic; travelled Europe; geologised in Africa; and then returned home to settle down in commonplace Australia and to "see Australia last"'. Madigan continued: 'My interest was drawn to Central Australia, the desert heart of the continent, as being remote and little known and likely to yield interesting results to the geologist ...' Madigan summarised the level of interest in and knowledge of the Australian desert just before the end of World War II. In short, it was remote and little known. The fact that so few shared Madigan's attitude helps explain why so little was known of our desert ecosystems until recently.

Madigan would have been raised on textbooks such as that in 1927, *The New South Wales intermediate geography text book*. This school text reflected the values of the time as much as it presented geographical information. In the section on Australia, under the heading of 'Deserts', the closing sentence reveals much about the shifting meaning of 'desert': 'As railways extend into the interior, land which was declared desert not long ago, is now being eagerly taken up for pastoral and agricultural purposes, with the result that some of the so-called deserts have vanished'. Land was classified according to its use, not its innate attributes. Eighty years later the United Nations (UN) declared 2006 'the year of the desert'. Where the 1927 school text wanted to transform the deserts into useful grazing lands, the UN recognised the environmental and economic value of deserts and moved to protect them worldwide.

By 1969, there was a revolution in thinking about Australia's arid ecosystems as witnessed in an Australian Academy of Science symposium, Australia's Arid Lands. In the opening sentences of the book there is the then-novel view that, 'History has shown that man's utilisation of the world's arid lands has, in general, caused progressive deterioration of the natural vegetation, of the animal resources, and finally of the landscape itself'. The author foreshadowed 'a conflict of interest between those whose primary concern is to conserve the arid land resource and those who wish to obtain an economic return from it'. These views were textbook material within two decades, as Steven Morton wrote: 'We must continually remind ourselves that two-thirds of our continent is arid, and that much of this huge area, with its rich diversity of organisms adapted to the uncertain rainfall, is under constant pressure from domestic stock and rabbits.' These are all modern sentiments, consistent with the UN's 'Millennium Assessment' report of 2005 and the UN's 'Global Deserts Outlook' of 2006. Let us turn now to the idea that other disciplines see the landscape differently.

To what extent do you identify with Roslynn Haynes, the English scholar who gave us a modern view of the Australian desert? She examined, in 1998, 'the changing reaction of white Australians to the desert – from fear to delight – and the contrast between these responses and the relationship that indigenous Australians have with their land'. She concluded that most Europeans of the 19th century, obsessed with their search for water and green, saw only monotony in the desert. 'It is the artists, writers and photographers of the last hundred years', she says, 'who have taught us to see the desert differently'. To her list of people providing fresh insight and a new perspective, we should add researchers, be they biologists, historians or archaeologists. It will need in the future to include at least some of you reading this textbook, and it could be in a range of disciplines, with the cultural component, rather than the geographical or biological elements, taking precedence.

The modern environmental biologist has an ever-increasing array of technological options to press into service that, in 1967, were not available. These include computers and everything that goes with them, such as modelling capacity (climate changes scenarios depend on it), radiotracking equipment for following animals, global positioning systems (GPS), a wealth of journals and computer-based searches for relevant material, better 4WD vehicles, genetic analyses for field studies, better microscopes, and a host of devices large and small that assist in the field work and the subsequent analyses. This continuing technological revolution will not, of itself, conserve our environment. It will be your outlook, and that of those making decisions to log a forest, clear the land, exploit the resources of this rich nation, establish a new national park, support a wildlife research program, manage urban wildlife, farm ecologically, or eat our native fauna as an alternative to introduced stock. Look again at the mob of sheep under the tree. Turn those sheep into the Australian population, and then think what will happen to the landscape. Is that where we are now? Imagine rewriting this textbook on a day in the future, the day of your retirement at age 65. What would you include, and how kindly will you look at that tree and mob of sheep, this textbook, and what has become of Australia?

An environmental ethic

Harry Recher came to Australia from the United States of America in 1967 to take up a lectureship at the University of Sydney. Since then he has held posts at the Australian Museum and the University of New England, and most recently was Professor of Environmental Management at Edith Cowan University. Harry has served as Chair of the National Biodiversity Council of Australia and President of the Royal Society of Western Australia and in 1994 he was awarded the Serventy Medal by the Royal Australasian Ornithologists Union for his contributions to ornithology. In 2004 he received an Order of Australia for 'service to ecological science, particularly through the development of ecosystem management in Australia, and as an educator, author and advocate for biodiversity'.

(Contributed by Harry Recher)

> Ecology may give us the scientific principles which we can use to describe the effects of a dam on the Franklin River, but it is only the development of an environmental ethic that gives us the moral and aesthetic foundation to defend [the Franklin River].[1] (Source: Irina Dunn, *A natural legacy*)

These were Irina Dunn's concluding words to *A natural legacy*, a book on ecological principles and environmental ethics for which she, Dan Lunney and I were co-editors. *A natural legacy* grew out of a series of lectures entitled 'Ecology for Conservationists' held at the Australian Museum, Sydney, in the 1970s. I initiated the lectures in frustration over what I saw as a lack of ecological understanding within Australia's rapidly expanding conservation movement. Not only did this weaken the argument for conservation and conservation planning, but it skewed the conservation movement towards decisions detrimental to Australia's environment.

The audience included business people, unionists, fellow scientists and the general public, as well as the conservationists at whom the lectures were aimed. Children as

[1] In the early 1980s, the Tasmanian government proposed to dam the Franklin River in the state's south-west wilderness. Conservationists saw this as both a threat to wilderness values and the destruction of one of Australia's last wild rivers. The dam was stopped in a classic confrontation between national and states' rights, with the High Court of Australia ruling that Australia's international obligations under the treaty to protect a World Heritage Area (the Southwest Wilderness) transcended the right of the state to build a dam.

young as 12 came, and the oldest to attend was in his eighties. I had completely misjudged the interest and concern for the environment among Australians of all backgrounds and ages. Moreover, *A natural legacy* became a widely used university text and was even adopted by schools in the United States of America, showing the need for a text that integrated ecology with conservation and conservation management. All of this was evidence of the growth of environmental awareness and responsibility in Australia, North America and Europe since the 1960s.

Figure 27.4 A youthful Harry Recher (right) surveying for marine invertebrates. (Source: D. Lunney)

Only a few years before organising the lectures and while a member of a scientific committee advising the New South Wales government on national parks, I remarked that I would be happy if government authorities just mentioned the word 'environment' when considering development proposals, much less do anything to protect the environment. So rapidly did the awareness of human impact on the Australian environment develop that by the 1980s New South Wales had enacted 'environmental impact' legislation requiring designated developments to assess the impacts the development would have on the environment and to modify the development if the impacts were judged unacceptable. A Land and Environment Court was created to adjudicate disputes, as well as the adequacy of environmental impact statements. Similar legislation followed in all the states and with the Commonwealth.

Proper assessment of environmental impacts requires applying ecological principles and detailed knowledge of the environment. At the same time that Australia developed environmental impact legislation and Australians expected more from government in the way of environmental management, universities across Australia began offering degrees in natural resource management.

The growth of environmental awareness over the past half-century among all people, not just Australians, is one of the great social and cultural changes to occur during my life. For more than 30 years, Australians have ranked the environment as one of their most important concerns and people routinely say they are willing to pay to protect and improve their environment. I cannot say this leaves me fully satisfied, feeling that the job is well done and the environment is secure for future generations to enjoy; too often environmental quality is still sacrificed in the interest of economic development. Australia still has a way to go in developing the 'environmental ethic' to which Irina Dunn referred.

Yes, there is greater awareness of and concern for the environment. Our knowledge and understanding is vastly greater in 2007 than it was in 1967 when I arrived in Australia from the United States of America. This means our capacity to manage the environment and to conserve Australia's flora and fauna is also vastly improved, but as a nation we still lack the will to make the sacrifices necessary to truly ensure that the Australian economy is financially secure and ecologically sustainable. This lack of will means two things to me. The first is that most Australians really do not understand what the environment is and how they fit into it. The second is that all the scientific knowledge and environment management skills in the world are meaningless unless people really care. Teaching people to care is the great challenge facing environmental biologists in the 21st century.

There is no better example of the gap between Australians' environmental concerns and actual achievement in environmental protection and management than the failure of successive governments to address the issue of human-induced global warming. When we published the second edition of *A natural legacy* in 1986, we referred to the threat to world climates posed by the rapid increase in greenhouse gases from human sources. This was more than 20 years ago and decisive action on global warming is barely beginning in 2008. In her chapter, Irina Dunn quoted a 1983 report from the United States Environmental Protection Agency: 'Temperature increases [from the increase in atmospheric greenhouse gases] are likely to be accompanied by dramatic changes in precipitation and storm patterns and a rise in global average sea level ... As a result, agricultural conditions will be significantly altered, environmental and economic systems potentially disrupted and political institutions stressed'. These were not idle comments, but were based on good science including a thorough knowledge of atmospheric physics and chemistry and a chronological record of climate and greenhouse gas concentrations in the atmosphere extending back thousands of years. The world's scientific community was united in seeing the threats posed by the accelerating emissions of greenhouse gases from the relentlessly expanding world economy and had warned against the threats of global warming since at least the 1960s, but the governments of major emitters of greenhouse gases, including Australia and the United States of America, denied the threat and took no action to curtail, much less reduce, emissions. Australian governments have taken global warming seriously only in the last 12 months as intense drought cut urban water supplies, threatened the total destruction of the Murray-Darling river system and meant there was no water for irrigation farming in the Murray-Darling in 2007.

It is hard for me to understand how Australian governments could ignore the threats to the nation's long-term sustainability by human-induced global warming. Partly the denial came because science could not make absolute predictions of climate change. It is still not certain exactly how high sea levels will rise, how much temperatures will increase, how these changes will affect rainfall and where the effects will be greatest. Denial was facilitated by a few contrarian scientists, supported by commercial interests and government, who argued that the evidence for accelerated global warming was imprecise or even wrong. In 1986, Australian government scientists prevented me from publishing an article in the Australian Museum magazine describing the potential threats to Australia from global warming. I was accused of being too extreme, yet all that I predicted in 1986 has happened and happened many times over. If anything, I should have been accused of being too conservative in my predictions.

The role that politics plays in the lives of environmental biologists is considerable and often unpleasant. I always believed that I had a responsibility not only to do research and publish the results for other scientists to read, but to make what I learned available to the people who paid for my education and who paid my wages – the taxpayers of America and Australia. Thus, a large part of my professional life has been dedicated to writing popular articles, giving public lectures, giving interviews to the media and appearing *pro bono* as an expert witness on behalf of community groups involved in litigation on environmental issues. It is an unfortunate fact of life that most community groups lack the financial resources to hire experts, whereas their opponents from government and industry not only have greater financial resources, but can call upon the scientists they support or employ to testify on their behalf.

As a public advocate for the environment, a scientist risks a great deal – research funds, promotions and even his or her job. There is also the loss of credibility with scientific peers, too many of whom believe scientists should be seen, not heard. For some reason, a prevailing view within the scientific community is that a scientist should present only facts, not opinions. I've always ignored that and for good reason. A scientist, such as an environmental biologist, has access to facts not known to the wider community and is trained to interpret them. Environmental biologists routinely interpret their knowledge of the environment to assess environmental impacts and develop plans of management. It is a small step to do the same when commenting on environmental issues of public concern.

Perhaps I was bold and my boldness helped fend off those who threatened me for my public commentary and courtroom testimony. However, I was also fortunate in working for 20 years in a public institution, the Australian Museum, with very liberal views on the role of science and scientists in public debates. From 1968, when I joined the museum, to 1988, when I took a position at the University of New England, the museum's scientists were free to comment publicly on issues within their area of expertise. My position as Head of Environmental Studies meant my area of expertise was directly concerned with the environment. Every environmental dispute within New South Wales, from sand mining for heavy metals along the coast, to woodchipping forests at Eden, to establishing pine plantations, to marina and canal estate proposals in estuaries, came across my desk. I commented on them so that the public would not remain in ignorance of their environmental impacts, and benefits, if any. A disappointment of my professional life has been how few scientists were willing to take a public stand, but instead stood behind the facade of scientific impartiality. No-one is impartial. No-one is without opinions and, when those opinions are founded on good science, there is a responsibility to make them available to the public.

As a young American, fresh to Australia when I arrived in 1967, I saw Australia, its landscape and its flora and fauna differently from most of my scientific colleagues. I definitely saw it differently from the politicians or the Australians who seemed only to see the continent's resources as economic wealth. Although I was still in my 20s when I left America, I'd already seen too much environmental destruction in the United States to believe it would be different in Australia. The two nations and their peoples were too similar in their democratic ideals of free enterprise and in their belief that the world was created for human benefit to expect otherwise. Maybe I was bold and outspoken because I grew up in the rough and tumble of New York City and its environs, but I also knew that I didn't want the Australia I found in 1967 to make the same environmental mistakes I had seen in the United States. I was fated to be disappointed.

After 40 years of environmental work in Australia, I can see where Australia repeated, and continues to repeat, many of the same mistakes made in the United States. I can also see where the growth of environmental concern and the excellent resource management education available at Australian universities for more than a generation now gives real promise of reversing many of those mistakes and much better environmental management in the future. I also see real changes within Australia's environment movement where peak bodies, such as the Wilderness Society, Birds Australia and Greenpeace, routinely employ articulate and well-trained scientists to provide scientific substance in their mission to achieve an ecologically sustainable Australia and world.

Caring for the environment means more than just your backyard and the air you breathe or the water you drink. It means more than protecting farmland from erosion and dryland salinity or protecting a few threatened species from extinction. Caring really means sharing. It means sharing the world's resources with all people regardless of nationality, race or religion. It means guaranteeing that future generations of people will have the same or greater opportunities to enjoy the natural world as we do. Sharing means allowing other species to use the world's resources and to have the opportunity to adapt and evolve as environments change. Sharing means using resources efficiently, recycling and developing technologies with minimal environmental impacts. Sharing may mean having fewer material goods, but it will also mean having more of other values. Sharing and caring means a richer, not a poorer world, but developing this ethic remains elusive. It means this generation of environmental biologists must be leaders and unafraid to take their knowledge, ideas and opinions to the public, for the public good. When that is done, we will have gone a long way towards building the environmental ethic Irina Dunn called for in *A Natural Legacy*.

Chapter summary

1. As knowledge of the biology of Leadbeater's possum increased, it was possible to perform population viability analyses. These led to a comprehensive set of management recommendations:
   - reserving all areas of old-growth forest from logging, as well as corridors of streamside vegetation that link suitable habitat into a network
   - reserving substantial areas within each timber production block to create a patchwork of developing, potential habitat across the landscape
   - reducing fire risk in all reserved areas and prohibiting any post-fire salvage logging in these areas if fires do occur, because fire-scarred surviving trees have a high chance of developing hollows.

   Significant new legislation passed in the 1980s and 1990s encouraged relevant biological research and the adoption of the findings.
2. Fossil deposits suggest, but do not confirm, that crown-of-thorns starfish outbreaks occurred in the past on the Great Barrier Reef. The two most likely human impacts that could contribute to contemporary outbreaks are overfishing of predators and pollution from nutrients, which could encourage survival of juvenile starfish. Protection measures, addressing overfishing and nutrient run-off, have been put in place but the effect of crown-of-thorns starfish on the Great Barrier Reef must still be monitored closely.
3. The Corrigin grevillea is scarce because of large-scale land clearing, which destroyed about 95% of its habitat. Conservation efforts centre on:
   - propagating plants from seeds and from tissue culture in the laboratory for re-establishment in suitable environments
   - using molecular biology techniques to determine the level of genetic diversity in surviving plants, so that new populations can be established with a stock of maximal genetic diversity
   - supporting newly established populations in the early years with weed control, drip irrigation and exclusion of rabbits.

   Collaboration between community groups, landholders and scientists is essential to sustaining these conservation efforts.
4. Professional scientists, as well as lay people, have their own perspectives and biases when interpreting their experiences in environmental research.

Test your knowledge

1. How does the conservation of Leadbeater's possum illustrate (a) the need to understand thoroughly an organism's basic biology before recommending conservation measures and (b) the importance of legislation to encourage and enforce conservation action?
2. What evidence is now available that outbreaks of crown-of-thorns starfish on the Great Barrier Reef are encouraged by human activities?
3. What actions are being undertaken to conserve the Corrigin grevillea?

Thinking scientifically

1. In this chapter Dan Lunney argued: 'The sequence of occupation and land use in the Western Division [of New South Wales] and the timing of the loss of native mammal species allows the conclusion to be drawn that it was sheep, and the way the land was managed for the export wool industry, that drove so many of the mammal species to extinction'. This contrasts with the view that feral animals were primarily responsible for the extinctions.
   (a) How might a scientist approach the problem of choosing between these two views?
   (b) What steps would be necessary to restore mammal communities in the Western Division of New South Wales?
2. In this chapter Harry Recher proposed: 'Caring really means sharing. It means sharing the world's resources with all people regardless of nationality, race or religion. It means guaranteeing that future generations of people will have the same or greater opportunities to enjoy the natural world as we do. Sharing means allowing other species to use the world's resources and to have the opportunity to adapt and evolve as environments change. Sharing means using resources efficiently, recycling and developing technologies with minimal environmental impacts'. Is this a scientifically defensible position?

Debate with friends

Debate the topic: 'Scientists should present facts, not opinions'.

Further reading

Ehrlich, P.R. and Ehrlich, A.H. (1996). *Betrayal of science and reason: how anti-environmental rhetoric threatens our future*. Island Press, Washington DC.

Jacobson, S.K. (1999). *Communication skills for conservation professionals*. Island Press, Washington DC.

Krauss, S.L., Dixon, B., Dixon, K.W. (2002). 'Rapid genetic decline in a translocated population of the endangered plant *Grevillea scapigera*.' *Conservation Biology* 16: 986–94.

Lindenmayer, D. (1996). *Wildlife and woodchips: Leadbeater's possum, a test case for sustainable forestry*. UNSW Press, Sydney.

Lindenmayer, D. (2007). *On borrowed time: Australia's environmental crisis and what we must do about it*. Penguin Group Australia, Camberwell, Victoria.

Recher, H.F., Lunney, D. and Dunn, I. (1986). *A natural legacy: ecology in Australia*. A.S. Wilson Inc.

Sapp, J. (1999). *What is natural? Coral reef crisis*. Oxford University Press, New York.

Wills, R. and Hobbs, R.J. (1998). *Ecology for everyone: communicating ecology to scientists, the public and the politicians*. Surrey Beatty and Sons, Chipping Norton, New South Wales.

Glossary

A horizon	Topsoil where most nutrient recycling occurs.
aboral surface	Surface opposite the mouth in an echinoderm.
acid rain	Precipitation of weak solutions of strong mineral acids derived from industrial pollutants such as sulfur dioxide. These react with oxygen and water vapour to give rain with a pH below 4.5.
acid sulfate soils	Soils rich in iron and sulfur, common in some wetlands, which, when exposed to air during clearing for agriculture or housing, can produce very acidic runoff.
active transport	Movement of dissolved substances across cell membranes into cells or organelles against a concentration gradient by using energy.
adaptation	A character that enhances the survival and reproductive success of an organism.
adaptive management	Treating different management actions in space and time as experimental treatments to increase understanding of the system being managed. Requires cycles of planning, action, reviewing and evaluating, and responding. Monitoring and researching management interventions are important components of these cycles.
aerobic	In the presence of air (oxygen).
aerobic respiration	Respiration occurring in the presence of oxygen in which ATP is formed from ADP (adenosine diphosphate). The key component, oxidative phosphorylation, occurs on the inner membrane of the mitochondrion.
aestivation	State of torpor used to avoid heat or drought.
agricultural ecosystem	An ecosystem where domesticated plants are the basis of the food chain supporting human populations.
allele	When a gene has more than one form, each is an allele.
allele frequency	Proportion of a particular allele in a population.
allopatric speciation	Formation of separate species through divergent evolution in populations that are geographically isolated. (Compare with sympatric speciation.)
alternation of generations	Life cycles of protists and plants with alternating gametophyte and sporophyte stages.
alternative stable state	A stable ecosystem structure which develops following successional change from a disturbance, but which is different from the ecosystem structure prior to disturbance.
amensalism	The interaction between two species in which one is disadvantaged and the other unaffected.
ammonium	The positive ion formed when ammonia is dissolved in water. A common form of inorganic nitrogen in water, from which it can be taken up and used in plant growth.
amniotic egg	Egg of reptiles, birds and mammals with a rigid or leathery shell providing support and protection, a food supply encased in a membrane continuous with the embryo's gut (yolk), a semipermeable chorion membrane

	beneath the shell for gaseous exchange, an allantoic membrane for storing waste products and respiration, and an inner membrane (the amnion) enclosing the embryo in a pool of fluid.
amoeboid	Cells without a firm wall that move by extensions of cytoplasm.
anaerobic	Not requiring oxygen to survive and grow.
anaerobic respiration	Respiration in the absence of oxygen leading to the production of ethanol (plants) or lactic acid (animals).
analogous	Pertaining to a character that is present in two or more taxa, but not in their common ancestor. (Compare with homologous character.)
ancestral	The form or state of a character that is presumed to be primitive because it is present in a group of taxa and also in other, more distantly related, taxa. (Compare with derived.)
angiosperms	Flowering plants with ovules (and seeds) enclosed in a carpel.
annual	A plant that germinates, grows, flowers and sets seed within 1 year.
anoxic	Without oxygen; used in relation to water and sediments.
anther	Part of a stamen bearing pollen sacs.
antheridia (singular: antheridium)	Sperm-producing structures. In plants there is an outer coat of sterile cells (cells that do not produce sperm), but no sterile layer in algae and fungi.
antheridiophore	A stalked structure supporting the antheridia in some liverworts.
apical meristem	Group of dividing cells at the tip of roots, shoots or some algal thalli.
archegonia (singular: archegonium)	Egg-producing structures. In algae and fungi there are no layers around the egg cell, but in plants there is a sterile layer around the egg extended into a neck.
archegoniophore	A stalked structure supporting the archegonia in some liverworts.
archival tag	Small tag containing a computer, attached to animals (typically marine) to record and store physiological data (body temperature, heart rate and swimming speed) and environmental data (water temperature, light level and salinity).
artificial selection	The differential reproductive success of different genotypes produced by human decisions about which organisms are used for breeding.
artificial substrates	Substrates that do not occur naturally in the environment, but are placed there by humans (e.g. jetty piles and the hulls of vessels) and that support growth of macroalgae, periphyton and mussels.
asexual reproduction	Reproduction that does not involve the union of two gametes. It results in clonal offspring that are genetically identical to the parent and each other.
asymmetric	Having no symmetry.
ATP (adenosine triphosphate)	The principal source of energy and phosphate groups for all reactions in cells.
atrium	1. Internal cavity of a sponge. 2. The chamber of the heart that receives blood from the veins.
attenuation coefficient	Calculation of the reduction in light intensity with water depth (= vertical attenuation coefficient, K_d).
autogamy	Form of sexual reproduction in ciliates.
autogenic recovery	Establishment of a plant and animal community in a degraded area through natural processes of colonisation.

autotroph	An organism with the capacity to make its own food. (See photoautotroph.)
B horizon	Soil layer beneath the A horizon. It grades into the weathered rock below and accumulates some materials washed down from the A horizon.
BACI	Before-after, control-impact. An approach to testing the effect of an environmental variable by assessing several control and impacted sites before and after the impact.
bacillus (plural: bacilli)	Rod-shaped prokaryote.
bacteriophage	A virus that infects bacteria.
bark	The outer layers of a shrub or tree trunk including the true bark, cork cambium, phelloderm and secondary phloem ending at the cambium.
barometric tides	Changes in water level in response to alterations in barometric pressure, for example in shallow estuaries.
basidium (plural basidia)	A club-shaped structure that bears spores of fungi in the basidiomycetes.
Batesian mimicry	Where one animal or plant species benefits by mimicking the appearance of another species; for example, a non-poisonous species has the colouration of a poisonous one. (Named after the naturalist H.W. Bates.)
benthos	Organisms living on or near the bottom of a sea, river or lake.
bicarbonate	See carbon dioxide.
bilateral symmetry	Characteristic of animals whereby one imaginary line could divide them along the long axis to produce two halves that are mirror images. Such animals have right and left sides, a distinct head end (anterior) and tail end (posterior), an upper surface (dorsal) and a lower surface (ventral). Also applied to flower shapes.
binary fission	Division of a cell into two equal halves; a means of asexual reproduction in unicellular organisms.
binomial system	Nomenclatural convention whereby the name of a species consists of two words: the genus name followed by the specific epithet (the binomial).
bioaccumulation	The increasing concentration of a substance in tissues, when it is absorbed at a faster rate than it is lost through excretion or degradation.
biodiversity	1. Genetic variation within organisms. 2. The range of species in an environment. 3. The full spectrum of the world's ecological communities and their associations with the physical environment.
biodiversity hotspots	Geographically defined areas containing many unique, threatened species of organisms.
biofilm	A layer of living bacteria and other organisms adhering to an inert support such as stone, gravel, wood or plastic, and which affects the properties of the water with which it is in contact (e.g. aquatic systems, septic tanks and sewage treatment plants).
biogeography	The study of the past and present geographical distributions of organisms, and the ecological and historical factors responsible for these distributions.
bioindicator	An organism whose growth or death is used as a test of the presence of a toxin in the environment.

biological control	The introduction of a predator or pathogen to control the numbers of a pest organism.
biological species concept	The definition of species which states that a species is a group of actually or potentially interbreeding populations that are reproductively isolated from other such populations.
biomagnification	The increasing concentration of a non-degraded substance (often a toxin) as it passes to organisms at higher levels in a food chain.
biomass	The mass of living material, usually of plants and/or animals, and typically expressed per unit area (e.g. kg per hectare).
biome	A climatically determined group of plants and animals spread over a wide area.
bioremediation	The use of living organisms to remove toxic chemicals from an environment.
biosphere	The collective ecosystems of the Earth.
bipedal	Walking upright on the hind limbs.
blade	The flat photosynthetic part of an algal thallus.
blastula	The hollow ball of cells arising after cell division in a zygote in an animal.
blood pigments	Coloured protein compounds that contain metal atoms; their function is to increase the oxygen-carrying capacity of blood.
bradymetabolic	Slowing the metabolism markedly at rest.
broadcast spawning	Release of large numbers of eggs into the environment (typically the ocean) for external fertilisation.
Calvin cycle	The cycle of reactions within chloroplasts in which the ATP and NADPH produced in the light-dependent reactions of photosynthesis power the synthesis of carbohydrate.
calyx	The outermost whorl of floral appendages.
CAM plants	Crassulacean acid metabolism plants, which open their stomates and take up carbon dioxide at night. The CO_2 is stored in the cell vacuoles at night, and fixed into carbon products during the day.
cambium	Sheet of meristematic cells that encircles the stems and roots of some plants. It divides, producing cells both internally and externally. The vascular cambium forms the secondary phloem and xylem. The cork cambium forms bark (externally) and an internal layer (phelloderm).
Cambrian	The earliest period of the Palaeozoic (545 million years ago), a time of dramatic increase in the diversity of organisms on Earth.
captive breeding	Breeding endangered species in captivity, with the long-term aim of reintroducing them to the wild.
carbon dioxide	A gas in the atmosphere; it dissolves in water, in large part then reacting to form carbonic acid, bicarbonate and carbonate, in an equilibrium that depends on pH and biological activity.
carbonate	See carbon dioxide.
carnivory	Consumption of animals as food by other animals (invertebrates or vertebrates) or plants.
carotenoid	Accessory photosynthetic pigments (yellow to reddish brown).

carpel	Central floral whorl that contains the ovules. One or more carpels are termed the gynoecium.
carrying capacity	The maximum population size an environment can sustain.
cartilage	Firm elastic tissue in vertebrates.
Casparian strip	The band of lignin and suberin that develops in primary walls of endodermal cells.
catch per unit effort	A population size estimation method that exploits falling capture rates over a series of trapping events when animals are removed after capture.
cell wall	The external, non-living, rigid structure enclosing the cell membrane of algal, plant, fungal and most prokaryotic cells. The primary constituent is cellulose (algae, plants and some fungi), chitin (some fungi) or polysaccharides (prokaryotes).
Cenozoic	An era in the geological time scale, from about 65 million years ago to the present day.
centrioles	Paired organelles occurring just outside the nuclear membrane of animal cells, many protists, and in cells of land plants that form motile sperm – their function is uncertain.
character	Any phenotypic feature or attribute of an organism that serves as a basis for comparison (= trait).
character state	Any of the range of values or measurements that a particular character (or trait) of an organism can take.
chemoautotroph	A prokaryote that obtains carbon from CO_2 without using light energy, and energy from oxidation of inorganic molecules such as ammonia or hydrogen sulfide.
chemoheterotroph	An organism that obtains energy and carbon from organic molecules.
chenopod	Salt-tolerant shrub from the family Chenopodiaceae.
chitin	The main structural component of the walls of fungi and the exoskeletons of insects and arthropods; also in molluscs.
chlorophyll	The main light-capturing pigment of plants and algae.
chloroplast	A membrane-bound organelle occurring in plant and algal cells; it contains the photosynthetic pigments and a small amount of DNA (cDNA).
choanocytes	In sponges, flagellated cells (collar cells) lining an internal cavity or atrium.
chromatid	One longitudinal half of a chromosome in which the DNA has been doubled, giving two chromatids.
chromatin	In a eukaryote, the combination of DNA and protein that constitutes the chromosomes.
chromosome	The strand of DNA and protein that carries the genetic information of a cell.
cilia (singular: cilium)	Hair-like extensions from the surface of some eukaryotic cells. In single-celled organisms, beating cilia move the whole cell; in multicellular organisms, cilia keep fluids in motion over the cell surface.
circinate vernation	The curled arrangement of young leaves and leaflets typical of a fern frond.

circular muscles	Muscles contracting in a radial plane of an animal's body. (Compare with longitudinal muscles.)
cladistics	Method of determining the phylogeny or evolutionary relationships among taxa based on the occurrence of shared, derived character states.
class	A rank within the hierarchy of taxonomic classification, between phylum (or division) and order.
classification	The hierarchy of groups of organisms, or the process of arranging organisms into such groups.
climax community	The end product of plant succession. Generally a community that is stable in the absence of disturbance.
clone	Progeny (cells or organisms) that are genetically identical to one another and to the parent.
closed circulatory system	A circulatory system where blood is confined within blood vessels and does not mix directly with fluids surrounding the cells.
closed forest	A forest with a closed canopy and an understorey of shade-tolerant shrubs, herbs and ferns. (Compare with open forest.)
clumped distribution	A distribution with individuals aggregated in patches.
coccus (plural: cocci)	A spherical prokaryote.
codes of nomenclature	Internationally agreed rules that govern the scientific naming of organisms.
co-dominant alleles	Alleles that, when present in a heterozygote, both produce products seen in the phenotype.
codon	The sequence of three nucleotides that specifies the position of a particular amino acid in a protein, or the beginning and end of a polypeptide.
coefficient of dispersion	The statistic calculated by dividing the variance of a sample by its mean. If the coefficient of dispersion is about 1, the population is distributed randomly. If the coefficient of dispersion is greater than 1 the population is clumped and if it is less than 1 the distribution is uniform.
coelom	A fluid-filled body cavity completely lined by tissue of mesodermal origin.
coenocytic	Describes a cell that has multiple nuclei.
cohort life table	Table made up of data obtained by following a group of individuals all born within a narrow time interval throughout their life spans, recording the probability of survival from one age class to the next and the number of young produced per female in each age class.
collenchyma	Living plant cells providing support by means of thickening deposited mostly in the corners of the cells.
colour	Electromagnetic radiation in the range visible to humans, from the red to violet ends of the spectrum (about 400–700 nm), which approximates to that used by plants in photosynthesis, attenuated in water with depth, in part because of water molecules, but especially because of chemicals dissolved in it (colour) and particles suspended in it (compare with turbidity).
commensalism	Interaction between two species where one species benefits and the other is unaffected.

community	A group of species that live together in a particular habitat.
companion cell	Cell type in phloem; living at maturity, lying adjacent to the sieve tube cell and (in angiosperms) formed from the same mother cell as the sieve tube.
compensation depth	Depth in a water body at which oxygen production through photosynthesis just balances oxygen consumption through respiration.
competition	An interaction in which both species are disadvantaged. In an ecological sense, refers to the competition between species to access limiting resources. See also interference competition and exploitation competition.
competitive exclusion principle	The concept that two species cannot coexist in a community if their requirements for life are identical.
conduction	The transfer of heat between bodies that are touching, such as an animal and a sun-warmed rock.
cone	A simple or complex elongated whorl of sporophylls (= strobilus).
cone scale	One of the microphylls or megaphylls making up a cone. May be a fusion between the megaphyll and a bract.
conjugation	The pairing of organisms (usually single-celled or filamentous) preceding the exchange of genetic material.
conservation biology	A new, integrated science that attempts to conserve biodiversity, developed in response to the human-induced, worldwide wave of habitat destruction and extinction.
conservation status	The likelihood that a species may become extinct over a given time.
consumer	An organism that obtains its food by eating other organisms. Primary consumers are animals that feed on plants (e.g. herbivores), secondary consumers are animals that eat herbivores (carnivores) and tertiary consumers eat carnivores, thus forming a simple food chain.
control (experimental)	A yardstick against which experimental treatments are compared. A control should differ from a treatment in only *one* variable, so any difference between the two can be attributed to that variable.
convection	The transfer of heat by a flow of air or water, such as a breeze or a stream current.
corals	Colonial marine invertebrates in the class Anthozoa (phylum Cnidaria). Within a coral colony, individual organisms live within a protective skeleton which they secrete. The calcium carbonate skeleton of hard corals forms the coral reefs of tropical seas.
coralloid root	The upward-growing branched root of a cycad that has a symbiotic nitrogen-fixing cyanobacterium. Also found in alders and she-oak roots infected with the nitrogen-fixing bacterium *Frankia*.
Coriolis effect	The effect on ocean currents of the spinning of the Earth, causing moving water to be deflected to the right in the Northern Hemisphere, and to the left in the Southern Hemisphere.
cork cambium	The sheet of lateral meristematic cells that divide to give rise to the bark.
corolla	The collective name for the whorl of petals in a flower.
corridors	Narrow strips of habitat that link isolated protected areas into a single system.

cortex	The tissue of parenchyma cells lying between the vascular tissue and the epidermis in roots and stems.
cotyledon	The first leaf (or leaves) formed on a gymnosperm or angiosperm embryo in the seed.
countercurrent	The flow of two fluids in opposite directions along two closely opposed vessels, allowing the transfer of heat, solutes, gases and so on from one fluid to the other.
coxal gland	The gland in some arthropods that collects and excretes metabolic waste. It takes its name from the coxa on the leg where it is located.
crassulacean acid metabolism	See CAM plants.
cristae	The folds within the inner membrane of the mitochondrion that increase the surface area within the matrix. Enzymes embedded in the membrane produce ATP.
cultural eutrophication	Enrichment of a water body by nutrient addition through human activity, to levels which greatly exceed those encountered in natural systems; characterised by blooms of algae, cyanobacteria and macroalgae. (Compare with eutrophication.)
culms	Upright stems arising from a rhizome.
cuticle	Secreted layer (cutin) on the outside of epidermal cells. In plants it is waxy, but in animals it may contain proteins and chitin. Makes walls more or less impervious to water.
cytochromes	Membrane-bound proteins in the mitochondrion that either carry out or catalyse the important series of redox reactions leading to the formation of ATP.
cytoskeleton	The network of submicroscopic fibres in the cytoplasm of a eukaryote cell.
dauciform roots	Short, parsnip-shaped roots producing large numbers of long, dense root hairs, increasing the surface area for nutrient uptake.
deciduous	Losing all leaves at one season of the year.
decomposers	Organisms responsible for the decomposition of dead plant and animal material.
decomposition	The process by which dead plant and animal material is broken down, predominantly by bacteria and fungi, to simple, inorganic soluble compounds, which can be used as respiratory substrates, and/or incorporated again into food webs.
definitive host	The host in which an adult parasite lives and reproduces.
density	The number of individuals per unit area or volume.
deoxyribonucleic acid (DNA)	The long, double-stranded, helical molecule in which the arrangement of nucleotide bases (adenine, cytosine, thymine and guanine) determines the inherited traits of an organism.
derived	The form or state of a character that is presumed to be advanced because it is present in a group of taxa but not in other, more distantly related, taxa. (Compare with ancestral.)
detrital food chain	A food chain in which bacteria of decay are consumed by other organisms, such as protozoa and zooplankton, which use the organic

	molecules found in detritus as a source of energy and nutrients; such organisms may in turn be consumed by higher-order consumers such as small fish, which may in turn be consumed by larger fish, constituting a food chain based on detritus.
detritivores	Animals that consume detritus, which they use as a source of energy and nutrients.
deuterostomes	Animals with radial cleavage of the egg, a coelom arising from outpocketings of the embryonic gut and where the embryonic anus forms before the embryonic mouth.
dichotomous key	A guide, constructed as a series of alternative choices, for identifying an organism as a member of a given taxon.
diffusion	The passive movement of dissolved molecules or ions from a region of high concentration to a region of low concentration.
dimictic stratification	A stratification event brought about by temperature, and occurring twice each year. (Compare with meromictic and monomictic stratification.)
dioecious	Having separate male and female individuals.
diploblastic	In animals, developing all tissues from one or the other of two cell layers in the developing embryo.
diploid	Having two sets of chromosomes, one from the male and one from the female parent.
directional selection	Selection that favours phenotypes towards one extreme of the normal character distribution, leading to a consistent directional change in the mean value of a character in a population.
dispersal	Dissemination of spores, gametes, offspring or organisms from their point of origin or release.
dispersion pattern	How individuals are spread across an area; may be clumped, dispersed or random.
disruptive selection	Selection that favours phenotypes towards both extremes of the normal character distribution, leading to a bimodal distribution of character values in a population.
dissolved oxygen	The amount of oxygen dissolved in water, usually recorded as mass per litre, but often as percentage saturation, expressed in relation to the concentration of oxygen in the water if it had been in equilibrium with air under the same temperature, pressure and salinity as the sample measured.
disturbance	Destruction or removal of any biotic or abiotic components of an ecosystem.
division	Taxonomic rank sometimes used instead of phylum for plants and fungi.
DNA	See deoxyribonucleic acid.
DNA fingerprinting	The analysis of the size of the fragments of DNA that result when the molecule is cut with specific enzymes.
DNA sequencing	The process of determining the order of nucleotides in a DNA molecule.
domain	An informal grouping of kingdoms in the hierarchy of taxonomic classification.
dominant allele	An allele the effect of which is seen in a diploid heterozygote organism.

dorsal	The upper side of a bilaterally symmetrical animal. Also applied to plant parts such as leaves.
double fertilisation	In angiosperms one sperm cell fuses with the egg cell to form the zygote; the second fuses with the central cell of the embryo sac and forms the endosperm.
ecological models	Models used to clarify the functioning of complex ecosystems. May be conceptual (e.g. taking the form of box-and-arrow diagrams depicting the flow of materials or energy being transferred between the different compartments of an ecosystem) or numerical, in which numerical values are assigned to each compartment and the rates of transfer between compartments are expressed as equations.
ecological zones	Zones of plant communities encountered when moving from one habitat to another, for example from a forest loam to the saturated soil of a wetland. Sometimes used in relation to zones encountered in different latitudes and altitudes.
ecosystem	A community of living organisms and the physical environment with which they interact.
ecosystem function	The flow of energy and nutrients through an ecosystem.
ecosystem services	The ecological services provided by individuals or species to all life on the planet, including clean oxygenated air, atmospheric moisture, clean water, pollination, dispersal of seeds, decomposition, nutrient cycling, flood control, and productive soils, lakes and oceans.
ecosystem structure	The physical state of an ecosystem, as measured by attributes such as species richness, species composition and total biomass.
ecotone	The area of overlap between different environments, such as between a terrestrial and an aquatic environment.
ectoderm	The cell layer in multicellular animals that gives rise to the epidermis and the nervous system.
ectoparasites	Parasites attached to the outside of their host.
ectotherm	An animal that regulates its body temperature behaviourally.
edge effect	1. The concept that conditions at the edge of an experimental plot will be different from those experienced by organisms in the centre of the plot. 2. The concept that the length of the perimeter of a reserve in relation to its area will affect weed and pest invasion. 3. The increased variety and density of plants at the junction of two communities.
El Niño	A climatic event caused by warm surface waters in the central Pacific Ocean, leading to reduced oceanic productivity and heavy summer rainfall along the Pacific coast of South America and drought in eastern Australia. (Compare with La Niña and see also ENSO.)
embryo sac	The female gametophyte of a flowering plant, usually of seven cells.
endemic	Occurring in one area and nowhere else.
endocrine gland	A ductless gland secreting specialist chemicals (hormones) to control a bodily function (together such organs form the endocrine system).
endocytosis	The taking up by eukaryotic cells of dissolved or particulate extracellular material by extending the plasma membrane around the external

	material to form an enclosed vesicle that releases its contents inside the cell.
endoderm	The cell layer in multicellular animals that develops into the digestive and respiratory tracts.
endodermis	The innermost layer of the cortex, adjacent to the pericycle in plant roots and some stems. Water and nutrient uptake into the xylem is controlled at this layer by Casparian strips and other wall thickenings.
endoparasites	Parasites living within the body of the host.
endoplasmic reticulum	The network of flattened tubes providing a large surface area of membrane within the cell. It may be smooth, or studded with ribosomes (rough endoplasmic reticulum).
endoskeleton	An internal skeleton with the muscles enclosing it.
endosperm	In flowering plants, a tissue that provides nutrients to the developing embryo, formed after fusion of a sperm cell with two nuclei of the central cell of the egg sac.
endostyle	The mucus-secreting organ characteristic of chordates and precursor to the thyroid gland in most vertebrates.
endosymbiont theory	The theory that chloroplasts and mitochondria originated as free-living prokaryotic cells that took up residence inside other eukaryotic cells. Supported by the fact that these organelles have their own loops of DNA.
endosymbiosis	A form of symbiosis in which one species (called the endosymbiont) lives within the body of another species. Usually restricted to symbiotic interactions where both species benefit and the endosymbiont is intracellular. (See symbiosis.)
endotherm	An animal regulating its body temperature internally by altering blood flow or muscular activity, or generating metabolic heat.
ENSO	El Niño Southern Oscillation: climatic pattern producing droughts in some parts of the Southern Hemisphere and devastating floods in others. Driven by the waters of the central Pacific Ocean called the Pacific warm pool.
environment	The set of external influences acting on an organism.
epicormic shoots	Dormant buds under the bark of trunks of trees or shrubs.
epicotyl	The portion of the stem of an embryo or young seedling between the cotyledons and first true leaves.
epidermis	The outer layer of cells of an animal, plant or alga.
epifauna	Benthic animals living on the sediment surface or on other animals.
epilimnion	Upper, mixed layer of a stratified water body.
epilithic	Attached to rock.
epipelic	On or in the sediment surface. (Compare with epipsammic.)
epiphytes	Plants or animals which live attached to plants, for example green algae, and bryozoans attached to the leaves of seagrass, and orchids attached to the stems of tropical trees.
epipsammic	Attached to sand grains. (Compare with epipelic.)
episodic wetland	Wetland filled seasonally or at longer intervals (= temporary wetland).

epizooic	Plants or animals which live attached to animals; examples are the algae attached to the carapace of marine crabs, and blue-green algae on the fur of sloths.
epoch	A subdivision of a period of geological time during the Cenozoic era.
era	A major subdividision of the geological time scale, marked by abrupt changes in the fossil record.
eukaryote	A cell containing discrete membrane-bound organelles and with a membrane-bound nucleus in which the DNA is organised into chromosomes.
euphotic zone	Upper part of the water column, sufficiently well illuminated that oxygen production by photosynthesis exceeds oxygen consumption in respiration during a 24-hour period.
euryhaline	Tolerating a wide variation in salinity. (Compare with stenohaline.)
eutrophication	Nutrient enrichment of a water body, leading to increased plant biomass. First used to describe the slow process of enrichment that leads under natural conditions to the invasion of water bodies by changing plant communities, and their infilling with organic materials such as peat. Now usually used almost exclusively to describe cultural eutrophication. (Compare with oligotrophic.)
evaporative cooling	The cooling that occurs when water evaporates from an organism. Energy is required to transform a liquid to a gas, and the energy comes from the organism's heat, so the organism is cooled.
evapotranspiration	The sum of the evaporation of water from a free water surface and that lost by evaporation from the leaves of plants. Used in the calculation of water budgets for wetlands and water catchments.
event-dependent systems	Communities in which regeneration is dependent on a disturbance such as a fire.
evolution	Genetic change in the characteristics of a population of organisms over generations.
ex situ conservation	Strategies for conserving a species away from the area of natural occurrence, such as in zoos, botanic gardens or new areas where the threats causing the problem in the original location are absent. (Compare with *in situ* conservation.)
exocytosis	Where substances for secretion from the cell are collected in vesicles that merge with plasma membrane and release their contents outside the cell.
exon	Stretch of the mRNA code of a eukaryote that remains after removal of the introns and code for protein structure.
exoskeleton	The skeleton on the outside of an animal, with the muscle attachments on the inside.
exploitation competition	Competition that does not involve direct contact between species (e.g. ground water use)
exponential growth	Growth that occurs when populations increase at their maximum rate under ideal conditions and the whole population multiplies by a constant factor in each generation.
extinction	The disappearance of a population or species from a given habitat.

extremophile	Micro-organisms that grow in environments that have physical or biological characteristics fatal to most species.
family	1. A rank within the hierarchy of taxonomic classification, between order and genus. 2. A group of organisms, consisting of parents, offspring and other closely associated relatives.
fen	A wetland, rich in organic matter, dominated by sedges and with the water level at the soil/peat surface.
feral	A domesticated plant or animal that has run wild.
fermentation	The breakdown of carbohydrates by living cells in the absence of oxygen.
fertilisation	The fusion of a sperm and egg cell.
fission	A type of asexual reproduction in which the parent divides into two or more parts. (See binary fission.)
fitness	The reproductive contribution of a given genotype to the subsequent generation, relative to that of other genotypes.
fjeldmark	A biome occurring in very exposed conditions in the alps, characterised by a combination of herbfield and very diminutive heath.
flagellae (singular: flagellum)	Long, whip-like structures used for movement in some eukaryotic and prokaryotic cells. Their internal structure is different in eukaryotes and prokaryotes.
flame cells	The excretory organs in flatworms.
food chain	The sequence of organisms along which energy and nutrients flow in an ecosystem, from primary producers to herbivores to carnivores.
food web	The system of linked or interacting food chains in an ecosystem.
founder effect	The reduction in genetic variation and the change in allele frequencies that occur when a small number of individuals from a larger population founds a new population.
frond	In a fern or palm, equivalent of a leaf (may be very large and divided).
fruiting body	The aggregation of fungal hyphae associated with sexual reproduction and spore dispersal.
fucoxanthin	A brown carotenoid pigment in brown algae.
functional feeding groups	Groups of organisms that carry out similar functions in an ecosystem.
fynbos	Heathland/low woodland occurring in the Cape region of South Africa.
gamete	A sex cell (usually haploid) that unites with another to form a zygote. Isogametes are of the same size. When one gamete is larger than the other, it is referred to as the female gamete or egg, and the smaller one is the male gamete or sperm.
gametophyte	The haploid stage of a protist or plant life cycle that produces the sexual gametes.
ganglion (plural: ganglia)	A cluster of nerve cells.
gastrula	The developmental stage in the embryos of multicellular animals following the blastula, in which one side of the blastula folds inwards, producing an internal sac that ultimately becomes the digestive tract.
gene	The sequence of nucleotide bases in DNA that codes for a functional product (RNA or protein).
gene flow	The exchange of genes among populations by migration and interbreeding.

genet	An individual arising from a single zygote.
genetic code	The sequence of three nucleotides in a codon that specifies a particular amino acid.
genetic drift	Random changes in allele frequencies from one generation to the next in a population.
genotype	The genetic constitution, or combination of alleles at one or more loci, of an organism.
genotype frequency	The proportion of a particular genotype in a population.
genus (plural: genera)	A group of related species. A rank within the hierarchy of taxonomic classification, between family and species, and designated by the first word of the species binomial.
germ cell	A diploid cell that produces gametes by meiosis.
glycolysis	The process occurring in the cell cytoplasm converting the six-carbon glucose molecule to two three-carbon pyruvate molecules.
gnathostome	A vertebrate with true jaws.
Golgi apparatus	The layer of flattened sacs receiving substances synthesised in the rough and smooth endoplasmic reticulum that are then modified and transported to various destinations.
Gondwana	The past supercontinent combining what are now separate Southern Hemisphere land masses (= Gondwanaland).
grana (singular: granum)	Stacks of thylakoid membranes in the chloroplast.
grazing food chain	A food chain based on the consumption of primary producers by grazing animals.
greenhouse effect	Warming created when electromagnetic energy, re-radiated from the Earth's surface, is absorbed by atoms of gases such as carbon dioxide and methane in the atmosphere, thus warming the molecules.
growth rings	Concentric rings of wood (xylem) of different density formed during periods of slow and fast growth. When such periods alternate annually, growth rings in trunks can be used to date trees.
guard cells	The pair of cells controlling the size of the aperture of a stomatal pore.
gymnosperm	A seed plant with ovules and seeds not enclosed in an ovary. Includes several formal phyla of the plant kingdom.
habitat	The environment of an organism; the place where it is usually found.
haemocoel	A body cavity consisting of blood-filled spaces.
halocline	The interface between the upper, mixed layer of a water body stratified by differences in salinity, and the lower, more saline, denser layer. Detected as a rapid change in salinity with depth. (Compare with thermocline.)
halophile	A micro-organism that grows in a highly saline environment.
haploid	Having a single set of chromosomes.
Hardy-Weinberg equilibrium	Maintenance of more or less constant allele and genotype frequencies from one generation to the next in a population that is very large, where mating is at random, and within which there is no mutation, migration or selection.
haustorium (plural: haustoria)	1. The modified root of a parasitic flowering plant that clasps the root or stem of the host and taps into its xylem and/or phloem. 2. Fungal hyphae that penetrate living cells and withdraw nutrient.

heathland	An area of nutrient-poor soils dominated by low shrubs.
hemimetabolous development	In insects: eggs hatch into nymphs resembling adults, but lacking wings. As they moult and grow, external wing buds develop until, at the final moult, the winged adult form emerges (= incomplete development).
hemiparasite	A plant that has green leaves and photosynthesises, but is wholly or partially dependent on a host for water and minerals.
herbfield	A vegetation type of mosses and small annuals.
herbivory	The consumption of plants or algae as food by invertebrate or vertebrate animals.
heritability	The extent to which phenotypic differences among individuals in a population for a particular character can be explained by genetic differences.
hermaphroditic	Producing both eggs and sperm.
heterodont	Having dissimilar teeth specialised for specific functions.
heterospory	Producing two types of spores (megaspores and microspores).
heterotroph	An organism that obtains energy by ingesting autotrophs or other heterotrophs and accessing the energy stored in the bonds of their organic molecules.
heterozygote	A diploid individual with a different allele for a gene on each chromosome. (Compare with homozygote.)
hibernation	A long-lasting state of inactivity, lowered metabolic activity and lowered body temperature. (Compare with torpor.)
holdfast	A basal structure of algae that attaches the thallus to the substratum.
holometabolous development	In insects: eggs hatch into larvae totally unlike the adult. After several moults they enter a non-feeding pupal stage and are extensively reorganised into the winged adult body (= complete development).
holoparasites	Plants lacking photosynthetic tissue and relying on other plants or the mycorrhizal fungi of other plants for food.
homeostasis	The ability to maintain a constant internal environment often different from the immediate surroundings.
homeothermic	Having the ability to regulate its body temperature at a relatively constant value, independent of fluctuations in the temperature of the environment (= homoiothermic).
homodont	Having similar teeth.
homoiothermic	See homeothermic.
homologous character	Pertaining to a character that is present in two or more taxa, and in their common ancestor. (Compare with analogous.)
homologous chromosome	In a diploid cell, one of a pair of matching chromosomes with the same sequence of genes.
homospory	Producing only one type (size) of spore.
homozygote	A diploid individual with the same allele for a gene on both chromosomes. (Compare with heterozygote.)
hormones	Substances produced by one tissue and transferred to another where they have a specific effect at extremely low concentration. In plants, hormones may have an effect in the tissue in which they are produced.

hummock grass	A characteristic grass shape that initially forms a very prickly hummock, but grows to a doughnut form with either bare ground or other plant species in the centre.
hydroids	The water-conducting cells of mosses: evolved independently of tracheids.
hydrologic budget	The balance between the inputs and losses of water for a system, calculated by working out the gains in water for a wetland from surface flow, groundwater input and rainfall, and comparing this with losses from evapotranspiration, groundwater outflow and surface flow (= water balance).
hydrolysis	The breaking of the bond to the third phosphate group in ATP in a chemical reaction involving water, producing a free phosphate group and a remaining molecule called ADP (adenosine diphosphate), with a release of energy from the broken chemical bond.
hydroperiod	The period during which a river flows in response to rainfall in its catchment.
hydrostatic skeleton	The fluid-filled, constant-volume cavity that permits muscles to be re-stretched after contraction.
hypertonic	Describes an external solution with solute concentration greater than that inside the cell. (See also hypotonic and isotonic.)
hyphae (singular: hypha)	Filamentous tubular strands that form the body of a fungus.
hypocotyl	The portion of the axis of an embryo or young seedling between the cotyledons and the root. The region in which the arrangement of vascular tissue changes from the root to the shoot pattern.
hypolimnion	The lower, unmixed layer of a stratified water body. (Compare with epilimnion.)
hypothesis	A testable prediction proposed to explain an observation.
hypothesis-testing approach	The approach to science that begins by making observations of an interesting pattern, event or puzzle, and then follows a logical sequence of constructing models and hypotheses that can be tested to explain the observations.
hypotonic	Describes an external solution with solute concentration less than that inside the cell. (See also hypertonic and isotonic.)
hysteresis	A phenomenon in which the recovery pathway of an ecosystem following a disturbance is different to the pathway followed during ecosystem degradation. The structure and function of the recovered ecosystem is thus likely to be different to the structure and function of the pre-disturbed ecosystem.
in situ conservation	Strategies for conserving a species within the area of natural occurrence.
inbreeding	Mating between related individuals, or more precisely mating between individuals that are more related than the population average.
inbreeding depression	Reduction in fitness of offspring arising from inbreeding.
independent assortment	In the first division of meiosis: the random allocation of the maternal or paternal chromosome of each pair of homologues to each pole of the cell

	(and consequently the independent assortment of the genes on those chromosomes).
infauna	Benthic animals living within the sediment.
information-theoretic approach	Approach to science that tests the consistency of data against multiple competing hypotheses, not just one. It assesses the strength of evidence for each competing hypothesis rather than comparing significance to non-significance at an arbitrary probability level.
instrumental	Conserving biodiversity for human use.
integument	The outer covering of cells and their secretions (e.g. hair and nails in mammals; cuticle in plants).
interference competition	Competition that involves direct contact such as fighting over prey.
intergenerational equity	Conserving biodiversity for the benefit of future generations.
intermediate disturbance hypothesis	The hypothesis that biodiversity in a community will be greatest when disturbances are intermediate in intensity and/or frequency.
intermediate host	The host in which a parasite passes a larval stage.
interspecific competition	Competition between species. (Compare with intraspecific competition.)
interspersion	In a field experiment, deliberately alternating control and treatment sites to ensure that all controls or all treatments are not clumped at one location.
intraspecific competition	Competition between individuals of the same species. (Compare with interspecific competition.)
introduced species	Plant or animal species that have become established in an area where they are not native (may be desirable agricultural or horticultural species or undesirable weeds and feral animals).
introduction	Moving captive-bred or wild-collected animals and plants to areas outside their historical range. (See introduced species.)
intron	Section of the mRNA strand that is transcribed from the DNA of a eukaryote organism, but is snipped out before the mRNA leaves the nucleus. (Compare with exon.)
isotonic	Describes an external solution with solute concentration equal to that inside the cell. (See also hypotonic and hypertonic.)
keystone species	Species present in small numbers, but essential for the persistence of many other species in the community.
kingdom	Highest formal category in the hierarchy of taxonomic classification, above phylum or division. Kingdoms may be grouped informally into domains.
Kjeldahl nitrogen	A measure of the amount of nitrogen in organic material. Involves digesting the organic material from an organism or water sample with strong acid to convert it to ammonia, which can then be analysed. Any ammonia present in the original sample will survive and be analysed, so Kjeldahl nitrogen includes ammonia. If a complete budget of nitrogen components is needed, as for a water sample, the amount of nitrogen in ammonia, nitrate, and nitrite should be analysed separately and added together.
Krebs cycle	The process within the mitochondrion that breaks down acetyl CoA (from pyruvate) to form carbon dioxide, 2 ATP, 6 NADH, and 2 $FADH_2$.

	It is called a cycle because it begins with the formation of a six-carbon compound (citrate) and finishes with a four-carbon compound (oxaloacetate) ready to accept two carbon atoms from acetyl CoA and begin the process anew.
K-selection	Maximising reproductive success by maturing later in life and producing fewer offspring that receive high levels of parental care or resources. *K*-selected organisms tend to be larger, longer-lived and good competitors, and maintain populations near carrying capacity. (Compare with *r*-selection.)
kwongan	Heathland/low woodland occurring in the south-west of Western Australia.
La Niña	Climatic event based on cooler water in the central Pacific Ocean leading to increasing marine productivity on the west coast of South America and good rainfall years on the east coast of Australia. (Compare with El Niño.)
lamina	A thin sheet applied to the photosynthetic blade of a leaf, the stacks of grana in a chloroplast, and so on.
larva (plural: larvae)	The juvenile stage in the life cycle of many animals.
lateral roots	Side roots arising from the pericycle region of a main root (= secondary roots).
Laurasia	Past supercontinent combining what are now separate Northern Hemisphere land masses.
Leeuwin current	Warm Southern Ocean flow, coming from the north between Australia and the Indonesian archipelago and flowing down the west coast, forcing the northward-flowing cold current offshore.
lentic	Non-flowing. (Compare with lotic.)
leptoids	The food-conducting cells of mosses: evolved independently of phloem.
life	Characterised by cellular structure, homeostasis, ability to respond to stimuli, sustained by a complex of internal biochemical reactions, able to reproduce, and growing and developing over time.
life history	The main events in an organism's life from birth (or germination in the case of a plant) to dispersal (if it occurs), feeding habits, reproduction and finally death.
light	Electromagnetic radiation that can be detected by the human eye (with a wavelength of about 400–700 nm), which approximates to that used by plants in photosynthesis. Radiation outside this range, such as ultraviolet and infra-red radiation, is not 'light'.
light-dependent reactions	Reactions that fix solar energy for use in producing sugars. They occur in two different light-collecting units, photosystem I and photosystem II, in different areas of the thylakoid membrane in the chloroplast.
light-independent reactions	Reactions within chloroplasts in which the ATP and NADPH produced in the light-dependent reactions of photosynthesis power the synthesis of carbohydrate.
lignin	A complex organic polymer that impregnates and strengthens the cellulose cell walls of plants.

lignotuber	The swelling of tissue produced in the cotyledon region of a seedling. It stores dormant buds that can produce new shoots if the tops of the plant are removed by grazing, drought or fire.
limiting factors	Factors that decrease birth rates and/or increase death rates as population densities rise.
limnetic zone	The open water of a shallow wetland (divided into euphotic and profundal).
limnology	Study of the physical and biological properties of fresh and saline inland waters.
littoral zone	The edge or shore region of a wetland where light penetrates to the substratum.
locus (plural: loci)	The location of a gene on a chromosome.
logistic growth	Growth occurring when exponentially growing populations eventually encounter one or more limiting factors constraining further growth.
longitudinal muscles	Muscles contracting in a longitudinal plane of an animal's body as opposed to a radial one.
lotic	Flowing systems, at least during parts of the annual hydrologic cycle (e.g. permanent and intermittently flowing rivers). (Compare with lentic.)
lysis	Breakdown of cell membranes resulting in leakage of cell contents and cell death.
lysosomes	Organelles arising from the rough endoplasmic reticulum and the Golgi apparatus, and containing enzymes that break down food particles and recycle damaged organelles. They also play a role in controlled cell death.
macroevolution	The appearance and extinction of species and higher taxa over evolutionary time.
macronucleus	A nucleus controlling metabolism and development in ciliate cells, which have several macronuclei and a single micronucleus.
macronutrient	One of nine key nutrients needed in relatively large amounts for healthy plant functioning.
macrophytes	Large plants associated with aquatic systems, including seagrasses, emergent sedges, water lilies, large complex algae and ferns.
mallee	A growth form of eucalypts with several thin trunks arising from a lignotuber.
Malpighian tubules	Excretory tubules in insects.
mangroves	Small trees growing rooted in mud subjected to tidal inundation, with a leaf canopy in the air above.
manipulative experiment	An experiment involving actively changing one part of the environment to determine the effects relative to other areas that are left unchanged.
mantle	The fold of soft tissue under the shell of molluscs.
mantle cavity	The space between the mantle and the body.
mark-release-recapture	The technique of population estimation involving capturing, marking and releasing animals and then estimating the population size from the proportion of marked animals retrieved on a later sampling occasion.
mass extinction	The extinction of a large number of species within a short interval of the geological time scale.

maximum sustainable yield	The largest harvest that can be taken while maintaining a harvested species at an intermediate population size, which will grow most rapidly under conditions of logistic growth.
medusa	A body form in cnidarians in which the main body floats in the water with the tentacles trailing beneath it.
megaphyll	A plant having large leaves with an extensive vascular system.
meiofauna	Small benthic invertebrates retained by a fine mesh.
meiosis	The sequence of two cell divisions by which a diploid cell gives rise to four haploid daughter cells.
mensurative experiment	An experiment in which the environment is not manipulated, but comparisons are made between groups already believed to differ in some way.
meristem	A region of undifferentiated, actively dividing cells.
meromictic stratification	Stratification established through a marked difference in salinity, and therefore density, between the upper and lower layers of a water body, so extreme that the water body remains permanently stratified, with anoxia at depth. (Compare with dimictic and monomictic stratification.)
mesoderm	The layer of cells that develops between the ectoderm and endoderm at the gastrula stage of an embryo.
mesoglea	The layer of gelatinous material between the epidermis and the gastrodermis of cnidarians.
mesohyl	The gelatinous, protein-rich matrix forming the middle layer in the bodies of sponges.
Mesozoic	An era in the geological time scale, about 248–65 million years ago.
metabolic rate	The rate of metabolism; usually measured as the rate at which a specific chemical reaction (e.g. respiration) is occurring.
metabolic wastes	The end products of metabolism that are not used further by the organism.
metabolism	The sum of all the chemical reactions occurring within the cells of a living organism.
metameric	Having segmentation of the body. Occurs in annelids, arthropods and chordates.
metamorphosis	Marked changes in structure between juvenile forms and the adult.
metanephridia	The excretory organ of molluscs.
metapopulation	A series of subpopulations the distributions of which do not overlap, but which exchange some individuals through migration.
metaxylem	A type of primary xylem formed from the procambium of a stem or root apex in regions where elongation has ceased.
microevolution	The changes in allele and genotype frequencies within a population over generations.
micronucleus	In ciliate cells that have several macronuclei and a single micronucleus, the micronucleus is involved with sexual reproduction.
micronutrient	Nutrient essential for healthy plant growth, but required in only small amounts.
microphyll	A small leaf derived from an outgrowth of the stem with little vascular system (e.g. in mosses).

microsatellite	Segments of DNA that have different numbers of short, repeated segments, in different individuals and can be used in DNA fingerprinting.
midrib	Large central vein of a leaf.
minimum viable population (MVP)	The smallest population size that can persist indefinitely.
mitochondrion (plural: mitochondria)	An organelle that converts the chemical energy of organic molecules such as sugars into the chemical energy of the molecule ATP (adenosine triphosphate) that then powers work within the cell.
mitosis	Cell division in eukaryotic cells where one cell gives rise to two daughter cells, each with the same chromosome number as the parent.
mixotrophic	An organism able to switch between heterotrophy and autotrophy.
monoecious	Producing both male and female gametes.
monomictic stratification	Stratification brought about by temperature and occurring once each year. (Compare with dimictic and meromictic stratification.)
monophyletic	Pertaining to a group of organisms or taxa that contains an ancestor and all of its descendants. (Compare with paraphyletic and polyphyletic.)
moss bog	A bog in alpine areas where sphagnum moss is the controlling life form of the vegetation.
Mullerian mimicry	Where two unrelated species that are harmful have similar colouration or appearance.
mutation	Heritable change due to alteration in the nucleotide sequence or arrangement of DNA in an organism.
mutualism	Interaction between species where both species benefit.
mycelium	A mass of fungal hyphae.
mycorrhiza (plural: mycorrhizae)	A symbiotic association of a plant root and a fungus. When the fungus coats the root and grows between the cells, it is termed ectomycorrhiza. When the fungus enters the living cells, it is termed endomycorrhiza.
NADH	Nicotinamide adenine dinucleotide, an electron carrier molecule used in cellular reactions.
NADPH	Nicotinamide adenine dinucleotide phosphate, an electron carrier molecule used in cellular reactions.
nasal glands	Glands found in marine birds and reptiles; they extract salt from ingested water and remove it from the animal's body.
natural selection	The differential reproductive success of different genotypes as a result of some organisms being better adapted to their environment than others.
nekton	Organisms that swim in open water.
nematocyst	Coiled, thread-like tube, often barbed, inverted within a capsule and used by cnidarians to immobilise prey. When triggered it discharges explosively, propelled by osmotic pressure.
nephridia	The excretory organs of annelids and larval echinoderms.
neural crest	The line of cells formed on the dorsal surface of the early embryo giving rise to parts of the nervous system, skeleton and some organs.
neuston	Organisms living on the surface of an aquatic habitat.
neutralism	Where the presence of two species has no effect on either one.

niche	The role played and the part of the habitat occupied by a particular species in a community. The fundamental niche is the environment in which it is possible for a species to grow; the realised niche is the part of the fundamental niche to which it is confined due to competition from other species.
nitrogen fixation	The process in which dissolved gaseous nitrogen is used by certain microbes as a substrate for the enzyme nitrogenase, which converts nitrogen to forms available for further metabolism within the microbe. The fixed nitrogen can then become available, in part, for other organisms.
nitrogenous wastes	Any metabolic waste product that contains nitrogen.
nomenclature	Naming of taxa.
notochord	In chordates, a stiff, flexible, internal rod that is the precursor of the vertebrate backbone.
nucleus	An organelle enclosed within its own double membrane and containing the chromosomes.
null hypothesis	The opposite of the hypothesis proposed to explain an observation – that is, that there will be no difference between control and treated groups.
nutrient cycling	The transfer of nutrient elements from one compartment of an ecosystem to another, until their eventual release to the environment.
nutrient spiralling	Nutrient uptake and release in flowing river systems; nutrients may be taken up by plants and animals at one place in a river, then released again, for example through the decay of organic material, carried downstream and taken up again.
nutrients	Compounds taken up and used by plants and animals; the term includes a suite of essential elements such as the macronutrients nitrogen, phosphorus and potassium. as well as 'trace elements' required in smaller amounts.
oligotrophic	Describes a usually deep water body, with clear, well-oxygenated water low in nitrogen and phosphorus.
open forest	Forest where the canopies of trees do not meet, and the understorey may be of diverse species.
operculum	1. Cap at the top of a moss capsule. 2. Cover over fish gills. 3. Cap on a eucalypt flower bud.
oral surface	The surface containing the mouth, for example in echinoderms.
order	A rank within the hierarchy of taxonomic classification, between class and family.
organ	A collection of tissues with a specific structure and function.
orographic relief	Rainfall related to altitude (orography deals with mountains).
orthophosphate	A dissolved ion of phosphate, the usual form present in water bodies, readily taken up by plants as the main source of phosphorus used in plant and animal metabolism.
osmoconformers	Organisms whose body fluids are the same osmotic concentration as the surrounding environment.
osmolarity	The solute concentration of a solution, measured as the number of moles of all dissolved solutes per kilogram of water.

osmoregulation	The process by which plants and animals balance the osmotic properties of the cells and tissues inside them against those of their external environment.
osmoregulator	An organism that maintains a constant internal osmotic concentration, despite changes in the external environment.
osmosis	Diffusion of water across a differentially permeable membrane; takes place from a region of greater water potential to one of lesser water potential.
osmotic potential	Change in free energy or chemical energy of water produced by solutes. Carries a negative sign (also called solute potential).
ovary	In an angiosperm, the ovule-bearing organ. In animals, the organ that produces eggs or ova.
oviparous	Laying fertilised eggs that hatch outside the body.
ovoviviparous	Carrying fertilised eggs that hatch inside the mother's body.
ovule	In gymnosperms and angiosperms, the female gametophyte with the egg cell, surrounded by an integument and other layers.
oxidation-reduction (redox) reactions	Reactions in the chloroplast that use the products of the energy-fixing, light-dependent reactions for the production of glucose from CO_2 in the Calvin cycle.
oxidative phosphorylation	The process occurring on the inner membrane of the mitochondrion that converts the potential energy of the high-energy electrons in NADH and $FADH_2$ into ATP.
Palaeozoic	An era in the geological time scale, about 543–248 million years ago.
Pangaea	The supercontinent that united Gondwana and Laurasia during the late Palaeozoic and early Mesozoic eras.
paramylon	The carbohydrate storage product of photosynthesis in euglenids.
paraphyletic	A group of organisms or taxa that contains an ancestor and some, but not all, of its descendants. (Compare with monophyletic and polyphyletic.)
parapodia	Thin, flattened outgrowths on the side of polychaete annelids used for locomotion.
parenchyma	Tissue type of protists and plants whereby cells remain joined by common walls after cell division. In vascular plants, large, thin-walled cells with a central vacuole, for example cells forming cortex, pith and leaf pallisade and mesophyll.
parthenogenesis	The development of a new individual from an unfertilised egg.
pathogens	Organisms causing a disease in another species.
particulates	Small particles suspended in water, including phytoplankton, bacteria and inorganic particles. Removed for analysis by filtration.
pedicellariae	Pincer-like structures on the body surface of echinoderms to keep it clear of settling organisms.
peduncle	The stalk of a flower.
pelagic	Occurring at or near the ocean surface, far from land.
peptidoglycan	Constituent of bacterial cell walls, rarely present in Archaea.

perennials	Long-lived plants (living longer than 2 years).
perianth	Calyx and corolla parts of a flower that are similar in appearance.
pericarp	Fruit wall; the outer, middle and inner layers are exocarp, mesocarp and endocarp.
pericycle	The outermost layer of parenchyma cells around the vascular cylinder of roots.
period	A subdivision of an era of geological time.
periphyton	Microscopic algae attached to hard substrata.
petiole	Leaf stalk.
pH	The measure of acidity, in which p stands for power and H stands for the concentration of hydrogen ions. Because the concentrations are so low, pH is expressed as a power of 10. Pure water has a hydrogen ion concentration of 10^{-7} moles of hydrogen per litre, written as pH 7. A pH below 7 is acidic; a pH above 7 is alkaline.
phagocytosis	The capture of food by the invagination of the external cell membrane.
pharynx	The first part of the digestive tract after the mouth (the throat in vertebrates).
phenotype	The physical appearance and biochemical properties of an organism, resulting from the interaction of heredity and the environment.
phloem	Tissue that conducts food both upwards and downwards in a plant. Composed of sieve cells (which transport sucrose), their adjacent companion cells and parenchyma fibres.
phosphorus	An essential element for plant and animal metabolism, taken up by plants and bacteria from dissolved forms of phosphate in the environment, and passing to animals through food webs.
photic zone	That part of the water column in which there is sufficient light for oxygen production through photosynthesis to exceed oxygen consumption through respiration (ideally calculated over a period of 24 hours).
photoautotroph	An organism (prokaryote or eukaryote) that obtains energy from sunlight and carbon from CO_2 by photosynthesis.
photoheterotroph	A prokaryote that obtains energy from sunlight, but carbon from organic molecules.
photosynthesis	The process in autotrophs by which solar energy is trapped in the chloroplasts and converted to chemical energy in the form of the bonds in ATP (adenosine triphosphate) molecules.
phylogeny	The evolutionary history or relationships of a group of organisms.
phylum (plural: phyla)	A rank within the hierarchy of taxonomic classification of animals and plants, between kingdom and class (= division, sometimes used in the classification of plants and fungi).
phytoplankton	Single-celled and filamentous algal protists and cyanobacteria suspended in the water column.
pinacocytes	Flattened cells making up the outer surface of a sponge.
pinnae (singular: pinna)	Lobes of a fern frond.
pioneer community	The initial community of colonising species on bare ground.
pith	Parenchymatous cells in the centre of a ring of vascular tissue in a root or stem.

pits	Small cavities in plant cell walls where the layer of cellulose of the primary cell wall is very thin (or absent) and the secondary cell wall does not form. Usually in matching positions in adjacent cells and associated with plasmodesmata (the thin strands of cytoplasm passing through small pores in plant cell walls connecting the cytoplasm of adjacent cells) in living cells and lateral water transport in dead xylem cells.
placenta	1. Spongy tissue that nourishes the embryo in mammals. 2. The region of the ovary in seed plants to which the ovule is attached.
plankton	Small plants (phytoplankton), animals (zooplankton) and protists suspended in open water.
plant growth form	The classification of a plant based on characteristics such as height, shape, foliage characteristics, and whether it is herbaceous or woody.
plant life form	The classification of a plant based mainly on the position of the apical buds, above or below ground.
planula	A ciliated, free-swimming larva.
plasma membrane	The outer membrane that consists of lipid and protein and encloses the contents of all living cells. It regulates the passage of dissolved substances into and out of the cell.
plastid	The membrane-bound organelle in eukaryotes that contains pigment.
plate tectonics	The theory of movements of plates of the Earth's crust resulting in changes of position and shape of continents, volcanoes, earthquakes, mountain building, and separation or combination of biota.
pleuston	Organisms of the surface film of the water, distributed by the wind.
plotless sampling	A method for determining the density of plants based on selecting random plants and measuring the distance between them and their nearest neighbours, or selecting random points and determining the distance between them and the nearest plants.
poikilothermic	Not regulating body temperature, so that it fluctuates with the temperature of the environment.
pollen	The male gametophyte of seed plants.
pollination	The transfer of pollen by abiotic or biotic means to the micropyle of the ovule (in gymnosperms) or the stigma (in angiosperms).
polymorphism	The presence of more than one allele at a locus within a population.
polyp	The body form of cnidarians living attached to the substratum, with the tentacles and the mouth pointing upwards.
polyphyletic	Pertaining to a group of organisms or taxa that are descended from more than one common ancestor. (Compare with monophyletic and paraphyletic.)
polyploid	Having more than two sets (the diploid number) of chromosomes, e.g. $3n$ = triploid, $4n$ = tetraploid.
population	A group of organisms of the same species, in a defined geographic area.
population ecology	The study of the interrelationships between individuals in a population, between the individuals of different populations, and between populations and their environment.
population viability analysis (PVA)	An analysis using computer simulations to predict what will happen to a population faced with real-world uncertainties.

pre-agricultural ecosystem	An ecosystem in which humans have no domesticated plants or animals and rely on plants and animals naturally occurring in their environment for food.
Pre-Cambrian	An era in the geological time scale, prior to about 543 million years ago.
precautionary principle	Where there are threats of serious or irreversible environmental damage, lack of full scientific certainty should not be used as a reason for postponing measures to prevent environmental degradation.
pressure flow hypothesis	The theory relating osmotically induced pressure gradients in phloem to sugar transport.
primary growth	A plant body resulting from growth from the apical meristems of the embryonic root and shoot apices, axillary buds and lateral roots. Composed of primary tissues.
primary root	The root that develops from the root apex of the embryo. In dicots and gymnosperms it forms the primary structure of the taproot.
primary succession	The sequence of communities that replace one another until a climax community is established.
prion	A pathogenic subviral particle consisting only of protein.
procambium	The strand of cells arising from a shoot or root apex that differentiates into the phloem and xylem of the vascular bundle. In dicots some procambial cells retain their meristematic ability and later form part of the vascular cambium.
productivity	The amount of organic material produced by plants expressed in carbon or energy produced per unit volume (or area) with time. Sometimes used for the generation of animal biomass.
profundal zone	The lowest layer of a water column, near the sediment surface and below the photic zone.
prokaryote	Unicellular or multicellular organism in the domains Bacteria and Archaea, comprising small cells with the plasma membrane encased within a cell wall. The DNA exists as a coiled, circular strand.
proteoid root	A globular cluster of lateral roots that increases surface area for nutrient uptake (= cluster root).
protoderm	The layer of the apical meristem that gives rise to the epidermis.
protostomes	Animals with spiral cleavage of the egg, a coelom arising from splitting of mesodermal cells and where the embryonic mouth forms before the embryonic anus.
protoxylem	The first xylem that forms from a root or shoot apex in regions where the organ is still elongating. The pattern of the position of these strands is referred to as protoxylem poles.
psammon	Organisms living between sand grains.
pseudocoelom	A fluid-filled body cavity not completely lined by tissue of mesodermal origin.
pseudopodia	Projections of cytoplasm seen in amoebozoans and rhizarians.
pseudoreplication	Treating subsamples as true independent replicates when they are in fact correlated (e.g. recording the response of one animal to the sound of a gunshot 10 times rather than testing 10 animals separately).

Q_{10}	The increase in rate of a physiological process or reaction produced by an increase in temperature of 10°C.
quadrat	Delimited area of ground, usually square or rectangular, used in population sampling.
rachis	Main axis of a divided fern leaf.
radial symmetry	Characteristic of an animal in which division in any longitudinal plane will divide the body into two equal halves. Also applied to flower shapes.
radiation	1. Radiation in the form of electromagnetic waves transfers heat between bodies that are not touching (e.g. an organism may gain heat from solar radiation and also radiate some of its own heat to the environment). 2. Evolution of several new species from an ancestral species.
radula	Protrusible, toothed, tongue-like membrane for feeding used by molluscs (except the filter-feeding bivalves).
ramet	An individual arising by asexual reproduction and therefore genetically identical to its parent.
random distribution	The distribution when the location of any individual is uninfluenced by the location of other individuals and all locations are equally likely to be occupied.
randomisation	Distribution of objects or treatments according to a random function, often using random number tables or computerised random number generators.
rank	The position or level of a taxon within the hierarchy of taxonomic classification.
rarity	Low levels of any one of the following: (1) the size of the local population, or the abundance of individuals (alpha rarity); (2) how specialised its habitat is (beta rarity); or (3) its geographic range or distribution (gamma rarity).
reallocation	Management actions that change the human land use of a site, with the aim of restoring or improving ecosystem function.
recessive allele	An allele the effect of which is not seen in a heterozygote individual.
reclamation	Management actions that stabilise a degraded site, usually aimed at increasing its use for human activity, not at restoring the pre-disturbance ecosystem.
recombination	Occurs during prophase of meiosis when paired strands of chromatids from homologous chromosomes break then rejoin, not with the original end, but with that from the homologous partner.
recovery plan	The collection of information and management documents that set out actions to assist threatened taxa and ecological communities within a planned and logical framework.
Red List of Threatened Species	International list from the IUCN of species classified according to perceived risk of extinction.
redox (reduction-oxidation)	Reactions involving the transfer of electrons in which one ion gains electrons (reduction) when another loses them (oxidation). The redox potential is the ability of a solution to oxidise or reduce, measured with an electrode.

redundant	Where two or more species perform similar roles in nutrient cycling or energy transfer in a community, the extra species are redundant, and extinction of one species need not lead to a change in ecosystem function.
rehabilitation	Management actions aimed at restoring ecosystem function, usually through the (partial) restoration of ecosystem structure.
reintroduction	The release of individuals of a species into an environment where they have become extinct.
replication	1. The use of multiple experimental subjects or sets of subjects. This ensures that the unique characteristics of one subject will not bias the experiment and allows calculation of a mean response and its variance. 2. To construct an exact copy of a DNA strand or chromatid.
resilience	The capacity of an ecosystem to recover from a disturbance and return to its original state.
respiration	The metabolic process in organisms by which organic molecules are oxidised to release energy, typically with carbon dioxide and water as by-products (= cellular respiration).
respiratory trees	The internal respiratory structures of sea cucumbers: paired, highly branched and muscular, extending from the hind-gut into the coelom.
resprouter	A tree or shrub that regrows from dormant epicormic buds, or buds in the lignotuber after defoliation by fire.
restoration	In the broad sense, management actions aimed at the full recovery of an ecosystem to its pre-disturbance structure and function. In the narrow sense, management actions aimed at reversing degradation and directing ecosystem recovery to one aspect of pre-disturbance structure and/or function.
rhizoids	Tubular outgrowths, one or several cells long, that anchor algae and lower plants to the substrate.
rhizome	An underground stem that grows horizontally rather than vertically.
ribonucleic acid (RNA)	A single-stranded nucleic acid with the bases adenine, cytosine, thyamine and uracil. Involved in transcribing information from DNA (messenger RNA; mRNA) and translating it to produce proteins (transfer RNA; tRNA) and ribosomal RNA; rRNA).
ribosome	The cell organelle that translates the DNA code into a linear sequence of amino acids, which collectively form proteins. Free ribosomes occur in the cytoplasm of prokaryotic or eukaryotic cells; bound ribosomes are on the endoplasmic reticulum in eukaryotic cells.
riparian	The habitat for plants and animals on the banks of a river or other water body.
river continuum concept	The concept that, as water flows rapidly from shady headwater streams with much coarse plant debris and passes through reaches of higher order to become sluggish lowland streams, there is a change in the functional feeding groups of invertebrates (shredders, collectors, scrapers, predators). (See also nutrient spiralling.)
RNA	See ribonucleic acid.

Rodinia	The supercontinent in the Pre-Cambrian before its breakup into Gondwana and Laurasia.
root	A multicellular, cylindrical organ formed from the root meristem in an embryo. Responsible for anchoring the plant, and for water and mineral nutrient uptake and transport through vascular tissues.
root hairs	Long tubular outgrowths from the epidermis just behind the root tip.
salinisation (primary)	Accumulation of soluble salts in the landscape, for example in salt lakes and salt deserts.
salinisation (secondary)	Accumulation of soluble salts in the landscape as a result of human activity, for example in response to clearing deep-rooted native plants and replacing them with shallow-rooted agricultural plants, and the irrigation of crops in areas that have saline groundwater.
salinity	The total amount of dissolved ions in water, dominated by sodium and chloride in marine and estuarine waters, and in Australian salt lakes; usually expressed as mg/L or parts per thousand (‰).
salt marsh	The community of plants rooted in sandy or muddy sediment, and subjected to tidal or longer-term inundation by saline water. The community is dominated by shrubby chenopods, or rhizomatous rushes or reeds.
satellites	Infectious subviral particles (similar to viroids).
saturation	1. The concentration of a solute at which no further solute will dissolve. Often used in relation to the salinity above which no further salt will dissolve, and the oxygen concentration above which no further oxygen will dissolve. 2. The presence of a factor at a level equal to or greater than that required for maximum response or activity.
savanna	A biome of grassland with scattered trees (also called woodland, especially in the Southern Hemisphere). The understorey may be diverse, with shrubs, herbs and grasses.
science	The process of solving questions raised by observations of nature, using reasoned evaluation of evidence and careful testing of ideas.
sclerenchyma	Plant cells, dead at maturity with thick lignified cell walls, providing strength to tissues. Includes sclereids (irregularly shaped cells), stone cells (rounded) and fibres (extremely long and thin with pointed ends).
sclerophylly	Pertaining to plant species having leaves with thick cuticles and reduced intercellular air spaces. It is a response to low soil nutrients and also confers some tolerance to arid conditions.
scolex	The organ at the anterior end of tapeworms for attaching to the host.
seagrass	Flowering plants rooted in sediment, and submerged in a marine or estuarine environment.
secondary growth	Growth from the vascular cambium (which increases girth in plants) and from the cork cambium (which results in a layer of bark) (= secondary thickening).
secondary roots	See lateral roots.
secondary succession	The succession of communities that occurs after a climax community is altered by a disturbance.

sedge	Plant from the family Cyperaceae. Often loosely also applied to restios (family Restionaceae), rushes (family Juncaceae) and reeds such as phragmites.
seed	The fertilised ovule comprising the zygote and nutritive tissue (endosperm in angiosperms or female gametophyte in gymnosperms) surrounded by a seed coat.
segmentation	Where the body is organised in a series of segments. Occurs in annelids, arthropods and chordates (= metamerism).
selection	The change in allele frequencies from one generation to the next in a population as a result of the differential reproductive success of different genotypes.
semipermeable membrane	A membrane through which solvent can move, but not solute; for example, one that allows water to move through, but not sugar molecules (= differentially permeable membrane).
septa (singular: septum)	1. Cross-walls in a fungal hypha. 2. Division between the two chambers of the ventricle of a heart. 3. Partition between the segments of annelid worms, isolating the coelomic fluid in adjacent segments.
serotinous seed banks	Stores of seeds held in persistent fruits and cones on plants (rather than in the soil).
sessile	1. Unable to move from one place. 2. A leaf with no petiole.
sexual reproduction	Reproduction involving the union of two haploid gametes to form a new diploid individual.
sieve cells	Long, thin cells with perforated plates at each end; components of phloem tissue.
silicate minerals	Combination of various elements and SiO_4^-, for example feldspars that contain magnesium, calcium, sodium and silicon, and mica that contains iron, magnesium, potassium and silicon.
SLOSS debate	The debate over whether species richness and biodiversity are maximised in one large reserve or in many smaller ones of an equal total area (single large or several small).
somatic cell	A body cell dividing by mitosis, as distinct from germ cells dividing by meiosis to produce gametes.
sorus (plural: sori)	A cluster of sporangia on a fern leaf.
speciation	The formation of new species.
species	1. A rank within the hierarchy of taxonomic classification, below genus and designated by a binomial name. 2. The smallest evolutionarily independent unit or group of organisms.
spiracles	External openings of the trachea of the respiratory system of insects.
sporangia (singular: sporangium)	A sac-like structure in which asexual spores form.
spores	Reproductive cells that can develop into a new individual without first fusing with another reproductive cell. (Compare with gamete.) In plants and algae, this includes the haploid cells produced from meiosis. In prokaryotes, protists and fungi, this includes resting spores resistant to unfavourable conditions.

sporophyll	A leaf-like structure bearing a sporangium.
sporophyte	The diploid stage of a protist or plant life cycle that produces haploid spores through meiosis.
stabilising selection	Selection that favours phenotypes near the mean of the normal character distribution, leading to a reduction in variance around intermediate character states.
stamen	Structure consisting of a filament and an anther, within which are the pollen grains (male gametophytes). Collectively known as the androecium.
static life table	Table recording data on age-specific survival and reproduction based on a cross-section of a population at a single time and constructed from sampling data.
stele	Vascular tissue (xylem, phloem and parenchyma) in a root or stem. It includes the pericycle in roots.
stem	The organ that develops from the shoot apex of an embryo. Usually cylindrical. Bears leaves at nodes. (See also stolon and rhizome.)
stenohaline	Aquatic animals that tolerate only small fluctuations in salinity. (Compare with euryhaline.)
stigma	1. The photoreceptor organ in protists. 2. The pollen receptor at the tip of a style in a flower.
stipe	The stalk between an algal holdfast and the photosynthetic blade.
stolon	A stem that grows horizontally above ground.
stomate (plural: stomates)	A pore in the surface of a leaf, bounded by two guard cells.
stratification	The separation of the water column into 'layers' of differing densities due to differences in salinity (chemical stratification) and/or temperature (thermal stratification). (Compare with monomictic, dimictic and meromictic stratification.)
stream order	The numbering of the reaches of a river in which each headwater stream is number one, and when it joins another the stream is designated number two, and so on.
stroma	The fluid enclosed within the inner membrane of the chloroplast where sugars are made from carbon dioxide and water during photosynthesis.
stromatolites	Dome-shaped rocks formed in saline waters, from layers of cyanobacteria, calcium carbonate and trapped sediments (both fossil and living).
stygofauna	Invertebrates living in ground water.
style	The structure between the stigma and the ovary of a flower.
suberin	A corky, waterproof substance.
succession	The chronological sequence of species or communities.
swim bladder	The gas-filled bladder in bony fish that gives adjustable neutral buoyancy, so the fish does not need to expend energy to hold its depth.
symbiosis	The living together of two organisms in a close relationship that may or may not be mutually beneficial.
sympatric speciation	The formation of separate species through divergent evolution in populations that are not geographically isolated. (Compare with allopatric speciation.)

synchronised spawning	The simultaneous release of eggs into the environment (typically the ocean) by a large number of individuals of the same species.
synergid cells	Part of the egg apparatus in the female gametophyte of a flowering plant. The egg cell has a synergid cell on each side.
systematics	The field of biology that aims to infer the phylogeny of organisms, classify them into a hierarchical series of groups and name the groups so classified.
tachymetabolic	Maintaining high levels of metabolism at rest.
taiga	Northern Hemisphere coniferous forest biome.
taproot	A root that develops from the extension of the radicle during seed germination. (See primary root.)
taxon (plural: taxa)	A group of organisms which is formally recognised at any rank within the hierarchy of taxonomic classification.
taxonomy	See systematics.
tegument	The outer layer of parasitic animals such as tapeworms and flukes.
temporary wetland	A wetland filled seasonally or at longer intervals (= episodic wetland).
terrigenous	Of, or produced by, the land, for example sediment derived from eroding land transported offshore by a river.
tetrapod	Having four legs.
thallus (plural: thalli)	A relatively undifferentiated body of certain protists and plants.
thermocline	The interface between the upper, mixed layer of a stratified water body and the lower, denser layer. Detected as a rapid change in temperature with depth. (Compare with halocline.)
thermogenesis	The production of heat within the tissues of an organism; it raises body temperature.
thermophiles	Micro-organisms that grow at extreme temperatures fatal to most species.
thermoregulation	The maintenance of an animal's body temperature within the range optimal for functioning.
threat	See threatening process.
threatened ecological community	An ecological community at risk of extinction.
threatening process	Any biological or environmental process (repeated or ongoing actions) that places a species at risk of extinction.
threshold	The level of some component of a system which, when reached, produces sudden or dramatic changes in other system components.
thylakoid	Interconnected hollow discs within the chloroplast which may be organised in stacks (grana). Their membranes contain chlorophyll, which traps solar energy in the light-dependent reactions of photosynthesis.
timber line	The altitude (or latitude) at which tree growth can no longer be supported by the climate.
tissue	An aggregation of cells with a specific appearance and function in a multicellular organism.

tissue culture	The method of propagating plants under sterile conditions and on (usually) defined media (= *in vitro* culture).
torpor	A state of inactivity, lowered metabolic activity and lowered body temperature for periods of hours or, at most, days.
torsion	In the larvae of some gastropods, the right and left muscles attaching the shell to the body grow unequally, twisting the viscera through a 180° rotation and bringing the mantle cavity, gills and the openings of the excretory, digestive and reproductive systems to the anterior.
total nitrogen	The amount of nitrogen in a given volume of water, calculated as the sum of nitrogen present in the organic and inorganic fractions – that is, organic nitrogen, oxidised nitrogen and ammonium nitrate.
total phosphorus	The amount of phosphorus in a given volume of water, calculated as the sum of phosphorus present in the organic and inorganic fractions – that is, organic phosphorus determined after acid digestion, plus dissolved phosphate.
trachea (plural: tracheae)	In insects, a system of branching tubes supplying oxygen to the cells. In vertebrates, the wind pipe.
tracheids	Water-conducting cells in the xylem. Long, thin cells with thick walls and tapering ends that do not have complete perforations. (Compare with vessel.)
trait	See character.
transcription	Copying of a DNA strand to produce a complementary RNA molecule.
transduction	The transfer by bacteriophage of fragments of prokaryote DNA to a new prokaryote cell.
transformation	The uptake and incorporation of new DNA from the environment into a cell's genome.
translation	Decoding of the genetic information in RNA molecules to make proteins.
translocation	Deliberate transfer of individuals of an endangered species from a breeding facility into the wild, or from one site where they occur naturally to another.
triploblastic	Developing all tissues from three cell layers in the embryo.
trochophore larva	Free-living larval stage of annelids and some molluscs characterised by a ring of cilia and a shape like a toy spinning top.
trophic level	A step in a food chain.
tube foot	In echinoderms, the external component of an extensive coelomic compartment forming the water vascular system. Each tube foot has a muscular internal sac (ampulla) at one end and often a sucker at the other (= podia).
turbidity	The measure of the suspended solids in the water.
upwelling	Regions where cold, nutrient-rich subsurface ocean currents are forced to the surface at continental shelves.
urban ecosystem	People living in a city and its suburbs and their surroundings.
urea	$(NH_2)_2.CO$. A nitrogen-rich, water-soluble organic waste formed during the respiration of protein and amino acids; an important excretory product in mammals and amphibians.

utilitarian	Conserving biodiversity for human use.
vacuole	A membranous sac in cytoplasm with a variety of functions.
vascular bundle	A strand of xylem and phloem separated by procambium in dicots, and often with a capping of fibres on the outer side. Xylem and phloem (the vascular tissues) together form the conducting tissue (the vascular system = stele).
vascular cambium	The ring of lateral meristematic cells derived from the procambium and de-differentiation of cortical cells. Divides to form xylem to the centre of the organ and phloem to the outside.
vegetation floristics	The classification of a biome on the basis of the main plant species present.
vegetation structure	The classification of a biome on the basis of the main growth forms of the plants.
veliger larva	The larval stage of molluscs that develops from trochophore larvae. In gastropods, it is the stage that undergoes torsion.
ventral	The lower side of a bilaterally symmetrical animal. Also applied to plant parts such as leaves.
ventricle	The chamber of a heart that pumps blood to body tissues or to gills/lungs.
vessel	A water-conducting cell in the xylem. Generally a shorter, wider cell than a tracheid with thick walls and perforations between the cells that are placed end to end.
viroids	Infectious small circles of single-stranded RNA.
virus	An infectious particle of protein and nucleic acid.
viviparous	1. Bearing live young. 2. Seeds that germinate while they are attached to the parent plant.
Walker circulation	Atmospheric circulation of air at the equatorial Pacific Ocean. It creates ocean upwelling off the South American coast.
wallum	Heathland/low woodland occurring in Queensland.
water balance	The difference between the rate of water intake and the rate of water loss by an organism.
water vascular system	The modification of the coelom in echinoderms to form a hydraulic system that powers many tiny tube feet used in movement, gas exchange, excretion and collecting food.
wood	Secondary xylem. Sapwood contains some living cells, but in heartwood all cells are dead.
woodland	Landscape characterised by scattered trees with undergrowth of shrubs and grasses (= savanna).
xylem	Tissue that conducts water and minerals upwards in a plant, comprised of tracheids, vessels (in angiosperms), fibres and parenchyma.
zooxanthellae (singular: zooxanthella)	Endosymbiont dinoflagellates in some corals, anemones and molluscs.
zygote	The diploid cell rising from the fusion of two haploid gametes.

Index